MECHANICAL ENGINEERING DESIGN

McGraw-Hill Series in Mechanical Engineering

Jack P. Holman, *Southern Methodist University*
Consulting Editor

Anderson: *Modern Compressible Flow: With Historical Perspective*
Dieter: *Engineering Design: A Materials and Processing Approach*
Eckert and Drake: *Analysis of Heat and Mass Transfer*
Ham, Crane, and Rogers: *Mechanics of Machinery*
Hartenberg and Denavit: *Kinematic Synthesis of Linkages*
Hinze: *Turbulence*
Hutton: *Applied Mechanical Vibrations*
Jacobsen and Ayre: *Engineering Vibrations*
Juvinall: *Engineering Considerations of Stress, Strain, and Strength*
Kays and Crawford: *Convective Heat and Mass Transfer*
Lichty: *Combustion Engine Processes*
Martin: *Kinematics and Dynamics of Machines*
Phelan: *Dynamics of Machinery*
Phelan: *Fundamentals of Mechanical Design*
Pierce: *Acoustics: An Introduction to Its Physical Principles and Applications*
Raven: *Automatic Control Engineering*
Rosenberg and Karnopp: *Introduction to Physical System Dynamics*
Schenck: *Theories of Engineering Experimentation*
Schlichting: *Boundary-Layer Theory*
Shames: *Mechanics of Fluids*
Shigley: *Dynamic Analysis of Machines*
Shigley: *Kinematic Analysis of Mechanisms*
Shigley: *Mechanical Engineering Design*
Shigley: *Simulation of Mechanical Systems*
Shigley and Uicker: *Theory of Machines and Mechanisms*
Stoecker and Jones: *Refrigeration and Air Conditioning*

MECHANICAL ENGINEERING DESIGN

Fourth Edition

Joseph Edward Shigley
Professor Emeritus
The University of Michigan

Larry D. Mitchell
Professor of Mechanical Engineering
Virginia Polytechnic Institute and State University

McGraw-Hill Book Company

New York   St. Louis   San Francisco   Auckland   Bogotá   Hamburg
Johannesburg   London   Madrid   Mexico   Montreal   New Delhi
Panama   Paris   São Paulo   Singapore   Sydney   Tokyo   Toronto

MECHANICAL ENGINEERING DESIGN

4567890 HALHAL 8987654

ISBN 0-07-056888-X

This book was set in Palatino by Santype-Byrd.
The editors were Rodger H. Klas and Madelaine Eichberg;
the designer was Robin Hessel;
the production supervisor was Dominick Petrellese.
New drawings were done by J & R Services, Inc.
Halliday Lithograph Corporation was printer and binder.

Library of Congress Cataloging in Publication Data

Shigley, Joseph Edward.
    Mechanical engineering design.

    (McGraw-Hill series in mechanical engineering)
    Includes index.
    1. Machinery—Design.   I. Mitchell, Larry D.
II. Title.
TJ230.S5   1983   621.8′15   82-10044
ISBN 0-07-056888-X

CONTENTS

7 DESIGN FOR FATIGUE STRENGTH 270

7-1 Introduction. 7-2 The *S-N* Diagram. 7-3 Low-Cycle Fatigue.
7-4 High-Cycle Fatigue. 7-5 Endurance-Limit Modifying Factors.
7-6 Surface Finish. 7-7 Size Effects. 7-8 Reliability.
7-9 Temperature Effects. 7-10 Stress-Concentration Effects.
7-11 Miscellaneous Effects. 7-12 Fluctuating Stresses. 7-13 Fatigue
Strength under Fluctuating Stresses. 7-14 Nonlinear Theories.
7-15 The Kimmelmann Factor of Safety. 7-16 Torsion. 7-17 Stresses
due to Combined Loading. 7-18 Cumulative Fatigue Damage.
7-19 Surface Strength.

Part 2 DESIGN OF MECHANICAL ELEMENTS

8 THE DESIGN OF SCREWS, FASTENERS, AND CONNECTIONS 357

8-1 Thread Standards and Definitions. 8-2 The Mechanics of Power
Screws. 8-3 Thread Stresses. 8-4 Threaded Fasteners. 8-5 Bolted
Joints in Tension. 8-6 Compression of Bolted Members. 8-7 Torque
Requirements. 8-8 Strength Specifications. 8-9 Bolt Preload: Static
Loading. 8-10 Selection of the Nut. 8-11 Bolt Preload: Fatigue Load-
ing. 8-12 Fatigue Loading. 8-13 Gasketed Joints. 8-14 Bolted and
Riveted Joints Loaded in Shear. 8-15 Centroids of Bolt Groups.
8-16 Shear of Bolts and Rivets Due to Eccentric Loading. 8-17 Keys,
Pins, and Retainers.

9 WELDED, BRAZED, AND BONDED JOINTS 412

9-1 Welding. 9-2 Butt and Fillet Welds. 9-3 Torsion in Welded
Joints. 9-4 Bending in Welded Joints. 9-5 The Strength of Welded
Joints. 9-6 Resistance Welding. 9-7 Bonded Joints.

10 MECHANICAL SPRINGS 442

10-1 Stresses in Helical Springs. 10-2 Deflection of Helical Springs.
10-3 Extension Springs. 10-4 Compression Springs. 10-5 Spring Materials. 10-6 Design of Helical Springs. 10-7 Critical Frequency of
Helical Springs. 10-8 Optimization. 10-9 Fatigue Loading.
10-10 Helical Torsion Springs. 10-11 Belleville Springs.
10-12 Miscellaneous Springs. 10-13 Energy-Storage Capacity.

PREFACE

This book has been written for engineering students who are beginning a course of study in mechanical engineering design. Such students will have acquired a set of engineering tools consisting, essentially, of mathematics, computer languages, and the ability to use the English language to express themselves in the spoken and written forms. Mechanical design involves a great deal of geometry, too; therefore, another useful tool is the ability to sketch and draw the various configurations which arise. Students will also have studied a number of basic engineering sciences, including physics, engineering mechanics, materials and processes, and the thermal-fluid sciences. These, the tools and sciences, constitute the foundation for the practice of engineering, and so, at this stage of undergraduate education, it is appropriate to introduce the professional aspects of engineering. These professional studies should integrate and use the tools and the sciences in the accomplishment of an engineering objective. The pressures upon the undergraduate curricula today require that we do this in the most efficient manner. Most engineering educators are agreed that mechanical design integrates and utilizes a greater number of the tools and the sciences than any other professional study. Mechanical design is also the very core of other professional and design types of studies in mechanical engineering. Thus studies in mechanical design seem to be the most effective method of starting the student in the practice of mechanical engineering.

One of the reasons for writing a new edition now is the recent increased emphasis on the creative aspects of design in so many colleges and universities. In the early 1950s a committee on evaluation of engineering education of the American Society for Engineering Education stated:

> Training for the creative and practical phases of economic design, involving analysis, synthesis, development and engineering research, is the most distinctive feature of professional engineering education.

> The technical goal of engineering education is preparation for performance of the functions of analysis and design or of the functions of construction, production or operation with full knowledge of analysis and design of the structure, machine, or process involved.

Though these goals were stated a generation ago, they are valid today. A way must be found to involve the engineering student in genuine design experiences.

In an academic atmosphere it is quite difficult, if not impossible, to duplicate all the constraints and options that face the engineering designer working in industry. A single comprehensive design situation can easily absorb a student's working time for weeks, or even months, using up valuable instructional time that would have been available for other, equally important, topics. For this reason our approach in this book is to present some short design problems or situations which illustrate the decision-making processes without demanding an inordinate amount of the student's time. We believe we have included a sufficient number of such problems. Many of them will suggest other possibilities to the instructor. In fact, many instructors will prefer to use situations encountered in their own experience.

Another reason for publishing this new edition is the now wide availability of machine computational devices, many of quite low cost. These have changed the way we design things quite drastically in a very short period of time. The programmable calculator and the personal computer have sufficient memory capacity for the solution of a great many design situations. Furthermore, a programmable calculator with a large memory can now be purchased for about three times the price of an average textbook. This is about the same cost ratio that once existed between a slide rule and a textbook. Thus there is very little excuse for an engineering student today to be deprived of such an important tool. The use of the programmable calculator, as well as of the personal computer, has been emphasized in this book by the inclusion of programming suggestions throughout.

Students and other readers of previous editions have requested that a larger number of worked-out examples be included. In response to this request our goal here has been to provide at least one worked-out example for all situations having some degree of difficulty. In addition, we have tried to eliminate the "busy work" that inevitably arises in the solution of some problems. Of course, this is impossible to do in the case of genuine design situations.

In Chap. 3 of this edition, the use of numerical integration to determine the deflection of beams by computer or programmable calculator has been added. The area-moment method for the analysis of beams is also included in response to many requests. And the material on columns has been expanded by the addition of design and computer approaches as well as by the inclusion of material on inelastic buckling.

A new chapter on materials is found in this edition. The content of Chap. 4 includes a discussion of plastics and elastomers, materials widely used in the products designed today.

Chapter 5 is a condensation of the discussion of statistics as it appeared in the previous editions. In this chapter there is a strong emphasis on the use of the statistics functions found on most calculators. A discussion of the Weibull distribution is also new to this edition.

The chapter on the strength of mechanical elements was too lengthy in the previous editions, so two chapters cover the subject here. Chapter 6 introduces the subject of designing for static strength. This chapter opens with detailed photographs of 14 actual part failures, some quite dramatic and most from equipment or machines familiar to most readers. It is our hope that these illustrations will give meaningful emphasis to the importance of strength to the designer. In addition to the usual material on failure theories, an expanded discussion of stress concentration has been added. Contributing to this is an illustration of the use of finite-element analysis to determine the stress-concentration factor in the root section of a gear tooth. Fracture mechanics is a subject we believe to be of considerable significance to the mechanical designer. Accordingly, we have concluded this chapter with what we believe is an adequate discussion of this important subject.

Chapter 7 is devoted entirely to the subject of designing to resist fatigue failure. New concepts included are the optional use of multiple factors of safety, the Kimmelmann factor of safety for combined mean and alternating stress situations, and a discussion of low-cycle fatigue. Significant additions have been made in the presentation of fatigue modification factors, especially those for size and surface finish.

In Chap. 8 there is expanded and up-to-date coverage of fastener standards, including standard bolt materials. The discussion of tension-loaded bolted joints has been completely revised and the recommendations updated in accordance with recent research results. New material on fatigue loading and bolt preload has been added, and there is a more complete coverage of gasketed joints.

In Chap. 9 the treatment of fillet welds has been improved by including the results from recent finite-element analysis papers. The inclusion of additional examples should serve to clarify some of the more difficult design situations encountered.

Chapter 10, on springs, now contains suggested computer routines, illustrative design examples and problems, theory and analysis of spring surge, hints on optimization techniques, and expanded material on torsion springs.

Charts for multi-viscosity lubricants have been added to Chap. 12 and the journal-bearing examples have been improved. Additions to Chap. 13 include a new table of Lewis form factors and four new tables of AGMA geometry factors which have just become available. Many of the sections have been rewritten and new examples and problems added.

Chapter 16, on brakes and clutches, has been simplified and new examples added. New sections on energy dissipation and temperature rise have also been included. A section on flywheel design has been added, too. This section takes advantage of the numerical integration routine commonly found in the program libraries of programmable calculators and personal computers.

Chapter 17 has been almost completely rewritten to take advantage of new materials, design methods, and rating factors for belt and chain drives. The section on wire rope has been expanded and now includes more information on factor of safety.

Slight revisions have also been made to the comprehensive Appendix in order to make this section more useful to the designer.

As in previous editions, this book is written for the practicing design engineer as well as for engineering students. In some cases we have addressed the design engineer separately from the student in recognition of the fact that the designer's needs may be somewhat different from those of the student and that different facilities may be available.

We are grateful to the persons listed on the following pages (and to those who have elected to remain anonymous) for the suggestions, opinions, ideas, and advice that we have received. In some cases their contributions were first used in previous editions, but the ideas were so valuable as to appear again in this edition.

We want particularly to express our gratitude to Professor Charles R. Mischke of the department of mechanical engineering at Iowa State University. Professor Mischke assiduously recorded his opinion of almost every page, problem, example, and illustration of the previous edition. Our only regret is that we were unable to take advantage of all his suggestions.

We also want to thank Professor David K. Felbeck of the department of mechanical engineering and applied mechanics of The University of Michigan for the material on fracture mechanics.

Diane D. Heiberg of the McGraw-Hill Book Company has been our editor during the period in which the planning and the writing of this book was done. We are very grateful to Diane for helping in the solution to the hundreds of problems that arose. Her prompt attention to each problem smoothed the way and brightened our day.

We also wish to express our gratitude to Ursula Smith of Bozeman, Montana, who edited the manuscript from beginning to end. Her diligence in examining the text word-by-word and line-by-line has contributed substantially to the readability and accuracy of the finished work.

ACKNOWLEDGMENTS

The authors express their gratitude for suggestions to:

Robert W. Adamson, California Polytechnic State University, San Luis Obispo, California; Charles W. Allen, California State University, Chico, California; Cemil Bagci, Tennessee Technological University, Cookeville, Tennessee; Rolin F. Barrett, North Carolina State University, Raleigh, North Carolina; Charles W. Beadle, University of California at Davis, Davis, California; Ian Begg, California Polytechnic State University, San Luis Obispo, California; W. Kenneth Bodger, California State University, Fresno, California; Arnold E. Carden, University of

Alabama, University, Alabama; Milton A. Chace, The University of Michigan, Ann Arbor, Michigan; Frederick A. Costello, University of Delaware, Newark, Delaware; Joseph Datsko, The University of Michigan, Ann Arbor, Michigan; Marvin W. Dixon, Clemson University, Clemson, South Carolina; Winston M. Dudley, California State University, Sacramento, California; Kenneth S. Edwards, Jr., University of Texas, El Paso, Texas; David K. Felbeck, The University of Michigan, Ann Arbor, Michigan; Ferdinand Freudenstein, Columbia University, New York, New York; Franklin D. Hart, North Carolina State University, Raleigh, North Carolina; Charles R. Hayleck, Jr., University of Maryland, College Park, Maryland; Cecil O. Huey, Jr., Clemson University, Clemson, South Carolina; Glen E. Johnson, University of Virginia, Charlottesville, Virginia; Robert C. Juvinall, The University of Michigan, Ann Arbor, Michigan; Geza Kardos, Carleton University, Ottawa, Ontario; William C. Kieling, University of Washington, Seattle, Washington; William A. Kleinhenz, University of Minnesota, Minneapolis, Minnesota; R. W. Landgraf, Ford Motor Company, Dearborn, Michigan; Charles Lipson, The University of Michigan, Ann Arbor, Michigan; Robert A. Lucas, Lehigh University, Bethlehem, Pennsylvania; Hamilton H. Mabie, Virginia Polytechnic Institute and State University, Blacksburg, Virginia; Charles R. Mischke, Iowa State University, Ames, Iowa; Charles Nuckolls, University of Central Florida, Orlando, Florida; Charles B. O'Toole, Pennsylvania State University, McKeesport, Pennsylvania; Frank W. Paul, Clemson University, Clemson, South Carolina; Melvin K. Richardson, Clemson University, Clemson, South Carolina; Donald R. Riley, University of Minnesota, Minneapolis, Minnesota; Fred P. J. Rimrott, University of Toronto, Toronto, Ontario; Stephen M. Ross, University of New Haven, West Haven, Connecticut; George N. Sandor, University of Florida, Gainesville, Florida; Arthur W. Sear, California State University, Los Angeles, California; Joseph F. Shelley, Trenton State College, Trenton, New Jersey; Walter L. Starkey, Ohio State University, Columbus, Ohio; Ralph I. Stevens, University of Iowa, Iowa City, Iowa; Edward N. Stevensen, Jr., University of Hartford, Hartford, Connecticut; and Ward O. Winer, Georgia Institute of Technology, Atlanta, Georgia.

Joseph Edward Shigley
Larry D. Mitchell

LIST OF SYMBOLS*

A	Area; constant
$\mathbf{a}, a$	Acceleration; constant; dimension; addendum
B	Constant
b	Weibull exponent; dedendum; section width; dimension; constant; fatigue-strength exponent
C	Coefficient; spring index; bearing load rating; column end-condition constant; gear factor; center distance; specific heat
c	Clearance; distance from neutral axis to outer surface in beam; fatigue-ductility exponent; coefficient of viscosity
D, d	Diameter
E	Modulus of elasticity; kinetic energy
e	Eccentricity; efficiency; strain value
$\mathbf{F}, F$	Force
F	Face width
f	Frequency; coefficient of friction
G	Shear modulus of elasticity
g	Acceleration due to gravity
H	Hardness number; pole distance; heat gained or lost; power
h	Section depth; bearing clearance
I	Moment of inertia; geometry factor
$\mathbf{i}$	Unit vector in x direction

* See Table A-27 for Greek alphabet

i	Integer
J	Polar moment of inertia; geometry factor; mechanical equivalent of heat
j	Unit vector in y direction
j	Integer
K	Stress-concentration factor; Buckingham wear factor; Wahl correction factor; gear factor; bearing rating ratio; strength coefficient; stress-intensity factor; bolt-torque coefficient
k	Unit vector in z direction
k	Spring rate; endurance-limit modification factor; radius of gyration
L	Length; life; lead
l	Length
M,M	Moment
m	Mass; margin of safety; slope; contact ratio; module
n	Factor of safety; strain-hardening exponent; speed in rpm
P,P	Force
P	Diametral pitch; bearing pressure
p	Pressure; linear or circular pitch
Q	First moment of area; flow volume
q	Load intensity; notch sensitivity; arc length
R,R	Reactive forces at supports and connections; position vector
R	Reliability; reduction in area
r	Radius; radial direction indicator; correlation coefficient
S	Strength; length scale; bearing characteristic number
s	Sample standard deviation; distance
T	Torque; temperature
t	Thickness; tangential direction; time
U	Energy; velocity; coefficient
u	Unit energy; velocity; coordinate
V	Shear force; velocity; rotation factor for bearings
v	Volume; kinematic viscosity
W	Weight; load; cold-work factor
w	Unit weight; width
X	Radial factor for bearings
x	Rectangular coordinate; distance
Y	Thrust factor for bearings; Lewis form factor
y	Rectangular coordinate; distance; Lewis form factor
Z	Section modulus; viscosity
z	Rectangular coordinate; distance; standard statistical variable

α	Coefficient of thermal expansion; thread angle; axial fatigue-stress correction factor; angle
β	Partial bearing angle
Γ	Pitch angle
γ	Shear strain; pitch angle; articulation angle
Δ	Increment or change
δ	Total deformation or elongation
ϵ	Unit engineering strain; eccentricity ratio
ε	True strain
η	Efficiency
θ	Twist angle; slope; a Weibull parameter
λ	Lead angle
μ	Poisson's ratio; population mean; coefficient of friction; absolute viscosity
ρ	Radius of curvature; density
Σ	Summation sign
σ	Normal stress
$\hat{\sigma}$	Population standard deviation
τ	Shear stress
ϕ	Pressure angle
ψ	Helix angle
ω	Angular velocity

MECHANICAL ENGINEERING DESIGN

FUNDAMENTALS OF MECHANICAL DESIGN

CHAPTER

1

INTRODUCTION

This book is a study of the decision-making processes which mechanical engineers use in the formulation of plans for the physical realization of machines, devices, and systems. These decision-making processes are applicable to the entire field of engineering design—not just to mechanical engineering design. To understand them, to apply them to practical situations, and to make them pay off, however, require a set of circumstances, a particular situation, or a vehicle, so to speak. In this book we have therefore chosen the field of mechanical engineering as the vehicle for the application of these decision-making processes.

The book is divided into two parts. Part 1 is concerned with the fundamentals of decision making, the mathematical and analytical tools we will require, and the actual subject matter to be employed in using these tools. Occasionally you may encounter a familiar subject. This is included so that you can review it, if necessary, but more importantly, to establish the nomenclature for use in the more advanced portions of the book, for continuity, and as a reference source.

In Part 2 the fundamentals are applied to many typical design situations which arise in the design or selection of the elements of mechanical systems. An attempt has been made to arrange Part 2 so that the basic or more common elements are studied first. In this manner, as you become familiar with the design of single elements, you can begin to put them together to form complete machines or systems. Thus, with respect to a whole system, Part 2 becomes progressively more comprehensive. The intent, therefore, is that Part 2 should be studied chapter by chapter in the order in which it is presented.

1-1 THE MEANING OF DESIGN

To design is to formulate a plan for the satisfaction of a human need. In the beginning the particular need to be satisfied may be quite well-defined. Here are two examples of well-defined needs.

  1 How can we obtain large quantities of power cleanly, safely, and eco-nomically without using fossil fuels and without damaging the surface of the earth?
  2 This gearshaft is giving trouble; there have been eight failures in the last six weeks. Do something about it.

On the other hand the particular need to be satisfied may be so nebulous and ill-defined that a considerable amount of thought and effort is necessary in order to state it clearly as a problem requiring a solution. Here are two examples.

  1 Lots of people are killed in airplane accidents.
  2 In big cities there are too many automobiles on the streets and highways.

This second type of design situation is characterized by the fact that neither the need nor the problem to be solved has been identified. Note, too, that the situation may contain many problems.

We can classify design too. For instance:

1	Clothing design	**7**	Bridge design
2	Interior design	**8**	Computer-aided design
3	Highway design	**9**	Heating system design
4	Landscape design	**10**	Machine design
5	Building design	**11**	Engineering design
6	Ship design	**12**	Process design

In fact there are an endless number, since we can classify design according to the particular article or product or according to the professional field.

In contrast to scientific or mathematical problems, design problems have no unique answers; it is absurd, for example, to request the "correct answer" to a design problem, because there is none. In fact a "good" answer today may well turn out to be a "poor" answer tomorrow, if there is a growth of knowledge during the period or if there are other structural or societal changes.

Almost everyone is involved with design in one way or another, even in daily living, because problems are posed and situations arise which must be solved. Consider the design of a family vacation. There may be seven different places to go, all at different distances from home. The costs of transportation are different for each, and some of the options require overnight stops on the way. The children would like to go to a lake or seashore resort. The wife would prefer to go to a large city with department store shopping, theatres, and nightclubs. The husband prefers a resort with a golf course and perhaps a nearby mountain trout stream. When these needs and desires are related to time and money, various solutions may be found. Of these there may or may not be one or more optimal solutions. But the solution chosen will include the travel route, the stops, the mode of transportation, and the names and locations of resorts, motels, camping sites, or other away-from-home facilities. It is not hard to see that there is really a rather large group of interrelated complex factors involved in arriving at one of the solutions to the vacation design problem.

A design is always subject to certain problem-solving constraints. For example, two of the constraints on the vacation design problem are the time and money available for the vacation. Note, too, that there are also constraints on the solution. In the case above some of those constraints are the desires and needs of each of the family members. Finally, the design solution found might well be optimal. In this case an optimal solution is obtained when each and every family member can say that he or she had a good time.

A design problem is *not* a hypothetical problem at all. Design has an authentic purpose—*the creation of an end result by taking definite action or the creation of something having physical reality.* In engineering the word "design" conveys different meanings to different persons. Some think of a designer as one who employs the drawing board to draft the details of a gear, clutch, or other machine member. Others think of design as the creation of a complex system, such as a communications network. In some areas of engineering the word *design* has been replaced by other terms such as *systems engineering* or *applied decision theory.* But no matter what words are used to describe the design function, in engineering it is still the process in which scientific principles and the tools of engineering—mathematics, computers, graphics, and English—are used to produce a plan which, when carried out, will satisfy a human need.

1-2 MECHANICAL ENGINEERING DESIGN

Mechanical design means the design of things and systems of a mechanical nature—machines, products, structures, devices, and instruments. For the most part, mechanical design utilizes mathematics, the materials sciences, and the engineering-mechanics sciences.

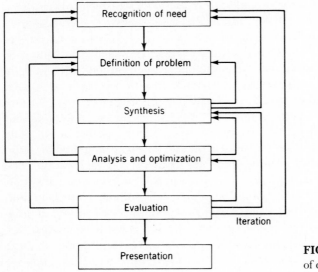

FIGURE 1-1 The phases
of design.

Mechanical engineering design includes all mechanical design, but it is a broader study because it includes all the disciplines of mechanical engineering, such as the thermal-fluids sciences, too. Aside from the fundamental sciences that are required, the first studies in mechanical engineering design are in mechanical design, and hence this is the approach taken in this book.

1-3 THE PHASES OF DESIGN

The total design process is of interest to us in this chapter. How does it begin? Does the engineer simply sit down at his or her desk with a blank sheet of paper and jot down some ideas? What happens next? What factors influence or control the decisions which have to be made? Finally, how does this design process end?

The complete process, from start to finish, is often outlined as in Fig. 1-1. The process begins with a recognition of a need and a decision to do something about it. After many iterations, the process ends with the presentation of the plans for satisfying the need. In the next several sections we shall examine these steps in the design process in detail.

1-4 RECOGNITION AND IDENTIFICATION

Sometimes, but not always, design begins when an engineer recognizes a need and decides to do something about it. *Recognition of the need* and phrasing the need often constitute a highly creative act, because the need may be only a vague discontent, a feeling of uneasiness, or a sensing that something is not right. The need is often not evident at all; recognition is usually triggered by a particular adverse circumstance or a set of random circumstances which arise almost simultaneously. For example, the need to do something about a food-packaging machine may be indicated by the noise level, by the variation in package weight, and by slight but perceptible variations in the quality of the packaging or wrap.

It is evident that a sensitive person, one who is easily disturbed by things, is more likely to recognize a need—and also more likely to do something about it. And for this reason sensitive people are more creative. A need is easily recognized after someone else has stated it. Thus the need in this country for cleaner air and water, for more parking facilities in the cities, for better public transportation systems, and for faster traffic flow has become quite evident.

There is a distinct difference between the statement of the need and the identification of the problem which follows this statement (Fig. 1-1). The problem is more specific. If the need is for cleaner air, the problem might be that of reducing the dust discharge from power-plant stacks, of reducing the quantity of irritants from automotive exhausts, or of quickly extinguishing forest fires.

Definition of the problem must include all the specifications for the thing that is to be designed. The specifications are the input and output quantities, the characteristics and dimensions of the space the thing must occupy, and all the limitations on these quantities. We can regard the thing to be designed as something in a black box. In this case we must specify the inputs and outputs of the box, together with their characteristics and limitations. The specifications define the cost, the number to be manufactured, the expected life, the range, the operating temperature, and the reliability. Obvious items in the specifications are the speeds, feeds, temperature limitations, maximum range, expected variations in the variables, and dimensional and weight limitations.

There are many implied specifications which result either from the designer's particular environment or from the nature of the problem itself. The manufacturing processes which are available, together with the facilities of a certain plant, constitute restrictions on a designer's freedom, and hence are a part of the implied specifications. A small plant, for instance, may not own cold-working machinery. Knowing this, the designer selects other metal-processing methods which can be performed in the plant. The labor skills available and the competitive situation also constitute implied specifications. Anything which limits the designer's freedom of choice is a specification. Many materials and sizes are listed in supplier's catalogs, for instance, but these are not all easily available and shortages frequently occur. Furthermore, inventory economics require that a manufacturer stock a minimum number of materials and sizes.

After the problem has been defined and a set of written and implied specifications has been obtained, the next step in design, as shown in Fig. 1-1, is the *synthesis* of the optimum solution. Now synthesis cannot take place without both *analysis and optimization*, because the system under design must be analyzed to determine whether the performance complies with the specifications. The analysis may reveal that the system is not an optimum one. If the design fails either or both of these tests, the synthesis procedure must begin again.

We have noted, and we shall do so again and again, that design is an iterative process in which we proceed through several steps, evaluate the results, and then return to an earlier phase of the procedure. Thus we may synthesize several components of a system, analyze and optimize them, and return to synthesis to see what effect this has on the remaining parts of the system. Both analysis and optimization require that we construct or devise abstract models of the system which will admit some form of mathematical analysis. We call these models *mathematical models*. In creating them it is our hope that we can find one which will simulate the real physical system very well.

1-5 EVALUATION AND PRESENTATION

As indicated in Fig. 1-1, *evaluation* is a significant phase of the total design process. Evaluation is the final proof of a successful design and usually involves the testing of a prototype in the laboratory. Here we wish to discover if the design really satisfies the need or needs. Is it reliable? Will it compete successfully with similar products? Is it economical to manufacture and to use? Is it easily maintained and adjusted? Can a profit be made from its sale or use?

Communicating the design to others is the final, vital step in the design process. Undoubtedly many great designs, inventions, and creative works have been lost to mankind simply because the originators were unable or unwilling to explain their accomplishments to others. Presentation is a selling job. The engineer, when presenting a new solution to administrative, management, or supervisory persons, is attempting to sell or to prove to them that this solution is a better one. Unless this can be done successfully, the time and effort spent on obtaining the solution have been largely wasted. When designers sell a new idea they also sell themselves. If they are repeatedly successful in selling ideas, designs, and new solutions to management, they begin to receive salary increases and promotions; in fact, this is how anyone succeeds in his or her profession.

Basically, there are only three means of communication. These are the *written*, the *oral*, and the *graphical* forms. Therefore the successful engineer will be technically competent and versatile in *all three forms* of communication. An otherwise competent person who lacks ability in any one of these forms is severely handicapped. If ability in all three forms is lacking, no one will ever know how competent that person is! The three forms of communication, writing, speaking, and drawing, are skills, that is, abilities which can be developed or acquired by any reasonably intelligent person. Skills are acquired only by practice—hour after monotonous hour of it. Musicians, athletes, surgeons, typists, writers, dancers, aerialists, and artists, for example, are skillful because of the number of hours, days, weeks, months, and years they have practiced. Nothing worthwhile in life can be achieved without work, often tedious, dull, and monotonous, and lots of it; and engineering is no exception.

Ability in writing can be acquired by writing letters, reports, memos, papers, and articles. It does not matter whether or not the articles are published—the practice is the important thing. Ability in speaking can be obtained by partici-

pating in fraternal, civic, church, and professional activities. This participation provides abundant opportunities for practice in speaking. To acquire drawing ability, pencil sketching should be employed to illustrate every idea possible. The written or spoken word often requires study for comprehension, but pictures are readily understood and should be used freely.

The competent engineer should not be afraid of the possibility of not succeeding in a presentation. In fact, occasional failure should be expected, because failure or criticism seems to accompany every really creative idea. There is a great deal to be learned from a failure, and the greatest gains are obtained by those willing to risk defeat. In the final analysis, the real failure would lie in deciding not to make the presentation at all.

The purpose of this section is to note the importance of *presentation* as the final step in the design process. No matter whether you are planning a presentation to your teacher or to your employer, you should communicate thoroughly and clearly, for this is the payoff. Helpful information on report writing, public speaking, and sketching or drafting is available from countless sources, and you should take advantage of these aids.

1-6 DESIGN CONSIDERATIONS

Sometimes the strength of an element is an important factor in the determination of the geometry and the dimensions of the element. In such a situation we say that *strength* is an important design consideration. When we use the expression *design consideration* we are referring to some characteristic which influences the design of the element or, perhaps, the entire system. Usually a number of these factors have to be considered in any given design situation.* Sometimes one of these will turn out to be critical, and when it is satisfied, the other factors no longer need to be considered. As an example, the following list of factors must often be considered:

1	Strength	12	Noise
2	Reliability	13	Styling
3	Thermal considerations	14	Shape
4	Corrosion	15	Size
5	Wear	16	Flexibility
6	Friction	17	Control
7	Processing	18	Stiffness
8	Utility	19	Surface finish
9	Cost	20	Lubrication
10	Safety	21	Maintenance
11	Weight	22	Volume

---

* In design literature the expression *design factor* is used by some writers to designate the ratio of the strength of an element to the internal stresses created by the external forces which act upon the element.

Some of these factors have to do directly with the dimensions, the material, the processing, and the joining of the elements of the system. Other factors affect the configuration of the total system. We shall be giving our attention to these factors, as well as many others, throughout the book.

In this book you will be faced with a great many design situations in which engineering fundamentals must be applied, usually in a mathematical approach, to resolve the problem or problems. This is completely correct and appropriate in an academic atmosphere where the need is actually to utilize these fundamentals in the resolution of professional problems. To keep the correct perspective, however, it should be observed that in many design situations the important design considerations are such that no calculations or experiments are necessary in order to define an element or a system. Students, especially, are often confounded when they run into situations in which it is virtually impossible to make a single calculation, and yet an important design decision must be made. These are not extraordinary occurrences at all; they happen every day. This point is made here so that you will not be misled into believing that there is a rational mathematical approach to every design decision.

1-7 FACTOR OF SAFETY

Strength is a *property* of a material or of a mechanical element. The strength of an element depends upon the choice, the treatment, and the processing of the material. Consider, for example, a shipment of 1000 springs. We can associate a strength S_i with the ith spring. When this spring is assembled into a machine, external forces are applied which result in stresses in the spring the magnitudes of which depend upon the geometry and are independent of the material and its processing. If the spring is removed from the machine unharmed, the stress due to the external forces will drop to zero as it was before assembly. But the strength S_i remains as one of the properties of the spring. So remember, *strength is an inherent property of a part*, a property built into the part because of the use of a particular material and process.

In this book we shall use the capital letter S to denote the strength, with appropriate subscripts to designate the kind of strength. Also, in accordance with accepted practice, we shall employ the Greek letters σ and τ in this book to designate the normal and shear stresses, respectively.*

The term *factor of safety* is applied to the factor used to evaluate the safeness of a member. Let a mechanical element be subjected to some effect which we will designate as F. We assume that F is a very general term and that it can be a force, a twisting moment, a bending moment, a slope, a deflection, or some kind of distortion. If F is increased, eventually it will become so large that any additional small increase would permanently impair the ability of the member to perform its

---

* See Sec. 2-1.

proper function. If we designate this limiting, or ultimate, value of F as F_u, then factor of safety is defined as

$$n = \frac{F_u}{F} \tag{1-1}$$

When F becomes equal to F_u, $n = 1$, and there is no safety at all. Consequently the term *margin of safety* is frequently used. Margin of safety is defined by the equation

$$m = n - 1 \tag{1-2}$$

The terms *factor of safety* and *margin of safety* are widely used in industrial practice, and the meaning and intent of these terms are clearly understood. However, the term F_u in Eq. (1-1), which is a very general term for any kind of strength, is a statistically varying quantity. In addition, the term F in Eq. (1-1) has a statistical variation of its own. For this reason a factor of safety $n > 1$ *does not preclude failure*. Because of this correlation between the degree of hazard and n, some authorities prefer to use the term *design factor* instead of "factor of safety." As long as you understand the intent, either term is correct. In this book *factor of safety* is the factor n of Eq. (1-1).

By far the greatest usage of factor of safety occurs when we are comparing stress with strength in order to estimate the amount of safety. Factor of safety is used to account for two separate, and usually unrelated, effects.

1 When many parts are to be manufactured from various shipments of materials, there occurs a variation in the strengths of the various parts for a variety of reasons, including processing, hot and cold working, and geometry.
2 When a part is assembled into, say, a machine and the machine is acquired by the ultimate user, there is a variation in the loading of the part and consequently the stresses induced by that loading, over which the manufacturer and designer have no control.

Thus factor of safety is used by engineers to account separately for the uncertainties that may occur in the strength of a part and the uncertainties that may occur with the loads acting on the part.

We shall designate as cases the three distinct procedures in which factor of safety is employed by engineers. These cases depend upon whether factor of safety is designated as a single quantity or is factored into components.

Case 1 The entire factor of safety is applied to the strength.

$$\sigma = \frac{S}{n} \quad \text{or} \quad \tau = \frac{S_s}{n} \tag{1-3}$$

Here the stresses σ and τ are called the *safe*, or the *design*, *stresses*. Note that S_s is a shear strength. Since only a single factor of safety is used in Eq. (1-3), n must include allowances for the uncertainties in strength and the uncertainties in loads. You should note that the relations of Eq. (1-3) imply that stress is linearly related to load. If there is any doubt about this proportionality, then Case 1 cannot be used.

When a part has already been designed and the geometry, loading, and strength are known, the factor of safety can be computed in order to evaluate the safety of the design. This approach is also used when a part has developed a history of failures and the engineer wishes to learn why some of the parts are failing. For this purpose, Eq. (1-3) is used in the form

$$n = \frac{S}{\sigma} \qquad \text{or} \qquad n = \frac{S_s}{\tau} \tag{1-4}$$

Case 2   The entire factor of safety is applied to the loading or to the stresses that result from that loading.

$$F_p = nF \qquad \text{or} \qquad \sigma_p = n\sigma \tag{1-5}$$

Here F_p is called the *allowable*, or the *permissible*, *load* and σ_p is the *allowable*, or *permissible stress*. The first relation in the set should always be used when the stress is not proportional to the load. The stress resulting from the permissible load can also be called the permissible stress. Equation (1-5) can then be used for design purposes by selecting a geometry such that the permissible stress is never greater than the strength S.

Case 3   The total or overall factor of safety is factored into components, and separate factors are used for the strength and for the loads, or the stresses produced by those loads. If there are two loads, say, then the overall or total factor of safety is

$$n = n_S n_1 n_2 \tag{a}$$

where n_S is used to account for all the variations or uncertainties concerned with the strength; n_1 accounts for all the uncertainties concerned with load 1; and n_2 accounts for all the uncertainties concerned with load 2.

When we apply a factor of safety, such as n_S, to the strength, we are really saying that under ordinary and reasonable circumstances the resulting strength will never be any smaller. Thus the smallest value of the strength is computed as

$$S(\text{min}) = \frac{S}{n_S} \tag{1-6}$$

When we apply a factor of safety, like n_1, to a load, or to the stress that results from that load, we are really saying that the resulting load or stress will never get any larger. Thus the largest stress or load, as the case may be, is

$$\sigma_p = n_j \sigma \qquad \text{or} \qquad F_p = n_j F \qquad\qquad (1\text{-}7)$$

where n_j is the component of the total factor of safety [n_1 or n_2 in Eq. (a)] used to account separately for the uncertainties concerning the stress or load. Again, it is appropriate to designate σ_p and F_p as *permissible* or *allowable values*.

For this case the stress-load relations can be represented by the general equation

$$\sigma_p = Cf(x_1, x_2, x_3, \ldots, x_i)F(n_1 F_1, n_2 F_2, n_3 F_3, \ldots, n_j F_j) \qquad (1\text{-}8)$$

where $C =$ a constant
    $f =$ a function of the geometry
    $F =$ a function of the loads, usually force and moment loads
    $x_i =$ dimensions of part to be designed
    $F_j =$ external loads applied to the part
    $n_j =$ factors of safety used to account for the individual variations in loads

Of course, an equation similar to Eq. (1-8) can be written for shear stresses. Equation (1-8) is used in design by solving it for all the unknown dimensions x_i.

Finally it is important to note that both the strength and the stresses in a machine part are likely to vary from point to point throughout the element. Various metalworking processes, such as forging, rolling, and cold forming, cause variations in the strength from point to point throughout the part. And we shall see in Chap. 2 that the stresses also vary from point to point. So remember that strength, stress, and safety are valid only at a particular point. In many cases they must be evaluated at several points in designing or analyzing a mechanical element.

For more on factor of safety see Sec. 6-2.

1-8 CODES AND STANDARDS

Once upon a time there were no standards for bolts, nuts, and screw threads. A $\frac{1}{2}$-in nut removed from a bolt on one machine would not mate with a $\frac{1}{2}$-in bolt removed from another. One manufacturer would produce, say, $\frac{1}{2}$-in bolts with 9 threads per inch; another used 12. Some fasteners had left-handed threads and sometimes the thread profiles differed. It wasn't unusual in the early days of the automobile to see a mechanic lay out the fasteners in a line as they were disassembled in order to avoid mixing them during reassembly. This lack of standards and uniformity was costly and inefficient for a great variety of reasons. It is no wonder that a person, disgusted with his or her inability to find a replacement for a damaged fastener, sometimes used bailing wire to fasten parts together.

A *standard* is a set of specifications for parts, materials, or processes intended to achieve uniformity, efficiency, and a specified quality. One of the important purposes of a standard is to place a limit on the number of items in the specifications so as to provide a reasonable inventory of tooling, sizes, shapes, and varieties.

A *code* is a set of specifications for the analysis, design, manufacture, and construction of something. The purpose of a code is to achieve a specified degree of safety, efficiency, and performance or quality. It is important to observe that safety codes *do not* imply *absolute safety*. In fact, absolute safety is impossible to obtain. Sometimes the unexpected event really does happen. Designing a building to withstand a 120-mph wind does not mean that the designer thinks a 150-mph wind is impossible; it simply means that he or she thinks it is highly improbable.

All of the organizations and societies listed below have established specifications for standards and safety or design codes. The name of the organization provides a clue to the nature of the standard or code. Some of the standards and codes, as well as addresses, can be obtained in most technical libraries. The organizations of interest to mechanical engineers are

Aluminum Association (AA)
American Gear Manufacturers Association (AGMA)
American Institute of Steel Construction (AISC)
American Iron and Steel Institute (AISI)
American National Standards Institute (ANSI)*
American Society of Mechanical Engineers (ASME)
American Society for Metals (ASM)
American Society of Testing and Materials (ASTM)
American Welding Society (AWS)
Anti-Friction Bearing Manufacturers Association (AFBMA)
Industrial Fasteners Institute (IFI)
National Bureau of Standards (NBS)
Society of Automotive Engineers (SAE)

1-9 ECONOMICS

The consideration of cost plays such an important role in the design decision process that one could spend as much time in studying the cost design factor as in the study of design itself. Here we shall introduce only a few general approaches and simple rules.

First, observe that nothing can be said in an absolute sense concerning costs. Materials and labor usually show an increasing cost from year to year. But the

---

* In 1966 the American Standards Association (ASA) changed its name to the United States of America Standards Institute (USAS). Then, in 1969, the name was again changed to American National Standards Institute, as shown above and as it is today. This means that you may occasionally find ANSI standards designated as ASA or USAS.

costs of processing the materials can be expected to exhibit a decreasing trend because of the use of automated machine tools. The cost of manufacturing a single product will vary from one city to another and from one plant to another because of overhead, labor, and freight differentials and slight manufacturing variations.

Standard Sizes

The use of standard sizes is a first principle of cost reduction. An engineer who specifies a G10350 bar of hot-rolled steel $2\frac{1}{8}$-in square, called a hot-rolled square, has added cost to the product, provided a 2- or $2\frac{1}{4}$-in square, both of which are standard, would do equally well. The $2\frac{1}{8}$-in size can be obtained by special order or by rolling or machining a $2\frac{1}{4}$-in bar, but these approaches add cost to the product. To assure that standard sizes are specified, the designer must have access to stock lists of the materials he or she employs. These are available in libraries or can be obtained directly from the suppliers.

A further word of caution regarding the selection of standard sizes of materials is necessary. Although a great many sizes are usually listed in catalogs, they are not all readily available. Some sizes are used so infrequently that they are not stocked. A rush order for such sizes may mean more expense and delay. Thus you should also have access to a list of preferred sizes. See Table A-13 for preferred inch and millimeter sizes.

There are many purchased parts, such as motors, pumps, bearings, and fasteners, which are specified by designers. In the case of these, too, the designer should make a special effort to specify parts which are readily available. If they are made and sold in large quantities, they usually cost considerably less than the odd sizes. The cost of ball bearings, for example, depends more upon the quantity of production by the bearing manufacturer than upon the size of the bearing.

Large Tolerances

Among the effects of design specifications on costs, those of tolerances are perhaps most significant. Tolerances in design influence the producibility of the end product in many ways, from necessitating additional steps in processing to rendering a part completely impractical to produce economically. Tolerances cover dimensional variation and surface-roughness range and also the variation in mechanical properties resulting from heat treatment and other processing operations.

Since parts having large tolerances can often be produced by machines with higher production rates, labor costs will be smaller than if skilled workers were required. Also, fewer of such parts will be rejected in the inspection process, and they are usually easier to assemble.

Breakeven Points

Sometimes it happens that, when two or more design approaches are compared for cost, the choice between the two depends upon another set of conditions, such as the quantity of production, the speed of the assembly lines, or some other condition. There then occurs a point corresponding to equal cost which is called the *breakeven point.*

As an example, consider a situation in which a certain part can be manufactured at the rate of 25 parts per hour on an automatic screw machine or 10 parts per hour on a hand screw machine. Let us suppose, too, that the setup time for the automatic is 3 h and that the labor cost for either machine is $20 per h, including overhead. Figure 1-2 is a graph of the cost versus production by the two methods. The breakeven point corresponds to 50 parts. If the desired production is greater than 50 parts, the automatic machine should be used.

Cost Estimates

There are many ways of obtaining relative cost figures so that two or more designs can be roughly compared. A certain amount of judgment may be required in some instances. For example, we can compare the relative value of two automobiles by comparing the dollar cost per pound of weight. Another way to compare the cost of one design with another is simply to count the number of parts. The design having the smallest number of parts is likely to cost less. Many other cost estimators can be used, depending upon the application, such as area, volume, horsepower, torque, capacity, speed, and various performance ratios.

Value Engineering

A method of evaluating several proposed designs using a systematic approach called *value analysis*, or *value engineering*, is often very useful and may even point the way to new design approaches. In this method* *value* is defined as a numerical ratio, the ratio of the *function*, or *performance*, to the *cost*.

As an elementary introduction to the methods of value analysis let us analyze the ball bearing of Fig. 1-3. The analysis is carried out in tabular form and is illustrated by Table 1-1. Since only bearing manufacturers have access to actual costs, the analysis which follows is a hypothetical situation. The steps proceed from left to right across the table. These are:

1 Name of part
2 Brief summary or summary of function
3 Estimate of percent of total performance contributed by this function
4 Cost of part
5 Percent of total cost
6 Value, the ratio of percent performance to percent cost

The reasoning behind this form of analysis is easily explained. As we shall learn in Chap. 11, the outer and inner rings provide raceways for the balls. Also,

---

* See D. Henry Edel, Jr., *Introduction to Creative Design*, Prentice-Hall, Englewood Cliffs, N.J., 1967, pp. 111–119.

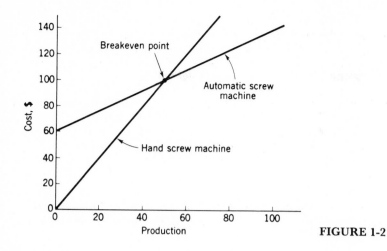

FIGURE 1-2

the bore of the inner ring and the outside diameter of the outer ring are very accurately ground so as to provide an excellent fit when the completed bearing is assembled into a machine. Further, the faces of these two rings often bear against other parts to permit the bearing to resist axial loads and to maintain alignment of rotating parts.

The function of the balls is to support the load and, by their rolling action, to reduce friction. If the balls were not kept apart by the separator, they would roll against one another and generate a larger frictional torque. The separator also assures the same number of balls in the load-carrying zone at all times. As noted

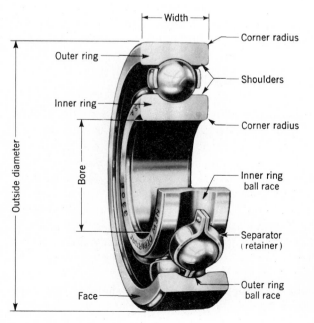

FIGURE 1-3 Nomenclature of a ball bearing. (*Courtesy of New-Departure-Hyatt Division, General Motors Corporation.*)

Table 1-1   VALUE ANALYSIS OF A BALL BEARING

Name of part	Function	% performance	Cost, $	% cost	Value
Outer ring	Provides ball race	20	2.35	47	0.425
Inner ring	Provides ball race	20	2.00	40	0.500
Balls	Supports load	50	0.50	10	5.000
Separator	Separates balls	10	0.10	2	5.000
Rivets	Fastens separator halves	0	0.05	1	0
Total		100	5.00	100	

in the table, the rivets secure the two separator halves together. Note that the value of the separator is equal to that of the balls even though the function of the separator is a secondary one. It would therefore be poor economy to attempt to eliminate the separator. A considerable saving might be made, however, if the value of the outer ring could be increased.

1-10   RELIABILITY

The statistical measure of the probability that a mechanical element will not fail in use is called the *reliability* of that element. The reliability R can be measured by a number having the range

$$0 \le R < 1 \tag{1-9}$$

A reliability of $R = 0.90$ means that there is a 90 percent chance that the part will perform its proper function without failure. A reliability of $R = 1$ cannot be obtained, since this means that failure is absolutely impossible.

1-11   SAFETY AND PRODUCT LIABILITY

The *strict liability* concept of product liability generally prevails in the United States. This concept states that the manufacturer of an article is liable for any damage or harm that results because of a defect. And it doesn't matter whether the manufacturer knew about the defect, or even could have known about it. For example, suppose an article was manufactured, say, 10 years ago. And suppose at that time the article could not have been considered defective based upon all technological knowledge then available. Ten years later, according to the concept of strict liability, the manufacturer is still liable. Thus, under this concept, the plaintiff only needs to prove that the article was defective and that the defect caused some damage or harm. Negligence of the manufacturer need not be proved.

It can be a very frightening experience for an engineer who has turned out what he or she considers a superlative design to be browbeaten on the witness stand by an experienced trial attorney and accused of killing children. A design book is not the place to dwell on this subject. You should encourage your local engineering chapters, such as Tau Beta Pi, to invite outside lecturers to discuss these topics with you. It is important that you learn how to protect yourself from being found guilty by a judge or jury for a "dangerous product design."

One of the touchy subjects that sometimes come up in the practice of engineering is what to do if you detect something that you consider poor engineering. If possible, of course, you should attempt to correct it or run sufficient tests to prove that your fears are groundless. If neither of these approaches is feasible then another approach is to place a memo in the design file and to keep a copy of the memo at home in case the original is "lost." While this approach may protect you, it may also result in your being passed over for promotion, or even in your being discharged. In the long run, if you really feel strongly about the fact that the engineering is poor, and you cannot live with it, then you should transfer or seek a new position.

The best approaches to the prevention of product liability are good engineering in both analysis and design, quality control, and comprehensive testing procedures. Advertising managers often make glowing promises in the warranties and sales literature for a product. These statements should be reviewed carefully by the engineering staff to eliminate excessive promises and to insert adequate warnings and instructions for use.

1-12 UNITS

In the symbolic equation of Newton's second law

$$F = MLT^{-2} \tag{1-10}$$

F stands for force, M for mass, L for length, and T for time. Units chosen for any three of these quantities are called *base units*. The first three having been chosen, the fourth unit is called a *derived unit*. When force, length, and time are chosen as base units, the mass is the derived unit and the system that results is called a *gravitational system of units*. When mass, length, and time are chosen as base units, force is the derived unit and the system that results is called an *absolute system of units*.

In the English-speaking countries, the *U. S. customary foot-pound-second system* (fps) and the *inch-pound-second system* (ips) are the two standard gravitational systems most used by engineers.* In the fps system the unit of mass is

$$M = \frac{FT^2}{L} = \frac{(\text{pound force})(\text{second})^2}{\text{foot}} = \text{lbf} \cdot \text{s}^2/\text{ft} = \text{slug} \tag{1-11}$$

---

* Most engineers prefer to use gravitational systems; this helps to explain some of the resistance to the use of SI units, since the International System (SI) is an absolute system.

Thus, length, time, and force are the three base units in the fps gravitational system.

The unit of time in the fps system is the second, abbreviated s. The unit of force in the fps system is the pound, more properly the *pound force*. We shall seldom abbreviate this unit as lbf; the abbreviation lb is permissible, since we shall be dealing only with the U. S. customary gravitational systems.* In some branches of engineering it is useful to represent 1000 lb as a kilopound and to abbreviate it as *kip*. Many writers add the letter s to *kip* to obtain the plural, but to be consistent with the practice of using only singular units we shall not do so here. Thus, 1 kip and 3 kip are used to designate 1000 and 3000 lb, respectively. Finally, we note in Eq. (1-11) that the derived unit of mass in the fps gravitational system is the lb·s²/ft, called a *slug*; there is no abbreviation for slug.

The unit of mass in the ips gravitational system is

$$M = \frac{FT^2}{L} = \frac{(\text{pound force})(\text{second})^2}{\text{inch}} = \text{lb} \cdot \text{s}^2/\text{in} \qquad (1\text{-}12)$$

This unit of mass has *not* been given a special name.

The International System of Units (SI) is an absolute system. The base units are the meter, the kilogram mass, and the second. The unit of force is derived using Newton's second law and is called the *newton* to distinguish it from the kilogram, which, as indicated, is the unit of mass. The units of the newton (N) are

$$F = \frac{ML}{T^2} = \frac{(\text{kilogram})(\text{meter})}{(\text{second})^2} = \text{kg} \cdot \text{m}/\text{s}^2 = \text{N} \qquad (1\text{-}13)$$

The weight of an object is the force exerted upon it by gravity. Designating the weight as W and the acceleration due to gravity as g, we have

$$W = mg \qquad (1\text{-}14)$$

In the fps system standard gravity is $g = 32.1740$ ft/s². For most cases this is rounded off to 32.2. Thus the weight of a mass of 1 slug in the fps system is

$$W = mg = (1 \text{ slug})(32.2 \text{ ft/s}^2) = 32.2 \text{ lb}$$

In the ips system, standard gravity is 386.088 or about 386 in/s². Thus, in this system, a unit mass weighs

$$W = (1 \text{ lb} \cdot \text{s}^2/\text{in})(386 \text{ in/s}^2) = 386 \text{ lb}$$

---

* The abbreviation *lb* for *pound* comes from Libra, the balance, the seventh sign of the zodiac, which is represented as a pair of scales.

Table 1-2 SI BASE UNITS

Quantity	Name	Symbol
Length	meter	m
Mass	kilogram	kg
Time	second	s
Electric current	ampere	A
Thermodynamic temperature	kelvin	K
Amount of matter	mole	mol
Luminous intensity	candela	cd

With SI units, standard gravity is 9.806 or about 9.80 m/s^2. Thus, the weight of a 1 kg mass is

$$W = (1 \text{ kg})(9.80 \text{ m/s}^2) = 9.80 \text{ N}$$

In view of the fact that weight is the force of gravity acting upon a mass, the following quotation is pertinent.

> The great advantage of SI units is that there is one, and only one unit for each physical quantity—the meter for length, the kilogram for mass, the newton for force, the second for time, etc. To be consistent with this unique feature, it follows that a given unit or word should not be used as an accepted technical name for two physical quantities. However, for generations the term "weight" has been used in both technical and nontechnical fields to mean either the force of gravity acting upon a body or the mass of the body itself. The reason for this double use of the term "weight" for two different physical quantities—force and mass—is attributed to the dual use of the pound units in our present customary gravitational system in which we often use weight to mean both force and mass.*

The seven SI base units, with their symbols, are shown in Table 1-2. These are dimensionally independent. Lowercase letters are used for the symbols unless they are derived from a proper name, then a capital is used for the first letter of the symbol. Note that the unit of mass uses the prefix kilo; this is the only base unit having a prefix.

Table 1-2 shows that the SI unit of temperature is the kelvin. The Celsius temperature scale (once called centigrade) is not a part of SI, but a difference of one degree on the Celsius scale equals one kelvin.

A second class of SI units comprises the derived units, many of which have special names. Table 1-3 is a list of those we shall find most useful in our work in this book.

---

* From "S.I., The Weight/Mass Controversy," *Mech. Eng.*, vol. 99, no. 9, September 1977, p. 40, and vol. 101, no. 3, March 1979, p. 42.

Table 1-3   EXAMPLES OF SI DERIVED UNITS*

Quantity	Unit	SI symbol	Formula
Acceleration	meter per second squared		$\text{m} \cdot \text{s}^{-2}$
Angular acceleration	radian per second squared		$\text{rad} \cdot \text{s}^{-2}$
Angular velocity	radian per second		$\text{rad} \cdot \text{s}^{-1}$
Area	square meter		m^2
Circular frequency	radian per second	ω	$\text{rad} \cdot \text{s}^{-1}$
Density	kilogram per cubic meter		$\text{kg} \cdot \text{m}^{-3}$
Energy	joule	J	$\text{N} \cdot \text{m}$
Force	newton	N	$\text{kg} \cdot \text{m} \cdot \text{s}^{-2}$
Force couple	newton meter		$\text{N} \cdot \text{m}$
Frequency	hertz	Hz	s^{-1}
Power	watt	W	$\text{J} \cdot \text{s}^{-1}$
Pressure	pascal	Pa	$\text{N} \cdot \text{m}^{-2}$
Quantity of heat	joule	J	$\text{N} \cdot \text{m}$
Speed	revolution per second		s^{-1}
Stress	pascal	Pa	$\text{N} \cdot \text{m}^{-2}$
Torque	newton meter		$\text{N} \cdot \text{m}$
Velocity	meter per second		$\text{m} \cdot \text{s}^{-1}$
Volume	cubic meter		m^3
Work	joule	J	$\text{N} \cdot \text{m}$

* In this book, negative exponents are seldom used; thus circular frequency, for example, would be expressed in rad/s.

The radian (symbol rad) is a supplemental unit in SI for a plane angle.

A series of names and symbols to form multiples and submultiples of SI units has been established to provide an alternative to the writing of powers of 10. Table A-1 includes these prefixes and symbols.

1-13   RULES FOR USE OF SI UNITS

The International Bureau of Weights and Measures (BIPM), the international standardizing agency for SI, has established certain rules and recommendations for the use of SI. These are intended to eliminate differences which occur among various countries of the world in scientific and technical practices.

Number Groups

Numbers having four or more digits are placed in groups of three and separated by a space instead of a comma. However, the space may be omitted for the special case of numbers having four digits. A period is used as a decimal point. These recommendations avoid the confusion caused by certain European countries in which a comma is used as a decimal point, and by the English use of a centered period. Examples of correct and incorrect usage are as follows:

1924 or 1 924 but not 1,924

0.1924 or 0.192 4 but not 0.192,4

192 423.618 50 but not 192,423.61850

The decimal point should always be preceded by a zero for numbers less than unity.

Use of Prefixes

The multiple and submultiple prefixes in steps of 1000 only are recommended (Table A-1). This means that length can be expressed in mm, m, or km, but not in cm, unless a valid reason exists.

When SI units are raised to a power the prefixes are also raised to the same power. This means that km^2 is

$$km^2 = (1000 \text{ m})^2 = (1000)^2 \text{ m}^2 = 10^6 \text{ m}^2$$

Similarly mm^2 is

$$(0.001 \text{ m})^2 = (0.001)^2 \text{ m}^2 = 10^{-6} \text{ m}^2$$

When raising prefixed units to a power, it is permissible, though not often convenient, to use the nonpreferred prefixes such as cm^2 or dm^3.

Except for the kilogram, which is a base unit, prefixes should not be used in the denominators of derived units. Thus the meganewton per square meter MN/m^2 is satisfactory, but the newton per square millimeter N/mm^2 is not to be used. Note that this recommendation avoids a proliferation of derived units.

Double prefixes should not be used. Thus, instead of millimillimeters mmm, use micrometers μm.

1-14 PRECISION AND ROUNDING OF QUANTITIES

The use of the electronic calculator having eight or ten digits or more in the display can easily lead to the appearance of great precision. Consider, for example, a computation which leads to the result $F = 142.047$ lb. This number contains six significant figures, and so the implication exists that the result is precise to all six figures. However, if the data entered into the computation consisted of numbers containing only three significant figures, then the answer $F = 142.047$ lb indicates a precision which does not exist. To express results with a nonexistent precision is misleading to say the least, but it is also dishonest, and may even be dangerous. So, in this section, we want to examine the means of learning the precision of a number, and then specify methods of rounding off a result to the correct precision.

When objects are counted, the result is always exact. When you purchase a dozen rolls at the bakery, you expect to receive exactly 12 rolls—no more, no less. But, as opposed to counting, measurements are always approximations. For example, in weighing yourself recently, you may have found your weight to be, say, 123 lb. This does not mean 123.000 lb because the three zeros after the decimal point imply that you were able to read the scales to a thousandth of a pound. Thus, a weight of 123 lb simply means that it is closer to 123 lb than it is to either 122 or 124.

The degree of precision obtained from a measurement depends upon a lot of factors, one of which is just how much precision is really needed. Usually great accuracy is expensive because such measurements must be taken with more expensive instruments and by more skillful workers. Unneeded precision is wasteful of both time and money.

The implied precision of any value is plus or minus one-half of the last significant digit used to state the value. Figure 1-4 shows a bar which measures 18.5 units in length. The length is closer to 18.5 units than it is to either 18.4 or to 18.6. Three significant figures were used to express this result. The precision of this measurement is ± 0.05 because the result would have been expressed as 18.5 units for all lengths between 18.45 and 18.55.

A value written as 18.50 contains four significant figures because the zero must be counted too. The value 18.50 implies that the measurement is closer to 18.50 than it is to either 18.49 or to 18.51. The precision is ± 0.005.

The measurement 0.0018 contains two significant figures because the zeros are only used to locate the decimal point. If the measurement is stated in the form 1.8 E-03 the two significant figures are evident. Then note that 1.80 E-03, which is the same as 0.001 80, has three significant figures.

The precision of a value such as 18 000 cannot be determined. The measurement could have been made to the nearest 1000, in which case there are only two significant figures. In such cases knowledge of how the measurement was made will usually reveal the precision. In this book ambiguous cases, such as this one, are arbitrarily assumed to have a precision equivalent to three significant figures. Note that if the value is stated as 1.80 $E + 04$, it is clear that there are three significant figures.

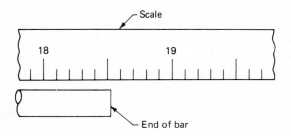

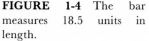

FIGURE 1-4 The bar measures 18.5 units in length.

Now suppose we have some value, say 142.507, which we know to be accurate only to three significant figures. We therefore wish to eliminate the insignificant figures. For this process, called *rounding off* a value, the following rules apply:

1 Retain the last digit unchanged if the first digit discarded is less than 5. For example 234.315 rounds to 234.3 if four significant figures are desired, or to 234 if three significant figures are desired.

2 Increase the last digit retained by one unit if the first digit discarded is greater than 5, or if it is a 5 followed by at least one digit other than zero. For example, 14.6 rounds off to 15; 14.50 rounds off to 15.

3 Retain the last digit unchanged if it is an even number and the first digit discarded is exactly 5 followed only by zeros. Thus 1.45 and 1.450 both round off to 1.4 if two significant figures are desired.

4 Increase the last digit by one unit if it is an odd number and the first digit discarded is exactly 5 followed only by zeros. Thus 1.55 rounds off to 1.6 and 15.50 rounds off to 16.

But note that most calculators with a FIX key do not follow rules 3 and 4 in rounding; instead, they always round up. Thus a calculator would round 1.45 to 1.5 and 1.55 to 1.6. Many engineers do the same.

Computations involving multiplication and division should be rounded after the computation has been performed. The answer should then be rounded off such that it contains no more significant digits than are contained in the least precise value. Thus, the product

$$(1.68)(104.2) = 175.056$$

should be rounded off to 175 because the value of 1.68 has three significant digits. Similarly

$$\frac{1.68}{104.2} = 0.016\ 122\ 8$$

should be rounded to 0.0161 for the same reason.

The rules are slightly different for computations involving addition and subtraction. Before performing the operation, round off all values to one significant digit farther to the right than that of the least accurate number. Then perform the addition or subtraction. Finally, round off the answer so that it has the same number of significant figures as are in the least accurate number. For example, suppose we wish to sum

$$A = 104.2 + 1.687 + 13.46$$

Of these, the number 104.2 is the least accurate. Therefore, using the round-off rules, we first compute

$$A = 104.2 + 1.69 + 13.46 = 119.35$$

Then we round off the result to $A = 119.4$ using rule 4. Note that the value of 1.687 was rounded off to 1.69 before the terms were summed.

1-15 ITERATION AND THE PROGRAMMABLE CALCULATOR

Many times in design a trial-and-error solution, that is, iteration, is the fastest way of solving a problem. Even when a solution is possible by graphic means or by expanding the function into an infinite series, a programmable-calculator solution may be more convenient. Usually the problem can be reduced in size so that only a very small number of program steps are required.

As an example, consider the problem of finding the wall thickness and the outside diameter (OD) of a tube when the area moment of inertia is given. From Table A-14, we find the formula to be

$$I = \frac{\pi}{64} (d^4 - d_i^4) \tag{a}$$

where I = area moment of inertia
d = outside diameter
d_i = inside diameter

Suppose the desired moment of inertia in SI units (see Example 3-3) is $I = 8.36(10)^5$ mm^4. Then the problem is to find an OD and a wall thickness having a moment of inertia equal to or greater than the given value. The program steps are as follows:

1 Enter a trial value of d in mm.
2 Enter a trial value of t, the wall thickness, in mm.
3 Compute d_i.
4 Compute I.
5 Return to step 1 and repeat this loop until a satisfactory solution is found.

One answer to the problem stated here is $d = 70$ mm and $t = 10$ mm.

In addition, most of the manuals that accompany programmable calculators contain programs or instructions for solving iteration-type problems. The Hewlett-Packard *HP-19C/HP-29C Applications Book* for example, presents a program which uses Newton's method to find a solution for $f(x) = 0$, where $f(x)$ is specified by the user. The Texas Instruments *TI 58/59 Master Library* contains a program that will evaluate a polynomial up to the 24th degree for the TI-58. Another program evaluates the roots of any function specified by the user. Both of these programs are available in a plug-in software module. Such programs usually require a first estimate of x and, sometimes, the interval to be explored.

PROBLEMS*

1-1* Devise a pole- and tree-climbing machine different from the belt and climbers used by electric and telephone linepersons.

1-2* Devise a powered toy saucer that will skip over water like a skipping stone.

1-3* Devise a cheap, lightweight, hand-held apple and peach picker.

1-4* Older persons, crippled by arthritis, often are unable to stoop low enough to put on their own sox and hosiery. See if you can design a cheap and simple device that will aid such people in accomplishing this task.

1-5* Develop a set of preliminary plans for a device to split logs up to 27 in long using an automobile jack.

1-6* Many times in the chapters to follow you will be converting from rectangular coordinates to polar coordinates. In order to save rethinking this problem each time it occurs, devise a routine for your calculator, or a subroutine for your programmable calculator or home computer, that will be readily available when needed. Be sure to record how the sign of the angle θ results for each of the four quadrants.

1-7* The same as Prob. 1-6, except do the polar-to-rectangular conversion.

1-8 Convert the following to appropriate SI units:

(*a*) The area of a $\frac{1}{2}$-acre lot

(*b*) 30 psi tire pressure

(*c*) 2400 rpm

(*d*) The mass of an automobile weighing 3200 lbf

(*e*) The displacement of a 425-in^3 engine

(*f*) 2 U.S. qt and 1 pt

(*g*) 55 mph

(*h*) 15 gal of gasoline

(*i*) 14.7 psi atmospheric pressure

1-9 A formula for bending stress is $\sigma = 32M/\pi d^3$. If $M = 300$ N·m and $d = 20$ mm what is the value of the stress σ in MPa? What is the stress in kpsi?

1-10 The formula for the deflection of a certain beam is $y = 64Fl^3/3\pi Ed^4$. If $F = 0.450$ kN, $l = 300$ mm, $E = 207$ GPa, and $d = 19$ mm, find the deflection y in μm.

1-11 In the formula of Prob. 1-10 what force F in kN is required to produce a deflection y of 1.50 mm if $l = 250$ mm, $E = 207$ GPa, and $d = 20$ mm?

1-12 In the formula of Prob. 1-9 find the diameter d in mm if $M = 200$ N·m and $\sigma = 150$ MPa.

1-13 The energy stored in a helical spring is given by the equation $U = 4F^2D^3N/d^4G$ where N is the number of coils. If $F = 4N$, $D = 11$ mm, $N = 36$, $G = 79.3$ GPa, and $d = 1.2$ mm, compute the energy stored in N·m.

1-14 Find the deflection of the spring of Prob. 1-13 in mm if the deflection formula is $y = 8FD^3N/d^4G$.

---

* The asterisk indicates a design-type problem or one having no unique result.

CHAPTER

2

STRESS ANALYSIS

The public today is clamoring for faster, cleaner, safer, and quieter machines. And design engineers, in their efforts to satisfy these demands with products at a lower cost, are being sued in the courts throughout the land for designing things that are unsafe, that are dangerous to human life. One of the real problems we shall face in this book is that of relating the strength of an element to the external loads which are imposed upon it. These external loads cause internal stresses in the element. To avoid product liability the design engineer must have positive assurance that these stresses will never exceed the strength. In this chapter we shall concern ourselves with the determination of these stresses. Then, in other chapters to follow, we shall introduce the concept of strength and begin the process of relating strength and stress to achieve safety.

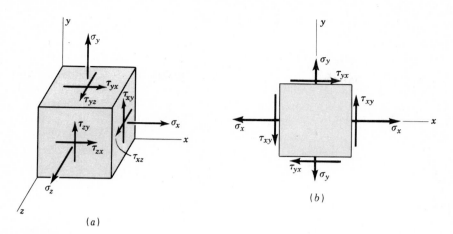

FIGURE 2-1

It is worth observing that the stresses which are calculated are only as reliable as are the external loads and geometries which are used in the determination. Once a product has been manufactured and placed in the hands of the consumer or user, the manufacturer has relinquished control over that product and, to some extent, over the external loads to which it may be subjected. It is for this reason, among others, that testing and experimental proving are so important in the design and development of a new machine or device.

2-1 STRESS

Figure 2-1a is a general three-dimensional stress element, showing three normal stresses σ_x, σ_y, and σ_z, all positive; and six shear stresses τ_{xy}, τ_{yx}, τ_{yz}, τ_{zy}, τ_{zx}, and τ_{xz}, also all positive. The element is in static equilibrium and hence

$$\tau_{xy} = \tau_{yx} \qquad \tau_{yz} = \tau_{zy} \qquad \tau_{zx} = \tau_{xz} \tag{2-1}$$

Outwardly directed normal stresses are considered as tension and positive. Shear stresses are positive if they act in the positive direction of a reference axis. The first subscript of a shear stress component is the coordinate normal to the element face. The shear stress component is parallel to the axis of the second subscript. The negative faces of the element will have shear stresses acting in the opposite direction; these are also considered as positive.

Figure 2-1b illustrates a state of *biaxial or plane stress*, which is the more usual case. For this state, only the normal stresses will be treated as positive or negative. The sense of the shear stress components will be specified by the clockwise (cw) or counterclockwise (ccw) convention. Thus, in Fig. 2-1b, τ_{xy} is ccw and τ_{yx} is cw.

2-2 MOHR'S CIRCLE

One of the most challenging problems in design is that of relating the strength of a mechanical element to the internal stresses which are produced by the external loads. Usually we have only a single value for the strength, such as the yield strength, but several stress components. One of the problems to be faced in a future chapter is how to take a stress element like Fig. 2-1b and relate it to a single strength magnitude to achieve safety. This section is the first step in the solution.

Suppose the element of Fig. 2-1b is cut by an oblique plane at angle ϕ to the x axis as shown in Fig. 2-2. This section is concerned with the stresses σ and τ which act upon this oblique plane. By summing the forces caused by all the stress components to zero, the stresses σ and τ are found to be*

$$\sigma = \frac{\sigma_x + \sigma_y}{2} + \frac{\sigma_x - \sigma_y}{2} \cos 2\phi + \tau_{xy} \sin 2\phi \qquad (2\text{-}2)$$

$$\tau = \frac{\sigma_x - \sigma_y}{2} \sin 2\phi + \tau_{xy} \cos 2\phi \qquad (2\text{-}3)$$

Differentiating the first equation with respect to ϕ and setting the result equal to zero gives

$$\tan 2\phi = \frac{2\tau_{xy}}{\sigma_x - \sigma_y} \qquad (2\text{-}4)$$

Equation (2-4) defines two particular values for the angle 2ϕ, one of which defines the maximum normal stress σ_1 and the other, the minimum normal stress σ_2. These two stresses are called the *principal stresses*, and their corresponding directions, the *principal directions*. The angle ϕ between the principal directions is 90°.

In a similar manner, we differentiate Eq. (2-3), set the result equal to zero, and obtain

$$\tan 2\phi = -\frac{\sigma_x - \sigma_y}{2\tau_{xy}} \qquad (2\text{-}5)$$

Equation (2-5) defines the two values of 2ϕ at which the shear stress τ reaches an extreme value.

It is interesting to note that Eq. (2-4) can be written in the form

$$2\tau_{xy} \cos 2\phi = (\sigma_x - \sigma_y) \sin 2\phi$$

---

* For the complete force analysis see Joseph E. Shigley, *Applied Mechanics of Materials*, McGraw-Hill, New York, 1976, pp. 259–261.

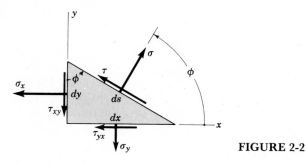

FIGURE 2-2

or

$$\sin 2\phi = \frac{2\tau_{xy} \cos 2\phi}{\sigma_x - \sigma_y} \tag{a}$$

Now substitute Eq. (a) for sin 2ϕ in Eq. (2-3). We obtain

$$\tau = -\frac{\sigma_x - \sigma_y}{2} \frac{2\tau_{xy} \cos 2\phi}{\sigma_x - \sigma_y} + \tau_{xy} \cos 2\phi = 0 \tag{2-6}$$

Equation (2-6) states that the shear stress associated with both principal directions is zero.

Solving Eq. (2-5) for sin 2ϕ, in a similar manner, and substituting the result in Eq. (2-2) yields

$$\sigma = \frac{\sigma_x + \sigma_y}{2} \tag{2-7}$$

Equation (2-7) tells us that the two normal stresses associated with the directions of the two maximum shear stresses are equal.

Formulas for the two principal stresses can be obtained by substituting the angle 2ϕ from Eq. (2-4) into Eq. (2-2). The result is

$$\sigma_1, \sigma_2 = \frac{\sigma_x + \sigma_y}{2} \pm \sqrt{\left(\frac{\sigma_x - \sigma_y}{2}\right)^2 + \tau_{xy}^2} \tag{2-8}$$

In a similar manner the two extreme-value shear stresses are found to be

$$\tau_1, \tau_2 = \pm \sqrt{\left(\frac{\sigma_x - \sigma_y}{2}\right)^2 + \tau_{xy}^2} \tag{2-9}$$

Your particular attention is called to the fact that an extreme value of the shear stress may not be the same as the maximum value. See Sec. 2-3.

A graphical method for expressing the relations developed in this section, called *Mohr's circle diagram*, is a very effective means of visualizing the stress state at

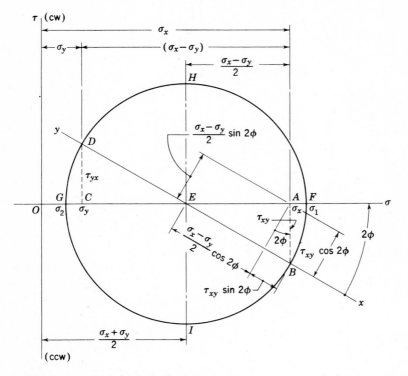

FIGURE 2-3 Mohr's circle diagram.

a point and keeping track of the directions of the various components associated with plane stress. In Fig. 2-3 we create a coordinate system with normal stresses plotted along the abscissa and shear stresses plotted as the ordinates. On the abscissa, tensile (positive) normal stresses are plotted to the right of the origin O and compressive (negative) normal stresses to the left. *On the ordinate clockwise (cw) shear stresses are plotted up; counterclockwise (ccw) shear stresses are plotted down.*

Using the stress state of Fig. 2-1b, we plot Mohr's circle diagram (Fig. 2-3) by laying off σ_x as OA, τ_{xy} as AB, σ_y as OC, and τ_{yx} as CD. The line DEB is the diameter of Mohr's circle with center at E on the σ axis. Point B represents the stress coordinates $\sigma_x \tau_{xy}$ on the x faces, and point D the stress coordinates $\sigma_y \tau_{yx}$ on the y faces. Thus EB corresponds to the x axis and ED to the y axis. The angle 2ϕ, measured counterclockwise from EB to ED, is $180°$, which corresponds to $\phi = 90°$, measured counterclockwise from x to y, on the stress element of Fig. 2-1b. The maximum principal normal stress σ_1 occurs at F, and the minimum principal normal stress σ_2 at G. The two extreme-value shear stresses, one clockwise and one counterclockwise, occur at H and I, respectively. You should demonstrate for yourself that the geometry of Fig. 2-3 satisfies all the relations developed in this section.

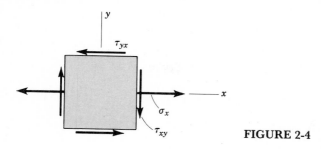

FIGURE 2-4

Though Eqs. (2-8) and (2-4) can be solved directly for the principal stresses and directions, a semigraphical approach is easier and quicker and offers fewer opportunities for error. This method is illustrated by the following example.

Example 2-1 The stress element shown in Fig. 2-4 has $\sigma_x = 80$ MPa and $\tau_{xy} = 50$ MPa cw.* Find the principal stresses and directions and show these on a stress element correctly aligned with respect to the xy system. Draw another stress element to show τ_1 and τ_2, find the corresponding normal stresses, and label the drawing completely.

Solution We shall construct Mohr's circle corresponding to the given data and then read the results directly from the diagram. You can construct this diagram with compass and scales and find the required information with the aid of scales and protractor, if you choose to do so. Such a complete graphical approach is perfectly satisfactory and sufficiently accurate for most purposes.

In the semigraphical approach used here, we first make an approximate freehand sketch of Mohr's circle and then use the geometry of the figure to obtain the desired information.

Draw the σ and τ axes first (Fig. 2-5a) and locate $\sigma_x = 80$ MPa along the σ axis. Then, from σ_x, locate $\tau_{xy} = 50$ MPa in the cw direction of the τ axis to establish point A. Corresponding to $\sigma_y = 0$, locate $\tau_{yx} = 50$ MPa in the ccw direction along the τ axis to obtain point D. The line AD forms the diameter of the required circle which can now be drawn. The intersection of the circle with the σ axis defines σ_1 and σ_2 as shown. The x axis is the line CA; the y axis is the line CD. Now, noting the triangle ABC, indicate on the sketch the length of the legs AB and BC, as 50 and 40 MPa, respectively. The length of the hypotenuse AC is

$$\tau_1 = \sqrt{(50)^2 + (40)^2} = 64.0 \text{ MPa}$$

and this should be labeled on the sketch too. Since intersection C is 40 MPa from

---

* Any stress components such as σ_y, τ_{zx}, etc., that are not given in a problem are always taken as zero.

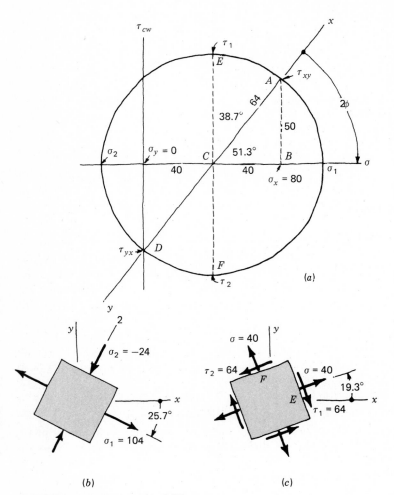

FIGURE 2-5 All stresses in MPa.

the origin, the principal stresses are now found to be

$$\sigma_1 = 40 + 64 = 104 \text{ MPa} \qquad \sigma_2 = 40 - 64 = -24 \text{ MPa}$$

The angle 2ϕ from the x axis cw to σ_1 is

$$2\phi = \tan^{-1} \frac{50}{40} = 51.3°$$

If your calculator has rectangular-to-polar conversion, you should always use it to obtain τ_1 and 2ϕ.

To draw the principal stress element (Fig. 2-5b), sketch the x and y axes parallel to the original axes (Fig. 2-4). The angle ϕ on the stress element must be measured in the *same* direction as is the angle 2ϕ on Mohr's circle. Thus, from x

measure 25.7° (half of 51.3°) clockwise to locate the σ_1 axis. The stress element can now be completed and labeled as shown.

The two maximum shear stresses occur at points E and F in Fig. 2-5a. The two normal stresses corresponding to these shear stresses are each 40 MPa, as indicated. Point E is 38.7° ccw from point A on Mohr's circle. Therefore, in Fig. 2-5c, draw a stress element oriented 19.3° (half of 38.7°) ccw from x. The element should then be labeled with magnitudes and directions as shown.

In constructing these stress elements it is important to indicate the x and y directions of the original reference system. This completes the link between the original machine element and the orientation of its principal stresses.   ////

Programming Hints

If you have a programmable calculator or access to a home or office computer you may decide to program the Mohr's circle problem for future use. The hints given here will help you to devise a very useful routine. However, you should first return to the problems in Chap. 1 and devise the rectangular-to-polar coordinate-conversions subroutine, unless you have already done so.

Now, in Fig. 2-3, notice that $(\sigma_x - \sigma_y)/2$ is the base of a right triangle, τ_{xy} is the ordinate, and the length of the hypotenuse corresponds to the extreme value of the shear stress. Thus, by entering the base and ordinate into your rectangular-to-polar conversion subroutine you can output the extreme value of the shear stress and the angle 2ϕ (see Fig. 2-3).

To help you decide what signs to use for the shear stresses in the routine, note that you will want the angle ϕ to indicate the angle *from* the x axis *to* the direction of the σ_1 stress.

As indicated in Fig. 2-3 the principal stresses can be found by adding and subtracting the extreme value of the shear stress to the term $(\sigma_x + \sigma_y)/2$. Table 2-1 displays some results to help in checking your program.

2-3   TRIAXIAL STRESS STATES

The general case of three-dimensional stress, or *triaxial stress*, is illustrated by Fig. 2-1a, as we have already learned. As in the case of plane or biaxial stress, a

Table 2-1   TYPICAL MOHR'S CIRCLE SOLUTIONS
(ROUNDED TO THREE PLACES)

σ_x	σ_y	τ_{xy}	σ_1	σ_2	τ_1	ϕ
8	2	4 ccw	10	0	5	26.6°
8	2	4 cw	10	0	5	−26.6°
2	−12	4 ccw	3.06	−13.06	8.06	14.9°
2	−8	4 cw	3.40	−9.40	6.40	−19.3°
−10	0	5 ccw	2.07	−12.07	7.07	67.5°

particular orientation of the stress element occurs in space for which all shear-stress components are zero. When an element has this particular orientation, the normals to the faces correspond to the principal directions and the normal stresses associated with these faces are the principal stresses. Since there are six faces, there are three principal directions and three principal stresses σ_1, σ_2, and σ_3.

In our studies of plane stress we were able to specify any stress state σ_x, σ_y, and τ_{xy} and find the principal stresses and principal directions. But six components of stress are required to specify a general state of stress in three dimensions, and the problem of determining the principal stresses and directions is much more difficult. It turns out that it is rarely necessary in design, and so we shall not investigate the problem in this book. The process involves finding the three roots to the cubic

$$\sigma^3 - (\sigma_x + \sigma_y + \sigma_z)\sigma^2 + (\sigma_x\sigma_y + \sigma_x\sigma_z + \sigma_y\sigma_z - \tau_{xy}^2 - \tau_{yz}^2 - \tau_{zx}^2)\sigma$$
$$- (\sigma_x\sigma_y\sigma_z + 2\tau_{xy}\tau_{yz}\tau_{zx} - \sigma_x\tau_{yz}^2 - \sigma_y\tau_{zx}^2 - \sigma_z\tau_{xy}^2) = 0 \qquad (2\text{-}10)$$

Having solved Eq. (2-10) for a given state of stress, one might obtain a principal stress element like that of Fig. 2-6a.

In plotting Mohr's circles for triaxial stress, the principal stresses are arranged so that $\sigma_1 > \sigma_2 > \sigma_3$. Then the result appears as in Fig. 2-6b. The stress coordinates $\sigma_N \tau_N$ for any arbitrarily located plane will always lie within the shaded area.

Examination of Fig. 2-6 shows that there will be three sets of extreme-value shear stresses, one for each circle. If the circles have different radii, the maximum shear stress will correspond to the largest radius. Thus the maximum shear stress is the largest of the three values

$$\tau = \frac{\sigma_1 - \sigma_2}{2} \qquad \text{or} \qquad \tau = \frac{\sigma_2 - \sigma_3}{2} \qquad \text{or} \qquad \tau = \frac{\sigma_1 - \sigma_3}{2}$$

But if we always arrange the principal stresses in the form $\sigma_1 > \sigma_2 > \sigma_3$ then it is

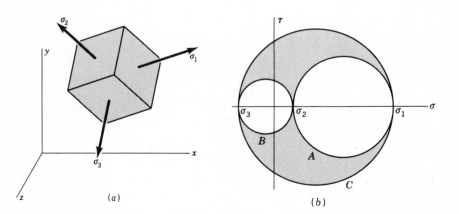

(a)   (b)

FIGURE 2-6   Mohr's circles for triaxial stress.

always true that

$$\tau_{max} = \frac{\sigma_1 - \sigma_3}{2}$$

(2-11)

Program for Two-Dimensional Stress

It is now expedient to develop a more comprehensive computer program for two-dimensional stress states. Such a program will be very useful in many of the chapters to follow. It is first desirable to define

$$\sigma_A = \frac{\sigma_x + \sigma_y}{2} + \tau_1 \qquad \sigma_B = \frac{\sigma_x + \sigma_y}{2} - \tau_1$$

(a)

where τ_1 is the solution to Eq. (2-9) using the positive sign. Now, use the program developed in the previous section to compute σ_A, σ_B, and ϕ, and use these results as input to the flow diagram of Fig. 2-7. If you are using a pro-

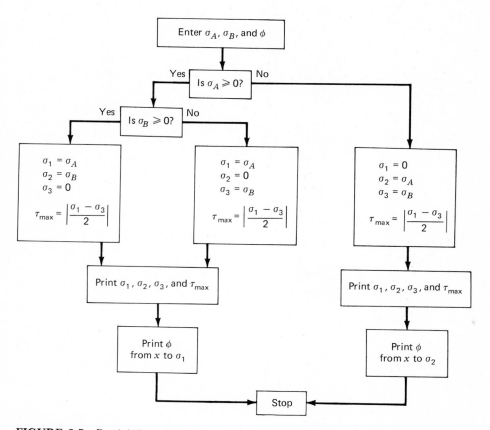

FIGURE 2-7 Partial flow diagram for a comprehensive solution to the two-dimensional stress problem.

Table 2-2 MOHR'S CIRCLE SOLUTIONS
(ROUNDED TO THREE PLACES)

σ_x	σ_y	τ_{xy}	σ_1	σ_2	σ_3	τ_{max}	ϕ
10	2	4 ccw	11.7	0.343	0	5.83	22.5°*
10	−2	4 cw	11.2	0	−3.21	7.21	−16.8°*
−2	−16	4 ccw	0	−0.938	−17.06	8.53	14.9°†

* From x to σ_1
† From x to σ_2

grammable calculator, this program will give you an opportunity to use the decision capability of the machine. The data of Table 2-2 can be used to check all three branches of the flow diagram.

2-4 UNIFORMLY DISTRIBUTED STRESSES

The assumption of a uniform distribution of stress is frequently made in design. The result is then often called *pure tension, pure compression,* or *pure shear,* depending upon how the external load is applied to the body under study. The word *simple* is sometimes used instead of "pure" to indicate that there are no other complicating effects. The tension rod is typical. Here a tension load F is applied through pins at the ends of the bar. The assumption of uniform stress means that if we cut the bar at a section remote from the ends, and remove the lower half, we can replace its effect by applying a uniformly distributed force of magnitude σA to the cut end. So the stress σ is said to be uniformly distributed. It is calculated from the equation

$$\sigma = \frac{F}{A} \tag{2-12}$$

This assumption of uniform stress distribution requires that:

1 The bar be straight and of a homogeneous material
2 The line of action of the force coincide with the centroid of the section
3 The section be taken remote from the ends and from any discontinuity or abrupt change in cross section

The same equation and assumptions hold for simple compression. A slender bar in compression may, however, fail by buckling, and this possibility must be eliminated from consideration before Eq. (2-12) is used.*

---

* See Sec. 3-12.

Use of the equation

$$\tau = \frac{F}{A} \tag{2-13}$$

for a body, say a bolt, in shear assumes a uniform stress distribution too. It is very difficult in practice to obtain a uniform distribution of shear stress; the equation is included because occasions do arise in which this assumption is made.

2-5   ELASTIC STRAIN

When a straight bar is subjected to a tensile load, the bar becomes longer. The amount of stretch, or elongation, is called *strain*. The elongation per unit length of the bar is called *unit strain*. In spite of these definitions, however, it is customary to use the word "strain" to mean "unit strain" and the expression "total strain" to mean total elongation, or deformation, of a member. Using this custom here, the expression for strain is

$$\epsilon = \frac{\delta}{l} \tag{2-14}$$

where δ is the total elongation (total strain) of the bar within the length l.

Shear strain γ is the change in a right angle of a stress element subjected to pure shear.

Elasticity is that property of a material which enables it to regain its original shape and dimensions when the load is removed. Hooke's law states that, within certain limits, the stress in a material is proportional to the strain which produced it. An elastic material does not necessarily obey Hooke's law, since it is possible for some materials to regain their original shape without the limiting condition that stress be proportional to strain. On the other hand, a material which obeys Hooke's law is elastic. For the condition that stress is proportional to strain, we can write

$$\sigma = E\epsilon \tag{2-15}$$

$$\tau = G\gamma \tag{2-16}$$

where E and G are the constants of proportionality. Since the strains are dimensionless numbers, the units of E and G are the same as the units of stress. The constant E is called the *modulus of elasticity*. The constant G is called the *shear modulus of elasticity*, or sometimes, the *modulus of rigidity*. Both E and G, however, are numbers which are indicative of the stiffness or rigidity of the materials. These two constants represent fundamental properties.

By substituting $\sigma = F/A$ and $\epsilon = \delta/l$ into Eq. (2-15) and rearranging, we obtain the equation for the total deformation of a bar loaded in axial tension or compression.

$$\delta = \frac{Fl}{AE} \tag{a}$$

Experiments demonstrate that when a material is placed in tension, there exists not only an axial strain, but also a lateral strain. Poisson demonstrated that these two strains were proportional to each other within the range of Hooke's law. This constant is expressed as

$$\mu = -\frac{\text{lateral strain}}{\text{axial strain}} \tag{2-17}$$

and is known as *Poisson's ratio*. These same relations apply for compression, except that a lateral expansion takes place instead.

The three elastic constants are related to each other as follows:

$$E = 2G(1 + \mu) \tag{2-18}$$

2-6 STRESS-STRAIN RELATIONS

There are many experimental techniques which can be used to measure strain. Thus, if the relationship between stress and strain is known, the stress state at a point can be calculated after the state of strain has been measured. We define the *principal strains* as the strains in the direction of the principal stresses. It is true that the shear strains are zero, just as the shear stresses are zero, on the faces of an element aligned in the principal directions. From Eq. (2-17) the three principal strains corresponding to a state of uniaxial stress are

$$\epsilon_1 = \frac{\sigma_1}{E} \qquad \epsilon_2 = -\mu\epsilon_1 \qquad \epsilon_3 = -\mu\epsilon_1 \tag{2-19}$$

The minus sign is used to indicate compressive strains. Note that, while the stress state is uniaxial, the strain state is triaxial.

Biaxial Stress

For the case of biaxial stress we give σ_1 and σ_2 prescribed values and let σ_3 be zero. The principal strains can be found from Eq. (2-17) if we imagine each principal stress to be acting separately and then combine the results by superposition. This gives

$$\epsilon_1 = \frac{\sigma_1}{E} - \frac{\mu\sigma_2}{E}$$

$$\epsilon_2 = \frac{\sigma_2}{E} - \frac{\mu\sigma_1}{E} \qquad (2\text{-}20)$$

$$\epsilon_3 = -\frac{\mu\sigma_1}{E} - \frac{\mu\sigma_2}{E}$$

Equations (2-20) give the principal strains in terms of the principal stresses. But the usual situation is the reverse. To solve for σ_1, multiply ϵ_2 by μ, and add the first two equations together. This produces

$$\epsilon_1 + \mu\epsilon_2 = \frac{\sigma_1}{E} - \frac{\mu\sigma_2}{E} + \frac{\mu\sigma_2}{E} - \frac{\mu^2\sigma_1}{E}$$

Solving for σ_1 gives

$$\sigma_1 = \frac{E(\epsilon_1 + \mu\epsilon_2)}{1 - \mu^2} \qquad (2\text{-}21)$$

Similarly,

$$\sigma_2 = \frac{E(\epsilon_2 + \mu\epsilon_1)}{1 - \mu^2} \qquad (2\text{-}22)$$

In solving these equations, remember that a tensile stress or strain is treated as positive and that compression is negative.

Triaxial Stress

The state of stress is said to be *triaxial* when none of the three principal stresses is zero. For this, the principal strains are

$$\epsilon_1 = \frac{\sigma_1}{E} - \frac{\mu\sigma_2}{E} - \frac{\mu\sigma_3}{E}$$

$$\epsilon_2 = \frac{\sigma_2}{E} - \frac{\mu\sigma_1}{E} - \frac{\mu\sigma_3}{E} \qquad (2\text{-}23)$$

$$\epsilon_3 = \frac{\sigma_3}{E} - \frac{\mu\sigma_1}{E} - \frac{\mu\sigma_2}{E}$$

In terms of the strains, the principal stresses are

$$\sigma_1 = \frac{E\epsilon_1(1 - \mu) + \mu E(\epsilon_2 + \epsilon_3)}{1 - \mu - 2\mu^2}$$

$$\sigma_2 = \frac{E\epsilon_2(1 - \mu) + \mu E(\epsilon_1 + \epsilon_3)}{1 - \mu - 2\mu^2} \qquad (2\text{-}24)$$

$$\sigma_3 = \frac{E\epsilon_3(1 - \mu) + \mu E(\epsilon_1 + \epsilon_2)}{1 - \mu - 2\mu^2}$$

Values of Poisson's ratio for various materials are given in Table A-7.

2-7 SHEAR AND MOMENT

Figure 2-8a *shows a beam supported by* reactions R_1 and R_2 and loaded by the concentrated forces F_1, F_2, and F_3. The direction chosen for the y axis is the clue to the sign convention for the forces. F_1, F_2, and F_3 are negative because they act in the negative y direction; R_1 and R_2 are positive.

If the beam is cut at some section location at $x = x_1$, and the left-hand portion removed as a free body, an *internal shear force* V and *bending moment* M must act on the cut surface to assure equilibrium. The shear force is obtained by summing the forces to the left of the cut section. The bending moment is the sum of the moments of the forces to the left of the section taken about an axis through the section. Shear force and bending moment are related by the equation

$$V = \frac{dM}{dx} \qquad (2\text{-}25)$$

Sometimes the bending is caused by a distributed load. Then, the relation between shear force and bending moment may be written

$$\frac{dV}{dx} = \frac{d^2 M}{dx^2} = -w \qquad (2\text{-}26)$$

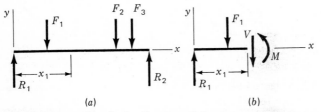

(a)   (b)

FIGURE 2-8 Free-body diagram of simply supported beam with V and M shown in positive directions.

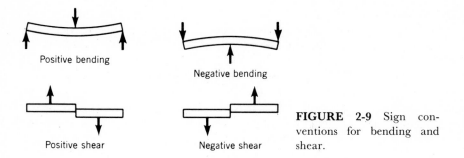

Positive bending

Negative bending

Positive shear

Negative shear

FIGURE 2-9 Sign conventions for bending and shear.

where w is a downward acting load of w units of force per unit length.

The sign conventions used for bending moment and shear force in this book are shown in Fig. 2-9.

The loading w of Eq. (2-26) is uniformly distributed. A more general distribution can be defined by the equation

$$q = \lim_{\Delta x \to 0} \frac{\Delta F}{\Delta x}$$

where q is called the *load intensity*; thus $q = -w$.

Equations (2-25) and (2-26) reveal additional relations if they are integrated. Thus, if we integrate between, say, x_A and x_B, we obtain

$$\int_{V_A}^{V_B} dV = \int_{x_A}^{x_B} q\ dx = V_B - V_A \tag{2-27}$$

which states that *the change in shear force from A to B is equal to the area of the loading diagram between x_A and x_B.*

In a similar manner

$$\int_{M_A}^{M_B} dM = \int_{x_A}^{x_B} V\ dx = M_B - M_A \tag{2-28}$$

which states that *the change in moment from A to B is equal to the area of the shear-force diagram between x_A and x_B.*

Example 2-2 Figure 2-10a shows a beam OC 18 in long loaded by concentrated forces of 50 lb at A and 70 lb at B. Find the reactions R_1 and R_2 and sketch the shear-force and moment diagrams.

Solution Summing moments about O gives $R_2 = 42.2$ lb. Then, summing forces in the y direction yields $R_1 = 77.8$ lb.

To plot the shear-force diagram (Fig. 2-10b) begin at the origin and work to the right. The shear force is $+77.8$ lb from O to A. At A it is reduced by 50 lb to $+27.8$ lb, and it continues with this value to B. At B it is reduced again by 70 lb to -42.2 lb. The positive reaction of 42.2 lb at C brings the shear force back to zero at the end.

To obtain the moment diagram (Fig. 2-10c), remember that the change in moment between any two points is equal to the area of the shear-force diagram between those points. As indicated by the loading diagram, which is a free-body diagram of the entire beam, the moment at O is zero. Therefore the moment at A is the area of the shear diagram between O and A. Thus

$$M_A = (77.8)(4) = 311.2 \text{ lb} \cdot \text{in}$$

Note that this could also be obtained by multiplying the reaction R_1 by the distance from O to A.

The change in moment from A to B is the area of the shear diagram between A and B. Thus

$$M_B - M_A = (27.8)(4) = 111.2 \text{ lb} \cdot \text{in}$$

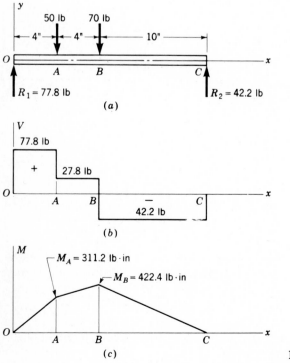

FIGURE 2-10

and

$$M_B = 111.2 + 311.2 = 422.4 \text{ lb} \cdot \text{in}$$

This result can also be obtained by taking moments about B as follows:

$$M_B = -(77.8)(8) + (50)(4) = -422.4 \text{ lb} \cdot \text{in}$$

Here we have obeyed the clockwise-counterclockwise rule for moments. Though the bending moment is negative using this rule, it is positive according to the convention of Fig. 2-9.   ////

2-8 SINGULARITY FUNCTIONS

The five singularity functions defined in Table 2-3 constitute a useful and easy means of integrating across discontinuities. By their use, general expressions for shear force and bending moment in beams can be written when the beam is loaded by concentrated forces or moments. As shown in the table, the concentrated moment and force functions are zero for all values of x except the particular value $x = a$. The unit step, ramp, and parabolic functions are zero only for values of x that are less than a. The integration properties shown in the table constitute a part of the mathematical definition too. The examples which follow show how these functions are used.*

Example 2-3 Derive expressions for the loading, shear-force, and bending-moment diagrams for the beam of Fig. 2-11.

Solution Using Table 2-3 and $q(x)$ for the loading function, we find

$$q = R_1 \langle x \rangle^{-1} - F_1 \langle x - a_1 \rangle^{-1} - F_2 \langle x - a_2 \rangle^{-1} + R_2 \langle x - l \rangle^{-1} \tag{1}$$

Next, we use Eq. (2-27) to get the shear force. Note that $V = 0$ at $x = -\infty$.

$$V = \int_{-\infty}^{x} q \, dx = R_1 \langle x \rangle^0 - F_1 \langle x - a_1 \rangle^0 - F_2 \langle x - a_2 \rangle^0 + R_2 \langle x - l \rangle^0 \tag{2}$$

* For additional examples see S. H. Crandall, Norman C. Dahl, and Thomas J. Lardner (eds.), *An Introduction to the Mechanics of Solids*, 2d ed., McGraw-Hill, New York, 1972, pp. 164–172.

Table 2-3 SINGULARITY FUNCTIONS

Function	Graph of $f_n(x)$	Meaning
Concentrated moment	$\langle x - a \rangle^{-2}$	$\langle x - a \rangle^{-2} = \begin{cases} 1 & x = a \\ 0 & x \neq a \end{cases}$ $\displaystyle\int_{-\infty}^{x} \langle x - a \rangle^{-2}\, dx = \langle x - a \rangle^{-1}$
Concentrated force	$\langle x - a \rangle^{-1}$	$\langle x - a \rangle^{-1} = \begin{cases} 1 & x = a \\ 0 & x \neq a \end{cases}$ $\displaystyle\int_{-\infty}^{x} \langle x - a \rangle^{-1}\, dx = \langle x - a \rangle^{0}$
Unit step	$\langle x - a \rangle^{0}$	$\langle x - a \rangle^{0} = \begin{cases} 0 & x < a \\ 1 & x \geq a \end{cases}$ $\displaystyle\int_{-\infty}^{x} \langle x - a \rangle^{0}\, dx = \langle x - a \rangle^{1}$
Ramp	$\langle x - a \rangle^{1}$	$\langle x - a \rangle^{1} = \begin{cases} 0 & x < a \\ x - a & x \geq a \end{cases}$ $\displaystyle\int_{-\infty}^{x} \langle x - a \rangle^{1}\, dx = \dfrac{\langle x - a \rangle^{2}}{2}$
Parabolic	$\langle x - a \rangle^{2}$	$\langle x - a \rangle^{2} = \begin{cases} 0 & x < a \\ (x - a)^{2} & x \geq a \end{cases}$ $\displaystyle\int_{-\infty}^{x} \langle x - a \rangle^{2}\, dx = \dfrac{\langle x - a \rangle^{3}}{3}$

A second integration, in accordance with Eq. (2-28), yields

$$M = \int_{-\infty}^{x} V\, dx = R_1 \langle x \rangle^{1} - F_1 \langle x - a_1 \rangle^{1} - F_2 \langle x - a_2 \rangle^{1} + R_2 \langle x - l \rangle^{1} \qquad (3)$$

The reactions R_1 and R_2 can be found by taking a summation of moments and forces as usual, or they can be found by noting that the shear force and bending moment must be zero everywhere except in the region $0 \leq x \leq l$. This means that

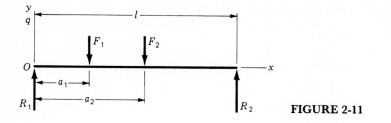

FIGURE 2-11

Eq. (2) should give $V = 0$ at x slightly larger than l. Thus

$$R_1 - F_1 - F_2 + R_2 = 0 \tag{4}$$

Since the bending moment should also be zero in the same region, we have from Eq. (3),

$$R_1 l - F_1(l - a_1) - F_2(l - a_2) = 0 \tag{5}$$

Equations (4) and (5) can now be solved for the reactions R_1 and R_2. ////

Example 2-4 Figure 2-12a shows the loading diagram for a beam cantilevered at O and having a uniform load w acting on the portion $a \le x \le l$. Derive the shear force and moment relations. M_1 and R_1 are the support reactions.

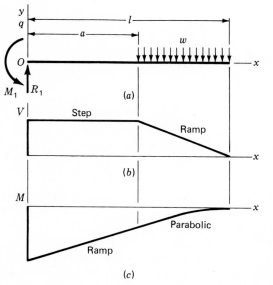

FIGURE 2-12 (a) Loading diagram for a beam cantilevered at O; (b) shear-force diagram; (c) bending-moment diagram.

Solution Following the procedure of Example 2-3 we find the loading function to be

$$q = -M_1\langle x \rangle^{-2} + R_1\langle x \rangle^{-1} - w\langle x - a \rangle^0 \tag{1}$$

Then integrating successively gives

$$V = \int_{-\infty}^{x} q\, dx = -M_1\langle x \rangle^{-1} + R_1\langle x \rangle^0 - w\langle x - a \rangle^1 \tag{2}$$

$$M = \int_{-\infty}^{x} V\, dx = -M_1\langle x \rangle^0 + R_1\langle x \rangle^1 - \frac{w}{2}\langle x - a \rangle^2 \tag{3}$$

The reactions are found by making x slightly larger than l because both V and M are zero in this region. Equation (2) will then give

$$-M_1(0) + R_1 - w(l - a) = 0 \tag{4}$$

which can be solved for R_1. From Eq. (3) we get

$$-M_1 + R_1 l - \frac{w}{2}(l - a)^2 = 0 \tag{5}$$

which can be solved for M_1. Figures 2-12*b* and 2-12*c* show the shear-force and bending-moment diagrams.   ////

2-9 NORMAL STRESSES IN BENDING

In deriving the relations for the normal bending stresses in beams, we make the following idealizations:

1   The beam is subjected to pure bending; this means that the shear force is zero, and that no torsion or axial loads are present.
2   The material is isotropic and homogeneous.
3   The material obeys Hooke's law.
4   The beam is initially straight with a cross section that is constant throughout the beam length.
5   The beam has an axis of symmetry in the plane of bending.
6   The proportions of the beam are such that it would fail by bending rather than by crushing, wrinkling, or sidewise buckling.
7   Cross sections of the beam remain plane during bending.

In Fig. 2-13*a* we visualize a portion of a beam acted upon by the positive bending moment M. The y axis is the axis of symmetry. The x axis is coincident

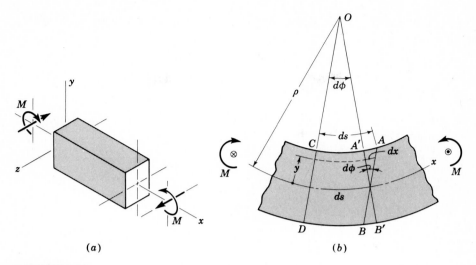

FIGURE 2-13

with the *neutral axis* of the section, and the xz plane, which contains the neutral axes of all cross sections, is called the *neutral plane*. Elements of the beam coincident with this plane have zero strain. The location of the neutral axis with respect to the cross section has not yet been defined.

Application of the positive moment will cause the upper surface of the beam to bend downward, and the neutral axis will then be curved, as shown in Fig. 2-13b. Because of the curvature, a section AB originally parallel to CD, since the beam was straight, will rotate through the angle $d\phi$ to $A'B'$. Since AB and $A'B'$ are both straight lines, we have utilized the assumption that plane sections remain plane during bending. If we now specify the radius of curvature of the neutral axis as ρ, the length of a differential element of the neutral axis as ds, and the angle subtended by the two adjacent sides CD and $A'B'$ as $d\phi$, then, from the definition of curvature, we have

$$\frac{1}{\rho} = \frac{d\phi}{ds} \qquad (a)$$

As shown in Fig. 2-13b, the deformation of a "fiber" at distance y from the neutral axis is

$$dx = y \, d\phi \qquad (b)$$

The strain is the deformation divided by the original length, or

$$\epsilon = -\frac{dx}{ds} \qquad (c)$$

where the negative sign indicates compression. Solving Eqs. (a), (b), and (c) simultaneously gives

$$\epsilon = -\frac{y}{\rho} \tag{d}$$

Thus the strain is proportional to the distance y from the neutral axis. Now, since $\sigma = E\epsilon$, we have for the stress

$$\sigma = -\frac{Ey}{\rho} \tag{e}$$

We are now dealing with pure bending, which means that there are no axial forces acting on the beam. We can state this in mathematical form by summing all the horizontal forces acting on the cross section and equating this sum to zero. The force acting on an element of area dA is $\sigma\, dA$; therefore

$$\int \sigma\, dA = -\frac{E}{\rho} \int y\, dA = 0 \tag{f}$$

Equation (f) defines the location of the neutral axis. The moment of the area about the neutral axis is zero, and hence the neutral axis passes through the centroid of the cross-sectional area.

Next we observe that equilibrium requires that the internal bending moment created by the stress σ must be the same as the external moment M. In other words,

$$M = \int y\sigma\, dA = \frac{E}{\rho} \int y^2\, dA \tag{g}$$

The second integral in Eq. (g) is the moment of inertia of the area about the z axis. That is,

$$I = \int y^2\, dA \tag{2-29}$$

This is called the *area moment of inertia*; since area cannot truly have inertia, it should not be confused with *mass moment of inertia*.

If we next solve Eqs. (g) and $(2-29)$ and rearrange them, we have

$$\frac{1}{\rho} = \frac{M}{EI} \tag{2-30}$$

This is an important equation in the determination of the deflection of beams, and

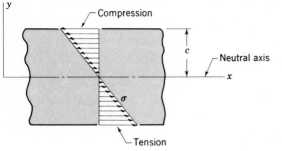

Compression

Neutral axis

σ

Tension

FIGURE 2-14

we shall employ it in Chap. 3. Finally, we eliminate ρ from Eqs. (e) and $(2\text{-}30)$ and obtain

$$\sigma = -\frac{My}{I} \tag{2-31}$$

Equation $(2\text{-}31)$ states that the bending stress σ is directly proportional to the distance y from the neutral axis and the bending moment M, as shown in Fig. 2-14. It is customary to designate $c = y_{max}$, to omit the negative sign, and to write

$$\sigma = \frac{Mc}{I} \tag{2-32}$$

where it is understood that Eq. $(2\text{-}32)$ gives the maximum stress. Then tensile or compressive maximum stresses are determined by inspection when the sense of the moment is known.

Equation $(2\text{-}32)$ is often written in the two alternative forms

$$\sigma = \frac{M}{I/c} \qquad \sigma = \frac{M}{Z} \tag{2-33}$$

where $Z = I/c$ is called the *section modulus*.

Example 2-5 A beam having a T section with the dimensions shown in Fig. 2-15 is subjected to a bending moment of 14 kip·in. Locate the neutral axis and find the maximum tensile and compressive bending stresses. The sign of the moment produces tension at the top surface.

Solution The area of the composite section is $A = 3.25$ in². Now divide the T section into two rectangles numbered 1 and 2 and sum the moments of these areas

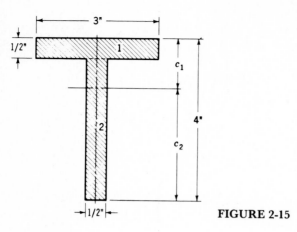

FIGURE 2-15

about the top edge. We then have

$$3.25c_1 = (\tfrac{1}{2})(3)(\tfrac{1}{4}) + (\tfrac{1}{2})(3\tfrac{1}{2})(2\tfrac{1}{4})$$

and hence $c_1 = 1.327$ in. Therefore $c_2 = 4 - 1.327 = 2.673$ in. Note that we have chosen not to round off these values.

Next we calculate the moment of inertia of each rectangle about its own centroidal axis. Using Table A-14, we find for the top rectangle

$$I_1 = \frac{bh^3}{12} = \frac{(3)(\tfrac{1}{2})^3}{12} = 0.0312 \text{ in}^4$$

Similarly, for the bottom rectangle,

$$I_2 = \frac{bh^3}{12} = \frac{(\tfrac{1}{2})(3\tfrac{1}{2})^3}{12} = 1.787 \text{ in}^4$$

We now employ the parallel-axis theorem to obtain the moment of inertia of the composite figure about its own centroidal axis. This theorem states

$$I_x = I_{cg} + Ad^2$$

where I_{cg} is the moment of inertia of the area about its own centroidal axis and I_x is the moment of inertia about any parallel axis a distance d removed. For the top rectangle, the distance is

$$d_1 = c_1 - \tfrac{1}{4} = 1.327 - 0.25 = 1.077 \text{ in}$$

And for the bottom rectangle,

$$d_2 = c_2 - 1\tfrac{3}{4} = 2.673 - 1.75 = 0.923 \text{ in}$$

Using the parallel-axis theorem twice, we now find

$$I = (I_1 + A_1 d_1^2) + (I_2 + A_2 d_2^2)$$
$$= [0.0312 + (1.5)(1.077)^2] + [1.787 + (1.75)(0.923)^2] = 5.048 \text{ in}^4$$

Finally then, the maximum tensile stress, which occurs at the top surface, is found to be

$$\sigma = \frac{Mc_1}{I} = \frac{14(10)^3(1.327)}{5.048} = 3680 \text{ psi}$$

Similarly, the maximum compressive stress at the lower surface is found to be

$$\sigma = -\frac{Mc_2}{I} = -\frac{14(10)^3(2.673)}{5.048} = -7400 \text{ psi} \qquad\qquad ////$$

Example 2-6 Determine a diameter for the solid round shaft, 18 in long, shown in Fig. 2-16. The shaft is supported by self-aligning bearings at the ends. Mounted upon the shaft are a V-belt sheave, which contributes a radial load of 400 lb to the shaft, and a gear, which contributes a radial load of 150 lb. Both loads are in the same plane and have the same directions. The bending stress is not to exceed 10 kpsi.

Solution As indicated in Chap. 1, certain assumptions are necessary. In this problem we decide on the following conditions:

1 The weight of the shaft is neglected.
2 Since the bearings are self-aligning, the shaft is assumed to be simply supported and the loads and bearing reactions to be concentrated.
3 The normal bending stress is assumed to govern the design.

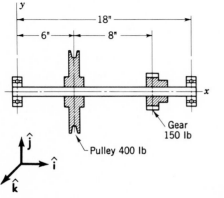

FIGURE 2-16

The assumed loading diagram is shown in Fig. 2-17a. Taking moments about C,

$$\sum M_c = -18R_1 + (12)(400) + (4)(150) = 0$$
$$R_1 = 300 \text{ lb}$$

Next

$$\sum M_0 = -(6)(400) - (14)(150) + 18R_2 = 0$$
$$R_2 = 250 \text{ lb}$$

This statics problem can also be solved by vectors. We define the unit vector triad **ijk** associated with the xyz axes, respectively. Then position vectors from the origin O to loading points A, B, and C are

$$\mathbf{r}_A = 6\mathbf{i} \qquad \mathbf{r}_B = 14\mathbf{i} \qquad \mathbf{r}_C = 18\mathbf{i}$$

Also, the force vectors are

$$\mathbf{R}_1 = R_1\mathbf{j} \qquad \mathbf{F}_A = -400\mathbf{j} \qquad \mathbf{F}_B = -150\mathbf{j} \qquad \mathbf{R}_2 = R_2\mathbf{j}$$

Then taking moments about O yields the vector equation

$$\mathbf{r}_A \times \mathbf{F}_A + \mathbf{r}_B \times \mathbf{F}_B + \mathbf{r}_C \times \mathbf{R}_2 = 0 \tag{1}$$

The terms in Eq. (1) are called *vector cross products*. The cross product of the two vectors

$$\mathbf{A} = x_A\mathbf{i} + y_A\mathbf{j} + z_A\mathbf{k} \qquad \mathbf{B} = x_B\mathbf{i} + y_B\mathbf{j} + z_B\mathbf{k}$$

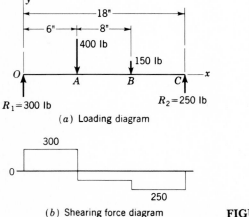

(a) Loading diagram

(b) Shearing force diagram

FIGURE 2-17

is

$$\mathbf{A} \times \mathbf{B} = (y_A z_B - z_A y_B)\mathbf{i} + (z_A x_B - x_A z_B)\mathbf{j} + (x_A y_B - y_A x_B)\mathbf{k}$$

But this relation is more conveniently viewed as the determinant

$$\mathbf{A} \times \mathbf{B} = \begin{vmatrix} \mathbf{i} & \mathbf{j} & \mathbf{k} \\ x_A & y_A & z_A \\ x_B & y_B & z_B \end{vmatrix}$$

Thus we compute the terms in Eq. (1) as follows:

$$\mathbf{r}_A \times \mathbf{F}_A = \begin{vmatrix} \mathbf{i} & \mathbf{j} & \mathbf{k} \\ 6 & 0 & 0 \\ 0 & -400 & 0 \end{vmatrix} = -2400\mathbf{k}$$

$$\mathbf{r}_B \times \mathbf{F}_B = \begin{vmatrix} \mathbf{i} & \mathbf{j} & \mathbf{k} \\ 14 & 0 & 0 \\ 0 & -150 & 0 \end{vmatrix} = -2100\mathbf{k}$$

$$\mathbf{r}_C \times \mathbf{R}_2 = \begin{vmatrix} \mathbf{i} & \mathbf{j} & \mathbf{k} \\ 18 & 0 & 0 \\ 0 & R_2 & 0 \end{vmatrix} = 18R_2 \mathbf{k}$$

Substituting these three terms into Eq. (1) and solving the resulting algebraic equation gives $R_2 = 250$ lb. Hence $\mathbf{R}_2 = 250\mathbf{j}$. Next we write

$$\mathbf{R}_1 = -\mathbf{F}_A - \mathbf{F}_B - \mathbf{R}_2 = -(-400\mathbf{j}) - (-150\mathbf{j}) - 250\mathbf{j} = 300\mathbf{j} \text{ lb}$$

In this instance the vector approach is more laborious. It is probably an advantage, however, for three-dimensional problems and also for programming the digital computer.

The next step is to draw the shear-force diagram (Fig. 2-17b). The maximum bending moment occurs at the point of zero shear force. Its value is

$$M = (300)(6) = 1800 \text{ lb} \cdot \text{in}$$

In view of the assumptions, Eq. (2-33) applies. We first compute the section modulus. From Table A-14,

$$\frac{I}{c} = \frac{\pi d^3}{32} = 0.0982 d^3 \tag{2}$$

Then, using Eq. (2-33),

$$\sigma = \frac{M}{I/c} = \frac{1800}{0.0982 d^3}$$

Substituting $\sigma = 10\ 000$ psi and solving for d yields

$$d = \sqrt[3]{\frac{1800}{(0.0982)(10\ 000)}} = 1.22 \text{ in}$$

Therefore we select $d = 1\frac{1}{4}$ in for the shaft diameter.   ////

2-10   VECTOR PROGRAMMING

If you have access to any kind of programmable computational facility, you should create the following programs for use. Since these problems are short, they can all be formed as subroutines and stored on a single magnetic card or other storage facility. Program flags or conditional transfers facilitate entering the various subroutines.   The following problems are suggested for inclusion:

1   Given $R\,\underline{/\theta}$, find $x\mathbf{i} + y\mathbf{j}$.
2   Given $x\mathbf{i} + y\mathbf{j}$, find $R\,\underline{/\theta}$.
3   Given θ, find $\hat{\mathbf{R}} = \bar{x}\mathbf{i} + \bar{y}\mathbf{j}$, where $\hat{\mathbf{R}}$ is a unit vector and $\bar{x}$ and $\bar{y}$ are the direction cosines.
4   Given $\mathbf{F}_1, \mathbf{F}_2, \mathbf{F}_3, \ldots$, in x, y, and z components, find $\sum \mathbf{F}$.
5   Given $\mathbf{C}$ and $\mathbf{C}'$ in x, y, and z components, find $\mathbf{C} \times \mathbf{C}'$.

Example 2-7   For the loading diagram shown in Fig. 2-18 find the reactions R_1 and R_2 and the value and location of the maximum bending moment.

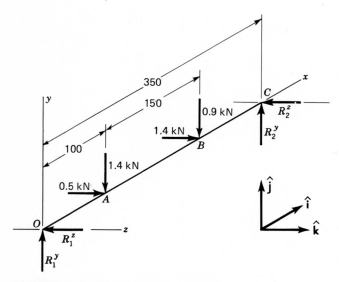

FIGURE 2-18   Dimensions in millimeters.

Solution Examination of the forces at A and B in Fig. 2-18 indicates that the reactions R_1^y and R_2^y will be in the positive y direction, and the reactions R_1^z and R_2^z will be in the negative z direction, and they are so shown. In writing equations, however, it is convenient to assume positive senses for all unknown quantities.

Using the vector approach, we first define the position vectors

$$\mathbf{r}_A = 100\mathbf{i} \qquad \mathbf{r}_B = 250\mathbf{i} \qquad \mathbf{r}_{BA} = 150\mathbf{i} \qquad \mathbf{r}_C = 350\mathbf{i}$$

where $\mathbf{r}_{BA}$ is the position of B as seen by an observer at A.

The forces and reactions expressed in vector form are

$$\mathbf{R}_1 = R_1^y\mathbf{j} + R_1^z\mathbf{k} \qquad \mathbf{F}_A = -1.4\mathbf{j} + 0.5\mathbf{k} \qquad \mathbf{R}_2 = R_2^y\mathbf{j} + R_2^z\mathbf{k}$$

$$\mathbf{F}_B = -0.9\mathbf{j} + 1.4\mathbf{k}$$

Then summing moments about the origin gives the vector expression

$$\sum \mathbf{M}_O = \mathbf{r}_A \times \mathbf{F}_A + \mathbf{r}_B \times \mathbf{F}_B + \mathbf{r}_C \times \mathbf{R}_2 = 0 \tag{1}$$

The cross products are computed as

$$\mathbf{r}_A \times \mathbf{F}_A = -50\mathbf{j} - 140\mathbf{k}$$

$$\mathbf{r}_B \times \mathbf{F}_B = -350\mathbf{j} - 225\mathbf{k}$$

$$\mathbf{r}_C \times \mathbf{R}_2 = -350R_2^z\mathbf{j} + 350R_2^y\mathbf{k}$$

Solving Eq. (1) for the unknown $\mathbf{R}_2$ gives

$$\mathbf{R}_2 = 1.043\mathbf{j} - 1.143\mathbf{k} \text{ kN} \qquad |\mathbf{R}_2| = 1.547 \text{ kN} \qquad\qquad \textit{Ans.}$$

Next, for equilibrium of forces, we have

$$\sum \mathbf{F} = \mathbf{R}_1 + \mathbf{F}_A + \mathbf{F}_B + \mathbf{R}_2 = 0 \tag{2}$$

When this equation is solved for $\mathbf{R}_1$ we get

$$\mathbf{R}_1 = 1.257\mathbf{j} - 0.757\mathbf{k} \text{ kN} \qquad |\mathbf{R}_1| = 1.467 \text{ kN} \qquad\qquad \textit{Ans.}$$

The maximum bending moment may occur at A or at B. The moment at A is

$$\mathbf{M}_A = -\mathbf{r}_A \times \mathbf{R}_1 = -75.7\mathbf{j} - 125.7\mathbf{k} \text{ N} \cdot \text{m} \qquad |\mathbf{M}_A| = 146.7 \text{ N} \cdot \text{m}$$

The moment at B is

$$\mathbf{M}_B = -\mathbf{r}_B \times \mathbf{R}_1 - \mathbf{r}_{BA} \times \mathbf{F}_A = -114.2\mathbf{j} - 104.2\mathbf{k} \; N \cdot m$$

And so the maximum moment is

$$|\mathbf{M}_B| = 154.7 \; N \cdot m \qquad\qquad\qquad\qquad\qquad\qquad\qquad Ans.$$

These computations can be verified by drawing two shear-force diagrams, one for the xz plane and one for the yz plane, and computing the two corresponding moment diagrams.   ////

2-11 BEAMS WITH UNSYMMETRICAL SECTIONS

The relations developed in Sec. 2-9 can also be applied to beams having unsymmetrical sections, provided that the plane of bending coincides with one of the two principal axes of the section.   We have found that the stress at a distance y from the neutral axis is

$$\sigma = -\frac{Ey}{\rho} \qquad\qquad\qquad\qquad\qquad\qquad\qquad\qquad (a)$$

Therefore, the force on the element of area dA in Fig. 2-19 is

$$dF = \sigma \, dA = -\frac{Ey}{\rho} \, dA$$

Taking moments of this force about the y axis and integrating across the section gives

$$M_y = \int z \, dF = \int \sigma z \, dA = -\frac{E}{\rho} \int yz \, dA \qquad\qquad\qquad (b)$$

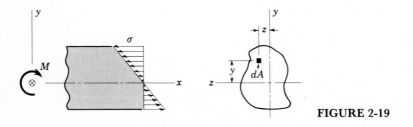

FIGURE 2-19

We recognize the last integral in Eq. (*b*) as the product of inertia I_{yz}. If the bending moment on the beam is in the plane of one of the principal axes, then

$$I_{yz} = \int yz \, dA = 0$$

$$(c)$$

With this restriction, the relations developed in Sec. 2-9 hold for any cross-sectional shape. Of course, this means that the designer has a special responsibility to assure that the bending loads do, in fact, come onto the beam in a principal plane.

2-12 SHEAR STRESSES IN BEAMS

We learned in solving the problems in Secs. 2-7 and 2-8 that most beams have *both* shear forces and bending moments present. It is only occasionally that we encounter beams subjected to pure bending, that is to say, beams having zero shear force. And yet, the flexure formula was developed using the assumption of pure bending. As a matter of fact, the reason for assuming pure bending was simply to eliminate the complicating effects of shear force in the development. For engineering purposes, the flexure formula is valid no matter whether a shear force is present or not. For this reason, we shall utilize the same normal bending-stress distribution [Eqs. (2-31) and (2-32)] when shear forces are present too.

In Fig. 2-20, we show a beam of constant cross section subjected to a shear

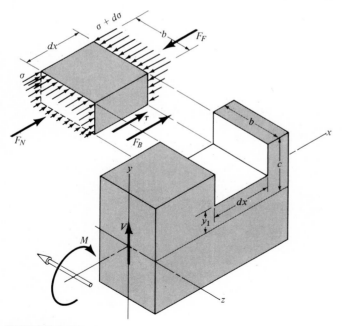

FIGURE 2-20

force V and a bending moment M. The direction of the bending moment is easier to visualize by associating the hollow vector with your right hand. The hollow vector points in the negative z direction. If you will place the thumb of your right hand in the negative z direction, then your fingers, when bent, will indicate the directon of the moment M. By Eq. (2-25), the relationship of V to M is

$$V = \frac{dM}{dx} \tag{a}$$

At some point along the beam, we cut a transverse section of length dx down to a distance y_1 above the neutral axis, as illustrated. We remove this section to study the forces which act. Because a shear force is present, the bending moment is changing as we move along the x axis. Thus, we can designate the bending moment as M on the near side of the section and as $M + dM$ on the far side. The moment M produces a normal stress σ and the moment $M + dM$ a normal stress $\sigma + d\sigma$, as shown. These normal stresses produce normal forces on the vertical faces of the element, the compressive force on the far side being greater than on the near side. The resultant of these two would cause the section to tend to slide in the $-x$ direction, and so this resultant must be balanced by a shear force acting in the $+x$ direction on the bottom of the section. This shear force results in a shear stress τ, as shown. Thus, there are three resultant forces acting on the element: F_N, due to σ, acts on the near face; F_F, due to $\sigma + d\sigma$, acts on the far face; and F_B, due to τ, acts on the bottom face. Let us evaluate these forces.

For the near face, select an element of area dA. The stress acting on this area is σ, and so the force is the stress times the area, or $\sigma\, dA$. The force acting on the entire near face is the sum of all the $\sigma\, dA$'s, or

$$F_N = \int_{y_1}^{c} \sigma\, dA \tag{b}$$

where the limits indicate that we integrate from the bottom $y = y_1$ to the top $y = c$. Using $\sigma = My/I$ [Eq. (2-31)], Eq. (b) becomes

$$F_N = \frac{M}{I} \int_{y_1}^{c} y\, dA \tag{c}$$

The force on the far face is found in a similar manner. It is

$$F_F = \int_{y_1}^{c} (\sigma + d\sigma)\, dA = \frac{M + dM}{I} \int_{y_1}^{c} y\, dA \tag{d}$$

The force on the bottom face is the shear stress τ times the area of the bottom face. Since this area is $b\, dx$, we have

$$F_B = \tau b\, dx \tag{e}$$

Summing these three forces in the x direction gives

$$\sum F_x = +F_N - F_F + F_B = 0 \tag{f}$$

If we substitute Eqs. (c) and (d) for F_N and F_F and solve the result for F_B, we get

$$F_B = F_F - F_N = \frac{M + dM}{I} \int_{y_1}^{c} y \, dA - \frac{M}{I} \int_{y_1}^{c} y \, dA = \frac{dM}{I} \int_{y_1}^{c} y \, dA \tag{g}$$

Next, using Eq. (e) for F_B and solving for the shear stress gives

$$\tau = \frac{dM}{dx} \frac{1}{Ib} \int_{y_1}^{c} y \, dA \tag{h}$$

Then, by the use of Eq. (a), we finally get the shear-stress formula as

$$\tau = \frac{V}{Ib} \int_{y_1}^{c} y \, dA \tag{2-34}$$

In this equation, the integral is the first moment of the area of the vertical face about the neutral axis. This moment is usually designated as Q. Thus,

$$Q = \int_{y_1}^{c} y \, dA \tag{2-35}$$

With this final simplification, Eq. (2-34) may be written as

$$\tau = \frac{VQ}{Ib} \tag{2-36}$$

In using this equation, note that b is the width of the section at the particular distance y_1 from the neutral axis. Also, I is the moment of inertia of the entire section about the neutral axis.

2-13   SHEAR STRESSES IN RECTANGULAR-SECTION BEAMS

The purpose of this section is to show how the equations of the preceding section are used to find the shear-stress distribution in a beam having a rectangular cross section. Figure 2-21 shows a portion of a beam subjected to a shear force V and a bending moment M. As a result of the bending moment, a normal stress σ is developed on a cross section such as A-A, which is in compression above the neutral axis and in tension below. To investigate the shear stress at a distance y_1 above the neutral axis, we select an element of area dA at a distance y above the

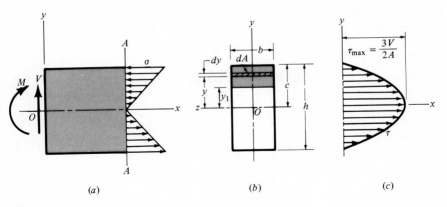

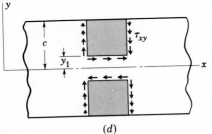

$$(d)$$

FIGURE 2-21

neutral axis. Then, $dA = b\ dy$, and so Eq. (2-35) becomes

$$Q = \int_{y_1}^{c} y\ dA = b \int_{y_1}^{c} y\ dy = \frac{by^2}{2} \Big|_{y_1}^{c} = \frac{b}{2}(c^2 - y_1^2) \qquad (a)$$

Substituting this value for Q into Eq. (2-36) gives

$$\tau = \frac{V}{2I}(c^2 - y_1^2) \qquad (2\text{-}37)$$

This is the general equation for the shear stress in a rectangular beam. To learn something about it, let us make some substitutions. From Table A-14, we learn that the moment of inertia for a rectangular section is $I = bh^3/12$; substituting $h = 2c$ and $A = bh = 2bc$ gives

$$I = \frac{Ac^2}{3} \qquad (b)$$

If we now use this value of I for Eq. (2-37), and rearrange, we get

$$\tau = \frac{3V}{2A}\left(1 - \frac{y_1^2}{c^2}\right) \qquad (2\text{-}38)$$

Table 2-4

y_1	τ
0	$1.500V/A$
$c/4$	$1.406V/A$
$c/2$	$1.125V/A$
$3c/4$	$1.656V/A$
c	0

Now let us substitute various values of y_1, beginning with $y_1 = 0$ and ending with $y_1 = c$. The results are displayed in Table 2-4. We note that the maximum shear stress exists with $y_1 = 0$, which is at the neutral axis. Thus,

$$\tau_{max} = \frac{3V}{2A} \tag{2-39}$$

for a rectangular section. As we move away from the neutral axis, the shear stress decreases until it is zero at the outer surface where $y_1 = c$. This is a parabolic distribution and is shown in Fig. 2-21c. It is particularly interesting and significant here to observe that the shear stress is maximum at the neutral axis where the normal stress, due to bending, is zero, and that the shear stress is zero at the outer surfaces where the bending stress is a maximum. Since horizontal shear stress is always accompanied by vertical shear stress, the distribution can be diagrammed as shown in Fig. 2-21d.

2-14 FLEXURAL SHEAR FOR OTHER SHAPES

The shear stress due to bending for a solid circular beam can be found from Eq. (2-34) as

$$\tau_{max} = \frac{4V}{3A} \tag{2-40}$$

and for a hollow circular section

$$\tau_{max} = \frac{2V}{A} \tag{2-41}$$

A good approximation for structural W and S shapes is given by

$$\tau_{max} = \frac{V}{A_w} \tag{2-42}$$

where A_w is the area of the web. The structural industry* uses various symbols to designate hot-rolled shapes. The symbol W designates a *wide-flange* shape. Thus a W10 × 45 is a hot-rolled steel shape, used for beams and columns, having a section depth of 10 in, a flange width of 8 in, and a weight of 45 lb/ft. An S is used to designate an *eye-beam*, which is similar to a wide-flange beam except it has a narrower flange. C designates a channel section.

2-15 SHEAR CENTER

When thin open sections are used for beams, a local failure, such as buckling, is always a possibility and should be investigated. Open sections have only one axis of symmetry, and if the plane of bending is perpendicular to this axis (Fig. 2-22), the possibility arises that the bending force may cause the beam to twist. If the plane of bending can be made to pass through the *shear center*, then twisting can be avoided.

The twisting effect can be explained by finding the shear center for a channel section. Figure 2-23 shows a channel section used as a cantilever and loaded on the end by a force P acting along the center line of the web. Point O is the shear center whose location e we wish to find. The force P produces a torque $T = Pe$, which is balanced by a countertorque $T = Hh$ at some section ab, at distance x from P. A bending moment $M = Px$ at section ab is resisted by the couple Fh. In order to simplify the analysis, we now make the following assumptions.

1   The flanges and web of the channel are assumed to be thin so that computations can be made based on the centerline dimensions.
2   The flexure formulas developed in the previous sections are assumed to apply even though there is no vertical axis of symmetry.
3   It is assumed that only the flanges resist the torsion through the development of the resisting couple Hh.
4   The vertical shear force $V = P$ is assumed to be resisted entirely by the web.
5   The bending stress σ is assumed to be resisted entirely by the couple Fh generated in the flanges.

Now consider an element of the top flange of length dx and width b, as shown in Fig. 2-23. A force F exists on the near face of this element and a force $F + dF$ on the far face, because the bending moment increases as x increases. The torsion Hh, remember we assumed this to be carried by the flanges, produces a shear stress τ on the near face of the element. This must be in equilibrium with another shear stress τ on the right face of the element. Summing forces in the x direction for the

* For an explanation and a list of standard sizes, see *Steel Construction Manual*, published by the American Institute of Steel Construction (AISC), New York; new editions of this reference are published frequently.

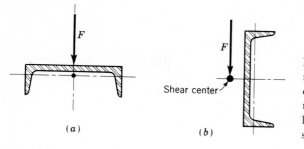

element then gives

$$\sum F_x = -F - \tau t_f \, dx + (F + dF) = 0 \qquad (a)$$

where t_f is the thickness of the flange. Solving gives

$$\tau = \frac{1}{t_f} \frac{dF}{dx} \qquad (b)$$

Now, the bending stress at a distance x from P is

$$\sigma = \frac{Mc}{I} = \frac{M(h/2)}{I} \qquad (c)$$

Remembering the fifth assumption, we write

$$F = \sigma b t_f \qquad (d)$$

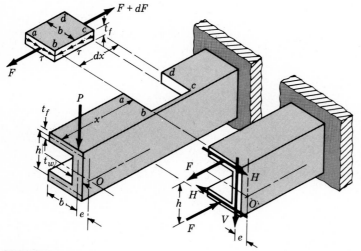

FIGURE 2-23

Combining Eqs. (c) and (d) yields

$$F = \frac{Mhbt_f}{2I} \qquad\qquad (e)$$

Taking the derivative of F with respect to M gives

$$dF = \frac{hbt_f}{2I}\, dM \qquad\qquad (f)$$

where dM is the change in moment from section ab to section cd. Now substitute dF from Eq. (f) into Eq. (b) and solve for the shear stress.

$$\tau = \frac{1}{t_f}\frac{dF}{dx} = \frac{hb}{2I}\frac{dM}{dx} \qquad\qquad (g)$$

From Eq. (2-25) we now find

$$\tau = V\frac{hb}{2I} \qquad\qquad (2\text{-}43)$$

Now, accept the fact that the shear stress τ on the face ab of the element changes from zero at a to a maximum at b. Therefore, the force H is

$$H = \frac{\tau}{2}\, bt_f = \frac{Vhb^2 t_f}{4I} \qquad\qquad (h)$$

Then the twisting moment is

$$T = Pe = Hh = \frac{Vh^2 b^2 t_f}{4I} \qquad\qquad (i)$$

and so

$$e = \frac{h^2 b^2 t_f}{4I} \qquad\qquad (2\text{-}44)$$

because $V = P$.

Note particularly that e is the distance from the *center of the web* to the shear center. Don't confuse this with the distance to the centroidal axis. They are different. See axis 2-2 in the figure for Table A-11.

2-16 TORSION

Any moment vector that is collinear with an axis of a mechanical element is called a *torque vector*, because the moment causes the element to be twisted about that axis. A bar subjected to such a moment is also said to be in *torsion*.

As shown in Fig. 2-24 the torque T applied to a bar can be designated by

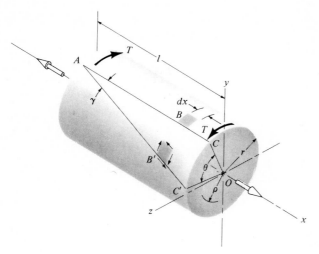

FIGURE 2-24

drawing arrows on the surface of the bar to indicate direction or by drawing torque vectors along the axes of twist of the bar. Torque vectors are the hollow arrows shown on the x axis in Fig. 2-24. Note that they conform to the right-hand rule.

The angle of twist for a round bar is

$$\theta = \frac{Tl}{GJ} \tag{2-45}$$

where T = torque
l = length
G = modulus of rigidity
J = polar area moment of inertia

For a solid round bar, the shear stress is zero at the center and maximum at the surface. The distribution is proportional to the radius ρ and is

$$\tau = \frac{T\rho}{J} \tag{2-46}$$

Designating r as the radius to the outer surface, we have

$$\tau_{\text{max}} = \frac{Tr}{J} \tag{2-47}$$

The assumptions used in the analysis are:

1 The bar is acted upon by a pure torque, and the sections under consideration are remote from the point of application of the load and from a change in diameter.

2 Adjacent cross sections originally plane and parallel remain plane and parallel after twisting, and any radial line remains straight.
3 The material obeys Hooke's law.

Equation (2-47) applies only to circular sections. For a solid round section,

$$J = \frac{\pi d^4}{32} \qquad in^4 \; lb/in^3 = \underline{in \cdot lb} \qquad\qquad T = J\ddot{\theta} \qquad (2\text{-}48)$$
$$s^2$$

where d is the diameter of the bar. For a hollow round section,

$$J = \frac{\pi}{32}\,(d^4 - d_i^4) \qquad\qquad\qquad (2\text{-}49)$$

where d_i is the inside diameter, often referred to as the ID.

In using Eq. (2-47) it is often necessary to obtain the torque T from a consideration of the horsepower and speed of a rotating shaft. For convenience, three forms of this relation are

$$H = \frac{2\pi T n}{(33\ 000)(12)} = \frac{FV}{33\ 000} = \frac{Tn}{63\ 000} \qquad\qquad (2\text{-}50)$$

$$T = \frac{63\ 000 H}{n} \qquad\qquad\qquad (2\text{-}51)$$

where H = horsepower
$\quad\quad\;\; T$ = torque, lb·in
$\quad\quad\;\; n$ = shaft speed, rpm
$\quad\quad\;\; F$ = force, lb
$\quad\quad\;\; V$ = velocity, fpm

If SI units are used, the applicable equation is

$$H = T\omega \qquad\qquad\qquad (2\text{-}52)$$

where H = power, W
$\quad\quad\;\; T$ = torque, N·m
$\quad\quad\;\; \omega$ = angular velocity, radians/s

Determination of the torsional stresses in noncircular members is a difficult problem, generally handled experimentally using a soap-film or membrane analogy, which we shall not consider here. Timoshenko and MacCullough,* how-

---

* S. Timoshenko and Gleason H. MacCullough, *Elements of Strength of Materials*, 3d ed., Van Nostrand, New York, 1949, p. 265. See also F. R. Shanley, *Strength of Materials*, McGraw-Hill, New York, 1957, p. 509.

ever, give the following approximate formula for the maximum torsional stress in a rectangular section.

$$\tau_{max} = \frac{T}{wt^2}\left(3 + 1.8\,\frac{t}{w}\right) \tag{2-53}$$

In this equation w and t are the width and thickness of the bar, respectively; they cannot be interchanged since t must be the shortest dimension. For thin plates in torsion, t/w is small and the second term may be neglected. The equation is also approximately valid for equal-sided angles; these can be considered as two rectangles, each of which is capable of carrying half the torque.

Example 2-8 The carbon-steel countershaft shown in Fig. 2-25 is 25 mm in diameter and supports two V-belt pulleys. The countershaft runs at 1100 rpm. The belt tension on the loose side of pulley A is 15 percent of the tension on the tight side. Except at the bearings, the shaft has a uniform diameter of 25 mm.

(*a*) What torque is transmitted between the pulleys?
(*b*) Compute the angle of twist of the shaft between the two pulleys.
(*c*) Find the maximum torsional stress in the shaft.

Solution (*a*) The torque transmitted is the difference between the two belt pulls

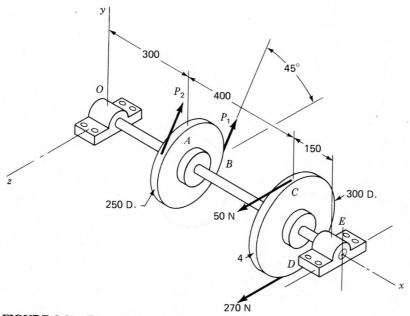

FIGURE 2-25 Dimensions in millimeters.

times the radius of the pulley. If we express the radius in meters, then

$$T = (P_2 - P_1)r = \frac{(270 - 50)(0.300)}{2} = 33 \text{ N} \cdot \text{m} \qquad \textit{Ans.}$$

The same result would have been obtained by expressing the force in kilonewtons and the radius in millimeters.

(*b*) To find the angle of twist, we have $l = 0.400$ m, and $G = 79.3$ GPa from Table A-7. The polar area moment of inertia is best expressed in millimeters. Thus

$$J = \frac{\pi d^4}{32} = \frac{\pi(25)^4}{32} = 38.35(10)^3 \text{ mm}^4$$

Now, since m $=$ mm$(10)^{-3}$, then m$^4 = [\text{mm}(10)^{-3}]^4 = \text{mm}(10)^{-12}$. Also, $G = 79.3(10)^9$ Pa. Noting that a pascal is a newton per square meter, we substitute these basic units into Eq. (2-45) to obtain

$$\theta = \frac{Tl}{GJ} = \frac{(33 \text{ N} \cdot \text{m})(0.400 \text{ m})}{[79.3(10)^9 \text{ Pa}][38.35(10)^3(10)^{-12} \text{ m}^4]}$$

$$= \frac{33(0.400)}{79.3(38.35)} = 0.004\ 34 \text{ rad} \qquad \textit{Ans.}$$

or

$$\theta = 0.249° \qquad\qquad\qquad\qquad\qquad\qquad\qquad\qquad \textit{Ans.}$$

(*c*) For the stress, we use Eq. (2-47) to get

$$\tau_{\text{max}} = \frac{Tr}{J} = \frac{[33(10)^3 \text{ N} \cdot \text{mm}](200 \text{ mm})}{38.35(10)^3 \text{ mm}^4} = 172 \frac{\text{N}}{\text{mm}^2} = 172 \text{ MPa} \qquad \textit{Ans.}$$

But see Table A-4 for other options.   ////

2-17 STRESSES IN CYLINDERS

Cylindrical pressure vessels, hydraulic cylinders, gun barrels, and pipes carrying fluids at high pressures develop both radial and tangential stresses with values which are dependent upon the radius of the element under consideration. In determining the radial stress σ_r and the tangential stress σ_t, we make use of the assumption that the longitudinal elongation is constant around the circumference of the cylinder. In other words, a right section of the cylinder remains plane after stressing.

Referring to Fig. 2-26, we designate the inside radius of the cylinder by *a*, the

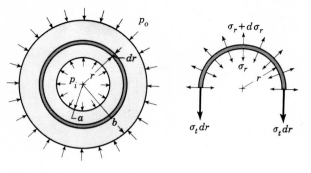

FIGURE 2-26 A cylinder subjected to both internal and external pressure.

outside radius by b, the internal pressure by p_i, and the external pressure by p_o. Let us consider the equilibrium of an infinitely thin semicircular ring cut from the cylinder at radius r and having a unit length. Taking a summation of forces in the vertical direction equal to zero, we have

$$2\sigma_t\, dr + 2\sigma_r r - 2(\sigma_r + d\sigma_r)(r + dr) = 0 \qquad (a)$$

Simplifying and neglecting higher-order quantities gives

$$\sigma_t - \sigma_r - r\frac{d\sigma_r}{dr} = 0 \qquad (b)$$

Equation (b) relates the two unknowns σ_t and σ_r, but we must obtain a second relation in order to evaluate them. The second equation is obtained from the assumption that the longitudinal deformation is constant. Since both σ_t and σ_r are positive for tension, Eq. (2-23) can be written

$$\epsilon_l = -\frac{\mu\sigma_t}{E} - \frac{\mu\sigma_r}{E} \qquad (c)$$

where μ is Poisson's ratio and ϵ_l is the longitudinal unit deformation. Both of these are constants, and so Eq. (c) can be rearranged in the form

$$-\frac{E\epsilon_l}{\mu} = \sigma_t + \sigma_r = 2C_1 \qquad (d)$$

Next, solving Eqs. (b) and (d) to eliminate σ_t produces

$$r\frac{d\sigma_r}{dr} + 2\sigma_r = 2C_1 \qquad (e)$$

Multiplying Eq. (e) by r gives

$$r^2\frac{d\sigma_r}{dr} + 2r\sigma_r = 2rC_1 \qquad (f)$$

We note that

$$\frac{d}{dr}(r^2\sigma_r) = r^2\frac{d\sigma_r}{dr} + 2r\sigma_r \tag{g}$$

Therefore

$$\frac{d}{dr}(r^2\sigma_r) = 2rC_1 \tag{h}$$

which, when integrated, gives

$$r^2\sigma_r = r^2C_1 + C_2 \tag{i}$$

where C_2 is a constant of integration. Solving for σ_r, we obtain

$$\sigma_r = C_1 + \frac{C_2}{r^2} \tag{j}$$

Substituting this value in Eq. (d), we find

$$\sigma_t = C_1 - \frac{C_2}{r^2} \tag{k}$$

To evaluate the constants of integration, note that, at the boundaries of the cylinder,

$$\sigma_r = \begin{cases} -p_i & \text{at } r = a \\ -p_o & \text{at } r = b \end{cases}$$

Substituting these values in Eq. (j) yields

$$-p_i = C_1 + \frac{C_2}{a^2} \qquad -p_o = C_1 + \frac{C_2}{b^2} \tag{l}$$

The constants are found by solving these two equations simultaneously. This gives

$$C_1 = \frac{p_i a^2 - p_o b^2}{b^2 - a^2} \qquad C_2 = \frac{a^2 b^2(p_o - p_i)}{b^2 - a^2} \tag{m}$$

Substituting these values into Eqs. (j) and (k) yields

$$\sigma_t = \frac{p_i a^2 - p_o b^2 - a^2 b^2(p_o - p_i)/r^2}{b^2 - a^2} \tag{2-54}$$

$$\sigma_r = \frac{p_i a^2 - p_o b^2 + a^2 b^2 (p_o - p_i)/r^2}{b^2 - a^2} \tag{2-55}$$

In the above equations positive stresses indicate tension and negative stresses compression.

Let us now determine the stresses when the external pressure is zero. Substitution of $p_o = 0$ in Eqs. (2-54) and (2-55) gives

$$\sigma_t = \frac{a^2 p_i}{b^2 - a^2} \left(1 + \frac{b^2}{r^2} \right) \tag{2-56}$$

$$\sigma_r = \frac{a^2 p_i}{b^2 - a^2} \left(1 - \frac{b^2}{r^2} \right) \tag{2-57}$$

These equations are plotted in Fig. 2-27 to show the distribution of stresses over the wall thickness. The maximum stresses occur at the inner surface, where $r = a$. Their magnitudes are

$$\sigma_t = p_i \frac{b^2 + a^2}{b^2 - a^2} \tag{2-58}$$

$$\sigma_r = -p_i \tag{2-59}$$

The stresses in the outer surface of a cylinder subjected only to external pressure are found similarly. They are

$$\sigma_t = -p_o \frac{b^2 + a^2}{b^2 - a^2} \tag{2-60}$$

$$\sigma_r = -p_o \tag{2-61}$$

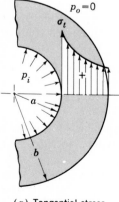

(a) Tangential stress distribution

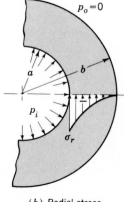

(b) Radial stress distribution

FIGURE 2-27 Distribution of stresses in a thick-walled cylinder subjected to internal pressure.

It should be realized that longitudinal stresses exist when the end reactions to pressures are taken by the pressure vessel itself. This stress is found to be

$$\sigma_l = \frac{p_i a^2}{b^2 - a^2} \tag{2-62}$$

We further note that Eqs. (2-58), (2-59), and (2-62) apply only to sections of the pressure vessel significantly distant from the ends, since various nonlinearities and stress concentration exist near these ends.

Thin-Walled Vessels

When the wall thickness of a cylindrical pressure vessel is about one-twentieth, or less, of its radius, the radial stress which results from pressurizing the vessel is quite small compared to the tangential stress. Under these conditions the tangential stress can be assumed to be uniformly distributed across the wall thickness. When this assumption is made, the vessel is called a *thin-walled pressure vessel*. The stress state in tanks, pipes, and hoops may also be determined using this assumption.

Let an internal pressure p be exerted on the wall of a cylinder of thickness t and inside diameter d_i. The force tending to separate two halves of a unit length of the cylinder is pd_i. This force is resisted by the tangential stress, also called the *hoop stress*, acting uniformly over the stressed area. We then have $pd_i = 2t\sigma_t$, or

$$\sigma_t = \frac{pd_i}{2t} \tag{2-63}$$

In a closed cylinder the longitudinal stress σ_l exists because of the pressure upon the ends of the vessel. If we assume this stress is also distributed uniformly over the wall thickness, we can easily find it to be

$$\sigma_l = \frac{pd_i}{4t} \tag{2-64}$$

Example 2-9 An aluminum alloy pressure vessel is made of tubing having an outside diameter of 8 in and a wall thickness of $\frac{1}{4}$ in.

    (*a*) What pressure can the cylinder carry if the permissible tangential stress is 12 kpsi and the theory for thin-walled vessels is assumed to apply?

    (*b*) Based on the pressure found in part (*a*), compute all of the stress components using thick-wall theory and show them on a three-dimensional Mohr's circle diagram. Is the thin-wall theory satisfactory for this analysis?

Solution (*a*) Here $d_i = 8 - 2(0.25) = 7.5$ in, $a = 7.5/2 = 3.75$ in, and $b = 8/2 = 4$ in. Then $t/a = 0.25/3.75 = 0.067$. Since this ratio is greater than one-twentieth, the thin-wall cylinder theory may not yield safe results.

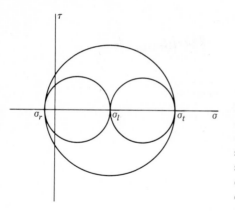

FIGURE 2-28 Mohr's circle diagram for pressure cylinder, drawn to scale, showing the principal stresses $\sigma_t = 12.4(10)^3$ psi, $\sigma_l = 5.81(10)^3$ psi, and $\sigma_r = -0.80(10)^3$ psi.

We first solve Eq. (2-63) to obtain the allowable pressure. This gives

$$p = \frac{2t\sigma_t}{d_i} = \frac{2(0.25)(12)(10)^3}{7.5} = 800 \text{ psi} \qquad Ans.$$

Then, from Eq. (2-64) we find the longitudinal stress to be

$$\sigma_l = \frac{pd_i}{4t} = \frac{800(7.5)}{4(0.25)} = 6000 \text{ psi} \qquad Ans.$$

(b) Using thick-wall cylinder theory and Eq. (2-58), we find the tangential stress at the inside radius to be

$$\sigma_t = p_i \frac{b^2 + a^2}{b^2 - a^2} = 800 \frac{(4)^2 + (3.75)^2}{(4)^2 - (3.75)^2} = 12\ 400 \text{ psi} \qquad Ans.$$

Then from Eqs. (2-59) and (2-62) we find the radial and longitudinal stresses to be

$$\sigma_r = -p_i = -800 \text{ psi} \qquad Ans.$$

$$\sigma_l = \frac{p_i a^2}{b^2 - a^2} = \frac{800(3.75)^2}{(4)^2 - (3.75)^2} = 5810 \text{ psi} \qquad Ans.$$

The three stresses σ_t, σ_r, and σ_l are principal stresses since there is no shear. The Mohr's circle diagram is shown in Fig. 2-28.

The tangential stresses in (a) and (b) differ by only about 3 percent. This is small enough such that we can consider the thin-wall theory to be satisfactory. But see Chap. 6 for additional information on the failure of parts subjected to three-dimensional stress states. ////

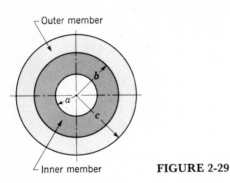

Outer member

Inner member   **FIGURE 2-29**

2-18   PRESS AND SHRINK FITS

When two cylindrical parts are assembled by shrinking or press-fitting one part
upon another, a contact pressure is created between the two parts. The stresses
resulting from this pressure may easily be determined with the equations of the
preceding section.

Figure 2-29 shows two cylindrical members which have been assembled with
a shrink fit. A contact pressure p exists between the members at radius b, causing
radial stresses $\sigma_r = -p$ in each member at the contacting surfaces. From Eq.
(2-60), the tangential stress at the outer surface of the inner member is

$$\sigma_{it} = -p\,\frac{b^2 + a^2}{b^2 - a^2} \tag{2-65}$$

In the same manner, from Eq. (2-58), the tangential stress at the inner surface of
the outer member is

$$\sigma_{ot} = p\,\frac{c^2 + b^2}{c^2 - b^2} \tag{2-66}$$

disgusting!

These equations cannot be solved until the contact pressure is known. In obtaining a shrink fit, the diameter of the male member is made larger than the diameter
of the female member. The difference in these dimensions is called the *interference*
and is the deformation which the two members must experience. Since these
dimensions are usually known, the deformation should be introduced in order to
evaluate the stresses. Let

δ_o = increase in radius of hole

and

δ_i = decrease in radius of inner cylinder

The tangential strain in the outer cylinder at the inner radius is

$$\epsilon_{ot} = \frac{\text{change in circumference}}{\text{original circumference}}$$

or

$$\epsilon_{ot} = \frac{2\pi(b + \delta_o) - 2\pi b}{2\pi b} = \frac{\delta_o}{b}$$

and so

$$\delta_o = b\epsilon_{ot} \tag{a}$$

but since

$$\epsilon_{ot} = \frac{\sigma_{ot}}{E_o} - \frac{\mu\sigma_{or}}{E_o}$$

then, from Eqs. (2-65) and (2-66), we have

$$\delta_o = \frac{bp}{E_o}\left(\frac{c^2 + b^2}{c^2 - b^2} + \mu\right) \tag{b}$$

This is the increase in radius of the outer cylinder. In a similar manner the decrease in radius of the inner cylinder is found to be

$$\delta_i = -\frac{bp}{E_i}\left(\frac{b^2 + a^2}{b^2 - a^2} - \mu\right) \tag{c}$$

Then the total deformation δ is

$$\delta = \delta_o - \delta_i = \frac{bp}{E_o}\left(\frac{c^2 + b^2}{c^2 - b^2} + \mu\right) + \frac{bp}{E_i}\left(\frac{b^2 + a^2}{b^2 - a^2} - \mu\right) \tag{2-67}$$

This equation can be solved for the pressure p when the interference δ is given. If the two members are of the same material, $E_o = E_i = E$ and the relation simplifies to

$$p = \frac{E\delta}{b}\left[\frac{(c^2 - b^2)(b^2 - a^2)}{2b^2(c^2 - a^2)}\right] \tag{2-68}$$

Substitution of this value of p in Eqs. (2-65) and (2-66) will then give the tangential stresses at the inner surface of the outer cylinder and at the outer surface of the inner cylinder. In addition, Eq. (2-67) or (2-68) can be employed to obtain the value of p for use in the general equations [Eqs. (2-54) or (2-55)] in order to obtain the stress at any point in either cylinder.

Assumptions

In addition to the assumptions both stated and implied by the development, it is necessary to assume that both members have the same length. In the case of a hub which has been press-fitted to a shaft, this assumption would not be true, and there would be an increased pressure at each end of the hub. It is customary to allow for this condition by the employment of a stress-concentration factor. The value of this factor depends upon the contact pressure and the design of the female member, but its theoretical value is seldom greater than 2.

2-19 THERMAL STRESSES AND STRAINS

When the temperature of an unrestrained body is uniformly increased, the body expands, and the normal strain is

$$\epsilon_x = \epsilon_y = \epsilon_z = \alpha(\Delta T) \tag{2-69}$$

where α is the coefficient of thermal expansion and ΔT is the temperature change, in degrees. In this action the body experiences a simple volume increase with the components of shear strain all zero.

If a straight bar is restrained at the ends so as to prevent lengthwise expansion and then is subjected to a uniform increase in temperature, a compressive stress will develop because of the axial constraint. The stress is

$$\sigma = \epsilon E = \alpha(\Delta T)E \tag{2-70}$$

In a similar manner, if a uniform flat plate is restrained at the edges and also subjected to a uniform temperature rise, the compressive stress developed is given by the equation

$$\sigma = \frac{\alpha(\Delta T)E}{1 - \mu} \tag{2-71}$$

The stresses represented by Eqs. (2-70) and (2-71), though due to temperature, are not thermal stresses inasmuch as they result from the fact that the edges were restrained. A *thermal stress* is one which arises because of the existence of a *temperature gradient* in a body.

Figure 2-30 shows the internal stresses within a slab of infinite dimensions

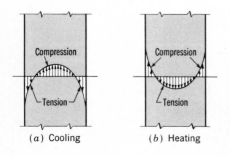

(a) Cooling   (b) Heating

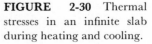

FIGURE   2-30 Thermal stresses in an infinite slab during heating and cooling.

Table 2-5 COEFFICIENTS OF THERMAL EXPAN-
SION (LINEAR MEAN COEFFICIENTS
FOR THE TEMPERATURE RANGE 0–
100°C)

Material	Celsius scale	Fahrenheit scale
Aluminum	$23.9(10)^{-6}$	$13.3(10)^{-6}$
Brass, cast	$18.7(10)^{-6}$	$10.4(10)^{-6}$
Carbon steel	$10.8(10)^{-6}$	$6.0(10)^{-6}$
Cast iron	$10.6(10)^{-6}$	$5.9(10)^{-6}$
Magnesium	$25.2(10)^{-6}$	$14.0(10)^{-6}$
Nickel steel	$13.1(10)^{-6}$	$7.3(10)^{-6}$
Stainless steel	$17.3(10)^{-6}$	$9.6(10)^{-6}$
Tungsten	$4.3(10)^{-6}$	$2.4(10)^{-6}$

during heating and cooling. During cooling, the maximum stress is the surface tension. At the same time, force equilibrium requires a compressive stress at the center of the slab. During heating, the external surfaces are hot and tend to expand but are restrained by the cooler center. This causes compression in the surface and tension in the center as shown.

Table 2-5 lists approximate values of the coefficient α for various engineering materials.

2-20 CURVED MEMBERS IN FLEXURE

The distribution of stress in a curved flexural member is determined by using the following assumptions.

1 The cross section has an axis of symmetry in a plane along the length of the beam.
2 Plane cross sections remain plane after bending.
3 The modulus of elasticity is the same in tension as in compression.

We shall find that the neutral axis and the centroidal axis of a curved beam, unlike a straight beam, are not coincident and also that the stress does not vary linearly from the neutral axis. The notation shown in Fig. 2-31 is defined as follows:

r_o = radius of outer fiber
r_i = radius of inner fiber
h = depth of section
c_o = distance from neutral axis to outer fiber
c_i = distance from neutral axis to inner fiber
r = radius of neutral axis

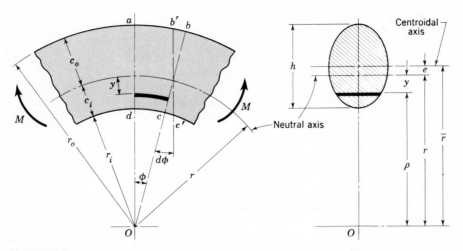

FIGURE 2-31   Note that y is positive in the direction toward point O.

$\bar{r}$ = radius of centroidal axis
e = distance from centroidal axis to neutral axis

To begin, we define the element $abcd$ by the angle ϕ. A bending moment M causes section bc to rotate through $d\phi$ to $b'c'$. The strain on any fiber at distance ρ from the center O is

$$\epsilon = \frac{\delta}{l} = \frac{(r - \rho)\, d\phi}{\rho\phi} \tag{a}$$

The normal stress corresponding to this strain is

$$\sigma = \epsilon E = \frac{E(r - \rho)\, d\phi}{\rho\phi} \tag{b}$$

Since there are no axial external forces acting on the beam, the sum of the normal forces acting on the section must be zero. Therefore

$$\int \sigma\, dA = E\, \frac{d\phi}{\phi} \int \frac{(r - \rho)\, dA}{\rho} = 0 \tag{c}$$

Now arrange Eq. (c) in the form

$$E\, \frac{d\phi}{\phi} \left(r \int \frac{dA}{\rho} - \int dA \right) = 0 \tag{d}$$

and solve the expression in parentheses. This gives

$$r \int \frac{dA}{\rho} - A = 0$$

or

$$r = \frac{A}{\int \frac{dA}{\rho}} \qquad\qquad (2\text{-}72)$$

This important equation is used to find the location of the neutral axis with respect to the center of curvature O of the cross section. The equation indicates that the neutral and the centroidal axes are not coincident.

Our next problem is to determine the stress distribution. We do this by balancing the external applied moment against the internal resisting moment. Thus, from Eq. (b),

$$\int (r - \rho)(\sigma\ dA) = E \frac{d\phi}{\phi} \int \frac{(r - \rho)^2\ dA}{\rho} = M \qquad\qquad (e)$$

Since $(r - \rho)^2 = r^2 - 2\rho r + \rho^2$, Eq. (e) can be written in the form

$$M = E \frac{d\phi}{\phi} \left(r^2 \int \frac{dA}{\rho} - r \int dA - r \int dA + \int \rho\ dA \right) \qquad\qquad (f)$$

Note that r is a constant; then compare the first two terms in parentheses with Eq. (d). These terms vanish, and we have left

$$M = E \frac{d\phi}{\phi} \left(-r \int dA + \int \rho\ dA \right) \qquad\qquad (g)$$

The first integral in this expression is the area A, and the second is the product $\bar{r}A$. Therefore

$$M = E \frac{d\phi}{\phi} (\bar{r} - r)A = E \frac{d\phi}{\phi} eA \qquad\qquad (h)$$

Now, using Eq. (b) once more, and rearranging, we finally obtain

$$\sigma = \frac{My}{Ae(r - y)} \qquad\qquad (2\text{-}73)$$

This equation shows that the stress distribution is hyperbolic. The maximum stresses occur at the inner and outer fibers and are

$$\sigma_i = \frac{Mc_i}{Aer_i} \qquad \sigma_o = \frac{Mc_o}{Aer_o} \tag{2-74}$$

These equations are valid for pure bending. In the usual and more general case, such as a crane hook, the U frame of a press, or the frame of a clamp, the bending moment is due to forces acting to one side of the cross section under consideration. In this case the bending moment is computed about the *centroidal axis*, *not* the neutral axis. Also, an additional axial tensile or compressive stress must be added to the bending stress given by Eqs. (2-73) and (2-74) to obtain the resultant stress acting on the section.

Example 2-10   A rectangular-section beam has the dimensions $b = 1$ in, $h = 3$ in and is subjected to a pure bending moment of 20 000 lb·in so as to produce compression of the inner fiber. Find the stresses for the following geometries:

(a)   The beam is straight.
(b)   The centroidal axis has a radius of curvature of 15 in.
(c)   The centroidal axis has a radius of curvature of 3 in.

Solution

(a) $\dfrac{I}{c} = \dfrac{bh^2}{6} = \dfrac{(1)(3)^2}{6} = 1.5$ in^3

$$\sigma_{max} = \frac{M}{I/c} = \frac{20\ 000}{1.5} = 13\ 300 \text{ psi}$$

(b)   We must integrate Eq. (2-72) first. Designating $dA = b\,d\rho$, we have

$$r = \frac{A}{\displaystyle\int \frac{1}{\rho}\,dA} = \frac{bh}{\displaystyle\int_{r_i}^{r_o} \frac{b}{\rho}\,d\rho} = \frac{h}{\ln \dfrac{r_o}{r_i}} \tag{2-75}$$

Now, $r_o = 15 + 1.5 = 16.5$ and $r_i = 15 - 1.5 = 13.5$ in. Solving Eq. (2-75) gives

$$r = \frac{3}{\ln (16.5/13.5)} = 14.950 \text{ in}$$

Then $e = \bar{r} - r = 15.00 - 14.950 = 0.050$ in.   Also $c_o = 1.550$ in and $c_i = 1.450$ in.

Equations (2-74) give

$$\sigma_i = -\frac{Mc_i}{Aer_i} = -\frac{(20\ 000)(1.450)}{(3)(0.050)(13.5)} = -14\ 330\ \text{psi}$$

$$\sigma_o = \frac{Mc_o}{Aer_o} = \frac{(20\ 000)(1.550)}{(3)(0.050)(16.5)} = 12\ 530\ \text{psi}$$

(c)   We have $r_o = 3 + 1.5 = 4.5$ in and $r_i = 3 - 1.5 = 1.5$ in.   Using Eq. (2-75) again, we obtain

$$r = \frac{3}{\ln\ (4.5/1.5)} = 2.731\ \text{in}$$

So, $e = \bar{r} - r = 3 - 2.731 = 0.269$ in, $c_o = 1.769$ in, and $c_i = 1.231$ in.   We have, from Eqs. (2-74),

$$\sigma_i = -\frac{Mc_i}{Aer_i} = -\frac{(20\ 000)(1.231)}{(3)(0.269)(1.5)} = -20\ 400\ \text{psi}$$

$$\sigma_o = \frac{Mc_o}{Aer_o} = \frac{(20\ 000)(1.769)}{(3)(0.269)(4.5)} = 9750\ \text{psi}$$

It is interesting to note that the difference in stresses for the straight beam and the beam having a 15-in radius of curvature is only about $7\frac{1}{2}$ percent.   Such an error may be acceptable in certain design situations.   It is here that you must utilize your own judgment in making the decision whether to treat the member as straight or as curved.   Only your own time will be saved by assuming the beam is straight.   If the radius of curvature is large compared with the section depth, and if your boss is nagging you for a quick result, you might well decide to increase the factor of safety by a modest amount, and compute the stress using the straight-beam assumption.   Of course, you should not risk this course of action in the classroom or on the job without being aware of the inaccuracies involved.   ////

Formulas for Some Common Sections

Sections most frequently encountered in the stress analysis of curved beams are shown in Fig. 2-32.   Formulas for the rectangular section were developed in Example 2-10, but they are repeated here for convenience.

$$\bar{r} = r_i + \frac{h}{2} \tag{2-76}$$

$$r = \frac{h}{\ln\ (r_o/r_i)} \tag{2-77}$$

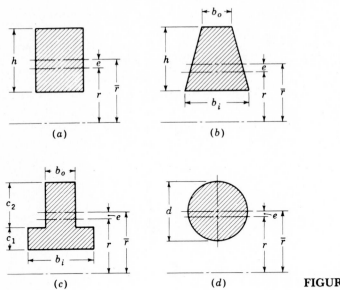

FIGURE 2-32

For the trapezoidal section in Fig. 2-32b, the formulas are

$$\bar{r} = r_i + \frac{h}{3} \frac{b_i + 2b_o}{b_i + b_o} \tag{2-78}$$

$$r = \frac{A}{b_o - b_i + [(b_i r_o - b_o r_i)/h] \ln (r_o/r_i)} \tag{2-79}$$

For the T section in Fig. 2-32c we have

$$\bar{r} = r_i + \frac{b_i c_1^2 + 2b_o c_1 c_2 + b_o c_2^2}{2(b_o c_2 + b_i c_1)} \tag{2-80}$$

$$r = \frac{b_i c_1 + b_o c_2}{b_i \ln [(r_i + c_1)/r_i] + b_o \ln [r_o/(r_i + c_1)]} \tag{2-81}$$

The equations for the solid round section of Fig. 2-32d are

$$\bar{r} = r_i + \frac{d}{2} \tag{2-82}$$

$$r = \frac{d^2}{4(2\bar{r} - \sqrt{4\bar{r}^2 - d^2})} \tag{2-83}$$

Note, very particularly, that Eq. (2-83) gives the radius of curvature of the neutral axis. Do not confuse the result with the radius of the section!

The formulas for other sections can be obtained by performing the integration indicated by Eq. (2-72). Cold forming of bars around a die to create a curved bar distorts the cross section, because the outer portion stretches and the inner portion compresses. The resulting cross section has a different geometry than the original, and it may not be possible to express it as an integrable function. In these cases the small, hand-held calculator should be used to perform a numerical integration using the equation

$$r = \frac{\sum \Delta A}{\sum \dfrac{\Delta A}{\rho}}$$

which is adapted from Eq. (2-72). Usually 15 to 20 ΔA segments will provide good results.

2-21   HERTZ CONTACT STRESSES*

A state of triaxial stress seldom arises in design. An exception to this occurs when two bodies having curved surfaces are pressed against one another. When this happens, point or line contact changes to area contact, and the stress developed in both bodies is three-dimensional. Contact-stress problems arise in the contact of a wheel and a rail, a cam and its follower, mating gear teeth, and in the action of rolling bearings. To guard against the possibility of surface failure in such cases, it is necessary to have means of computing the stress states which result from loading one body against another.

When two solid spheres of diameters d_1 and d_2 are pressed together with a force F, a circular area of contact of radius a is obtained. Specifying E_1, μ_1 and E_2, μ_2 as the respective elastic constants of the two spheres, the radius a is given by the equation

$$a = \sqrt[3]{\frac{3F}{8} \frac{[(1 - \mu_1^2)/E_1] + [(1 - \mu_2^2)/E_2]}{(1/d_1) + (1/d_2)}} \tag{2-84}$$

The pressure within each sphere has a semi-elliptical distribution, as shown in Fig. 2-33. The maximum pressure occurs at the center of the contact area and is

$$p_{max} = \frac{3F}{2\pi a^2} \tag{2-85}$$

---

* See in W. Flügge (ed.), J. L. Lubkin, "Contact Problems" *Handbook of Engineering Mechanics*, McGraw-Hill, New York, 1962, sec. 42-1. This section contains some history and many references, including two books in Russian devoted entirely to contact stresses.

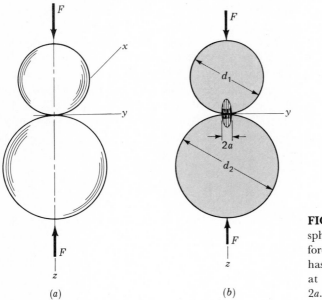

FIGURE 2-33 (a) Two spheres held in contact by force F; (b) contact stress has an elliptical distribution at face of contact of width $2a$.

Equations (2-84) and (2-85) are perfectly general and also apply to the contact of a sphere and a plane surface or of a sphere and an internal spherical surface. For a plane surface use $d = \infty$. For an internal surface the diameter is expressed as a negative quantity.

The maximum stresses occur on the z axis, and these are principal stresses. Their values are

$$\sigma_x = \sigma_y = -p_{max} \left\{ \left[1 - \frac{z}{a} \tan^{-1} \left(\frac{1}{z/a} \right) \right] (1 + \mu) - \frac{1}{2 \left(1 + \dfrac{z^2}{a^2} \right)} \right\} \qquad (2\text{-}86)$$

$$\sigma_z = \frac{-p_{max}}{1 + \dfrac{z^2}{a^2}} \qquad (2\text{-}87)$$

These equations are valid for either sphere, but the value used for Poisson's ratio must correspond with the sphere under consideration. The equations are even more complicated when stress states off the z axis are to be determined, because here the x and y coordinates must also be included. But these are not required for design purposes because the maximums occur on the z axis.

Mohr's circles for the stress state described by Eqs. (2-86) and (2-87) are a point and two coincident circles. Since $\sigma_x = \sigma_y$, $\tau_{xy} = 0$, and

$$\tau_{xz} = \tau_{yz} = \frac{\sigma_x - \sigma_z}{2} = \frac{\sigma_y - \sigma_z}{2} \qquad (2\text{-}88)$$

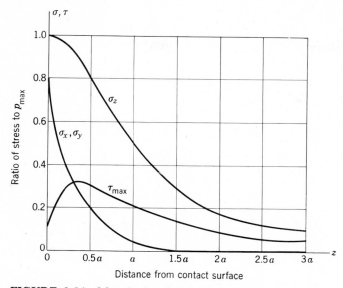

FIGURE 2-34 Magnitude of the stress components below the surface as a function of the maximum pressure for contacting spheres. Note that the maximum shear stress is slightly below the surface and is approximately $0.3p_{max}$. The chart is based on a Poisson's ratio of 0.30

Figure 2-34 is a plot of Eqs. (2-86) and (2-87) for a distance of $3a$ below the surface. Note that the shear stress reaches a maximum value slightly below the surface. It is the opinion of many authorities that this maximum shear stress is responsible for the surface fatigue failure of contacting elements. The explanation is that a crack originates at the point of maximum shear stress below the surface and progresses to the surface and that the pressure of the lubricant wedges the chip loose.

Figure 2-35 illustrates a similar situation in which the contacting elements are two cylinders of length l and diameters d_1 and d_2. As shown in Fig. 2-35b the area of contact is a narrow rectangle of width $2b$ and length l, and the pressure distribution is elliptical. The half-width b is given by the equation

$$b = \sqrt{\frac{2F}{\pi l} \frac{[(1 - \mu_1^2)/E_1] + [(1 - \mu_2^2)/E_2]}{(1/d_1) + (1/d_2)}}$$ (2-89)

The maximum pressure is

$$p_{max} = \frac{2F}{\pi bl}$$ (2-90)

Equations (2-89) and (2-90) apply to a cylinder and a plane surface, such as a

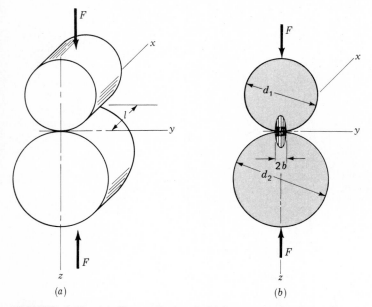

FIGURE 2-35 (*a*) Two cylinders held in contact by force F uniformly distributed along cylinder length l; (*b*) contact stress has an elliptical distribution at face of contact of width $2b$.

rail, by making $d = \infty$ for the plane surface. The equations also apply to the contact of a cylinder and an internal cylindrical surface; in this case d is made negative.

The stress state on the z axis is given by the equations

$$\sigma_x = -2\mu p_{\text{max}}\left(\sqrt{1 + \frac{z^2}{b^2}} - \frac{z}{b}\right) \tag{2-91}$$

$$\sigma_y = -p_{\text{max}}\left\{\left[2 - \frac{1}{1 + \dfrac{z^2}{b^2}}\right]\sqrt{1 + \frac{z^2}{b^2}} - 2\frac{z}{b}\right\} \tag{2-92}$$

$$\sigma_z = \frac{-p_{\text{max}}}{\sqrt{1 + \dfrac{z^2}{b^2}}} \tag{2-93}$$

These three equations are plotted in Fig. 2-36 up to a distance of $3b$ below the surface. Though τ_{zy} is not the largest of the three shear stresses for all values of z/b, it is a maximum at about $z/b = 0.75$ and is larger at that point than either of the other two shear stresses for any value of z/b.

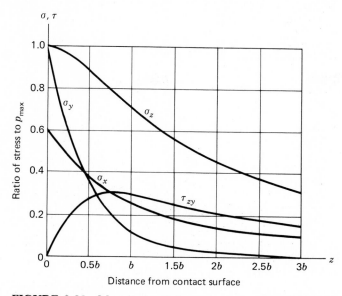

FIGURE 2-36 Magnitude of the stress components below the surface as a function of the maximum pressure for contacting cylinders. τ_{zy} becomes the largest of the three shear stresses at about $z/b = 0.75$; its maximum value is $0.30P_{max}$. The chart is based on a Poisson's ratio of 0.30. Can you tell which two principal stresses are used to determine τ_{max} when $z/b < 0.75$?

PROBLEMS*

Sections 2-1 to 2-3

2-1 to 2-8 For each of the problems listed in the table below and their stress states, construct Mohr's circle diagram and find the principal stresses and the extremes of the shear stresses. Draw a stress element properly oriented from the x axis to show the principal stresses and their directions. Calculate the angle from the x axis to σ_1.

Prob.	σ_x, kpsi	σ_y, kpsi	τ_{xy}, kpsi
2-1	10	−4	0
2-2	10	0	4 cw
2-3	−2	−8	4 ccw
2-4	10	5	1 cw
2-5	−17	−4	7 cw
2-6	−10	−12	6 ccw
2-7	4	0	8 cw
2-8	4	12	7 ccw

* The asterisk indicates a problem that may not have a unique result.

2-9 to 2-16   The same as Prob. 2-1 for the stress states shown in the following table:

Prob.	σ_x, MPa	σ_y, MPa	τ_{xy}, MPa
2-9	-30	-60	30 ccw
2-10	-90	10	40 cw
2-11	75	15	32 ccw
2-12	120	0	60 ccw
2-13	84	-42	96 cw
2-14	-50	60	80 ccw
2-15	-60	0	40 cw
2-16	20	-60	40 cw

2-17   If $\sigma_1 = 12$ kpsi and $\sigma_2 = 2$ kpsi, what are the values of σ_x, σ_y, and τ_{xy} if the angle from the x axis to the direction of σ_1 is 30° cw?   30° ccw?

2-18   Suppose the largest principal stress is 14 MPa and σ_2 is 6 MPa.   What are the values of σ_x, σ_y, and τ_{xy} if the angle from the x axis to the direction of the largest principal stress is 100° ccw?   80° cw?

2-19   Find the values of all three principal stresses and the maximum shear stress for all the odd-numbered problems in the table containing Prob. 2-1.

2-20   The same as Prob. 2-19 for the even-numbered problems.

2-21   The same as Prob. 2-19 for the odd-numbered problems in the table containing Prob. 2-9.

2-22   The same as Prob. 2-21 for the even-numbered problems.

Sections 2-4 to 2-6

2-23   A steel rod 80 mm long and 15 mm in diameter is acted upon by a compressive load of 175 kN.   The material is carbon steel.   Find:
   (a)   The compressive stress
   (b)   The axial strain
   (c)   The deformation
   (d)   The increase in diameter of the rod

2-24   Draw Mohr's three-circle diagram and find the three principal stresses and the maximum shear stress for the following conditions:
   (a)   Pure tension
   (b)   Pure compression
   (c)   Pure torsion
   (d)   Equal tension in the x and y directions

2-25   A steel rod $\frac{1}{2}$ in in diameter and 40 in long carries a tensile load of 3000 lb.
   (a)   Find the stress and elongation of the rod.
   (b)   If this rod is to be replaced by an aluminum rod which is to have the same elongation, what should be its diameter?
   (c)   Calculate the stress in the aluminum rod.

2-26*   A press is to be designed so that the elongation of the four steel tension members A, as shown, does not exceed 0.0625 in and so that the tensile stress does not exceed

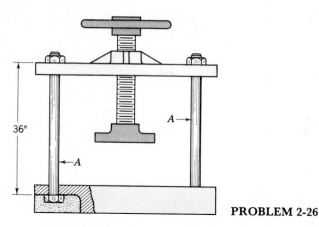

PROBLEM 2-26

15 kpsi.   Find the diameters of the tension members if the press is to be rated at 5 tons.   Use $E = 30$ Mpsi.

2-27   A plastic ball is inflated enough to produce tangential stresses $\sigma_x = \sigma_y = 300$ psi. The radial thickness of the material is 0.05 in before inflation.   Find the thickness after inflation if the tensile modulus of elasticity is 500 kpsi and the shear modulus is 200 kpsi.   Compute the unit tangential strain in the x direction.

2-28   Assuming that the compression members of the truss shown in the figure will not buckle, find the cross-sectional area of the tension members using an allowable stress of 50 MPa.   The dimensions are $OB = BC = 5$ m, $AB = 6$ m.   The force is $F = 500$ kN.   Find also the total deformation of each tension member using $E = 207$ GPa.

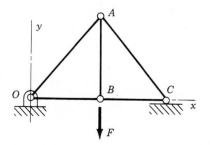

PROBLEM 2-28

2-29   Suppose the side stay BC on the catamaran shown in the figure has an effective modulus of elasticity of 9.5 Mpsi and an effective cross-sectional area of 0.006 in^2. The distance BC is 20 ft.   A turnbuckle is used to stretch the wire rope BC by 2 in. If point B remains stationary, what is the resulting tension in the stay?

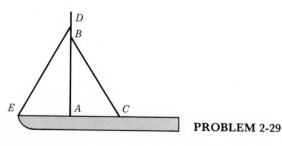

PROBLEM 2-29

2-30 A built-up connecting rod consists of three steel bars, each $\frac{1}{4}$ in thick $\times$ $1\frac{1}{4}$ in wide as shown in the figure. During assembly it was found that one of the bars measured only 31.997 in between pin centers, the other two bars measuring exactly 32.000 in. Determine the stress in each bar after assembly but before an external load is applied.

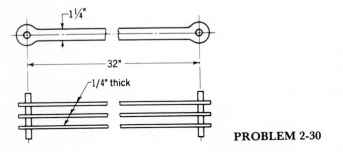

PROBLEM 2-30

2-31 An aluminum rod, $\frac{3}{4}$ in in diameter $\times$ 48 in long, and a steel rod, $\frac{1}{2}$ in in diameter $\times$ 32 in long, are spaced 60 in apart and fastened to a horizontal beam which carries a 2000-lb load, as shown in the figure. The beam is to remain horizontal after the load is applied. For the purposes of this problem we assume the beam is weightless and absolutely rigid.
(*a*) Find the location X of the load.
(*b*) Find the stress in each rod.

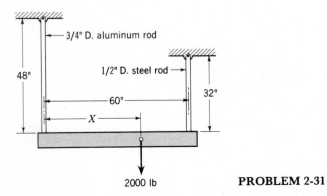

PROBLEM 2-31

2-32 A steel bolt, having a nominal diameter of 20 mm and a pitch of 2.5 mm (pitch is the distance from thread to thread in the axial direction), and an aluminum tube,

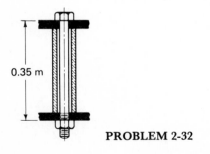

PROBLEM 2-32

40 mm OD by 22 mm ID, act as a spacer for two plates, as shown in the figure. The nut is pulled snug and then given a one-third additional turn. Find the resulting stress in the bolt and in the tube, neglecting the deformation of the plates.

2-33 Develop expressions for the strain in the x and y directions when the tensile stresses σ_x and σ_y are equal.

2-34 Electrical strain gauges mounted on the surface of a loaded machine part were aligned in the direction of the principal stresses and gave the following values for the strains:

$$\epsilon_1 = 0.0021 \text{ m/m} \qquad \epsilon_2 = -0.000\ 76 \text{ m/m}$$

Determine the three principal stresses and the unknown normal strain. What assumptions were necessary? Use $E = 207$ GPa and $\mu = 0.292$.

Sections 2-7, 2-8

2-35, 2-36 Find the reactions at the supports and plot the shear-force and bending-moment diagrams for each of the beams shown in the figure.

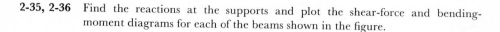

PROBLEM 2-35

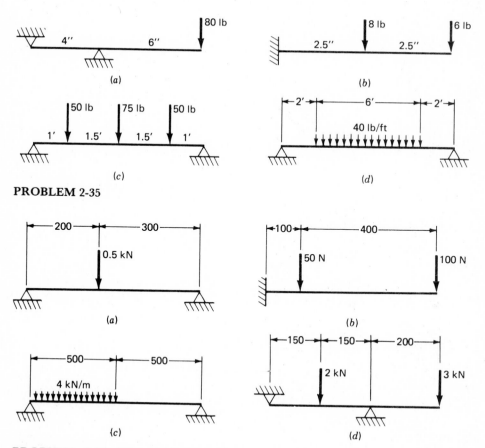

PROBLEM 2-36 Dimensions in millimeters.

2-37* The formed sheet-metal bracket shown in the figure is to be fastened to a smooth wall using round-head wood screws and washers. Determine the magnitude and location of the reactions based on the total distributed load W and the dimensions of the bracket.

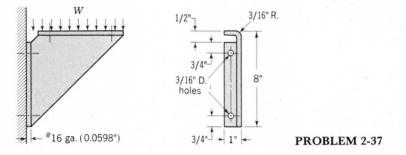

PROBLEM 2-37

2-38 to 2-41 Using singularity functions, find general expressions for the loading, shear force, and moment for each beam shown in the figure and compute the reactions at the supports. Where possible, check your results with Table A-12.

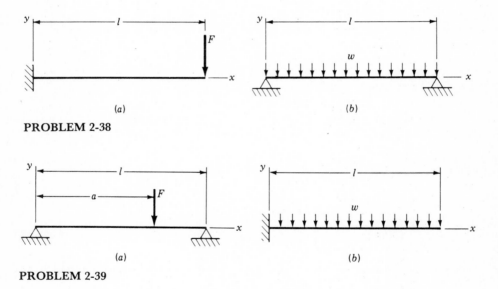

(a)   (b)

PROBLEM 2-38

(a)   (b)

PROBLEM 2-39

(a)   (b)

PROBLEM 2-40

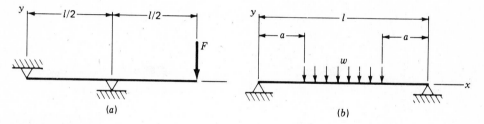

(a)

(b)

PROBLEM 2-41

Sections 2-9 and 2-10

2-42* The figure shows a rubber roll used for sizing cloth. Bearings A and B support the rotating shaft, which is loaded in bending by another similar roll. Crowning of the rolls causes both of them to load their shafts with a uniformly distributed load of 125 lb/in of length. Select a hollow steel tubing from Table A-10 to be used for the shaft such that the normal bending stress will not exceed 16 kpsi.

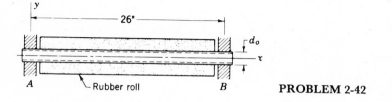

Rubber roll

PROBLEM 2-42

2-43* The figure shows a shaft mounted in bearings A and B. The forces acting on the gear and pulley that are keyed to the shaft cause bending. Assume the loads are concentrated, use an allowable bending stress of 24 kpsi, and find the dimensions of a standard-size hollow round shaft from Table A-10 to safely resist the loads.

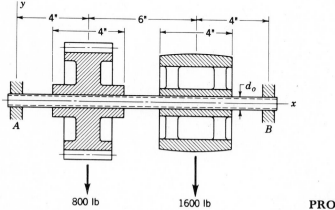

800 lb   1600 lb

PROBLEM 2-43

2-44* Cast iron has a strength in compression of about three to four times the strength in tension, depending upon the grade.

(a) Using the stress as the prime consideration, design an optimum T section for a

cast-iron beam using a uniform section thickness such that the compressive
stress will be related to the tensile stress by a factor of 4.

(b) Would it make any difference in the design if weight, rather than stress, is the
prime consideration?

2-45 The shaft shown in the figure is supported in bearings at O and C and is subjected to
bending loads due to the force components acting at A and B.

(a) Sketch two moment diagrams, one for each plane of bending, and compute the
moment components at A and B. Also compute the maximum bending
moment.

(b) The shaft is to be made of cold-drawn UNS G10180 steel (see Table A-17). If
a factor of safety of 2.3 based on yield strength is to be used, find the allowable
normal bending stress in SI units.

(c) Find the safe shaft diameter at the point of maximum bending moment to the
nearest millimeter.

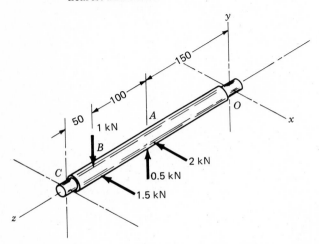

PROBLEM 2-45 Dimensions in millimeters.

2-46 The figure illustrates an overhanging shaft supported in bearings, assumed to be
self-aligning, at O and B. The shaft is loaded by external forces at A and C. If the

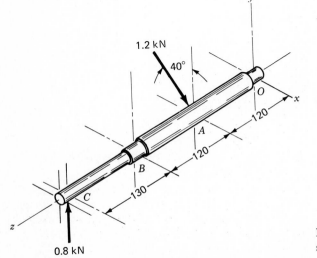

PROBLEM 2-46 Dimensions in millimeters.

diameter of the shaft section at B is 20 mm, what is the bending stress at this section? Is the stress likely to be higher at section A than at B?

Sections 2-11 to 2-15

2-47 The figure shows a round shaft of diameter d and length l loaded in bending by a force F acting in the center of the span. Three stress elements identified in section A-A are B, on top and in the xy plane; C, on the front side and in the xz plane; and D, on the bottom and in the xy plane. Draw each of these stress elements properly oriented with respect to xyz, show the stresses which act upon them, and write the stress formulas in terms of F and the geometry of the bar.

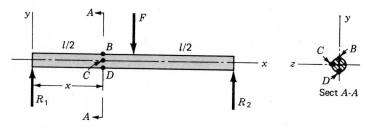

PROBLEM 2-47

2-48 A steel shaft $\frac{7}{8}$ in in diameter and 22 in long between bearings supports two loads as shown.
   (a)   Sketch the shear-force and bending-moment diagrams and compute the shear forces and bending moments at points where the loads are applied.
   (b)   Draw a stress element corresponding to point A on the shaft and compute and show on this element all stresses that act.
   (c)   The same as in b for a stress element at B.

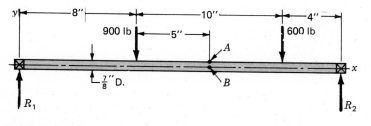

PROBLEM 2-48

2-49 The figure shows a shaft mounted in bearings at A and D and having pulleys at B and C. The forces shown acting on the pulley surfaces represent the belt pulls on the tight and loose sides of the pulleys. Compute the torque applied to the shaft by the pulley at C and by the pulley at B. Find an appropriate diameter for the shaft using a permissible normal stress of 16 kpsi and/or a permissible shear stress of 12 kpsi.

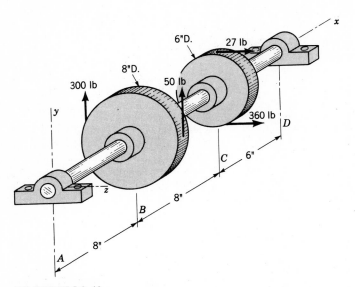

PROBLEM 2-49

2-50   Illustrated in the figure is a shaft having two pulleys and with bearings (not shown)
at A and B.   The belt pulls and their directions are shown for each pulley.   The
coordinate system may be identified by the superscripts on the bearing reaction
vectors.   Find the safe diameter of the shaft using a permissible normal stress of
24 kpsi and a permissible shear stress of 12 kpsi.

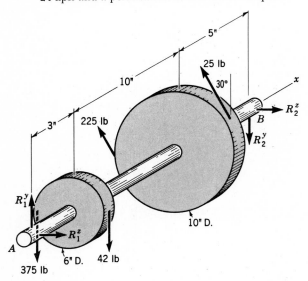

PROBLEM   2-50   The
partially   concealed   375-lb
force vector is the belt pull
on the 6-in pulley.

2-51   The figure is a free-body diagram of a steel shaft supported in bearings at O and B
resulting in bearing reactions R_1 and R_2.   The shaft is loaded by the forces $F_A =$
4.5 kN and $F_C = 1.8$ kN and by the equal and opposite torques $T_A = -T_C = 600$
N · m.
    (*a*)   Draw the shear-force and bending-moment diagrams and compute values cor-
responding to points O, A, B, and C.

(b) Locate a stress element on the bottom (minus y side) of the shaft at A. Find all the stress components which act. Do not look through the shaft to see this element; look from underneath!

(c) Using the stress element of b, draw Mohr's circle diagram, find both principal stresses, and label the locations of σ_1, σ_2, τ_{max}, both reference axes, and the angle from the x axis to the axis containing σ_1.

(d) Draw the principal stress element correctly oriented from a horizontal x axis and label it completely, with angles and stresses.

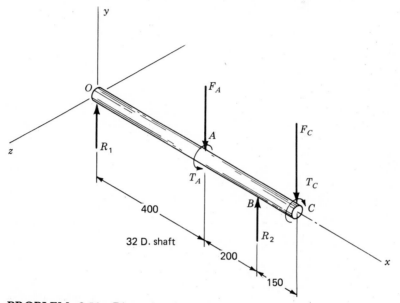

PROBLEM 2-51 Dimensions in millimeters.

2-52 The stresses are to be computed at two points on the cantilever shown in the figure. These are the stress element which is shown at A on top of the bar and parallel to the

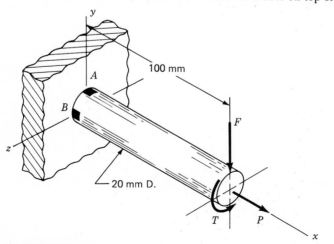

PROBLEM 2-52

xz plane and the one at *B* that is on the front side of the bar and parallel to the *xy* plane. The forces are $F = 0.55$ kN, $P = 8$ kN, and $T = 30$ N·m. Draw both stress elements, label the axes and the stress components using proper magnitudes and directions.

2-53   The figure shows the crankshaft and flywheel of a one-cylinder air compressor. In use, a part of the energy stored in the flywheel is used to produce a portion of the piston force *P*. In this problem you are to assume that the entire piston force *P* results from the torque of 600 N·m delivered to the crankshaft by the flywheel. A stress element is to be located on the top surface of the crankshaft at *A*, 100 mm from the left bearing. The sides of the element are parallel to the *xz* axes.

(*a*)   Compute the stress components that act at *A*.

(*b*)   Find the principal stresses and their directions for the element at *A*.

(*c*)   Make a sketch of the principal stress element and orient it correctly with reference to the *x* and *z* axes. Label completely.

(*d*)   Sketch another stress element correctly oriented to show the maximum shear stress and the corresponding normal stresses. Label this element too.

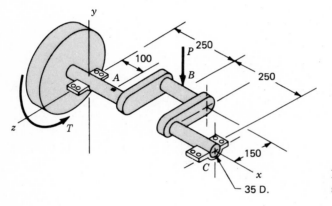

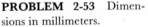

PROBLEM 2-53   Dimensions in millimeters.

2-54   A 100-mm-diameter solid circular shaft can carry a torsion of T_s N·m without exceeding a certain maximum shear stress. What proportion of this torque can a hollow shaft having a wall thickness of 10 mm and the same OD carry? Both shafts are to have the same maximum shear stress. Compute the ratio of the mass of the solid shaft to that of the hollow shaft.

2-55   The figure illustrates a bevel gear keyed to a 1-in-diameter shaft. The shaft is supported by a combined radial and thrust bearing at *D* and a bearing which takes pure radial load at *F*. Torque to the shaft is applied through the spur gear at *E*. In a force analysis of a bevel gear, the equivalent forces acting at the large end of the teeth are often used. In this problem the resultant force at the pitch line and at the large end of the teeth is specified in three components: a thrust force $W_T = 129$ lb, acting in the negative *x* direction; a tangential force $W = 500$ lb, acting in the negative *z* direction (the coordinate system is right-handed, with the *z* direction coming out of the paper); and a radial force $W_R = 129$ lb, acting in the negative *y* direction. The portion of the shaft to the right of the bearing *F* may be treated as a cantilever. Then the effect of these forces is to cause bending, compression, and torsion of the shaft. Neglect direct shear, and calculate all the stresses which act on an element at A, an element at B, and an element at C. Draw each stress element, show the stresses acting on it, and calculate the principal stresses and the maximum shear stress in each case.

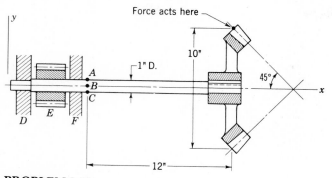

PROBLEM 2-55

2-56 The bit brace shown in the figure is used to apply a torque of 150 lb·in and a thrust of 50 lb to the bit, in boring a hole.

(a) Make a three-dimensional abstract drawing of the brace and calculate and show all the reactions.

(b) Select stress elements at the top, bottom, and two sides at section A-A and compute all the stresses acting on these elements.

(c) The bit used has a shank diameter of $\frac{1}{4}$ in where it projects from the chuck at section B-B. If the distance from B-B to the surface of the wood is 3 in, find the largest of the normal stresses at this section when the plane of the brace is horizontal.

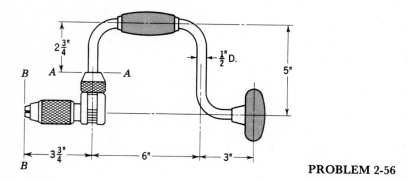

PROBLEM 2-56

2-57 The figure shows a crank loaded by a force $F = 300$ lb which causes twisting and bending of a $\frac{3}{4}$-in-diameter shaft fixed to a support at the origin of the reference system. In actuality, the support may be an inertia which we wish to rotate, but for the purposes of a stress analysis we can consider this as a statics problem.

(a) Draw separate free-body diagrams of the shaft AB and the arm BC, and compute the values of all forces, moments, and torques which act. Label the directions of the coordinate axes on these diagrams.

(b)   Compute the maximum torsional stress and the maximum bending stress in the
arm BC and indicate where these act.

(c)   Locate a stress element on the top surface of the shaft at A, and calculate all the
stress components which act upon this element.

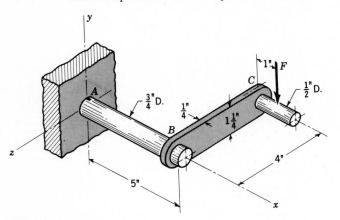

PROBLEM 2-57

Sections 2-17 to 2-19

2-58 A light pressure cylinder is made of an aluminum alloy. This cylinder has a $3\frac{1}{2}$-in
OD, a 0.065-in wall thickness, and material properties $E = 10.3$ Mpsi and
$\mu = 0.334$. If the internal pressure is 1860 psi, what are the principal normal strains
on an element on the circumference?

2-59 A magnesium tube is 5 in in OD and has a wall thickness of $\frac{1}{2}$ in. The tubing is used
as a pressure vessel to hold a fluid at an internal pressure of 4 kpsi. Calculate the
radial- and tangential-stress components and the three principal normal strains at
the outer and inner radii.

2-60 A cylinder is 300 mm OD by 200 mm ID and is subjected to an external pressure of
140 MPa. The longitudinal stress is zero. What is the maximum shear stress and at
what radius does it occur?

2-61 Derive the relations for the stress components in a thin-wall spherical pressure vessel.

2-62 A gun barrel is assembled by shrinking an outer barrel over an inner barrel so that
the maximum principal stress equals 70 percent of the yield strength of the material.
Both members are of steel and have as their properties a yield strength $S_y = 78$ kpsi,
$E = 30$ Mpsi, and $\mu = 0.292$. The nominal radii of the barrels are $\frac{3}{16}$, $\frac{3}{8}$, and $\frac{9}{16}$ in.

(a)   Plot the stress distribution for both parts of the barrels.

(b)   What value of interference should be used in assembly?

(c)   When the gun is fired, an internal pressure of 40 kpsi is created. Plot the
resulting stress distributions.

2-63 A cylinder is 25 mm ID by 50 mm OD and is subjected to an internal pressure of
150 MPa. Find the tangential stress at the inner and outer surfaces.

2-64 Find the tangential stress at the inner and outer surfaces of the cylinder of Prob. 2-63
when there is an external pressure of 150 MPa and the internal pressure is zero.

2-65 A steel tire $\frac{3}{8}$ in thick is to be shrunk over a cast-iron rim 16 in in diameter and 1 in
thick. Calculate the inside diameter to which the tire must be bored in order to

induce a hoop stress of 20 kpsi in the tire. The rim is made of grade No. 30 cast iron having $E = 13$ Mpsi and $\mu = 0.211$.

2-66 A carbon-steel gear hub having a nominal hole diameter of 1 in is to be shrink-fitted to a carbon-steel shaft using a class FN4 fit. The hub has a nominal thickness of $\frac{1}{2}$ in.

    (*a*)   Find the maximum tangential and radial stresses in the hub and shaft when the loosest fit is obtained.

    (*b*)   The same as (*a*), except for the tightest fit.

2-67 An alloy steel gear has a hub $\frac{3}{4}$ in thick and a nominal hole diameter of $1\frac{1}{2}$ in. This hub is to be shrink-fitted over a steel shaft using a class FN5 fit. Select suitable steels from Table A-17 such that the hub will not be stressed beyond the tensile yield strength.

Sections 2-20 and 2-21

2-68 Plot the distribution of stresses across section *A-A* of the crane hook shown in part (*a*) of the figure. The load is $F = 5$ kip.

2-69 Find the stress at the inner and outer surfaces at section *A-A* of the frame shown in part (*b*) of the figure if $F = 500$ lb.

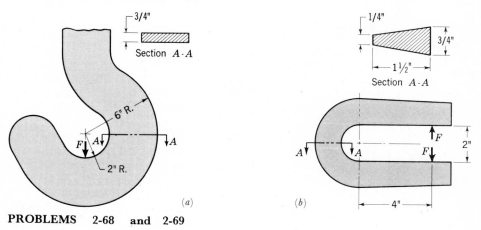

(*a*)   (*b*)

PROBLEMS   2-68   and   2-69

2-70 The eye bolt shown in the figure is made of a cold-drawn low-carbon steel having a yield strength of 62 kpsi. In a complete analysis of the strength of this fastener, the stripping strength of the threads would have to be considered, but in this problem you are to consider only the possibility of failure in the eye. Therefore, based on the yield strength, find the maximum axial pull that can be applied. Next find the maximum bending load that can be applied at *A* if the bolt is supported at *B*.

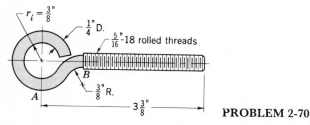

PROBLEM 2-70

2-71   The figure shows a section of a square tube which has been distorted by cold forming it to a right-angle bend with a radius of $r_i = 50$ mm to the inner surface. Using numerical integration find the distance to the centroidal axis measured from the bottom of the section. Also, find the eccentricity e. The grid spacing is 2 mm.

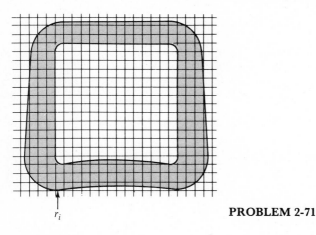

r_i

PROBLEM 2-71

General Problems

2-72*   Part (a) of the figure is an illustration of the No. 3029T mayfly hook No. 14, 2X long, TDE. It is made of forged Sheffield 0.95 carbon wire. The first number is simply the catalog number. The size is No. 14, and the designation 2X long means that the shank is longer than standard and corresponds to that of a No. 12 hook (two sizes larger). TDE means turned-down eye; if the leader is properly attached, this provides a means of giving a straight pull (no eccentricity) from the leader into the shank. This wire has a tensile strength of 375 kpsi and a yield strength of approximately 340 kpsi. What pull would have to be applied to the hook to cause yielding?

2-73*   In this problem you are to find the stresses in the U frame of the coping saw shown in part (b) of the figure. To solve the problem you need only know that the saw blade is assembled with a tension of 65 lb and that the material of the frame is cold-drawn UNS G10180 steel. However, you may be interested to know that the

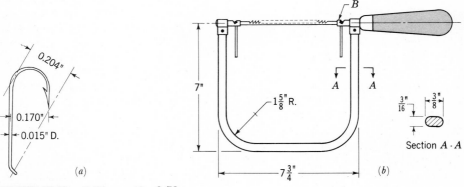

(a)

(b)

PROBLEMS 2-72 and 2-73

blade has a small pin at each end which engages the slotted jaws. Jaw *B* has a long threaded shank which mates with internal threads in the handle of the saw. Thus, unscrewing the handle permits the two arms of the U to spread apart, relieves the tension on the blade, and allows the blade to be removed and replaced with a new one or rotated about its axis to a more favorable cutting position relative to the work.

2-74* The disk sander shown in part (*a*) of the figure uses a $\frac{1}{3}$-hp motor with a full-load speed of 3440 rpm.

- (*a*) What force *F* should be applied to the wood block to develop the full capacity of the motor?
- (*b*) Bearing *A* takes both thrust and radial loads; bearing *B* will take only a radial load. Find the bearing reactions due to the force *F* found in part (*a*).
- (*c*) Determine the magnitude and location of the maximum principal stress in the shaft.
- (*d*) List all assumptions made in solving this problem.

2-75* Part (*b*) of the figure shows the exterior dimensions of a 1-gal paint container. You are to design a wire handle to snap into the lugs on centerline *A-A*. The bending stress in the wire should not exceed 60 kpsi if a permanent set is to be avoided. Determine the diameter and geometric configuration of the wire. The paint weighs 13 lb. Remember, the handle should fold out of the way when the buckets are nested in a shipping carton.

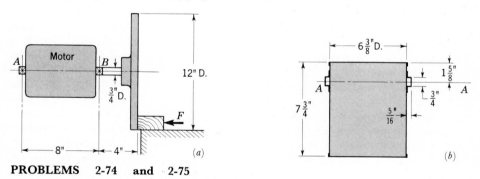

PROBLEMS   2-74   and   2-75

2-76* The tentative design of a welded hand-operated lever for use on an agricultural machine is illustrated in the figure. It consists of a tapered steel handle, fitted with a rubber grip, welded to a thick-walled tube on centerline *A-A* which fits on a stud attached to the machine frame. The operator applies a force *F* to the grip while in a seated position with the handle at his right side and approximately at his waistline. Reactions to this force are the output force at *B*, which is always perpendicular to the 3-in dimension, and the stud reaction at *A*. Regardless of the output force required, the lever should be designed so that it will take any load *F* that the operator might choose to apply because the forward lever motion is arrested by a stop.

- (*a*) Draw free-body diagrams of each element of the lever assembly and find the magnitude and direction of all forces which act.
- (*b*) Determine a suitable set of dimensions *t* and *w* for the tapered handle, and *b* and *h* for the two webs. Use a low-carbon hot-rolled steel having a yield strength of 45 kpsi and a generous factor of safety. (*Note:* The design of welded

joints will be considered in a later chapter; so assume the welds will be adequate.)

(c) Find the maximum principal stresses in the thick-walled tubes and their locations. Are these stresses significant?

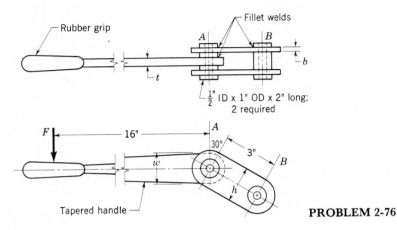

PROBLEM 2-76

2-77* A torsion tube is made by drilling a hole through the side of a steel tube $1\frac{1}{8}$ in OD $\times \frac{7}{8}$ in ID and inserting a handle as shown in part (a). Both the tube and the handle are of cold-drawn steel having a yield strength of 60 kpsi. An optimum length and diameter for the handle are desired such that a person, using his or her full strength, will neither bend the handle nor crush or distort the wall of the tube where the handle bears against it. Use the smallest factor of safety possible consistent with your assumptions.

2-78* A clamp is to be designed according to the dimensions shown in part (b) of the figure. The material to be used is cast iron having a tensile strength of 20 kpsi and a compressive strength of 60 kpsi. The clamp is to be designed for a limiting normal load of 1200 lb. Referring to section A-A in the figure, the dimension t cannot be less than $\frac{3}{16}$ in because of the limitations of the casting process; the final size selected for t should also be an integral number of sixteenths. Determine a satisfactory set of dimensions for the cross section.

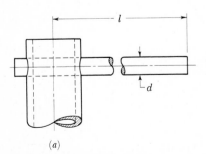

(a)

PROBLEMS 2-77 and 2-78

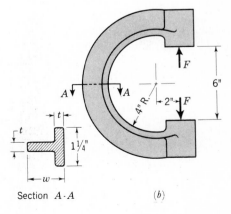

Section A-A (b)

2-79* A steel wheel $\frac{1}{4}$ in wide and 3 in in diameter supports a load of 400 lb while rolling without slipping on a flat surface of steel. Using Fig. 2-36, find the distance in inches below the wheel surface at which the maximum shear stress occurs. Draw a stress element for this point, aligned in the direction of the principal stresses, and compute σ_x, σ_y, and σ_z using Fig. 2-36. Draw Mohr's three-circle diagram and find all three maximum shear stresses. Values of the elastic constants are given in Table A-7.

2-80* Design the horizontal members *IJ* and *IK* of the hydraulic floor crane shown in the figure for a $\frac{1}{2}$-ton load. The members are to be made of hot-rolled welded mechani-

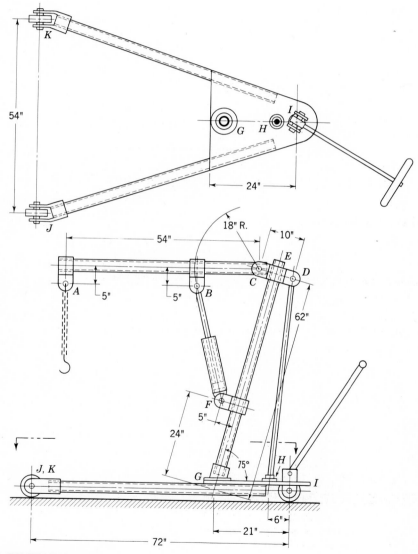

PROBLEM 2-80

cal tubing having a yield strength of 46 kpsi. The factor of safety should not be less than 3. The tubing is continuously welded to the plate at the rear of the crane. Select tubing from the standard sizes in Table A-10 or other source.

2-81* Design the boom of the crane in Prob. 2-80 using the same conditions and tubing specified for that problem.

2-82* Design the tension member *DH* of the crane of Prob. 2-80. Use a low-carbon steel round having a yield strength of 46 kpsi and a factor of safety of at least 1.5. The crane load is $\frac{1}{2}$ ton as in Prob. 2-80.

2-83* Design the mast of the crane in Prob. 2-80 using $S_y = 46$ kpsi and $n = 3$.

2-84* Design the boom of the pulpwood loader of Prob. 3-48. Use two low-carbon hot-rolled steel plates separated by spacers to provide out-of-plane stiffness. Select generous factors of safety and show locations of critical loads and stresses.

CHAPTER

3

DEFLECTION AND STIFFNESS CONSIDERATIONS

A structure or mechanical element is said to be *rigid* when it does not bend, or deflect, or twist too much when an external force, moment, or torque is applied. But if the movement due to the external disturbance is large, the member is said to be *flexible*. The words *rigidity* and *flexibility* are qualitative terms which depend upon the situation. Thus the floor of a building which bends only 0.1 in because of the weight of a machine placed upon it would be considered very rigid if the machine is heavy. But a surface plate which bends 0.01 in because of its own weight would be considered too flexible.

Deflection analysis enters into design situations in many ways. A snap ring, or retaining ring, must be flexible enough so that it can be bent, without permanent deformation, and assembled; and then it must be rigid enough to hold the assembled parts together. In a transmission the gears must be supported by a rigid shaft. If the shaft bends too much, that is, if it is too flexible, the teeth will not mesh properly, resulting in excessive impact, noise, wear, and early failure. In rolling sheet or strip steel to prescribed thicknesses, the rolls must be crowned, that is, curved, so that the finished product will be of uniform thickness. Thus, to design the rolls it is necessary to know exactly how much they will bend when a sheet of steel is rolled between them. Sometimes mechanical elements must be designed to have a particular force-deflection characteristic. The suspension system of an automobile, for example, must be designed within a very narrow range to achieve an optimum bouncing frequency for all conditions of vehicle loading because the human body is comfortable only within a limited range of frequencies.

3-1 SPRING RATES

Elasticity is that property of a material which enables it to regain its original configuration after having been deformed. A *spring* is a mechanical element which exerts a force when deformed. Figure 3-1a shows a straight beam of length l simply supported at the ends and loaded by the transverse force F. The deflection y is linearly related to the force, as long as the elastic limit of the material is not exceeded, as indicated by the graph. This beam can be described as a *linear spring*.

In Fig. 3-1b a straight beam is supported on two cylinders such that the length between supports decreases as the beam is deflected by the force F. A larger force is required to deflect a short beam than a long one, and hence the more this beam is deflected, the stiffer it becomes. Also, the force is not linearly related to the deflection, and hence this beam can be described as a *nonlinear stiffening spring*.

Figure 3-1c is a dish-shaped round disk. The force necessary to flatten the disk increases at first and then decreases as the disk approaches a flat configuration, as shown by the graph. Any mechanical element having such a characteristic is called a *nonlinear softening spring*.

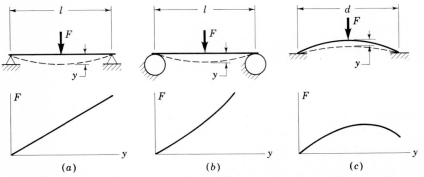

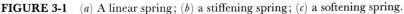

FIGURE 3-1   (a) A linear spring; (b) a stiffening spring; (c) a softening spring.

If we designate the general relationship between force and deflection by the equation

$$F = F(y) \qquad (a)$$

then *spring rate* is defined as

$$k(y) = \lim_{\Delta y \to 0} \frac{\Delta F}{\Delta y} = \frac{dF}{dy} \qquad (3\text{-}1)$$

where y must be measured in the direction of F and at the point of application of F. Most of the force-deflection problems encountered in this book are linear, as in Fig. 3-1a. For these, k is a constant, also called the *spring constant*; consequently Eq. (3-1) is written

$$k = \frac{F}{y} \qquad (3\text{-}2)$$

We might note that Eqs. (3-1) and (3-2) are quite general and apply equally well for torques and moments, provided angular measurements are used for y. For linear displacements the units of k are often lb per in or N per m, and for angular displacements lb $\cdot$ in per radian or N $\cdot$ m per radian.

3-2 TENSION, COMPRESSION, AND TORSION

The relation for the total extension or deformation of a uniform bar has already been developed in Sec. 2-5, Eq. (a). It is repeated here for convenience.

$$\delta = \frac{Fl}{AE} \qquad (3\text{-}3)$$

This equation does not apply to a long bar loaded in compression if there is a possibility of buckling (Sec. 3-12). Using Eqs. (3-2) and (3-3), we see that the spring constant of an axially loaded bar is

$$k = \frac{AE}{l} \qquad (3\text{-}4)$$

The angular deflection of a uniform round bar subjected to a twisting moment T was given in Eq. (2-45), and is

$$\theta = \frac{Tl}{GJ} \qquad (3\text{-}5)$$

where θ is in radians. If we multiply Eq. (3-5) by $180/\pi$ and substitute $J = \pi d^4/32$ for a solid round bar, we obtain

$$\theta = \frac{585 Tl}{Gd^4} \tag{3-6}$$

where θ is in degrees.

Equation (3-5) can be rearranged to give the torsional spring rate as

$$k = \frac{T}{\theta} = \frac{GJ}{l} \tag{3-7}$$

When the word "simple" is used to describe the loading, the meaning is that no other load is present and that no geometric complexities are present. Thus, a bar loaded in *simple tension* is a uniform bar acted upon by a tensile load directed along the centroidal axis, and no other loads are acting on the bar.

3-3 DEFLECTION DUE TO BENDING

Beams deflect a great deal more than axially loaded members, and the problem of bending probably occurs more often than any other loading problem in design. Shafts, axles, cranks, levers, springs, brackets, and wheels, as well as many other elements, must often be treated as beams in the design and analysis of mechanical structures and systems. The subject of bending, however, is one which you should have studied as preparation for reading this book. It is for this reason that we include here only a brief review to establish the nomenclature and conventions to be used throughout this book.

In Sec. 2-9 we developed the relation for the curvature of a beam subjected to a bending moment M [Eq. (2-30)]. The relation is

$$\frac{1}{\rho} = \frac{M}{EI} \tag{3-8}$$

where ρ is the radius of curvature. From studies in mathematics we also learn that the curvature of a plane curve is given by the equation

$$\frac{1}{\rho} = \frac{d^2y/dx^2}{[1 + (dy/dx)^2]^{3/2}} \tag{3-9}$$

where the interpretation is that y is the deflection of the beam at any point x along its length. The slope of the beam at any point x is

$$\theta = \frac{dy}{dx} \tag{a}$$

For many problems in bending, the slope is very small, and for these the denominator of Eq. (3-9) can be taken as unity. Equation (3-8) can then be written

$$\frac{M}{EI} = \frac{d^2y}{dx^2} \tag{b}$$

Noting Eqs. (2-25) and (2-26) and successively differentiating Eq. (b) yields

$$\frac{V}{EI} = \frac{d^3y}{dx^3} \tag{c}$$

$$\frac{q}{EI} = \frac{d^4y}{dx^4} \tag{d}$$

It is convenient to display these relations in a group as follows:

$$\frac{q}{EI} = \frac{d^4y}{dx^4} \tag{3-10}$$

$$\frac{V}{EI} = \frac{d^3y}{dx^3} \tag{3-11}$$

$$\frac{M}{EI} = \frac{d^2y}{dx^2} \tag{3-12}$$

$$\theta = \frac{dy}{dx} \tag{3-13}$$

$$y = f(x) \tag{3-14}$$

The nomenclature and conventions are illustrated by the beam of Fig. 3-2. Here, a beam of length $l = 20$ in is loaded by the uniform load $w = 80$ lb per in of beam length. The x axis is positive to the right, and the y axis positive upward. All quantities, loading, shear, moment, slope, and deflection, have the same sense as y; they are positive if upward, negative if downward.

The values of the quantities at the ends of the beam, where $x = 0$ and $x = l$, are called their *boundary values*. For this reason the beam problem is often called a *boundary-value problem*. The reactions $R_1 = R_2 = +800$ lb and the shear forces $V_0 = +800$ lb and $V_l = -800$ lb are easily computed using the methods of Chap. 2. The bending moment is zero at each end because the beam is simply supported. Note that the beam-deflection curve must have a negative slope at the left boundary and a positive slope at the right boundary. This is easily seen by examining the deflection of Fig. 3-2. The magnitude of the slope at the bound-

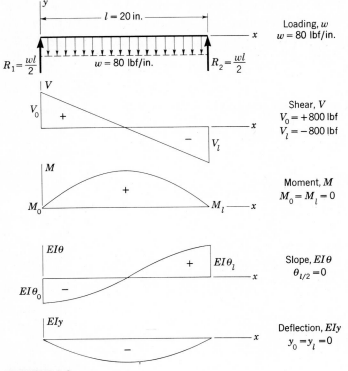

FIGURE 3-2

aries is, as yet, unknown; because of symmetry, however, the slope is known to be zero at the center of the beam. We note, finally, that the deflection is zero at each end.

3-4 DEFLECTIONS BY USE OF SINGULARITY FUNCTIONS

The following examples are illustrations of the use of singularity functions.

Example 3-1   As a first example, we choose the beam of Table A-12-6, which is a simply supported beam having a concentrated load F not in the center. Writing Eq. (3-10) for this loading gives

$$EI \frac{d^4y}{dx^4} = q = -F\langle x - a \rangle^{-1} \qquad 0 < x < l \tag{1}$$

Note that the reactions R_1 and R_2 do not appear in this equation, as they did in Chap. 2, because of the range chosen for x. If we now integrate from 0 to x, not

$-\infty$ to x, according to Eq. (3-11) we get

$$EI \frac{d^3y}{dx^3} = V = -F\langle x - a\rangle^0 + C_1 \tag{2}$$

Using Eq. (3-12) this time we integrate again and obtain

$$EI \frac{d^2y}{dx^2} = M = -F\langle x - a\rangle^1 + C_1 x + C_2 \tag{3}$$

At $x = 0$, $M = 0$, and Eq. (3) gives $C_2 = 0$. At $x = l$, $M = 0$, and Eq. (3) gives

$$C_1 = \frac{F(l - a)}{l} = \frac{Fb}{l}$$

Substituting C_1 and C_2 into Eq. (3) gives

$$EI \frac{d^2y}{dx^2} = M = \frac{Fbx}{l} - F\langle x - a\rangle^1 \tag{4}$$

Note that we could have obtained this equation by summing moments about a section a distance x from the origin. Next, integrate Eq. (4) twice in accordance with Eqs. (3-13) and (3-14). This yields

$$EI \frac{dy}{dx} = EI\theta = \frac{Fbx^2}{2l} - \frac{F\langle x - a\rangle^2}{2} + C_3 \tag{5}$$

$$EIy = \frac{Fbx^3}{6l} - \frac{F\langle x - a\rangle^3}{6} + C_3 x + C_4 \tag{6}$$

The constants of integration C_3 and C_4 are evaluated using the two boundary conditions $y = 0$ at $x = 0$, and $y = 0$ at $x = l$. The first condition, substituted into Eq. (6), gives $C_4 = 0$. The second conditon, substituted into Eq. (6), yields

$$0 = \frac{Fbl^2}{6} - \frac{Fb^3}{6} + C_3 l$$

whence

$$C_3 = -\frac{Fb}{6l}(l^2 - b^2)$$

Upon substituting C_3 and C_4 into Eq. (6) we obtain the deflection relation as

$$y = \frac{F}{6EIl} \left[bx(x^2 + b^2 - l^2) - l\langle x - a \rangle^3 \right] \tag{7}$$

Compare Eq. (7) with the two deflection equations in Table A-12-6 and note the use of singularity function enables us to express the entire relation with a single equation.   ////

Example 3-2 Find the deflection relation, the maximum deflection, and the reactions for the statically indeterminate beam shown in Fig. 3-3a using singularity functions.

Solution The loading diagram and approximate deflection curve are shown in Fig. 3-3b. Based upon the range $0 < x < l$, the loading equation is

$$q = R_2 \langle x - a \rangle^{-1} - w \langle x - a \rangle^0 \tag{1}$$

We now integrate this equation four times in accordance with Eqs. (3-10) to (3-14). The results are

$$V = R_2 \langle x - a \rangle^0 - w \langle x - a \rangle^1 + C_1 \tag{2}$$

$$M = R_2 \langle x - a \rangle^1 - \frac{w}{2} \langle x - a \rangle^2 + C_1 x + C_2 \tag{3}$$

$$EI\theta = \frac{R_2}{2} \langle x - a \rangle^2 - \frac{w}{6} \langle x - a \rangle^3 + \frac{C_1}{2} x^2 + C_2 x + C_3 \tag{4}$$

$$EIy = \frac{R_2}{6} \langle x - a \rangle^3 - \frac{w}{24} \langle x - a \rangle^4 + \frac{C_1}{6} x^3 + \frac{C_2}{2} x^2 + C_3 x + C_4 \tag{5}$$

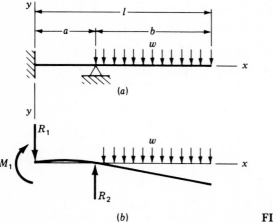

(a)

(b)   FIGURE 3-3

As in the previous example, the constants are evaluated by applying appropriate boundary conditions. First we note from Fig. 3-3b that both $EI\theta = 0$ and $EIy = 0$ at $x = 0$. This gives $C_3 = 0$ and $C_4 = 0$. Next we observe that at $x = 0$ the shear force is the same as the reaction. So Eq. (2) gives $V(0) = R_1 = C_1$. Since the deflection must be zero at the reaction R_2 where $x = a$, we have from Eq. (5) that

$$\frac{C_1}{6} a^3 + \frac{C_2}{2} a^2 = 0 \qquad \text{or} \qquad C_1 \frac{a}{3} + C_2 = 0 \tag{6}$$

Also the moment must be zero at the overhanging end where $x = l$; Eq. (3) then gives

$$R_2(l - a) - \frac{w}{2}(l - a)^2 + C_1 l + C_2 = 0$$

Simplifying and noting that $l - a = b$ gives

$$C_1 a + C_2 = -\frac{wb^2}{2} \tag{7}$$

Equations (6) and (7) are now solved simultaneously for C_1 and C_2. The results are

$$C_1 = R_1 = -\frac{3wb^2}{4a} \qquad C_2 = \frac{wb^2}{4}$$

With R_1 known, we can sum the forces in the y direction to zero to get R_2. Thus

$$R_2 = -R_1 + wb = \frac{3wb^2}{4a} + wb = \frac{wb}{4a}(4a + 3b)$$

The moment reaction M_1 is obtained from Eq. (3) with $x = 0$. This gives

$$M(0) = M_1 = C_2 = \frac{wb^2}{4}$$

The complete equation of the deflection curve is obtained by substituting the known terms for R_2 and the constants C_i into Eq. (5). The result is

$$EIy = \frac{wb}{24a}(4a + 3b)\langle x - a \rangle^3 - \frac{w}{24}\langle x - a \rangle^4 - \frac{wb^2 x^3}{8a} + \frac{wb^2 x^2}{8} \tag{8}$$

The maximum deflection occurs at the free end of the beam at $x = l$. By making this substitution into Eq. (8) and manipulating the resulting expression, one finally obtains

$$y_{max} = -\frac{wb^3l}{8EI} \tag{9}$$

Examination of the deflection curve of Fig. 3-3b reveals that the curve will have a zero slope at some point between R_1 and R_2. By replacing the constants in Eq. (4), setting $\theta = 0$, and solving the result for x we readily find that the slope is zero at $x = 2a/3$. The corresponding deflection at this point can be obtained by using this value of x in Eq. (8). The result is

$$y(x = 2a/3) = \frac{wa^2b^2}{54EI} \tag{10}$$

////

3-5   THE METHOD OF SUPERPOSITION

While an engineer may enjoy the challenge of solving beam problems and it may sharpen his or her problem-solving capability, a designer has more important and more urgent things to do. If someone else has already solved the problem, the practicing engineer can save an employer money and increase the flow of cash to the stockholders by using worked-out solutions. The solutions to bending problems which occur most frequently in design are included in Table-A-12. Solutions to still other beam-deflection problems may be found in various handbooks. If the solution to a particular problem cannot be found, it still may be possible to obtain one using the method of superposition. This method can be employed for all linear force-deflection problems, that is, those in which the force and deflection are linearly related. The *method of superposition* uses the principle that the deflection at any point in a beam is equal to the sum of the deflections caused by each load acting separately. Thus, if a beam is bent by three separate forces, the deflection at a particular point is the sum of three deflections, one for each force.

Example 3-3   Figure 3-4a shows a member loaded in bending by the forces F_1 and F_2. The deflection at B must not exceed 200 μm. Using the method of superposition, determine the size of a steel tube to support these loads safely.

Solution   In Fig. 3-4b the beam has been redrawn with only the force F_1 acting. And in Fig. 3-4c only the force F_2 is shown acting. The total deflection, by superposition, at B is the sum of the deflections due separately to F_1 and F_2 acting.

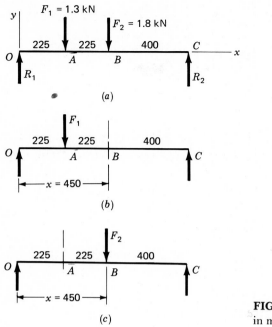

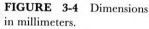

FIGURE 3-4 Dimensions in millimeters.

From Table A-12-6, for Fig. 3-4b we have

$$y'_B = \frac{Fa(l-x)}{6EIl} \, (x^2 + a^2 - 2lx) \tag{1}$$

The values are $F = 1300$ N, $a = 225$ mm, $(l - x) = 400$ mm, $E = 207$ GPa, $l = 850$ mm, and $x = 450$ mm. Using Table A-6, we have

$$y'_B = \frac{1300(225)(400)}{6(207)(850)I} \, [(450)^2 + (225)^2 - 2(850)(450)]$$

$$= \frac{-5.673(10)^7}{I} \, \mu m$$

For Fig. 3-4c, we have also from Table A-12-6 that

$$y''_B = \frac{Fbx}{6EIl} \, (x^2 + b^2 - l^2) \tag{2}$$

Here $F = 1800$ N, $b = 400$ mm, $x = 450$ mm, and the rest of the values are the same as before. Solving gives

$$y_B'' = \frac{1800(400)(450)}{6(207)(850)I} [(450)^2 + (400)^2 - (850)^2]$$

$$= \frac{-1.1049(10)^8}{I} \mu m$$

Then the total deflection at B is

$$y_B = y_B' + y_B'' = -\frac{1.672}{I} (10)^8 \mu m$$

Since y_B is not to exceed 200 μm, we have that

$$I \geq \frac{1.672(10)^8}{200} = 8.36(10)^5 \text{ mm}^4$$

From Table A-14, the area moment of inertia of a hollow section is

$$I = \frac{\pi}{64} (d^4 - d_i^4)$$

By programming this equation on, say, the programmable calculator, it is easy to try various values of the diameter and wall thickness. By this method we find that a tube having a 70 mm OD and a wall thickness of 10 mm has a moment of inertia of $I = 8.72(10)^5$ mm^4, which is satisfactory.   ////

3-6 THE GRAPHICAL-INTEGRATION METHOD

It often happens that the geometry, or the method of loading a beam, makes the deflection problem so difficult that it is impractical to solve it by classical methods. Under such conditions one can employ numerical integration, using desk calculators or electronic computers, or graphical integration. Though of limited accuracy, the graphical approach is fast, and it provides a good physical understanding of what is happening. For many purposes the accuracy is completely satisfactory.

Graphical integration can be explained by reference to Fig. 3-5; the function is plotted in a and the integral in b. Three graphical scales are necessary, one for the dependent variable, one for the independent variable, and one for the integral. Let us denote x as the independent variable and y as the dependent variable. Then the integral of the function $y = f(x)$ between, say, $x = a$ and $x = b$ is simply the area below the curve $y = f(x)$ between ordinates erected to x at a and b. In

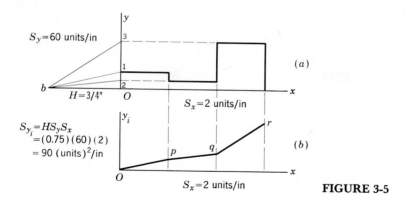

$S_y = 60$ units/in

$H = 3/4"$

b

(a)

$S_x = 2$ units/in

$S_{y_i} = HS_yS_x$
    $= (0.75)(60)(2)$
    $= 90$ (units)^2/in

(b)

p q r

$S_x = 2$ units/in

FIGURE 3-5

graphical integration we choose the two ordinates close together and average them so that the procedure consists merely in finding a graphical scheme which will enable us to draw lines whose slopes are proportional to the average ordinate in each interval.

To integrate the function of Fig. 3-5a choose the line Ob, called the *pole distance* and designated as H, any convenient length, say $\frac{3}{4}$ in. Project the heights of the various rectangles to the y axis and draw lines connecting these intersections with the pole b. Each line, b-1, b-2, and b-3, has a slope proportional to the height of the corresponding rectangle. The integral in Fig. 3-5b is obtained by drawing lines Op parallel to b-1, pq parallel to b-2, and qr parallel to b-3. The scale of the integral is

$$S_{yi} = HS_x S_y \qquad (3\text{-}15)$$

where H = pole distance, in
    S_x = scale of x in units of x per inch
    S_y = scale of y in units of y per inch
    S_{yi} = scale of the integral in units of x times units of y per inch

---

Example 3-4 This method of solution is now applied to the problem of obtaining the deflection of a beam (Fig. 3-6). In this figure is a stepped shaft (1), originally drawn one-fourth size, which is further reduced for reproduction reasons. Loading diagram (2) shows a load of 50 lb per in extending over a distance of 10 in of the shaft length. The reactions at the bearings have been calculated and are indicated as R_1 and R_2. The shearing-force diagram (3) has been obtained from the loading diagram, using the conditions of static equilibrium. The moment diagram (4) was obtained by graphically integrating the shearing-force diagram. The construction is shown, and also the calculation for the scale of the moment diagram.

The next step is to obtain the numerical values of the moment at selected points along the shaft. This is done by scaling the diagram. The values of the

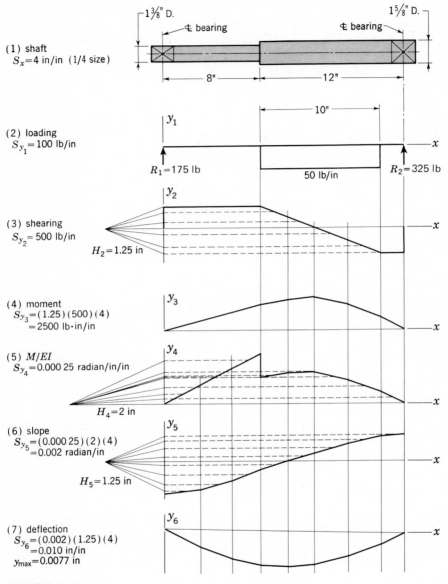

FIGURE 3-6

moment of inertia are then calculated for each diameter. The moments are next divided by the products of the modulus of elasticity ($E = 30$ Mpsi) and the moments of inertia, and these values are plotted (5) to obtain the M/EI diagram. (If the moment of inertia is constant, this operation may be performed after the deflection curve is obtained. In case this is done, note that the deflection curve then becomes the yEI curve.) The M/EI diagram is now integrated twice to obtain the deflection curve (7). In integrating the slope diagram (6) it is necessary to guess at the location of zero slope, that is, the position at which to place the x axis. Should this guess be wrong, and it usually is, the deflection curve will not close with a horizontal line. The line should be drawn so as to close the deflection curve, and measurements of deflection made in the vertical direction. (Do not measure perpendicular to the closing line unless it is horizontal.) The correct location of zero slope is found as follows. Draw a line parallel to the closing line and tangent to the deflection curve. The point of tangency is the point of zero slope, and this is also the location of the maximum deflection. ////

3-7 NUMERICAL INTEGRATION

Instead of using graphical methods to determine beam deflections, we can employ numerical integration using a programmable calculator or any home computer. An example that is included in this section requires only 30 data locations, and the program is quite short.

The *trapezoidal method* of numerical integration can be applied to any function $f = f(x)$, no matter whether it is continuous or discontinuous. For example, in Fig. 3-7a we have computed the values $y_0, y_1, y_2, \ldots, y_N$ of a function $y = f(x)$ corresponding to $x_0, x_1, x_2, \ldots, x_N$, respectively. Note that each pair of x values is separated by the same amount Δx.

The integral of a function $y = f(x)$ is simply the area between the function $f(x)$ and the x axis. Thus an approximate method of integrating the function y in Fig. 3-7a is to find the areas enclosed in each segment and add them to get the total, which is then the approximate integral.

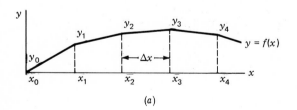

(a)

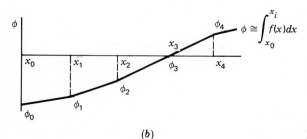

(b)

FIGURE 3-7

In Fig. 3-7a the areas of the first two segments are, respectively,

$$\frac{y_0 + y_1}{2}\,\Delta x \qquad \text{and} \qquad \frac{y_1 + y_2}{2}\,\Delta x$$

Thus, for each segment, the area is the mean value of the two ordinates times the increment Δx. If we designate the integral as ϕ, then for the first segment, we have

$$\phi_1 = \int_{x_0}^{x_1} f(x)\,dx \cong \frac{y_0 + y_1}{2}\,\Delta x + \phi_0 \tag{3-16}$$

The integral at the end of the second segment is

$$\phi_2 = \int_{x_0}^{x_2} f(x)\,dx = \frac{y_1 + y_2}{2}\,\Delta x + \phi_1 \tag{3-17}$$

Thus, in general

$$\phi_i = \frac{y_{i-1} + y_i}{2}\,\Delta x + \phi_{i-1} \tag{3-18}$$

Note that the approximation sign has been replaced by an equals sign merely for convenience.

The value of the term ϕ_0 in Eq. (3-16) is dictated by the constraints on the system, as will be shown.

Sometimes a discontinuity like that of Fig. 3-8a occurs in the function to be integrated. This would cause no problem in a step-by-step arithmetic hand integration. But both $y_{i,\,max}$ and $y_{i,\,min}$ have to be stored at two x_i storage locations when a programmable calculator is used. Then the integration algorithms require a special means for indentifying these double data points. This difficulty can be avoided by using the average

$$y_{i,\,av} = \frac{y_{i,\,max} + y_{i,\,min}}{2} \tag{3-19}$$

at this point, because the area between the stations x_{i-1} and x_{i+1} is the same.

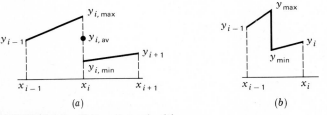

(a)   (b)

FIGURE 3-8   Other discontinuities.

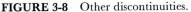

A similar type of discontinuity sometimes occurs inside the increment Δx as shown in Fig. 3-8b. In this case, disregard the discontinuity and use Eq. (3-18) as before; these errors tend to average out.

When a beam is statically determinant, you can compute the bending moment at a variety of points. Since the dimensions of the beam and its material are known, the moment values can be divided by the product EI to get a set of M/EI values. Figure 3-9 is a sketch of an M/EI diagram and a corresponding slope and deflection diagram, both of which could have been obtained by integrating the M/EI function using Eq. (3-18).

If we substitute the moment and slope functions into Eq. (3-18) for the first increment, we get

$$\theta_1 = \left[\left(\frac{M}{EI} \right)_0 + \left(\frac{M}{EI} \right)_1 \right] \left(\frac{\Delta x}{2} \right) + \theta_0 \tag{3-20}$$

and for the deflection

$$y_1 = (\theta_0 + \theta_1) \left(\frac{\Delta x}{2} \right) + y_0 \tag{3-21}$$

When the integrations are first performed it is necessary to estimate a value to use for the starting value of the slope θ_0. Designate this estimated value as θ_0^*, as shown in Fig. 3-9b. The corresponding values of y obtained from this integration

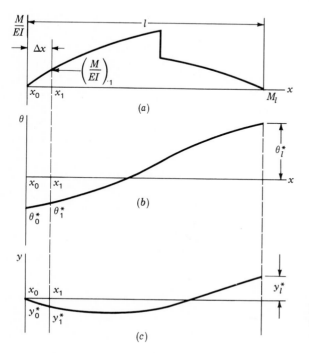

FIGURE 3-9 (a) Moment function; (b) slope function; (c) deflection function.

are also designated as y^*, as shown in Fig. 3-9c. If the beam has simple supports at $x = 0$ and at $x = l$, then $y_0^* = 0$, but incorrect results for y will be obtained because θ_0^* was estimated.

If the estimate of θ_0^* is too small, then, at the other end of the diagram θ_l^* will be too large and, as a result y_l^* will be positive. If the estimate of θ_0^* is too large, then θ_l^* will be too small and y_l^* will be negative.

Since the beam has simple supports, y_l^* should be zero. Thus, our problem is to find a starting value for the slope that will cause the deflection to be zero at $x = l$. Since even the programmable calculator will integrate from $x = 0$ to $x = l$ in a fairly short time, it is possible to perform the double integration twice, each with a different starting value of θ_0^*. This will result in two corresponding values of y_l^*. From these two runs we can find the correct starting value, here called θ_0.

Let $\theta_0^* = \theta_A$ and $y_l^* = y_A$ be the results of the first run, and $\theta_0^* = \theta_B$ and $y_l^* = y_B$ the data for the second run. For good results try to choose these such that y_A and y_B have opposite signs and are fairly close to zero. Then from the equation of a straight line we can write

$$y_A = a + b\theta_A \qquad \text{and} \qquad y_B = a + b\theta_B$$

Solving for a and b gives

$$a = \frac{y_B \theta_A - y_A \theta_B}{\theta_A - \theta_B} \qquad b = \frac{y_A - y_B}{\theta_A - \theta_B} \tag{a}$$

With $y_l = 0$, we have that $a + b\theta_0 = 0$. Thus

$$\theta_0 = -\frac{a}{b} = \frac{y_B \theta_A - y_A \theta_B}{y_B - y_A} \tag{3-22}$$

is the correct value to use because it makes $y_l = 0$.

The number of stations or segments used depends upon the accuracy desired and the storage limitations of the calculator or computer. Very good results can be obtained using $N = 20$ and the results will be quite acceptable with $N = 10$. The flow chart shown in Fig. 3-10 was used to create the program for the example in this section. The following data storage locations are required: $\Delta x/2$, i, N, address of $(M/EI)_{i-1}$ and $(M/EI)_i$ for indirect addressing, θ_{i-1}, θ_i, y_{i-1}, y_i, $(M/EI)_0$, $(M/EI)_1$, $(M/EI)_2$, ..., $(M/EI)_N$.

Note, in the step $i = i + 1$ in Fig. 3-10 that the addresses for $(M/EI)_{i-1}$ and $(M/EI)_i$ must also be incremented because of the use of indirect addressing. Setting a flag when the correct value of θ_0 is entered permits each θ and y value to be recorded in each loop during the final run.

If you have program room left, it is desirable to add two subprograms. The first of these is used to enter the M/EI values. Again, indirect addressing can be used to shorten the entry time. The second subprogram is used to solve Eq.

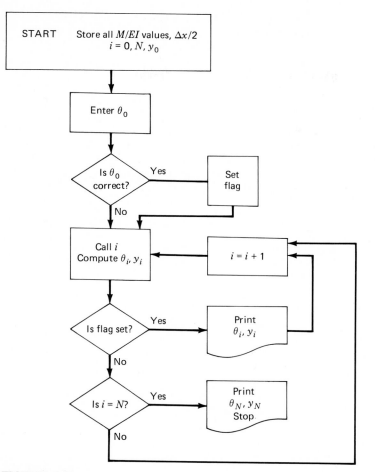

FIGURE 3-10

(3-22). This equation is not difficult to solve, but the numbers to be substituted may consist of eight or ten digits and be in exponential notation, making for an unwieldable computation.

Example 3-5 Using 20 steps, solve Example 3-4 by numerical integration.

Solution A tabulation of M and M/EI values is shown in Table 3-1. The table also lists the values of θ and y obtained in the final run.

At station 8 there are two M/EI values because of the change in the moment of inertia. These are averaged to obtain a single value for use in the integration.

The two trial runs are not tabulated. The first run gave $\theta_A = -1.36(10)^{-3}$ rad and $y_A = -1.407(10)^{-2}$ in. The second run gave $\theta_B = 5.0(10)^{-4}$ rad and

Table 3-1 NUMERICAL INTEGRATION OF THE BEAM OF FIG. 3-6*

x_i in	M_i lb·in	$(M/EI)_i \times 10^{-4}$ rad/in	$\theta_i \times 10^{-4}$ rad	$y_i \times 10^{-3}$ in
0	0	0	−12.96	0
1	175	0.3324	−12.80	−1.29
2	350	0.6649	−12.30	−2.54
3	525	0.9973	−11.47	−3.73
4	700	1.2398	−10.35	−4.82
5	875	1.6622	−8.90	−5.79
6	1050	1.9947	−7.07	−6.58
7	1225	2.3271	−4.91	−7.18
8	1400	$\left\{\begin{matrix} 2.6596 \\ 1.3635 \end{matrix}\right\}$†	−2.74	−7.57
9	1550	1.5095	−0.980	−7.75
10	1650	1.6069	0.578	−7.77
11	1700	1.6556	2.21	−7.63
12	1700	1.6556	3.86	−7.33
13	1650	1.6069	5.50	−6.86
14	1550	1.5095	7.05	−6.23
15	1400	1.3635	8.49	−5.46
16	1200	1.1687	9.76	−4.54
17	950	0.9252	10.80	−3.52
18	650	0.6330	11.58	−2.40
19	325	0.3165	12.06	−1.21
20	0	0	12.22	0

  * All values except M_i are rounded off.
  † $(M/EI)_8 = 2.0116$ rad/in averaged.

$y_B = 1.593(10)^{-2}$ in. Substituting these into Eq. (3-22) gave $\theta_0 = -1.296(10)^{-3}$ rad as the correct starting value, as shown in Table 3-1.   ////

Professor Mischke of Iowa State University points out that there are certain situations in which the use of trapezoidal integration may introduce serious errors. He states:

> Because of the nature of stepped shafts, M/EI is piecewise linear. This means that the first integration of piecewise linear function can be carried out using a trapezoidal rule, and the result is exact. The second integration is of a piecewise quadratic function, and the repeated use of the trapezoidal rule is inexact.*

  * Charles R. Mischke, "An Exact Numerical Method for Determining the Bending Deflection and Slope of Stepped Shafts," *Advances in Reliability and Stress Analysis*, ASME, 1979, pp. 101–115.

Mischke further observes that this error only occurs when the moment diagram crosses the abscissa. Thus we can safely employ trapezoidal integration for both slope and deflection, provided the moment does not change sign. If the moment does change sign, then the trapezoidal rule can still be used to obtain the slope, but now Simpson's rule must be used to derive the deflection function from the slope. Simpson's rule is based on fitting a parabola to three of the points of the function to be integrated. With the slope given, the equation is

$$y_i = \frac{\Delta x}{6} \left(\theta_{i-2} + 4\theta_{i-1} + \theta_i\right) + y_{i-2} \tag{3-23}$$

Note that this is an integration over two steps instead of one as for trapezoidal integration. This means that if you choose 10 steps for y, then you must have 20 steps for θ. In the case of trapezoidal integration it was possible to integrate for θ and y successively in each step. This cannot be done using Simpson's rule because three points on the slope function must be obtained before the corresponding point on the deflection function can be found. Though this requires more program memory, the problem can be handled quite nicely on the calculators with large numbers of data registers.

3-8   THE AREA-MOMENT METHOD

In Sec. 2-7, we learned methods for deriving the moment diagram directly from the loading diagram by employing the principles of statics. We have seen in this chapter that the moment diagram can also be obtained from the loading diagram by integrating twice. But the deflection diagram can be obtained from the moment diagram by integrating twice also. Thus it should be possible to derive the deflection diagram from the moment diagram by applying the principles of statics.

This method of deflection determination is called the *area-moment method*. It can be stated as follows: *the vertical distance between any point A on a deflection curve and a tangent through point B on the curve is the moment with respect to A of the area of the moment diagram between A and B divided by the stiffness EI.* To use this statement, first find the area of the parts of the moment diagram or, if preferred, the M/EI diagram. Then, second, multiply the areas by their centroidal distances from the axis of moments. Areas and centroidal distances for typical portions of moment diagrams are shown in Fig. 3-11.

Example 3-6   Figure 3-12a shows a $1\frac{1}{4}$-in-diameter steel shaft upon which are mounted two gears. If the shaft bends excessively, the gears will mesh improperly and an early failure can be expected. In this example the gear forces, shown in Fig. 3-12b, are assumed to be in the same plane and of magnitude $F_1 = 120$ lb and

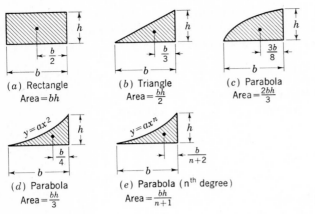

(a) Rectangle
Area = bh

(b) Triangle
Area = $\frac{bh}{2}$

(c) Parabola
Area = $\frac{2bh}{3}$

(d) Parabola
Area = $\frac{bh}{3}$

(e) Parabola (n^{th} degree)
Area = $\frac{bh}{n+1}$

FIGURE 3-11

$F_2 = 90$ lb.   Find the shaft deflection at each gear.   The loading diagram indicates that the bearings are self-aligning.

Solution   This problem can be solved by employing the area-moment method only once, but we shall apply it twice, to illustrate the method of superposition again.

The moment of inertia is $I = \pi d^4/64 = \pi(1.25)^4/64 = 0.120$ in^4.   Therefore $EI = 30(10)^6(0.120) = 3.6(10)^6$ lb·in^2.   Our first step is to calculate the deflections due only to F_1 acting.   In Fig. 3-13 the loading and moment diagrams have been constructed.   By statics the two reactions are found, and then the moments at A and B.   An exaggerated deflection curve is drawn on the loading diagram, and a tangent constructed at the left-hand reaction.   Then, by the area-moment method, the distance $C'C$ is equal to the moment of the area of the moment diagram about C, divided by EI. Thus

$$EI(C'C) = \overset{\text{area}}{[(200/2)(2)]} \; \overset{\text{arm}}{(10\tfrac{2}{3})} + \overset{\text{area}}{[(200/2)(10)]} \; \overset{\text{arm}}{(6\tfrac{2}{3})} = 8800$$

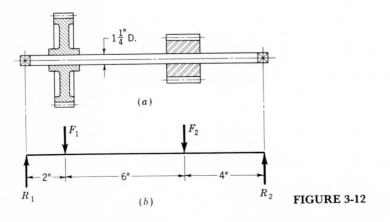

FIGURE 3-12

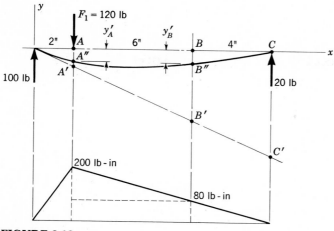

FIGURE 3-13

So

$$C'C = \frac{8800}{3.6(10)^6} = 2.44(10)^{-3} \text{ in}$$

In this approach, the areas may be either positive or negative, but the moment arms are *always* positive. The distance $C'C$ is a positive quantity, measured from C' to C. However, it is easier to keep track of the signs in this approach using carefully made sketches of the moment and deflection diagrams. By similar triangles we also find

$$A'A = 4.074(10)^{-4} \text{ in} \qquad B'B = 1.630(10)^{-3} \text{ in}$$

Next, we find the distance $A'A''$ by taking moments about A.

$$EI(A'A'') = \underset{\text{area}}{[(200/2)(2)]} \underset{\text{arm}}{(\tfrac{2}{3})} = 133$$

$$A'A'' = \frac{133}{3.6(10)^6} = 3.704(10)^{-5} \text{ in}$$

The distance $B'B''$ is found by taking the moments of two triangular areas and one rectangular area about B.

$$EI(B'B'') = \underset{\text{area}}{[(200/2)(2)]} \underset{\text{arm}}{(6\tfrac{2}{3})} + \underset{\text{area}}{[(80)(6)]} \underset{\text{arm}}{(3)} + \underset{\text{area}}{[(120/2)(6)]} \underset{\text{arm}}{(4)} = 4213$$

$$B'B'' = \frac{4213}{3.6(10)^6} = 1.17(10)^{-3} \text{ in}$$

The deflection at A due only to F_1 is now found to be

$$y'_A = A'A'' - A'A = 3.704(10)^{-4} \text{ in}$$

In a similar manner,

$$y'_B = B'B'' - B'B = -4.60(10)^{-4} \text{ in}$$

The next step is to calculate the deflections at A and B due only to F_2. This is done in a similar manner, and the results are

$$y''_A = -3.45(10)^{-4} \text{ in} \qquad y''_B = -7.11(10)^{-4} \text{ in}$$

By superposition, the total deflection is the sum of the deflections caused by each load acting separately. Hence

$$y_A = -7.15(10)^{-4} \text{ in} \qquad y_B = -11.71(10)^{-4} \text{ in} \qquad\qquad ////$$

3-9 STRAIN ENERGY

The external work done on an elastic member in deforming it is transformed into *strain*, or *potential*, *energy*. The potential energy stored by a member when it is deformed through a distance y is the average force times the deflection, or

$$U = \frac{F}{2}y = \frac{F^2}{2k} \tag{3-24}$$

Equation (3-24) is general in the sense that the force F also means torque, or moment, provided, of course, that consistent units are used for k. By substituting appropriate expressions for k, formulas for strain energy for various simple loadings may be obtained. For simple tension and compression, for example, we employ Eq. (3-4) and obtain

$$U = \frac{F^2 l}{2AE} \tag{3-25}$$

For torsion we use Eq. (3-7); the torsional strain energy is then found to be

$$U = \frac{T^2 l}{2GJ} \tag{3-26}$$

To obtain an expression for the strain energy due to direct shear, consider the

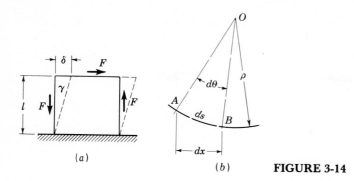

(a)

(b) **FIGURE 3-14**

element with one side fixed in Fig. 3-14a.   The force F places the element in pure shear, and the work done is

$$U = \frac{F\delta}{2}$$ (a)

Now, since the shear strain is

$$\gamma = \frac{\delta}{l} = \frac{\tau}{G} = \frac{F}{AG}$$

then

$$U = \frac{F^2 l}{2AG}$$ $(3\text{-}27)$

The strain energy stored in a beam or lever by bending may be obtained by referring to Fig. 3-14b.   Here AB is a section of the elastic curve of length ds having a radius of curvature ρ.   The strain energy stored in this section of the beam is

$$dU = \frac{M}{2} d\theta$$ (b)

Now, since $\rho \, d\theta = ds$, we have

$$dU = \frac{M \, ds}{2\rho}$$ (c)

Next, using Eq. (3-8) to eliminate ρ,

$$dU = \frac{M^2 \, ds}{2EI}$$ (d)

The strain energy in an entire beam is obtained by adding the energies in all the elemental sections. For small deflections $ds \approx dx$, and so

$$U = \int \frac{M^2 \, dx}{2EI} \qquad (3\text{-}28)$$

A useful relation can be obtained by dividing Eqs. (3-25) and (3-26) by the volume lA. We then obtain the two following expressions for the strain energy per unit volume:

$$u = \frac{\sigma^2}{2E} \qquad u = \frac{\tau^2}{2G} \qquad (3\text{-}29)$$

Suppose we wish to design a member to store a large amount of energy (this problem often arises in the design of springs). Then these expressions tell us that the material ought to have a high strength, because σ appears in the numerator, and amazingly, a *low* modulus of elasticity, because E appears in the denominator.

Equation (3-28) gives the strain energy due to pure bending. Most problems encountered in design are not pure bending, but the shear is so small that it is neglected. However, you ought to be able to calculate the shear strain energy for yourself and then decide whether to neglect it or not. So let us select a rectangular-section beam of width b and depth h subjected to a vertical shear force F. Using Eq. (2-34) with $I = bh^3/12$ and $dA = b \, dy$, we find the shear stress to be

$$\tau = \frac{V}{Ib} \int_y^{h/2} y \, dA = \frac{3V}{2bh^3} (h^2 - 4y^2) \qquad (e)$$

Taking an element of volume $dv = b \, dy \, dx$ and using Eq. (3-29) for the shear energy gives

$$dU = \frac{b\tau^2}{2G} \, dy \, dx \qquad (f)$$

If we now substitute τ from Eq. (e) into Eq. (f) and integrate with respect to y, we find

$$dU = \frac{9V^2 \, dx}{8Gbh^6} \int_{-h/2}^{+h/2} (h^4 - 8h^2y^2 + 16y^4) \, dy = \frac{3V^2 \, dx}{5Gbh}$$

and hence

$$U = \frac{3}{5} \int_0^l \frac{V^2 \, dx}{Gbh} \qquad (3\text{-}30)$$

Note that this expression holds only for a rectangular cross section and gives the strain energy due to transverse shear. The total energy must include that due to bending too.

A cantilever having a concentrated load F on the free end has a constant shear force $V = -F$, and Eq. (3-30) yields

$$U = \frac{3F^2l}{5Gbh} \tag{3-31}$$

Juvinall* states that the constant $\frac{3}{5}$ should be replaced by the approximate value of $\frac{1}{2}$ for other cross sections. Popov† shows that the strain energy due to shear in a cantilever is less than 1 percent of the total when the beam length is ten or more times the beam depth. Thus, except for very short beams, the strain energy given by Eqs. (3-30) and (3-31) is negligible.

As another example of the use of Eq. (3-30), take a simply supported beam having a uniformly distributed load w. The equation for the shear force is

$$V = \frac{wl}{2} - wx \tag{g}$$

Solving Eq. (3-30) gives

$$U = \frac{3}{5Gbh} \int_0^l \left(\frac{wl}{2} - wx \right)^2 dx = \frac{w^2l^3}{20Gbh} \tag{3-32}$$

3-10 THE THEOREM OF CASTIGLIANO

Castigliano's theorem states that *when forces operate on elastic systems, the displacement corresponding to any force may be found by obtaining the partial derivative of the total strain energy with respect to that force.* As in our studies of spring rates, the terms *force* and *displacement* should be broadly interpreted, since they apply equally to moments and to angular displacements. Mathematically, the theorem of Castigliano is

$$\delta_i = \frac{\partial U}{\partial F_i} \tag{3-33}$$

where δ_i is the displacement of the point of application of the ith force F_i in the direction of F_i.

---

* Robert C. Juvinall, *Engineering Considerations of Stress, Strain, and Strength*, McGraw-Hill, New York, 1967, p. 147.

† Egor P. Popov, *Introduction to Mechanics of Solids*, Prentice-Hall, Englewood Cliffs, N.J., 1968, p. 487.

Sometimes the deflection of a structure is required at a point where no force or moment is acting. In this case we can place an imaginary force Q_i at that point, develop the expression for δ_i, and then set Q_i equal to zero. The remaining terms give the deflection at the point of application of the imaginary force Q_i in the direction in which it was imagined to be acting.

Equation (3-33) can also be used to determine the reactions in indeterminate structures. The deflection is zero at these reactions, and so we merely solve the equation

$$\frac{\partial U}{\partial R_j} = 0 \tag{3-34}$$

to obtain the reaction force R_j. If there are several indeterminate reactions, Eq. (3-34) is written once for each to obtain a set of equations, which are then solved simultaneously.

Castigliano's theorem, of course, is valid only for the condition in which the displacement is proportional to the force which produced it.

Example 3-7 Find the maximum deflection of a simply supported beam with a uniformly distributed load.

Solution The maximum deflection will occur at the center of the beam, and so we place an imaginary force Q acting downward at this point. The end reactions are

$$R_1 = R_2 = \frac{wl}{2} + \frac{Q}{2}$$

Between $x = 0$ and $x = l/2$ the moment is

$$M = \left(\frac{wl}{2} + \frac{Q}{2}\right)x - \frac{wx^2}{2}$$

The strain energy for the whole beam is twice as much as for one-half; neglecting direct shear, this is

$$U = 2 \int_0^{l/2} \frac{M^2 \, dx}{2EI}$$

Therefore the deflection at the center is

$$y_{max} = \frac{\partial U}{\partial Q} = 2 \int_0^{l/2} \frac{2M}{2EI} \frac{\partial M}{\partial Q} \, dx = \frac{2}{EI} \int_0^{l/2} \left(\frac{wlx}{2} + \frac{Qx}{2} - \frac{wx^2}{2}\right)\frac{x}{2} \, dx$$

Since Q is imaginary, we can now set it equal to zero. Integration yields

$$y_{max} = \frac{2}{EI}\left[\frac{wlx^3}{12} - \frac{wx^4}{16}\right]_0^{l/2} = \frac{5wl^4}{384EI}$$

It is worth noting that it doesn't matter whether you differentiate first, and then integrate, or whether you integrate first and then differentiate.* ////

Example 3-8   Use Castigliano's theorem to find the downward deflection of point C, the downward deflection of point B, and the angular deflection of AB of the cantilevered crank shown in Fig. 3-15.

Solution   The arm BC acts as a cantilever with a support at B. The strain energy is

$$U_{BC} = \int_0^r \frac{M_{BC}^2\, dz}{2EI_2} \tag{1}$$

where $M_{BC} = Fz$.

The reaction at B, due to F, is a torque $T = Fr$, and a force F producing a bending moment in AB of $M_{AB} = Fx$. Thus the strain energy stored in AB is

$$U_{AB} = \frac{T^2 l}{2GJ} + \int_0^l \frac{M_{AB}^2\, dx}{2EI_1} \tag{2}$$

Therefore the strain energy for the entire crank is

$$U = \int_0^r \frac{M_{BC}^2\, dz}{2EI_2} + \frac{T^2 l}{2GJ} + \int_0^l \frac{M_{AB}^2\, dx}{2EI_1} \tag{3}$$

Taking the derivative of U with respect to F then gives

$$y = \frac{\partial U}{\partial F} = \frac{1}{EI_2}\int_0^r M_{BC}\frac{\partial M_{BC}}{\partial F}\, dz + \frac{Tl}{GJ}\frac{\partial T}{\partial F} + \frac{1}{EI_1}\int_0^l M_{AB}\frac{\partial M_{AB}}{\partial F}\, dx \tag{4}$$

---

* One of my Purdue math professors, whose name has long been forgotten, explained this as follows: "It doesn't matter whether you take your clothes off first, and then get into the bathtub, or whether you get into the bathtub first, and then take your clothes off." J.E.S.

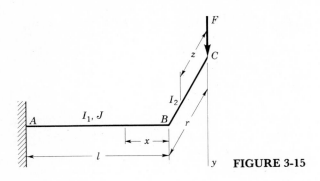

FIGURE 3-15

Now, since $M_{BC} = Fz$, $T = Fr$, and $M_{AB} = Fx$, we have

$$\frac{\partial M_{BC}}{\partial F} = z \qquad \frac{\partial T}{\partial F} = r \qquad \frac{\partial M_{AB}}{\partial F} = x$$

Substituting these values in Eq. (4) gives

$$y = \frac{1}{EI_2} \int_0^r Fz^2 \, dz + \frac{Fr^2 l}{GJ} + \frac{1}{EI_1} \int_0^l Fx^2 \, dx = \frac{Fr^3}{3EI_2} + \frac{Fr^2 l}{GJ} + \frac{Fl^3}{3EI_1} \qquad Ans.$$

To find the downward deflection at B, place an imaginary force Q at B, acting downward. Then the moment M_{AB} is

$$M_{AB} = (F + Q)x$$

According to Castigliano's theorem, the deflection at B is

$$y_B = \frac{\partial U}{\partial Q} = \frac{1}{EI_2} \int_0^r M_{BC} \frac{\partial M_{BC}}{\partial Q} \, dz + \frac{Tl}{GJ} \frac{\partial T}{\partial Q} + \frac{1}{EI_1} \int_0^l M_{AB} \frac{\partial M_{AB}}{\partial Q} \, dx \qquad (5)$$

Here

$$\frac{\partial M_{BC}}{\partial Q} = \frac{\partial}{\partial Q} (Fz) = 0 \qquad \frac{\partial T}{\partial Q} = \frac{\partial}{\partial Q} (Fr) = 0 \qquad \frac{\partial M_{AB}}{\partial Q} = \frac{\partial}{\partial Q} [(F + Q)x] = x$$

Thus Eq. (5) becomes

$$y_B = \frac{1}{EI_1} \int_0^l (F + Q)x^2 \, dx$$

But, since Q is imaginary, we set it equal to zero, integrate, and obtain

$$y_B = \frac{Fl^3}{3EI_1} \qquad Ans.$$

To find the rotation of AB, place an imaginary torque Q_T acting at B. Then the strain energy stored in AB is

$$U_{AB} = \frac{(T + Q_T)^2 l}{2GJ} + \int_0^l \frac{M_{AB}^2}{2EI_1} \, dx \qquad (6)$$

The strain energy stored in BC is unchanged. The angular deflection at B is now given by the expression

$$\theta_B = \frac{\partial U}{\partial Q_T} \qquad (7)$$

Differentiating, setting $Q_T = 0$ as before, and solving, gives

$$\theta_B = \frac{Frl}{GJ} \qquad \qquad Ans.$$

////

3-11 DEFLECTION OF CURVED MEMBERS

Machine frames, springs, clips, fasteners, and the like, frequently occur as curved shapes. These members are easily analyzed for deflection by using Castigliano's theorem. Consider, for example, the curved frame of Fig. 3-16a. This frame is loaded by the force F and we wish the deflection in the x direction. If we cut the ring section at some position θ and show the force components on the cut section (Fig. 3-16b), then it can be seen that the force F causes bending, the component F_θ causes tension, and the component F_r causes direct shear. The total strain energy results from each of these effects and is

$$U = \int \frac{M^2 \, ds}{2EI} + \int \frac{F_\theta^2 \, ds}{2AE} + \frac{1}{2} \int \frac{F_r^2 \, ds}{GA} \qquad (a)$$

where, by using the Juvinall approximation of $\frac{1}{2}$ in the third term, a nonrectangu-

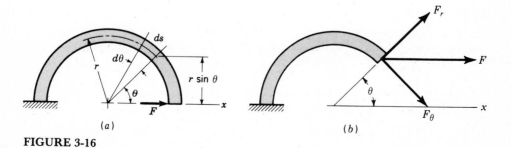

(a)   (b)

FIGURE 3-16

lar cross section has been assumed (see Sec. 3-10). This is an application of Eq. (3-30) and a factor of $\frac{3}{2}$ would be used for a rectangular section. According to Castigliano's theorem, the deflection produced by the force F is

$$\delta_x = \frac{\partial U}{\partial F} = \int_0^s \frac{M}{EI} \frac{\partial M}{\partial F} \, ds + \int_0^s \frac{F_\theta}{AE} \frac{\partial F_\theta}{\partial F} \, ds + \int_0^s \frac{F_r}{GA} \frac{\partial F_r}{\partial F} \, ds \tag{b}$$

The factors for the first term in this equation are

$$M = Fr \sin \theta \qquad \frac{\partial M}{\partial F} = r \sin \theta \qquad ds = r \, d\theta$$

Using Fig. 3-16b we see that the factors for the second and third terms are

$$F_\theta = F \sin \theta \qquad \frac{\partial F_\theta}{\partial F} = \sin \theta$$

$$F_r = F \cos \theta \qquad \frac{\partial F_r}{\partial F} = \cos \theta$$

Substituting all these into Eq. (b) and factoring the result gives

$$U = \frac{Fr^3}{EI} \int_0^\pi \sin^2 \theta \, d\theta + \frac{Fr}{AE} \int_0^\pi \sin^2 \theta \, d\theta + \frac{Fr}{GA} \int_0^\pi \cos^2 \theta \, d\theta$$

$$= \frac{Fr^3}{2EI} + \frac{Fr}{2AE} + \frac{Fr}{2GA} \tag{3-34}$$

This is the deflection of the free end of the frame in the direction of F. Because the radius is cubed in the first term, the second two terms will be negligible for large-radius frames.

Example 3-9 A thin ring is loaded by the two equal and opposite forces F in Fig. 3-17a. Find the maximum bending moment in the ring.

Solution A free-body diagram of one quadrant is shown in Fig. 3-17b. The moment M_A cannot be found by statics and so we shall employ Castigliano's theorem. Since the cross section at A does not rotate, due to the load F, we have

$$\frac{\partial U}{\partial M_A} = 0 \tag{1}$$

where U is the strain energy for a single quadrant. Now consider the differential

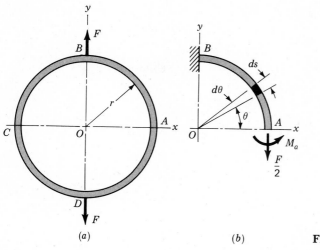

(a)   (b)   **FIGURE 3-17**

cross section at angle θ to the x axis in Fig. 3-17b.   The bending moment at this section is

$$M = M_A - \frac{F}{2}\,(r - x) = M_A - \frac{Fr}{2}\,(1 - \cos\theta) \tag{2}$$

since $x = r\cos\theta$.   The strain energy is

$$U = \int \frac{M^2\,ds}{2EI} = \int_0^{\pi/2} \frac{M^2 r\,d\theta}{2EI} \tag{3}$$

since $ds = r\,d\theta$.   The partial derivative of U with respect to M_A must be zero by Eq. (1).   Therefore

$$\frac{\partial U}{\partial M_A} = \frac{r}{EI} \int_0^{\pi/2} M\,\frac{\partial M}{\partial M_A}\,d\theta = 0 \tag{4}$$

From Eq. (2) we see that $\partial M/\partial M_A = 1$.   Therefore

$$\int_0^{\pi/2} M\,d\theta = \int_0^{\pi/2} \left[M_A - \frac{Fr}{2}\,(1 - \cos\theta) \right] d\theta = 0 \tag{5}$$

Integrating and solving for M_A gives

$$M_A = \frac{Fr}{2}\left(\frac{1}{2} - \frac{1}{\pi} \right) \tag{6}$$

If we substitute Eq. (6) back into (2) we obtain

$$M = \frac{Fr}{2}\left(\cos\theta - \frac{2}{\pi}\right)$$

The maximum occurs at B where $\theta = \pi/2$, and is

$$M_B = -\frac{Fr}{\pi} \qquad\qquad\qquad Ans.$$

////

3-12 EULER COLUMNS

A short bar loaded in pure compression by a force P acting along the centroidal axis will shorten, in accordance with Hooke's law, until the stress reaches the elastic limit of the material. If P is increased still more, the material bulges and is squeezed into a flat disk or fractures.

Now visualize a long, thin, straight bar, such as a yardstick, loaded in pure compression by another force P acting along the centroidal axis. As P is increased from zero, the member shortens according to Hooke's law, as before. However, if the member is sufficiently long, as P increases a critical value will be reached designated P_{cr}, corresponding to a condition of unstable equilibrium. At this point any little crookedness of the member or slight movement of the load or support will cause the member to collapse by buckling.

If the compression member is long enough to fail by buckling, it is called a *column*; otherwise it is a simple compression member. Unfortunately, there is no line of demarcation which clearly distinguishes a column from a simple compression member. Then, too, a column failure can be a very dangerous failure because there is no warning that P_{cr} has been exceeded. In the case of a beam, an increase in the bending load causes an increase in the beam deflection, and the excessive deflection is a visible indication of the overload. But a column remains straight until the critical load is reached, after which there is sudden and total collapse. Depending upon the length, the actual stresses in a column at the instant of buckling may be quite low. For this reason the criterion of safety consists in a comparison of the actual load with the critical load.

The relationship between the critical load and the column material and geometry is developed with reference to Fig. 3-18a. We assume a bar of length l loaded by a force P acting along the centroidal axis on rounded or pinned ends. The figure shows that the bar is bent in the positive y direction. This requires a negative moment, and hence

$$M = -Py \qquad\qquad\qquad (a)$$

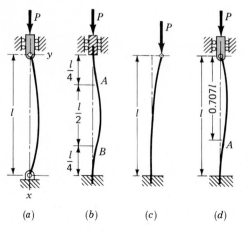

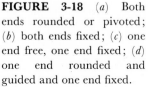

FIGURE 3-18 (*a*) Both ends rounded or pivoted; (*b*) both ends fixed; (*c*) one end free, one end fixed; (*d*) one end rounded and guided and one end fixed.

If the bar should happen to bend in the negative *y* direction, a positive moment would result, and so $M = -Py$, as before. Using Eq. (3-12), we write

$$\frac{d^2y}{dx^2} = -\frac{P}{EI}y \tag{b}$$

or

$$\frac{d^2y}{dx^2} + \frac{P}{EI}y = 0 \tag{3-35}$$

This resembles the well-known differential equation for simple harmonic motion. The solution is

$$y = A \sin \sqrt{\frac{P}{EI}}\,x + B \cos \sqrt{\frac{P}{EI}}\,x \tag{c}$$

where *A* and *B* are constants of integration and must be determined from the boundary conditions of the problem. We evaluate them using the conditions that $y = 0$ at $x = 0$ and at $x = l$. This gives $B = 0$, and

$$0 = A \sin \sqrt{\frac{P}{EI}}\,l \tag{d}$$

The trivial solution of no buckling occurs with $A = 0$. However, if $A \neq 0$, then

$$\sin \sqrt{\frac{P}{EI}}\,l = 0 \tag{e}$$

Equation (e) is satisfied by $\sqrt{P/EI}\, l = N\pi$, where N is an integer. Solving for $N = 1$ gives the critical load

$$P_{cr} = \frac{\pi^2 EI}{l^2} \tag{3-36}$$

which is called the *Euler column formula*; it applies only to rounded-end columns. If we substitute these results back into Eq. (c), we get the equation of the deflection curve as

$$y = A \sin \frac{\pi x}{l} \tag{3-37}$$

which indicates that the deflection curve is a half-wave sine. We are only interested in the minimum critical load, which occurs with $N = 1$. However, though it is not of any importance here, values of N greater than 1 result in deflection curves which cross the axis at points of inflection and are multiples of half-wave sines.

Using the relation $I = Ak^2$, where A is the area and k the radius of gyration enables us to rearrange Eq. (3-36) into the more convenient form

$$\frac{P_{cr}}{A} = \frac{\pi^2 E}{(l/k)^2} \tag{3-38}$$

with l/k designated as the *slenderness ratio*. The solution to Eq. (3-38) is called the *critical unit load*. And though the unit load has the dimensions of stress, you are cautioned very particularly not to call it a stress! To do so might lead you into the error of comparing it with a strength, the yield strength, for example, and coming to the false conclusion that a margin of safety exists. Equation (3-38) shows that the critical unit load depends *only* upon the modulus of elasticity and the slenderness ratio. Thus a column obeying the Euler formula made of high-strength alloy steel is no better than one made of low-carbon steel, since E is the same for both.

The critical loads for columns with different end conditions can be obtained by solving the differential equation or by comparison. Figure 3-18b shows a column with both ends fixed. The inflection points are at A and B, a distance $l/4$ from the ends. The distance AB is the same curve as a rounded-end column. Substituting the length $l/2$ for l in Eq. (3-36), we obtain

$$P_{cr} = \frac{\pi^2 EI}{(l/2)^2} = \frac{4\pi^2 EI}{l^2} \tag{3-39}$$

In Fig. 3-18c is shown a column with one end free and one end fixed. This curve is equivalent to half the curve for columns with rounded ends, so that if a length of $2l$ is substituted into Eq. (3-36), the critical load becomes

$$P_{cr} = \frac{\pi^2 EI}{(2l)^2} = \frac{\pi^2 EI}{4l^2} \qquad (3\text{-}40)$$

A column with one end fixed and one end rounded, as in Fig. 3-18d, occurs frequently. The inflection point is at A, a distance of $0.707l$ from the rounded end. Therefore

$$P_{cr} = \frac{\pi^2 EI}{(0.707l)^2} = \frac{2\pi^2 EI}{l^2} \qquad (3\text{-}41)$$

We can account for these various end conditions by writing the Euler equation in the two following forms:

$$P_{cr} = \frac{C\pi^2 EI}{l^2} \qquad \frac{P_{cr}}{A} = \frac{C\pi^2 E}{(l/k)^2} \qquad (3\text{-}42)$$

Here, the factor C is called the *end-condition constant*, and it may have any one of the theoretical values $\frac{1}{4}$, 1, 2, or 4, depending upon the manner in which the load is applied. In practice it is difficult, if not impossible, to fix the column ends so that the factors $C = 2$ or $C = 4$ would apply. Even if the ends are welded, some deflection will occur. Because of this, some designers never use a value of C greater than unity. However, if liberal factors of safety are employed, and if the column load is accurately known, then a value of C not exceeding 1.2 for both ends fixed, or for one end rounded and one end fixed, is not unreasonable, since it supposes only partial fixation. Of course, the value $C = \frac{1}{4}$ must always be used for a column having one end fixed and one end free. These recommendations are summarized in Table 3-2.

Table 3-2 END-CONDITION CONSTANTS FOR EULER COLUMNS
[TO BE USED WITH EQ. (3-42)]

Column end conditions	End-condition constant C		
	Theoretical value	Conservative value	Recommended value*
Fixed-free	$\frac{1}{4}$	$\frac{1}{4}$	$\frac{1}{4}$
Rounded-rounded	1	1	1
Fixed-rounded	2	1	1.2
Fixed-fixed	4	1	1.2

* To be used only with liberal factors of safety when the column load is accurately known.

3-13 EULER VS. JOHNSON COLUMNS

We have previously noted the absence of any clear distinction between a simple compression member and a column. To illustrate the problem, Fig. 3-19a is a plot of the criteria of failure of both simple compression members and of Euler columns. If a member is short, it will fail by yielding; if it is long, it will fail by buckling. Consequently, the graph has the slenderness ratio l/k as the abscissa, and the unit load P/A as the ordinate. Then an ordinate through any desired slenderness ratio will either intersect the line AB and define a simple compression member, since failure would be by yielding, or the ordinate will intersect the line BD and define a column, because failure would occur by buckling. Unfortunately, this theory does not work out quite so beautifully in practice.

The results of a large number of experiments indicate that, within a broad region around point B (Fig. 3-19a), column failure begins before the unit load reaches a point represented by the graph ABD. Furthermore, the test points obtained from these experiments are scattered. Researchers have surmised that the failure of experiment to verify the theory in the vicinity of point B is explained by the fact that it is virtually impossible to construct an ideal column. Very small deviations can have an enormous effect upon the value of the critical load. Such factors as built-in stresses, initial crookedness, and very slight load eccentricities must all contribute to the scatter and to the deviation from theory.

Many different column formulas, most of them empirical, have been devised to overcome some of the disadvantages of the Euler equation. The *parabolic*, or *J. B. Johnson, formula* is widely used in the machine, automotive, aircraft, and structural-steel construction fields. This formula often appears in the form

$$\frac{P_{cr}}{A} = a - b\left(\frac{l}{k}\right)^2 \tag{3-43}$$

where a and b are constants that are adjusted to cause the formula to fit experimental data. Figure 3-19b is a graph of two formulas, the Euler equation and the

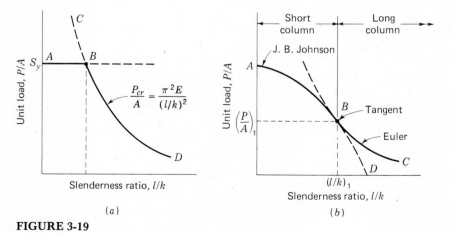

(a)   (b)

FIGURE 3-19

parabolic equation. Notice that curve ABD is the graph of the parabolic formula, while curve BC represents the Euler equation. In analyzing a column to determine the critical buckling load, only part AB of the parabolic graph and part BC of the Euler graph should be used.

The constants a and b in Eq. (3-43) are evaluated by deciding where the intercept A, in Fig. 3-19b, is to be located, and where the tangent point B is desired. Notice that the coordinates of point B are specified as $(P/A)_1$ and $(l/k)_1$.

One of the most widely used versions of the parabolic formula is obtained by making the intercept A correspond to the yield strength S_y of the material, and making the parabola tangent to the Euler curve at $(P/A)_1 = S_y/2$. Thus, the first constant in Eq. (3-43) is $a = S_y$. To get the second constant, substitute $S_y/2$ for P_{cr}/A and solve for $(l/k)_1$. Using Eq. (3-42) we get

$$\left(\frac{l}{k}\right)_1 = \sqrt{\frac{2\pi^2 CE}{S_y}} \tag{3-44}$$

Then, substituting all this into Eq. (3-43) yields

$$\frac{S_y}{2} = S_y - b\,\frac{2\pi^2 CE}{S_y}$$

or

$$b = \left(\frac{S_y}{2\pi}\right)^2 \frac{1}{CE} \tag{3-45}$$

which is to be used for the constant in

$$\frac{P_{cr}}{A} = S_y - b\left(\frac{l}{k}\right)^2 \tag{3-46}$$

Of course, this equation should only be used for slenderness ratios up to $(l/k)_1$. Then the Euler equation is used when l/k is greater than $(l/k)_1$.

3-14 INELASTIC BUCKLING

The *tangent-modulus theory* is a modification to the Euler equation suggested by Engesser in 1899. Engesser proposed that the modulus of elasticity E in the Euler equation be replaced by the tangent modulus E_t when the critical unit load exceeds the elastic limit of the material. The tangent modulus is the rate of change of stress with strain and is given by the equation

$$E_t = \frac{d\sigma}{d\epsilon} \tag{3-47}$$

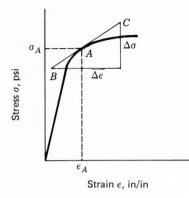

FIGURE 3-20 Typical stress-strain diagram. To find the tangent modulus corresponding to σ_A, construct tangent BAC. Then $E_t = \Delta\sigma/\Delta\epsilon$.

where $d\sigma/d\epsilon$ is the local slope of the engineering stress-strain diagram. It is true that this slope can be estimated using the strain-hardening exponent and other relations developed in Chap. 4. However, this approach is not recommended because there are so many materials that do not conform to this empirical approximation.

The engineering stress-strain diagram in Fig. 3-20 is used to illustrate the graphical method of obtaining the tangent modulus. Note that there is a particular value of E_t corresponding to every value of σ after passing the elastic limit.

We obtain the *tangent-modulus equation* by substituting E_t for E in Eq. (3-42). This gives

$$\frac{P_{cr}}{A} = \frac{C\pi^2 E_t}{(l/k)^2} \tag{3-48}$$

This equation is also called the *Euler-Engesser equation.** We note, once more, that for every value of the unit load P_{cr}/A there exists a particular value of E_t.

Examination of most stress-strain diagrams will show that the tangent modulus decreases very rapidly with increase in stress. This means that the Engesser unit load should be considered as the ultimate value of the column strength.

3-15 THE SECANT FORMULA

An eccentric column load is one in which the line of action of the column forces is not coincident with the centroidal axis of the cross section. The distance between the two axes is called the eccentricity e. The product of the force and the eccentricity produces an initial moment Pe. When this moment is introduced into the analysis, a rational formula can be deduced, valid for any slenderness ratio.

---

* A very good discussion of this equation may be found in F. R. Shanley, *Strength of Materials*, McGraw-Hill, New York, 1957, pp. 581–590.

The result is called the *secant formula*; it is usually expressed as

$$\frac{P}{A} = \frac{S_y}{1 + (ec/k^2) \sec\left[(1/k)\sqrt{P/4AE}\right]}$$

(3-49)

In this equation c is the distance from the neutral plane of bending to the outer surface. The term ec/k^2 is called the *eccentricity ratio*. Figure 3-21 is a plot of Eq. (3-49) for various values of the eccentricity ratio and for a steel having a yield strength of 40 kpsi. Euler's equation is shown for comparison purposes.

3-16 COLUMN DESIGN

We have learned that a column failure can be very dangerous because there is no visible evidence of impending buckling. For this reason very generous factors of safety should be chosen. Factors of safety in the range $2 \leq n \leq 8$ should be chosen if the uncertainties in the material, geometry, and loading are not too large.

Equation (1-1) should always be used in defining factor of safety for columns. The equation then becomes

$$n = \frac{P_{cr}}{P}$$

(3-50)

where P is the actual column load. You are cautioned very particularly *not* to evaluate factor of safety by dividing the yield strength by the unit load. Column failures can occur when the unit load is many times less than the yield strength.

Euler-Johnson Columns

Depending upon the cross section, Euler and Johnson columns can usually be designed without using iteration techniques. For a solid-round-section column,

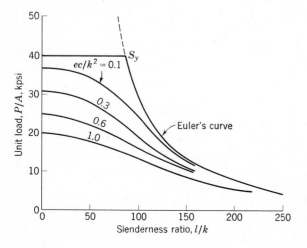

FIGURE 3-21 Comparison of the secant and Euler formulas.

the moment of inertia is

$$I = \frac{\pi d^4}{64} \tag{a}$$

and the radius of gyration is $d/4$. Solve Eq. (3-50) for P_{cr} first. Then, with P_{cr}, l, C, E, and S_y given, use the Euler equation [Eq. (3-42)] to find I as follows:

$$I = \frac{P_{cr} l^2}{C\pi^2 E} \tag{3-51}$$

Now use Eq. (a) to compute the column diameter. Thus

$$d = \left(\frac{64I}{\pi}\right)^{1/4} \tag{b}$$

Now we check to see if the result is truly an Euler column. So use Eq. (3-44) to find $(l/k)_1$. If $4l/d - (l/k)_1 \geq 0$, the column is an Euler column and Eq. (b) gives the correct diameter. If the inequality is less than zero, we must use the Johnson equation. Solving Eqs. (3-45) and (3-46) for the diameter gives

$$d = 2\left(\sqrt{\frac{P_{cr}}{\pi S_y} + \frac{S_y l^2}{C\pi^2 E}}\right) \tag{3-52}$$

These steps are summarized in the flow diagram of Fig. 3-22.

A procedure for designing a rectangular-section column is given in the following example.

Example 3-10   Design a rectangular-section steel column in which the width is about three times the thickness. Use a column design load of 5000 lb and a factor of safety of 4. The material has a yield strength of 75 kpsi and a modulus of elasticity of 30 Mpsi. Assume the end-condition constant $C = 1$ for buckling in the weakest direction.

(a)   Find a suitable set of dimensions for $l = 15$ in.
(b)   Find the dimensions for $l = 8$ in.

Solution   (a) Using Eq. (3-50), we find $P_{cr} = nP = 4(5) = 20$ kip. Then solving Eq. (3-42) for I gives

$$I = \frac{P_{cr} l^2}{C\pi^2 E} = \frac{20(10)^3(15)^2}{1\pi^2(30)(10)^6} = 0.0152 \text{ in}^4 \tag{1}$$

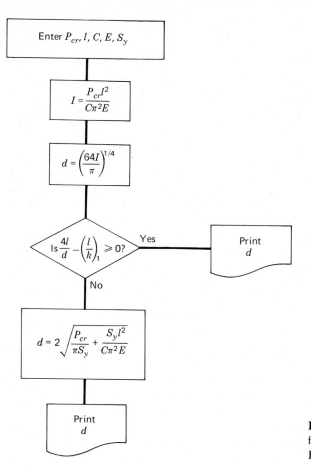

FIGURE 3-22 Flow chart for designing a Johnson or Euler round column.

The moment of inertia of a rectangular section about the weakest axis is

$$I = \frac{wt^3}{12} = \frac{t^4}{4} \tag{2}$$

because $w = 3t$. Using Eqs. (1) and (2) and solving for t gives $t = 0.497$ in. The radius of gyration corresponding to this thickness, from Table A-14, is $k = 0.289t = 0.289(0.497) = 0.144$ in. Thus the slenderness ratio is

$$\frac{l}{k} = \frac{15}{0.144} = 104$$

We next check to see if this is an Euler column. Equation (3-44) gives

$$\left(\frac{l}{k}\right)_1 = \sqrt{\frac{2C\pi^2 E}{S_y}} = \sqrt{\frac{2(1)(\pi)^2(30)(10)^6}{75(10)^3}} = 88.8 \tag{3}$$

So this is indeed an Euler column. The dimensions selected are rounded up to a $\frac{1}{2} \times 1\frac{1}{2}$-in bar.

(b) For the 8-in column, Eq. (1) becomes

$$I = \frac{P_{cr}\, l^2}{C\pi^2 E} = \frac{20(10)^3(8)^2}{1\pi^2(30)(10)^6} = 4.32(10)^{-3} \text{ in}^4 \tag{4}$$

Combining this result with Eq. (2) gives a thickness of 0.363 in. Then $k = 0.289(0.363) = 0.105$ in, so $l/k = 8/0.105 = 76$. Comparing this value with the result of Eq. (3) shows that this is a Johnson column, and so the thickness just found is incorrect. Thus we must use Eqs. (3-45) and (3-46) to obtain a set of dimensions. The constant b is found to be

$$b = \left(\frac{S_y}{2\pi}\right)^2 \frac{1}{CE} = \left[\frac{75(10)^3}{2}\right]^2 \frac{1}{1(30)(10)^6} = 4.75$$

Using $A = wt = 3t^2$ and $k = 0.289t$, we rearrange Eq. (3-46) and solve for t as follows:

$$t = \sqrt{\frac{1}{S_y}\left[\frac{P_{cr}}{3} + \frac{bl^2}{(0.289)^2}\right]} = \sqrt{\frac{1}{75(10)^3}\left[\frac{20(10)^3}{3} + \frac{4.75(8)^2}{(0.289)^2}\right]} = 0.371 \text{ in}$$

So we choose a rectangular section $\frac{3}{8} \times 1\frac{1}{8}$ in.   *Ans.*

////

Euler-Engesser Columns

Rather extensive experimental and graphical procedures must be undertaken in order to design columns using the Engesser proposal. Since the method yields the practical limiting column load, the procedure in many cases is completely justified.

The first step is to obtain the compressive stress-strain diagram. Shanley states that the tensile test will be sufficiently accurate unless the material has been cold-worked.[*] The stress-strain diagram shown in Fig. 3-23 is for an AISI 4142 steel, austenized at 1500°F, quenched in agitated oil at 180°F, and tempered at 400°F. The mechanical properties are $S_y = 245$ kpsi by 0.2 percent offset, $S_u = 325$ kpsi, $E = 30$ Mpsi, and $H_B = 560$ Bhn.[†]

Having plotted the stress-strain diagram, use the method previously described

---

[*] Ibid., p. 584.

[†] Data source: R. W. Landgraf, *Cyclic Deformation and Fatigue Behavior of Hardened Steels*, Report no. 320, Department of Theoretical and Applied Mechanics, University of Illinois, Urbana, 1968.

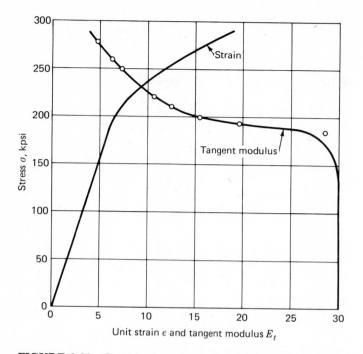

FIGURE 3-23   Combined stress vs. strain and stress vs. tangent-modulus diagrams.   The unit strain scale is in milli-inches per inch.   The tangent modulus scale is in Mpsi.

to find an assortment of values of the tangent modulus, corresponding to specified stress values.   This has been done in Fig. 3-23 and the results plotted on the same graph.

The next step is to compute a series of slenderness ratios so that a design chart like that of Fig. 3-24 can be prepared.   The easiest way to do this is to arrange Eq. (3-48) in the form

$$\frac{l}{k} = \pi \sqrt{\frac{CE_t}{P_{cr}/A}} \tag{3-53}$$

Then use Fig. 3-23 to get values of σ, used for P_{cr}/A, and E_t for substitution. When enough points are obtained the design chart can be plotted.

Note, in Fig. 3-24, that the Engesser curve begins at a point substantially below the yield strength.   The reason for this apparent discrepancy is that the yield strength was obtained by the 0.2-percent-offset method.

It is quite simple to obtain column dimensions using these design charts, so they should be filed for future use.   It may also be desirable to plot curves for other end conditions on the same chart.

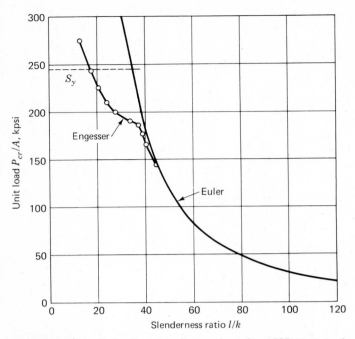

FIGURE 3-24 Euler-Engesser design chart for AISI 4142 steel, heat-treated to 560 Bhn.   Yield strength shown for reference purposes.

Secant Columns

Equation (3-49) cannot be solved explicitly for the unit load.   For this reason a design chart should be prepared similar to Fig. 3-21.   Solving Eq. (3-49) for the slenderness ratio gives

$$\frac{l}{k} = \frac{1}{\sqrt{\dfrac{P/A}{4E}}} \cos^{-1} \frac{ec/k^2}{\dfrac{S_y}{P/A} - 1} \tag{3-54}$$

Now a variety of slenderness ratios can be found for each value of ec/k^2 by substituting various values of P/A in this equation; then the results can be used to plot the design chart.   The equation requires only a few program steps.

PROBLEMS*

Sections 3-1 and 3-2

3-1   The stepped shaft shown in the figure is steel.   Find the torsional spring rate.

---

* The asterisk indicates a design-type problem, one with no unique result, or a challenging problem.

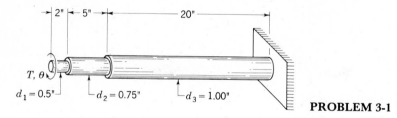

$$T, \theta$$

$$d_1 = 0.5"$$ $$d_2 = 0.75"$$ $$d_3 = 1.00"$$

PROBLEM 3-1

3-2 A steel bar 250 mm long is acted upon by a simple tensile load of 18 kN. The diameter of the bar is 5 mm. Find the elongation, the spring rate, and the stress in the bar in appropriate SI units.

3-3 A 1.50-mm-diameter steel wire 8 m in length is subjected to a tensile force of 1.25 kN. Find the resulting tensile stress, the total elongation, and the spring rate of the wire.

3-4 A steel torsion spring is made from a round bar of steel 15 mm in diameter and 2.5 m long. Based on a maximum torque of 300 N · m find the torsional stress, the angular deflection in degrees, and the spring rate.

3-5 The two gears shown in the figure have the respective tooth numbers N_1 and N_2. An input torque T_1 is applied to the end of a shaft, having a torsional spring constant k_1. This torque is resisted by an output torque T_2 at B, exerted at the end of the driven shaft, whose torsional stiffness is k_2.

(a) Assume the shaft is fixed at B and find an expression for the torsional spring rate of the entire system based upon a deflection measured at A.

(b) Assume the shaft is fixed at A and find an expression for the torsional spring rate corresponding to the deflection at B.

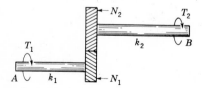

PROBLEM 3-5

3-6 Find the diameter of a solid round steel shaft which is to transmit $\frac{1}{20}$ hp at 1 rpm if the angular deflection is not to exceed $1°$ in a length of 30 diameters. Find the corresponding torsional stress.

3-7 Find the diameter of a solid round steel shaft to transmit 15 kW at 60 s^{-1} if the angular deflection is not to exceed $1°$ in a length of 30 diameters. Compute the corresponding torsional stress, and comment on the result.

3-8* A clothesline 50 ft long is made from No. 10 gauge W & M steel wire (see Table A-25). The line is installed with an initial tension of 50 lb. Suppose a 10-lb weight is suspended from the middle of the line. How much will the line sag? How much will the line stretch? What is the resulting tensile stress in the wire? Is the sag linearly related to the magnitude of the weight? If not, is the response similar to that of a softening spring or a stiffening spring?

Sections 3-3 to 3-5

3-9* A cantilever as in Table A-12-2 has an intermediate load $F = 300$ lb. The dimensions are $a = 4$ in and $b = 16$ in. The beam is to be designed such that the maximum deflection is not more than $\frac{1}{16}$ in and the maximum stress is not over 24 kpsi.

The cross section is to be a square tube to be formed of sheet steel using a U.S. Standard even-numbered gauge size (see Table A-25). After forming, the edges of the sheet are to be welded together and ground smooth.

3-10 Determine a set of cross-sectional dimensions for a steel straightedge 1 m long such that, when it is optimally supported, the deflection due to its own weight will be less than 12.5 μm.

3-11 A $2 \times 2 \times \frac{3}{8}$-in steel angle supports the load shown in the figure. Find the maximum deflection.

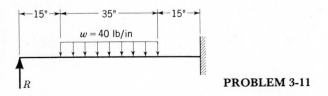

PROBLEM 3-11

3-12 A round steel shaft supports the loads shown in the figure. The shaft has a diameter of 30 mm and is supported by preloaded antifriction (ball) bearings that introduce some end constraint. Calculate the deflection at the center by assuming, first, that the ends are fixed as shown in the figure, and by assuming, second, that the shaft is simply supported at the ends. What is the ratio of these two deflections?

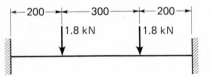

PROBLEM 3-12 Dimensions in millimeters.

3-13 Select a standard steel angle with equal legs from Table A-8 to support the load in the figure such that the deflection will not exceed $\frac{1}{16}$ in.

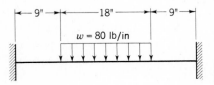

PROBLEM 3-13

3-14 Select a round steel bar to support the load shown in the figure such that the maximum deflection will not exceed 0.40 mm.

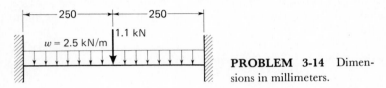

PROBLEM 3-14 Dimensions in millimeters.

3-15 A torsion spring, as shown in part *a* of the figure, can be treated as a cantilever for analysis purposes.

    (*a*) Determine the dimensions *b* and *h* of a rectangular steel section such that a force $F = 20$ N will produce a deflection of exactly 75 mm. Use $b = 0.10h$.

(*b*)   The yield strength of the steel used is 400 MPa. Will this yield strength be exceeded by the stress when the force is 20 N?

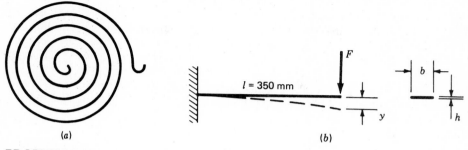

(*a*)   (*b*)

PROBLEM 3-15

3-16   The figure shows a cantilever steel spring of rectangular cross section. Determine the dimensions of the spring so that it has a scale of 140 lb/in. What is the maximum stress if the operating range is $\frac{1}{4}$ in?

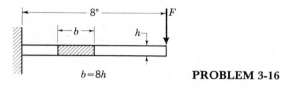

$b=8h$ **PROBLEM 3-16**

3-17   The figure is a structural-engineering drawing of a beam spanning two columns. The beam is made up of two channel sections mounted back to back and spaced $\frac{1}{4}$ in

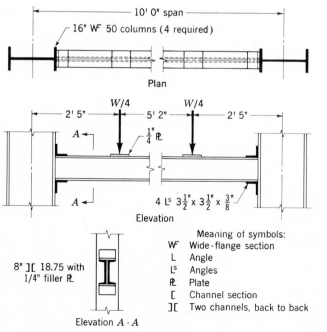

PROBLEM 3-17

apart, using short sections of $\frac{1}{4}$-in plates. Four short angles are used to secure the beam to the columns by riveting, bolting, or welding. Structural shapes are specified by stating their major dimension or dimensions, together with the weight per foot. Thus the wide-flange beam, used here as a column, weighs 50 lb/ft.† Dimensions and properties of shapes frequently used in mechanical design are listed in Tables A-8 to A-11. Examination of Table A-11, for example, indicates that the 8-in back-to-back channels have a flange width of 2.527 in and a web thickness of 0.487 in.

In this problem two of these beams are used to support a machine whose total weight is W. The figure shows that this weight is to be transferred into the beams at four points, two on each beam. Find:

(a) The shear and moment diagrams for the beams
(b) The weight W that can be supported by the two beams, allowing a maximum bending stress of 16 kpsi
(c) The maximum deflection caused by this weight

Sections 3-6 and 3-7

3-18 The figure shows the drawing of a countershaft and its loading diagram. Bearings A and B are self-aligning; find the maximum deflection.

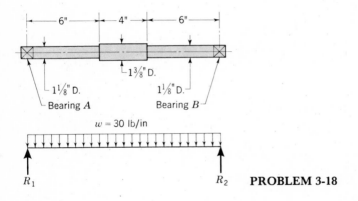

PROBLEM 3-18

3-19 Determine the maximum deflection of the shaft shown in the figure. The material is steel.

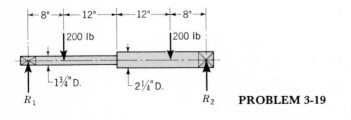

PROBLEM 3-19

† See AISC Handbook.

Sections 3-9 to 3-11

3-20 Using the theorem of Castigliano, find the maximum deflection of the cantilever shown in the figure. Neglect direct shear.

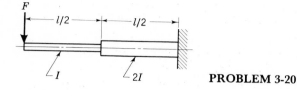

PROBLEM 3-20

3-21 In the figure, a is a drawing of a shaft and b is the loading diagram. Determine the diameter d so that the maximum deflection does not exceed 0.010 in. The material is an alloy steel with a modulus of elasticity of 30 Mpsi. Use Castigliano's theorem.

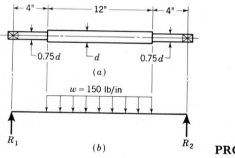

PROBLEM 3-21

3-22 Find the deflection at the point of application of the force F to the cantilever in the figure and the deflection at the free end. Use Castigliano's theorem.

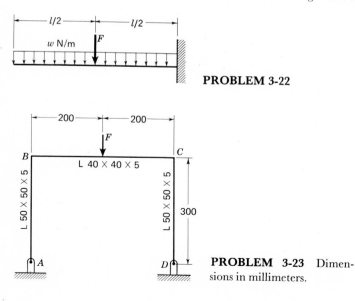

PROBLEM 3-22

PROBLEM 3-23 Dimensions in millimeters.

3-23 The frame shown in the figure is composed of three structural aluminum angles welded at points B and C and bolted to a supporting structure at points A and D. The moments of inertia about the bending axes are $I_{AB} = I_{CD} = 112.5(10)^3$ mm^4 and $I_{BC} = 55.6(10)^3$ mm^4. Determine the magnitude of the load F in kilonewtons such that the maximum deflection of BC is not greater than 1.5 mm.

3-24 The figure illustrates an X frame in the xz plane. The frame is made of two members of length $2l$ welded at the center of each at an X angle of θ. Each member has a moment of inertia I. The ends A, B, and D are simply supported. Find an expression for the downward deflection at C due to the force F.

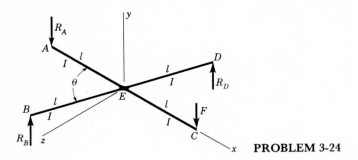

PROBLEM 3-24

3-25 The L frame shown in the figure has one end A twisted by the torque T and the other end C is simply supported by the reaction R. The rectangular and polar moments of inertia are I_1, J_1, and I_2, J_2, as shown, and both legs are made of the same material. Use Castigliano's theorem and find the angular rotation of the frame at A in the direction of the applied torque.

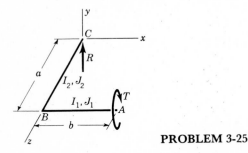

PROBLEM 3-25

3-26 Find the total deformation of the ring of Example 3-9 in the y direction.

3-27 The bell-crank lever shown in the figure is pivoted to a fixed frame at B. Using Castigliano's theorem, develop an expression for the deflection at A. Assume that point C does not move.

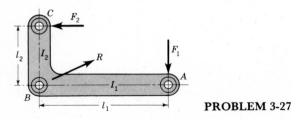

PROBLEM 3-27

3-28 Find the total deformation of the ring of Example 3-9 in the x direction.

3-29 The figure shows a welded steel bracket loaded by a force $F = 5$ kN. Using the assumed loading diagram and Castigliano's method, find the maximum deflection of the end.

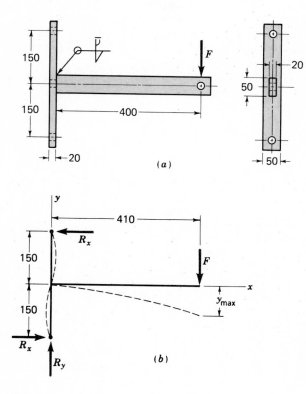

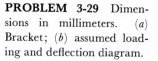

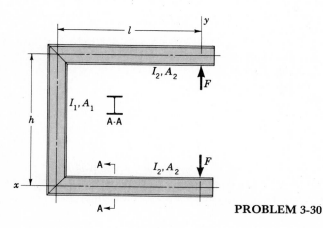

PROBLEM 3-29 Dimensions in millimeters. (*a*) Bracket; (*b*) assumed loading and deflection diagram.

3-30 The rectangular-shaped C frame shown in the figure is welded of three wide-flange beams. Find an expression for the deflection of the frame at the force F and in the direction of F.

PROBLEM 3-30

Sections 3-12 to 3-16

3-31 A column with both ends rounded is made of hot-rolled UNS G10150 steel with a 10×25-mm rectangular cross section. Find the buckling load in kilonewtons for the following column lengths: 85, 175, 400, and 600 mm.

3-32 A column with one end fixed and one end rounded is made of hot-rolled UNS G10100 steel. The member is a rectangular-section bar $\frac{1}{2} \times 1\frac{1}{2}$ in. Use the theoretical end-condition constant and find the buckling load for the following column lengths: 0.50, 2, and 4 ft.

3-33* Find the safe compressive load for a $4 \times 4 \times \frac{1}{2}$-in structural steel angle 5 ft long. Use $n = 4$.

3-34 An Euler column with one end fixed and one end free is to be made of an aluminum alloy. The cross-sectional area of the column is to be 600 mm^2 and it is to have a length of 2.5 m. Determine the column buckling load corresponding to the following shapes:

  (*a*) A solid round bar
  (*b*) A round tube with a 50 mm OD
  (*c*) A 50-mm square tube
  (*d*) A square bar

3-35 A steel tube having a $\frac{3}{16}$-in wall thickness is to be designed to safely support a column load of 3.60 kip. The column will have both ends rounded and will be made of UNS G10350 cold-drawn steel. Use a factor of safety of 4 and find the outside diameter to the nearest $\frac{1}{8}$ in for the following column lengths: 3, 15, and 45 in.

3-36 A steel tube having a 5-mm wall thickness is to be designed to safely support a column load of 15 kN. The column will have both ends rounded and will be made of UNS G10180 cold-drawn steel. Find an appropriate outside diameter to the nearest 2.5 mm for the following column lengths: 50 mm, 400 mm, and 1 m. Use $n = 3$.

3-37 The figure shows a hydraulic cylinder having a clevis mount. Because of the bearing and seal and because of the stiffness of the cylinder itself, the piston end of the rod may be regarded as fixed. The outboard end of the rod may be either a free end or rounded and guided, depending upon the application. Of course, the column length is taken as the distance l when fully extended.

  Consider a typical application with a hydraulic pressure of 3500 psi, a 3-in cylinder bore, a factor of safety of 3, and a medium-carbon-steel-rod material having a yield strength of 70 kpsi. Use the recommended value for end-condition constant, based on one end fixed and the other end pinned and guided, to find a safe diameter d for piston-rod lengths of 96, 48, and 24 in.

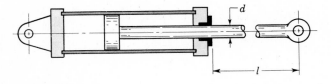

PROBLEM 3-37

3-38 Bars OA and AB in the figure are made of UNS G10100 hot-rolled steel and have a cross section of $1 \times \frac{1}{4}$ in, as shown. Based on recommended values of end-condition constants, what weight W would cause a column failure?

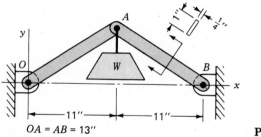

$OA = AB = 13''$

PROBLEM 3-38

3-39* A column having one end fixed and the other end pivoted and guided is to be made of either UNS G10100 hot-rolled steel or the more expensive UNS G10500 cold-drawn steel, depending upon the designer's decision. The nominal column load is 1400 lb and a factor of safety of 3.25 is to be used. If this column is to be round, find a safe diameter to the nearest $\frac{1}{16}$ in for 40-, 20-, and 7.5-in lengths.

3-40* A rectangular-section column having both ends fixed for buckling about the weakest axis and both ends rounded for buckling about the strongest axis is to be designed of UNS G10180 hot-rolled steel or of UNS G10350 steel, heat-treated and drawn to 1000°F. The column design load (nominal load) is 15 kN and a factor of safety of 3.5 is to be used. The width of the section is to be four times the thickness. Find the dimensions to the nearest millimeter for 500-, 250-, and 125-mm lengths.

General Problems

3-41* The figure is a schematic drawing of a space frame to be designed to support a dead load of 5000 lb at D. Use standard or preferred sizes with tubing for the compression members and round bars for the tension members. Make sketches to show exactly how the ends are designed.

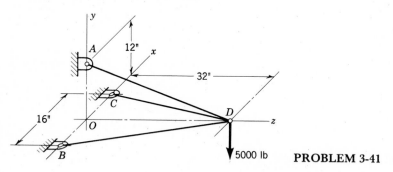

5000 lb   **PROBLEM 3-41**

3-42* A tripod 1.5 m high when erected is made of aluminum tubing having an OD of 10 mm and a wall thickness of 1.5 mm. The material used has a yield strength of 135 MPa. If the upper ends of the legs are pinned together, what load would cause failure?

3-43* The bracket shown in the figure consists of a base plate to which are welded four short $1\frac{1}{2} \times 1\frac{1}{2} \times \frac{1}{4}$-in angles and three steel bars $\frac{3}{8} \times 1$ in. All parts are of steel having $S_y = 43$ kpsi. Find the maximum load F that this bracket will carry, basing your calculations on the strength of the three bars. If necessary, the two compres-

sion members can also be secured to each other at some intermediate point, using a
bolt and spacer. Can this bracket be redesigned for more efficient use of the material? If so, state what you would do.

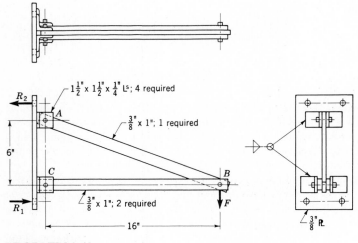

PROBLEM 3-43

3-44 Refer to the figure of the coping saw, which has a 7-in throat, in Prob. 2-73. You
are to design the frame for a coping saw having a 4-in throat using the same frame
cross section, $\frac{3}{16} \times \frac{3}{8}$ in. Use cold-drawn steel having a yield strength of 80 kpsi and
a factor of safety of 1.20. The frame is to be formed by cold bending, which builds a
helpful residual stress into the curve. Assume that this stress is 20 kpsi, a not
unreasonable assumption. Because of this helpful residual stress, the strength in the
curved portions is effectively 100 kpsi. Find the exact geometry of the frame prior to
assembly of the blade and find the tension in the blade after it has been assembled.
Use same blade length.

3-45* Design the piston rod of the cylinder at *FB* of the hydraulic floor crane of Prob.
2-80. Select both the material and the factor of safety. The rated crane load is $\frac{1}{2}$
ton.

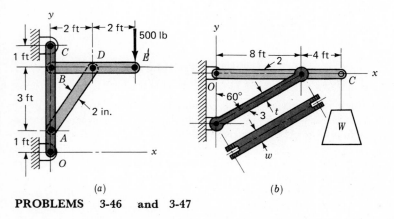

(a)   (b)

PROBLEMS   3-46   and   3-47

3-46* Determine appropriate cross-section shapes, dimensions, and end configurations for the three members of the frame shown in part (*a*) of the figure. Choose materials and appropriate factors of safety for each mode of loading. The 500-lb force is a dead load acting on a pin at *E*.

3-47* Design the compression member shown in part (*b*) of the figure. Use a factor of safety of 5 to find appropriate cross-section dimensions *w* and *t*. Choose a material and a method of manufacture for very small production quantities. The weight *W* is 4000 lb.

3-48* The figure shows a tentative design for a pulpwood loader. The machine is mounted directly on the flatbed of the truck and is used to load and unload the pulpwood. Though it might run higher, the average weight of logs handled by the loader is about 10 kN. Determine the stroke required for each of the hydraulic cylinders. Choose materials and factors of safety and specify the piston-rod diameters.

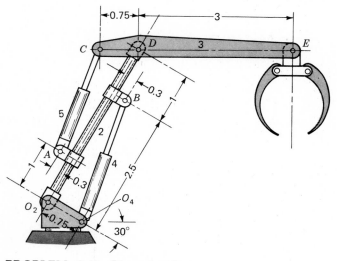

PROBLEM 3-48   Dimensions in meters.

CHAPTER

4

MATERIALS AND
THEIR PROPERTIES

In chaps. 2 and 3 methods of determining the stresses and deflections of machine members were discussed. It was found that in some methods the elastic properties of the material are used. The stress existing in a machine member has no meaning unless the strength of the material is known; this strength is a property of the particular material in use.

The selection of a material for a machine or structural member is one of the decisions the designer is called upon to make. This decision is usually made before the dimensions of the part are determined. After choosing the material and process (the two cannot be divorced), the designer can then proportion the member so that the internal stresses and strains have reasonable and satisfactory values compared with the properties associated with failure of the material.

As important as the stress and deflection of mechanical parts are, the selection of a material is not always based upon these factors. There are many parts which have no loads on them whatever. These parts may be designed merely to fill up space. Members must frequently be designed to resist corrosion. Sometimes temperature effects are more important in design than stress and strain. So many other factors besides stress and strain may govern the design of parts that a versatile background in materials and processes is necessary.

4-1 STATIC STRENGTH

The standard tensile test is used to obtain a variety of characteristics and strengths that are used in design. Figure 4-1 illustrates a typical tension-test specimen and some of the dimensions that are often employed. The original diameter d_0 and length of the gauge l_0, used to measure the strains, are recorded before the test is begun. The specimen is then mounted in the test machine and slowly loaded in tension while the load and strain are observed. At the conclusion of, or during, the test the results are plotted as a *stress-strain diagram* (Fig. 4-2).

Point P in Fig. 4-2 is called the *proportional limit*. This is the point at which the curve first begins to deviate from a straight line. Point E is called the *elastic limit*. No permanent set will be observable in the specimen if the load is removed at this point. Between P and E the diagram is not a perfectly straight line, even though the specimen is elastic. Thus Hooke's law, which states that stress is proportional to strain, applies only up to the proportional limit.

During the tension test, many materials reach a point at which the strain begins to increase very rapidly without a corresponding increase in stress. This point is called the *yield point*. Not all materials have a yield point that is so easy to find. For this reason, *yield strength S_y* is often defined by an *offset method* as shown in Fig. 4-2. Such a yield strength corresponds to a definite or stated amount of permanent set, usually 0.2 or 0.5 percent of the original gauge length.

The *ultimate*, or *tensile*, *strength S_u* or S_{ut} corresponds to point U in Fig. 4-2 and is the maximum stress reached on the stress-strain diagram. Some materials exhibit a downward trend after the maximum stress is reached. These fracture at point F on the diagram in Fig. 4-2. Others, such as some of the cast irons and high-strength steels, fracture when the diagram is still rising.

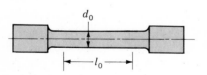

FIGURE 4-1 A typical tension-test specimen. Some values used for d_0 are 0.1, 0.25, and 0.50 in or 2.5, 6.25, and 12.5 mm. Common gauge lengths l_0 used are 10, 25, and 50 mm in SI and 1.0 or 2.0 in.

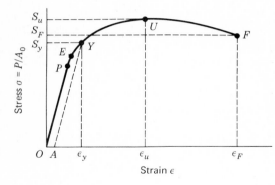

FIGURE 4-2 Stress-strain diagram obtained from the standard tensile test of a ductile material. P marks the proportional limit; E, the elastic limit; Y, the offset yield strength as defined by offset distance OA; U, the maximum or ultimate strength; and F, the fracture strength.

To determine the strain relations for the stress-strain test, let

l_0 = original gauge length
l_i = gauge length corresponding to any load P_i
A_0 = original cross-sectional area
A_i = area of smallest cross section under load P_i

The unit strain from Chap. 2 is

$$\epsilon = \frac{l_i - l_0}{l_0} \tag{4-1}$$

The term *true stress* is used to indicate the result obtained when any load used in the tension test is divided by the *true* or *actual* cross-sectional area of the specimen. This means that both the load and the cross-sectional area must be measured simultaneously during the test. If the specimen has necked, especial care must be taken to measure the area at the smallest part.

In plotting the true stress-strain diagram it is customary to use a term called *true strain*, sometimes called *logarithmic strain*. True strain is the sum of each incremental elongation divided by the current length of the filament, or

$$\varepsilon = \int_{l_0}^{l_i} \frac{dl}{l} = ln\, \frac{l_i}{l_0} \tag{4-2}$$

where l_0 is the original gauge length and l_i is the gauge length corresponding to load P_i.

The most important characteristic of a true stress-strain diagram (Fig. 4-3) is that the true stress increases all the way to fracture. Thus, as shown in Fig. 4-3, the true fracture stress σ_F is greater than the true ultimate stress σ_u. Contrast this with Fig. 4-2 where the fracture strength S_F is less than the ultimate strength S_u.

Bridgman has pointed out that the true stress-strain diagram of Fig. 4-3 should be corrected because of the triaxial stress state that exists in the neck of the

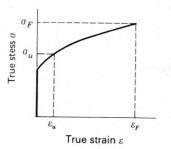

FIGURE 4-3 True stress-strain diagram plotted using cartesian coordinates.

specimen.* He observes that the tension is greatest on the axis and smallest on the periphery and that the stress state consists of an axial tension uniform all the way across the neck, plus a hydrostatic tension, which is zero on the periphery and increases to a maximum value on the axis.

Bridgman's correction for the true stress during necking is particularly significant. Designating σ_C as the computed true stress and σ_{ACT} as the corrected or actual stress, R as the radius of the neck (Fig. 4-4), and D as the smallest neck diameter, the equation is

$$\sigma_{ACT} = \frac{\sigma_C}{\left(1 + \dfrac{4R}{D}\right)\left[ln\left(1 + \dfrac{D}{4R}\right)\right]} \tag{4-3}$$

When necking occurs, the engineering strain given by Eq. (4-1) will not be the same at all points within the gauge length. A more satisfactory relation can be obtained by using areas. Since the volume of material remains the same during the test $A_0 l_0 = A_i l_i$. Consequently $l_i = l_0(A_0/A_i)$. Substituting this value of l_i into Eq. (4-1) and canceling terms gives

$$\epsilon = \frac{A_0 - A_i}{A_i} \tag{4-4}$$

But see also Eq. (4-9).

Compression tests are more difficult to make, and the geometry of the test specimens differs from the geometry of those used in tension tests. The reason for this is that the specimen may buckle during testing or it may be difficult to get the stresses distributed evenly. Other difficulties occur because ductile materials will bulge after yielding. However, the results can be plotted on a stress-strain diagram, too, and the same strength definitions can be applied. For many materials the compressive strengths are about the same as the tensile strengths. When

---

* P. W. Bridgman, "The Stress Distribution at the Neck of a Tension Specimen," *ASM*, vol. 32, 1944, p. 553.

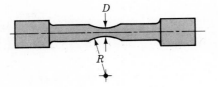

FIGURE 4-4 Tension specimen after necking. Radius of neck is R; diameter of smallest portion of neck is D.

substantial differences occur, however, as is the case with the cast irons, the tensile and compressive strengths should be stated separately.

Torsional strengths are found by twisting bars and recording the torque and the twist angle. The results are then plotted as a *torque-twist diagram*. By using the equations in Chap. 2 for torsional stress, both the elastic limit and the *torsional yield strength* S_{sy} may be found. The maximum point on a torque-twist diagram, corresponding to point U on Fig. 4-1, is T_u. The equation

$$S_{su} = \frac{T_u r}{J} \qquad (a)$$

where r is the radius of the bar and J is the polar area moment of inertia, defines the *modulus of rupture* for the torsion test. Note that the use of Eq. (a) implies that Hooke's law applies to this case. This is not true, because the outermost area of the bar is in a plastic state at the torque T_u. For this reason the quantity S_{su} is called the modulus of rupture. It is incorrect to call S_{su} the ultimate torsional strength.

All of the stresses and strengths defined by the stress-strain diagram of Fig. 4-2 and other similar diagrams are specifically known as *engineering stresses* and *strengths* and/or *nominal stresses* and *strengths*. These are the values normally used in all engineering design. The adjectives "engineering" or "nominal" are used here to emphasize that the stresses are computed using the *original* or *unstressed cross-sectional area* of the specimen. In this book we shall only use these modifiers when we specifically wish to call attention to this distinction.

4-2 PLASTIC DEFORMATION

Explanations of the relationships between stress and strain in the plastic region were first proposed by Ludwik[*] in 1909. Additions to the theory were then made by Hollomon[†] in 1945, by Nelson and Winlock[‡] in 1949, and later by Datsko.[§]

* P. Ludwik, *Elemente der technologischen Mechanik*, Springer, Berlin, 1909.

† John H. Hollomon, "Tensile Deformation," *Trans. Am. Inst. Mining and Metallurgical Engrs.* (Iron and Steel Division), vol. 162, 1945, pp. 268–290.

‡ Paul G. Nelson and Joseph Winlock, "A Method of Determining the Percentage Elongation at Maximum Load in the Tension Test," *ASTM Bulletin* TP15, January 1949, pp. 53–55.

§ Joseph Datsko, *Materials in Design and Manufacture*, University of Michigan, Dept. of Mech. Eng. and Appl. Mech., Ann Arbor, 1977, chap. 5; see also Joseph Datsko, *Material Properties and Manufacturing Processes*, Wiley, New York, 1966, chap. 1.

Hertzberg refers to the plastic region as a type II stress-strain behavior, consisting of an irreversible homogeneous plastic flow.* We have seen, in Fig. 4-3, that the true stress-strain curve rises all the way to a maximum stress. This increasing value describes a process called *strain hardening*.

Hollomon suggested that the plastic region can be described approximately by the equation

$$\sigma = K\varepsilon^n \qquad (4\text{-}5)$$

where σ = true stress
K = a strength coefficient
ε = true plastic strain
n = strain-hardening exponent

A graph of this equation is a straight line when plotted on log-log paper, as shown in Fig. 4-5. The graph contains three zones of interest: the elastic zone, on line AB, called type I behavior by Hertzberg; the plastic zone on line $Y_2 C$ defining type II behavior; and the intermediate zone.

According to Hooke's law the equation of the elastic portion is

$$\sigma = E\varepsilon \qquad (4\text{-}6)$$

where E is the modulus of elasticty. Taking the logarithm of both sides of Eq. (4-6) and recognizing the equation of a straight line, we get

$$\log \sigma = \log E + 1 \log \varepsilon = b + mx$$

And from Eq. (4-5) we obtain

$$\log \sigma = \log K + n \log \varepsilon$$

From this we conclude that the elastic portion of the line is the same for all materials, having a slope of unity (45° angle) and passing through the point $\sigma = E$ and $\varepsilon = 1$. We also see that the elastic portion has an exponent $n = 1$ and an intercept $K = E$ relative to the true stress-strain equation [Eq. (4-5)].

The constant K in Eq. (4-5) is the true stress corresponding to a true strain of unity. This constant can be obtained by extending the plastic stress-strain line until it intersects an ordinate through $\varepsilon = 1$ ($\log \varepsilon = 0$). The height of this ordinate is $\log K$, as measured parallel to the $\log \sigma$ axis of Fig. 4-5.

The shape of the elastic-plastic zone between the two straight lines varies from one material to another. The three possible yield points Y_1, Y_2, and Y_3 describe the various possibilities that might be observed. An extension of the plastic line would intersect the elastic line at Y_2 and describe an ideal material.

---

* Richard W. Hertzberg, *Deformation and Fracture Mechanics of Engineering Materials*, Wiley, New York, 1976, pp. 6–26.

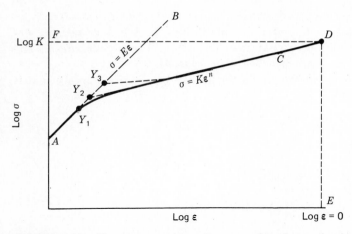

FIGURE 4-5 True stress-strain diagram plotted on log-log paper. Since the values of ε are less than unity, their logarithms are negative. At point $E \varepsilon = 1$, $\log \varepsilon = 0$ and the ordinate through E locates D and defines the logarithm of the constant K at F.

Most engineering materials are said to *overyield* to Y_3 because they have a yield strength greater than the ideal value. The alloys of the steels, coppers, brasses, and nickels all have this characteristic. The point Y_1 describes what might be called *under yielding*. Only a few engineering materials have this characteristic, a fully annealed aluminum alloy being one of them.

The relationship between logarithmic strain and unit strain can be obtained by rearranging Eq. (4-1) to

$$\epsilon = \frac{l_i - l_0}{l_0} = \frac{l_i}{l_0} - 1 \tag{4-7}$$

Then we see that

$$\frac{l_i}{l_0} = \epsilon + 1$$

and so, from Eq. (4-2), we get

$$\varepsilon = ln(\epsilon + 1) \tag{4-8}$$

Equations (4-7) and (4-8) are used in the Datsko approach to plot the elastic portion of the diagram. Thus, for this portion of the experiment, the data are acquired in the conventional manner, using an extensometer to obtain the elongation. The extensometer is not used for the plastic portion of the true stress-strain diagram because the average strain is no longer useful. Its use would also expose an expensive precision instrument to damage at fracture.

The approach for the plastic region consists in measuring the area of the specimen, being particularly careful to obtain this value at the smallest cross section between the gauge points. Sometimes it is necessary to measure "diameters" of the specimen in two directions, perpendicular to each other, in case the cross section becomes oval-shaped.

From Eq. (4-4) we have

$$\frac{A_0}{A_i} = \epsilon + 1$$

Thus, from Eq. (4-8), we find the logarithmic strain for areas to be

$$\varepsilon = ln \frac{A_0}{A_i} \tag{4-9}$$

If the diagram is to be corrected, then the necking radius R should also be measured and the stress corrected using the Bridgman equation [Eq. (4-3)].

The exponent n represents the slope of the plastic line, as we have seen. This slope is easily obtained after the plastic line has been drawn through the points in the plastic region of the diagram. Another, and easier, method of obtaining the exponent is possible for materials having an ultimate strength greater than the nominal stress at fracture. For these materials the exponent is the same as the logarithmic strain corresponding to the ultimate strength. The proof is as follows:

$$P_i = \sigma A_i = K A_i(\varepsilon)^n \tag{a}$$

where we have used Eq. (4.5). Now, from Eq. (4-9), we have

$$A_i = \frac{A_0}{e^\varepsilon} \tag{b}$$

and so Eq. (a) becomes

$$P_i = \frac{K A_0(\varepsilon)^n}{e^\varepsilon} \tag{c}$$

Now the maximum point on the load-deformation diagram, or nominal stress-strain diagram, at least for some materials, is coincident with a zero slope. Thus, for these materials, the derivative of the load with respect to the logarithmic strain must be zero. Thus

$$\frac{dP_i}{d\varepsilon} = K A_0 \frac{d}{d\varepsilon} \left(\varepsilon^n e^{-\varepsilon}\right) = K A_0 (n\varepsilon^{n-1} e^{-\varepsilon} - \varepsilon^n e^{-\varepsilon}) = 0 \tag{d}$$

Solving gives $\varepsilon = n$; but this corresponds to the ultimate load, and so

$$n = \varepsilon_u \tag{4-10}$$

Note again that this relation is only valid if the load-deformation diagram has a point of zero slope.

4-3   STRENGTH AND HARDNESS

It is possible for two metals to have exactly the same strength, yet for one of these metals to have a superior ability to absorb overloads, because of the property called *ductility*.   A useful measure of ductility is the percent *reduction in area* that occurs in the standard tension-test specimen.   The equation is

$$R = \frac{A_0 - A_F}{A_0}\,(100) \tag{4-11}$$

where A_0 is the original area, A_F is the area at fracture, and R is the reduction in area in percent.   This figure is tabulated along with the strengths in the material-property tables in the appendix.

   The characteristics of a ductile material which permit it to absorb large overloads provide an additional safety factor in design.   Ductility is also import-ant because it is a measure of that property of a material which permits it to be cold-worked.   Such operations as bending, drawing, heading, and stretch forming are metal-processing operations which require ductile materials.

   Consider next the results of a tension test such as the one diagrammed in Fig. 4-6a. Both Marin[*] and Juvinall[†] show that if the load is released at some point I the total strain is made up of two parts, the permanent plastic deformation ϵ_p and the elastic deformation ϵ_e.   Thus

$$\epsilon = \epsilon_p + \epsilon_e \tag{a}$$

If the cold-worked specimen is now reloaded, point I will be reached again, and the stress-strain diagram traced out will be the same as it would have been if the specimen had not been unloaded in the first place.

   Both Marin and Juvinall show that the unloading and reloading occurs along straight lines that are approximately parallel to the initial elastic line OY.   Thus

$$\epsilon_e = \frac{\sigma_i}{E} \tag{b}$$

   [*] Joseph Marin, *Mechanical Behavior of Engineering Materials*, Prentice-Hall, Englewood Cliffs, N. J., 1962, p. 10.
   [†] Robert C. Juvinall, *Engineering Considerations of Stress, Strain, and Strength*, McGraw-Hill, New York, 1967, p. 99.

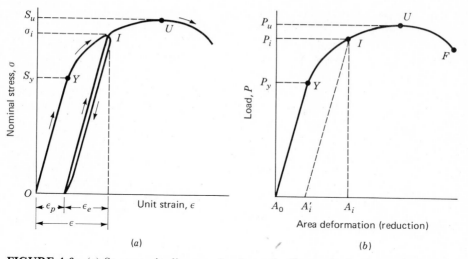

FIGURE 4-6 (*a*) Stress-strain diagram showing unloading and reloading at point *I* in the plastic region; (*b*) analogous load-deformation diagram.

And so

$$\epsilon = \frac{\sigma_i}{E} + \epsilon_p \tag{c}$$

It is possible to construct a similar diagram, as in Fig. 4-6*b*, where the abscissa is the reduction in area and the ordinate is the applied load. Using this diagram we imagine a specimen loaded by a force P_i to the point *I* on the diagram. If P_i is now released, the specimen relaxes and the cross-sectional area *increases* from A_i to A_i'. The *cold-work factor* is defined as

$$W = \frac{A_0 - A_i'}{A_0} \cong \frac{A_0 - A_i}{A_0} \tag{4-12}$$

Notice that W can also be expressed as a percentage by multiplying by 100. We note also that the approximate sign in Eq. (4-12) is usually replaced by an equals sign.

Suppose we load a specimen once more by P_i to point *I* on Fig. 4-6*b*. Since the specimen has been plastically deformed, it has been cold-worked, and so it now has a new yield strength. This new yield strength must be based upon the area A_i' instead of A_0. Therefore, the new yield strength is

$$S_y' = \frac{P_i}{A_i'} = K(\varepsilon_i)^n \tag{4-13}$$

where ε_i is the logarithmic strain corresponding to point *I*.

The ultimate strength also changes because of the reduced area and, after cold work, becomes

$$S_u' = \frac{P_u}{A_i'} \tag{d}$$

Now $P_u = S_u A_0$, where S_u is the engineering ultimate strength. With the help of Eq. (4-12), Eq. (d) becomes

$$S_u' = \frac{S_u A_0}{A_0(1 - W)} = \frac{S_u}{1 - W} \tag{4-14}$$

which is valid only if point I is to the left of point U in Fig. 4-6b, that is, if $\varepsilon_i \le \varepsilon_u$. Datsko points out that for points to the right of U the ultimate strength and the yield strength are equal and are

$$S_u' = S_y' = K(\varepsilon_i)^n \qquad \varepsilon_i > \varepsilon_u \tag{4-15}$$

Hardness

Most hardness-testing systems employ a standard load which is applied to a ball or pyramid in contact with the material to be tested. The hardness is then expressed as a function of the size of the resulting indentation. This means that hardness is an easy property to measure, because the test is nondestructive and test specimens are not required. Usually the test can be conducted directly on an actual machine element.

In the *Brinell* hardness test* a force is applied to a ball and the hardness number H_B is equal to the applied load divided by the spherical area of the indentation. The fact that the Brinell hardness number can be used to obtain a good estimate of the ultimate tensile strength of steel is of particular value. The relation is

$$S_u = 500H_B \tag{4-16}$$

where S_u is in psi. In SI units the corresponding relation is

$$S_u = 3.45H_B \tag{4-17}$$

where S_u is in MPa. The relationship is not quite so simple for other materials.†

---

* The word *Brinell* is one of the most misspelled words in materials engineering. Remember, it has two l's, as in bell.

† See Datsko, *Materials in Design and Manufacture*, op. cit., pp. 5-34–5-37.

4-4 IMPACT PROPERTIES

An external force applied to a structure or part is called an *impact load* if the time of application is less than one-third the lowest natural period of vibration of the part or structure. Otherwise it is called simply a *static load*.

The *Charpy* and *Izod notched-bar tests* utilize bars of specified geometries to determine brittleness and impact strength. These tests are helpful in comparing several materials and in the determination of low-temperature brittleness. In both tests the specimen is struck by a pendulum released from a fixed height, and the energy absorbed by the specimen, called the *impact value*, is computed from the height of swing after fracture.

The effect of temperature on impact values is shown in Fig. 4-7. Notice the narrow region of critical temperatures where the impact value increases very rapidly. In the low-temperature region the fracture appears as a brittle, shattering type, whereas the appearance is a tough, tearing type above the critical-temperature region. The critical temperature seems to be dependent on both the material and the geometry of the notch. For this reason designers should not rely too heavily on the results of notched-bar tests.

The average strain rate used in obtaining the stress-strain diagram is about 0.001 in·s/in or less. When the strain rate is increased, as it is under impact conditions, the strengths increase as shown in Fig. 4-8. In fact, at very high strain rates the yield strength seems to approach the ultimate strength as a limit. But note that the curves show little change in the elongation. This means that the ductility remains about the same. Also, in view of the sharp increase in yield strength, a mild steel could be expected to behave elastically throughout practically its entire strength range under impact conditions.

These conclusions seem to be justified for both ferrous and nonferrous materials. Thus, except for fatigue loading, the use of static properties in designing to resist impact loads is on the conservative side.

4-5 CREEP AND TEMPERATURE PROPERTIES

Machine and structural members are often required to endure temperatures different from those at which the mechanical properties are usually obtained. Here we wish to explore the changes that occur in material properties due to temperature.

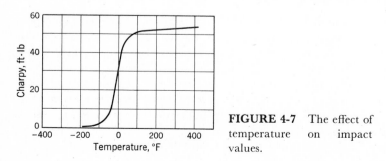

FIGURE 4-7 The effect of temperature on impact values.

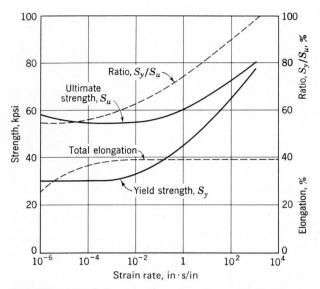

FIGURE 4-8 Influence of strain rate on tensile properties. (*By M. J. Manjoine.*)

Figure 4-9 shows the effects of temperature on the tensile properties of steels. Designers should always consult manufacturers' bulletins for exact properties, but the figure shows typical characteristics. Note that there is only a small change in the tensile strength until a critical temperature is reached, then it falls off rapidly. However, too many engineers remember this fact and forget that the yield strength continually decreases with temperature increase. For example, the material would have a yield strength of about 70 percent of room-temperature yield strength at 500°F. Ignoring this fact has led designers into errors that have resulted in a number of steam-turbine component failures in service.

Many tests have been made of ferrous metals subjected to constant loads for long periods of time at elevated temperatures. The specimens were found to be permanently deformed during the tests, even though at times the actual stresses were less than the yield strength of the material obtained from short-time tests made at the same temperature. This continuous deformation under load is called *creep*.

The usual problem in the design of members subjected to elevated temperatures is to select a material and stress such that, for the life of the part, a certain limiting value of the creep will not be exceeded. In the case of parts which have a very short life, it is not difficult to devise tests which will provide the necessary information. On the other hand, some steam-turbine parts are expected to have a life of 20 years or more. Since it is not practical to run such long tests, experimental results must be extrapolated in order to provide the necessary design information.

One of the most useful tests which have been devised is the long-time creep

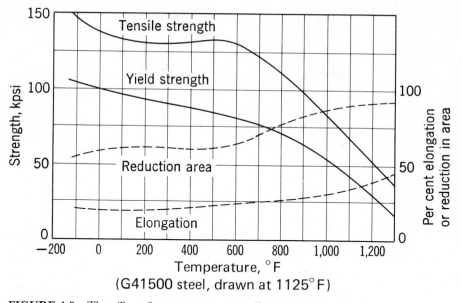

FIGURE 4-9   The effect of temperature on tensile properties of steel.

test under constant load.   Figure 4-10 illustrates a curve which is typical of this kind of test.   The curve is obtained at a constant stated temperature.   A number of tests are usually run simultaneously at different stress intensities.   The curve exhibits three distinct regions.   In the first stage are included both the elastic and the plastic deformation. This stage shows a decreasing creep rate which is due to the strain hardening.   The second stage shows a constant minimum creep rate caused by the annealing effect.   In the third stage the specimen shows a considerable reduction in area, the unit stress is increased, and a higher creep eventually leads to fracture.

A stress which would cause fracture within the time usually available for testing would normally be too high, except for members with a very short life. For this reason creep-time tests are usually extended only long enough to establish

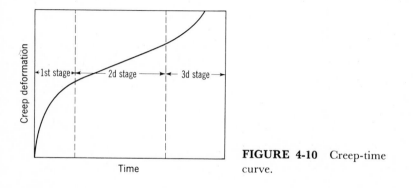

FIGURE 4-10   Creep-time curve.

the first stage, or the first and a portion of the second stage. Investigators must be cautious in not ending the tests too soon. The shape of the curve is influenced by such factors as crystal-grain size, corrosion, and age hardening. Some of these factors may not show up during the early portions of the test, and so it is preferable to extend the tests until at least 10 percent of the expected life has been exceeded.

4-6 NUMBERING SYSTEMS

The Society of Automotive Engineers (SAE) was the first to recognize the need, and to adopt a system, for the numbering of steels. Later the American Iron and Steel Institute (AISI) adopted a similar system. In 1975 the SAE published the Unified Numbering System for Metals and Alloys (UNS); this system also contains cross-reference numbers for other material specifications.* The UNS uses a letter prefix to designate the material, as, for example, G for the carbon and alloy steels, A for the aluminum alloys, C for the copper-base alloys, and S for the stainless or corrosion-resistant steels. For some materials not enough agreement has as yet developed in the industry to warrant the establishment of a designation.

For the steels, the first two numbers following the letter prefix indicate the composition, excluding the carbon content. The various compositions used are as follows:

G10	Plain carbon	G46	Nickel-molybdenum
G11	Free-cutting carbon steel with more sulfur or phosphorus	G48	Nickel-molybdenum
		G50	Chromium
G13	Manganese	G51	Chromium
G23	Nickel	G52	Chromium
G25	Nickel	G61	Chromium-vanadium
G31	Nickel-chromium	G86	Chromium-nickel-molybdenum
G33	Nickel-chromium	G87	Chromium-nickel-molybdenum
G40	Molybdenum	G92	Manganese-silicon
G41	Chromium-molybdenum	G94	Nickel-chromium-molybdenum
G43	Nickel-chromium-molybdenum		

The second number pair refers to the approximate carbon content. Thus, G10400 is a plain carbon steel with a carbon content of 0.37 to 0.44 percent. The fifth number following the prefix is used for special situations. For example, the old designation AISI 52100 represents a chromium alloy with about 100 points of carbon. The UNS designation is G51216.

The UNS designations for the stainless steels, prefix S, utilize the older AISI designations for the first three numbers following the prefix. The next two num-

---

* Many of the materials discussed in the balance of this chapter are listed in the appendix tables. Be sure to review these.

Table 4-1 ALUMINUM ALLOY DES-IGNATIONS

Aluminum 99.00% pure and greater	Ax1xxx
Copper alloys	Ax2xxx
Manganese alloys	Ax3xxx
Silicon alloys	Ax4xxx
Magnesium alloys	Ax5xxx
Magnesium-silicon alloys	Ax6xxx
Zinc alloys	Ax7xxx

bers are reserved for special purposes. The first number of the group indicates the approximate composition. Thus 2 is a chromium-nickel-manganese steel, 3 is a chromium-nickel steel, and 4 is a chromium alloy steel. Sometimes stainless steels are referred to by their alloy content. Thus S30200 is often called an 18-8 stainless steel, meaning 18 percent chromium and 8 percent nickel.

The prefix for the aluminum group is the letter A. The first number following the prefix indicates the processing. For example, A9 is a wrought aluminum while A0 is a casting alloy. The second number designates the main alloy group as shown in Table 4-1. The third number in the group is used to modify the original alloy or to designate the impurity limits. The last two numbers refer to other alloys used with the basic group.

The American Society for Testing and Materials (ASTM) numbering system for cast iron is in widespread use. This system is based on the tensile strength. Thus ASTM No. 30 cast iron has a minimum tensile strength of 30 kpsi. Note from the appendix however that the *typical* tensile strength is 31 kpsi. Readers should be careful to designate which of the two values is used in design and problem work because of the significance of factor of safety.

4-7 METAL PROCESSES

The designer must be familiar with all of the metalworking processes in order to produce the most satisfactory design. These processes include casting, machining, hot and cold working, and the various heat treatments. A detailed discussion of these is beyond the scope of this book.*

4-8 ALLOY STEELS

While a plain carbon steel is an alloy of iron and carbon with small amounts of manganese, silicon, sulfur, and phosphorus, the term *alloy steel* is applied when one or more elements other than carbon are introduced in sufficient quantities to

---

* The following two books are typical references: Benjamin W. Neibel and Alan B. Draper, *Product Design and Process Engineering*, McGraw-Hill, New York, 1974; and Joseph Datsko, *Material Properties and Manufacturing Processes*, op. cit.

modify its properties substantially. The alloy steels not only possess more desirable physical properties but also permit a greater latitude in the heat-treating process.

Chromium The addition of chromium results in the formation of various carbides of chromium which are very hard, yet the resulting steel is more ductile than a steel of the same hardness produced by a simple increase in carbon content. Chromium also refines the grain structure so that these two combined effects result in both increased toughness and increased hardness. The addition of chromium increases the critical range of temperatures and moves the eutectoid point to the left. Chromium is thus a very useful alloying element.

Nickel The addition of nickel to steel also causes the eutectoid point to move to the left and increases the critical range of temperatures. Nickel is soluble in ferrite and does not form carbides or oxides. This increases the strength without decreasing the ductility. Case hardening of nickel steels results in a better core than can be obtained with plain carbon steels. Chromium is frequently used in combination with nickel to obtain the toughness and ductility provided by the nickel and the wear resistance and hardness contributed by the chromium.

Manganese Manganese is added to all steels as a deoxidizing and desulfurizing agent, but if the sulfur content is low and the manganese content is over 1 percent, the steel is classified as a manganese alloy. Manganese dissolves in the ferrite and also forms carbides. It causes the eutectoid point to move to the left and lowers the critical range of temperatures. It increases the time required for transformation so that oil quenching becomes practicable.

Silicon Silicon is added to all steels as a deoxidizing agent. When added to very-low-carbon steels it produces a brittle material with a low hysteresis loss and a high magnetic permeability. The principal use of silicon is with other alloying elements, such as manganese, chromium, and vanadium, to stabilize the carbides.

Molybdenum While molybdenum is used alone in a few steels, it finds its greatest use when combined with other alloying elements, such as nickel, chromium, or both. Molybdenum forms carbides and also dissolves in ferrite to some extent, so that it adds both hardness and toughness. Molybdenum increases the critical range of temperatures and substantially lowers the transformation point. Because of this lowering of the transformation point, molybdenum is most effective in producing desirable oil-hardening and air-hardening properties. Except for carbon, it has the greatest hardening effect, and because it also contributes to a fine grain size, this results in the retention of a great deal of toughness.

Vanadium Vanadium has a very strong tendency to form carbides; hence it is used only in small amounts. It is a strong deoxidizing agent and promotes a fine grain size. Since some vanadium is dissolved in the ferrite, it also toughens the steel. Vanadium gives a wide hardening range to steel, and the alloy can be hardened from a higher temperature. It is very difficult to soften vanadium steel by tempering; hence it is widely used in tool steels.

Tungsten Tungsten is widely used in tool steels because the tool will maintain its hardness even at red heat. Tungsten produces a fine, dense structure and adds both toughness and hardness. Its effect is similar to that of molybdenum, except that it must be added in greater quantities.

4-9 CORROSION-RESISTING STEELS

Iron-base alloys containing at least 12 percent chromium are called *stainless steels*. The most important characteristic of these steels is their resistance to many, but not all, corrosive conditions. The four types available are the ferritic chromium steels, the austenitic chromium-nickel steels, and the martensitic and precipitation-hardenable stainless steels.

The ferritic chromium steels have a chromium content ranging from 12 to 27 percent. Their corrosion resistance is a function of the chromium content, so that alloys containing less than 12 percent still exhibit some corrosion resistance, although they may rust. The quench-hardenability of these steels is a function of both the chromium and the carbon content. The very-high-carbon steels have good quench-hardenability up to about 18 percent chromium, while in the lower carbon ranges it ceases at about 13 percent. If a little nickel is added, these steels retain some degree of hardenability up to 20 percent chromium. If the chromium content exceeds 18 percent, they become difficult to weld, and at the very high chromium levels the hardness becomes so great that very careful attention must be paid to the service conditions. Since chromium is expensive, the designer will choose the lowest chromium content consistent with the corrosive conditions.

The chromium-nickel stainless steels retain the austenitic structure at room temperature; hence they are not amenable to heat treatment. The strength of these steels can be greatly improved by cold working. They are not magnetic unless cold-worked. Their work-hardenability properties also cause them to be difficult to machine. All the chromium-nickel steels may be welded. They have greater corrosion-resistant properties than the plain chromium steels. When more chromium is added for greater corrosion resistance, more nickel must also be added if the austenitic properties are to be retained.

4-10 CASTING MATERIALS

Gray cast iron Of all the cast materials, gray cast iron is the most widely used. This is because it has a very low cost, is easily cast in large quantities, and is easy to machine. The principal objections to the use of gray cast iron are that it is brittle and that it is weak in tension. In addition to a high carbon content (over 1.7 percent and usually greater than 2 percent), cast iron also has a high silicon content, with low percentages of sulfur, manganese, and phosphorus. The result-ant alloy is composed of pearlite, ferrite, and graphite, and under certain con-ditions the pearlite may decompose into graphite and ferrite. The resulting prod-uct then contains all ferrite and graphite. The graphite, in the form of thin flakes distributed evenly throughout the structure, darkens it, hence, the name *gray cast iron*.

Gray cast iron is not readily welded because it may crack, but this tendency may be reduced if the part is carefully preheated. While the castings are gener-ally used in the as-cast condition, a mild anneal reduces cooling stresses and improves the machinability. The tensile strength of gray cast iron varies from 15 to 60 kpsi, and the compressive strengths are three to four times the tensile strengths. The modulus of elasticity varies widely, with values extending all the way from 11 to 22 Mpsi.

White cast iron   If all the carbon in cast iron is in the form of cementite and pearlite, with no graphite present, the resulting structure is white and is known as *white cast iron*. This may be produced in two ways. The composition may be adjusted by keeping the carbon and silicon content low, or the gray-cast-iron composition may be cast against chills in order to promote rapid cooling. By either method a casting with large amounts of cementite is produced, and as a result the product is very brittle and hard to machine but also very resistant to wear. A chill is usually used in the production of gray-iron castings in order to provide a very hard surface within a particular area of the casting, while at the same time retaining the more desirable gray structure within the remaining por-tion. This produces a relatively tough casting with a wear-resistant area.

Malleable cast iron   If white cast iron within a certain composition range is annealed, a product called *malleable cast iron* is formed. The annealing process frees the carbon so that it is present as graphite, just as in gray cast iron but in a different form. In gray cast iron the graphite is present in a thin flake form, while in malleable cast iron it has a nodular form and is known as *temper carbon*. A good grade of malleable cast iron may have a tensile strength of over 50 kpsi, with an elongation of as much as 18 percent. The percentage elongation of a gray cast iron, on the other hand, is seldom over 1 percent. Because of the time required for annealing (up to 6 days for large and heavy castings), malleable iron is necessarily somewhat more expensive than gray cast iron.

Ductile and nodular cast iron   Because of the lengthy heat treatment required to produce malleable cast iron, a cast iron has long been desired which would combine the ductile properties of malleable iron with the ease of casting and machining of gray iron and at the same time would possess these properties in the as-cast conditions. A process for producing such a material using cesium with magnesium seems to fulfill these requirements.

Ductile cast iron, or *nodular cast iron*, as it is sometimes called, is essentially the same as malleable cast iron, because both contain graphite in the form of spheroids. However, ductile cast iron in the as-cast condition exhibits properties very close to those of malleable iron, and if a simple 1-h anneal is given and is followed by a slow cool, it exhibits even more ductility than the malleable product. Ductile iron is made by adding magnesium to the melt; since magnesium boils at this temperature, it is necessary to alloy it with other elements before it is introduced.

Ductile iron has a high modulus of elasticity (25 Mpsi) as compared with gray cast iron, and it is elastic in the sense that a portion of the stress-strain curve is a straight line. Gray cast iron, on the other hand, does not obey Hooke's law, because the modulus of elasticity steadily decreases with increase in stress. Like gray cast iron, however, nodular iron has a compressive strength which is higher than the tensile strength, although the difference is not as great. Since this is a new product, its full range of application has not as yet developed, but an obvious area of application is for those castings requiring shock and impact resistance.

Alloy cast irons Nickel, chromium, and molybdenum are the most common alloying elements used in cast iron. Nickel is a general-purpose alloying element, usually added in amounts up to 5 percent. Nickel increases the strength and density, improves the wearing qualities, and raises the machinability. If the nickel content is raised to 10 to 18 percent, an austenitic structure with valuable heat- and corrosion-resistant properties results. Chromium increases the hardness and wear resistance and, when used with a chill, increases the tendency to form white iron. When chromium and nickel are both added, the hardness and strength are improved without a reduction in the machinability rating. Molybdenum added in quantities up to 1.25 percent increases the stiffness, hardness, tensile strength, and impact resistance. It is a widely used alloying element.

Cast steels The advantage of the casting process is that parts having complex shapes can be manufactured at costs less than fabrication by other means, such as welding. Thus the choice of steel castings is logical when the part is complex and when it must also have a high strength. The higher melting temperatures for steels do aggravate the casting problems and require closer attention to such details as core design, section thicknesses, fillets, and the progress of cooling. The same alloying elements used for the wrought steels can be used for cast steels to improve the strength and other mechanical properties. Cast-steel parts can also be heat-treated to alter the mechanical properties, and, unlike the cast irons, they can be welded.

4-11 NONFERROUS METALS

Aluminum The outstanding characteristics of aluminum and its alloys are their strength-weight ratio, their resistance to corrosion, and their high thermal and electrical conductivity. The density of aluminum is about 0.10 lb/in³ (2770 kg/m³), compared with 0.28 lb/in³ (7750 kg/m³) for steel. Pure aluminum has a tensile strength of about 13 kpsi, but this can be improved considerably by cold working and also by alloying with other materials. The modulus of elasticity of aluminum, as well as of its alloys, is 10.3 Mpsi, which means that it has about one-third the stiffness of steel.

Considering the cost and strength of aluminum and its alloys, they are among the most versatile materials from the standpoint of fabrication. Aluminum can be processed by sand casting, die casting, hot or cold working, or extruding. These alloys can be machined, press-worked, soldered, brazed, or welded. Aluminum melts at 1215°F, which makes it very desirable for the production of either permanent or sand-mold castings. It is commercially available in the form of plate, bar, sheet, foil, rod, and tube and in structural and extruded shapes. Certain precautions must be taken in joining aluminum by soldering, brazing, or welding; these joining methods are not recommended for all alloys.

The corrosion resistance of the aluminum alloys depends upon the formation of a thin oxide coating. This film forms spontaneously because aluminum is inherently very reactive. Constant erosion or abrasion removes this film and allows corrosion to take place. An extra-heavy oxide film may be produced by the process called *anodizing*. In this process the specimen is made to become the anode in an electrolyte, which may be chromic acid, oxalic acid, or sulfuric acid. It is possible in this process to control the color of the resulting film very accurately.

The most useful alloying elements for aluminum are copper, silicon, manganese, magnesium, and iron. Aluminum alloys are classified as *casting alloys* or *wrought alloys*. The casting alloys have greater percentages of alloying elements to facilitate casting, but this makes cold working difficult. Many of the casting alloys, and some of the wrought alloys, cannot be hardened by heat treatment. The alloys that are heat-treatable use an alloying element which dissolves in the aluminum. The heat treatment consists of heating the specimen to a temperature which permits the alloying element to pass into solution, then quenching so rapidly that the alloying element is not precipitated. The aging process may be accelerated by heating slightly, which results in even greater hardness and strength. One of the better-known heat-treatable alloys is duraluminum, or A92017 (4 percent Cu, 0.5 percent Mg, 0.5 percent Mn). This alloy hardens in 4 days at room temperature. Because of this rapid aging, the alloy must be stored under refrigeration after quenching and before forming, or it must be formed immediately after quenching. Other alloys (such as A95053) have been developed which age-harden much more slowly, so that only mild refrigeration is required before forming. After forming, they are artificially aged in a furnace and possess approximately the same strength and hardness as the A92024 (245) alloys. Those alloys of aluminum which cannot be heat-treated can be hardened only by cold working. Both work hardening and the hardening produced by heat treatment may be removed by an annealing process.

Magnesium   The density of magnesium is about 0.065 lb/in^3 (1800 kg/m^3), which is two-thirds that of aluminum and one-fourth that of steel. Since it is the lightest of all commercial metals, its greatest use is in the aircraft industry, but uses are now being found for it in other applications. Although the magnesium alloys do not have great strength, because of their light weight the strength-weight ratio compares favorably with the stronger aluminum and steel alloys. Even so, magnesium alloys find their greatest use in applications where strength is not an important consideration. Magnesium will not withstand elevated temperatures; the yield point is definitely reduced when the temperature is raised to that of boiling water.

Magnesium and its alloys have a modulus of elasticity of 6.5 Mpsi in tension and in compression, although some alloys are not as strong in compression as in tension. Curiously enough, cold working reduces the modulus of elasticity.

Copper-base alloys   When copper is alloyed with zinc, it is usually called *brass*. If it is alloyed with another element, it is often called *bronze*. Sometimes the other element is specified too, as, for example, *tin bronze* or *phosphor bronze*. There are hundreds of variations in each category.

Brass with 5 to 15 percent zinc   These brasses are easy to cold-work, especially those with the higher zinc content. They are ductile but often hard to machine. The corrosion resistance is good. Alloys included in this group are *gilding brass* (5 percent Zn), *commercial bronze* (10 percent Zn), and *red brass* (15 percent Zn). Gilding brass is used mostly for jewelry and articles to be gold-plated; it has the same ductility as copper but greater strength, accompanied by poor machining characteristics. Commercial bronze is used for jewelry and for forgings and stampings because of its ductility. Its machining properties are poor, but it has excellent cold-working properties. Red brass has good corrosion resistance as well as high-temperature strength. Because of this it is used a great deal in the form of tubing or piping to carry hot water in such applications as radiators or condensers.

Brass with 20 to 36 percent zinc   Included in this group are *low brass* (20 percent Zn), *cartridge brass* (30 percent Zn), and *yellow brass* (35 percent Zn). Since zinc is cheaper than copper, these alloys cost less than those with more copper and less zinc. They also have better machinability and slightly greater strength; this is offset, however, by poor corrosion resistance and the possibility of season cracking at points of residual stresses. Low brass is very similar to red brass and is used for articles requiring deep-drawing operations. Of the copper-zinc alloys, cartridge brass has the best combination of ductility and strength. Cartridge cases were originally manufactured entirely by cold working; the process consisted in a series of deep draws, each draw being followed by an anneal to place the material in condition for the next draw; hence, the name cartridge brass. While the hot-working ability of yellow brass is poor, it can be used in practically any other fabricating process and is therefore employed in a large variety of products.

When small amounts of lead are added to the brasses, their machinability is greatly improved and there is some improvement in their abilities to be hot-worked.   The addition of lead impairs both the cold-working and welding proper-ties.   In this group are *low-leaded brass* ($32\frac{1}{2}$ percent Zn, $\frac{1}{2}$ percent Pb), *high-leaded brass* (34 percent Zn, 2 percent Pb), and *free-cutting brass* ($35\frac{1}{2}$ percent Zn, 3 percent Pb).   The low-leaded brass is not only easy to machine but has good cold-working properties.   It is used for various screw-machine parts.   High-leaded brass, sometimes called *engraver's brass*, is used for instrument, lock, and watch parts.   Free-cutting brass is also used for screw-machine parts and has good corrosion resistance with excellent mechanical properties.

Admiralty metal (28 percent Zn) contains 1 percent tin, which imparts excel-lent corrosion resistance, especially to saltwater.   It has good strength and duc-tility but only fair machining and working characteristics.   Because of its cor-rosion resistance it is used in power-plant and chemical equipment.   *Aluminum brass* (22 percent Zn) contains 2 percent aluminum and is used for the same purposes as admiralty metal because it has nearly the same properties and charac-teristics.   In the form of tubing or piping, it is favored over admiralty metal, because it has better resistance to erosion caused by high-velocity water.

Brass with 36 to 40 percent zinc   Brasses with more than 38 percent zinc are less ductile than cartridge brass and cannot be cold-worked as severely.   They are frequently hot-worked and extruded.   *Muntz metal* (40 percent Zn) is low in cost and mildly corrosion-resistant.   *Naval brass* has the same composition as Muntz metal except for the addition of 0.75 percent tin, which contributes to the cor-rosion resistance.

Bronze   *Silicon bronze*, containing 3 percent silicon and 1 percent manganese in addition to the copper, has mechanical properties equal to those of mild steel, as well as good corrosion resistance.   It can be hot- or cold-worked, machined, or welded.   It is useful wherever corrosion resistance combined with strength is required.

Phosphor bronze, made with up to 11 percent tin and containing small amounts of phosphorus, is especially resistant to fatigue and corrosion.   It has a high tensile strength and a high capacity to absorb energy, and it is also resistant to wear. These properties make it very useful as a spring material.

Aluminum bronze is a heat-treatable alloy containing up to 12 percent alu-minum.   This alloy has strength and corrosion-resistance properties which are better than brass, and in addition, its properties may be varied over a wide range by cold working, heat treating, or changing the composition.   When iron is added in amounts up to 4 percent, the alloy has a high endurance limit, a high shock resistance, and excellent wear resistance.

Beryllium bronze is another heat-treatable alloy, containing about 2 percent beryllium.   This alloy is very corrosion-resistant and has high strength, hardness, and resistance to wear.   Although it is expensive, it is used for springs and other parts subjected to fatigue loading where corrosion resistance is required.

4-12 PLASTICS

The term *thermoplastics* is used to mean any plastic that flows or is moldable when heat is applied to it; the term is sometimes applied to plastics moldable under pressure. Such plastics can be remolded when heated.

A *thermoset* is a plastic for which the polymerization process is finished in a hot molding press where the plastic is liquified under pressure. These plastics cannot be remolded.

Table 4-2 lists some of the most widely used thermoplastics, together with some of their characteristics and the range of their properties. Table 4-3, listing some of the thermosets, is similar. These tables are presented for information only and should not be used to make a final design decision. The range of properties and characteristics that can be obtained with plastics is very great. The influence of many factors, such as cost, moldability, coefficient of friction, weathering, impact strength, and the effect of fillers and reinforcements, must be considered. Manufacturers' catalogs will be found quite helpful in making possible selections.

4-13 ELASTOMERS

Table 4-4 lists some of the characteristics of the various thermoset or vulcanized elastomers, including natural rubber. These characteristics are presented only as a general guide and to show the range of properties that are available. These elastomers vary widely in cost and other factors. Unless the designer has had experience in the design and selection of elastomers, it is good practice to specify the requirements of the particular application and depend upon the manufacturer to specify the most suitable material.

Thermoplastic elastomers are also becoming available. These generally require shorter processing times.

4-14 VISCOELASTICITY

Plastics have the characteristic that when they are loaded beyond the elastic limit, the stress-strain relations are dependent on time. Materials that exhibit this characteristic are said to be *viscoelastic*. You can observe the viscoelastic effect yourself by folding a 3 × 5 index card and watching it unfold itself.

Since the actual stress-strain-time relations for many materials may be quite complex, it is often necessary to idealize a material by assuming linearity. Such a material is then called a *Maxwell material**, and the stress-strain relations are

---

* E. H. Lee, "Viscoelasticity," in W. Flügge (ed.), *Handbook of Engineering Mechanics*, McGraw-Hill, New York, 1962, chap. 53.

Table 4-2 THE THERMOPLASTICS

Name	S_u kpsi	E Mpsi	Hardness, Rockwell	Elongation, %	Dimensional stability	Heat resistance	Chemical resistance	Processing
ABS group	2–8	0.10–0.37	60–110R	3–50	Good	*	Fair	EMST
Acetal group	8–10	0.41–0.52	80–94M	40–60	Excellent	Good	High	M
Acrylic	5–10	0.20–0.47	92–100M	3–75	High	*	Fair	EMS
Fluoroplastic group	0.50–7	...	50–80D	100–300	High	Excellent	Excellent	MPR†
Nylon	8–14	0.18–0.45	112–120R	10–200	Poor	Poor	Good	CEM
Phenylene oxide	7–18	0.35–0.92	115R, 106L	5–60	Excellent	Good	Fair	EFM
Polycarbonate	8–16	0.34–0.86	62–91M	10–125	Excellent	Excellent	Fair	EMS
Polyester	8–18	0.28–1.6	65–90M	1–300	Excellent	Poor	Excellent	CLMR
Polyimide	6–50	...	88–120M	Very low	Excellent	Excellent	Excellent†	CLMP
Polyphenylene sulfide	14–19	0.11	122R	1.0	Good	Excellent	Excellent	M
Polystyrene group	1.5–12	0.14–0.60	10–90M	0.5–60	...	Poor	Poor	EM
Polysulfone	10	0.36	120R	50–100	Excellent	Excellent†	Excellent†	EFM
Polyvinyl chloride	1.5–7.5	0.35–0.60	65–85D	40–450	...	Poor	Poor	EFM

* Heat-resistant grades available.

† With exceptions

C	Coatings	L	Laminates
E	Extrusions	M	Moldings
F	Foams	P	Press and sinter methods
R	Resins		
S	Sheet		
T	Tubing		

Source: These data have been obtained from the *Machine Design Materials Reference Issue*, published by Penton/IPC, Cleveland, Ohio. These reference issues are published about every 2 years and constitute an excellent source of data on a great variety of materials.

Table 4-3 THE THERMOSETS

Name	S_u kpsi	E Mpsi	Hardness, Rockwell	Elongation, %	Dimensional stability	Heat resistance	Chemical resistance	Processing
Alkyd	3–9	0.05–0.30	99M*	...	Excellent	Good	Fair	M
Allylic	4–10	...	105–120M	...	Excellent	Excellent	Excellent	CM
Amino group	5–8	0.13–0.24	110–120M	0.30–0.90	Good	Excellent*	Excellent*	LR
Epoxy	5–20	0.03–0.30*	80–120M	1–10	Excellent	Excellent	Excellent	CMR
Phenolics	5–9	0.10–0.25	70–95E	...	Excellent	Excellent	Good	EMR
Silicones	5–6	...	80–90M	...	...	Excellent	Excellent	CLMR

* With exceptions.

C	Coatings	L Laminates	R Resins
E	Extrusions	M Moldings	S Sheet
F	Foams	P Press and sinter methods	T Tubing

Source: These data have been obtained from the *Machine Design Materials Reference Issue,* published by Penton/IPC, Cleveland, Ohio. These reference issues are published about every 2 years and constitute an excellent source of data on a great variety of materials.

Table 4-4 SOME CHARACTERISTICS AND PROPERTIES OF THERMOSET ELASTOMERS

Name	S_u kpsi	Elonga-tion %	Hard-ness Shore A	Brittle point °F	Resistance to					
					Flame	Weather	Water	Acids*	Alkalies*	Oils
Natural rubber	4	700	30–90	−80	D	D	A	A	A	NR
Polyisoprene	4	700	30–90	−80	D	D	A	A	A	NR
Styrene butadiene	3.5	600	40–90	−80	D	D	B–A	C	C	NR
Butadiene	3	600	40–80	−100	D	D	B	C	C	NR
Isobutane isoprene	3	800	40–80	−80	D	A	C–A	A	A	NR
Chlorinated isobutene isoprene	3	700	40–80	−80	D	A	B–A	A	A	NR
Ethylene propylene copolymer	3	600	30–90	−90	D	A	A	A	A	NR
Chloro-sulfonated polyethylene	3	500	50–90	−70	B–A	A	B–A	A	A	A
Chloroprene (neoprene)	4	600	30–90	−80	B–A	A	B	A	A	A
Chlorinated polyethylene	4	500	40–90	−40	A	A	B	A	A	A
Nitrile butadiene (high nitrile)	4	600	40–90	−40	D	D	A	B	B	A
Nitrile butadiene (low nitrile)	3.5	600	40–90	−90	D	D	B	B	B	A
Epichloro-hydrins	2.5	400	40–90	−10 to −50	B–D	B	B	B–C	B–D	A
Polyacrylate	2.5	400	40–85	−40	D	A	D	D–C	D–C	A
Silicones	1.2	700	30–85	−90 to −180	C–A	A	B–A	B	A	B
Urethanes	10	700	40–100	−60 to −90	D–A	A	D–C	C	C	A
Fluoro-silicone	1.2	400	60–80	−85	A	A	A	A	A	A
Fluoro-carbon	2.5	300	60–95	−40	A	A	A	A	A	A

* Diluted

NOTE: A = excellent, B = good, C = fair, D = poor, NR = not recommended.

Source: These data have been obtained from *Machine Design Materials Reference Issue*, published by Penton/IPC, Cleveland, Ohio. These reference issues are published about every 2 years and constitute an excellent source of detailed data on a great variety of materials.

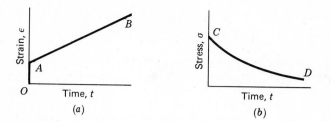

FIGURE 4-11 Characteristics of an ideal viscoelastic material. (a) Creep-time relations; (b) relaxation curve.

expressed by the differential equation

$$\frac{1}{E}\frac{d\sigma}{dt} + \frac{\sigma}{c} = \frac{d\epsilon}{dt} \tag{4-18}$$

where E is the modulus of elasticity and c is the coefficient of viscosity. A typical solution to this equation is shown in Fig. 4-11. In Fig. 4-11a a constant load is applied which results in the linear creep relation AB. The distance OA is the elastic response, which takes place without time delay, while AB is the viscoelastic response. When the load is removed, the relaxation or stress-time response of Fig. 4-11b is obtained. Note that the curve CD corresponds to the unfolding of the 3 × 5 card.

PROBLEMS

4-1 A specimen of medium-carbon steel having an initial diameter of 0.503 in was tested in tension using a gauge length of 2 in. The following data were obtained for the elastic and plastic states:

Elastic state		Plastic state	
Load P lb	Elongation δ in	Load P lb	Area A_i in^2
1000	0.0004	8 800	0.1984
2000	0.0006	9 200	0.1978
3000	0.0010	9 100	0.1963
4000	0.0013	13 200	0.1924
7000	0.0023	15 200	0.1875
8400	0.0028	17 000	0.1563
8800	0.0036	16 400	0.1307
9200	0.0089	14 800	0.1077

Note that there is some overlap in the data. Plot the engineering or nominal stress-strain diagram using two scales for the unit strain ϵ, one from zero to about 0.02 in/in and the other from zero to maximum strain. From this diagram find the modulus of elasticity, the 0.2-percent-offset yield strength, the ultimate strength, and the percent reduction in area.

4-2 Compute the true stress and the logarithmic strain using the data of Prob. 4-1. Plot the results on log-log paper or take the logarithms of σ and ε and plot a diagram using cartesian coordinates.

(a) Find the modulus K and the strain-hardening exponent n.

(b) Determine the yield strength and the ultimate strength after the specimen has had 20 percent cold work.

4-3 The same as Prob. 4-1 except the initial diameter is 12.83 mm, the gauge length is 50 mm, and the following data apply.

Elastic state		Plastic state	
Load P kN	Elongation δ mm	Load P kN	Area A_i mm^2
5.5	0.005	37.0	128.9
7.8	0.015	37.1	128.8
13.6	0.025	37.0	127.3
18.9	0.035	37.0	125.4
26.9	0.050	42.0	120.1
33.2	0.065	46.4	114.3
35.6	0.080	48.0	107.8
36.8	0.110	48.1	103.3
37.0	0.145	47.9	94.1
37.1	0.200	47.3	83.4
		46.5	76.0
		40.3	68.8

4-4 The same as Prob. 4-2 but use the data of Prob. 4-3 and 5 percent cold work.

CHAPTER

5

STATISTICAL CONSIDERATIONS

The use of statistics in mechanical design provides a method of dealing with characteristics which have a variability. Statistics is used in this book so that we can express in numerical terms such things as the reliability and the life of a mechanical part. In these days, when there are so many product-liability suits in the courts, it is no longer satisfactory to say that a product is expected to have a long and trouble-free life. We must now find a way to express such things as product life and product reliability in numerical form.

A side benefit that you will derive from this chapter is the ability to use the statistical keys on your calculator. These provide quite useful information, even on everyday problems. It is assumed that you are familiar with such topics as permutations, combinations, probability, and probability theorems. These subjects are usually studied in basic college algebra.

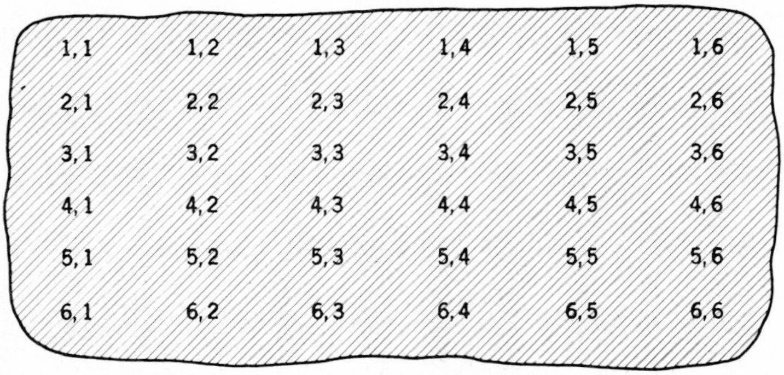

1,1	1,2	1,3	1,4	1,5	1,6
2,1	2,2	2,3	2,4	2,5	2,6
3,1	3,2	3,3	3,4	3,5	3,6
4,1	4,2	4,3	4,4	4,5	4,6
5,1	5,2	5,3	5,4	5,5	5,6
6,1	6,2	6,3	6,4	6,5	6,6

FIGURE 5-1   Sample space showing all possible outcomes of the toss of two dice.

5-1 RANDOM VARIABLES

Consider a collection of 20 tensile-test specimens that have been machined from a like number of samples selected at random from a carload shipment of, say, UNS G10200 cold-drawn steel. It is reasonable to expect that there will be differences in the ultimate tensile strengths S of each of these test specimens. Such differences may occur because of differences in the sizes of the specimens, in the strength of the material itself, or both. Such an experiment is called a *random experiment*, because the specimens were selected at random. The strength S determined by this experiment is called a *random,* or a *stochastic, variable.* So a random variable is a variable quantity, such as strength, size, or weight, whose value depends upon the outcome of a random experiment.

Let us define a random variable x as the sum of the numbers obtained when two dice are tossed. Either die can display any number from 1 to 6; Fig. 5-1, called the *sample space*, displays all possible outcomes. Note that x has a specific value for each possible outcome; for the event 5, 4, $x = 5 + 4 = 9$. It is useful to form a table showing the values of x and the corresponding values of the probability of x, called $p = f(x)$. This is easily done from Fig. 5-1 merely by counting, since there are 36 points, each having a weight $w = \frac{1}{36}$. The results are shown in Table 5-1. Any table, such as this, listing all possible values of a random variable, together with the corresponding probabilities, is called a *probability distribution*.

The values of Table 5-1 are plotted in graphical form in Fig. 5-2. Here it is clear that the probability is a function of x. This *probability function* $p = f(x)$ is often called the *frequency function* or, sometimes, the *probability density*. The probability that x is less than or equal to a certain value x_i can be obtained from the

Table 5-1   A PROBABILITY DISTRIBUTION

x	2	3	4	5	6	7	8	9	10	11	12
$f(x)$	$\frac{1}{36}$	$\frac{2}{36}$	$\frac{3}{36}$	$\frac{4}{36}$	$\frac{5}{36}$	$\frac{6}{36}$	$\frac{5}{36}$	$\frac{4}{36}$	$\frac{3}{36}$	$\frac{2}{36}$	$\frac{1}{36}$

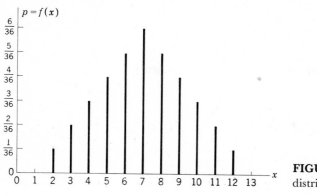

FIGURE 5-2 Frequency distribution.

probability function by summing the probability of all x's up to and including x_i. If we do this with Table 5-1, letting x_i equal 2, then 3, and so on, up to 12, we get Table 5-2, which is called a *cumulative probability distribution*. The function $F(x)$ in Table 5-2 is called a *cumulative probability function*. In terms of $f(x)$ it may be expressed mathematically in the general form

$$F(x_i) = \sum_{x_j \leqslant x_i} f(x_j) \tag{5-1}$$

where $F(x_i)$ is properly called the *distribution function*. The cumulative distribution may also be plotted as a graph (Fig. 5-3).

The variable x of this example is called a *discrete random variable* because x has only discrete values. A *continuous random variable* is one that can take on any value in a specified interval; for such variables, graphs like Figs. 5-2 and 5-3 would be plotted as continuous curves.

5-2 THE ARITHMETIC MEAN, VARIANCE, AND STANDARD DEVIATION

In studying the variations in the mechanical properties and characteristics of mechanical elements, we shall generally be dealing with a finite number of elements. The total number of elements, called the *population*, may in some cases be quite large. In such cases it is usually impractical to measure the characteristics of each member of the population, because this involves destructive testing in some cases, and so we select a small part of the group, called a *sample*, for these determi-

Table 5-2 A CUMULATIVE PROBABILITY DISTRIBUTION

x	2	3	4	5	6	7	8	9	10	11	12
$F(x)$	$\frac{1}{36}$	$\frac{3}{36}$	$\frac{6}{36}$	$\frac{10}{36}$	$\frac{15}{36}$	$\frac{21}{36}$	$\frac{26}{36}$	$\frac{30}{36}$	$\frac{33}{36}$	$\frac{35}{36}$	$\frac{36}{36}$

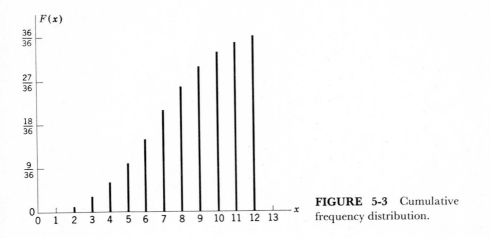

FIGURE 5-3 Cumulative frequency distribution.

nations. Thus the *population* is the entire group, and the *sample* is a part of the population.

The arithmetic mean of a sample, called the *sample mean*, consisting of N elements, is defined by the equation

$$\bar{x} = \frac{x_1 + x_2 + x_3 + \cdots + x_N}{N} = \frac{1}{N} \sum_{1}^{N} x_j \tag{5-2}$$

In a similar manner a population consisting of N elements has a *population mean* defined by the equation

$$\mu = \frac{x_1 + x_2 + x_3 + \cdots + x_N}{N} = \frac{1}{N} \sum_{1}^{N} x_j \tag{5-3}$$

The *mode* and the *median* are also used as measures of central value. The *mode* is the value that occurs most frequently. The *median* is the middle value if there are an odd number of cases; it is the mean of the two middle values if there are an even number.

Besides the arithmetic mean, it is useful to have another kind of measure which will tell us something about the spread, or dispersion, of the distribution. For any random variable x, the deviation of the ith observation from the mean is $x_i - \bar{x}$. But since the sum of these is always zero, we square them, and define *sample variance* as

$$s_x^2 = \frac{(x_1 - \bar{x})^2 + (x_2 - \bar{x})^2 + \cdots + (x_N - \bar{x})^2}{N - 1} = \frac{1}{N - 1} \sum_{1}^{N} (x_j - \bar{x})^2 \tag{5-4}$$

The *sample standard deviation*, defined as the square root of the variance, is

$$s_x = \left[\frac{1}{N - 1} \sum_{1}^{N} (x_j - \bar{x})^2 \right]^{1/2} \tag{5-5}$$

Equations (5-4) and (5-5) are not well suited for use in a calculator. For such purposes, use the alternate form

$$s_x = \left[\frac{\sum x^2 - \dfrac{(\sum x)^2}{N}}{N-1} \right]^{1/2} \tag{5-6}$$

for the standard deviation, and

$$s_x^2 = \frac{\sum x^2}{N} - \bar{x}^2 \tag{5-7}$$

for the variance.

It should be observed that some authors define the variance and the standard

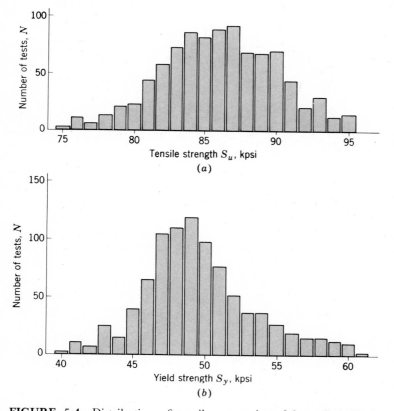

FIGURE 5-4 Distribution of tensile properties of hot-rolled UNS G10350 steel, as rolled. These tests were made from round bars varying in diameter from 1 to 9 in. (*a*) Tensile-strength distributions from 930 heats; $\bar{S}_u = 86.0$ kpsi, $s_{S_u} = 4.04$ kpsi. (*b*) Yield-strength distribution from 899 heats; $\bar{S}_y = 49.5$ kpsi, $s_{S_y} = 5.36$ kpsi. (*By permission, Metals Handbook, vol. 1, 8th ed., American Society for Metals, Metals Park, Ohio, 1961, p. 64.*)

deviation using N instead of $N - 1$ in the denominator. For large values of N there is very little difference. For small values, the denominator $N - 1$ actually gives a better estimate of the variance of the population from which the sample is taken.

Sometimes we are going to be dealing with the standard deviation of the strength of an element. So you must be careful not to be confused by the notation. Note that we are using the *capital letter S* for *strength* and the *lowercase letter s* for *standard deviation*. Figure 5-4 will be useful in visualizing these ideas.

Equations (5-4) to (5-7) apply specifically to the *sample* of a population. When an entire population is considered, the same equations apply, but $\bar{x}$ is replaced with μ, and N weighting is used in the denominators instead of $N - 1$. Also, the resulting standard deviation and variance are designated as $\hat{\sigma}_x$ and $\hat{\sigma}_x^2$; the caret, or hat, is used to avoid confusion with normal stress.

Using the Calculator

Most of the small, everyday pocket calculators have statistics keys to be used for the special purpose of computing the mean and the standard deviation. Does your calculator have a $\sum +$ key? If so, you have statistical capability. When you enter a number and press $\sum +$, the calculator adds this quantity to the total already contained in a memory register. So be sure to clear the memory registers before you start. At the same time you press $\sum +$, the calculator also displays and stores N; the quantity $\sum x^2$ is also computed and stored in a third register. When all the data are entered, press $\bar{x}$, the mean-value key, to display the mean. Depending upon the make of your calculator, other keys can be pressed to recall $\sum x$ and $\sum x^2$ from memory. See your user's manual. The manual will also tell you which keys to press to obtain the standard deviation and, with some calculators, the variance.

If you have even a small programmable calculator, the chances are that you have the capability of assimilating data-point pairs. On these, the data are entered in pairs, using an exchange key or an **ENTER** key to separate the terms of each pair. As each data point is keyed in, it is stored in the memory registers as shown in Table 5-3.

Table 5-3

Register		Contents
1	$\sum y$	Dependent variable
2	$\sum y^2$	
3	N	
4	$\sum x$	Independent variable
5	$\sum x^2$	
6	$\sum xy$	

The following example can be used to get acquainted with the statistics keys on your calculator.

Example 5-1 This example is used to determine the mean and the standard deviation for the grades of a sample of 10 students selected alphabetically. In the table below, x is the final examination grade and y is the course grade. Use your calculator to verify the results shown.

N	1	2	3	4	5	6	7	8	9	10
x	85	74	69	78	93	79	46	75	85	84
y	72.7	52.7	73.2	83.2	78.2	82.3	63.0	82.3	85.7	84.9

$$\sum y = 758.20 \qquad \bar{y} = 75.82 \qquad s_y = 10.75$$

$$\sum x = 768.00 \qquad \bar{x} = 76.80 \qquad s_x = 12.80$$

The standard deviations shown here use $N - 1$ weighting. Some calculators provide the option of using either N or $N - 1$. Also, the results shown above were recorded with the display in the **FIX 2** mode. ////

5-3 REGRESSION

Statisticians use a process of analysis called *regression* to obtain a curve which best fits a set of data points. The process is called *linear regression* when the best-fitting straight line is to be found. The meaning of the word "best" is open to argument because there can be many meanings. The usual method, and the one employed here, is that of choosing a best line based on minimizing the squares of the deviations of the data points from the line.

Figure 5-5 shows a set of data points approximated by the line *AB*. The

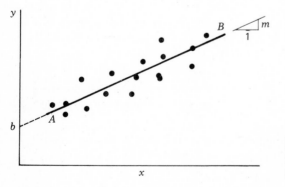

FIGURE 5-5 Set of data points approximated by regression line *AB*.

equation of the straight line is

$$y = mx + b \qquad\qquad (a)$$

where m is the slope and b is the y intercept.   The regression equations are

$$m = \frac{\sum xy - \dfrac{\sum x \sum y}{N}}{\sum x^2 - \dfrac{(\sum x)^2}{N}} \qquad\qquad (5\text{-}8)$$

$$b = \frac{\sum y - m \sum x}{N} \qquad\qquad (5\text{-}9)$$

Examination of Table 5-3 shows that the terms of these equations are available in calculator memory, or can be easily computed.

Having established a slope and intercept, the next point of interest is to discover how well x and y correlate with each other.   If the data points are scattered all over the xy plane, there is obviously no correlation.   But, if all the data points coincide with the regression line, then there is perfect correlation. Most statistical data will be in between these extremes.   A *correlation coefficient r*, having the range $-1 \le r \le +1$, has been devised to answer these questions.   The formula is

$$r = \frac{m s_x}{s_y} \qquad\qquad (5\text{-}10)$$

A negative r indicates that the regression line has a negative slope.   If $r = 0$, there is no correlation; if $r = 1$, there is perfect correlation.

Equations (5-8) to (5-10) can be solved directly on your calculator if it has the $\sum +$ key.   There are special calculators for statistics problems in which the constants m, b, and r can be obtained by pressing appropriate keys after the statistical data have been entered.   If you have a programmable calculator the regression equations should be in the calculator library; they can be solved using a software module or a magnetic card or can be programmed directly for solution. See your user's manual.   The following example can be used to familiarize yourself with the capability of your own calculator.

Example 5-2 *   Find the slope, intercept, and correlation coefficient for the following data:

---

* From *HP-19C/HP-29C Applications Book*, Hewlett-Packard, Corvallis, Ore., 1977, p. 105.

x_i	40.5	38.6	37.9	36.2	35.1	34.6
y_i	104.5	102	100	97.5	95.5	94

What value of y corresponds to $x = 37$?

Solution $y = 33.53 + 1.76x$ *Ans.*

$r = 0.995$ *Ans.*

At

$x = 37,$ $y = 98.65$ *Ans.*

////

Nonlinear regression can be performed merely by operating on one or both of the elements of the data pair with any mathematical function. For example, using natural logarithms, exponentials, and reciprocals, you can obtain logarithmic and exponential curves, as well as others.

5-4   THE NORMAL DISTRIBUTION

One of the most important of the many distributions which occur in the study of statistics is the *normal*, or *gaussian*, *distribution*. The equation of the normal curve is

$$f(x) = \frac{1}{\hat{\sigma}\sqrt{2\pi}}\, e^{-(x-\mu)^2/2\hat{\sigma}^2} \tag{5-11}$$

where $f(x)$ is the *frequency function*. The total area under this curve, from $x = -\infty$ to $x = +\infty$, is one square unit. Therefore the area between any two points, say, from $x = a$ to $x = b$, is the proportion of cases which lie between the two points. A plot of Eq. (5-11) yields a bell-shaped curve having a maximum ordinate at $x = \mu$ and decreasing to zero at both $+\infty$ and $-\infty$. If the standard deviation $\hat{\sigma}$ is small, the curve is tall and thin; if $\hat{\sigma}$ is large, the curve is short and broad; see Fig. 5-6.

In order to analyze different statistical cases, it is convenient to replace x in Eq. (5-11) with the *standardized variable* z defined as

$$z = \frac{x - \mu}{\hat{\sigma}} \tag{5-12}$$

where now z is to be normally distributed with a mean of zero. Equation (5-11)

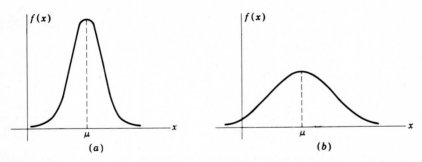

FIGURE 5-6   The shape of the normal distribution curve, (a) small $\hat{\sigma}$; (b) large $\hat{\sigma}$.

then becomes a *unit normal distribution* and is written

$$f(z) = \frac{1}{\sqrt{2\pi}} \, e^{-z^2/2} \qquad\qquad (5\text{-}13)$$

The distribution is illustrated in Fig. 5-7. The ordinate at z is $f(z)$, obtained using Eq. (5-13). The areas $Q(z)$ and $P(z)$ under the curve can be found in statistical tables or approximated. Since the total area under the curve is unity

$$P(z) = 1 - Q(z) \qquad\qquad (5\text{-}14)$$

The term $Q(z)$ is obtained by integrating the area under the curve. It can be approximated and then programmed for calculation using a polynomial expansion.* The results are found to be

$$Q(z) = f(z)(b_1 t + b_2 t^2 + b_3 t^3 + b_4 t^4 + b_5 t^5) + \varepsilon(z) \qquad\qquad (5\text{-}15)$$

where

$$t = \frac{1}{1 + p|z|}, \qquad\qquad |\varepsilon(z)| < 7.5(10)^{-8}$$

$$p = 0.231\,641\,9 \qquad\qquad b_1 = 0.319\,381\,530$$
$$b_2 = -0.356\,563\,782 \qquad\qquad b_3 = 1.781\,477\,937$$
$$b_4 = -1.821\,255\,978 \qquad\qquad b_5 = 1.330\,274\,429$$

The equations in this section can be evaluated by using statistical tabulations, by programming on any computer or calculator, or by using programmable-calculator software. The following example illustrates the use of the above equations in a design situation.

---

* Hewlett-Packard *Applications Book*, op. cit., p. 107.

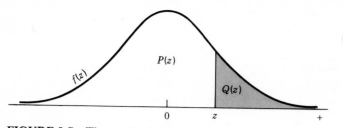

FIGURE 5-7   The standard normal distribution.

Example 5-3   In a shipment of 250 connecting rods, the mean tensile strength is found to be 45 kpsi and the standard deviation 5 kpsi.   Assume the strengths have a normal distribution.

(*a*)   How many have a strength less than 39.5 kpsi?
(*b*)   How many are expected to have a strength between 39.5 and 59.5 kpsi?

Solution   (*a*) Substituting in Eq. (5-12) gives the standard variable as

$$z_{39.5} = \frac{x - \mu}{\hat{\sigma}} = \frac{S - \bar{S}}{\hat{\sigma}} = \frac{39.5 - 45}{5} = -1.10$$

Since the distribution is symmetrical, the area to the left of $z = -1.10$ (Fig. 5-7) is the same as the area to the right of $z = 1.10$.   Therefore, we enter $z = 1.10$ into the calculator program.   This gives the ordinate of the curve, after rounding, $f(z) = 0.2179$.   Also found is $Q(z) = 0.135\ 666\ 101$, which is the area under the curve to the right of z.   By rounding, we find that 13.57 percent of the specimens have a strength less than 39.5 kpsi, and so

$$N = 250(0.1357) = 33.925$$

and, by rounding off, we find that 34 connecting rods are likely to have strengths less than 39.5 kpsi.

(*b*) Corresponding to $S = 59.5$ kpsi, we have

$$z_{59.5} = \frac{59.5 - 45}{5} = 2.90$$

Entering   this   value   into   the   program   yields $f(z) = 0.005\ 953$ and $Q(z) = 0.001\ 865\ 880$.   Examination of Fig. 5-8 shows that $z = 2.90$ is quite far out on the forward tail.   Since there is unit area beneath the entire curve, the area between $z_{39.5}$ and $z_{59.5}$ is, from Fig. 5-8,

$$P(z) = 1 - [Q(z_{39.5}) + Q(z_{59.5})]$$
$$= 1 - (0.135\ 666\ 101 + 0.001\ 865\ 880) = 0.862\ 468$$

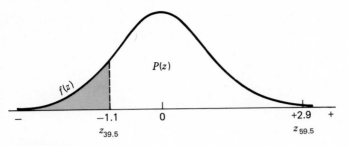

FIGURE 5-8

which is the proportion having strengths between 39.5 and 59.5 kpsi.   The number of connecting rods in this range is

$$N = 250(0.862) = 215.5$$

or about 216 rods when rounded off.   ////

The inverse of the previous analysis is necessary if $Q(z)$ is given and z is to be found.   The following approximation can be used for programming:

$$y = t - \frac{c_0 + c_1 t + c_2 t^2}{1 + d_1 t + d_2 t^2 + d_3 t^3} + \varepsilon(Q)$$

$$\varepsilon(Q) < 4.5(10)^{-4}, \qquad t = \begin{cases} \sqrt{ln(1/Q^2)} & 0 < Q \le 0.5 \\ \sqrt{ln[1/(1-Q)^2]} & 0.5 < Q < 1 \end{cases}$$

$$c_0 = 2.515\ 517 \qquad c_1 = 0.802\ 853$$
$$c_2 = 0.010\ 328 \qquad d_1 = 1.432\ 788$$
$$d_2 = 0.189\ 269 \qquad d_3 = 0.001\ 308$$

$$z = \begin{cases} y & 0 < Q \le 0.5 \\ -y & 0.5 < Q < 1 \end{cases} \qquad (5\text{-}16)$$

Example 5-4   A group of shafts are to be machined to 25.500 mm in diameter with a tolerance of ± 0.030 mm.   If a sample of 200 shafts is taken and found to have a mean of 25.500 mm, what must be the standard deviation in order to assure that 95 percent of the shafts are within acceptable dimensions?   Assume the diameters are normally distributed.

Solution   Adding and subtracting the tolerance indicates that 95 percent of the shafts must be in the range

$$d_{min} = 25.470 \text{ mm} \qquad to \qquad d_{max} = 25.530 \text{ mm}$$

Thus, 2.5 percent of them will be over d_{max} and 2.5 percent under d_{min}. So $Q(z) = 0.025$. Entering this value into a calculator subroutine gives $z = 1.9604$. Rearranging Eq. (5-12) gives

$$\hat{\sigma} = \frac{x - \mu}{z}$$

Using $x = d_{max} = 25.530$, $\mu = \bar{d} = 25.500$ mm, and the value of z just found, gives

$$\hat{\sigma} = \frac{25.530 - 25.500}{1.9604} = 0.0153 \text{ mm}$$ *Ans.*

////

5-5 POPULATION COMBINATIONS

Cases often arise in design in which we must make a statistical analysis of the result of combining two or more populations in some specified manner. For example, we may wish to assemble a population of cranks into a population of machines. The cranks all have a population of strengths S_i, and the machines subject them to a population of stresses σ_j. For safety, the strengths must be greater than the stresses, but since we are dealing with distributions, it may happen that some of the cranks will be subjected to stresses greater than their strengths. Thus we are interested in determining the reliability of these assemblies.

Another problem involving several populations occurs in the dimensioning of mating parts. We may wish, say, to specify a range of shaft diameters and another range of bearing sizes such that the shafts will assemble into the bearings with a specified clearance range. Thus we have a population of shaft diameters and a population of bearing diameters which result in a population of clearances. Our problem might be to specify the shaft and bearing sizes such that 99 percent of the assemblies have satisfactory clearances.

One of the very interesting problems arising in design is called *tolerance stacking*. Suppose eight parts having mean lengths $\bar{x}_1, \bar{x}_2, \ldots, \bar{x}_8$ and tolerances of $\pm x_1, \pm x_2, \ldots, \pm x_8$ are to be assembled end to end into a space whose mean length and tolerances are $\bar{y}$ and $\pm y$, respectively. And suppose, furthermore, that we require that these parts fit with a clearance $\bar{c}$ within the limits $\pm c$. What is the probability of getting a satisfactory assembly when the elements are selected at random from each of the several populations?

The means of two or more populations may be either added or subtracted to obtain a resultant mean. Thus, for addition,

$$\mu = \mu_1 + \mu_2 \tag{5-17}$$

and for subtraction,

$$\mu = \mu_1 - \mu_2 \tag{5-18}$$

where in each case μ is the mean of the resulting distribution.

The standard deviations follow the Pythagorean theorem. Thus the standard deviation for both addition and subtraction is

$$\hat{\sigma} = \sqrt{\hat{\sigma}_1^2 + \hat{\sigma}_2^2} \tag{5-19}$$

5-6 DIMENSIONING—DEFINITIONS AND STANDARDS

Though many decisions relating to the dimensioning of parts and assemblies can be made quite satisfactorily by the design draftsman, the engineer must be familiar with standard practices in dimensioning in order to retain control of the design for which he or she is directly responsible. Furthermore, the tolerancing of mating parts and assemblies has a significant effect on the cost and reliability of the finished product, and hence some of these decisions cannot be left to the judgment of technicians.

The following terms are used generally in dimensioning.

Nominal size The size we use in speaking of an element. For example, we may specify a $\frac{1}{2}$-in bolt or a $1\frac{1}{2}$-in pipe. Either the theoretical size or a measured size may be quite different. The bolt, say, may actually measure 0.492 in. And the theoretical size of a $1\frac{1}{2}$-in standard-weight pipe is 1.61 in for the outside diameter.

Basic size The exact theoretical size. Limiting variations in either the plus or minus direction begin from the basic dimension.

Limits The stated maximum and minimum permissible dimensions.

Tolerance The difference between the two limits.

Bilateral tolerance The variation in both directions from the basic dimension. That is, the basic size is between the two limits; for example, 1.005 ± 0.002 in.

Unilateral tolerance The basic dimension is taken as one of the limits, and variation is permitted in only one direction; for example,

$$1.005 \begin{array}{c} +0.004 \\ -0.000 \end{array} \text{in}$$

Natural tolerance A tolerance equal to plus and minus three standard deviations from the mean. For a normal distribution this assures that 99.73 percent of production is within the tolerance limits.

Clearance A general term which refers to the mating of cylindrical parts such as a bolt and a hole, or a journal and a bearing. The word *clearance* is used only when the internal member is smaller than the external member. The *diametral clearance* is the measured difference in the two diameters. The *radial clearance* is the difference in the two radii.

Interference The opposite of clearance for the mating of cylindrical parts in which the internal member is larger than the external member.

Allowance The minimum stated clearance or the maximum stated interference for mating parts.

The USASI (United States of America Standards Institute) has adopted a useful set of standards for the mating of cylindrical parts. A portion of these standards has been reproduced as Table A-15 for use in design. This table gives the shaft and hole tolerances for both clearance and interference fits ranging from very loose running fits to heavy force and shrink interference fits. These fits are tabulated for the *basic hole system* in which the lower limit of the hole dimension is made equal to the theoretical dimension and all other limits are taken from this starting point. If the holes are finished by reaming, the basic dimension should be the size of a standard reamer. If the holes are finished by grinding, any suitable dimension may be used as the basic size.

There is, of course, no law which requires the engineer to employ these standards. They do, however, furnish a good starting point for the development of other tolerances.

5-7 STATISTICAL ANALYSIS OF TOLERANCING

Variations in dimensions of parts in a production process may occur purely by chance as well as for specific reasons. For example, the operating temperature of a machine tool changes during start-up, and this may have an effect on the dimensions of the parts produced during the first 30 min of running time. When all such causes of variations in part dimensions have been eliminated, the production process is said to be *in statistical control*. Under these conditions all variations in dimensions occur at random.

We will assume that production is in statistical control for our investigations in this section and that the variations in part dimensions have a normal distribution.

Problems involving the tolerancing of mating cylindrical parts occur frequently. If the two parts must mate with a clearance fit, the problem is to discover the percentage of rejected assemblies due either to a fit that is too tight or to one that is too loose. A similar problem occurs when the parts must mate with an interference fit. Both of these problems can be treated using the statistical approaches already developed. An example will serve to illustrate the notation and approach.

Example 5-5 A shaft and hole are to mate with a class RC8 loose running fit. The nominal diameter is 1 in. What percentage of assemblies is likely to be rejected because the clearance is less than 0.006 in? What percentage is likely to be rejected for having clearances greater than 0.009 in? Assume natural tolerances.

Solution Using Table A-15, we find the hole and shaft dimensions shown in Fig. 5-9. Using c for clearance and the subscripts H and S for hole and shaft, respectively, we find the maximum and minimum clearances to be

$$c_{max} = d_{H, max} - d_{S, min} = 1.0035 - 0.9935 = 0.0100 \text{ in}$$

$$c_{min} = d_{H, min} - d_{S, max} = 1.0000 - 0.9955 = 0.0045 \text{ in}$$

Therefore the mean clearance is

$$\mu_c = \frac{c_{max} + c_{min}}{2} = \frac{0.0100 + 0.0045}{2} = 0.007\ 25 \text{ in}$$

The means of the hole and shaft diameters are easily found. They are

$$\bar{d}_H = 1.001\ 75 \text{ in} \qquad \bar{d}_S = 0.9945 \text{ in}$$

Subtracting these from the maximums and minimums shown in Fig. 5-9 gives the tolerances, respectively, as

$$t_H = \pm 0.001\ 75 \text{ in} \qquad t_S = \pm 0.001\ 0 \text{ in}$$

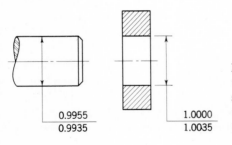

0.9955
0.9935

1.0000
1.0035

FIGURE 5-9 Note that the limits are dimensioned so that the upper figure is the one the machinist gets to first.

Since these are assumed to be natural tolerances, the standard deviations are

$$\hat{\sigma}_H = \frac{0.001\ 75}{3} = 0.000\ 55 \text{ in} \qquad \hat{\sigma}_S = \frac{0.001\ 0}{3} = 0.000\ 33 \text{ in}$$

Then, using Eq. (5-19), the standard deviation of the clearance is

$$\hat{\sigma}_c = \sqrt{\hat{\sigma}_H^2 + \hat{\sigma}_S^2} = \sqrt{(0.000\ 55)^2 + (0.000\ 33)^2} = 0.000\ 644 \text{ in}$$

The standardized variable is to correspond to a clearance not less than 0.006 in. Therefore, from Eq. (5-12), we have

$$z = \frac{c - \mu_c}{\hat{\sigma}_c} = \frac{0.006 - 0.007\ 25}{0.000\ 644} = -1.945$$

From our program we find

$$Q(z) = 0.0259$$

or 2.59 percent.

For clearances greater than 0.009 in we have

$$z = \frac{c - \mu_c}{\hat{\sigma}_c} = \frac{0.009 - 0.007\ 25}{0.000\ 644} = 2.717$$

Then

$$Q(z) = 0.003\ 294$$

or 0.33 percent.   ////

Another problem which arises in dimensioning is illustrated in Fig. 5-10. Here parts A and B are to be placed end to end and then assembled into C. One of the questions that might be asked is: What percentage of the assemblies is

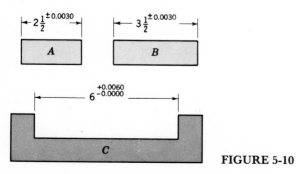

FIGURE 5-10

likely to be rejected because of interference? The solution consists in summing the population of A's to the population of B's to obtain a resultant population of X's. Then the X's and the C's may be treated in exactly the same manner as the shaft-and-hole problem in Example 5-5.   For the assembly $X = A + B$ we have, from Eqs. (5-18) and (5-19),

$$\mu_X = \mu_A + \mu_B \tag{5-20}$$

$$\hat{\sigma}_X = \sqrt{\hat{\sigma}_A^2 + \hat{\sigma}_B^2} \tag{5-21}$$

Since A and B could in themselves be subassemblies, the process described by these equations can be extended indefinitely.

Next, we combine Eqs. (5-18) and (5-19) to predict interference between X and C. The result is

$$z = \frac{\mu_C - \mu_X}{\sqrt{\hat{\sigma}_C^2 + \hat{\sigma}_X^2}} \tag{5-22}$$

Example 5-6 Suppose the standard deviations for the assembly of Fig. 5-10 are $\hat{\sigma}_A = 0.0010$ in, $\hat{\sigma}_B = 0.0010$ in, and $\hat{\sigma}_C = 0.0015$ in. Find the percentage of assemblies in which interference between $X = A + B$ and C is likely to occur.

Solution   From Fig. 5-10 we find

$$\mu_A = 2.5000 \text{ in} \qquad \mu_B = 3.5000 \text{ in} \qquad \mu_C = 6.0030 \text{ in}$$

Then

$$\mu_X = \mu_A + \mu_B = 2.5000 + 3.5000 = 6.0000 \text{ in}$$

Also

$$\hat{\sigma}_X = \sqrt{\hat{\sigma}_A^2 + \hat{\sigma}_B^2} = \sqrt{(0.0010)^2 + (0.0010)^2} = 0.001\ 41 \text{ in}$$

Then, using Eq. (5-22),

$$z = \frac{\mu_C - \mu_X}{\sqrt{\hat{\sigma}_C^2 + \hat{\sigma}_X^2}} = \frac{6.0030 - 6.0000}{\sqrt{(0.0015)^2 + (0.001\ 41)^2}} = 1.453$$

Entering the program with this value of z we find the probability of interference to be

$$Q(z) = 0.073\ 112$$

or 7.31 percent.

/////

5-8   THE WEIBULL DISTRIBUTION*

A distribution particularly useful for predicting life is called the Weibull distribution. As shown in Fig. 5-11, it is really a family of distributions because of the variety of density functions that can be obtained by changing parameters.

The equation of the *frequency*, or *density, function* is

$$f(x) = \frac{b}{\theta - x_0}\left(\frac{x - x_0}{\theta - x_0}\right)^{b-1}\left\{\exp\left[-\left(\frac{x - x_0}{\theta - x_0}\right)^b\right]\right\} \tag{5-23}$$

where x_0 = minimum expected value of x
θ = a characteristic or scale value
b = Weibull slope

These three parameters are determined from experimental data by plotting failure points on specially prepared Weibull probability paper. If the data obeys the Weibull distribution, the line so plotted is a straight line having a slope b; hence it is called the Weibull slope. See Fig. 11-6.

The cumulative distribution function is

$$F(x) = \int_{x_0}^{x} f(x)\ dx = 1 - \exp\left[-\left(\frac{x - x_0}{\theta - x_0}\right)^b\right] \tag{5-24}$$

Equation (5-24) simplifies to

$$F(x) = 1 - \exp\left[-\left(\frac{x}{\theta}\right)^b\right] \tag{5-25}$$

when $x_0 = 0$. This occurs quite often since it is reasonable to expect a zero minimum life at times.

* A comprehensive discussion of this topic with many practical applications is contained in Charles Lipson and Narenda J. Sheth, *Statistical Design and Analysis of Engineering Experiments*, McGraw-Hill, New York, 1973.

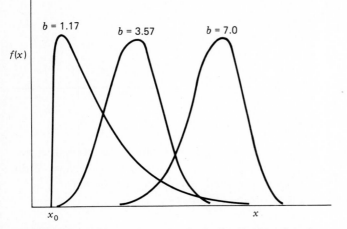

FIGURE 5-11 Plot of the Weibull distribution for three values of the slope b. The value $b = 1.17$ is often used to predict the life of rolling bearings. A close approximation to the normal distribution is obtained by using $b = 3.57$.

PROBLEMS

Sections 5-1 to 5-3

5-1   A series of fatigue tests was run on some steel test specimens to determine the bending endurance limit S_e and the torsional or shear endurance limit S_{se} (see Chap. 7). The results are shown in the table that follows in kpsi. Use this table to determine whether it is possible to predict the torsional endurance limit when only the bending endurance limit is given. If so, what is the relation?

S_e	98	94	69	65	63	67	57	55	58	57	49
S_{se}	52	49	39	37	35	34	33	31	29	29	28
S_e	49	50	35	34	31	32	25				
S_{se}	26	25	21	20	17	16	12				

5-2   A group of steels were tested for the Brinell hardness H_B, the yield strength S_y, and the ultimate strength S_u. The results, with the strengths in kpsi, are shown below.*
      (a)   Using the hardness as x and the yield strength as y, determine if it is possible to predict the yield strength by measuring only the hardness.

---

* The data shown are from actual tests of high-strength steels.

(b) The same as (a) for the ultimate strength.

(c) What is the value of S_u if $H_B = 300$? Use the results of (b).

H_B	705	595	500	450	390	670	560	475	450
S_y	265	270	245	220	185	235	245	250	230
S_u	300	325	265	230	195	355	325	280	255
H_B	380	475	450	400	660	480	460	405	
S_y	200	275	270	210	295	280	260	215	
S_u	205	295	280	225	375	290	270	220	

5-3   The process of winding compression coil springs results in variations in some of the characteristics. In attempting to meet the required tolerances it would be interesting to learn if it is necessary to control all of the characteristics, independently of each other, in the winding process. In an attempt to answer this question, a sample of 12 springs has been selected at random and the spring rate and free length of each measured. The results are shown below.

(a) Compute the mean and the standard deviation for each characteristic.

(b) Compute the correlation coefficient using l_F as x and k as y. In your opinion can we hold the spring rate to close limits by controlling only the free length?

l_F, mm	79.68	80.13	80.32	79.81	79.92	80.21
k, N/m	1749	1769	1784	1757	1761	1773
l_F, mm	79.97	79.84	80.03	80.11	79.97	80.00
k, N/m	1770	1760	1767	1770	1760	1763

5-4   The table below shows the grades in percent received by 10 students in Econ 301 and Math 316.

(a) Find the mean and the standard deviation for the grades in each course.

(b) Is there any correlation between the grades received in the two courses?

Econ 301	96	93	90	86	86	80	78	76	68	84
Math 316	78	90	74	82	96	68	94	86	78	90

5-5   The following data are given for the lives in hours of the filament of a standard household light bulb: 2098, 1720, 1922, 1877, 2050, 2900, 1978, 1620, and 2010. Compute the sample mean and the sample standard deviation.

Section 5-4

5-6   A population of 800 springs has been manufactured and the mean spring rate was found to be 1200 N/m with a standard deviation of 42 N/m.
   (a)   How many have a rate less than 1100 N/m?
   (b)   How many have a rate between 1150 and 1250 N/m?

5-7   A class of 30 students received a mean grade on an examination of 82. The standard deviation was 7.5.
   (a)   How many students received 95 or more?
   (b)   How many students received less than 60?
   (c)   What percentage of the class received grades within one standard deviation of the mean?

5-8   The standard deviation of the fatigue strength for a certain group of steels was found to be 6 percent. If a part is to be designed to have a probability of non-failure of 99 percent, by what factor should the mean fatigue strength be reduced to obtain the design strength? Assume the normal distribution holds.

5-9   A shipment of 150 steel pins has been checked for diameter using "Go" and "Not Go" gauges. The pins were to be machined to 0.375 in in diameter with a tolerance of ± 0.003 in. Four pins were rejected as being under. Six pins were rejected as being over. If the dimensions are assumed to have a normal distribution, what can you learn from these test results?

Sections 5-5 to 5-7

5-10   The commercial diametral tolerance of music wire up to 0.026 in in diameter is ± 0.0003 in. A 0.010-in-diameter music wire is used to manufacture coil springs in which the wire is stressed in torsion. If the diametral tolerance is the same as the natural tolerance, determine the standard deviation of the stress, assuming all other factors are constant. The applied torque is 0.25 oz · in.

5-11   The standard manufacturing tolerance of a cold-drawn steel flat* up to $\frac{3}{4}$ in in width is 0.003 in. A $\frac{1}{8} \times \frac{5}{8}$-steel flat is subjected to a bending load such that the neutral plane is parallel to the $\frac{5}{8}$-in dimension. Suppose both the thickness and width have natural tolerances of $+0.000$ and -0.003 in. What is the standard deviation of the bending stress if the moment is 64 lb · in and all other factors are constant?

5-12   Based on the tolerances shown in Fig. 5-10 and the dimensions of A and B, find the mean dimension for C such that only 1 percent interference is obtained. Use the information from Example 5-6.

5-13   A shaft and hole have the dimensions 0.6230 ± 0.0010 and 0.6260 ± 0.0010 in with standard deviations of 0.0004 and 0.0005 in, respectively. Compute the allowance. What percentage of assemblies can be expected to have a smaller allowance?

5-14   A shaft and hole are to have natural tolerances of 0.002 and 0.003 in, respectively. Determine the mean clearance such that not more than 5 percent of the assemblies will have clearances less than 0.002 in. What percentage of these will have clearances greater than 0.007 in?

---

* A *flat* is any steel bar having a rectangular cross section. Flats generally range in sizes from $\frac{1}{8}$ in thick $\times \frac{3}{16}$ in wide up to 3 in thick $\times$ 12 in wide. Sizes $\frac{3}{16}$ in and thinner are called *strip*. The overlap is intentional.

CHAPTER

6

DESIGN FOR STATIC STRENGTH

In Chap. 1 we learned that *strength is a property or characteristic of a material or of a mechanical element.* This property may be inherent in a material or may result from the treatment and processing of a material. The strength of a mechanical part is a property completely independent of whether or not that part is subjected to load or force. In fact, this strength property is a characteristic of the element even before it is assembled into a machine or system.

A *static load* is a stationary force or moment acting on a member. To be stationary the force or moment must have an unchanging magnitude, unchanging point or points of application, and an unchanging direction. A static load can be axial tension or compression, a shear load, a bending load, a torsional load, or any combination of these. But the load cannot change in any manner if it is to be considered as static. Sometimes a load is *assumed* to be static when it is known that some variation is to be expected. This assumption is usually made to get a rough idea of the component dimensions and to simplify design computations when variations in loads are few or minor in nature.

The purpose of this chapter is to develop the relations between strength and loads to achieve optimum component dimensions, with the requirement that the part will not fail in service. This means that we must learn how parts fail in service and how to specify a margin of safety to assure that they will not fail. The part failures illustrated in Figs. 6-1 to 6-14 exemplify the need for the designer to be well versed in the knowledge needed to assure satisfactory performance of the parts he or she designs. The broken parts illustrated in these figures were presumably designed by experienced people armed with all the knowledge and the background data needed. And yet the parts failed. The probable reasons for the failures are described in the figure captions.

It is worth noting here that nearly all of the numerical data that we use in design have some uncertainties. Published data on material strengths, such as those in the Appendix of this book, are always uncertain. Tabulated strengths may be the minimum expected values, or typical values, but the strength of an actual specimen will almost certainly turn out to be different from the published values. This uncertainty concerning strength even extends to the parts themselves. If one connecting rod from every 100 manufactured, for example, is tested to failure, the strength of each of the remaining 99 rods still has some degree of uncertainty.

In addition to the uncertainties concerning strengths, a similar uncertainty exists with regard to the magnitudes of the loads. For example, the bolt in a

FIGURE 6-1   Fatigue failure of an automotive cooling fan due to vibrations caused by a defective water pump.

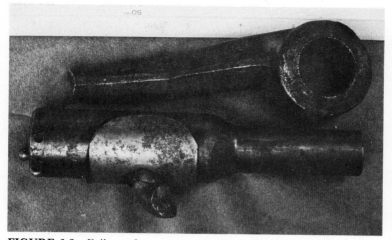

FIGURE 6-2 Failure of an automotive drag link. This failure occurred at 140 000 mi. Fortunately the car was in park and against a curb. Such a failure results in total disconnect of the steering wheel from the steering mechanism.

bolted connection is loaded in axial tension and in torsion when the nut is tightened. The amount of this tension and torsion depends upon the fit of the threads, the friction between the threads, the friction between the nut face and the washer, and the wrench torque. Thus there will always remain some doubt or uncertainty about the actual tension and torsion remaining in the bolt after the wrench torque has been applied. Similar uncertainties will exist for almost any kind of loading you can think of.

FIGURE 6-3 Typical failure of a stamped steel alternator bracket at about 25 000 mi. Failure probably due to residual stresses caused by the cold-forming operation. The high failure rate prompted the manufacturer to redesign the bracket as a die casting.

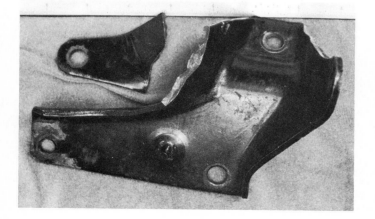

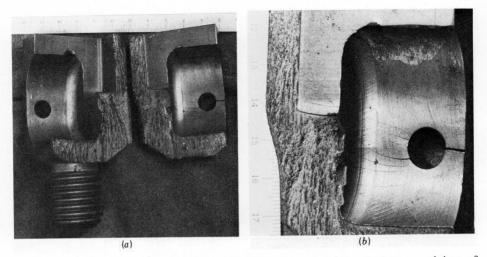

(a)   (b)

FIGURE 6-4 Chain test fixture that failed in one cycle. To alleviate complaints of excessive wear the manufacturer decided to case-harden the material. (*a*) Two halves showing fracture; this is an excellent example of brittle fracture initiated by stress concentration. (*b*) Enlarged view of one portion to show cracks induced by stress concentration at the support-pin holes.

To account for all the uncertainties in design we employ the concept of *factor of safety*. This assures that we can relate the loads acting on a mechanical part, or the stresses resulting from those loads, to the strength of the part in order to achieve a safe and trouble-free design.

6-1 STATIC STRENGTH

Ideally, in designing any machine element, the engineer should have at his or her disposal the results of a great many strength tests of the particular material chosen. These tests should have been made on specimens having the same heat treatment, surface finish, and size as the element the engineer proposes to design; and the tests should be made under exactly the same loading conditions as the part will experience in service. This means that, if the part is to experience a

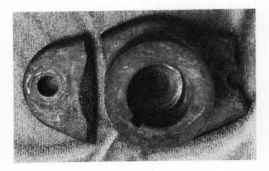

FIGURE 6-5 Impact failure of a lawnmower-blade driver hub. The blade impacted a surveying pipe marker.

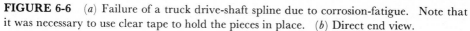

FIGURE 6-6 (*a*) Failure of a truck drive-shaft spline due to corrosion-fatigue. Note that it was necessary to use clear tape to hold the pieces in place. (*b*) Direct end view.

bending load, it should be tested with a bending load. If it is to be subjected to combined bending and torsion, it should be tested under combined bending and torsion. If it is made of heat-treated UNS G10400 steel drawn at 900°F with a ground finish, the specimens tested should be of the same material prepared in the same manner. Such tests will provide very useful and precise information. Whenever such data are available for design purposes, the engineer can be assured that he or she is doing the best possible job of engineering.

The cost of gathering such extensive data prior to design is justified if failure of the part may endanger human life or if the part is manufactured in sufficiently large quantities. Refrigerators and other appliances, for example, have very good

FIGURE 6-7 Failure of an interior die-cast car-door handle. Failure occurred at about every 45 000 mi. Probable causes were the electroplating material, stress concentration, the long lever arm required to operate a "sticky" door-release mechanism, and the high actuation forces.

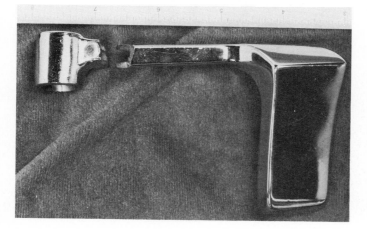

FIGURE 6-8 Automotive rocker-arm articulation-joint fatigue failure.

FIGURE 6-9 Valve-spring failure caused by spring surge in an overrevved engine. The fractures exhibit the classic 45° shear failure.

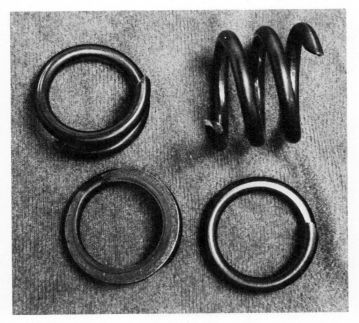

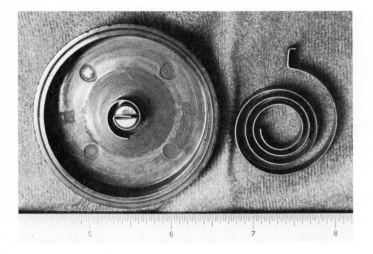

FIGURE 6-10 Torsional spring failure. This is a Corvair choke spring; a very rare failure.

FIGURE 6-11 Brittle fracture of a lock washer in one cycle. The washer failed when it was installed.

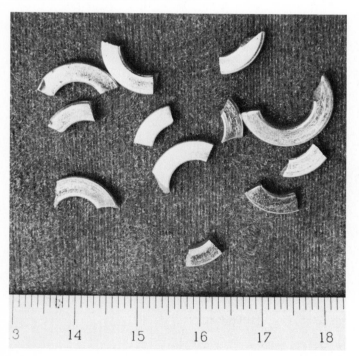

FIGURE 6-12 Failure of an overhead pulley retaining bolt on a
weight-lifting machine. A manufacturing error caused a gap that
forced the bolt to take the entire moment load.

reliabilities because the parts are made in such large quantities that they can be
thoroughly tested in advance of manufacture. The cost of making these tests is
very low when it is divided by the total number of parts manufactured.

You can now appreciate the following four design categories.

1 Failure of the part would endanger human life, or the part is made in
extremely large quantities; consequently, an elaborate testing program is
justified during design.
2 The part is made in large enough quantities so that a moderate series of
tests is feasible.

FIGURE 6-13 Fatigue failure of a die-cast residence door
bumper.   This bumper is installed on the door hinge to prevent
the doorknob from impacting the wall.

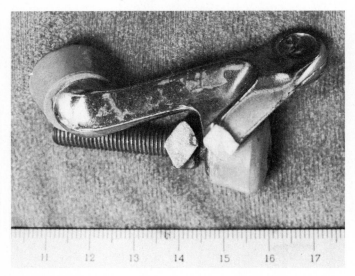

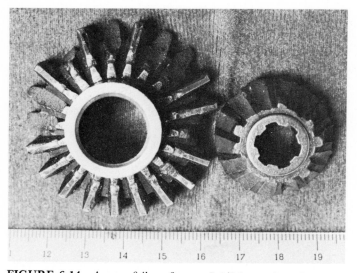

FIGURE 6-14 A gear failure from a 7 1/2-hp outboard motor. The large gear is 1 7/8 in outside diameter and had 21 teeth; 6 are broken. The pinion had 14 teeth; all are broken. Failure occurred when the propeller struck a steel auger placed in the lake bottom as an anchorage. Owner had replaced the shear pin with a substitute pin.

3 The part is made in such small quantities that testing is not justified at all; or the design must be completed so rapidly that there is not enough time for testing.

4 The part has already been designed, manufactured, and tested and found to be unsatisfactory. Analysis is required to understand why the part is unsatisfactory and what to do to improve it.

It is with the last three categories that we shall be mostly concerned in this book. This means that the designer will usually have only published values of yield strength, ultimate strength, and percentage elongation, such as those listed in the Appendix, and little else. With this meager information the engineer is expected to design against static and dynamic loads, biaxial and triaxial stress states, high and low temperatures, and large and small parts! The data usually available for design have been obtained from the simple tension test, where the load was applied gradually and the strain given time to develop. Yet these same data must be used in designing parts with complicated dynamic loads applied thousands of times per minute. No wonder machine parts sometimes fail.

The data provided by the simple tension test provide an enormous amount of information concerning the probable behavior of a mechanical part when placed in service. As we have learned in Chap. 4, the yield strength, the ultimate strength, and the true fracture stress can be obtained from the test. The test, if carried out in sufficient detail, also reveals the strain at fracture, the percent reduction in area, and the strain-hardening exponent. These data provide sufficient information so that we can estimate the yield strength and ultimate strength corresponding to any amount of cold work.

To sum up, *the fundamental problem of the designer is to use the simple tension-test data and relate them to the strength of the part, regardless of the stress state or the loading situation.* The balance of this chapter is devoted to the solution of this problem for the case of static loads.

6-2 STATIC LOADS AND FACTOR OF SAFETY

In Chap. 1 we learned that *factor of safety n* is defined by either of the equations

$$n = \frac{F_u}{F} \quad \text{or} \quad n = \frac{S}{\sigma} \tag{6-1}$$

where F_u represents the maximum load that will still enable the part to perform its proper function. Thus F_u is the limiting value of F. In the second term of the equation, S is the strength and σ is the stress. In this case, too, S is the limiting value of σ. Of course, if S is a shear strength then σ must be a shear stress; that is, the two must be consistent.

It is sometimes convenient to define two factors of safety. One of them n_S is used to account for the uncertainties in the strength. The other n_1 accounts for the uncertainties with regard to the load. Thus the total factor of safety will be

$$n = n_S n_1 \tag{6-2}$$

When n_S is applied to the strength, we have

$$\sigma_p = \frac{S}{n_S} \tag{6-3}$$

where σ_p is called the *permissible stress*. When n_1 is applied to the limiting load F_u, we have

$$F_p = \frac{F_u}{n_1} \tag{6-4}$$

where F_p is called the *permissible load*.

A few special cases occur in which either form of Eq. (6-1) can be used directly in design, thus bypassing the need to employ both Eqs. (6-3) and (6-4). One of these is the case of pure tension (see Sec. 2-4), where the second form of Eq. (6-1) can be used directly to compute the dimensions of the part to be designed or to obtain the factor of safety when the geometry is known.

Example 6-1 A connecting rod having a rectangular cross section is loaded in pure tension by a force $F = 4.8$ kN. The yield strength of the material is 320 MPa. Using $n_S = 1.2$ and $n_1 = 2.0$, find the cross-sectional dimensions of the rod if the width is to be six times the thickness.

Solution The area of the bar is $A = wt = 6t^2$. The tensile stress is

$$\sigma = \frac{F}{A} = \frac{4.8(1000)}{6t^2} = \frac{800}{t^2}$$

Now $n = n_S n_1 = 1.2(2.0) = 2.40$. From Eq. (6-1), $\sigma = S/n$. Upon substitution we have

$$\frac{800}{t^2} = \frac{320(10)^6}{2.40} \tag{1}$$

If we solve Eq. (1) for the thickness t and change the result to millimeters, we have

$$t = \sqrt{\frac{800(2.40)}{320(10)^6}} = 2.45 \text{ mm}$$

Thus a good set of dimensions is 2.5 mm thick by 15 mm wide.   ////

The example above illustrates how the second form of Eq. (6-1) can be used to design or analyze a component. However, there is some danger involved in its use. For example, a long compression member, such as a column, is a case in which the stress is not linearly related to the load. In this case use of the strength formula for factor of safety could give very misleading results.

In other cases the critical stress may result from several loads. The use of the second form of Eq. (6-1) would then imply that the uncertainties regarding each load are identical. This is rarely true. Consider the implications. Suppose the critical stress in an element is caused by three separate loads F_1, F_2, and F_3. Suppose also that the uncertainties in these loads are to be accounted for by the three load safety factors n_1, n_2, and n_3, respectively. If n_2, say, happens to be the largest of these three factors, then it would be very inefficient design to use only n_2 in designing the part. This difference in the two approaches is demonstrated in the following example.

Example 6-2 A round axle is loaded in static bending by the three forces shown in Fig. 6-15. The bar is to be made of cold-drawn UNS G10100 steel having a tabulated yield strength of 44 kpsi. The strength safety factor is to be $n_S = 1.20$.

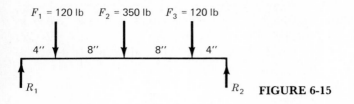

$F_1 = 120$ lb $F_2 = 350$ lb $F_3 = 120$ lb

4''   8''   8''   4''

R_1 R_2 **FIGURE 6-15**

The nature of the loads F_1, F_2, and F_3 are such that they can be expected to overload differently from each other if the axle is misused in service. Because of this, we choose $n_1 = n_3 = 1.25$, and $n_2 = 2.40$.

(a) Find a suitable diameter for the axle using n_1, n_2, n_3, and n_s.
(b) Find a suitable diameter using a single factor of safety.

Solution   (a) Using $F_p = n_i F_i$, we get

$$F_{1,p} = 1.25(120) = 150 \text{ lb}$$

$$F_{2,p} = 2.40(350) = 840 \text{ lb}$$

$$F_{3,p} = 1.25(120) = 150 \text{ lb}$$

The maximum bending moment occurs in the center of the beam. Its magnitude is

$$M = 12\left(\frac{F_1 + F_2 + F_3}{2}\right) - 8F_1 \tag{1}$$

and so the permissible moment is

$$M_p = 12\left(\frac{150 + 840 + 150}{2}\right) - 8(150) = 5640 \text{ lb} \cdot \text{in}$$

The permissible stress is

$$\sigma_p = \frac{M_p}{I/c} = \frac{32M_p}{\pi d^3}$$

Then, using Eq. (6-3), we have

$$\frac{S}{n_S} = \frac{32M_p}{\pi d^3} \qquad \text{or} \qquad \frac{44(10)^3}{1.20} = \frac{32(5640)}{\pi d^3}$$

Solving for the diameter gives $d = 1.161$ in.   *Ans.*

(b) Using the largest of n_1, n_2, n_3, we find the total factor of safety to be

$$n = n_S n_2 = 1.20(2.40) = 2.88$$

Using the specified design loads we next find the moment from Eq. (1) to be

$$M = 12\left(\frac{120 + 350 + 120}{2}\right) - 8(120) = 2580 \text{ lb} \cdot \text{in}$$

Since $\sigma = M/(I/c)$, we have, from Eq. (6-1)

$$\frac{S}{n} = \frac{M}{I/c} = \frac{32M}{\pi d^3} \quad \text{or} \quad \frac{44(10)^3}{2.88} = \frac{32(2580)}{\pi d^3}$$

When this equation is solved, we get $d = 1.198$ in.   *Ans.*

////

Based upon yield-strength distributions, such as the one in Fig. 5-4b, a factor of safety n_S should be about 1.20 for a reliability of 90 percent and about 1.40 for a 99 percent reliability. Of course, these factors can be decreased significantly using inspection and other quality-control techniques, since Fig. 5-4b is probably about the worst conceivable distribution. On the other hand, if the part must operate in an adverse environment, for example, under unusual temperatures or corrosive conditions, these factors may have to be adjusted to much higher values.

It is more difficult to provide a set of guidelines for choosing the load safety factors. Almost every situation will have to be investigated separately. Knowing the source of the load, that is, how it is produced, will usually provide some insight. The following questions might be asked.

1 Is it a preload? That is, does it come about because of the assembly operation? If so, what conditions or factors might affect its magnitude?
2 Does the load result from a power source? If it is a motor, have you considered both the stalling torque and the starting torque? If the power source is an engine, how might this affect the maximum load?
3 Does the load result from driven machinery? If so, can it be stalled? Is there anything else that can go wrong?
4 Does the load result from human activity or control? If so, what might happen if control is lost? Or, is it possible to apply an overload?
5 Does the load result from natural phenomena such as earthquakes, wind, or flood?
6 In your numerical evaluation of the load, which of the following words is most descriptive?

(a) Typical   (b) Average   (c) Maximum

(d)	Minimum	(e)	Expected	(f)	Rated
(g)	Usual	(h)	They say	(i)	Limiting
(j)	Steady	(k)	Nonsteady	(l)	Estimated

7   What are the consequences of a part failure?

Most of the time the load safety factors may be selected based on past experience with similar designs. But thoughtful answers to the above questions will help you in selecting appropriate factors.

6-3   FAILURE THEORIES

In designing parts to resist failure we usually assure ourselves that the internal stresses do not exceed the strength of the material. If the material to be used is ductile, then it is the yield strength that we are usually interested in, because a permanent deformation would constitute failure. There are exceptions to this rule, however.

Many of the more brittle materials, such as the cast irons, do not have a yield point, and so we must utilize the ultimate strength as the criterion of failure in design. In designing parts of brittle materials it is also necessary to remember that the ultimate compressive strength is much greater than the ultimate tensile strength.

The strengths of ductile materials are about the same in tension and compression. We usually assume this to be so in design unless we have information to the contrary.

Now consider the general two-dimensional stress element of Fig. 6-16. In Chap. 2 we studied similar stress situations and learned of the various loading patterns that might produce such a stress state. Our problem now is how to relate a stress state such as that of Fig. 6-16 to a single strength, such as a tensile yield, to achieve safety. The solution of this problem is the subject of the balance of this chapter.

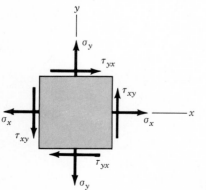

FIGURE 6-16

6-4 THE MAXIMUM-NORMAL-STRESS THEORY

This theory is presented only for its historical interest. Its predictions do not agree with experiment and, in fact, often give results on the unsafe side.

The maximum-normal-stress theory states that *failure occurs whenever the largest principal stress equals the strength.*

Suppose we arrange the three principal stresses for any stress state in the form

$$\sigma_1 > \sigma_2 > \sigma_3$$

Then if yielding is the criterion of failure, this theory predicts that failure occurs whenever

$$\sigma_1 = S_{yt} \quad \text{or} \quad \sigma_3 = -S_{yc} \tag{6-5}$$

where S_{yt} and S_{yc} are the tensile and compressive yield strengths, respectively. (Note that the subscripts t and c are usually dropped when these two strengths are equal.) If the ultimate strength is used, as it would be for brittle materials, then failure occurs whenever

$$\sigma_1 = S_{ut} \quad \text{or} \quad \sigma_3 = -S_{uc} \tag{6-6}$$

where S_{ut} and S_{uc} are the ultimate tensile and compressive strengths, respectively.

For pure torsion (see Fig. 6-17b) $\sigma_1 = \tau = -\sigma_3$ and $\sigma_2 = 0$. Thus, the maximum-normal-stress theory predicts that a part would fail in torsion when $\tau = S_y$. But experiments show that parts loaded in torsion will permanently deform when the maximum torsional stress is about 60 percent of the yield strength. This is one of the reasons the maximum-normal-stress theory is not recommended.

FIGURE 6-17 (*a*) Mohr's circle for simple tension; (*b*) Mohr's circle for pure torsion.

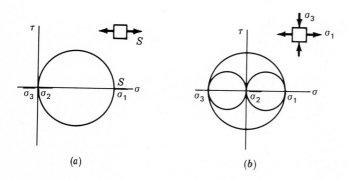

(*a*)   (*b*)

6-5   THE MAXIMUM-SHEAR-STRESS THEORY

This is an easy theory to use, it is always on the safe side of test results, and it has been used in many design codes. It is used only to predict *yielding*, and hence it applies only to *ductile* materials.

The maximum-shear-stress theory states that *yielding begins whenever the maximum shear stress in any mechanical element becomes equal to the maximum shear stress in a tension-test specimen of the same material when that specimen begins to yield.*

In Fig. 6-17a is shown Mohr's circle for the simple tension test. The maximum shear stress is seen to be

$$\tau_{max} = \frac{\sigma_1}{2} = \frac{S}{2} \qquad\qquad (a)$$

Mohr's circle for pure torsion is shown in Fig. 6-17b. The maximum shear stress is

$$\tau_{max} = \frac{\sigma_1 - \sigma_3}{2} \qquad\qquad (b)$$

because we have assumed the principal stresses to be arranged in the form $\sigma_1 > \sigma_2 > \sigma_3$. Thus the maximum-shear-stress theory predicts that failure will occur whenever

$$\tau_{max} = \frac{S_y}{2} \qquad \text{or} \qquad \sigma_1 - \sigma_3 = S_y \qquad\qquad (6\text{-}6)$$

Note that this theory also states that the yield strength in shear is given by the equation

$$S_{sy} = 0.50 S_y \qquad\qquad (6\text{-}7)$$

6-6   THE DISTORTION-ENERGY THEORY

This failure theory is also called the *shear-energy theory* and the *von Mises-Hencky theory*. It is only slightly more difficult to use than the maximum-shear-stress theory, and it is the best theory to use for *ductile* materials. Like the maximum-shear-stress theory, it is employed to define only the beginning of *yield*.

The distortion-energy theory originated because of the observation that ductile materials stressed hydrostatically (equal tension or compression) had yield strengths greatly in excess of the values given by the simple tension test. It was postulated that yielding was not a simple tensile or compressive phenomenon at all, but rather, that it was related somehow to the angular distortion of the stressed element. Now, one of the earlier theories of failure predicted that yield-ing would begin whenever the total strain energy stored in the stressed element became equal to the strain energy stored in an element of the tension-test speci-men at the yield point. This theory, called the *maximum-strain-energy theory*, is no longer used, but it was a forerunner of the distortion-energy theory. It was argued, Why not take the total strain energy and subtract from it whatever energy is used only to produce a volume change? Then whatever energy is left will be that which produces angular distortion alone. Let us see how this works.

Figure 6-18a shows an element acted upon by stresses arranged so that $\sigma_1 > \sigma_2 > \sigma_3$. For a unit cube the work done in any one of the principal directions is

$$u_n = \frac{\sigma_n \epsilon_n}{2} \tag{a}$$

where $n = 1$, 2, or 3. Therefore, from Eq. (2-23), the total strain energy is

$$u = u_1 + u_2 + u_3 = [1/(2E)][\sigma_1^2 + \sigma_2^2 + \sigma_3^2 - 2\mu(\sigma_1\sigma_2 + \sigma_2\sigma_3 + \sigma_3\sigma_1)] \tag{b}$$

We next define a stress

$$\sigma_{av} = \frac{\sigma_1 + \sigma_2 + \sigma_3}{3} \tag{c}$$

and apply this stress to each of the principal directions of a unit cube (Fig. 6-18b). The remaining stresses $\sigma_1 - \sigma_{av}$, $\sigma_2 - \sigma_{av}$, and $\sigma_3 - \sigma_{av}$, shown in Fig. 6-18c, will produce only angular distortion. Substituting σ_{av} for σ_1, σ_2, and σ_3 in Eq. (b) gives the amount of strain energy producing only volume change.

$$u_v = \frac{1}{2E} [3\sigma_{av}^2 - 2\mu(3)\sigma_{av}^2] = \frac{3\sigma_{av}^2}{2E} (1 - 2\mu) \tag{d}$$

FIGURE 6-18 (a) Element with triaxial stresses; this element undergoes both volume change and angular distortion. (b) Element under hydro-static tension undergoes only volume change. (c) Element has angular distortion without volume change.

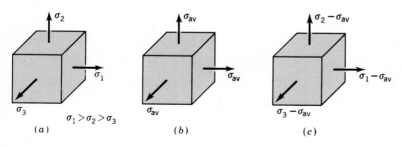

If we now substitute $\sigma_{av}^2 = [(\sigma_1 + \sigma_2 + \sigma_3)/3]^2$ into Eq. (d) and simplify the expression, we get

$$u_v = \frac{1 - 2\mu}{6E}(\sigma_1^2 + \sigma_2^2 + \sigma_3^2 + 2\sigma_1\sigma_2 + 2\sigma_2\sigma_3 + 2\sigma_3\sigma_1) \qquad (e)$$

Then the distortion energy is obtained by subtracting Eq. (e) from Eq. (b). This yields

$$u_d = u - u_v = \frac{1 + \mu}{3E}\left[\frac{(\sigma_1 - \sigma_2)^2 + (\sigma_2 - \sigma_3)^2 + (\sigma_3 - \sigma_1)^2}{2}\right] \qquad (6\text{-}8)$$

Note that the distortion energy is zero if $\sigma_1 = \sigma_2 = \sigma_3$.

For the simple tension test, $\sigma_1 = S_y$ and $\sigma_2 = \sigma_3 = 0$. Therefore the distortion energy is

$$u_d = \frac{1 + \mu}{3E}S_y^2 \qquad (6\text{-}9)$$

The criterion is obtained by equating Eqs. (6-8) and (6-9).

$$2S_y^2 = (\sigma_1 - \sigma_2)^2 + (\sigma_2 - \sigma_3)^2 + (\sigma_3 - \sigma_1)^2 \qquad (6\text{-}10)$$

which defines the beginning of yield for a triaxial stress state.

If σ_1, σ_2, or σ_3 is zero, the stress state is biaxial. Then, let σ_A be the larger of the two nonzero stresses, and let σ_B be the smaller. Equation (6-10) then reduces to

$$S_y^2 = \sigma_A^2 - \sigma_A\sigma_B + \sigma_B^2 \qquad (6\text{-}11)$$

For pure torsion $\sigma_B = -\sigma_A$ and $\tau = \sigma_A$; consequently

$$S_{sy} = 0.577S_y \qquad (6\text{-}12)$$

Comparison of Eq. (6-12) with (6-7) shows that the distortion-energy criterion predicts a yield strength in shear appreciably higher than that predicted by the maximum-shear-stress theory. How does it compare with the yield strength in shear as predicted by the maximum-normal-stress theory?

For analysis and design purposes it is convenient to define a *von Mises* stress, from Eq. (6-11), as

$$\sigma' = \sqrt{\sigma_A^2 - \sigma_A\sigma_B + \sigma_B^2} \qquad (6\text{-}13)$$

The corresponding equation for a triaxial stress state is

$$\sigma' = \sqrt{\frac{(\sigma_1 - \sigma_2)^2 + (\sigma_2 - \sigma_3)^2 + (\sigma_3 - \sigma_1)^2}{2}} \tag{6-14}$$

It is possible to bypass a Mohr's circle analysis for the special case of combined bending and torsion when finding the von Mises stress. A Mohr's circle for this stress state will reveal the two nonzero principal stresses to be

$$\sigma_A = \frac{\sigma_x}{2} + \tau_{xy} \qquad \sigma_B = \frac{\sigma_x}{2} - \tau_{xy} \tag{f}$$

When these two stresses are substituted into Eq. (6-13), we learn

$$\sigma' = \sqrt{\sigma_x^2 + 3\tau_{xy}^2} \tag{6-15}$$

6-7 FAILURE OF DUCTILE MATERIALS

It is now time to summarize the results of the three preceding sections and relate them to experimental results. The maximum-normal-stress theory is included for comparison purposes, even though it is only of historical interest. For a biaxial stress state one of the three principal stresses will be zero. Let the remaining nonzero stresses be σ_A and σ_B, as in the previous section. A plot of all three failure theories on a σ_A, σ_B coordinate system gives the graph of Fig. 6-19. Well-documented experiments indicate that the distortion-energy theory predicts yielding with greatest accuracy in all four quadrants. Thus, accepting the distortion-energy theory as the correct one, we see that the maximum-shear-stress theory gives results on the conservative side, because its graph is inside the distortion-energy ellipse.

Note that the maximum-normal-stress theory is the same as the maximum-shear-stress theory in the first and third quadrants. However, the graph of the maximum-normal-stress theory is outside the distortion-energy ellipse in the second and fourth quadrants. Thus it would be very dangerous to use the maximum-normal-stress theory, since it might predict safety when in fact no safety exists.

Generally a designer will employ the maximum-shear-stress theory if the dimensions need not be held too closely, if a quick size estimate is needed, or if the factors of safety are known to be generous. The distortion-energy theory predicts failure most accurately, and so it would be used when the margin of safety is to be held to very close limits or when the cause of an actual part failure is being investigated.

It is important to note that there are instances in which parts may fail, not by yielding, but by a brittle fracture. For more on this mode of failure see Sec. 6-13.

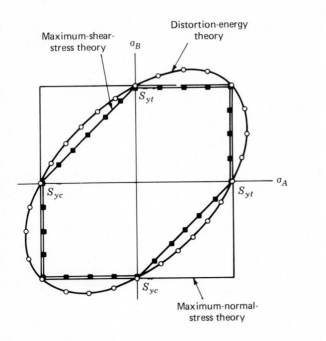

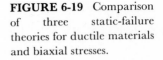

FIGURE 6-19 Comparison of three static-failure theories for ductile materials and biaxial stresses.

For machine computation of the factor of safety n_S or just n, use the program in Sec. 2-3 first to obtain the three principal stresses. Then use the following steps:

1 Store σ_1, σ_2, σ_3, and S_y using consistent units.
2 Is $\sigma_1 \geq |\sigma_3|$? If YES, go to step 3. If NO, go to step 4.
3 For maximum-normal-stress theory $n_S = S_y/\sigma_1$; go to step 5.
4 For maximum-normal-stress theory $n_S = -S_y/\sigma_3$; go to step 5.
5 For maximum-shear-stress theory $n_S = S_y/(\sigma_1 - \sigma_3)$.
6 Use Eq. (6-14) to compute σ'. Then for the distortion-energy theory $n_S = S_y/\sigma'$.

Example 6-3 A material has a yield strength of $S_y = 100$ kpsi. Compute the factor of safety for each of the failure theories for ductile materials. Plot the results on a graph like Fig. 6-19. Use the following stress states:

(a) $\sigma_1 = 70$ kpsi, $\sigma_2 = 70$ kpsi, $\sigma_3 = 0$
(b) $\sigma_1 = 70$ kpsi, $\sigma_2 = 30$ kpsi, $\sigma_3 = 0$
(c) $\sigma_1 = 70$ kpsi, $\sigma_2 = 0$, $\sigma_3 = -30$ kpsi
(d) $\sigma_1 = 0$, $\sigma_2 = -30$ kpsi, $\sigma_3 = -70$ kpsi

Solution The results, rounded to three significant figures, are shown in Table 6-1. The results of part (c) are interesting. The maximum-normal-stress theory

Table 6-1 FACTORS OF SAFETY n_S

Part	a	b	c	d
Maximum-normal-stress theory	1.43	1.43	1.43	1.43
Maximum-shear-stress theory	1.43	1.43	1.00	1.43
Distortion-energy theory	1.43	1.64	1.13	1.64

predicts safety with $n_S = 1.43$. The maximum-shear-stress theory predicts failure, since $n_S = 1$. But the distortion-energy theory predicts some safety with $n_S = 1.13$. The graph of Fig. 6-20 is also helpful in comparing the results.

The following procedure is useful in obtaining points for the distortion-energy ellipse. Let $\sigma_B = \sigma_A$ and $f(\sigma) = S_y$ in Eq. (6-11). Then, by some manipulation, we can write

$$\sigma_A = \frac{f(\sigma)}{\sqrt{1 - a + a^2}} \tag{6-16}$$

With $f(\sigma)$ given, select positive and negative values for a and solve for σ_A and σ_B. About six to eight points in each quadrant will yield a smooth curve. A programmable calculator will be useful in making the calculations. ////

Example 6-4 Whenever it is suspected that a part might be overloaded during misuse, an overload factor should be applied to each suspicious load in addition to the factor of safety used to account for other uncertainties. This example is

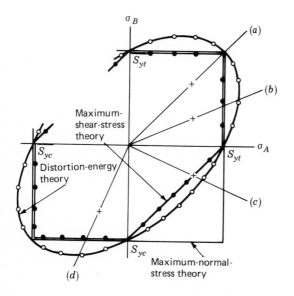

FIGURE 6-20 Graph of failure theories for ductile materials with $S_y = 100$ kpsi. The stress coordinates are marked by the plus signs.

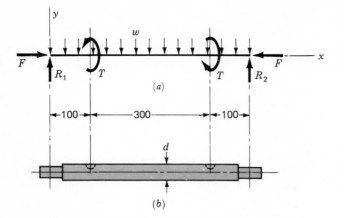

FIGURE 6-21   (*a*) Assumed loading diagram; (*b*) shape diagram.

intended to show how factors of safety can be employed in design when the possibility of multiple overloading occurs.

The member illustrated in Fig. 6-21 has a uniform bending load $w = 0.72$ kN/m, an axial preload $F = 2$ kN, and a torsional load $T = 24$ N·m. We wish to determine the diameter of the member using a material having a yield strength $S_y = 385$ MPa.

An examination of the particular application of the hypothetical member of Fig. 6-21, let us say, indicates no prediction for overloading of the preload F. However, misuse could increase the torsional load by as much as 43 percent and the bending load by as much as 80 percent.

We choose a factor of safety $n_S = 1.40$ to account for uncertainties in the strength. We also choose a general or overall factor of safety for loading of $n_L = 1.40$ to account for differences in assembly, variations in fits, and speed variations. Thus the total factor of safety for the load F is

$$n_F = n_L = 1.40$$

But, because of the possibility of overloads, for torsion and bending we have

$$n_T = 1.43n_L = 1.43(1.40) = 2.00$$

$$n_w = 1.80n_L = 1.80(1.40) = 2.52$$

Thus the effect of overloading is to increase the factors of safety that correspond to the particular loads that might be affected.

Solution   The critical point will be at the top surface and in the middle of the member since, at this point, the compressive bending stress adds to the compressive axial stress. Column action is neglected. A stress element at the critical point would be aligned in the xz directions. At this point the permissible loads are:

$$M_p = n_w \frac{wl^2}{8} = \frac{2.52(0.72)(500)^2}{8(10)^3} = 56.7 \text{ N} \cdot \text{m}$$

$$T_p = n_T T = 2.00(24) = 48 \text{ N} \cdot \text{m}$$

$$F_p = n_F F = 1.40(2) = 2.8 \text{ kN}$$

The permissible stress, corresponding to the yield strength, is

$$\sigma_p = \frac{S_y}{n_S} = \frac{385}{1.40} = 275 \text{ MPa}$$

The stress equations will contain both d^3 and d^2 terms, and so it is probably quicker to use a trial-and-error approach. As the first trial we select $d = 20$ mm. The following stresses result:

$$\sigma_{x, p} = -\frac{M_p}{I/c} - \frac{F_p}{A} = -\frac{32M_p}{\pi d^3} - \frac{4F}{\pi d^2}$$

Thus,

$$\sigma_{x, p} = -\left[\frac{32(56.7)}{\pi(0.02)^3} + \frac{4(2800)}{\pi(0.02)^2}\right](10)^{-6} = -81.1 \text{ MPa}$$

Also

$$\tau_{xz, p} = \frac{T_p r}{J} = \frac{16T_p}{d^3} = \left[\frac{16(48)}{\pi(0.02)^3}\right](10)^{-6} = 30.6 \text{ MPa cw}$$

When these values are entered into a machine computation program we obtain $\sigma_1 = 10.25$ MPa, $\sigma_2 = 0$, and $\sigma_3 = -91.35$ MPa. The von Mises stress is found to be $\sigma' = 96.9$ MPa which is considerably less than the permissible stress of 275 MPa.

A second trial using $d = 15$ mm gives $\sigma_{x, p} = -187$ MPa, $\tau_{xz, p} = 72.4$ MPa cw, $\sigma_1 = 24.8$ MPa, $\sigma_2 = 0$, and $\sigma_3 = -212$ MPa. The von Mises stress for $d = 15$ mm is 225 MPa. Thus, compared to $\sigma_p = 275$ MPa, we still have a margin of safety of 50 MPa.

A third trial using $d = 14$ mm gives $\sigma_{x, p} = -229$ MPa, $\tau_{xz, p} = 89.1$ MPa cw, $\sigma_1 = 30.6$ MPa, $\sigma_2 = 0$, and $\sigma_3 = -259$ MPa. The von Mises stress is found to be 276 MPa, barely over the permissible value of 275 MPa. While this is acceptable accuracy, still we choose to use $d = 15$ mm as the final result. As a check on this decision you should find the safety margin in MPa for $d = 14$ mm and $d = 15$ mm, using the maximum-shear-stress theory.

Note, however, from Table A-13, that 14 mm and 16 mm are preferred sizes, but 15 mm is not.   ////

6-8 FAILURE OF BRITTLE MATERIALS

In selecting a failure theory for use with brittle materials we first note the following characteristics of these materials:

1 A graph of stress versus strain is a smooth continuous line to the failure point; failure occurs by fracture, and hence these materials do not have a yield strength.
2 The compressive strength is usually many times greater than the tensile strength.
3 The ultimate torsional strength S_{su}, that is, the modulus of rupture, is approximately the same as the tensile strength.

The maximum-normal-stress theory and the Coulomb-Mohr theory have both been used to predict fracture of brittle materials. The maximum-normal-stress theory has already been investigated. In using this theory the test points shown in Fig. 6-19 would be changed to the values S_{ut} and S_{uc}.

The *Coulomb-Mohr theory*, sometimes called the *internal-friction theory*, is based upon the results of two tests, the tensile test and the compression test. On the σ, τ coordinate system, plot both circles, one for S_{ut} and one for S_{uc}. Then the Coulomb-Mohr theory states that fracture occurs for any stress situation which produces a circle tangent to the envelope of the two test circles. If we arrange the principal stresses so that $\sigma_1 > \sigma_2 > \sigma_3$, then the critical stresses are σ_1 and σ_3. These two stresses and the strengths are related by the equation

$$\frac{\sigma_1}{S_{ut}} - \frac{\sigma_3}{S_{uc}} = 1 \qquad \begin{array}{l} \sigma_1 \geq 0 \\ \sigma_3 \leq 0 \end{array} \qquad\qquad (6\text{-}17)$$

which defines fracture by the Coulomb-Mohr theory for a stress state in the fourth quadrant of Fig. 6-22. In this equation S_{uc} is a positive quantity.

Figure 6-22 also exhibits a number of test points to enable us to compare the maximum-normal-stress theory predictions with those of the Coulomb-Mohr theory.* In the first quadrant, where σ_A and σ_B both have the same sense, we

---

* L. F. Coffin, "The Flow and Fracture of a Brittle Material," *Trans. ASME*, vol. 72, *J. Appl. Mech.*, vol. 17, 1950; pp. 233–248; R. C. Grassi and I. Cornet, "Fracture of Gray Cast Iron Tubes under Biaxial Stresses," *Trans. ASME*, vol. 71, *J. Appl. Mech.*, vol. 16, 1949, pp. 178–182.

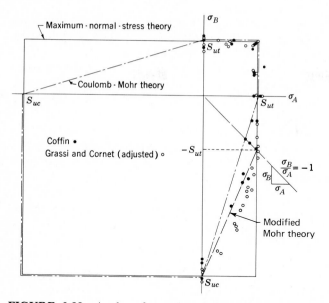

FIGURE 6-22 A plot of experimental data points from tests of gray cast iron subjected to biaxial stresses. The data were adjusted to correspond to $S_{ut} = 32$ kpsi and $S_{uc} = 105$ kpsi. Superimposed on the plot are graphs of the maximum-normal-stress theory, the Coulomb-Mohr theory, and the modified Mohr theory.

note that the two theories are identical, and hence that either theory can be used satisfactorily to predict failure. It is in the fourth quadrant, where σ_A and σ_B have opposite senses that the two theories differ. We note first that the Coulomb-Mohr theory is on the conservative side, since all data points fall outside.† Note also, in the figure, the line having the slope $\sigma_B/\sigma_A = -1$. For pure torsion, $\sigma_B = -\sigma_A$, and hence the intersection of this line with the graph of a failure theory yields the value of S_{su}, as predicted by that theory. Note that its intersection with the maximum-normal-stress theory yields $S_{su} = S_{ut}$, which we have already noted is one of the characteristics of brittle materials. But the Coulomb-Mohr theory predicts a value of S_{su} somewhat less than S_{ut}.

By manipulating Eq. (6-17) it is possible to deduce that

$$S_3 = \frac{S_{uc}}{\dfrac{S_{uc}}{S_{ut}} \dfrac{\sigma_1}{\sigma_3} - 1} \tag{6-18}$$

---

† A conservative theory is perfectly satisfactory for design purposes where the object is to determine a set of dimensions such that the part will not fail, but it is completely useless in an analysis if the object is to learn why something failed.

where S_3 is the limiting value of σ_3 corresponding to any specified ratio σ_1/σ_3 within the limits imposed by Eq. (6-17). Equation (6-18) is especially convenient in design and analysis, and it avoids the need for graphical solutions.

The *modified Mohr theory*, shown in the fourth quadrant of Fig. 6-22, is not as conservative as the Coulomb-Mohr theory, but it does a better job of predicting fracture. Burton Paul* proposed a similar theory, though slightly different, which we shall not include here. Note that the modified Mohr theory does not differ from the maximum-normal-stress theory until σ_3 becomes less than $-S_{ut}$. For this region of the fourth quadrant of Fig. 6-22, the modified Mohr theory is expressed by the equation

$$ S_3 = \frac{S_{uc}}{\dfrac{S_{uc} - S_{ut}}{S_{ut}} \dfrac{\sigma_1}{\sigma_3} - 1} \qquad \begin{array}{l} \sigma_3 \leq -S_{ut} \\[4pt] \sigma_1 \geq 0 \end{array} \qquad\qquad (6\text{-}19) $$

Computer or Calculator Program

The program which follows is designed to compute the factor of safety for each of the three failure theories for brittle materials when the strengths and stresses are given. It is a tricky but interesting program. The steps are as follows:

1   Enter and print S_{ut}, S_{uc}, σ_1, σ_2, and σ_3. Be sure to enter S_{uc} as a positive number. Be sure the stresses are ordered $\sigma_1 > \sigma_2 > \sigma_3$.

2   Is $\sigma_3 = 0$? If NO, go to step 4. If YES, go on.

3   Compute $n = S_{ut}/\sigma_1$. This result is valid for all three theories. Print results and stop.

4   Is $\sigma_1 = 0$? If YES, go to step 6. If NO, go on.

5   Is $\sigma_1 \geq 0$? If NO, go to step 6. If YES, go to step 7.

6   Compute $n = -S_{uc}/\sigma_3$. This is valid for all three theories. Print results and stop.

7   Solve the Coulomb-Mohr equation [Eq. (6-18)] for S_3. Compute $n = S_3/\sigma_3$ and print the result.

8   Is $[(\sigma_3/-\sigma_1) - 1] \geq 0$? If YES, go to step 10. If NO, go on.

9   Compute $n = S_{ut}/\sigma_1$. This result is valid for the maximum-normal-stress theory and for the modified Mohr theory. Print both results and stop.

10   Solve the equation for the modified Mohr theory [Eq. (6-19)] for S_3. Compute $n = S_3/\sigma_3$ and print the result.

11   Compute and store $n_1 = S_{ut}/\sigma_1$ and $n_2 = -S_{uc}/\sigma_3$.

12   Is $(n_2 - n_1) \geq 0$? If YES, go to step 14. If NO, go on.

13 $n = n_2$ for the maximum-normal-stress theory. Print and stop.

14 $n = n_1$ for the maximum-normal-stress theory. Print and stop.

---

* Burton Paul, "A Modification of the Coulomb-Mohr Theory of Fracture," *Trans. ASME, J. Appl. Mech.*, ser. E, vol. 28, no. 2, June 1961, pp. 259–268.

Example 6-5 A small, 6-mm-diameter pin was designed of ASTM No. 40 cast iron. The pin was designed to take an axial compressive load of 3.5 kN combined with a torsional load of 9.8 N·m after the load safety factors had been applied. Compute the strength factor of safety remaining for each of the three failure theories.

Solution   We shall solve using the graphical approach first. The axial compressive stress is

$$\sigma_x = \frac{F}{A} = \frac{4F}{\pi d^2} = -\frac{4(3.5)(10)^3}{\pi(6)^2} = -124 \text{ MPa}$$

The torsional shear stress is

$$\tau_{xy} = \frac{16T}{\pi d^3} = \frac{16(9.8)(10)^3}{\pi(6)^3} = 231 \text{ MPa}$$

When Mohr's circle diagram is constructed, the principal stresses are found to be $\sigma_1 = 177$ MPa, $\sigma_2 = 0$, and $\sigma_3 = -301$ MPa.

Using typical values of the strengths, instead of the minimums, we find from the Appendix that $S_{ut} = 293$ MPa and $S_{uc} = 965$ MPa.

The next step is to plot a graph to scale corresponding to the fourth quadrant of Fig. 6-22, using the strengths and stresses just found. This has been done in Fig. 6-23. Point A represents the coordinates σ_1, σ_3 of the actual stress state. If σ_1 and σ_3 increased in magnitude but retained the same ratio to each other, then points B, C, and D would represent failure by each theory. Thus, if OA represents the stress state, then AB, AC, and AD represent the respective margins of safety. The corresponding factors of safety are equal to OB divided by OA, OC divided by OA, and OD divided by OA.

Another way to obtain the factors of safety is to project points B, C, and D to the σ_1 or σ_3 axis. The resulting intersections define corresponding strengths S_1 or S_3, if a mnemonic notation is used. Thus, in Fig. 6-23 we can read out the strength S_3 for each failure theory. The factor of safety for the maximum-normal-stress theory is found to be

$$n_S = \frac{-S_3}{\sigma_3} = \frac{-500}{-301} = 1.6 \qquad\qquad Ans.$$

For the modified Mohr theory we get

$$n_S = \frac{-S_3}{\sigma_3} = \frac{-405}{-301} = 1.35 \qquad\qquad Ans.$$

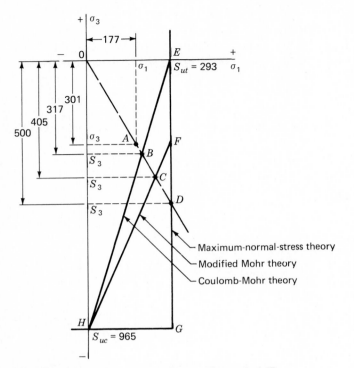

FIGURE 6-23   *All stresses and strengths are in MPa.*

The result for the Coulomb-Mohr theory is

$$n_S = \frac{-S_3}{\sigma_3} = \frac{-317}{-301} = 1.05$$ *Ans.*

The results of a graphical solution are completely satisfactory in every way, as no additional accuracy is meaningful. However, a programmable-calculator solution, using the program presented previously, gave 1.655, 1.092, and 1.365, respectively. The results are given to four significant figures in case you might wish to check your own program. Two or three significant figures give all the accuracy that is normally needed.   ////

6-9   STRESS CONCENTRATION

In the development of the basic stress equations for tension, compression, bending, and torsion, it was assumed that no irregularities occurred in the member under consideration. But it is quite difficult to design a machine without permitting some changes in the cross sections of the members. Rotating shafts must have shoulders designed on them so that the bearings can be properly seated and so that they will take thrust loads; and the shafts must have key slots machined into them for securing pulleys and gears. A bolt has a head on one end and screw threads on the other end, both of which are abrupt changes in the cross section. Other parts require holes, oil grooves, and notches of various kinds. Any discontinuity in a machine part alters the stress distribution in the neighborhood of the discontinuity so that the elementary stress equations no longer describe the state of stress in the part. Such discontinuities are called *stress raisers*, and the regions in which they occur are called areas of *stress concentration*.

A *theoretical*, or *geometric, stress-concentration factor* K_t or K_{ts} is used to relate the actual maximum stress at the discontinuity to the nominal stress. The factors are defined by the equations

$$K_t = \frac{\sigma_{max}}{\sigma_0} \qquad K_{ts} = \frac{\tau_{max}}{\tau_0} \tag{6-20}$$

where K_t is used for normal stresses and K_{ts} for shear stresses. The nominal stress σ_0 or τ_0 is more difficult to define. Generally it is the stress calculated by using the elementary stress equations and the net area, or net cross section. But sometimes the gross cross section is used instead, and so it is always wise to check before calculating the maximum stress.

The subscript t in K_t means that this stress-concentration factor depends for its value only on the *geometry* of the part. That is, the particular material used has no effect on the value of K_t. This is why it is called a *theoretical* stress-concentration factor.

6-10 DETERMINATION OF STRESS-CONCENTRATION FACTORS

It is possible to analyze certain geometrical shapes by using the methods of the theory of elasticity to determine the values of stress-concentration factors. Figure 6-24, for example, represents an infinite plate stressed uniformly in tension by σ_0.

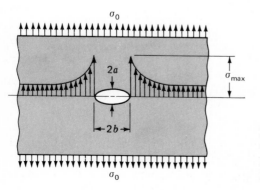

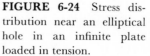
FIGURE 6-24 Stress distribution near an elliptical hole in an infinite plate loaded in tension.

A small elliptical hole in the plate will have a stress at the edge of

$$\sigma_{max} = \sigma_0\left(1 + \frac{2b}{a}\right) \qquad (6\text{-}21)$$

If a and b are equal, the ellipse becomes a circle and Eq. (6-21) reduces to

$$\sigma_{max} = 3\sigma_0 \qquad (6\text{-}22)$$

so that $K_t = 3$. In this example note that the plate is infinite and the nominal stress σ_0 is the tensile stress at a point remote from the discontinuity.

Equation (6-21) can be used to determine the stress at the edge of a transverse crack by making a very small compared with b. In this case K_t is seen to become a very large number. On the other hand, if b is made small compared with a, the equation gives the stress at the edge of a longitudinal crack and it is seen that K_t approaches unity.

Photoelasticity

A very dependable and widely used method of determining stresses at a point is the method of *photoelasticity*. A transparent material having double-refraction properties, when stressed, is cut into the same shape as the part whose stresses are desired. The model is placed in a loading frame, and a beam of polarized light is directed through it onto a photographic plate or screen. When the model is loaded, fringes of colored light originate at the points of maximum stress and, when the load is increased, move from the edges of the image toward the center. A certain stress is associated with each fringe so that one can determine the stresses at the edges merely by counting the fringes as they originate.

Figure 6-25 is a picture, taken by photoelastic methods, of the fringes on a gear tooth loaded by the force w_r. Stress concentration exists at the point of application of the force as well as at both fillets at the root of the tooth.

FIGURE 6-25 Stress distribution in a gear tooth determined by photoelastic procedures.

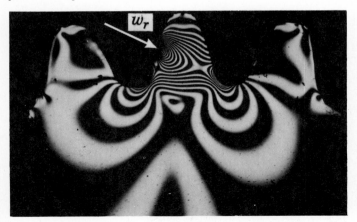

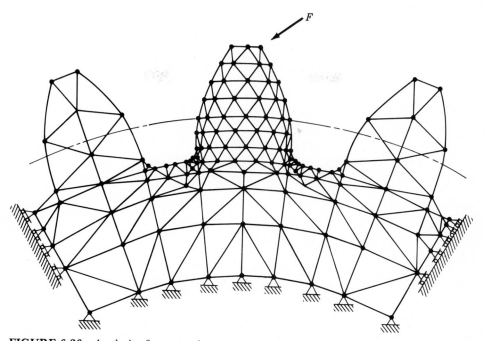

FIGURE 6-26 Analysis of gear-tooth stresses by the finite-element method. (*By permission of Cemil Bagci of Tennessee Technological University; the figure was included in the paper "Finite Stress Elements and Applications in Machine Design" presented at the 6th Applied Mechanisms Conference, Denver, 1979.*)

Finite-Element Techniques

This is a very powerful and new approach made possible by advances in computers and computer-aided design methods in recent years. The finite elements may be lines, triangles, or any convenient geometric shapes. The member to be analyzed is first divided into a large number of finite elements, which may be of different sizes. Starting from the known loading and boundary configuration and constraints, a computer analysis is made and iterated until all conditions are satisfied. Use of the method requires knowledge of matrix mathematics and linear elasticity and familiarity with a computer language.

Figure 6-26 is an illustration showing how the method is applied for the analysis of gear-tooth stresses. A force F acts at a junction of elements 1 and 2 at the tip of the tooth. Triangular finite-elements are particularly useful for two-dimensional analysis. Note how the constraints and boundary conditions are shown.*

---

* The following books are recommended for additional information.

J. L. Meek, *Matrix Structural Analysis*, McGraw-Hill, New York, 1971.

Harold C. Martin and Graham F. Carey, *Introduction to Finite Element Analysis*, McGraw-Hill, New York, 1973.

O. C. Zienkiewicz, *The Finite Element Method*, 3d ed., McGraw-Hill, New York, 1977.

Other Methods

Also used for experimentally determining stress-concentration factors are the grid, brittle-coating, brittle-model, and strain-gauge methods.

The *grid method* consists in scratching or drawing a grid of lines on the part or on a model of it. Two sets of equally spaced lines at right angles to each other may be used, or sometimes a series of concentric circles intersected by radial lines is employed to form the grid. The part or model is placed under a known load, and the change in the spacing of the lines is used to map the strain. Since the strains may be quite small, some form of magnification must be employed in making the measurements. After the strains are mapped, the stresses are calculated by the biaxial stress-strain relations of Chap. 2. Of course the lines must be scribed accurately, and the strains measured accurately, if good results are to be obtained with this method.

The *brittle-coating* method employs a lacquer developed by the Magnaflux Corporation under the name Stresscoat. A uniform layer of Stresscoat is sprayed onto the part under carefully controlled temperature and humidity conditions. After the lacquer has dried, a load is applied to the part, causing tiny cracks to form at all areas of high tensile stress. The directions of the cracks are always perpendicular to the direction of the tensile stresses, because the lacquer is brittle and fails in tension. The first crack formed on the part is, of course, the region of highest tensile stress. As more load is applied to the part, other cracks form, indicating the areas of lower stress. The principal advantage of the brittle-coating method is that it can be applied to very irregular surfaces. Except when used by a very skillful investigator, this method will probably not yield an accuracy of better than 15 percent.

The *brittle-model* method employs a very brittle material, such as plaster, for a calibration specimen and a model of the part to be investigated. Each specimen is loaded until fracture occurs, and the respective loads are then compared to obtain K_t.

Strain-gauge methods use either electrical or mechanical means to measure the strains. As in the case of the grid methods, the strains are averaged over the gauge length, and consequently will not give the true stresses at a point.

Intuitive Methods

It is very important for the designer to develop a "feel" for stress concentration in order to know intuitively when it exists and what to do about it. Such a sense will also assist greatly in extrapolating K_t from charts or in estimating the stress-concentration factor when experimental means cannot be used.

The flow analogy is probably the best single means of visualizing stress concentration. Figure 6-27a shows a uniform bar in tension, and we visualize the force as flowing through the bar. Thus each flow line in the figure represents a certain amount of force, and since the bar is uniform, the flow lines are uniformly spaced. Now, in Fig. 6-27b, we have cut notches in the bar to represent stress

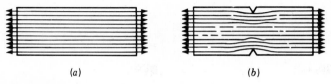

FIGURE 6-27

raisers. At sections remote from the notches the flow lines will be uniformly spaced, as they are in (*a*). As these lines of force approach the notch, those closest to it will have to bend the most in order to pass through the restricted opening. The severity of the stress concentration is proportional to the amount of bending of the flow lines. This can be explained in another manner. Figure 6-28 illustrates two bars in tension; one is straight, and the other is curved. For the same elongation the straight bar will carry much more load than the curved one. Or, to put it another way, the curved bar will elongate many times more than the straight one if the same load is applied to each. Thus introduction of the notch in Fig. 6-27*b* causes a deterioration of the load-carrying capacity, and this deterioration is greatest for the material immediately adjacent to the notch.

The flow analogy is important because it gives us a physical picture of why stress concentration exists. The flow analogy can also be used as a qualitative tool to learn what to do to decrease stress concentration. A few examples will suffice for purposes of illustration.

Without doubt the failure which occurs most frequently in rotating machinery is the fatigue failure of a shaft. Bearings, gears, and other parts have to be seated against a shoulder, and it is at the base of this shoulder that failure occurs. Figure 6-29*a* shows such a part, together with the flow lines. For the lowest K_t the fillet radius *r* must be as large as possible, but usually *r* is limited by the design of the mating part. A larger fillet can be used and a lower K_t obtained by undercutting the shoulder as in (*b*). The method illustrated in Fig. 6-29*c* is sometimes quite useful. The first notch should be deeper than the second, but the diameter at the bottom of the notch should be slightly larger than (*d*). Surprisingly enough, the cutting of additional notches is often a very effective way of reducing stress concentration. Many difficult problems involving stress concentration have been solved in this way by removing material instead of adding it. When used with members subjected to bending, the bending-moment diagram should be examined to make sure that the additional notches are not cut at a point of higher bending moment. In fact, the improvement shown in Fig. 6-29*d* is probably only an improvement if the additional cut is taken at a point of lower bending moment.

A great many of the discontinuities found in practice cannot be avoided. But

FIGURE 6-28 For the same elongation, the straight bar takes more load than the curved bar.

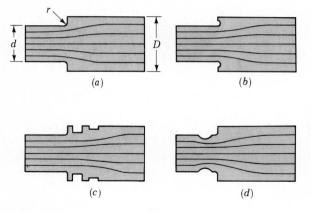

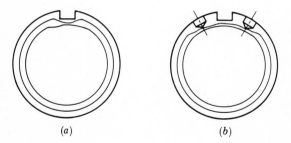

FIGURE 6-29 A shouldered shaft in bending. (*a*) The usual design; (*b-d*) improved designs to reduce the amount of stress concentration.

a good designer will often go to extreme lengths in an effort to locate the discontinuity in a region of low nominal stress.

Figure 6-30 shows how the flow analogy can be used for shear and illustrates one means of improving the cross section. It is always good practice in design to sketch and study each stress raiser when it occurs and investigate it by drawing flow lines to see whether or not improvement can be obtained.

6-11 STRESS-CONCENTRATION CHARTS

For convenience, all stress-concentration charts and data are included in Table A-26. You are cautioned to look at each chart before calculating the nominal stress, since some of the stresses are based on the net area, while others are calculated by using the gross area or cross section.

6-12 STRESS CONCENTRATION AND STATIC LOADS

It is always somewhat disconcerting to a beginner in stress analysis to study geometric stress-concentration factors in considerable detail and then learn, to his or her amazement, that the full value of these factors is seldom used in actual design. The reason for this is that the geometric stress-concentration factors depend for their value only on the geometry of the part, as the name implies. The actual material of which the part is made does not enter at all into the problem of

FIGURE 6-30 (*a*) Section of a torsionally loaded shaft, showing the key slot. The corners at the bottom of the slot are very nearly square. (*b*) Improvement obtained by drilling holes on each side of the key slot.

determining K_t. When one begins to study the materials used to form the parts, it is found that the stress-concentration effect really depends upon whether the material is brittle or ductile.

An ideal brittle material is probably one that has a steep stress-strain diagram completely up to fracture. The full value of K_t should probably be used with such materials. But there is no use in discussing ideal brittle materials, since no sane designer would ever specify one for a load-carrying member. Many of the cast irons, especially those with low tensile strengths, are considered brittle; yet they are far from being ideally brittle. Cast irons, generally, are not homogeneous; they often have slag inclusions, pockets, and graphite. In addition, the sand-casting process always produces a nonsmooth surface. Thus cast irons inherently have a great many discontinuities. Now, if to all these discontinuities we add one more in the form of a notch or hole, this is really not going to make very much difference in the stress distribution in the part. That is, the strength of the part was determined in the first place when many discontinuities existed. Consequently it is never necessary to apply the full value of K_t, even to such a brittle material as cast iron. In the next chapter we shall be more specific and state exactly how much K_t should be reduced for various cast irons.

Stress concentration is a highly localized effect. The high stresses exist only in a very small region in the vicinity of the notch. In the case of ductile materials the statement is usually made that the first load applied to the member causes the material to yield at the discontinuity and relieves the stress concentration. Consequently, for ductile materials under static loads, it is not necessary to apply a stress-concentration factor at all! It will be of value to explore this statement further.

The discussion to follow is completely qualitative; that is, the methods presented are not intended to be used for determining numerical values of stress or strength. Our purpose is only to explain why yielding relieves stress concentration in ductile materials.

We begin by defining a hypothetical material of high ductility having the properties described by the stress-strain diagram of Fig. 6-31a. Many of the more ductile steels, incidentally, have stress-strain diagrams very nearly like this material. We also visualize a notched rectangular bar (Fig. 6-31b) loaded in bending by the moment M and having a geometric stress-concentration factor $K_t = 3$. Since $I = bh^3/12$ and $h = 2c$, the nominal stress at the root of the notch is

$$\sigma_0 = \frac{Mc}{I} = \frac{Mc}{bh^3/12} = \frac{3M}{2bc^2}$$

The maximum stress is

$$\sigma_{max} = K_t \sigma_0 = \frac{9M}{2bc^2}$$

These two stresses are shown on the stress-distribution diagrams of Fig. 6-31c and d, and σ_{max} is shown as point A on the stress-strain diagram.

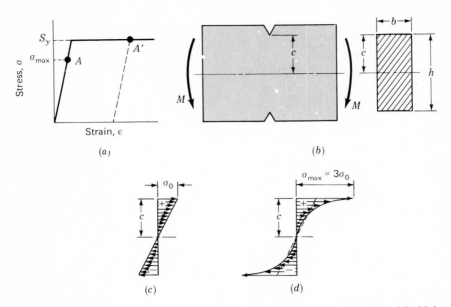

FIGURE 6-31 (a) Stress-strain diagram for a hypothetical material with high ductility; (b) notched beam loaded by moments M; (c) nominal stress distribution; (d) stress distribution at notched section due to stress concentration.

Now suppose that the moment M is gradually increased. Point A will move up the stress-strain diagram until σ_{max} becomes equal to S_y. At this point yielding will begin at the bottom of the notch. This yielding will move inward toward the neutral axis as more and more moment is applied, and point A will move out on the horizontal portion of the stress-strain diagram. At some time or another in the life of the beam we might suppose that a slight overload had been applied which caused yielding to progress to a point halfway between the bottom of the notch and the neutral axis. Designate the bending moment corresponding to this overload as M' and the new position of point A as A'. The new stress distribution under this overload will probably be somewhat like that of Fig. 6-32a. The elastic portion will be very nearly triangular-shaped, as shown in the figure, because this region is not too close to the bottom of the notch. Our next problem is to find out what happens when the bending moment M' is removed.

We can find the bending moment required to yield 50 percent of the beam by multiplying the stresses by their areas and then finding the moment of the resulting force about the neutral axis. Thus, from Fig. 6-32a, we find

$$M' = 2\left[\left(S_y b \frac{c}{2}\right)\left(\frac{3c}{4}\right) + \left(\frac{S_y}{2}\frac{bc}{2}\right)\left(\frac{2}{3}\frac{c}{2}\right)\right] = \frac{11}{12}S_y bc^2$$

To find the stress distribution that remains after the beam is unloaded, we apply another moment to the beam, equal to M' in magnitude but opposite in direction.

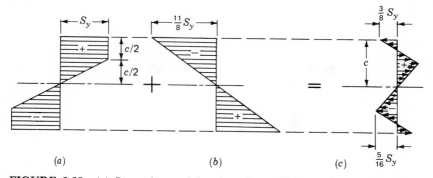

FIGURE 6-32 (*a*) Stress due to M' acting alone; (*b*) stress due to M'' acting alone; (*c*) resultant stress created by summing M' and M'' is a residual stress.

Call this moment M''. The resultant moment will then be

$$M' - M'' = 0$$

The stress at the bottom of the notch created by this unloading moment is

$$\sigma = \frac{M''c}{I} = -\frac{M'c}{I} = -\frac{11}{12}\,S_y bc^2 \frac{c}{bh^3/12} = -\frac{11}{8}\,S_y$$

and the stress distribution due to this moment acting alone is shown in Fig. 6-32*b*. The final stress distribution with zero moment on the beam is then obtained by adding together the distributions of Fig. 6-32*a* and *b*. The result is shown in Fig. 6-32*c*. This is called a *residual stress*. It is a stress that has been permanently built into the beam by the overload. If we now apply a bending moment large enough to cause the original beam to yield, the stress at the bottom of the notch will be less than S_y by the amount $3S_y/8$. Thus, yielding has relieved the stress concentration. This effect will be even more pronounced with most materials because the stress-strain diagram continues to climb upward after yielding begins.

We can conclude, then, that yielding in the vicinity of a stress raiser is beneficial in improving the strength of the part and that stress-concentration factors need not be employed when the material is ductile and the loads are static.

6-13 INTRODUCTION TO FRACTURE MECHANICS*

The use of elastic stress-concentration factors provides an indication of the average load required on a part for the onset of plastic deformation, or yielding; these factors are also useful for analysis of the loads on a part that will cause fatigue

* We are very grateful to Professor David K. Felbeck of the Department of Mechanical Engineering and Engineering Mechanics at the University of Michigan for writing the material on fracture mechanics in this chapter. J.E.S., L.D.M.

fracture. However, stress-concentration factors are limited to structures for which all dimensions are precisely known, particularly the radius of curvature in regions of high stress concentration. When there exists a crack, flaw, inclusion, or defect of unknown small radius in a part, the elastic stress-concentration factor approaches infinity as the root radius approaches zero, thus rendering the stress-concentration factor useless. Furthermore, even if the radius of curvature of the flaw tip is known, the high local stresses there will lead to local plastic deformation surrounded by a region of elastic deformation. Elastic stress-concentration factors are no longer valid for this situation, so analysis from the point of view of stress-concentration factors does not lead to criteria useful for design when very sharp cracks are present.

By combining analysis of the gross elastic changes in a structure or part that occur as a sharp brittle crack grows with measurements of the energy required to produce new fracture surfaces, it is possible to calculate the average stress (if no crack were present) which will cause crack growth in a part. Such calculation is possible only for parts with cracks for which the elastic analysis has been completed, and for materials that crack in a relatively brittle manner and for which the fracture energy has been carefully measured. The term *relatively brittle* is rigorously defined in the test procedures,[*] but it means, roughly, *fracture without yielding occurring throughout the fractured cross section.*

Thus glass, hard steels, strong aluminum alloys, and even low-carbon steel below the ductile-to-brittle transition temperature can be analyzed in this way. Fortunately, ductile materials blunt sharp cracks, as we have previously discovered, so that fracture occurs at average stresses of the order of the yield strength, and the designer is prepared for this condition. The middle ground of materials that lie between "relatively brittle" and "ductile" is now being actively analyzed, but exact design criteria for these materials are not yet available and will not be covered here.

6-14 STRESS STATE IN A CRACK

Suppose a sharp, transverse, full-thickness crack of length $2a$ is located in the center of a rectangular plate of material, as in Fig. 6-33. An average axial tensile stress σ is applied to both ends of the plate. If the plate length $2h$ is large compared with the width $2b$, and width $2b$ is large compared with the crack length $2a$, elastic analysis shows that the conditions for crack growth are controlled by the magnitude of the elastic stress intensity factor K, and that for this case

$$K_0 = \sigma\sqrt{\pi a} \tag{6-23}$$

We shall employ the SI units of MPa $\sqrt{m}$ and the U.S. customary units of

---

[*] ASTM Standard E 399-72, *Annual Book of ASTM Standards*, 1972, or the latest version of this standard.

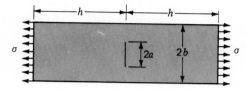

FIGURE 6-33 Plate of length $2h$, width $2b$, containing a central crack of length $2a$; tensile stress σ acts in longitudinal direction.

kpsi $\sqrt{\text{in}}$ for the factor K_0. If, for example, $h/b = 1$ and $a/b = 0.5$, then the magnitude of K_0 must be modified, in this case by a factor of 1.32, so then

$$K_1 = 1.32\sigma\sqrt{\pi a} \tag{6-24}$$

Thus it can be seen that K_1 is a function of the average axial stress and the geometry of the part. Solutions for this particular problem over a wide range of ratios of h/b and a/b have been calculated and are given graphically in Fig. 6-34, where K_1 is the desired value and K_0 is the base value calculated from Eq. (6-23).

6-15   CRITICAL STRESS-INTENSITY FACTOR

The previous section describes the conditions in a part under a given applied stress by calculation of the stress-intensity factor. This value is, from the designer's point of view, a condition analogous to *stress*. This section discusses the other half of the design equation, the value analogous to *strength* of the material, that is, the *critical stress-intensity factor*, also called *fracture toughness*, and designated by the symbol K_c.

Through carefully controlled testing of a given material, the stress-intensity

FIGURE 6-34   Plate containing a central crack loaded in longitudinal tension.

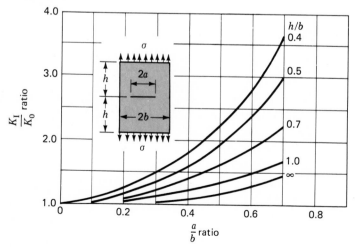

factor at which a crack will propagate is measured. This is the critical stress-intensity factor K_c. Thus for a known applied stress σ acting on a part of known or assumed crack length $2a$, when the magnitude of K reaches K_c, the crack will propagate. For the designer, the factor of safety n is thus

$$n = \frac{K_c}{K} \tag{6-25}$$

The enormous power of this method of analysis is that it enables the designer to use the value of K_c (usually measured in a single edge-notch specimen) for a given material in the design of a part that may be much more complex than the original test specimen.

Example 6-6   A ship steel deck that is 30 mm thick, 12 m wide and 20 m long (in the tensile-stress direction) is operated below its ductile-to-brittle transition temperature (with $K_c = 28.3$ MPa $\sqrt{m}$). If a 65-mm-long central transverse crack is present, calculate the tensile stress for catastrophic failure. Compare this stress with the yield strength for this steel.

Solution   From Fig. 6-33, $2a = 65$ mm, $2b = 12$ m, and $2h = 20$ m. Thus $a/b = 32.5/6(10)^3 = 0.005$ and $h/b = 10/6 = 1.67$. Since a/b is so small this may be considered as an infinite plate, and so Eq. (6-23) need not be modified. Solving for the stress then gives $\sigma = K_I/\sqrt{\pi a}$. Since fracture will occur when $K_I = K_{Ic}$, we have

$$\sigma = \frac{K_{Ic}}{\sqrt{\pi a}} = \frac{28.3\sqrt{(10)^3}}{\sqrt{\pi(32.5)}} = 88.6 \text{ MPa}$$

Thus catastrophic fracture will occur at a strength-stress ratio of

$$\frac{S_y}{\sigma} = \frac{240}{88.6} = 2.71$$

The designer who ignored the crack and designed to a factor of safety against yielding of

$$n = \frac{S_y}{\sigma} = 3.0$$

would thus experience unexpected and premature failure.   ////

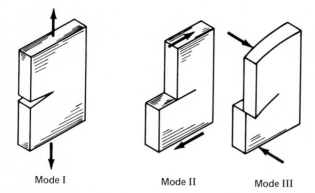

Mode I   Mode II   Mode III

FIGURE 6-35   Deformation modes: Mode I is tension; Modes II and III are both shear modes.

Deformation Modes

The three possible ways of separating a plate are shown in Fig. 6-35. Note that Modes II and III are fundamentally shear modes of fracture, but Mode II stresses and deformation stay within the plane of the plate. Mode III is out-of-plane shear.

Some stress analyses and fewer critical stress-intensity factor measurements have been made for Modes I and III, but they are still limited in scope. The nomenclature K_I for stress-intensity factor and K_{Ic} for critical stress-intensity factor under Mode I conditions are in general use, so for clarity the subscript I will be added here.

When analysis of K and measurements of K_c are made generally available for Modes II and III, then more design can be extended to these modes. The procedure is exactly the same as for Mode I analysis.

Effect of Thickness

In general, increasing the thickness of a part leads to a decrease in K_{Ic}. The value of K_{Ic} becomes asymptotic to a minimum value with increasing thickness; this minimum value is called the *plane strain critical stress-intensity factor*, because deformation in the thickness direction at the tip of a crack is constrained by the surrounding elastic material so that most of the strain occurs in the two directions that lie in the plane of the plate. The test requirements* for measuring K_{Ic} provide for essentially plane strain conditions, so the published values of K_{Ic} are usually plane strain values. Since use of the minimum (plane strain) value of K_{Ic} will lead to a conservative design for thinner parts, the plane strain K_{Ic} is usually used. However, if the designer has available a reliable value of K_{Ic} for the thickness of the part to be designed, this value should be used.

---

* ASTM Standard E 399-72, op. cit.

Flaw Size

In practice, the crack length and location assumed in design are the worst combination of crack size and location, leading to the weakest structure. Thus $2a$ is the longest crack that will not be discovered by the crack detection methods used in manufacture and in service for the part. The location and orientation of this crack or cracks must be selected as the worst conceivable. Sometimes more than one location might be critical, so analysis of the part with cracks in either or both locations must be made.

The failure analyst often has a simpler task if fractography can establish accurately the location and size of the crack that led to the final fracture. It is then a matter of determining K_I as a function of stress (which may not be known) and comparing K_I with K_{Ic} measured for the material to obtain an estimate of the stress at the time of final fracture.

6-16   FRACTURE TOUGHNESS FACTORS

A substantial number of geometries for stress-intensity factors have been compiled in recent years.* Some of these are included here as Figs. 6-36 to 6-41. If K_I is needed for a configuration not included in the literature, the designer's only recourse is to carry out the complete analysis himself. A large body of literature on this subject is summarized in a form useful to the designer,† and typical values of K_{Ic} are listed in Table 6-2.‡

Note carefully in Table 6-2 the general inverse relationship between yield strength and K_{Ic}. This often leads to the choice of a material of lower yield strength and higher K_{Ic}, as is shown in the example that follows.

Example 6-7 A plate of width 1.4 m and length 2.8 m is required to support a tensile force in the 2.8-m direction of 4.0 MN. Inspection procedures will only detect through-thickness edge cracks larger than 2.7 mm. The two Ti-6AL-4V alloys in Table 6-2 are being considered for this application, for which the safety factor must be 1.3 and minimum weight is important. Which alloy should be used?

---

* H. Tada, P. C. Paris, and G. R. Irwin, *The Stress Analysis of Cracks Handbook*, Del Research, Hellertown, Pa., 1973; G. C. M. Sih, *Handbook of Stress Intensity Factors*, Lehigh University, Bethlehem, Pa., 1973; D. P. Rooke and D. J. Cartwright, *Compendium of Stress Intensity Factors*, H.M.S.O., Hillingdon Press, Uxbridge, England, 1976.

† S. T. Rolfe and J. M. Barsom, *Fracture and Fatigue Control in Structures*, Prentice-Hall, Englewood Cliffs, N. J., 1977; R. W. Hertzberg, *Deformation and Fracture Mechanics of Engineering Materials*, Wiley, New York, 1976.

‡ For an extensive compilation of K_c values, see *Damage Tolerant Design Handbook*, MCIC-HB-01, Air Force Materials Laboratory, Wright-Patterson Air Force Base, Ohio, December 1972 and supplements.

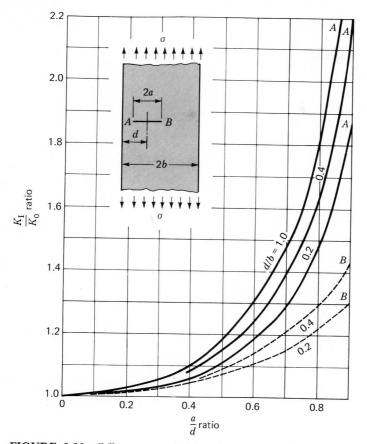

FIGURE 6-36 Off-center crack in a plate in longitudinal tension; solid curves are for the crack tip at A; dashed curves for tip at B.

Solution   (*a*) We elect first to determine the thickness required to resist yielding. Since $\sigma = P/wt$, we have $t = P/w\sigma$.   But

$$\sigma_p = \frac{S_y}{n} = \frac{910}{1.3} = 700 \text{ MPa}$$

Thus

$$t = \frac{P}{w\sigma_p} = \frac{4.0(10)^3}{1.4(700)} = 4.08 \text{ mm}$$ *Ans.*

where we have $S_y = 910$ MPa for the weaker titanium alloy.   For the stronger

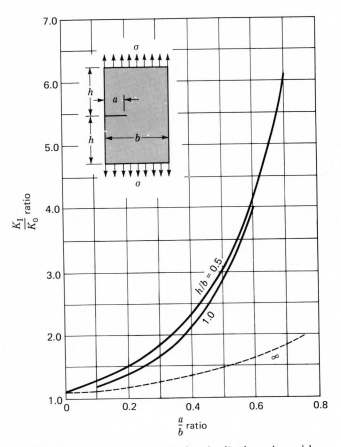

FIGURE 6-37 Plate loaded in longitudinal tension with a crack at the edge; for the solid curve there are no constraints to bending; the dashed curve was obtained with bending constraints added.

alloy we have, from Table 6-2,

$$\sigma_p = \frac{1035}{1.3} = 796 \text{ MPa}$$

and so the thickness is

$$t = \frac{P}{w\sigma_p} = \frac{4.0(10)^3}{1.4(796)} = 3.59 \text{ mm} \qquad\qquad Ans.$$

(*b*) Now let us find the thickness required to prevent crack growth. Using

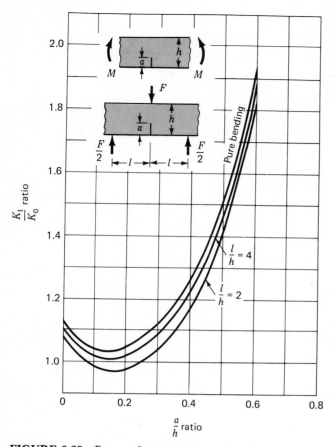

FIGURE 6-38 Beams of rectangular cross section having an edge crack.

Fig. 6-37, we have

$$\frac{h}{b} = \frac{2.8/2}{1.4} = 1 \qquad \frac{a}{b} = \frac{2.7}{1.4(10)^3} = 0.001\ 93$$

Corresponding to these ratios we find from Fig. 6-37 that $K_I/K_0 = 1.1$. Thus $K_I = 1.1\ \sigma\sqrt{\pi a}$. From Table 6-2 we next find $K_{Ic} = 115$ MPa $\sqrt{m}$ for the weaker of the two alloys. The stress at fracture will be

$$\sigma = \frac{K_{Ic}}{1.1\ \sqrt{\pi a}} = \frac{115\ \sqrt{(10)^3}}{1.1\ \sqrt{\pi(2.7)}} = 1135\ \text{MPa}$$

This stress is larger than the yield strength, and so yielding governs the design when the weaker of the two alloys is used.

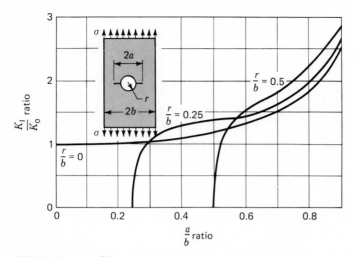

FIGURE 6-39 Plate in tension containing a circular hole with two cracks.

For the stronger alloy we see from Table 6-2 that $K_{Ic} = 55$. Thus

$$\sigma = \frac{K_{Ic}}{1.1\sqrt{\pi a}} = \frac{55\sqrt{(10)^3}}{1.1\sqrt{\pi(2.7)}} = 543 \text{ MPa}$$

Then the permissible stress is $\sigma_p = \sigma/n = 543/1.3 = 418$ MPa. Thus, the required

FIGURE 6-40 A cylinder loaded in axial tension having a radial crack of depth a extending completely around the circumference of the cylinder.

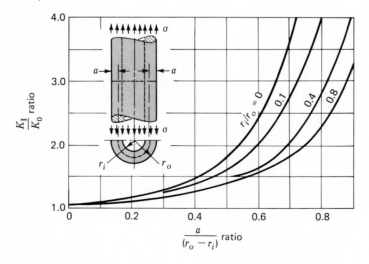

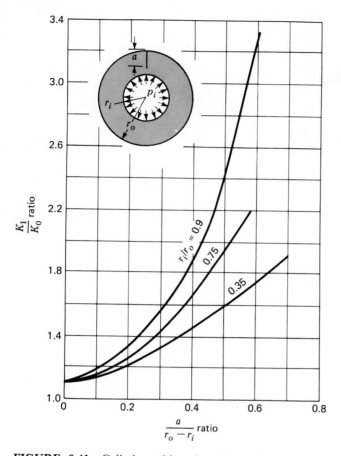

FIGURE 6-41 Cylinder subjected to internal pressure p_i having a radial crack in the longitudinal direction of depth a. Use Eq. (2-56) for the tangential stress at $r = r_o$.

thickness is

$$t = \frac{P}{w\sigma_p} = \frac{4.0(10)^3}{1.4(418)} = 6.84 \text{ mm}$$ *Ans.*

This example shows that the fracture toughness K_{1c} limits the design when the stronger alloy is used and so a thickness of 6.84 mm is required. When a weaker alloy is used, the design is limited by its yield strength, giving a thickness of only 4.08 mm. Thus the weaker alloy leads to a thinner and lighter-weight choice.   ////

Table 6-2   VALUES OF K_{Ic} FOR SOME ENGINEERING MATERIALS

Material		K_{Ic}		Yield strength	
Previous designation	UNS designation	MPa $\sqrt{m}$	kpsi $\sqrt{in}$	MPa	kpsi
Aluminum					
2024-T851	A92024-T851	26	24	455	66
7075-T651	A97075-T651	24	22	495	72
Titanium					
Ti-6AL-4V	R56401	115	105	910	132
Ti-6AL-4V*	R56401*	55	50	1035	150
Steel					
4340	G43400	99	90	860	125
4340*	G43400*	60	55	1515	220
52100	G52986	14	13	2070	300

* Heat-treated to a higher strength.

6-17   STRESS CORROSION

Parts subjected to continuous static loads in certain corrosive environments may, over a period of time, develop serious cracks. This phenomena is known as *stress-corrosion cracking*. Examples of such cracking are door-lock springs, watch springs, lock washers, marine and bridge cables, and other highly stressed parts subject to atmospheric or other corrosive surroundings. The stress, environment, time, and alloy structure of the part all seem to have an influence on the cracking, with each factor speeding up the influence of the other (Fig. 6-6).

Stress-time tests* can be made on specimens in a corrosive environment in order to determine the limiting value of the fracture toughness. The curve shown in Fig. 6-42 typifies the results of many of these experiments. The tests must be run on a number of specimens, each subjected to a constant but different load and each having the same size initial crack. It will then be found that the rate of crack growth depends both upon stress and upon time. When the times to fracture corresponding to each value of K_1 are noted and plotted, a curve like that of Fig. 6-42 will be obtained. The limiting value of the stress-intensity factor is here designated as K'_{Ic}, corresponding to point C on the curve. Crack growth will not be obtained for stress-intensity factors less than this value, no matter how long the loaded specimen remains in the environment. Unfortunately these tests require a great deal of time for completion, usually not less than 1000 h.†

---

* See H. O. Fuchs and R. I. Stephens, *Metal Fatigue in Engineering*, Wiley, New York, 1980, p. 218.

† For some values of the stress-intensity factors K'_{Ic}, see *Damage Tolerant Handbook*, Metals and Ceramics Information Center, Battelle, Columbus, Ohio, 1975.

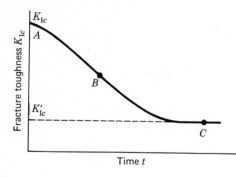

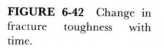

FIGURE 6-42 Change in fracture toughness with time.

PROBLEMS*

Sections 6-1 to 6-7

6-1 A ductile steel has a yield strength of 40 kpsi. Find factors of safety corresponding to failure by the maximum-normal-stress theory, the maximum-shear-stress theory, and the distortion-energy theory, respectively, for each of the following stress states:

   (a) $\sigma_x = 10$ kpsi, $\sigma_y = -4$ kpsi

   (b) $\sigma_x = 10$ kpsi, $\tau_{xy} = 4$ kpsi cw

   (c) $\sigma_x = -2$ kpsi, $\sigma_y = -8$ kpsi, $\tau_{xy} = 4$ kpsi ccw

   (d) $\sigma_x = 10$ kpsi, $\sigma_y = 5$ kpsi, $\tau_{xy} = 1$ kpsi cw

6-2 A machine element is loaded so that $\sigma_1 = 20$ kpsi, $\sigma_2 = 0$ kpsi, and $\sigma_3 = -15$ kpsi; the material has a minimum yield strength in tension and compression of 60 kpsi. Find the factor of safety for each of the following failure theories:

   (a)   Maximum-normal-stress theory

   (b)   Maximum-shear-stress theory

   (c)   Distortion-energy theory

6-3 A machine part is statically loaded and has a yield strength of 350 MPa. For each stress state indicated below, find the factor of safety using each of the three static-failure theories:

   (a) $\sigma_A = 70$ MPa, $\sigma_B = 70$ MPa

   (b) $\sigma_A = 70$ MPa, $\sigma_B = 35$ MPa

   (c) $\sigma_A = 70$ MPa, $\sigma_B = -70$ MPa

   (d) $\sigma_A = -70$ MPa, $\sigma_B = 0$ MPa

6-4 Based on the use of UNS C27000 hard yellow brass rod as the material, find factors of safety for each of the three static-failure theories for the following stress states:

   (a) $\sigma_x = 70$ MPa, $\sigma_y = 30$ MPa

   (b) $\sigma_x = 70$ MPa, $\tau_{xy} = 30$ MPa cw

   (c) $\sigma_x = -10$ MPa, $\sigma_y = -60$ MPa, $\tau_{xy} = 30$ MPa ccw

   (d) $\sigma_x = 50$ MPa, $\sigma_y = 20$ MPa, $\tau_{xy} = 40$ MPa cw

6-5 A force F applied at D near the end of a 15-in lever shown in the figure results in certain stresses in the cantilevered bar $OABC$. The bar $(OABC)$ is made of UNS G10350 steel which is forged, machined, and heat-treated and tempered to 800°F. What force F would cause the cantilevered bar to yield?

---

* The asterisk indicates a problem that may not have a unique result.

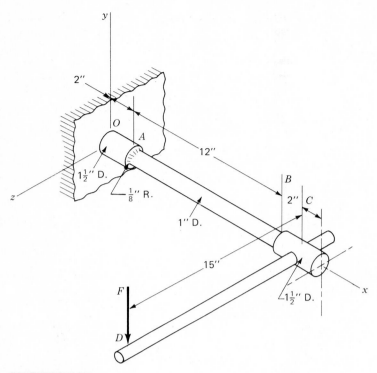

PROBLEM 6-5

6-6 The figure shows a round bar subjected to the vector moment $\mathbf{M} = 1.75\mathbf{i} + 1.10\mathbf{k}$ kN·m. The material is UNS A95056-H38 aluminum alloy. A stress element A located on top of the bar is oriented in the xz plane as shown. Using the stresses on this element, determine the factor of safety guarding against a static failure by using the maximum-shear-stress theory and the distortion-energy theory.

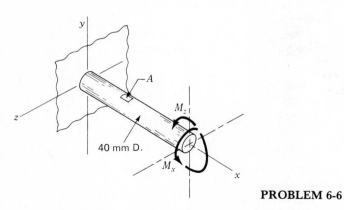

PROBLEM 6-6

6-7 A lever subjected to a downward static force of 400 lb is keyed to a 1-in round bar as shown in the figure.

(*a*) Find the critical stresses in the round bar.

(b) The round bar is made of UNS G46200 steel, heat-treated and drawn to 800°F. Based on static loading, find the factor of safety by using the distortion-energy theory.

(c) As a check on (b), find the factor of safety using the maximum-shear-stress theory. Should the result be greater or less than that obtained in (b)? Why?

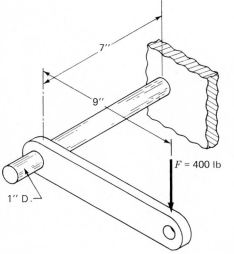

PROBLEM 6-7

6-8 Develop a computer program or calculator program to find the diameter of a shaft subject to static bending and torsion using the maximum-shear-stress theory of failure. Given data are the allowable bending and torsional moments, the yield strength, and the strength factor of safety.

6-9 The same as Prob. 6-8, but use the distortion-energy theory.

6-10* The cantilevered round bar of Prob. 6-7 is to be made of UNS G10150 cold-drawn seamless tubing. Use Table A-10 to find a suitable size with a load safety factor of 1.80 and a strength factor of 1.30. The yield strength is 47 kpsi.

6-11 The figure shows a cantilevered tube to be made of UNS A92014-T4 aluminum alloy. We wish to find a set of cross-section dimensions for the tube based upon a bending load $F = 0.80$ kN, an axial tension $P = 7.2$ kN, and a torsional load $T = 38$

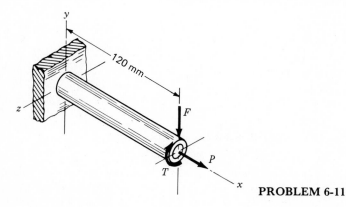

PROBLEM 6-11

$N \cdot m$. The load factors of safety are to be $n_{LF} = 2.20$, $n_{LP} = 1.30$, and $n_{LT} = 1.90$. Use a strength safety factor $n_S = 1.50$.

6-12 A UNS A92024-T3 aluminum tube has a 3-in OD and a 0.049-in wall. It is subjected to an internal pressure of 1200 psi. Find the factor of safety guarding against yielding by each of the three theories for ductile materials.

6-13 A thin-walled pressure vessel is made of UNS A93003-H14 aluminum alloy tubing. The vessel has an OD of 60 mm and a wall thickness of 1.50 mm. What internal pressure would cause the material to yield?

6-14* A thick-walled cylinder is to have an inside diameter of 0.500 in; it will be made of UNS G41400 cold-drawn steel, and it must resist an internal pressure of 5 kpsi based on a factor of safety of at least 4. Specify a satisfactory outside diameter, basing your decision on yielding as predicted by the maximum-shear-stress theory.

6-15 A $1\frac{1}{2}$-in-diameter UNS G10350 cold-drawn steel shaft has a forged gear mounted on it with a class FN 4 fit. This fit has a mean interference of 0.0023 in. The gear hub is 3 in long and $2\frac{1}{2}$ in in diameter. Find the critical value of the von Mises stress based on a mean fit.

6-16 A gun barrel is assembled by shrinking an outer barrel over an inner barrel so that the maximum principal stress is 70 percent of the yield strength of the material. The material of both members is steel, $S_y = 78$ kpsi, $E = 30$ Mpsi, and $\mu = 0.292$. The nominal radii of the barrels are $\frac{3}{16}$, $\frac{3}{8}$, and $\frac{9}{16}$ in. Calculate the critical von Mises stress for both members.

6-17 Suppose the gun of Prob. 6-16 is fired with an internal pressure of 40 kpsi. What are the new values of the critical von Mises stress?

Section 6-8

6-18 Using typical values for the strengths of ASTM No. 40 cast iron, find the strength safety factors corresponding to fracture by the maximum-normal-stress theory, the Coulomb-Mohr theory, and the modified Mohr theory, respectively, for each of the following stress states:
(a) $\sigma_x = 10$ kpsi, $\sigma_y = -4$ kpsi
(b) $\sigma_x = 10$ kpsi, $\tau_{xy} = 4$ kpsi cw
(c) $\sigma_x = -2$ kpsi, $\sigma_y = -8$ kpsi, $\tau_{xy} = 4$ kpsi ccw
(d) $\sigma_x = 10$ kpsi, $\sigma_y = -30$ kpsi, $\tau_{xy} = 10$ kpsi cw

6-19 Tests on a particular melt of ASTM No. 20 cast iron gave $S_{ut} = 150$ MPa and $S_{uc} = 600$ MPa. Find the strength factor of safety for each of the three failure theories for brittle materials for the following stress states:
(a) $\sigma_x = 50$ MPa, $\tau_{xy} = 30$ MPa cw
(b) $\sigma_x = -80$ MPa, $\sigma_y = -40$ MPa, $\tau_{xy} = 20$ MPa ccw
(c) $\sigma_x = 40$ MPa, $\sigma_y = 30$ MPa, $\tau_{xy} = 10$ MPa ccw
(d) $\sigma_x = 30$ MPa, $\sigma_y = -60$ MPa, $\tau_{xy} = 30$ MPa cw

6-20 Due to a heavy shrink fit, a hollow ASTM No. 40 cast-iron member, having a 1-in-diameter hole and a $1\frac{1}{2}$-in OD, is subjected to an external pressure of 38 kpsi. Has the part failed?

Sections 6-15 and 6-16

6-21* Determine the dimensions to the nearest $\frac{1}{8}$ in of the rectangular steel cantilever spring shown in the figure. The spring is to be designed to have a spring rate of about 100 lb/in. The material has $S_y = 180$ kpsi, $E = 30$ Mpsi, and $K_{Ic} = 62$

kpsi $\sqrt{\text{in.}}$ Use $a = \frac{1}{16}$ in as the smallest visible crack depth and in the worst location on the beam.   Compare the maximum spring deflections under the load F based on crack propagation with the deflection based on yielding if the overall factor of safety is to be 3.00.

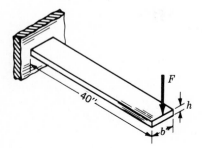

PROBLEM 6-21 $b = 6h.$

6-22   A nominal 3-in aluminum pipe is 3.500 in OD by 3.068 in ID.   It is proposed to use the alloy A92024-T851 for such a pipe.   Use an overall factor of safety of 2.5 and a crack depth of $\frac{1}{16}$ in and find the internal pressure such a pipe can carry based on yielding and based on crack propagation.

CHAPTER
7

DESIGN FOR
FATIGUE
STRENGTH

In Chap. 6 we were concerned with design and analysis of parts subjected to static loads. It is an entirely different matter when the parts are subjected to time-varying, or nonstatic, loads. Thus in this chapter, we want to learn how parts fail under nonstatic conditions and how to design them to safely resist such conditions.

7-1 INTRODUCTION

In obtaining the properties of materials relating to the stress-strain diagram, the load is applied gradually, giving sufficient time for the strain to develop. With the usual conditions, the specimen is tested to destruction so that the stresses are applied only once. These conditions are known as *static conditions* and are closely approximated in many structural and machine members.

The condition frequently arises, however, in which the stresses vary or fluctuate between values. For example, a particular fiber on the surface of a rotating shaft, subjected to the action of bending loads, undergoes both tension and compression for each revolution of the shaft. If the shaft is a part of an electric motor rotating at 1725 rpm, the fiber is stressed in tension and in compression 1725 times each minute. If, in addition, the shaft is also axially loaded (caused, for example, by a helical or worm gear), an axial component of stress is superimposed upon the bending component. This results in a stress, in any one fiber, which is still fluctuating but which is fluctuating between different values. These and other kinds of loads occurring in machine members produce stresses which are called *repeated*, *alternating*, or *fluctuating stresses*.

Machine members are often found to have failed under the action of repeated or fluctuating stresses, and yet the most careful analysis reveals that the actual maximum stresses were below the ultimate strength of the material and quite frequently even below the yield strength. The most distinguishing characteristic of these failures has been that the stresses have been repeated a very large number of times. Hence the failure is called a *fatigue failure*.

A fatigue failure begins with a small crack. The initial crack is so minute that it cannot be detected by the naked eye and is even quite difficult to locate in a Magnaflux or x-ray inspection. The crack will develop at a point of discontinuity in the material, such as a change in cross section, a keyway, or a hole. Less obvious points at which fatigue failures are likely to begin are inspection or stamp marks, internal cracks, or even irregularities caused by machining. Once a crack has developed, the stress-concentration effect becomes greater and the crack progresses more rapidly. As the stressed area decreases in size, the stress increases in magnitude until, finally, the remaining area fails suddenly. A fatigue failure, therefore, is characterized by two distinct areas of failure (Fig. 7-1). The first of

FIGURE 7-1 A fatigue failure of a $7\frac{1}{2}$-in-diameter forging at a press fit. The specimen is UNS G10450 steel, normalized and tempered, and has been subjected to rotating bending. (*Courtesy of The Timken Company*).

these is that due to the progressive development of the crack, while the second is due to the sudden fracture. The zone of the sudden fracture is very similar in appearance to the fracture of a brittle material, such as cast iron, which has failed in tension.

When machine parts fail statically, they usually develop a very large deflection, because the stress has exceeded the yield strength, and the part is replaced before fracture actually occurs. Thus many static failures are visible ones and give warning in advance. But a fatigue failure gives no warning; it is sudden and total, and hence dangerous. It is a relatively simple matter to design against a static failure because our knowledge is quite complete. But fatigue is a much more complicated phenomenon, only partially understood, and the engineer seeking to rise to the top of the profession must acquire as much knowledge of the subject as possible. Anyone who lacks knowledge of fatigue can double or triple factors of safety and get a design that will not fail. But such designs will not compete in today's market, and neither will the engineers who produce them.

7-2 THE *S-N* DIAGRAM

To determine the strength of materials under the action of fatigue loads, specimens are subjected to repeated or varying forces of specified magnitudes while the cycles or stress reversals are counted to destruction. The most widely used fatigue-testing device is the R. R. Moore high-speed rotating-beam machine. This machine subjects the specimen to pure bending (no transverse shear) by means of weights. The specimen, shown in Fig. 7-2, is very carefully machined and polished, with a final polishing in an axial direction to avoid circumferential scratches. Other fatigue-testing machines are available for applying fluctuating or reversed axial stresses, torsional stresses, or combined stresses to the test specimens.

To establish the fatigue strength of a material, quite a number of tests are necessary because of the statistical nature of fatigue. For the rotating-beam test a constant bending load is applied, and the number of revolutions (stress reversals) of the beam required for failure is recorded. The first test is made at a stress which is somewhat under the ultimate strength of the material. The second test is made with a stress which is less than that used in the first. This process is continued, and the results plotted as an *S-N* diagram (Fig. 7-3). This chart may be plotted on semilog paper or on log-log paper. In the case of ferrous metals and

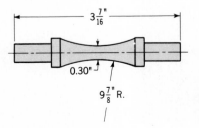

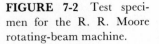

FIGURE 7-2 Test specimen for the R. R. Moore rotating-beam machine.

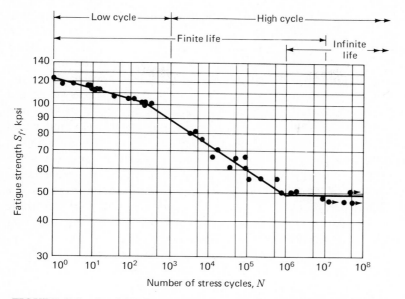

FIGURE 7-3 An *S-N* diagram plotted from the results of completely reversed axial fatigue tests. Material: UNS G41300 steel normalized; $S_{ut} = 116$ kpsi; maximum $S_{ut} = 125$ kpsi. (*Data from NACA Technical Note 3866, December 1966.*)

alloys, the graph becomes horizontal after the material has been stressed for a certain number of cycles. Plotting on log paper emphasizes the bend in the curve, which might not be apparent if the results were plotted by using cartesian coordinates.

The ordinate of the *S-N* diagram is called the *fatigue strength* S_f; a statement of this strength must always be accompanied by a statement of the number of cycles *N* to which it corresponds.

Soon we shall learn that these *S-N* diagrams can be determined either for the actual test specimen or for an actual mechanical element. Even when the material of the test specimen and that of the mechanical element are identical, there will be significant differences between the two diagrams.

In the case of the steels, a knee occurs in the graph, and beyond this knee failure will not occur, no matter how great the number of cycles. The strength corresponding to the knee is called the *endurance limit* S_e, or the *fatigue limit*. The graph of Fig. 7-3 never does become horizontal for nonferrous metals and alloys, and hence these materials do not have an endurance limit.

We note that a stress cycle ($N = 1$) constitutes a single application and removal of a load and then another application and removal of the load in the opposite direction. Thus $N = \frac{1}{2}$ means the load is applied once and then removed, which is the case with the simple tension test.

The body of knowledge available on fatigue failure from $N = \frac{1}{2}$ to $N = 1000$ cycles is generally classified as *low-cycle fatigue*, as indicated in Fig. 7-3. *High-cycle fatigue*, then, is concerned with failure corresponding to stress cycles greater than 10^3 cycles.

We also distinguish a *finite-life region* and an *infinite-life region* in Fig. 7-3. The boundary between these regions cannot be clearly defined except for a specific material; but it lies somewhere between 10^6 and 10^7 cycles for steels, as shown in Fig. 7-3.

As noted previously, it is always good engineering to conduct a testing program on the materials to be employed in design and manufacture. This, in fact, is a requirement, not an option, in guarding against the possibility of a fatigue failure. *Because of this necessity for testing it would really be unnecessary for us to proceed any further in the study of fatigue failure except for one important reason: the desire to know why fatigue failures occur so that the most effective method or methods can be used to improve fatigue strength.* Thus our primary purpose in studying fatigue is to understand why failures occur so that we can guard against them in an optimum manner. For this reason, the analytical and design approaches presented in this book, or in any other book for that matter, do not yield absolutely precise results. The results should be taken as a guide, as something which indicates what is important and what is not important in designing against fatigue failure.

The methods of fatigue-failure analysis represent a combination of engineering and science. Often science fails to provide the answers which are needed. But the airplane must still be made to fly—safely. And the automobile must be manufactured with a reliability that will ensure a long and trouble-free life and at the same time produce profits for the stockholders of the industry. Thus, while science has not yet completely explained the actual mechanism of fatigue, the engineer must still design things that will not fail. In a sense this is a classic example of the true meaning of engineering as contrasted with science. Engineers use science to solve their problems *if* the science is available. But available or not, the problem must be solved, and whatever form the solution takes under these conditions is called engineering.

The determination of endurance limits by fatigue testing is quite lengthy and expensive. For preliminary and prototype design especially, but for some failure analysis as well, a quick method of estimating the endurance limits is needed. There are great quantities of data in the literature on the results of rotating-beam tests and simple tension tests. By plotting these as in Fig. 7-4 it is possible to see whether there is any correlation between the two sets of results. The graph appears to suggest that the endurance limit ranges from about 35 to 60 percent of the tensile strength for steels up to about $S_{ut} = 200$ kpsi (1400 MPa). Beginning at about $S_{ut} = 200$ kpsi the scatter appears to increase, but the trend seems to level off, as suggested by the dashed horizontal line at $S'_e = 100$ kpsi.

Another series of tests for various microstructures is shown in Table 7-1. In this table the endurance limits vary from about 23 to 63 percent of the tensile strength.*

---

* But see H. O. Fuchs and R. I. Stephens, *Metal Fatigue in Engineering*, Wiley, New York, 1980, pp. 69–71, which reports a range of 35 to 60 percent for steels having $S_{ut} < 1400$ MPa and as low as 20 percent for high-strength steels.

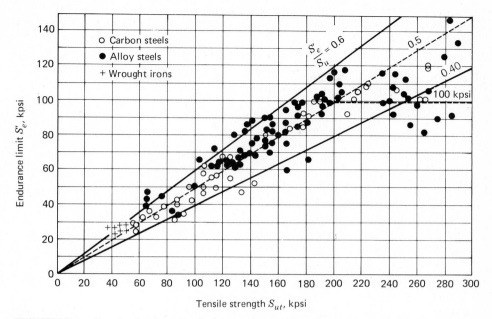

FIGURE 7-4 Graph of endurance limits versus tensile strengths from actual test results for a large number of wrought irons and steels. Ratios of S'_e/S_{ut} = 0.60, 0.50, and 0.40 are shown by the solid and dashed lines. Note also the horizontal dashed line for $S'_e = 100$ kpsi. Points shown having a tensile strength greater than 200 kpsi have a mean endurance limit $\bar{S}_e = 107$ kpsi and a standard deviation of 13.5 kpsi. (*Collated from data compiled by H. J. Grover, S. A. Gordon, and L. R. Jackson in Fatigue of Metals and Structures, Bureau of Naval Weapons Document NAVWEPS 00-25-534, 1960 rev.; and from Fatigue Design Handbook, SAE, 1968, p. 42.*)

Now it is important here to observe that the dispersion of the endurance limit is *not* due to a dispersion, or spread, in the tensile strengths of the specimen. This spread will occur even when the tensile strengths of a large number of specimens remain exactly the same.

Students who are solving problems for practice purposes need a uniform approach for estimating endurance limits so that all members of the class will obtain the same "correct" answer. It is for this reason that we decide on the

Table 7-1 ENDURANCE-LIMIT RATIO S'_e/S_{ut} FOR VARIOUS STEEL MICRO-STRUCTURES

	Ferrite		Pearlite		Martensite	
	Range	**Average**	**Range**	**Average**	**Range**	**Average**
Carbon steel	0.57–.63	0.60	0.38–0.41	0.40	...	0.25
Alloy steel	...	...	...	...	0.23–0.47	0.35

Source: Adapted from L. Sors, *Fatigue Design of Machine Components*, Pergamon Press, Oxford, England, 1971.

following relation (*for students only*) for estimating the *mean endurance limit* $\bar{S}'_e$ of the rotating-beam specimen:

$$\bar{S}'_e = 0.5S_{ut} \qquad\qquad S_{ut} \leq 200 \text{ kpsi } (1400 \text{ MPa})$$

$$\text{(7-1)}$$

$$\bar{S}'_e = 100 \text{ kpsi } (700 \text{ MPa}) \qquad S_{ut} > 200 \text{ kpsi } (1400 \text{ MPa})$$

Practicing engineers have several options available. If the cost of the project justifies it, experimental procedures should be employed to obtain the mean endurance limit and the standard deviation. This is good sound engineering practice and should be employed whenever possible. A second approach for practicing engineers is to use Eq. (7-1) but allow a generous standard deviation, say 15 percent. (See Sec. 7-8.) A third approach would be to use the line $S'_e/S_{ut} = 0.40$ in Fig. 7-4 and then to use $S'_e = 80$ kpsi when $S_{ut} > 200$ kpsi.

The prime mark on $\bar{S}'_e$ in Eq. (7-1) refers to the rotating-beam specimen itself, because we wish to reserve the symbol S_e for the endurance limit of a particular machine element. Soon we shall learn that these two strengths may be quite different. The overhead bar on the symbol $\bar{S}'_e$ is intended to draw particular attention to the fact that the strength being specified is a *mean value*, and hence actual results may vary in either direction from this mean value.

The data of Table 7-1 emphasize the difficulty of attempting to provide a single rule for deriving the endurance limit from the tensile strength. The table also shows a part of the cause of this difficulty. The table shows that steels treated to give various microstructures have various S'_e/S_{ut} ratios. In general the more ductile microstructures have a higher ratio. Martensite has a very brittle nature and is highly susceptible to fatigue-induced cracking; thus the ratio is low. When designs include detailed heat-treating specifications to obtain specific microstructures, it is possible to use a better estimate of the endurance limit based on test data; such estimates are much more reliable and indeed should be used.

A graph like that of Fig. 7-4 can also be plotted for the cast irons and cast steels. When this is done, we find the following approximations to be appropriate *for student use only*:

$$\bar{S}'_e = 0.45S_{ut} \qquad\qquad S_{ut} \leq 88 \text{ kpsi } (600 \text{ MPa})$$

$$\text{(7-2)}$$

$$\bar{S}'_e = 40 \text{ kpsi } (275 \text{ MPa}) \qquad S_{ut} > 88 \text{ kpsi } (600 \text{ MPa})$$

Practicing engineers should use actual test data. Equation (7-2) is valid for both cast iron and cast steel. The results differ only slightly from the values shown in Table A-21.

Processors of aluminum and magnesium alloys publish very complete tabulations of the properties of these materials, including the fatigue strengths, which ordinarily run from about 30 to 40 percent of the tensile strength, depending upon whether the material is cast or wrought. These materials do not have an endurance limit, and the fatigue strength is usually based on 10^8 or $5(10)^8$ cycles of stress reversal.

The *S-N* curves for plastics have a continually decreasing fatigue strength with life N, as do most aluminums. Because of this and because of the diversity of compositions of plastics and of their composite structural uses, it is necessary to develop test data for their *S-N* curves or to obtain data from the manufacturer or vendor. The ratio S_f'/S_{ut} for lives of 10^7 cycles or more varies from 0.18 to 0.43 for various plastics and composite structures.

Having now outlined methods of assessing the mean fatigue properties of various materials, our next problem is to develop guidelines for the scatter in such data. We can do this by finding the standard deviation of the endurance limit.

As indicated by the data displayed in Table 7-2, the standard deviation of the endurance limit varies from 4 to 10 percent. Using reasonable care, inspection, and quality-control techniques, it is probable that a standard deviation of 8 percent of the endurance limit can be held, and in this book we shall use this figure in our reliability analyses.

The bar on the symbol $\bar{S}_e$, to indicate the mean, is an inconvenient notation which we shall now abandon. Instead we shall use S_e' for the endurance limit of the rotating-beam specimen corresponding to a reliability of 50 percent. That is, this shall indicate the mean endurance limit. Reliabilities different from 50 percent will be accounted for later.

Table 7-2 STANDARD DEVIATION OF THE ENDURANCE LIMIT*

Material† (UNS No.)	Tensile strength		Endurance limit		Standard deviation	
	MPa	kpsi	MPa	kpsi	kpsi	%
G41300 steel (axial)	730	106	276	40	1.1	2.7
G43400 steel (axial)‡	799	116	338	49	4.4	9.1
G43400 steel	965	140	489	71	3.5	4.9
	1310	190	586	85	6.7	7.8
	1580	230	620	90	5.3	5.9
	1790	260	668	97	6.3	6.5
G43500 steel	2070	300	689	100	4.4	4.4
R50001-series titanium alloy	1000	145	579	84	5.4	6.4
A97076 aluminum alloy	524	76	186	27	1.6	6.0
C63000 aluminum bronze	806	117	331	48	4.5	9.4
C17200 beryllium copper	1210	175	248	36	2.7	7.5

* Reported by F. B. Stulen, H. N. Cummings, and W. C. Schulte, "Preventing Fatigue Failures, part 5," *Machine Design*, vol. 33, June 22, 1961 p. 161.

† Alloys are heat-treated, hot-worked; specimens are smooth, subjected to long-life rotating-beam tests except as noted.

‡ Reported from axial fatigue tests by D. B. Kececioglu and L. B. Chester, "Combined Axial Stress Fatigue Reliability for AISI 4130 and 4340 Steels," ASME Paper, No. 75-WA/DE-17.

7-3   LOW-CYCLE FATIGUE*

In this section we shall study the fatigue resistance of test specimens and machine elements when they are subjected to stress reversals up to about 10^3 cycles. A knowledge of fatigue resistance in this low-cycle region is desirable for the design of short-lived devices, such as missiles, and for the design of other machines when the possibility of some very large overloads during the lifetime of the machine must be considered.

A fatigue failure almost always begins at a local discontinuity such as a notch, crack, or other area of high stress concentration. When the stress at the discontinuity exceeds the elastic limit, plastic strain occurs. If a fatigue fracture is to occur, there must exist cyclic plastic strains. Thus, in the study of low-cycle fatigue it is necessary to investigate the behavior of materials subject to cyclic deformation.

In 1910 Bairstow verified by experiment Bauschinger's theory that the elastic limits of iron and steel can be changed, either up or down, by the cyclic variations of stress.[†] In general, the elastic limits of annealed steels are likely to increase when subjected to cycles of stress reversals, while cold-drawn steels exhibit a decreasing elastic limit.

Test specimens subjected to reversed bending are not suitable for strain cycling because of the difficulty of measuring plastic strains. Consequently, most of the research has been done using axial specimens. By using electrical transducers it is possible to generate signals that are proportional to the stress and strain, respectively.[‡] These signals can then be displayed on an oscilloscope or plotted on an XY plotter. R. W. Landgraf has investigated the low-cycle fatigue behavior of a large number of very high strength steels and during his research he made many cyclic stress-strain plots.[§] Figure 7-5 has been constructed to show the general appearance of these plots for the first few cycles of controlled cyclic strain. In this case the strength decreases with stress repetitions, as evidenced by the fact that the reversals occur at ever smaller stress levels. As previously noted, other materials may be strengthened, instead, by cyclic stress reversals.

Slightly different results may be obtained if the first reversal occurs in the compressive region; this is probably due to the fatigue-strengthening effect of compression.

---

* An understanding of the balance of the chapter is not dependent on a knowledge of this section.

† L. Bairstow, "The Elastic Limits of Iron and Steel Under Cyclic Variations of Stress," *Philosophical Transactions* Series A, vol. 210, Royal Society of London, 1910, pp. 35–55.

‡ Methods of dynamic force and strain measurements may be found in Joseph E. Shigley and John J. Uicker, Jr., *Theory of Machines and Mechanisms*, McGraw-Hill, New York, 1980, pp. 545–550.

§ R. W. Landgraf, *Cyclic Deformation and Fatigue Behavior of Hardened Steels*, Report no. 320, Department of Theoretical and Applied Mechanics, University of Illinois, Urbana, 1968, pp. 84-90.

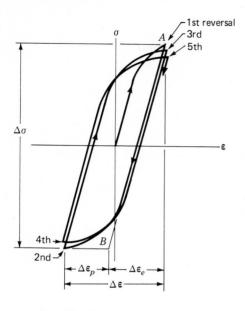

FIGURE 7-5 True stress, true strain hysteresis loops showing first five stress reversals of a cyclic softening material. The graph is slightly exaggerated for clarity. Note that the slope of line AB is the modulus of elasticity E. The stress range is $\Delta\sigma$; $\Delta\varepsilon_p$ is the plastic strain range and $\Delta\varepsilon_e$ the elastic strain range. The total strain range is $\Delta\varepsilon = \Delta\varepsilon_p + \Delta\varepsilon_e$.

Landgraf's paper contains a number of plots that compare the monotonic stress-strain relations in both tension and compression with the cyclic stress-strain curve.* Two of these have been redrawn and are shown in Fig. 7-6. The importance of these is that they emphasize the difficulty of attempting to predict the fatigue strength of a material from known values of monotonic yield or ultimate strengths in the low-cycle region.

The SAE Fatigue Design and Evaluation Steering Committee released a report in 1975 in which the life in reversals to failure is related to the strain amplitude.† The report contains a plot of this relationship for SAE 1020 hot-rolled steel; the graph has been reproduced as Fig. 7-7. To explain the graph we first define the following terms:

Fatigue ductility coefficient ε_F' is the true strain corresponding to fracture in one reversal (point A in Fig. 7-5). The plastic strain line begins at this point in Fig. 7-7.

Fatigue strength coefficient σ_F' is the true stress corresponding to fracture in one reversal (point A in Fig. 7-5). Note in Fig. 7-7 that the elastic strain line begins at σ_F'/E.

Fatigue ductility exponent c is the slope of the plastic strain line in Fig. 7-7 and is the power to which the life $2N$ must be raised to be proportional to the true plastic strain amplitude.

Fatigue strength exponent b is the slope of the elastic strain line, and is the power to which the life $2N$ must be raised to be proportional to the true stress amplitude.

Now, from Fig. 7-5, we see that the total strain is the sum of the elastic and

---

* Ibid., pp. 58–62.

† *Technical Report on Fatigue Properties*, SAE J1099, 1975.

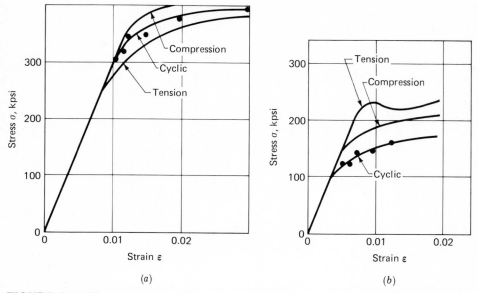

FIGURE 7-6 Monotonic and cyclic stress-strain results.   (*a*) Ausformed H-11 steel, 660 Bhn; (*b*) SAE 4142 steel, 400 Bhn.

plastic components. Therefore the total strain amplitude is

$$\frac{\Delta\varepsilon}{2} = \frac{\Delta\varepsilon_e}{2} + \frac{\Delta\varepsilon_p}{2} \tag{a}$$

The equation of the plastic strain line in Fig. 7-7 is

$$\frac{\Delta\varepsilon_p}{2} = \varepsilon_F'(2N)^c \tag{7-3}$$

The equation of the elastic strain line is

$$\frac{\Delta\varepsilon_e}{2} = \frac{\sigma_F'}{E}(2N)^b \tag{7-4}$$

Therefore, from Eq. (*a*), we have for the total strain amplitude

$$\frac{\Delta\varepsilon}{2} = \frac{\sigma_F'}{E}(2N)^b + \varepsilon_F'(2N)^c \tag{7-5}$$

which is the Manson-Coffin relationship between fatigue life and total strain.[*] Some values of the coefficients and exponents are listed in Table 7-3. Many more are included in the SAE J1099 report.

---

[*] J. F. Tavernelli and L. F. Coffin Jr., "Experimental Support for Generalized Equation Predicting Low Cycle Fatigue," and S. S. Manson, discussion, *Trans. ASME, J. Basic Eng.*, vol. 84, no. 4, pp. 533–537.

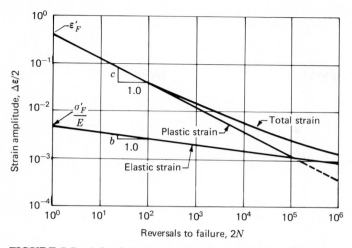

FIGURE 7-7 A log-log plot showing how the fatigue life is related to the true strain amplitude for hot-rolled SAE 1020 steel. (*Reproduced from Tech. Report SAE J 1099, by permission.*)

Manson* has proposed a simplification to Eq. (7-5) which Fuchs and Stephens† recommend for first approximations. The equation is

$$\Delta\varepsilon = \frac{3.5S_u}{EN^{0.12}} + \left(\frac{\varepsilon_F}{N}\right)^{0.6} \tag{7-6}$$

which Manson calls the method of universal slopes. The equation does simplify the job of estimating fatigue life, or failure, because the three terms S_u, ε_F, and E can all be obtained from the simple monotonic tension test.

Though Eq. (7-5) is a perfectly legitimate equation for obtaining the fatigue life of a part when the strain and other cyclic characteristics are given, it appears to be of little use to the designer. The question of how to determine the total strain at the bottom of a notch or discontinuity has not been answered. There are no tables or charts of strain concentration factors in the literature. It is possible that strain concentration factors will become available in research literature very soon because of the increase in the use of finite-element analysis. Moreover, finite-element analysis can of itself approximate the strains that will occur at all points in the subject structure. Until these appear, or until finite-element analysis becomes an everyday tool for design engineers, another, very empirical approach can be employed.‡ If we rewrite Eq. (4-5) in terms of the cyclic stress-strain

* S. S. Manson, "Fatigue: A Complex Subject—Some Simple Approximations," *Exp. Mech.*, vol. 5, no. 7, July 1965, p. 193.

† Fuchs and Stephens, op. cit., p. 78.

‡ See Eqs. (12), (13), and (14) in SAE J1099 *Technical Report*, op. cit.

Table 7-3 CYCLIC PROPERTIES OF SOME HIGH-STRENGTH STEELS*

AISI number	Processing	Brinell hardness H_B	Cyclic yield strength S_y', kpsi	Fatigue strength coefficient σ_F', kpsi	Fatigue ductility coefficient ε_F'	Fatigue strength exponent b	Fatigue ductility exponent c	Fatigue strain-hardening exponent n'
1045	Q & T 80°F	705	...	310	...	−0.065	−1.0	0.10
1045	Q & T 360°F	595	250	395	0.07	−0.055	−0.60	0.13
1045	Q & T 500°F	500	185	330	0.25	−0.08	−0.68	0.12
1045	Q & T 600°F	450	140	260	0.35	−0.07	−0.69	0.12
1045	Q & T 720°F	390	110	230	0.45	−0.074	−0.68	0.14
4142	Q & T 80°F	670	300	375	...	−0.075	−1.0	0.05
4142	Q & T 400°F	560	250	385	0.07	−0.076	−0.76	0.11
4142	Q & T 600°F	475	195	315	0.09	−0.081	−0.66	0.14
4142	Q & T 700°F	450	155	290	0.40	−0.080	−0.73	0.12
4142	Q & T 840°F	380	120	265	0.45	−0.080	−0.75	0.14
4142†	Q & D 550°F	475	160	300	0.20	−0.082	−0.77	0.12
4142	Q & D 650°F	450	155	305	0.60	−0.090	−0.76	0.13
4142	Q & D 800°F	400	130	275	0.50	−0.090	−0.75	0.14

* The steels listed in this table are the same as those in Table A-18.
† Deformed 14 percent.

Source: Data from R. W. Landgraf, *Cyclic Deformation and Fatigue Behavior of Hardened Steels*, Report no. 320, Department of Theoretical and Applied Mechanics, University of Illinois, Urbana, 1968.

relations, we have

$$\sigma_a = \frac{\Delta\sigma}{2} = K' \left(\frac{\Delta\varepsilon_p}{2}\right)^{n'} \tag{7-7}$$

where σ_a is the *true stress amplitude*, K' is the *cyclic strength coefficient*, and n' is the *cyclic strain-hardening exponent*. If we solve Eq. (7-7) for the amplitude of the plastic strain, we have

$$\frac{\Delta\varepsilon_p}{2} = \left(\frac{\sigma_a}{K'}\right)^{1/n'} \tag{b}$$

The elastic strain amplitude can be written as

$$\frac{\Delta\varepsilon_e}{2} = \frac{\sigma_a}{E} \tag{c}$$

Substituting Eqs. (b) and (c) into (a) gives

$$\frac{\Delta\varepsilon}{2} = \frac{\sigma_a}{E} + \left(\frac{\sigma_a}{K'}\right)^{1/n'} \tag{7-8}$$

Using an iteration technique, such as the one suggested in Sec. 1-15, Eq. (7-8) can be solved with Eq. (7-5) for the stress amplitude σ_a when the desired life is specified. You are warned, however, that Eq. (7-7) is an idealization of the cyclic stress-strain curve in the plastic region and may be highly approximate. Therefore a great deal of caution should be observed in using this approach.

If the value of the strength coefficient K' is not available, it can be obtained from the equation

$$K' = \frac{\sigma_F'}{(\varepsilon_F')^{n'}} \tag{7-9}$$

where σ_F' and ε_F' have been substituted for σ_a and $\Delta\varepsilon_p/2$, respectively, in Eq. (7-7).

7-4 HIGH-CYCLE FATIGUE

As indicated in Fig. 7-3, high-cycle fatigue is the region beyond $N = 10^3$ cycles of stress. In many cases the possibility of yielding will govern design decisions at the lower end of this region; yet we must always be alert to the possibility of a fatigue failure.

The purpose of this section is to develop methods of approximating the *S-N* diagram for steels when the results of the simple tension test are known. With sufficient data available, a similar approach may be justified for other materials.

In this section we shall use a log S-log N coordinate system similar to that shown in Fig. 7-3. It is true, however, that a semilog system using a linear S vs log N may have a better statistical fit.*

We have already learned [Eq. (7-1)] that a good estimate of the mean endurance limit for steels is either half of the tensile strength or 100 kpsi (700 MPa), whichever is the least, at 10^6 cycles. In order to get an approximation of the S-N diagram using only results obtained from the simple tension test, we now require an estimate of the mean fatigue strength at 10^3 cycles. Figures 7-8 and 7-9 will help in arriving at this estimate.

In Fig. 7-8 the data are plotted using an abscissa scaled to the number of stress reversals to failure $2N_f$, because a complete cycle consists of two stress reversals. Thus the origin at $2N_f = 10^0$ corresponds to one stress reversal which is the same as $\frac{1}{2}$ cycle. So it is logical to begin the S-N lines in Fig. 7-8 with the true fracture stress at $N_f = \frac{1}{2}$ cycle. However, we are only interested now in obtaining the fatigue strength at $N = 1000$ cycles ($2N_f = 2000$). When we do this for all seven S-N lines and divide the results by the tensile strengths, we obtain the following strength ratios: 0.89, 0.76, 0.75, 0.76, 0.79, 0.79, and 0.83. The mean is found to be $S_f' = 0.80S_u$. Note that this is the same as the mean found by extrapolation of the test results shown in Fig. 7-9.

We also observe that the dispersions of the strengths appear to be constant over the life range. In an earlier edition of this book it was thought that there was less scatter at 10^3 cycles than at 10^6, but Fig. 7-9 seems to refute this observation.

In accordance with the observations of Figs. 7-8 and 7-9 and Eq. (7-1), *it is recommended that a line on the log S-log N chart joining $0.8S_{ut}$ at 10^3 cycles and S_e' at 10^6 cycles be used to define the mean fatigue strength S_f' corresponding to any life N.*

Practicing engineers should employ this recommendation only when it is verified by extensive testing. In the absence of such tests it might be prudent to substitute a minimum strength line for the mean strength. Based on the data presented here such a line would pass through $S_f' = 0.65S_{ut}$ at 10^3 cycles and $S_e' = 0.35S_{ut}$ at 10^6 cycles. In addition, remember that some steels may not even possess a fatigue limit.

An easy way to obtain the fatigue strength S_f' corresponding to a given number of cycles N is to plot the S-N diagram on 2×3 log-log paper. The values are then easy to read off. A disadvantage of this approach is that the slope of the S-N line is so small that accurate results are difficult to obtain.

The use of log-log paper can be avoided if a scientific calculator is at hand. First, write the equation of the S-N line as

$$\log S_f' = b \log N + C \tag{7-10}$$

---

* Larry D. Mitchell, *An Investigation of the Correlation of the Acoustic Emission Phenomenon with the Scatter in Fatigue Data*, Pub. no. IP-719, University of Michigan, Oct. 1965, p. 88.

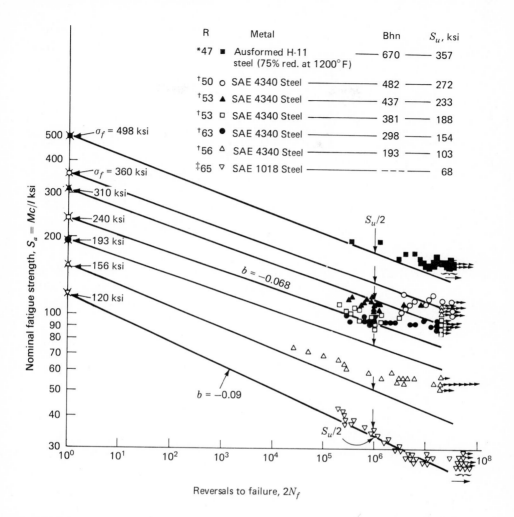

FIGURE 7-8 Bending fatigue results showing the effect of heat treatment on fatigue strength of steel. In this figure σ_f is the true fracture stress, designated as σ_F in Chap. 4, and b is the fatigue-strength exponent. The references are

* F. Borik, W. M. Justusson, and V. F. Zackay, "Fatigue Properties of an Ausformed Steel," *Trans. ASM*, vol. 56, 1963, pp. 327–338.

† J. R. Shure and G. W. Brock, *A Comparison of the Fatigue Behavior of Leaded and Non-leaded AISI 4340 Steel at High Hardness Levels*, TAM Report no. 87, University of Illinois, Theoretical and Applied Mechanics Dept., June 1955.

‡ F. C. Rally and G. M. Sinclair, *Influence of Strain Aging on the Shape of the S-N Diagram*, TAM Report no. 87, University of Illinois, Theoretical and Applied Mechanics Dept., June 1955.

(*This figure is from Fatigue Design Handbook, SAE, 1968, p. 27, and is reprinted with permission © 1968, Society of Automotive Engineers, Inc.*)

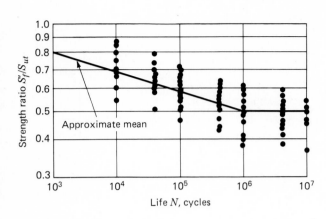

FIGURE 7-9 A compilation of the average rotating-beam, normalized S-N curves for 25 different steels. Notice that the actual S-N lines have been removed for clarity. Only the points on the curves are shown. (*Collated from data compiled by H. J. Grover, S. A. Gordon, and L. R. Jackson in Fatigue of Metals and Structures, Bureau of Naval Weapons Document NAVWEPS 00-25-534, 1960.*)

This line must intersect 10^6 cycles at S'_e and 10^3 cycles at $0.8S_{ut}$. When these are substituted into Eq. (7-10), the resulting equations can be solved for b and C. The results are

$$b = -\frac{1}{3} \log \frac{0.8S_{ut}}{S'_e} \tag{7-11}$$

$$C = \log \frac{(0.8S_{ut})^2}{S'_e} \tag{7-12}$$

Units of kpsi or MPa are most convenient for use in these equations. Having solved them for the two constants, we can find S'_f when N is given by solving Eq. (7-10).

$$S'_f = 10^C N^b \qquad 10^3 \le N \le 10^6 \tag{7-13}$$

Alternately, if S'_f is given and N is desired, then Eq. (7-10) yields

$$N = 10^{-C/b} S'^{1/b}_f \qquad 10^3 \le N \le 10^6 \tag{7-14}$$

Of course, you should demonstrate for yourself that these relations are correct.

The terms S'_f and S'_e in Eqs. (7-11) and (7-14) designate, respectively, the fatigue strength and endurance limit of a rotating-beam specimen. When these equations are to be used for an actual machine or structural element, the terms should be replaced by S_f and S_e, as in the example which follows.

Example 7-1 The endurance limit of a steel member is 112 MPa and the tensile strength is 385 MPa. What is the fatigue strength corresponding to a life of $70(10)^3$ cycles?

Solution   Since $0.8S_{ut} = 0.8(385) = 308$ MPa, Eqs. (7-11) and (7-12) give, respectively

$$b = -\frac{1}{3} \log \frac{0.8S_{ut}}{S_e} = -\frac{1}{3} \log \frac{308}{112} = -0.146$$

$$C = \log \frac{(0.8S_{ut})^2}{S_e} = \log \frac{(308)^2}{112} = 2.928$$

Then, from Eq. (7-13), we find the finite-life strength to be

$$S_f = 10^{2.928}[70(10)^3]^{-0.146} = 166 \text{ MPa} \qquad\qquad Ans.$$

////

7-5   ENDURANCE-LIMIT MODIFYING FACTORS

We have seen that the rotating-beam specimen used in the laboratory to determine endurance limits is prepared very carefully and tested under closely controlled conditions.   It is unrealistic to expect the endurance limit of a mechanical or structural member to match values obtained in the laboratory.

Marin* classifies some of the factors that modify the endurance limit, and these are shown in Table 7-4.   To account for the most important of these conditions we employ a variety of modifying factors, each of which is intended to account for a single effect.   Using this idea we may write

$$S_e = k_a k_b k_c k_d k_e k_f S'_e \qquad\qquad (7-15)$$

where S_e = endurance limit of mechanical element
$\quad\quad S'_e$ = endurance limit of rotating-beam specimen
$\quad\quad k_a$ = surface factor
$\quad\quad k_b$ = size factor
$\quad\quad k_c$ = reliability factor
$\quad\quad k_d$ = temperature factor
$\quad\quad k_e$ = modifying factor for stress concentration
$\quad\quad k_f$ = miscellaneous-effects factor

We shall learn soon that some of these factors also have an effect on the low-cycle or static strengths.

---

* Joseph Marin, *Mechanical Behavior of Materials*, Prentice-Hall, Englewood Cliffs, N. J., 1962, p. 224.

Table 7-4 CONDITIONS AFFECTING THE ENDURANCE LIMIT

Material: Chemical composition, basis of failure, variability
Manufacturing: Method of manufacture, heat treatment, fretting-corrosion, surface condition, stress
        concentration
Environment: Corrosion, temperature, stress state, relaxation times
Design: Size, shape, life, stress state, stress concentration, speed, fretting, galling

The multiplicative nature of Eq. (7-15) had been assumed for years but had not been proven. In the mid-60s an extensive statistical analysis[*] of a UNS G43400 (electric furnace, aircraft quality) steel resulted in a correlation coefficient for multiplication models to reach 0.85 and additive models to reach 0.40. We therefore conclude that Eq. (7-15) is a valid model for predicting endurance limits of mechanical elements.

7-6 SURFACE FINISH

The surface of the rotating-beam specimen is highly polished, with a final polishing in the axial direction to smooth out any circumferential scratches. Obviously, most machine elements do not have such a high-quality finish. The modification factors, k_a, shown in Fig. 7-10, depend upon the quality of the finish

FIGURE 7-10 Surface-finish modification factors for steel. These are the k_a factors for use in Eq. (7-15).

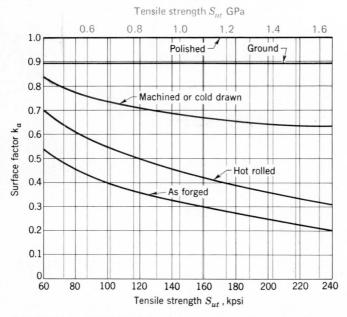

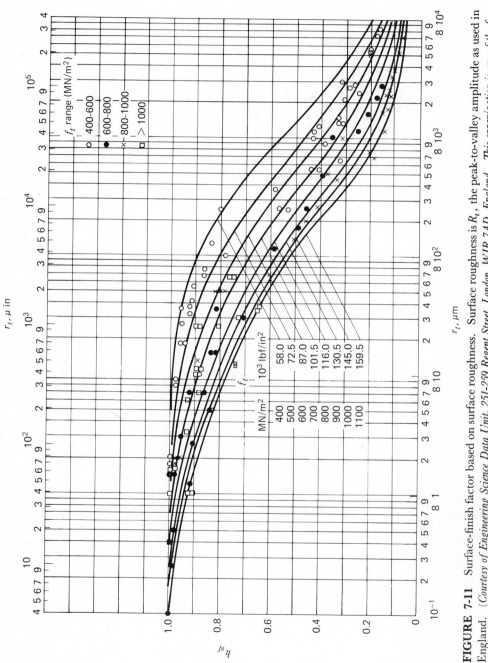

FIGURE 7-11 Surface-finish factor based on surface roughness. Surface roughness is R_t, the peak-to-valley amplitude as used in England. (*Courtesy of Engineering Science Data Unit, 251-259 Regent Street, London, WIR 7AD, England. This organization is one of the few that continue compilation and distribution of current fatigue data.*)

and upon the tensile strength. You should examine this chart carefully. It emphasizes the great importance of having a good surface finish when fatigue failure is a possibility. The factors shown on this chart have been obtained by condensing large compilations of data from tests of wrought steels and are probably also valid for cast steels and the better grades of cast iron.

More recent research on the surface-finish effect has concentrated on the surface roughness. The surface roughness, R_a, in this country, is the numerical value of the mean roughness height or amplitude measured from the roughness centerline. It is approximately half of the average distance from peak to valley. Surface roughness is measured in microinches (μin) or in micrometers (μm). Since the research indicates a correlation between the surface-finish factor and the existence of surface notches, the surface factor k_a is plotted as a function of the surface roughness. Figure 7-11 is such a graph showing the most recent data available. Note from the caption that the data is plotted using the English or European method of surface-finish measurement.

We might note that the data of Fig. 7-11 should not be used for parts that are hot-rolled or forged. The reason for this is that these processes modify, decarburize, or reorient the surface material, and change the surface finish.

Table 7-5 has been prepared to show the range of surface roughness R_a for some common manufacturing processes. Some of the data of Fig. 7-11 has been converted to R_a roughness values and replotted as Fig. 7-12 to make it easier to find surface factors when the processing is known. You should be very cautious

Table 7-5 TYPICAL RANGE OF SURFACE ROUGHNESS NUMBERS R_a

Process	Roughness μin Usual	Extreme	Roughness μm Usual	Extreme
Shaping	500–63	1000–16	15–1.5	25–0.4
Drilling	250–63	1000–16	6–1.5	25–0.4
Milling	250–32	1000–8	6–0.8	25–0.2
Broaching	125–32	250–16	3–0.8	6–0.4
Reaming	125–32	250–16	3–0.8	6–0.4
Lathe work	250–16	1000–2	6–0.4	25–0.05
Grinding	63–4	250–1	1.5–0.1	6–0.02
Polishing	16–4	32–0.5	0.4–0.1	0.8–0.01
Lapping	16–2	32–0.5	0.4–0.05	0.8–0.01
Sand casting	1000–500	2000–250	15–13	50–6
Investment casting	125–63	250–16	3–1.5	6–0.4
Extruding	125–32	500–16	3–0.8	13–0.4
Cold drawing	125–32	250–8	3–0.8	6–0.2
Die casting	63–32	125–16	1.5–0.8	3–0.4

Source: *Machinery's Handbook*, 20th ed., Industrial Press, 1975, p. 2395.

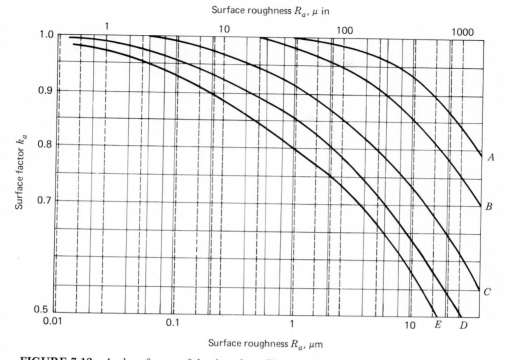

FIGURE 7-12 A plot of some of the data from Fig. 7-11 for the most-used range of surface roughness converted to U. S. roughness measure R_a. Note that the dashed lines are for roughness in μin. Spacing is 1, 2, 4, 6, 8, 10, etc., for both sets of lines. The letters designate the following ultimate strengths: A, 400 MPa, 58 kpsi; B, 500 MPa, 73 kpsi; C, 700 MPa, 102 kpsi; D, 900 MPa, 130 kpsi; E, 1100 MPa, 160 kpsi.

about relying too much upon the graph of Fig. 7-12. The data upon which the equation and graph are based were determined only for ground surfaces. There is no reason to believe that it can be applied to other surface finishes. Engineers who wish to use Fig. 7-11 must use the European method of surface measurement.

7-7 SIZE EFFECTS

The round rotating-beam test gives the endurance limit for a specimen usually 0.30 in in diameter. In SI 7.5, 10, or 12.5 mm are commonly used. It turns out that the endurance limits of machine elements having a larger size or a different cross section seldom approach the values found from the standard rotating-beam tests. This effect, due to the dimensions, the shape, and the method of loading, is called the *size effect*. Unfortunately, the testing of large specimens having different geometries is very costly, requiring expensive laboratory facilities. For this reason there are only limited data available.

Bending and Torsion

Kuguel* has proposed a theory based on the idea that failure is related to the probability of a high stress interacting with a critical flaw within a certain volume. When the volume of material subjected to a high stress is large there is a larger probability of failure. Kuguel uses a volume of material that is stressed to 95 percent of maximum or more and compares this to the equivalent rotating-beam volume to obtain a size factor. Unfortunately it is difficult to determine this volume in the vicinity of a discontinuity or area of stress concentration.

Mischke† has developed a statistical approach which does yield a very reliable value of the size factor. This method requires the designer to run tests on the specific material used.

Sors‡ states that the most modern theory is that of von Phillipp§ and that it shows good agreement with test results. The von Philipp theory is based on the assumption that the stress distribution in a rotating-beam specimen does not obey the basic theory. He reasons that the elastic deformations are delayed by the grains in the inner zones which have a lower stress and provide some support to the outer grains. Thus the fatigue-stress distribution resembles the solid line in Fig. 7-13a. The rather amazing discovery made by von Philipp is that the distance s in Fig. 7-13a, within which the deformations take place, is a constant for each material and is

$$s = \begin{cases} 3.1 \text{ mm} & \text{for steel} \\ 1.0 \text{ mm} & \text{for light alloys} \end{cases}$$

Thus a plot of the size factor was found to consist of a horizontal line for small thicknesses followed by a parabolic decrease as shown in Fig. 7-13b. The approach is quite detailed and very conservative when compared to actual test results.

Figure 7-14 is a plot of actual test data for specimens less than 0.30 in up to 2.2 in in diameter. The generally decreasing value is quite apparent even though the data exhibit a great deal of scatter. For reference purposes the results of the Kuguel theory are plotted as a dashed line. The solid line is plotted from the

---

* R. Kuguel, "A Relation Between Theoretical Stress Concentration Factor and Fatigue Notch Factor Deduced from the Concept of Highly Stressed Volume," *Proc. ASTM*, vol. 61, 1961, pp. 732–748.

† Charles R. Mischke, "A Probabilistic Model of Size Effect in the Fatigue Strength of Rounds in Bending and Torsion," ASME Paper no. 79-DE-16, 1979.

‡ L. Sors, *Fatigue Design of Machine Components*, pt. I, Pergamon Press, Oxford, 1971, pp. 42–44.

§ H. von Philipp, «Einfluss von Querschnittgrösse und Querschnittform auf die Dauerfestigkeit bei ungleichmässig verteilen Spannungen», thesis, Munich, 1941 [Forschg. a. d. Geb. d. Ing.—Wesens (1942)].

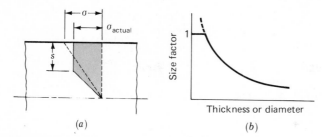

(a) (b)

FIGURE 7-13 (a) Elastic and actual bending-stress distribution according to von Phillipp; (b) size-factor variation with diameter.

results of the relation

$$
k_b = \begin{cases} 0.869d^{-0.097} & 0.3 \text{ in} < d \le 10 \text{ in} \\ 1 & d \le 0.3 \text{ in} \quad \text{or} \quad d \le 8 \text{ mm} \\ 1.189d^{-0.097} & 8 \text{ mm} < d \le 250 \text{ mm} \end{cases} \tag{7-16}
$$

which we shall employ in this book for round bars in bending and torsion.

One of the problems that arise in using Eq. (7-16) is what to do when a

FIGURE 7-14 Effect of specimen size on the endurance limit in reversed bending and torsion. (*Compiled from R. B. Heywood, Design Against Fatigue of Metals, Reinhold, New York, 1962; and from H. J. Grover, S. A. Gordon, and L. R. Jackson, Fatigue of Metals and Structures, Bureau of Naval Weapons Document NAVEWPS 00-25-534, 1960 rev.*)

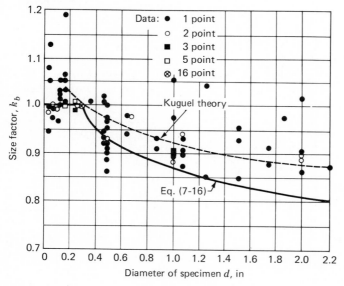

noncircular section is used. That is, what is the size factor for a bar $\frac{1}{4}$ in thick and 2 in wide? Professor Charles R. Mischke of Iowa State University has suggested the following approach. In Eq. (7-16) use an *effective dimension d*, obtained by equating the volume of material stressed at and above 95 percent of maximum stress to the same volume in the rotating-beam specimen. It turns out that when the 95-percent-stress volumes of the rotating-beam section and the section under study are equated, the lengths cancel, and so we need only consider the areas. Thus, Fig. 7-15 shows the 95-percent-stress areas for some typical sections. For a rotating-round section the 95-percent-stress area is the area in a ring having an outside diameter d and an inside diameter of $0.95d$. For the nonrotating-round section the 95-percent-stress area is the area outside of two parallel chords having a spacing of $0.95d$.

The effective dimension d for a rectangular section $\frac{1}{4}$ in thick by 2 in wide is obtained by equating the two 95-percent-stress areas, one for the round or rotating-beam section and the other for the rectangular section. From Fig. 7-15, we get

$$0.95A = 0.0766d^2 = 0.05hb$$

Substituting h and b gives

$$d = \sqrt{\frac{0.05hb}{0.0766}} = \sqrt{\frac{0.05(\frac{1}{4})(2)}{0.0766}} = 0.571 \text{ in}$$

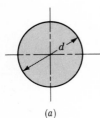

(a)

(b)

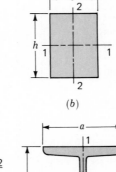

(c)

(d)

FIGURE 7-15 The 95-percent-stress areas for typical sections. Subscripts indicate axis of bending. Average flange thickness is t_f. (a) Solid or hollow round, $0.95A$ (rotating) = $0.0766d^2$, $0.95A$ (nonrotating) = $0.0107d^2$; (b) rectangle, $0.95A_{1\text{-}1}$ = $0.95A_{2\text{-}2} = 0.05hb$; (c) channel, $0.95A_{1\text{-}1} = 0.05ab$. $t_f > 0.025a$, $0.95A_{2\text{-}2}$ = $0.05xa + 0.1t_f$ $(b - x)$; (d) wide-flange or eye-beam, $0.95A_{1\text{-}1} = 0.10at_f$, $0.95A_{2\text{-}2} = 0.05ba$, $t_f > 0.025a$.

Table 7-6 AXIAL STRENGTHS IN kpsi

S_{uc}	S'_e	S_{uc}	S'_e	S_{uc}	S'_e
65	29.5	325	114	280	96
60	30	238	109	295	99
82	45	130	67	120	48
64	48	207	87	180	84
101	51	205	96	213	75
119	50	255	99	242	106
105	43	280	114	292	105
195	78	325	117	134	60
210	87	355	122	145	64
230	105	225	87	227	116
265	105				

Axial Loading

Repeated testing has shown that there is no apparent size effect for specimens tested in axial or push-pull fatigue. This means that the endurance limit, say, of a group of specimens having a diameter of $\frac{1}{4}$ in will be the same as another group having a 2-in diameter in push-pull fatigue. There does, however, appear to be a definite difference between the endurance limit in push-pull or axial fatigue and the endurance limit of the rotating-beam specimen.

A very extensive collection of data has been made by R. W. Landgraf, now of Ford Motor Company, on axial fatigue.* These results were recorded in terms of the true strain, instead of the stress, but, by applying Eqs. (7-5) and (7-8), the results can be interpreted in terms of stress. Table 7-6 is a tabulation of the ultimate compressive strength and the corresponding value of the axial endurance limit for a comprehensive series of 31 tests.

When a regression analysis of this data is made we find the y intercept at 19.2 kpsi and a slope of 0.314. The correlation coefficient is 0.944. This gives the equation

$$S'_e = 19.2 + 0.314 S_{uc} \qquad S_{uc} \geq 60 \tag{7-17}$$

where the strengths are in kpsi. The limitation on S_{uc} is necessary to account for the nonzero intercept. Of course, if the value of S'_e from Eq. (7-17) is used in Eq. (7-15) then $k_b = 1$.

We observe that Eq. (7-17) gives endurance limits of about 50 percent of the ultimate strength for the low-strength steels, but less than this for the high-strength steels.

---

* Landgraf, op cit., and by personal communication.

Others* have found the axial endurance limit to be significantly less than the rotating-beam endurance limit for tensile strengths from 54 to 170 kpsi. However, the data have considerable scatter resulting in size factors from 0.57 to 0.78. For this reason two axial size factors are recommended

$$k_b = \begin{cases} 0.71 & \text{when used with testing} \\ 0.60 & \text{when used without testing} \end{cases} \tag{7-18}$$

The higher value is an average that can be used when the mechanical properties are available from actual test results of the material to be used. The smaller value is a low-boundary value to be used when test results are not available.

7-8 RELIABILITY†

In this section we shall outline an analytical approach which will enable you to design a mechanical element subjected to fatigue loads such that it will last for any desired life at any stated reliability. In many ways life and reliability may constitute a more effective method of measuring design efficiency than the use of a factor of safety because both life and reliability are easily measured. The approach presented here is a logical one, but additional testing is needed before it can be recommended for general use. In particular, do not expect the method to yield

---

* Sors, op. cit., pp. 9–13; and H. J. Grover, S. A. Gordon, and L. R. Jackson, *Fatigue of Metals and Structures*, Bureau of Naval Weapons Document NAVWEPS 00-25-534, 1960 rev., pp. 282–314.

† The following books and papers on reliability are recommended for additional study and information:

Jack R. Benjamin and C. Allin Cornell, *Probability, Statistics, and Decision for Civil Engineers*, McGraw-Hill, New York, 1970.

Charles Lipson and Narendra J. Sheth, *Statistical Design and Analysis of Engineering Experiments*, McGraw-Hill, New York, 1973.

Edward B. Haugen and Paul H. Wirsching, "Probabilistic Design," *Machine Design*, vol. 47, nos. 10–14, 1975.

Dimitri Kececioglu, Louie B. Chester, and Thomas M. Dodge, "Combined Bending-Torsion Fatigue Reliability of AISI 4340 Steel Shafting with $K_t = 2.34$," *ASME* Paper no. 74-WA/DE-12, 1974.

C. Mischke, "A Method of Relating Factor of Safety and Reliability," *ASME* Paper no. 69-WA/DE-6, 1969.

C. Mischke, "Designing to a Reliability Specification," *SAE* Paper no. 740643, 1974.

C. R. Mischke, "A Rationale for Mechanical Design to a Reliability Specification, Implementing Mechanical Design to a Reliability Specification, and Organizing the Computer for Mechanical Design," *ASME* paper, Design Engineering Technical Conference, New York, Oct. 5–9, 1974.

My Dao-Thien and M. Massoud, "On the Probabilistic Distributions of Stress and Strength in Design Problems," *ASME* Paper no. 74-WA/DE-7, 1974.

absolute values. Its greatest use will be as a guide to reveal the most effective thing to do to improve life and reliability of actual parts.

In order to define the exact meaning of reliability let us suppose that we have a large group or population of mechanical parts. We can associate a certain strength S and a certain stress σ with each part. But since there are a large number of them, we have a population of strengths and a population of stresses. These two populations might be distributed somewhat as shown in Fig. 7-16. Using the notation of Chap. 5 we designate μ_σ and $\hat{\sigma}_\sigma$ as the mean and the standard deviation of the stress, and μ_S and $\hat{\sigma}_S$ as the mean and standard deviation of the strength. Although strength is generally greater than the stress, Fig. 7-16 shows that the forward tail of the stress distribution may overlap the rearward tail of the strength distribution and result in some failures. To determine the reliability, we combine both populations using Eqs. (5-18) and (5-19). The combined population then has a mean value and standard deviation of

$$r = S - \sigma \qquad \mu_r = \mu_S - \mu_\sigma \qquad \hat{\sigma}_r = \sqrt{\hat{\sigma}_S^2 + \hat{\sigma}_\sigma^2}$$

The corresponding standardized variable z_r is

$$z_r = \frac{r - \mu_r}{\hat{\sigma}_r} = \frac{r - (\mu_S - \mu_\sigma)}{\sqrt{\hat{\sigma}_S^2 + \hat{\sigma}_\sigma^2}} \qquad (7\text{-}19)$$

Then, using Eq. (5-14), the reliability R is

$$R = P(z_r) = 1 - Q(z_r) \qquad (7\text{-}20)$$

Equations (7-19) and (7-20) enable us to determine the reliability of any given material-stress situation, or, conversely, with Eq. (5-15), to determine the standardized variable z_r for any desired reliability.

Example 7-2 Determine the mean strength needed to assure a 90 percent probability of not yielding (reliability) if the standard deviation of the mean yield strength is $0.03S_y$ and that of the stress is 0.12σ.

FIGURE 7-16 Graph of stress and strength distribution showing the mean stress μ_σ and the mean strength μ_S.

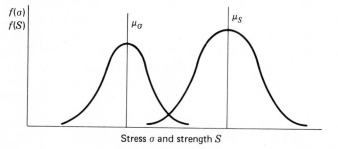

Stress σ and strength S

Solution Using the calculator program for Eq. (5-15), or a normal distribution table, we find $z_r = 1.2817$ for $Q(z_r) = 0.10$. Using $r = 0$, we have, from Eq. (7-19), that

$$z_r = \frac{-(\mu_S - \mu_\sigma)}{\sqrt{\hat{\sigma}_S^2 + \hat{\sigma}_\sigma^2}} = -1.282$$

So

$$\mu_S - \mu_\sigma = 1.282 \sqrt{(0.03\mu_S)^2 + (0.12\mu_\sigma)^2}$$

For a first estimate

$$\mu_S - \mu_\sigma \cong 1.282(0.12\mu_\sigma) = 0.1538\mu_\sigma$$

and

$$\mu_S = 1.154\,\mu_\sigma$$

An iterative solution yields

$$\mu_S = 1.1595\,\mu_\sigma$$

Notice that the approximation is only 0.4 percent in error. The above result should be interpreted to mean that the yield strength must be at least 116 percent of the mean stress to provide a 90 percent reliability.   ////

Rarely is all the information available to carry out such a detailed interference analysis. For the most part you will not know the scatter or the distribution of the applied load. This means that an interference theory calculation is not feasible.

We must, however, take care of the known scatter in the fatigue data. This means that we will not carry out a true reliability analysis but will actually answer the question, What is the probability that a *known* (assumed) stress will exceed the strength of a randomly selected component made from this material population?

An examination of Table 7-2 shows that the standard deviation of the endurance limit for steels is not likely to exceed 8 percent. In fact, data presented by Haugen and Wirsching* also show standard deviations of less than 8 percent. This means that we can obtain the endurance limit corresponding to any specified reliability R merely by subtracting a number of standard deviations from the

* Op. cit., no. 12, May 15, 1975.

Table 7-7 RELIABILITY FACTORS k_c CORRESPONDING TO AN 8 PERCENT STANDARD DEVIATION OF THE ENDURANCE LIMIT

Reliability R	Standardized variable z_r	Reliability factor k_c
0.50	0	1.000
0.90	1.288	0.897
0.95	1.645	0.868
0.99	2.326	0.814
0.999	3.091	0.753
0.999 9	3.719	0.702
0.999 99	4.265	0.659
0.999 999	4.753	0.620
0.999 999 9	5.199	0.584
0.999 999 99	5.612	0.551
0.999 999 999	5.997	0.520

mean endurance limit. Thus, the reliability factor k_c is

$$k_c = 1 - 0.08z_r \tag{7-21}$$

Table 7-7 shows the standardized variable z_r corresponding to the various reliabilities required in design together with the corresponding reliability factor k_c computed from Eq. (7-21).*

In using the approach suggested here, it is important to remember that the actual distribution of fatigue strengths can be better approximated by the Weibull distribution†,‡ than by the normal distribution. The normal distribution is preferred here because of the convenience of matching stress with strength.

In using Table 7-7 you should observe carefully the conditions applying to the use of Eq. (7-1) to find S'_e. Briefly, these conditions require that the ultimate tensile strength S_{ut} be known with certainty, unless S'_e is found by some other method.

Students who are solving problems for practice purposes require a standard approach for selecting S_{ut} so that the "correct" answer will be the same for all members of the class. You may have noted in referring to Table A-17 that the strengths are estimated minimum values for hot-rolled and cold-drawn bars, and that these are typical values when the steels are heat-treated. Thus these values can be obtained provided one adheres to the specifications and takes sufficient care in inspection and in processing. In this book, therefore, we shall assume that

---

* We are very grateful to Professor Charles Mischke of Iowa State University for working out the higher values in this table and permitting us to include them here. J.E.S., L.D.M.

† Lipson and Sheth, op. cit., p. 324.

‡ See Sec. 5-8.

quality-control procedures have been established which assure that the tensile strength S_{ut} is always equal to or greater than the values shown in Table A-17 when fatigue failure must be guarded against.

You are cautioned very particularly that these recommendations are for students only. Corresponding guidelines in the practice of engineering must be obtained from in-plant experience.

7-9   TEMPERATURE EFFECTS

In Figs. 4-9 and 4-10 we learned that temperature changes all of the mechanical properties of a material and that the existence of a static or mean stress also induces creep in the material. It is also probably true that there is no fatigue limit for materials operating at high temperatures. This means that the *S-N* diagram for steels, say, would lose the bend, or knee, in the curve when used at sufficiently high temperatures.

The significance of the factor k_d in Eq. (7-15) is to remind the designer that the temperature effects must be considered. Even though much testing has been done, values of the factor k_d are very elusive. It is for this reason that laboratory testing is strongly recommended when new and untried materials are under consideration.

High temperatures mobilize dislocations and reduce the fatigue resistance in many materials. This mobilization results in the conversion of an essentially time-independent failure process into a time-dependent process. There are also complicated interactions involving creep due to static or mean stress, environmental atmosphere, and the fatigue process. Forest[*] collected a great deal of data, some of which are plotted in Fig. 7-17, so that a general idea of the effect of temperature can be obtained. Note the large difference in the effects for a low-carbon steel and for a strong alloy steel. The mild steel endurance limit increases up to about 350°C. This effect is caused by cyclic-strain aging. Figure 4-9 shows that temperature has about the same effect on the ultimate strength. But Fig. 4-9 also shows that the yield strength falls continuously with temperature. Overlooking of this fact has caused some disastrous failures.

Next, note from Fig. 7-17 that the alloy steel, UNS G43400, retains about the same fatigue strength up to 450°C, and that the mild steel returns to its room-temperature strength at the same temperature. We can approximate these observations with the relation

$$k_d = \begin{cases} 1.0 & T \leq 450°C \ (840°F) \\ 1 - 5.8(10)^{-3}(T-450) & 450°C < T \leq 550°C \\ 1 - 3.2(10)^{-3}(T-840) & 840°F < T \leq 1020°F \end{cases} \tag{7-22}$$

---

[*] P. G. Forest, *Fatigue of Metals*, Pergamon Press, London, 1962.

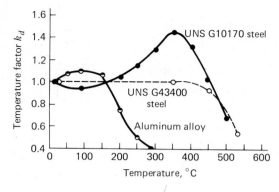

FIGURE 7-17 Graph of temperature factor for two wrought steels and a wrought aluminum alloy.

for non-high-temperature steels and for *problem solution only*. At the maximum temperature permitted by this relation, k_d is about 0.40. It is likely that a part would fail by yielding or by creep at such a high temperature, so there is little value in attempting to predict this factor for higher temperatures.

With sufficient data, similar relations can be expressed for the cast irons and for the nonferrous alloys.

It is important to remember that the temperature factor must be applied to both ends of the *S-N* diagram because the fatigue strength is reduced by the same factor at 10^3 cycles as at 10^6 cycles.

7-10 STRESS-CONCENTRATION EFFECTS

Most mechanical parts have holes, grooves, notches, or other kinds of discontinuities which alter the stress distribution as discussed in Secs. 6-9 to 6-12.

Several points must be reiterated here before we apply stress-concentration-factor effects to predict the endurance limit.

- $\sigma_{\max} = K_t \sigma_0 \qquad \tau_{\max} = K_{ts} \tau_0$
- σ_0 or τ_0 *must* be calculated by the stress equation used in the development of the stress-concentration-factor table.
  - Abbreviated charts of stress-concentration factors are given in Table A-26.
  - K_t and K_{ts} are theoretical values.
  - Stress concentration is a highly localized effect.
- K_t and K_{ts} need *not* be applied to static stresses in ductile materials but *must* be used on the static stresses in high-strength, low-ductility, case-hardened, and/or heavily cold-worked materials.

Stress concentration does have to be considered when parts are made of brittle materials or when they are subject to fatigue loading. Even under these conditions, however, it turns out that some materials are not very sensitive to the existence of notches or discontinuities, and hence the full values of the theoretical stress concentration factors need not be used. For these materials it is convenient to use a reduced value of K_t. The resulting factor is defined by the equation

$$K_f = \frac{\text{endurance limit of notch-free specimens}}{\text{endurance limit of notched specimens}} \qquad (a)$$

This factor is usually called a *fatigue stress-concentration factor*, although it is used for brittle materials under static loads, too.

Now, in using K_f, it does not matter, algebraically, whether it is used as a factor for *increasing the stress* or whether it is used for *decreasing the strength*. This simply means that it may be placed on one side of the equation or the other. But a great many troublesome problems can be avoided if K_f is treated as a factor which reduces the strength of a member. Therefore we shall call K_f a *fatigue-strength reduction factor* and shall nearly always use it in this sense. This means that the modifying factor for stress concentration k_e of Eq. (7-15) and K_f have the relation

$$k_e = \frac{1}{K_f} \qquad (7\text{-}23)$$

Notch sensitivity q is defined by the equation

$$q = \frac{K_f - 1}{K_t - 1} \qquad (7\text{-}24)$$

where q is usually between zero and unity. Equation (7-24) shows that, if $q = 0$, $K_f = 1$, and the material has no sensitivity to notches at all. On the other hand, if $q = 1$, then $K_f = K_t$, and the material has full sensitivity. In analysis or design work one first determines K_t from the geometry of the part. Then, the material having been specified, q can be found, and the equation is solved for K_f.

$$K_f = 1 + q(K_t - 1) \qquad (7\text{-}25)$$

For steels and UNS A92024 aluminum alloys use Fig. 7-18 to find q when the parts are subjected either to rotating-beam action or to reversed axial loading. Use Fig. 7-19 for parts subjected to reversed shear.

Both charts (Figs. 7-18 and 7-19) show that for large notch radii, and especially for high-strength materials, the sensitivity index approaches unity. This means that whenever there is any doubt, one can always make $K_f = K_t$ and err on the safe side. Also, if the notch radius is quite large—and it should be designed this way if at all possible—then q is not far from unity and the error of assuming K_f equal to K_t will be quite small.

It does happen sometimes that two discontinuities, each having a particular stress-concentration factor, occur at the same point. When this situation arises,

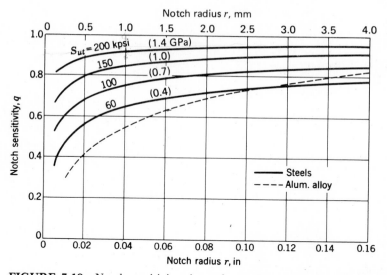

FIGURE 7-18 Notch-sensitivity charts for steels and UNS A92024-T wrought aluminum alloys subjected to reversed bending or reversed axial loads. For larger notch radii use the values of q corresponding to $r = 0.16$ in (4 mm). (*Reproduced by permission from George Sines and J. L. Waisman (eds.), Metal Fatigue, McGraw-Hill, New York, 1959, pp. 296, 298.*

find each fatigue-strength reduction factor and then multiply them together to get an equivalent factor.*

Brittle Materials

The notch sensitivity of the cast irons is very low, varying from about zero to 0.20, depending upon the tensile strength. To be on the conservative side, it is recommended that a notch sensitivity $q = 0.20$ be used for all grades of cast iron.

Since brittle materials have no yield strength, the stress-concentration factor K_f must be applied to the static strength S_{ut} or S_{uc} as well as to the endurance limit as observed earlier. This means that both ends of the *S-N* diagram for cast iron would be lowered the same amount since K_f must be used to reduce the strength at each end.

7-11 MISCELLANEOUS EFFECTS

Though the factor k_f is intended to account for the reduction in endurance limit due to all other effects, it is really intended as a reminder that these must be accounted for, because actual values of k_f are not available.

---

* A more accurate approach to this problem is described by Sors, op. cit., pt. I, pp. 57–61, 65–66.

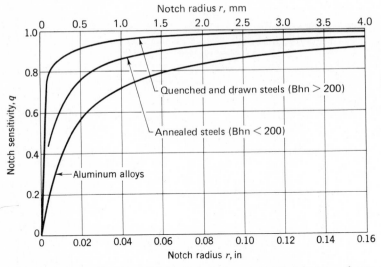

FIGURE 7-19 Notch-sensitivity curves for materials in reversed torsion. For larger notch radii use the values of q corresponding to $r = 0.16$ in (4 mm).

Residual stresses may either improve the endurance limit or affect it adversely. Generally, if the residual stress in the surface of the part is compression, the endurance limit is improved. Fatigue failures appear to be tensile failures, or at least to be caused by tensile stress, and so anything which reduces tensile stress will also reduce the possibility of a fatigue failure. Operations such as shot peening, hammering, and cold rolling build compressive stresses into the surface of the part and improve the endurance limit significantly. Of course, the material must not be worked to exhaustion.

The endurance limits of parts which are made from rolled or drawn sheets or bars, as well as parts which are forged, may be affected by the so-called *directional characteristics* of the operation. Rolled or drawn parts, for example, have an endurance limit in the transverse direction which may be 10 to 20 percent less than the endurance limit in the longitudinal direction.

Parts which are case-hardened may fail at the surface or at the maximum core radius, depending upon the stress gradient. Figure 7-20 shows the typical triangular stress distribution of a bar under bending or torsion. Also plotted as a heavy line on this figure are the endurance limits S_e for the case and core. For this example the endurance limit of the core rules the design because the figure shows that the stress σ or τ, whichever applies, at the outer core radius, is appreciably larger than the core endurance limit.

Of course, if stress concentration is also present, the stress gradient is much steeper, and hence failure in the core is unlikely.

Corrosion   It is to be expected that parts which operate in a corrosive atmosphere will have a lowered fatigue resistance. This is, of course, true, and it

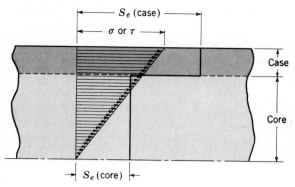

FIGURE 7-20 The failure of a case-hardened part in bending or torsion. In this example failure occurs in the core.

is due to the roughening or pitting of the surface by the corrosive material. But the problem is not so simple as the one of finding the endurance limit of a specimen which has been corroded. The reason for this is that the corrosion and the stressing occur at the same time. Basically, this means that in time any part will fail when subjected to repeated stressing in a corrosive atmosphere. There is no fatigue limit. Thus the designer's problem is to attempt to minimize the factors that affect the fatigue life. These are:

- Mean or static stress
- Alternating stress
- Electrolyte concentration
- Dissolved oxygen in electrolyte
- Material properties and composition
- Temperature
- Cyclic frequency
- Fluid flow rate around specimen
- Local crevices

Electrolytic plating   Metallic coatings, such as chromium plating, nickel plating, or cadmium plating, reduce the endurance limit by as much as 50 percent. In some cases the reduction by coatings has been so severe that it has been necessary to eliminate the plating process. Zinc plating does not affect the fatigue strength. Anodic oxidation of light alloys reduces bending endurance limits by as much as 39 percent but has no effect on the torsional endurance limit.

Metal spraying   Metal spraying results in surface imperfections that can initiate cracks. Limited tests show reductions of 14 percent in the fatigue strength.

Cyclic frequency.   If, for any reason, the fatigue process becomes time-dependent then it also becomes frequency-dependent. Under normal conditions fatigue failure is independent of frequency. But when corrosion or high temperatures, or both, are encountered, the cyclic rate becomes important. The slower the frequency and the higher the temperature, the higher the crack propagation rate and the shorter the life at a given stress level.

Frettage corrosion This phenomena is the result of microscopic motions of tightly fitting parts or structures. Bolted joints, bearing-race fits, wheel hubs, and any set of tightly fitted parts are examples. The process involves surface discoloration, pitting, and eventual fatigue. The frettage factor k_f depends upon the material of the mating pairs and ranges from 0.24 to 0.90. A detailed list of typical values is given by Fuchs and Stephens.*

Example 7-3 Corresponding to a reliability of 99 percent, estimate the endurance limit of a 1-in round UNS G10150 cold-drawn steel bar.

Solution From Table A-17, we find $S_{ut} = 56$ kpsi and $H_B = 111$. From Fig. 7-10, $k_a = 0.84$. Since the endurance limit means the bending endurance limit unless specified differently, we use Eq. (7-16) for k_b. This gives

$$k_b = 0.869d^{-0.097} = 0.869(1)^{-0.097} = 0.869$$

From Table 7-7, $k_c = 0.814$. Since nothing is stated, we assume $k_d = k_e = k_f = 1$. From Eq. (7-1)

$$S'_e = 0.5S_{ut} = 0.5(56) = 28 \text{ kpsi}$$

Then, from Eq. (7-15)

$$S_e = (0.84)(0.869)(0.814)(28) = 16.6 \text{ kpsi} \qquad\qquad Ans.$$

////

Example 7-4 Figure 7-21a shows a rotating shaft supported in ball bearings at A and D and loaded by the nonrotating force F. Using the methods of the preceding sections, estimate the life of the part.

Solution From Fig. 7-21b we learn that failure will probably occur at B rather than at C. Point B has a smaller cross section, a higher bending moment, and a higher stress-concentration factor. It is unlikely that failure would occur under the load F, even though the maximum moment occurs there, because there is no stress concentration and the section is larger.

---

* Op. cit., p. 229.

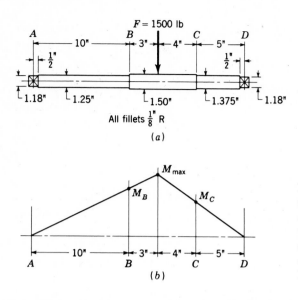

FIGURE 7-21 (a) Shaft drawing. The shaft rotates and the load is stationary; the material is UNS G10350 steel, drawn at 1000°F, with a machined finish. (b) Bending-moment diagram.

We shall solve the problem by finding the strength at point B, since it will probably be different at other points, and comparing this strength with the stress at point B.

From Table A-17 find that $S_{ut} = 103$ kpsi and $S_{yt} = 72$ kpsi. Therefore

$$S'_e = (0.5)(103) = 51.5 \text{ kpsi}$$

The surface factor is found to be $k_a = 0.73$. The size factor for point B is found from Eq. (7-16) as

$$k_b = 0.869(1.25)^{-0.097} = 0.850$$

The reliability is always taken as 50 percent unless a specific value is given; therefore $k_c = 1$. Also, $k_d = 1$, since nothing concerning temperature is stated. Then, using Fig. A-26-9, we calculate for point B

$$\frac{D}{d} = \frac{1.5}{1.25} = 1.20 \qquad \frac{r}{d} = \frac{0.125}{1.25} = 0.10$$

and find $K_t = 1.60$. Then, entering Fig. 7-18 with $S_{ut} = 103$ kpsi, find $q = 0.82$. The fatigue-strength reduction factor is found to be

$$K_f = 1 + q(K_t - 1) = 1 + 0.82(1.60 - 1) = 1.49$$

Consequently, the modifying factor for stress concentration is

$$k_e = \frac{1}{K_f} = \frac{1}{1.49} = 0.671$$

The endurance limit at point B is therefore

$$S_e \text{ (at } B) = k_a k_b k_e S'_e = (0.73)(0.850)(0.671)(51.5) = 21.4 \text{ kpsi}$$

Now, to determine the stress at B; the bending moment is

$$M_B = 10 \frac{9F}{22} = (10) \frac{(9)(1500)}{22} = 6140 \text{ lb} \cdot \text{in}$$

The section modulus is $I/c = \pi d^3/32 = \pi(1.25)^3/32 = 0.192 \text{ in}^3$. Therefore the stress is

$$\sigma = \frac{M}{I/c} = \frac{6140}{0.192} = 32(10)^3 \text{ psi}$$

Since this stress is greater than the endurance limit, the part will have only a finite life.

Following the procedure of Example 7-1, we first find $0.8S_{ut} = 0.8(103) = 82.4$ kpsi. Then

$$b = -\frac{1}{3} \log \frac{0.8S_{ut}}{S_e} = -\frac{1}{3} \log \frac{82.4}{21.4} = -0.195$$

$$C = \log \frac{(0.8S_{ut})^2}{S_e} = \log \frac{(82.4)^2}{21.4} = 2.501$$

Next, for use in Eq. (7-14), we compute $-C/b = -2.501/(-0.195) = 12.83$ and $1/b = 1/-0.195 = -5.13$. Thus

$$N = 10^{-C/b}S_f'^{1/b} = 10^{12.83}(32)^{-5.13} = 128(10)^3 \text{ cycles} \qquad \textit{Ans.}$$

$$////$$

7-12 FLUCTUATING STRESSES

Quite frequently it is necessary to determine the strength of parts corresponding to stress situations other than complete reversals. Many times in design the stresses fluctuate without passing through zero. Figure 7-22 illustrates some of the various stress-time relationships which may occur. The components of stress with which we must deal, some of which are shown in Fig. 7-22d, are

$$\sigma_{min} = \text{minimum stress} \qquad \sigma_m = \text{mean stress}$$

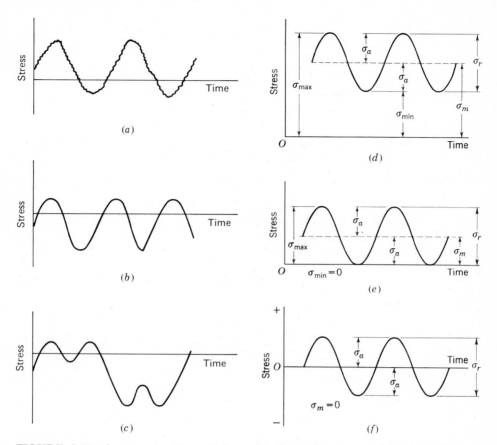

FIGURE 7-22 Some stress-time relations: (a) fluctuating stress with high-frequency ripple; (b) and (c) nonsinusoidal fluctuating stress; (d) sinusoidal fluctuating stress; (e) repeated stress; (f) completely reversed sinusoidal stress.

σ_{max} = maximum stress σ_r = stress range

σ_a = stress amplitude σ_s = steady, or static, stress

The steady, or static, stress is *not* the same as the mean stress; in fact, it may have any value between σ_{min} and σ_{max}. The steady stress exists because of a fixed load or preload applied to the part, and it is usually independent of the varying portion of the load. A helical compression spring, for example, is always loaded into a space shorter than the free length of the spring. The stress created by this initial compression is called the steady, or static, component of the stress. It is not the same as the mean stress.

We shall have occasion to apply the subscripts of these components to shear stresses as well as normal stresses.

The following relations are evident from Fig. 7-22:

$$\sigma_m = \frac{\sigma_{max} + \sigma_{min}}{2} \qquad\qquad\qquad (7\text{-}26)$$

$$\sigma_a = \frac{\sigma_{max} - \sigma_{min}}{2} \qquad\qquad\qquad (7\text{-}27)$$

Although some of the stress components have been defined by using a sine stress-time relation, the exact shape of the curve does not appear to be of particular significance.

7-13   FATIGUE STRENGTH UNDER FLUCTUATING STRESSES

Now that we have defined the various components of stress associated with a part subjected to fluctuating stress, we want to vary both the mean stress and the stress amplitude, to learn something about the fatigue resistance of parts when subjected to such situations. Two methods of plotting the results of such tests are in general use and are both shown in Fig. 7-23.

The *modified Goodman diagram* of Fig. 7-23a has the mean stress plotted along the abscissa and all other components of stress plotted on the ordinate, with tension in the positive direction. The endurance limit, fatigue strength, or finite-life strength, whichever applies, is plotted on the ordinate above and below the origin. The mean-stress line is a 45° line from the origin to point A, representing the tensile strength of the part. The modified Goodman diagram consists of the lines constructed from point A to S_e (or S_f) above and below the origin. A better average through the points of failure would be obtained by constructing curved lines from point A. Note that the yield strength is also plotted on both axes, because yielding would be the criterion of failure if σ_{max} exceeded S_y.

Another fatigue diagram which is often used is that of Fig. 7-23b. Here the mean stress is also plotted on the abscissa, tension to the right and compression to the left. But the ordinate has only the stress amplitude σ_a plotted along it. Thus, with this diagram, only two of the stress components are used. The endurance limit, fatigue strength, or finite-life strength, whichever applies to the particular problem, is the limiting value of the stress amplitude, and so it is plotted on the ordinate. A straight line from S_e to S_u on the abscissa is also the modified Goodman criterion of failure. Note that when the mean stress is tension, most of the failure points fall above this line. On the compression side, however, the failure points show that the magnitude of the mean stress has no effect. The Soderberg line, drawn from S_e to S_y, has also been proposed as a criterion for design, because yielding is also used to define failure. Note, however, that the modified Goodman line errs on the safe side, and so the Soderberg line is even more conservative.

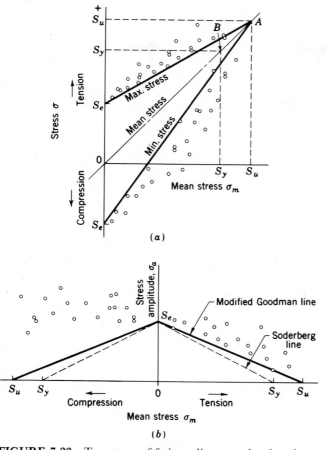

FIGURE 7-23 Two types of fatigue diagrams showing simulated points of failure: (*a*) modified Goodman diagram; (*b*) diagram showing the modified Goodman line.

Now that we have used these two diagrams to learn how parts fail, we can use them to construct a well-defined criterion of fatigue or static failure.

As noted, the modified Goodman diagram is particularly useful because it contains all the stress components. In Fig. 7-24 the diagram has been redrawn to show all these stress components and also the manner in which it will be used to define failure. When the mean stress is compression, failure is defined by the two heavy parallel lines originating at $+S_e$ and $-S_e$ and drawn downward and to the left. When the mean stress is tension, failure is defined by the maximum-stress line or by the yield strength as indicated by the heavy outline to the right of the ordinate. The modified Goodman diagram is particularly useful for analysis when all the dimensions of the part are known and the stress components can be easily calculated. But it is rather difficult to use for design, that is, when the dimensions are unknown.

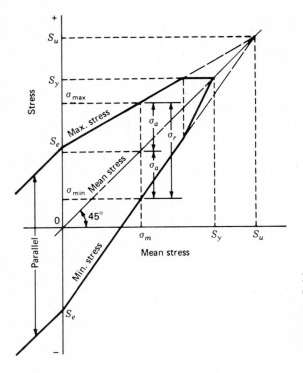

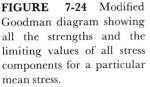

FIGURE 7-24 Modified Goodman diagram showing all the strengths and the limiting values of all stress components for a particular mean stress.

The fatigue diagram of Fig. 7-25* is the one we shall employ frequently for design purposes, where the heavy lines will be taken as the criterion of failure. To explain, note that the yield strength has been plotted on the mean-stress axis, both for tension and for compression, and also on the stress-amplitude axis. A line from S_y to S_{yc} defines failure by compressive yielding; from S_y to S_{yt}, failure by tensile yielding. We construct the modified Goodman line for tensile mean stress and a horizontal line from S_e to the left for compressive mean stress. The intersections of the two lines in each quadrant are the transition points between a failure by fatigue and a failure by yielding. The heavy outline therefore specifies when failure by either method will occur.

---

* A justifiable confusion exists with respect to the correct terminology for the fatigue diagrams of Figs. 7-23 to 7-25. Figure 7-23a uses the axes originally employed by Goodman, but the diagram is a modification of his proposal. This modified Goodman diagram is modified again in Fig. 7-23b because a new set of axes has been employed. Still another modification is introduced in Fig. 7-24, where the yield strength has been added as a limiting factor and the diagram has been extended into the compressive-stress region. The fourth modification is obtained by converting this to the axes of Fig. 7-25. The original Goodman diagram, which is not shown, has not been employed for many years. For these reasons many people, quite correctly, simplify the whole situation and call Fig. 7-24 the *Goodman diagram*, and Fig. 7-25 the *modified Goodman diagram*.

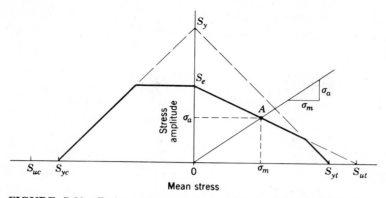

FIGURE 7-25 Fatigue diagram showing how to find the limiting values of σ_a and σ_m when the ratio of the two is given.

In design work the force amplitude and the mean force can usually be calculated or determined. Sometimes we are working with bending moments or torsional moments. In these cases one can usually calculate the mean moment and the amplitude of the moment. The stress amplitude and the mean stress are related to these through the dimensions which are to be found. The ratio σ_a/σ_m is the same as F_a/F_m or M_a/M_m. So a line from the origin through point A can be drawn and the limiting values of σ_a or σ_m found as the projections of this point on the two axes.

The simplicity of the approach described here is worth noting. The value of the fatigue strength plotted on the ordinate has previously been corrected for size, surface finish, reliability, stress concentration, and other effects. Consequently one need not worry about which stress components these factors should have been applied to.*

---

* More sophisticated approaches are available, however. Some of these include a factor of safety, some are directed specifically to design, in contrast to analysis, and others are quite analytical. The following references are highly authoritative and are recommended:

W. N. Findley, J. J. Coleman, and B. C. Hanley, "Theory for Combined Bending and Torsion Fatigue," *Proceedings of the International Conference on the Fatigue of Metals*, London, 1956.

Robert C. Juvinall, *Engineering Considerations of Stress, Strain, and Strength*, McGraw-Hill, New York, 1967, pp. 268–314.

Robert E. Little, "Analysis of the Effect of Mean Stress on Fatigue Strength of Notched Steel Specimens," Publ. IP-63Q, The University of Michigan, Ann Arbor, 1963.

L. D. Mitchell and D. T. Vaughn, "A General Method for the Fatigue-Resistant Design of Mechanical Components, Part 1 Graphical," *ASME* Paper no. 74-WA/DE-4, "Part 2 Analytical," *ASME* Paper no. 74WA/DE-5, 1974.

George Sines and J. L. Waisman (eds.), *Metal Fatigue*, McGraw-Hill, New York, 1959, chap. 7.

The modified Goodman criterion applies for the cast irons when the mean stress is tension.* The existence of a compressive mean stress has no effect on the endurance limit, however.

Example 7-5 It is desired to determine the size of a UNS G10500 cold-drawn steel bar to withstand a tensile preload of 8 kip and a fluctuating tensile load varying from 0 to 16 kip. Owing to the design of the ends, the bar will have a geometric stress-concentration factor of 2.02 corresponding to a fillet whose radius is $\frac{3}{16}$ in. Determine a suitable diameter for an infinite life and a factor of safety of at least 2.0.

Solution From Table A-17, $S_y = 84$ kpsi and $S_{ut} = 100$ kpsi. So $S'_e = 0.5 S_{ut} = 0.5(100) = 50$ kpsi. We next find $k_a = 0.73$ from Fig. 7-10. For axial loads, $k_b = 0.60$. From Fig. 7-18, $q = 0.86$, and so

$$K_f = 1 + q(K_t - 1) = 1 + 0.86(2.02 - 1) = 1.87$$

so that $k_e = 1/K_f = 1/1.87 = 0.535$. These are all the necessary corrections, so that

$$S_e = (0.73)(0.60)(0.535)(50) = 11.7 \text{ kpsi}$$

Next we determine the stresses in terms of their dimensions. The static stress is

$$\sigma_s = \frac{F_s}{A} = \frac{8}{\pi d^2/4} = \frac{10.2}{d^2} \text{ kpsi}$$

The stress range is

$$\sigma_r = \frac{F_r}{A} = \frac{16}{\pi d^2/4} = \frac{20.4}{d^2} \text{ kpsi}$$

Then

$$\sigma_a = \frac{\sigma_r}{2} = (10.2/d^2) \text{ kpsi}$$

and in this case,

$$\sigma_m = \sigma_s + \sigma_a = (20.4/d^2) \text{ kpsi}$$

---

* See *Metals Handbook*, 8th ed., American Society for Metals, Metals Park, Ohio, 1961, pp. 356 and 357.

Therefore

$$\sigma_a/\sigma_m = 0.50$$

To relate the stresses and strengths, we plot a fatigue diagram (Fig. 7-26). Note that only the tensile side is needed. Note that the line through S_y does not make a 45° angle because the axes have different scales. The intersection of the modified Goodman line with another line at a slope of $\sigma_a/\sigma_m = 0.50$ defines *two* values of *strength*. Using a mnemonic notation for these, S_a is a strength corresponding to the stress σ_a, and S_m is a strength corresponding to the stress σ_m. For a factor of safety of 2.0 we have

$$\sigma_a \le S_a/2.0$$

Consequently

$$10.2/d^2 \le 9.5/2.0 \qquad \text{or} \qquad d \ge 1.47 \text{ in}$$

We therefore choose $d = 1\frac{1}{2}$ in to get it in fractional-inch or stock size.   ////

Example 7-6   We wish to size the crankshaft of Fig. 7-27b to support a moment $M_{\max} = 480$ N · m which occurs at point C on the shaft and varies according to the graph shown. The material is to be UNS G10350 steel, heat-treated and drawn to 1200°F, with a machined finish. The reliability is to be 95 percent and the stress-concentration factor is 1.6 corresponding to a 3-mm fillet at point C. We have elected to choose the factors of safety separately for each set of uncertainties and so have chosen $n_{yS} = 1.3$ for the yield strength, $n_{uS} = 1.10$ for the ultimate strength, and $n_{eS} = 1.8$ for the endurance limit. For the loading we have chosen $n_1 = 2$. While the torsional stress should never be neglected, in this case we wish to do so in order to illustrate the use of multiple factors of safety when fluctuating stresses are present. To add the complicating effects of torsion would

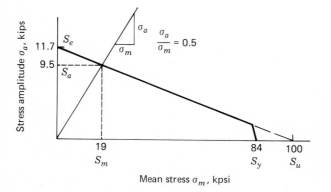

FIGURE 7-26 Corresponding to $\sigma_a/\sigma_m = 0.50$, S_a is the *strength amplitude* and S_m is the *mean strength*. To enlarge the ordinate, different scales were used for the axes.

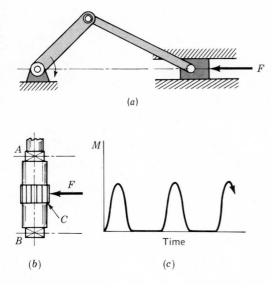

FIGURE 7-27 (*a*) Slider-crank mechanism used on assembly line to exert the pushing force F; (*b*) crank-shaft supported in bearings at A and B and splined in the center for crank; (*c*) moment-time relation.

obscure some of the points to be made, and besides we shall be covering the problem of torsional fatigue soon.

Solution From Table A-17 we find $S_y = 62(6.89) = 427$ MPa and $S_{ut} = 91(6.89) = 627$ MPa. Applying factors of safety to these strengths yields safe values of

$$S_y(\text{min}) = \frac{427}{1.3} = 328 \text{ MPa} \qquad S_u(\text{min}) = \frac{627}{1.10} = 570 \text{ MPa}$$

Let us first size the shaft based on the possibility of a static failure. Since $I/c = \pi d^3/32$, the permissible stress is, from Eq. (2-32),

$$\sigma_p = \frac{M}{I/c} = \frac{32 M_{\text{max}} n_1}{\pi d^3} \tag{1}$$

where the permissible stress is obtained when the maximum moment is increased by the amount of the load factor of safety. Solving Eq. (1) for d gives

$$d = \sqrt[3]{\frac{32 M_{\text{max}} n_1}{\pi \sigma_p}} = \sqrt[3]{\frac{32(480)(2)}{\pi(328)(10)^{-3}}} = 31.0 \text{ mm}$$

The next problem is to size the shaft based on the possibility of a fatigue failure. So we enter Fig. 7-10 with $S_{ut} = 627$ MPa and find $k_a = 0.74$. Using the diameter of 31 mm found based on yielding, we use Eq. (7-16) for the size factor.

$$k_b = 1.189 d^{-0.097} = 1.189(31)^{-0.097} = 0.852$$

However, we choose $k_b = 0.80$ to be on the safe side. Next from Table 7-7 we find $k_c = 0.868$ for 95 percent reliability. Then, from Fig. 7-18, $q = 0.84$ based on $r = 3$ mm and $S_{ut} = 627$ MPa. Therefore

$$K_f = 1 + q(K_t - 1) = 1 + 0.84(1.6 - 1) = 1.50$$

and so $k_e = 1/1.5 = 0.667$. Now we find $S'_e = 0.5S_{ut} = 0.5(627) = 313.5$ MPa. Then the corrected value is

$$S_e = k_a k_b k_c k_e S'_e = (0.74)(0.80)(0.868)(0.667)(313.5) = 107.4 \text{ MPa}$$

Applying the factor of safety n_{es} gives

$$S_e(\text{min}) = \frac{107.4}{1.8} = 59.7 \text{ MPa}$$

An examination of Fig. 7-27c shows that the mean moment and the moment amplitude are equal to each other and are half the maximum moment. Consequently $\sigma_a/\sigma_m = 1$. Using Eq. (1) we find the two permissible fatigue-stress components to be

$$\sigma_{a,p} = \sigma_{m,p} = \frac{32(M_{max}/2)(n_1)}{\pi d^3} = \frac{32(480/2)(2)}{\pi d^3} = \frac{4890}{d^3} \text{ MPa} \tag{2}$$

The modified Goodman line has been plotted in Fig. 7-28 using strengths that have been reduced to minimums by the respective factors of safety. Note that the yield strength has been omitted because the size based on yielding has already been obtained. Plotting the line $\sigma_a/\sigma_m = 1$ on the diagram yields the two strengths $S_a = S_m = 54$ MPa. Replacing $\sigma_{a,p}$ with S_a enables us to solve Eq. (2)

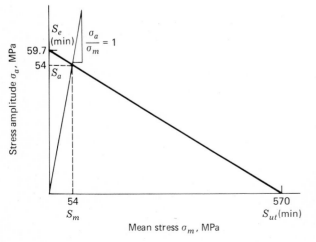

FIGURE 7-28 Fatigue diagram plotted using minimum strength values; axes have different scales.

for the diameter. The result is

$$d = \sqrt[3]{\frac{4890}{54(10)^{-3}}} = 44.9 \text{ mm}$$

Using Table A-13 we therefore choose $d = 45$ mm.

The size factor was assumed to be $k_b = 0.80$. The actual value should have been

$$k_b = 1.189(45)^{-0.097} = 0.822$$

This result shows that we are on the safe side.   ////

Example 7-7   While it is possible to use multiple factors of safety in design, as in the previous example, the reverse is not true. This means that when we analyze a part that has already been designed to find out whether it is safe or unsafe, only a single factor of safety can be found. To illustrate this point, suppose the shaft of Example 7-6 had been machined to a diameter of 50 mm from a bar of hot-rolled UNS G10150 steel. Find the factors of safety guarding against a static failure and a fatigue failure.

Solution   From Table A-17 we find $S_y = 27(6.89) = 186$ MPa and $S_{ut} = 50(6.89) = 344$ MPa. Now substitute S_y for σ_p in Eq. (1) of the previous example and solve for the factor of safety. Thus

$$n = \frac{\pi d^3 S_y}{32 M_{max}} = \frac{\pi (50)^3 (186)(10)^{-3}}{32(480)} = 4.76 \qquad\qquad Ans.$$

So there is an ample factor of safety guarding against a static failure.

We must next find the endurance limit. Entering Fig. 7-10 with $S_{ut} = 344$ MPa we find $k_a = 0.85$. The size factor is

$$k_b = 1.189(50)^{-0.097} = 0.814$$

Since there may be no reliability at all, we use $k_c = 1$ for analysis; and $k_e = 0.667$ as in Example 7-6. Next $S'_e = 0.5 S_{ut} = 0.5(344) = 172$ MPa; so

$$S_e = k_a k_b k_c k_e S'_e = (0.85)(0.814)(1)(0.667)(172) = 79.4 \text{ MPa}$$

The next step is to construct a fatigue diagram like Fig. 7-28 using $S_e = 79.4$ MPa, $S_{ut} = 344$ MPa, and $\sigma_a/\sigma_m = 1$. From such a diagram we find $S_a = 64.5$ MPa.

We now write Eq. (2) of the previous example in the form

$$\frac{S_a}{n} = \frac{32(M_{max}/2)}{\pi d^3} \tag{1}$$

Solving for the factor of safety gives

$$n = \frac{\pi d^3 S_a}{32(M_{max}/2)} = \frac{\pi(50)^3(64.5)(10)^{-3}}{32(480/2)} = 1.65 \qquad Ans.$$

While this might indicate a small margin of safety, it may not be large enough to cover all the uncertainties involved. The importance of this kind of analysis to the designer is that it reveals weaknesses in design and shows where improvements are most likely to pay the biggest dividends in safety.   ////

Machine Computation

The modified Goodman line can be placed in equation form for machine computation by writing the equation of a straight line in intercept form. The equation is

$$\frac{x}{a} + \frac{y}{b} = 1 \tag{a}$$

where a and b are the x and y intercepts, respectively. This equation for the modified Goodman line is thus

$$\frac{\sigma_m}{S_{ut}} + \frac{\sigma_a}{S_e} = 1 \tag{7-28}$$

For machine computation it is convenient to manipulate this equation to the form

$$S_m = \frac{S_e}{\dfrac{\sigma_a}{\sigma_m} + \dfrac{S_e}{S_{ut}}} \tag{7-29}$$

which is easier to program. In using Eq. (7-29) and, in fact, the methods of this section, be very careful that the ratio σ_a/σ_m remains a constant no matter how the forces or moments may vary. See Sec. 7-15 for instances in which σ_a/σ_m is not a constant. Note in Example 7-5 that σ_a and σ_m both result from a single force and, hence, must remain constant.

7-14 NONLINEAR THEORIES

The modified Goodman criterion is a very conservative theory of the effect of mean stress on the fatigue resistance, as indicated in Fig. 7-23. The various nonlinear theories do a better job of predicting failure since they appear to take a mean path through test data points. It is probably not worth going to the extra trouble to use such theories, though, unless all the strengths are known with some accuracy.

One of the most widely used theories is the *Gerber parabolic relation;* it is expressed in either of the following forms:

$$\frac{S_a}{S_e} + \left(\frac{S_m}{S_{ut}}\right)^2 = 1$$

$$S_a = S_e\left[1 - \left(\frac{S_m}{S_{ut}}\right)^2\right]$$

$$(7\text{-}30)$$

Most of the nonlinear theories are empirical, but Marin* states that a relation having a theoretical basis can be obtained by equating the elastic strain energy absorbed by the rotating-beam specimen to the corresponding strain energy obtained from a fluctuating-stress state. The result is called the *quadratic,* or *elliptic, equation.* It is

$$\left(\frac{S_a}{S_e}\right)^2 + \left(\frac{S_m}{S_{ut}}\right)^2 = 1$$

$$S_a = S_e\left[1 - \left(\frac{S_m}{S_{ut}}\right)^2\right]^{1/2}$$

$$(7\text{-}31)$$

Kececioglu, Chester, and Dodge† have proposed a solution which has been verified by many experiments. The equation is

$$\left(\frac{S_a}{S_e}\right)^a + \left(\frac{S_m}{S_{ut}}\right)^2 = 1$$

$$S_a = S_e\left[1 - \left(\frac{S_m}{S_{ut}}\right)^2\right]^{1/a}$$

$$(7\text{-}32)$$

---

* Marin, op. cit., p. 189.
† Op. cit.

The constant a was found to be 2.606 for AISI 4340 steel having H_B in the range of 320 to 370 Bhn. But at a reliability higher than 50 percent the exponent was found to be $a = 2.750$.

Professor Cemil Bagci of Tennessee Technological University states that the Kececioglu relation predicts failure better than the Gerber equation.* However, tests must be carried out on each proposed material to determine the exponent a. Bagci also feels that a good criterion should include the possibility of a failure by yielding. His equation is

$$\frac{S_a}{S_e} + \left(\frac{S_m}{S_y}\right)^4 = 1$$

$$S_a = S_e\left[1 - \left(\frac{S_m}{S_y}\right)^4\right]$$

$$(7\text{-}33)$$

The above four nonlinear theories are plotted in Fig. 7-29, together with the modified Goodman line. To complete the set of all theories we repeat the modified Goodman relations and add the Soderberg relations. The modified Goodman equation is

$$\frac{S_a}{S_e} + \frac{S_m}{S_{ut}} = 1$$

$$S_a = S_e\left(\frac{S_m}{S_{ut}}\right)$$

$$(7\text{-}34)$$

The Soderberg relation is

$$\frac{S_a}{S_e} + \frac{S_m}{S_y} = 1$$

$$S_a = S_e\left(1 - \frac{S_m}{S_y}\right)$$

$$(7\text{-}35)$$

There are two principal ways in which any of the relations in the collection of fluctuating fatigue-strength equations are used. They are used (1) for analysis, i.e., when a part has already been designed the relations can be used to estimate the factor of safety or to predict the life of the part; and (2) for design, i.e., either a single factor of safety or multiple factors of safety can be used to determine the dimensions or geometry of a part. The examples that follow illustrate these uses.

---

* Cemil Bagci, "Computer-Aided Fatigue Design of Power Transmission Shafts with Strength Constraints Using a Finite Line Element Technique and a Proposed Fatigue Failure Criterion," ASME Paper no. 79-DET-103.

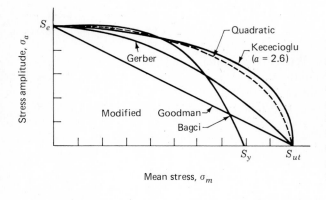

FIGURE 7-29 An accurate plot of all the nonlinear theories, with the modified Goodman line added for comparison purposes. You may wish to complete the diagram by adding the Soderberg theory as a red or green straight line from S_e to S_y.

Example 7-8 A part is made of a steel having $S_{ut} = 100$ kpsi, $S_y = 80$ kpsi, and a fully corrected endurance limit $S_e = 25$ kpsi. The stress components are $\sigma_a = 16/d^2$ and $\sigma_m = 30/d^2$ in kpsi. Find the factor of safety using the Gerber theory if $d = 2$ in. Find the factor of safety guarding against a static failure.

Solution $\sigma_a = 16/(2)^2 = 4$ kpsi, $\sigma_m = 30/(2)^2 = 7.5$ kpsi. Then $S_a = \sigma_a n = 4n$ and $S_m = \sigma_m n = 7.5n$. Substituting in the first form of Eq. (7-30) gives

$$\frac{4n}{25} + \left(\frac{7.5n}{100}\right)^2 = 1$$

Solving this equation for the factor of safety yields $n = 5.28$ *Ans.*

To compute the factor of safety guarding against yielding, we compute

$$\sigma_{max} = \sigma_a + \sigma_m = 4 + 7.5 = 11.5 \text{ kpsi}$$

Then

$$n = \frac{S_y}{\sigma_{max}} = \frac{80}{11.5} = 6.96$$ *Ans.*

///

Example 7-9 Find the diameter of the part in Example 7-8 if the factor of safety is to be 3. Use the Gerber equation.

Solution $S_a = n\sigma_a = 3(16/d^2) = 48/d^2$, $S_m = n\sigma_m = 3(30/d^2) = 90/d^2$. Substituting into the first form of Eq. (7-30) gives

$$\frac{48}{25d^2} + \left(\frac{90}{100d^2}\right)^2 = 1$$

This equation simplifies to

$$d^4 - 1.92d^2 - 0.81 = 0$$

Many calculators have programs to evaluate the zeros of any function specified by the user. Using such a program we find only one root in the interval $1 \leq d \leq 2$. This result is $d = 1.51$ in

Ans.

////

Example 7-10 We wish to determine the size of the part in Example 7-8 using multiple factors of safety. These are $n_1 = 2$ for the stress amplitude, $n_2 = 1.4$ for the mean stress, $n_{eS} = 1.35$ for the endurance limit, $n_{yS} = 1.20$ for the yield strength, and $n_{uS} = 1.25$ for the ultimate strength. Find a safe diameter using the Soderberg theory.

Solution Applying the factors of safety to the strengths first, we get

$$S_e(\text{min}) = \frac{S_e}{n_{eS}} = \frac{25}{1.35} = 18.5 \text{ kpsi}$$

$$S_y(\text{min}) = \frac{S_y}{n_{yS}} = \frac{80}{1.20} = 66.7 \text{ kpsi}$$

$$S_u(\text{min}) = \frac{S_u}{n_{uS}} = \frac{100}{1.25} = 80 \text{ kpsi}$$

Next, the permissible stresses are

$$\sigma_{a,p} = \sigma_a n_1 = \frac{16(2)}{d^2} = \frac{32}{d^2} \text{ kpsi}$$

$$\sigma_{m,p} = \sigma_m n_2 = \frac{40(1.4)}{d^2} = \frac{56}{d^2} \text{ kpsi}$$

We now let $S_a = \sigma_{a,p}$ and $S_m = \sigma_{m,p}$ and substitute into Eq. (7-35).

$$\frac{32}{18.5d^2} + \frac{56}{66.7d^2} = 1$$

Solving gives $d = 1.60$ in

Ans.

////

7-15 THE KIMMELMANN FACTOR OF SAFETY

The assumption that the alternating stress and the mean stress vary in exactly the same ratio when overloads occur has been shown by Kimmelmann to be arbitrary.* He argues that the factor of safety should be computed as the ratio of the stresses that cause failure to the maximum value under the operating conditions.

Consider a situation in which the alternating stress remains relatively constant, but the mean stress component depends upon whether a rotating part has been properly balanced. Failure to balance the part or carelessness resulting in a poor balance could cause the mean stress to increase to three or more times the normal value.

Many other stress states may occur in which the ratio σ_a/σ_m does not remain constant with overload, and consequently the methods of Sec. 7-13 cannot be used.

The general situation as illustrated by Kimmelmann is seen in Fig. 7-30. Line *ABC* is called a *load line*, or a *Kimmelmann line*. This line shows points P_1, P_2, P_3, P_4 representing successive stress states as the loading increases. Point *B*, of course, represents failure and defines the strength S_a corresponding to the alternating component and the strength S_m corresponding to the mean component.

If the stress state at P_2 represents the normal or design stress situation, then two factors of safety can be found. These are

$$n_a = \frac{S_a}{\sigma_a} \qquad n_m = \frac{S_m}{\sigma_m} \tag{7-36}$$

These factors of safety are only equal to each other for the special situation in which the Kimmelmann line is a straight line passing through the origin, that is, when σ_a/σ_m is a constant ratio. In general you can expect to find both straight and curved load lines, depending upon the particular loading situation encountered.

Example 7-11 A flat leaf spring is used to retain a roller-follower mechanism in contact with a plate cam. The range of motion of the follower is fixed and so the alternating component of the bending moment in the spring is fixed, too. However, the spring preload can be adjusted to accommodate various cam speeds. When the cam is run at a high speed, the preload must be increased so as to prevent follower float or jump. For lower speeds, the preload should be decreased to obtain a longer life for the cam and roller contact surfaces.

The spring is a steel cantilever 16 in long, 3 in wide, and $\frac{1}{16}$ in thick, as shown

---

* Kimmelmann, «Gépalkatrészek szilárdsági számitásai ismételt igénybevételeknél» ("Mechanical Design of Machine Components Subjected to Repeated Loading"), *Nehézipari Könyvkiadó*, 1951.

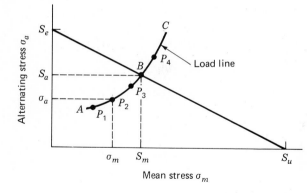

FIGURE 7-30 Fatigue diagram showing the load line *ABC* intersecting the modified Goodman line.

in Fig. 7-31. The strengths are $S_u = 150$ kpsi, $S_y = 127$ kpsi, and $S_e = 28$ kpsi fully corrected. The total cam motion is 2 in and we wish to preload the spring by deflecting it 2 in for slow speeds and up to 5 in for high speeds. Determine factors of safety for each of these conditions and also check for the possibility of a static failure by yielding. Use the modified Goodman criterion.

Solution   The moment of inertia is

$$I = \frac{bh^3}{12} = \frac{3\left(\frac{1}{16}\right)^3}{12} = 6.10(10)^{-5} \text{ in}^4$$

The deflection is related to the force by the equation $y = Fl^3/3EI$. Therefore the spring rate is

$$k = \frac{F}{y} = \frac{3EI}{l^3} = \frac{3(30)(10)^6(6.10)(10)^{-5}}{(16)^3} = 1.34 \text{ lb/in}$$

The bending stress caused by a 1-in deflection is

$$\sigma = \frac{Mc}{I} = \frac{1.34(16)\left(\frac{1}{32}\right)}{6.10(10)^{-5}} (10)^{-3} = 11 \text{ kpsi}$$

Thus, for slow speeds, the stress components are

$$\sigma_a = 11 \text{ kpsi} \qquad \sigma_m = 2(11) = 22 \text{ kpsi}$$

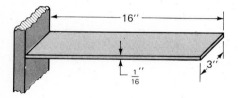

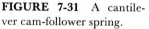

FIGURE 7-31 A cantilever cam-follower spring.

And, for high speeds

$$\sigma_a = 11 \text{ kpsi} \qquad \sigma_m = 5(11) = 55 \text{ kpsi}$$

These sets of points are plotted as P_1 and P_2, respectively, in Fig. 7-32. Since σ_a does not change, the load line is the horizontal line through $\sigma_a = 11$, P_1, P_2, and point A on the modified Goodman line. Using $S_m = 91$ kpsi, the factor of safety for each condition is

$$n_m = \frac{S_m}{\sigma_m} = \frac{91}{22} = 4.14 \text{ for slow speeds} \qquad\qquad Ans.$$

$$n_m = \frac{91}{55} = 1.65 \text{ for high speeds} \qquad\qquad Ans.$$

The maximum stress will be

$$\sigma_{\max} = \sigma_a + \sigma_m = 11 + 55 = 66 \text{ kpsi}$$

Therefore the factor of safety guarding against a failure by yielding is

$$n_y = \frac{S_y}{\sigma_{\max}} = \frac{127}{66} = 1.92 \qquad\qquad Ans.$$

$$////$$

Example 7-12 A machine weighing 13.5 tons is to be supported on two steel channels, each 6 ft in length. It is estimated that a vibrating vertical force from 20 to 40 percent of the machine weight could result if the rotating parts of the machine are not properly balanced. We need to determine the size of channel sections to be used. The material is a mild, hot-rolled steel having $S_u = 58$ kpsi,

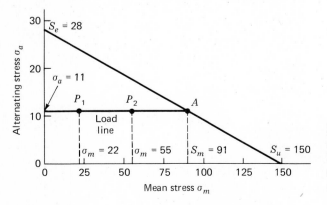

FIGURE 7-32 Fatigue diagram with stresses in kpsi. Note that different scales are used for σ_a and σ_m.

$S_y = 32$ kpsi, and $S_e = 28$ kpsi fully corrected. The channels are to be connected to other members at their ends, but, for safety, it is assumed that the ends are simply supported, resulting in the loading diagram shown in Fig. 7-33.

(a) Based on an overall factor of safety of 2, determine the size of channels to be used.

(b) Some designers prefer to choose factors of safety to account for each uncertainty separately (see Sec. 1-7). Find an appropriate channel size using a factor of safety for the yield strength of 1.3, for the ultimate strength of 1.2, for the endurance limit of 1.4, for the mean stress of 1.25, and for the alternating stress of 1.00.

Solution (a) Since $R_1 = W/2$ in Fig. 7-33, the maximum bending moment due to the weight alone, is

$$M = 12 \frac{W}{2} = 12 \frac{(13.5)(2000)}{2} = 162\,000 \text{ lb} \cdot \text{in}$$

This results in a mean stress whose magnitude is

$$\sigma_m = \frac{M}{Z} = \frac{162\,000}{Z} \text{ psi}$$

where Z is the section modulus. The alternating stress is

$$\sigma_a = 0.40 \frac{M}{Z} = \frac{0.40(162\,000)}{Z} = \frac{64\,800}{Z} \text{ psi}$$

We really don't know at this stage whether the design is governed by the static strength or by the fatigue strength. We can assume either one to find a possible size, then check the possibility of failure by the other mode. Assuming that static failure will govern the design, we compute the maximum stress to be

$$\sigma_{\max} = \sigma_a + \sigma_m = \frac{1}{Z}(64\,800 + 162\,000) = \frac{226\,800}{Z} \text{ psi}$$

Then, since $S_y/n = \sigma_{\max}$, we have

$$\frac{32\,000}{2} = \frac{226\,800}{Z}$$

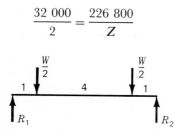

FIGURE 7-33 The 13.5-ton machine is supported at the two points as shown. Dimensions in feet.

from which we find $Z = 14.17$ in^3. For one channel $Z = 14.17/2 = 7.09$ in^3. From Table A-11 we find that an 8-in $\times$ 11.50-lb channel has a section modulus of $Z_{1-1} = 8.10$ in^3. So this is our tentative selection.

To check for a fatigue failure, we now compute

$$\sigma_a = \frac{64\ 800}{Z} = \frac{64\ 800(10)^{-3}}{2(8.10)} = 4.00 \text{ kpsi}$$

$$\sigma_m = \frac{162\ 000}{Z} = \frac{162\ 000(10)^{-3}}{2(8.10)} = 10.00 \text{ kpsi}$$

Since the mean stress cannot vary, the load line is the vertical line through $\sigma_m = 10$ kpsi on Fig. 7-34a. At point A we see $S_a = 23.2$ kpsi. Therefore, the

FIGURE 7-34 Fatigue diagrams showing load lines for two stress states

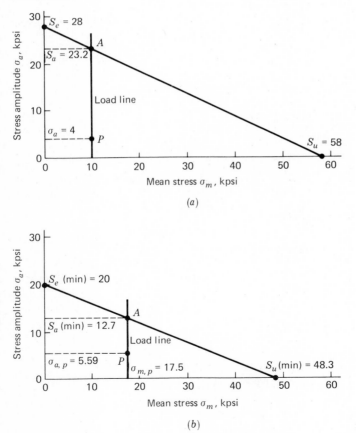

(a)

(b)

factor of safety guarding against a fatigue failure is

$$n_a = \frac{S_a}{\sigma_a} = \frac{23.2}{4.00} = 5.8$$

Thus the possibility of a static failure governs the design and the 8-in, 11.5-lb channel is made the final selection.

(b) Applying the factors of safety to the strengths gives

$$S_y(\text{min}) = \frac{S_y}{n_{ys}} = \frac{32}{1.3} = 24.6 \text{ kpsi}$$

$$S_u(\text{min}) = \frac{S_u}{n_{us}} = \frac{58}{1.2} = 48.3 \text{ kpsi}$$

$$S_e(\text{min}) = \frac{S_e}{n_{es}} = \frac{28}{1.4} = 20 \text{ kpsi}$$

Designating n_1 as the factor of safety for the stress amplitude and n_2 as the factor of safety for the mean stress, we find the permissible stresses to be

$$\sigma_{a,\,p} = n_1 \sigma_a = (1.0)\left(\frac{64\ 800}{Z}\right) = \frac{64\ 800}{Z} \text{ psi}$$

$$\sigma_{m,\,p} = n_2 \sigma_m = (1.25)\left(\frac{162\ 000}{Z}\right) = \frac{202\ 500}{Z} \text{ psi}$$

The maximum permissible stress is found to be

$$\sigma_{\text{max},\,p} = \frac{1}{Z}(64\ 800 + 202\ 500) = \frac{267\ 300}{Z} \text{ psi}$$

Equating this to $S_y(\text{min})$ and solving, gives $Z = 10.87 \text{ in}^3$ for two channels. Using Table A-11, we find that two C6 × 5.80 members (6-in, 5.80-lb channels) provide $Z = 11.6 \text{ in}^3$, which is safe.

Next we must check for the possibility of a fatigue failure. Using $Z = 11.6$ in^3, the permissible stresses are found to be

$$\sigma_{a,\,p} = \frac{64\ 800}{11.6}(10)^{-3} = 5.59 \text{ kpsi}$$

$$\sigma_{m,\,p} = \frac{202\ 500}{11.6}(10)^{-3} = 17.5 \text{ kpsi}$$

Plotting these values and the strengths on the fatigue diagram of Fig. 7-34*b*, we find the load point P to be in the safe region. In fact, $S_a(\min) = 12.7$ kpsi, giving an additional factor of safety of

$$\frac{S_a(\min)}{\sigma_{a,\,p}} = \frac{12.7}{5.59} = 2.27$$

as protection against a fatigue failure.   ////

7-16   TORSION

In Sec. 6-5 we learned that the maximum-shear-stress theory predicted the yield strength in shear to be

$$S_{sy} = 0.50S_y \tag{a}$$

and that this relation gives conservative values. Thus Eq. (a) is useful for design, because it is easy to apply and to remember, but not for the analysis of failure. A more accurate prediction of failure, we learned, is given by the distortion-energy theory, which predicts the yield strength in shear to be

$$S_{sy} = 0.577S_y \tag{b}$$

Interestingly enough, as indicated by experiments whose results are shown in Fig. 7-35, these two theories are also useful in predicting the endurance limit in shear S_{se} when the bending endurance limit S_e is known. Thus the maximum-shear-stress theory conservatively predicts

$$S_{se} = 0.50S_e \tag{7-37}$$

and the distortion-energy theory yields

$$S_{se} = 0.577S_e \tag{7-38}$$

We shall employ only Eq. (7-38) in this book because, as shown in Fig. 7-35, it predicts failure more accurately.

Let us now consider the case in which there is a torsional stress amplitude τ_a and a torsional mean stress τ_m. Corresponding strengths are the torsional or shear endurance limit S_{se}, the yield strength in shear S_{sy}, and the torsional modulus of rupture S_{su}. Using these strengths, it should be possible to construct a torsional fatigue diagram corresponding to that of Fig. 7-25. When we do this, and also plot a number of experimental observations of failures on it, we get the diagram of Fig. 7-36.

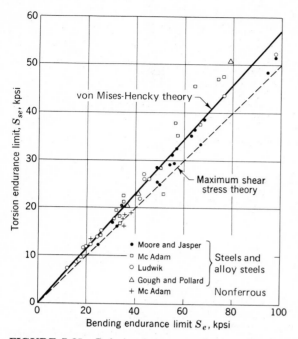

FIGURE 7-35 Relation between endurance limits in torsion and in bending. (*From Thomas J. Dolan, "Stress Range," in Oscar J. Horger (ed.), ASME Handbook-Metals Engineering—Design, McGraw-Hill, New York, 1953, sec. 6-2, p. 97; reproduced by permission of the publishers.*)

The interesting thing about Fig. 7-36 is that, up to a certain point, torsional mean stress has no effect on the torsional endurance limit. Thus it is not necessary to construct such a diagram for torsion at all! Instead, a fatigue failure is indicated if

$$\tau_a = S_{se} \qquad (7\text{-}39)$$

FIGURE 7-36 Fatigue diagram for combined alternating and mean torsional stress showing failure points.

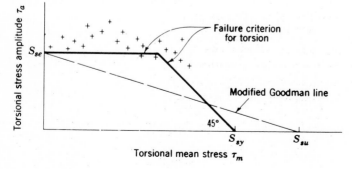

and a static failure if

$$\tau_{max} = \tau_a + \tau_m = S_{sy} \tag{7-40}$$

as indicated by the solid line of Fig. 7-36. Of course, these equations can be used for design too if a factor of safety is used.

7-17 STRESSES DUE TO COMBINED LOADING

One of the most frequently encountered problems in design is that of a rotating shaft subjected to a constant torque and a stationary bending load. An element on the surface of the shaft has a torsional stress $\tau = Tc/J$ which is constant in magnitude and direction when referred to a mark made on the shaft surface. But, owing to the bending moment, the same element will have a normal stress $\sigma = \pm Mc/I$ varying from tension to compression and back again, as the shaft rotates. If the stresses on the element are analyzed using a Mohr's circle diagram, it will be found that the principal stresses do not maintain the same orientation, relative to a mark on the surface, as the shaft rotates.

The problem is even more complicated when it is realized that the normal stresses σ_x and σ_y as well as the shear stress τ_{xy}, in the general two-dimensional stress state, may have both mean and alternating components. In this book we shall present a method of using the distortion-energy theory applied to fatigue to solve this problem, because all available experimental evidence shows the theory to be conservative and because the method employs the basic theory already developed in this chapter.*

The problem may also be complicated when axial loading exists in combination with other loads because the size factors are different. Still another complication arises when separate stress-concentration factors must be used for each type of loading. Each of these problems will be considered in the discussion that follows.

Basic Theory

To apply the theory, construct two stress elements, one for the mean stresses and one for the alternating stresses. Two Mohr's circles are then drawn, one for each element, and the principal mean stresses obtained from one circle and the principal alternating stresses obtained from the other. We can then define mean and

---

* The following papers discuss this subject in much greater detail:

R. E. Little, "Fatigue Stresses from Complex Loadings," *Machine Design*, Jan. 6, 1966, pp. 145–149.

W. R. Miller, K. Ohji, and J. Marin, "Rotating Principal Stress Axes in High-Cycle Fatigue," ASME Paper no. 66-WA/Met-9, 1966.

Dimitri Kececioglu et al., op. cit.

alternating von Mises stresses as

$$\sigma'_m = \sqrt{\sigma_{1m}^2 - \sigma_{1m}\sigma_{2m} + \sigma_{2m}^2}$$

$$\sigma'_a = \sqrt{\sigma_{1a}^2 - \sigma_{1a}\sigma_{2a} + \sigma_{2a}^2} \tag{7-41}$$

for the biaxial stress state. Alternatively the need for Mohr's circles can be eliminated by using the equations

$$\sigma'_m = \sqrt{\sigma_{xm}^2 - \sigma_{xm}\sigma_{ym} + \sigma_{ym}^2 + 3\tau_{xym}^2}$$

$$\sigma'_a = \sqrt{\sigma_{xa}^2 - \sigma_{xa}\sigma_{ya} + \sigma_{ya}^2 + 3\tau_{xya}^2} \tag{7-42}$$

For the uniaxial stress state, Eq. (7-42) reduces to

$$\sigma'_m = \sqrt{\sigma_{xm}^2 + 3\tau_{xym}^2} \qquad \sigma'_a = \sqrt{\sigma_{xa}^2 + 3\tau_{xya}^2} \tag{7-43}$$

Though seldom needed, the corresponding von Mises stresses for a triaxial stress state are

$$\sigma'_m = \tfrac{1}{2}\sqrt{(\sigma_{1m} - \sigma_{2m})^2 + (\sigma_{2m} - \sigma_{3m})^2 + (\sigma_{3m} - \sigma_{1m})^2}$$

$$\sigma'_a = \tfrac{1}{2}\sqrt{(\sigma_{1a} - \sigma_{2a})^2 + (\sigma_{2a} - \sigma_{3a})^2 + (\sigma_{3a} - \sigma_{1a})^2} \tag{7-44}$$

It is apparent from Eqs. (7-41) to (7-44) that we now have a means of combining any set of shear, normal, and bending stresses, all of which may have both mean- and alternating-stress components. The resulting two von Mises or equivalent stress components can then be applied in design or analysis exactly as the simple $\sigma_a - \sigma_m$ stress state is applied in Secs. 7-12 to 7-15.

Several words of caution are appropriate. Formulas like Eqs. (7-41) to (7-44) can also be written by using the maximum and minimum stress components, but these will not produce the same results. Also, be sure to use the method of Sec. 7-16 when only shear stresses exist, that is, when both σ_x and σ_y are zero. While Eqs. (7-42) and (7-43) could be used for such cases, the results will be quite conservative.

Example 7-13 A bar of steel has $S_{ut} = 700$ MPa, $S_y = 500$ MPa, and a fully corrected endurance limit $S_e = 200$ MPa. For each of the cases below find the factors of safety which guard against static and fatigue failures.

(a) $\tau_m = 140$ MPa

(b) $\tau_m = 140$ MPa, $\tau_a = 70$ MPa

(c) $\tau_{xym} = 100$ MPa, $\sigma_{xa} = 80$ MPa

(d) $\sigma_{xm} = 60$ MPa, $\sigma_{xa} = 80$ MPa, $\tau_{xym} = 70$ MPa, $\tau_{xya} = 35$ MPa

Solution   (*a*) The yield strength in shear is

$$S_{sy} = 0.577S_y = 0.577(500) = 288 \text{ MPa}$$

Therefore

$$n(\text{static}) = S_{sy}/\tau_{\max} = 288/140 = 2.06 \qquad\qquad\qquad Ans.$$

Since τ_m is a steady stress, there is no fatigue.
    (*b*) The maximum shear stress, from Fig. 7-22, is

$$\tau_{\max} = \tau_m + \tau_a = 140 + 70 = 210 \text{ MPa}$$

Therefore

$$n(\text{static}) = S_{sy}/\tau_{\max} = 288/210 = 1.37 \qquad\qquad\qquad Ans.$$

From Eq. (7-38) we find the endurance limit in shear to be

$$S_{se} = 0.577S_e = 0.577(200) = 115 \text{ MPa}$$

Based on Eq. (7-39) we therefore have

$$n(\text{fatigue}) = S_{se}/\tau_a = 115/70 = 1.64 \qquad\qquad\qquad Ans.$$

(*c*) The maximum von Mises stress occurs when the alternating component is summed with the mean component. Using Eq. (7-43), to bypass the use of a Mohr's circle, gives

$$\sigma'_{\max} = \sqrt{\sigma_{xa}^2 + 3\tau_{xym}^2} = \sqrt{(80)^2 + 3(100)^2} = 191 \text{ MPa}$$

Therefore the factor of safety guarding against a static failure is

$$n(\text{static}) = \frac{S_y}{\sigma'_{\max}} = \frac{500}{191} = 2.62 \qquad\qquad\qquad Ans.$$

The distortion-energy theory must also be used to find the possibility of a fatigue failure. Using Eq. (7-43) again we obtain

$$\sigma'_m = \sqrt{\sigma_{xm}^2 + 3\tau_{xym}^2} = \sqrt{3(100)^2} = 173 \text{ MPa}$$

$$\sigma'_a = \sigma_{xa} = 80 \text{ MPa}$$

Next we plot these two components on the stress axes of the fatigue diagram of

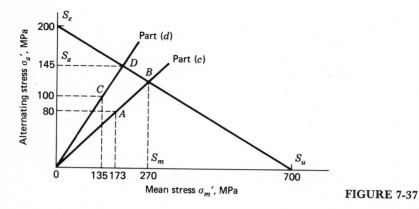

FIGURE 7-37

Fig. 7-37. A line drawn from the origin to point A, determined by these two stress coordinates, intersects the modified Goodman line at B, giving a mean strength $S_m = 270$ MPa as shown. Therefore the factor of safety guarding against fatigue failure is

$$n(\text{fatigue}) = S_m/\sigma'_m = 270/173 = 1.56 \qquad \qquad Ans.$$

(d) To determine the possibility of a static failure we first compute the maximum normal stresses and maximum shear stresses, assuming that eventually the maximums might occur simultaneously. This gives

$$\sigma_{x\,max} = \sigma_{xm} + \sigma_{xa} = 60 + 80 = 140 \text{ MPa}$$

$$\tau_{xy\,max} = \tau_{xym} + \tau_{xya} = 70 + 35 = 105 \text{ MPa}$$

Bypassing the use of Mohr's circle again, we find the maximum von Mises stress to be

$$\sigma'_{max} = \sqrt{(140)^2 + 3(105)^2} = 229 \text{ MPa}$$

Then the factor of safety is found to be

$$n(\text{static}) = S_y/\sigma'_{max} = 500/229 = 2.18 \qquad \qquad Ans.$$

To determine the possibility of a fatigue failure we use Eq. (7-43) to get

$$\sigma'_m = \sqrt{\sigma^2_{xm} + 3\tau^2_{xym}} = \sqrt{(60)^2 + 3(70)^2} = 135 \text{ MPa}$$

$$\sigma'_a = \sqrt{\sigma^2_{xa} + 3\tau^2_{xya}} = \sqrt{(80)^2 + 3(35)^2} = 100 \text{ MPa}$$

These two stress components are plotted on Fig. 7-37 as before, giving point C. A line through the origin and point C intersects the modified Goodman line at D and yields the alternating strength as $S_a = 145$ MPa. Note that we could just as well have used the mean strength, as before. The factor of safety is

$$n(\text{fatigue}) = S_a/\sigma_a' = 145/100 = 1.45 \qquad\qquad Ans.$$

////

Stress Concentration

In the development of Eq. (7-15) the factor k_e is used to reduce the endurance limit to account for stress concentration. It is obtained as the reciprocal of the fatigue-strength reduction factor K_f. When several of the stress components in Eqs. (7-42) or (7-43) have stress raisers, the several stress-concentration factors differ from each other and, consequently, a unique value of k_e does not exist.

The solution to this problem is to use $k_e = 1$ in Eq. (7-15) and to increase the various alternating stress components by their respective fatigue-strength reduction factors K_f. Thus K_f is used to increase the stress instead of decreasing the strength.

Of course, for brittle materials, the mean stresses must be increased by their respective K_f values too.

Axial Loads

In Sec. 7-7 we learned that the size factor k_b for axial loading is often different from k_b for bending and torsion. Thus a correction must be made when a part is subjected to any load combinations in which axial loading is present. The procedure we shall take is to use k_b in Eq. (7-15) for bending and torsion. Then multiply all of the axial alternating stress components in Eqs. (7-42) or (7-43) by the *axial correction factor* defined as

$$\alpha = \frac{k_b(\text{bending})}{k_b(\text{axial})} \qquad\qquad (7-45)$$

If the analysis is carried out by use of a Mohr's circle diagram, then the axial-stress components should be multiplied by the factor α before the principal stresses are found.

7-18   CUMULATIVE FATIGUE DAMAGE

Instead of a single reversed stress σ for n cycles, suppose a part is subjected to σ_1 for n_1 cycles, σ_2 for n_2 cycles, etc. Under these conditions our problem is to estimate the fatigue life of a part subjected to these reversed stresses, or to estimate the factor of safety if the part has an infinite life. A search of the literature reveals that this problem has not been solved completely. Therefore the results obtained using either of the approaches presented here should be employed as guides to indicate how you might seek improvement. They should *never* be used to obtain absolute values unless your own experiments indicate the feasibility of doing so. An approach consistently in agreement with experiment has not yet been reported in the literature of the subject.

The theory which is in greatest use at the present time to explain cumulative fatigue damage is the *Palmgren-Miner cycle-ratio summation theory,* also called *Miner's rule.** Mathematically, this theory is stated as

$$\frac{n_1}{N_1} + \frac{n_2}{N_2} + \cdots + \frac{n_i}{N_i} = C \tag{7-46}$$

where n is the number of cycles of stress σ applied to the specimen and N is the life corresponding to σ. The constant C is determined by experiment and is usually found in the range

$$0.7 \leq C \leq 2.2$$

Many authorities recommend using $C = 1$, and then Eq. (7-46) may be written

$$\sum \frac{n}{N} = 1 \tag{7-47}$$

To illustrate the use of Miner's rule, let us choose a steel having the properties $S_{ut} = 90$ kpsi and $S'_{e,0} = 40$ kpsi, where we have used the designation $S'_{e,0}$ instead of the more usual S'_e, to indicate the endurance limit of the *virgin,* or *undamaged, material.* The log S–log N diagram for this material is shown in Fig. 7-38 by the heavy solid line. Now apply, say, a reversed stress $\sigma_1 = 60$ kpsi for $n_1 = 3000$ cycles. Since $\sigma_1 > S'_{e,0}$, the endurance limit will be damaged, and we wish to find the new endurance limit $S'_{e,1}$ of the damaged material using Miner's rule. The figure shows that the material has a life $N_1 = 8320$ cycles, and consequently, after the application of σ_1 for 3000 cycles, there are $N_1 - n_1 = 5320$ cycles of life remaining. This locates the finite-life strength $S_{f,1}$ of the damaged material, as shown on Fig. 7-38. To get a second point, we ask the question, With n_1 and N_1 given, how many cycles of stress $\sigma_2 = S'_{e,0}$ can be applied before the damaged material fails? This corresponds to n_2 cycles of stress reversal, and hence, from Eq. (7-47), we have

$$\frac{n_1}{N_1} + \frac{n_2}{N_2} = 1 \tag{a}$$

---

* A. Palmgren «Die Lebensdauer von Kugellagern», *ZVDI,* vol. 68, 1924, pp. 339–341; M. A. Miner, "Cumulative Damage in Fatigue," *J. Appl. Mech.,* vol. 12, *Trans. ASME,* vol. 67, 1945 pp. A159–A164.

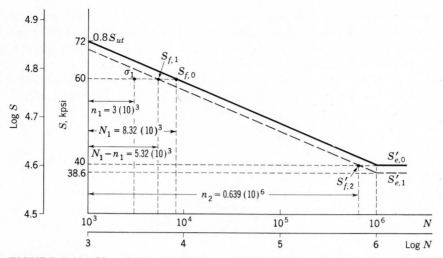

FIGURE 7-38 Use of Miner's rule to predict the endurance limit of a material that has been overstressed for a finite number of cycles.

or

$$n_2 = \left(1 - \frac{n_1}{N_1}\right)N_2 \tag{b}$$

Then

$$n_2 = \left[1 - \frac{3(10)^3}{8.32(10)^3}\right](10)^6 = 0.639(10)^6 \text{ cycles}$$

This corresponds to the finite-life strength $S_{f,2}$ in Fig. 7-38. A line through $S_{f,1}$ and $S_{f,2}$ is the log S–log N diagram of the damaged material according to Miner's rule. The new endurance limit is $S_{e,1} = 38.6$ kpsi.

Though Miner's rule is quite generally used, it fails in two ways to agree with experiment. First, note that this theory states that the static strength S_{ut} is damaged, that is, decreased, because of the application of σ_1; see Fig. 7-38 at $N = 10^3$ cycles. Experiments fail to verify this prediction.

Miner's rule, as given by Eq. (7-47), does not account for the order in which the stresses are applied, and hence ignores any stresses less than $S'_{e,0}$. But it can be seen in Fig. 7-38 that a stress σ_3 in the range $S'_{e,1} < \sigma_3 < S'_{e,0}$ would cause damage if applied after the endurance limit had been damaged by the application of σ_1.

*Manson's** approach overcomes both of the deficiencies noted for the

_____

* S. S. Manson, A. J. Nachtigall, C. R. Ensign, and J. C. Freche, "Further Investigation of a Relation for Cumulative Fatigue Damage in Bending," *Trans. ASME, J. Eng. Ind.*, ser. B, vol. 87, no. 1, February 1965, pp. 25–35.

Palmgren-Miner method; historically it is a much more recent approach, and it is just as easy to use. Except for a slight change, we shall use and recommend the Manson method in this book. Manson plotted the S–log N diagram instead of a log S–log N plot as is recommended here. Manson also resorted to experiment to find the point of convergence of the S–log N lines corresponding to the static strength, instead of arbitrarily selecting the intersection of $N = 10^3$ cycles with $S = 0.8S_{ut}$ as is done here. Of course, it is always better to use experiment, but our purpose in this book has been to use the simple tension-test data to learn as much as possible about fatigue failure.

The method of Manson, as presented here, consists in having all log S–log N lines, that is, lines for both the damaged as well as the virgin material, converge to the same point, $0.8S_{ut}$ at 10^3 cycles. In addition, the log S–log N lines must be constructed in the same historical order in which the stresses occur.

The data from the preceding example are used for illustrative purposes. The results are shown in Fig. 7-39. Note that the strength $S_{f,1}$ corresponding to $N_1 - n_1 = 5.32(10)^3$ cycles is found in the same manner as before. Through this point and through $0.8S_{ut}$ at 10^3 cycles draw the heavy dashed line to meet $N = 10^6$ cycles and define the endurance limit $S'_{e,1}$ of the damaged material. In this case the new endurance limit is 34.1 kpsi, somewhat less than that found by Miner's method.

It is now easy to see from Fig. 7-39 that a reversed stress $\sigma = 36$ kpsi, say, would not harm the endurance limit of the virgin material, no matter how many cycles it might be applied. However, if $\sigma = 36$ kpsi should be applied *after* the material had been damaged by $\sigma_1 = 60$ kpsi, then additional damage would be done.

FIGURE 7-39 Use of Manson's method to predict the endurance limit of a material that has been overstressed for a finite number of cycles.

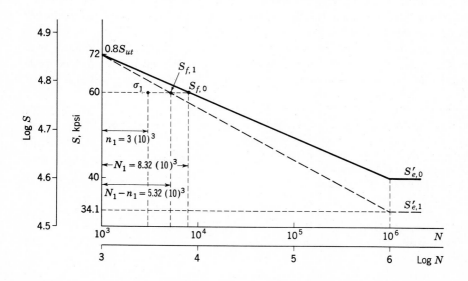

7-19 SURFACE STRENGTH

Our studies thus far have dealt with the failure of a machine element by yielding, by fracture, and by fatigue. The endurance limit obtained by the rotating-beam test is frequently called the *flexural endurance limit* because it is a test of a rotating beam. In this section we shall study a property of *mating materials* called the *surface endurance shear*. The design engineer must frequently solve problems in which two machine elements mate with one another by rolling, sliding, or a combination of rolling and sliding contact. Obvious examples of such combinations are the mating teeth of a pair of gears, a cam and follower, a wheel and rail, or a chain and sprocket. A knowledge of the surface strength of materials is necessary if the designer is to create machines having a long and satisfactory life.

When two surfaces roll or roll and slide against one another with sufficient force, a pitting failure will occur after a certain number of cycles of operation. Authorities are not in complete agreement on the exact mechanism of the pitting;* although the subject is quite complicated, they do agree that the Hertz stresses, the number of cycles, the surface finish, the hardness, the degree of lubrication, and the temperature all influence the strength. In Sec. 2-21 it was learned that, when two surfaces are pressed together, a maximum shear stress is developed slightly below the contacting surface. It is postulated by some authorities that a surface fatigue failure is initiated by this maximum shear stress and then is propagated rapidly to the surface. The lubricant then enters the crack which is formed and, under pressure, eventually wedges the chip loose.

Professor Charles Mischke of Iowa State University has observed that the pressure acting upon both contacting surfaces is repeated and hence would resemble a half cycle of Fig. 7-22a for each rotation.† This may not be true at all for a stress element below the surface where the shear stress is maximum. Mischke states that it is necessary to follow the interior point from approach to recession to see how the loading changes. He theorizes that the stresses at the depth of the maximum shear stress may be completely reversed.

To determine the surface fatigue strength of mating materials, Buckingham‡ designed a simple machine for testing a pair of contacting rolling surfaces in connection with his investigation of the wear of gear teeth. Buckingham and, later, Talbourdet§ gathered large numbers of data from many tests so that considerable design information is now available. To make the results useful for

---

* See, for example, Stewart Way, "Pitting Due to Rolling Contact," *Trans. ASME,* vol. 57, 1935, pp. A-49–A-58; and Charles Lipson and L. V. Colwell (eds.), *Handbook of Mechanical Wear,* University of Michigan Press, Ann Arbor, 1961, p. 95.

† By personal communication.

‡ Earl Buckingham, *Analytical Mechanics of Gears,* McGraw-Hill, New York, 1949, chap. 23.

§ As reported by W. D. Cram, 'Experimental Load-Stress Factors,' Charles Lipson and L. V. Colwell (eds.), *Engineering Approach to Surface Damage,* University of Michigan, Ann Arbor, 1958.

designers, Buckingham defined a *load-stress factor*, also called a *wear factor*, which is derived from the Hertz equations. Equations (2-89) and (2-90) for contacting cylinders are found to be

$$b = \sqrt{\frac{2F}{\pi l} \frac{(1 - \mu_1^2)/E_1 + (1 - \mu_2^2)/E_2}{(1/d_1) + (1/d_2)}} \tag{7-48}$$

$$p_{\text{max}} = \frac{2F}{\pi b l} \tag{7-49}$$

where b = half-width of rectangular contact area
$\quad F$ = contact force
$\quad l$ = width of cylinders
$\quad \mu$ = Poisson's ratio
$\quad E$ = modulus of elasticity
$\quad d$ = cylinder diameter

On the average, $\mu = 0.30$ for engineering materials. Thus, let $\mu = \mu_1 = \mu_2 = 0.30$. Also, it is more convenient to use the cylinder radius; so let $2r = d$. If we then designate the width of the cylinders as w instead of l and remove the square-root sign, Eq. (7-48) becomes

$$b^2 = 1.16 \frac{F}{w} \frac{(1/E_1 + (1/E_2)}{(1/r_1) + (1/r_2)} \tag{7-50}$$

Next, define a new kind of endurance property called *surface endurance strength*, which is qualified by the number of cycles at which the first tangible evidence of fatigue is observed. Using Eq. (7-49), this strength is

$$S_C = \frac{2F}{\pi b w} \tag{7-51}$$

which may also be called the *contact endurance strength*, the *contact fatigue strength*, or the *Hertzian endurance strength*. This strength is the contacting pressure which, after a specified number of cycles, will cause failure of the surface. Such failures are often called *wear* because they occur after a very long time. They should not be confused with abrasive wear, however. By substituting the value of b from Eq. (7-50) into (7-51) and rearranging, we obtain

$$2.857 S_C^2 \left(\frac{1}{E_1} + \frac{1}{E_2} \right) = \frac{F}{w} \left(\frac{1}{r_1} + \frac{1}{r_2} \right) \tag{7-52}$$

The left side of this equation contains E_1, E_2, and S_C, constants which come about because of the selection of a certain material for each element of the pair.

We call this K_1, *Buckingham's load-stress factor*. Having selected the two materials, K_1 is computed from the equation

$$K_1 = 2.857 S_C^2 \left(\frac{1}{E_1} + \frac{1}{E_2} \right)$$ (7-53)

With K_1 known, we now write the design equation as

$$K_1 = \frac{F}{w} \left(\frac{1}{r_1} + \frac{1}{r_2} \right)$$ (7-54)

which, if satisfied, defines a surface fatigue failure in 10^8 cycles of repeated stress according to Talbourdet's experiments. Since we usually want to define safety n instead of failure, we would write Eq. (7-54) in the form

$$\frac{K_1}{n} = \frac{F}{w} \left(\frac{1}{r_1} + \frac{1}{r_2} \right)$$ (7-55)

Values of the surface fatigue strength for steels can be obtained from the equation

$$S_C = \begin{cases} 0.4 H_B - 10 \text{ kpsi} \\ 2.76 H_B - 70 \text{ MPa} \end{cases}$$ (7-56)

where H_B is the Brinell hardness number and where it is understood that these strengths are only valid up to 10^8 cycles of repeated contact stress. If the two materials have different hardnesses the lesser value is generally, though not always, used. The results of this procedure agree with the values of the load-stress factors recommended by Buckingham.

PROBLEMS*

Sections 7-1 to 7-4

7-1   What is the fatigue strength of a rotating-beam specimen made of UNS G10180 hot-rolled steel corresponding to a life of $250(10)^3$ cycles of stress reversal? What would be the life of the specimen if the alternating stress were 40 kpsi?

7-2   Derive Eqs. (7-11) to (7-14).

7-3   Write a calculator program for solving Eqs. (7-11) to (7-14).

Sections 7-5 to 7-11

7-4   A $\frac{3}{16}$-in drill rod was heat-treated and ground and the measured hardness was found to be $H_B = 490$. What is the endurance limit?

---

* The asterisk indicates a design type problem or one which may have no unique result.

7-5   Find the endurance limit of a 1-in bar of UNS G10350 steel, heat-treated and tempered to 1000°F, if the bar has a machined finish.

7-6   Find the endurance limit corresponding to a reliability of 99.9 percent for a UNS G43400 steel bar, heat-treated and tempered to 1000°F, if the bar is about $1\frac{1}{2}$ in in diameter and has no points of stress concentration. The material is ground and polished.

7-7   Two connecting rods having approximately $\frac{3}{4}$ in diameters are made as forgings. They are to have reliabilities of 99.99 percent. One is made of the very best and most expensive steel available at the time, a UNS G43400 steel, heat-treated and tempered to 600°F. The other is made of an ordinary and easily obtainable medium-carbon steel, a UNS G10400 steel, heat-treated and tempered to 1000°F. Find both endurance limits. Is there any advantage in using the high-priced steel? Why?

7-8   A portion of a machine member is shown in the figure. It is loaded by completely reversed axial forces F which are uniformly distributed across the width. The material is a UNS G10180 cold-drawn steel flat. For 90 percent reliability and infinite life determine the maximum force F that can be applied.

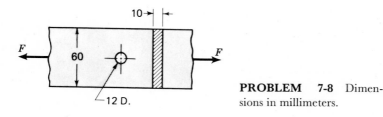

PROBLEM   7-8   Dimensions in millimeters.

7-9   The figure is an idealized representation of a machine member subjected to the action of an alternating force F which places the member in completely reversed bending. The material is UNS G10500 steel, heat-treated and tempered to 600°F, with a ground finish. Based on 50 percent reliability, infinite life, and no margin of safety, determine the maximum value of the alternating force F which can probably be applied.

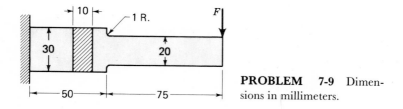

PROBLEM   7-9   Dimensions in millimeters.

7-10   The shaft shown in the figure rotates at 1720 rpm and is to have a life of 3 min at 50 percent reliability. The steel used has the following properties: $S_{ut} = 89$ kpsi, $E = 30$ Mpsi, 22.5 percent elongation in 2 in, $H_B = 178$. The shaft is finished by grinding. It is simply supported in antifriction bearings at A and B and is loaded by the static forces $F_1 = 2.0$ kip and $F_2 = 3.0$ kip. Progressing from A to B, the stress-concentration factors for the shoulders in bending are $K_t = 2.00$, 1.94, 2.08, and 2.02. Find the factor of safety guarding against failure.

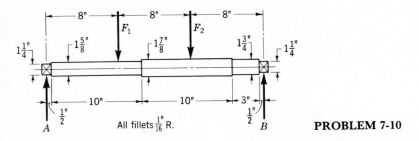

PROBLEM 7-10

7-11 The bar shown in the figure is of UNS A92017-T4 wrought aluminum alloy, and it has a fatigue strength of 18 kpsi at $5(10)^8$ cycles for reversed axial loading. Find the factor of safety if the axial load shown is completely reversed.

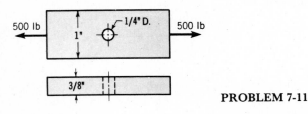

PROBLEM 7-11

7-12 The bar shown in the figure is machined from a UNS G10350 cold-drawn steel flat. The axial load shown is completely reversed. Find the factor of safety.

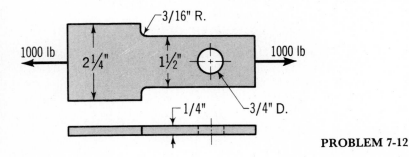

PROBLEM 7-12

7-13 The figure is a drawing of a rotating shaft loaded in completely reversed bending by the 400-lb force. The shaft is to be made of UNS G41400 steel, heat-treated to a

PROBLEM 7-13

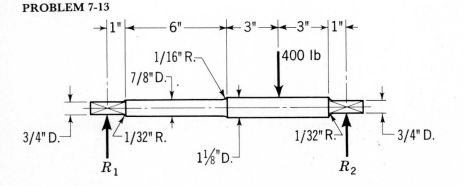

hardness of $H_B = 376$, with a ground finish in critical places. Based on an infinite life, find the factor of safety.

7-14   The rotating shaft shown in the figure is machined from a 50-mm bar of cold-drawn UNS G10350 steel. The shaft is designed for an infinite life and a reliability of 99.99 percent. What factor of safety guards against a fatigue failure if the force F is 3.0 kN?

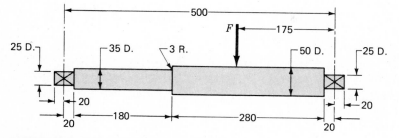

PROBLEM 7-14   Dimensions in millimeters.

7-15   The connecting rod shown in the figure is machined from a 10-mm thick bar of UNS G10350 cold-drawn steel. The bar is axially loaded by completely reversed forces acting on pins through the two drilled holes. Theoretical stress-concentration factors for the fillets may be obtained from Fig. A-26-5, and for the pin-loaded holes from Fig. A-26-12. Based on infinite life, 99 percent reliability, and a factor of safety of 2, determine the maximum safe reversed axial load that can be employed.

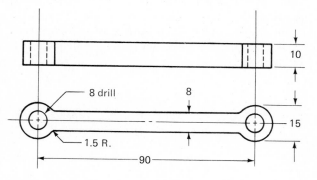

PROBLEM 7-15   Dimensions in millimeters.

7-16   The shaft shown in the figure has bearing reactions R_1 and R_2, rotates at 1150 rpm, and supports the 10-kip bending force. The specifications call for a ductile steel having $S_{ut} = 120$ kpsi and $S_y = 90$ kpsi. The shaft is to be machined and is to have a life of $80(10^3)$ cycles corresponding to 90 percent reliability. Find the safe diameter d based on a factor of safety of 1.60.

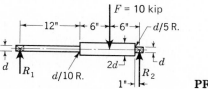

PROBLEM 7-16

7-17   The figure shows a rotating shaft loaded by two bending forces having the bearing reactions R_1 and R_2.   Point A is a shaft shoulder which is required for positioning the left-hand bearing.   The grinding-relief groove at B is 2.5 mm deep (see Fig. A-26-14).   The surface AB is ground, but the groove is machined.   The material of the shaft is UNS G43400 steel, heat-treated and tempered to 1000°F under conditions such that the ultimate tensile strength is 1.30 GPa.   Determine the factor of safety corresponding to a life of $0.35(10)^6$ revolutions of the shaft.

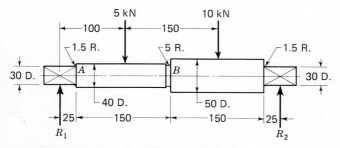

PROBLEM 7-17   Dimensions in millimeters.

7-18*   Shown in the figure is a bell-crank lever used on job-shop printing presses.   The lever is shown in its central position, and it oscillates through a total angle of 24 deg, or 12 deg each way.   The normal speed is $n = 160$ oscillations per minute, and $F_1 = 250 \cos 2\pi nt$ lb, where t is the time in minutes.   Hardened-steel pins $\frac{5}{8}$ in in diameter are used to transfer the load through nylon bushings A and B, which have an outside diameter of $\frac{7}{8}$ in and a length of $1\frac{1}{4}$ in.   They are press-fitted into holes in the lever.   The nylon bushing C has an outside diameter of $1\frac{1}{8}$ in and a length of $1\frac{1}{4}$ in.   It is also press-fitted into the lever and works on a hardened-steel stud fastened to the frame of the press.   Make a complete design of the lever, considering manufacture in quantities of 50 to 100 pieces.

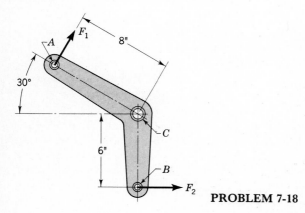

PROBLEM 7-18

7-19*   A countershaft is to be designed to support the two pulleys shown in the figure.   The countershaft rotates at 900 rpm, and 15 hp is to be transmitted.   The forces exerted on the shaft due to the belt pull are in the same plane.   The bearings mounted on the two supporting steel angles can be unfastened to assemble the pulleys and belts.

Make a complete dimensioned drawing of the shaft, specify all machining operations, and prove by calculation that the design will be satisfactory.

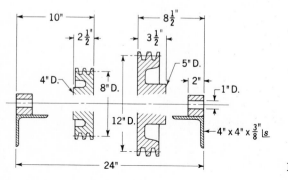

PROBLEM 7-19

Sections 7-12 and 7-13

7-20 A mechanical part is made of steel with the properties $S_u = 600$ MPa, $S_y = 480$ MPa, and $S_e = 200$ MPa. Determine the factor of safety for the following stress states:
(a) A bending stress alternating between 40 and 100 MPa
(b) A bending stress alternating between 0 and 200 MPa
(c) A pure axial compressive stress which fluctuates between 0 and 200 MPa

7-21 The figure shows a formed round-wire cantilever spring subjected to a varying force. A hardness test made on 25 springs gave a minimum hardness of 380 Bhn. It is apparent from the mounting details that there is no stress concentration. A visual inspection of the springs indicates that the surface finish corresponds closely to a hot-rolled finish. Based on a 50 percent reliability, find the number of cycles of load application likely to cause failure.

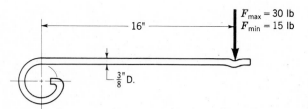

PROBLEM 7-21

7-22 The figure is a drawing of a 12-gauge (0.1094-in) $\times \frac{3}{4}$-in latching spring. The spring is assembled by deflecting it 0.075 in initially, and then deflecting it an additional 0.15 in during each latching operation. The material is machined high-carbon steel heat-treated to 490 Bhn. The stress concentration at the bend is 1.70, corresponding to a fillet radius of $\frac{1}{8}$ in and uncorrected for notch sensitivity.
(a) Find the maximum and minimum latching forces F.
(b) Will the spring fail in fatigue? Why?

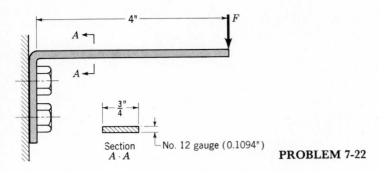

Section
A - A
No. 12 gauge (0.1094")

PROBLEM 7-22

7-23 The figure shows a short, rectangular link-rod which is loaded by the forces F acting upon the pins at each end. The forces F vary so as to produce axial tension ranging from 70 to 30 kN. A UNS G41300 steel has been selected for the link which is to be heat-treated and tempered to 1000°F after which all surfaces will be finished by grinding. Determine the dimension t to the nearest millimeter, basing the design on an infinite life at 99.999 percent reliability and a factor of safety of 1.35.

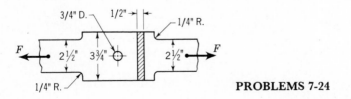

10 D. pins   18 R.

36

PROBLEM 7-23 Dimensions in millimeters.

7-24 The figure shows the free-body diagram of a connecting rod portion having stress concentration at two places. The forces F fluctuate between a tension of 4 kip and a compression of 16 kip. The material of the rod is cold-drawn UNS G10180 steel. Neglecting column action, find the factors of safety which guard against both static and fatigue failure, based on infinite life and $R = 0.50$.

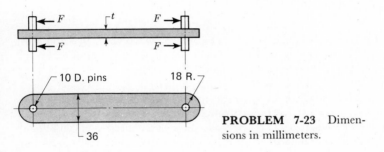

3/4" D.   1/2"   1/4" R.

F $2\frac{1}{2}$" $3\frac{3}{4}$" $2\frac{1}{2}$" F

1/4" R.

PROBLEMS 7-24

7-25 The figure shows two views of a flat steel spring loaded in bending by the force F. The spring is assembled so as to produce a preload $F_{\min} = 0.90$ kN. The force then varies from this minimum to a maximum of 3.0 kN. The spring is forged of a 95-point carbon steel and has the following properties after a suitable heat treatment: $S_u = 1400$ MPa, $S_y = 950$ MPa, $H_B = 399$, 12 percent elongation in 50 mm. Find the thickness t if $K_t = 2.50$, the factors of safety are 1.30 for the yield strength, 1.15 for the ultimate strength, 1.25 for the endurance limit, and 1.40 for the loading.

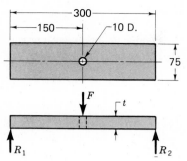

PROBLEM 7-25 Dimensions in millimeters.

7-26* The figure shows the tentative design of a cantilever screw-machine part which is to resist a load P varying from 0 to 3400 lb for an infinite number of cycles. The part is to be manufactured of AISI 1018 cold-drawn steel. Determine a suitable diameter d for the member. List all assumptions and decisions made in the computations.

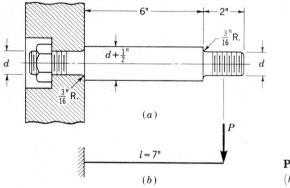

PROBLEM 7-26 (*a*) Part; (*b*) loading diagram.

7-27* The figure illustrates a table-model pneumatic impact press. A power cylinder mounted in a guide bracket A may slide vertically on a base-supported column C. In use, a chuck on the piston rod holds a tool which may be a marking, riveting, staking, stamping, or swaging tool. Not shown in the figure is a mechanism in which a foot lever is pressed which causes the entire power cylinder to move downward in the bracket until the tool is in contact with the work. A slight additional

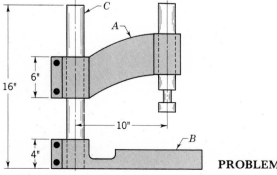

PROBLEM 7-27

pressure on the foot lever then "triggers" a valve so that sufficient air pressure is admitted to deliver the impact. The force of the impact is adjustable up to 7000 lb by regulating the volume of air admitted. Design column C for manufacture in quantities of 1000 units.

Sections 7-14 and 7-15

7-28 A part is made of a steel having $S_{ut} = 600$ MPa, $S_y = 480$ MPa, and a fully corrected endurance limit of $S_e = 160$ MPa. The stress components are $\sigma_r = 150/d^3$ and $\sigma_{max} = 240/d^3$ GPa where d is millimeters. Factors of safety for the strengths and the mean stress are all 1.20; factor of safety for the stress amplitude is 2.00. Find a safe diameter using the Gerber theory.

7-29 A diameter of 15 mm has been selected for the part of Prob. 7-28. If the factors of safety for the strengths and for the mean stress are all 1.20, what is the factor of safety for the stress amplitude based on the Gerber theory?

7-30 A diameter of 15 mm has been tentatively selected for the part of Prob. 7-28. Expected uncertainties in the strengths and the stress range are quite small, and factors of safety of 1.2 are satisfactory for all these. But there is reason to believe that the maximum stress may be subject to severe overloading. If the other factors of safety are maintained, what is the factor of safety corresponding to the maximum stress? Use the modified Goodman relation.

7-31 A part is made of a steel having $S_{ut} = 80$ kpsi, $S_y = 68$ kpsi, and fully corrected endurance limit of $S_e = 22$ kpsi. The part is subjected to a static bending stress of $\sigma_s = 84/d^3$ kpsi, where d is in inches. A stress range $\sigma_r = 50/d^3$ kpsi, also in bending, is superimposed on the static stress. We wish to find a safe diameter d for the part using a factor of safety of 1.30 for all the strengths and for the static stress. But the stress range may be subject to severe overloads, and so we wish to allow a factor of safety of 2.40 for the stress range. Use the Gerber theory and find a safe diameter d for the member.

7-32 It has been proposed that the part in Prob. 7-31 be made of a steel having $S_{ut} = 120$ kpsi, $S_y = 96$ kpsi, and a fully corrected endurance limit of $S_e = 28$ kpsi. A diameter $d = 1\frac{3}{4}$ in has been selected. Expected uncertainties in the strengths and in the static stress are normal, and factors of safety of 1.30 are satisfactory for all these. But a generous factor of safety is needed for the stress range. Use the modified Goodman relation to find the factor of safety for the stress range, assuming the other factors are held to 1.30.

7-33 Solve Prob. 7-29 graphically using the modified Goodman line with the Kimmelmann load line. Overloading occurs with the stress amplitude.

7-34 Use the modified Goodman line and the Kimmelmann load line to solve Prob. 7-32 by graphics. As indicated in the problem statement, overloading occurs with the stress range.

Section 7-16

7-35 A rotating steel shaft has the following properties: $S_u = 90$ kpsi, $S_y = 70$ kpsi, $S_e = 30$ kpsi, $S_{su} = 67$ kpsi, $S_{sy} = 40$ kpsi, and $S_{se} = 17$ kpsi. This shaft is subjected to a steady torsional stress of 9 kpsi. In addition, due to torsional vibration, the shaft is

subjected to completely reversed torsional stress of magnitude 6 kpsi. Find the factor
of safety for a fatigue failure and a static failure.

7-36   A 20-mm-diameter shaft is made of cold-drawn UNS G10350 steel and has a 6-mm-
diameter hole drilled transversely through it. Determine the factor of safety guarding against both fatigue and static failures for the following loads:
   (a)   The shaft is subject to a torque which fluctuates between 0 and 90 N · m.
   (b)   The shaft is subject to a completely reversed torque of 40 N · m.
   (c)   The shaft is loaded by a steady torque of 50 N · m together with an alternating
          component of 35 N · m.

Section 7-17†

7-37   A bar of steel has the properties $S_e = 40$ kpsi, $S_y = 60$ kpsi, and $S_u = 80$ kpsi. For
each of the cases below find the factor of safety guarding against a static failure and
either the factor of safety guarding against a fatigue failure or the expected life of the
part.
   (a)   A steady torsional stress of 15 kpsi and an alternating bending stress of 25 kpsi
   (b)   A steady torsional stress of 20 kpsi and an alternating torsional stress of 10 kpsi
   (c)   A steady torsional stress of 15 kpsi, an alternating torsional stress of 10 kpsi, and
          an alternating bending stress of 12 kpsi
   (d)   An alternating torsional stress of 30 kpsi
   (e)   An alternating torsional stress of 15 kpsi and a steady tensile stress of 15 kpsi

7-38   A spherical pressure vessel 24 in in diameter is made of cold-drawn UNS G10180
steel No. 10 gauge (0.1345 in). The vessel is to withstand an infinite number of
pressure fluctuations from 0 to $p_{\max}$.
   (a)   What maximum pressure will cause static yielding?
   (b)   What maximum pressure will eventually cause a fatigue failure? In any case,
          the joints and connections are adequately reinforced and do not weaken the
          vessel.

7-39   The figure illustrates a steel shaft supported in bearings at R_1 and R_2 and loaded by
completely reversed bending forces and by static (nonmoving) torsion. The shaft is
of UNS G10500 steel, heat-treated and tempered to 600°F, and has a ground finish.
Find the factor of safety based on the possibility of a fatigue failure for an infinite life
at 90 percent reliability.

PROBLEM 7-39

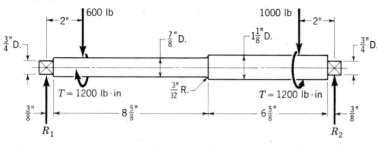

† Unless specified otherwise, use the modified Goodman relation for this set of
problems.

7-40 The figure shows a stationary torsion-bar spring loaded statically by the forces $F = 35$ N and by a torque T which varies from 0 to 8 N·m. The material is UNS G61500 steel, heat-treated and tempered to 1000°F. The ends of the spring are ground up to the shoulders. The 2.5-m body of the spring has a hot-rolled surface finish. The geometric stress-concentration factors at the shoulders are 1.68 for bending and 1.42 for torsion. Determine a suitable diameter d to the nearest millimeter, using a factor of safety of at least 1.80.

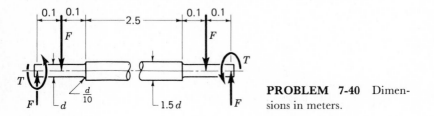

PROBLEM 7-40 Dimensions in meters.

7-41 A rotating shaft is loaded in static torsion and reversed bending by the stresses $\sigma_x = 10$ kpsi and $\tau_{xy} = 6$ kpsi referred to a mark on the shaft surface. Find the magnitude of the principal stresses and the direction of σ_1 from the x axis as the shaft makes one-half a revolution. Use 15° increments of shaft angle.

General Problems

7-42* The figure illustrates a Geneva wheel mounted on one end of a shaft of diameter d. The shaft, mounted on bearings A, B, C, and D, carries an inertial load on the other end which is to be indexed, that is, given an intermittent motion. It is well known that the angular acceleration of a Geneva, like that of a cam follower, has both a positive and a negative peak during cycle. In this example, bending is practically eliminated by the four bearings, and so a fatigue failure of the shaft by reversed torsion is the principal design problem. The material selected for the shaft is AISI 1112 cold-drawn steel. Corresponding to the maximum indexing speed, a torque amplitude of 4000 lb·in was found. The bearing mountings produce a geometric stress-concentration factor in shear of 1.90. If the shaft is to have a long and

PROBLEM 7-42

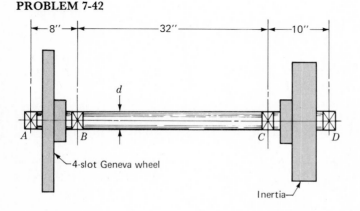

trouble-free life, what must be the diameter? Size the shaft so that the strength is at least 125 percent greater than the stress.

7-43 In this problem you are to design a countershaft on which the pinion and gear shown in the figure can be appropriately mounted. You should begin by sketching the shaft outline and locating the bearings. Mount the gears and bearings so that no axial motion is possible and so as to get the lowest bending or torsional moments wherever discontinuities exist. The radial bending forces exerted on the shaft are 4.2 kN for the pinion and 1.5 kN for the gear. These forces are stationary and in the same plane and same direction and will therefore produce completely reversed bending stresses on the rotating shaft. The torque transmitted is 200 N · m and steady. Select an appropriate material, finish, heat treatment, and margin of safety. Determine all the dimensions so that the shaft will not fail in fatigue, and record these on a freehand sketch.

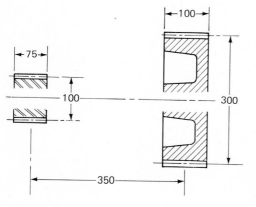

PROBLEM 7-43 Dimensions in millimeters.

Design of Mechanical Elements

CHAPTER

8

THE DESIGN
OF SCREWS,
FASTENERS, AND
CONNECTIONS

This book presupposes a knowledge of the elementary methods of fastening. Typical methods of fastening or joining parts include the use of such items as bolts, nuts, cap screws, setscrews, rivets, spring retainers, locking devices, and keys. Studies in engineering graphics and in metal processes often include instruction on various joining methods, and the curiosity of any person interested in engineering naturally results in the acquistion of a good background knowledge of fastening methods. Consequently, the purpose of this chapter is not to describe the various fasteners or tabulate available sizes, but rather to select and specify suitable ones in the design of machines and devices.

The subject is one of the most interesting in the entire field of mechanical design. The number of new inventions in the fastener field over any period you might care to mention has been tremendous. There is an overwhelming variety of fasteners available for the designer's selection. Another thing: Did you know that a good bolt material should be strong and tough, but a good nut material should be soft and ductile? Or did you know that there are certain applications where you should tighten the bolt as tightly as possible and, if it does not fail by twisting in two during tightening, there is a very good possibility that the bolt never will fail? In the material to follow you will discover the why of these questions. You will learn why a nut or bolt loosens and what you must do to keep it tight. Methods of joining parts are extremely important in the engineering of a quality design, and it is necessary to have a thorough understanding of the performance of fasteners and joints under all conditions of use and design.

Jumbo jets such as Boeing's 747 and Lockheed's L1011 require as many as 2.5 million fasteners, some of which cost several dollars apiece. The 747, for example, needs about 70 000 titanium fasteners, costing about $150,000 in all; 400 000 other close-tolerance fasteners, costing about $250,000; and 30 000 squeeze rivets priced at 50 cents each, installed. To keep costs down, Boeing, Lockheed, and their subcontractors constantly review new fastener designs, installation techniques, and tooling. Cost-saving designs and tooling will find a ready market, which will grow in value as jumbo jets proliferate.*

8-1 THREAD STANDARDS AND DEFINITIONS

The terminology of screw threads, illustrated in Fig. 8-1, is explained as follows:

The *pitch* is the distance between adjacent thread forms measured parallel to the thread axis. The pitch is the reciprocal of the number of thread forms per inch N.

The *major diameter d* is the largest diameter of a screw thread.

The *minor diameter d_r* is the smallest diameter of a screw thread.

The lead l, not shown, is the distance the nut moves parallel to the screw axis when the nut is given one turn. For a single thread, as in Fig. 8-1, the lead is the same as the pitch.

A *multiple-threaded* product is one having two or more threads cut beside each other (imagine two or more strings wound side by side around a pencil). Standardized products such as screws, bolts, nuts etc., all have single threads; a *double-threaded* screw has a lead equal to twice the pitch; a *triple-threaded* screw has a lead equal to three times the pitch, etc.

All threads are made according to the *right-hand rule* unless otherwise noted.

Figure 8-2 shows the thread geometry for the three English thread standards in most general use. Figure 8-2a represents the *American National (Unified)* thread standard which has been approved in this country and in Great Britain for use on all standard threaded products. The thread angle is 60° and the crests and roots of the thread may be either flat or rounded.

---

* *Product Engineering*, vol. 41, no. 8, Apr. 13, 1970, p. 9.

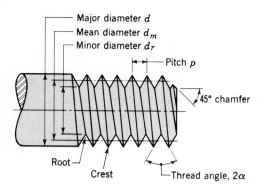

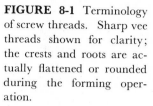

FIGURE 8-1 Terminology of screw threads. Sharp vee threads shown for clarity; the crests and roots are actually flattened or rounded during the forming operation.

Tables 8-1 and 8-2 will be useful in specifying and designing threaded parts. Note that the thread size is specified by giving the pitch p for metric sizes and by giving the number of threads per inch N for the unified sizes. The screw sizes in Table 8-2 under $\frac{1}{4}$ in in diameter are numbered or gauge sizes. The second column in Table 8-2 shows that a No. 8 screw has a nominal diameter of 0.1640 in.

A great many tensile tests of threaded rods have shown that an unthreaded rod having a diameter equal to the mean of the pitch and minor diameters will have the same tensile strength as the threaded rod. The area of this unthreaded rod is called the tensile-stress area A_t of the threaded rod; values of A_t are listed in both tables.

Two major Unified thread series are in common use, UN and UNR. The difference between these is simply that a root radius must be used in the UNR series. Because of reduced thread stress-concentration factors, UNR series have improved fatigue strengths. Unified threads are specified by stating the nominal diameter, the number of threads per inch, and the thread series, for example, $\frac{5}{8}$-18 UNRF or 0.625-18 UNRF.

Metric threads are specified by writing the diameter and pitch in millimeters, in that order. Thus, M12 × 1.75 is a thread having a nominal major diameter of

FIGURE 8-2 (a) American National or Unified thread; width of root flat is $p/4$; (b) square thread; (c) Acme thread.

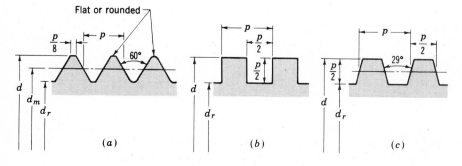

Table 8-1 DIAMETERS AND AREAS OF COARSE-PITCH AND FINE-PITCH METRIC THREADS (ALL DIMENSIONS IN MILLIMETERS)*

Nominal major diameter, d	Coarse-pitch series			Fine-pitch series		
	Pitch p	Tensile-stress area, A_t	Minor-diameter area, A_r	Pitch p	Tensile-stress area, A_t	Minor-diameter area, A_r
1.6	0.35	1.27	1.07			
2	0.04	2.07	1.79			
2.5	0.45	3.39	2.98			
3	0.5	5.03	4.47			
3.5	0.6	6.78	6.00			
4	0.7	8.78	7.75			
5	0.8	14.2	12.7			
6	1	20.1	17.9			
8	1.25	36.6	32.8	1	39.2	36.0
10	1.5	58.0	52.3	1.25	61.2	56.3
12	1.75	84.3	76.3	1.25	92.1	86.0
14	2	115	104	1.5	125	116
16	2	157	144	1.5	167	157
20	2.5	245	225	1.5	272	259
24	3	353	324	2	384	365
30	3.5	561	519	2	621	596
36	4	817	759	2	915	884
42	4.5	1120	1050	2	1260	1230
48	5	1470	1380	2	1670	1630
56	5.5	2030	1910	2	2300	2250
64	6	2680	2520	2	3030	2980
72	6	3460	3280	2	3860	3800
80	6	4340	4140	1.5	4850	4800
90	6	5590	5360	2	6100	6020
100	6	6990	6740	2	7560	7470
110				2	9180	9080

* The equations and data used to develop this table have been obtained from ANSI B1.1-1974 and B18.3.1-1978. The minor diameter was found from the equation $d_r = d - 1.226\ 869p$, and the pitch diameter from $d_m = d - 0.649\ 519p$. The mean of the pitch diameter and the minor diameter was used to compute the tensile-stress area.

12 mm and a pitch of 1.75 mm. Note the letter M, which precedes the diameter, is the clue to the metric designation.

Square and Acme threads are used on screws when power is to be transmitted. Since each application is a special one, there is really no need for a standard relating the diameter to the number of threads per inch.

Modifications are frequently made to both Acme and square threads. For instance, the square thread is sometimes modified by cutting the space between the teeth so as to have an included thread angle of 10–15°. This is not difficult, since these threads are usually cut with a single-point tool anyhow; the modifi-

Table 8-2 DIAMETERS AND AREAS OF UNIFIED SCREW THREADS UNC AND UNF*

Size designation	Nominal major diameter, in	Coarse series—UNC			Fine series—UNF		
		Threads per inch, N	Tensile-stress area A_t, in^2	Minor-diameter area A_r, in^2	Threads per inch, N	Tensile-stress area A_t, in^2	Minor-diameter area A_r, in^2
0	0.0600				80	0.001 80	0.001 51
1	0.0730	64	0.002 63	0.002 18	72	0.002 78	0.002 37
2	0.0860	56	0.003 70	0.003 10	64	0.003 94	0.003 39
3	0.0990	48	0.004 87	0.004 06	56	0.005 23	0.004 51
4	0.1120	40	0.006 04	0.004 96	48	0.006 61	0.005 66
5	0.1250	40	0.007 96	0.006 72	44	0.008 80	0.007 16
6	0.1380	32	0.009 09	0.007 45	40	0.010 15	0.008 74
8	0.1640	32	0014 0	0.011 96	36	0.014 74	0.012 85
10	0.1900	24	0.017 5	0.014 50	32	0.020 0	0.017 5
12	0.2160	24	0.024 2	0.020 6	28	0.025 8	0.022 6
$\frac{1}{4}$	0.2500	20	0.031 8	0.026 9	28	0.036 4	0.032 6
$\frac{5}{16}$	0.3125	18	0.052 4	0.045 4	24	0.058 0	0.052 4
$\frac{3}{8}$	0.3750	16	0.077 5	0.067 8	24	0.087 8	0.080 9
$\frac{7}{16}$	0.4375	14	0.106 3	0.093 3	20	0.118 7	0.109 0
$\frac{1}{2}$	0.5000	13	0.141 9	0.125 7	20	0.159 9	0.148 6
$\frac{9}{16}$	0.5625	12	0.182	0.162	18	0.203	0.189
$\frac{5}{8}$	0.6250	11	0.226	0.202	18	0.256	0.240
$\frac{3}{4}$	0.7500	10	0.334	0.302	16	0.373	0.351
$\frac{7}{8}$	0.8750	9	0.462	0.419	14	0.509	0.480
1	1.0000	8	0.606	0.551	12	0.663	0.625
$1\frac{1}{4}$	1.2500	7	0.969	0.890	12	1.073	1.024
$1\frac{1}{2}$	1.5000	6	1.405	1.294	12	1.315	1.260

* This table was compiled from ANSI B1.1-1974. The minor diameter was found from the equation $d_r = d - 1.299\,038p$, and the pitch diameter from $d_m = d - 0.649\,519p$. The mean of the pitch diameter and the minor diameter was used to compute the tensile-stress area.

cation retains most of the high efficiency inherent in square threads and makes the cutting simpler. Acme threads are sometimes modified to a stub form by making the tooth shorter. This results in a larger minor diameter, and consequently a stronger screw.

8-2 THE MECHANICS OF POWER SCREWS

A power screw is a device used in machinery to change angular motion into linear motion and, usually, to transmit power. Familiar applications include the lead screws of lathes and the screws for vises, presses, and jacks.

A schematic representation of the application of power screws to a power-driven press is shown in Fig. 8-3. In use, a torque T is applied to the ends of the screws through a set of gears, thus driving the head of the press downward against the load.

In Fig. 8-4 a square-threaded power screw with single thread having a mean diameter d_m, a pitch p, a lead angle λ, and a helix angle ψ is loaded by the axial compressive force F. We wish to find an expression for the torque required to raise this load, and another expression for the torque required to lower the load.

First, imagine that a single thread of the screw is unrolled or developed (Fig. 8-5) for exactly one turn. Then one edge of the thread will form the hypotenuse of a right triangle whose base is the circumference of the mean-thread-diameter circle and whose height is the lead. The angle λ, in Figs. 8-4 and 8-5, is the lead angle of the thread. We represent the summation of all the unit axial forces acting upon the normal thread area by F. To raise the load, a force P acts to the right (Fig. 8-5a), and to lower the load, P acts to the left (Fig. 8-5b). The friction force is the product of the coefficient of friction μ with the normal force N, and

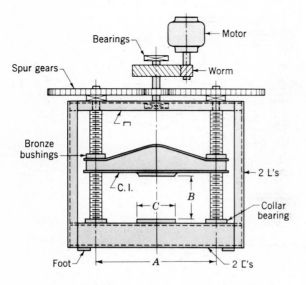

FIGURE 8-3 Small press operated by power screws.

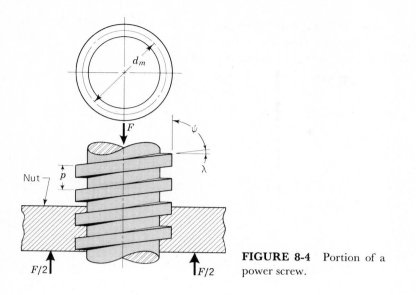

FIGURE 8-4 Portion of a power screw.

acts to oppose the motion. The system is in equilibrium under the action of these forces, and hence, for raising the load, we have

$$\sum F_H = P - N \sin \lambda - \mu N \cos \lambda = 0$$

$$\sum F_V = F + \mu N \sin \lambda - N \cos \lambda = 0$$

$$(a)$$

In a similar manner, for lowering the load, we have

$$\sum F_H = -P - N \sin \lambda + \mu N \cos \lambda = 0$$

$$\sum F_V = F - \mu N \sin \lambda - N \cos \lambda = 0$$

$$(b)$$

Since we are not interested in the normal force N, we eliminate it from each of

FIGURE 8-5 Force diagrams: (a) lifting the load; (b) lowering the load.

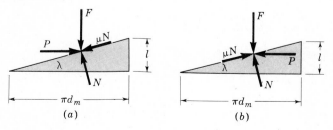

these sets of equations and solve the result for P. For raising the load this gives

$$P = \frac{F(\sin \lambda + \mu \cos \lambda)}{\cos \lambda - \mu \sin \lambda} \tag{c}$$

and for lowering the load,

$$P = \frac{F(\mu \cos \lambda - \sin \lambda)}{\cos \lambda + \mu \sin \lambda} \tag{d}$$

Next, divide numerator and denominator of these equations by $\cos \lambda$ and use the relation $\tan \lambda = l/\pi d_m$ (Fig. 8-5). We then have, respectively,

$$P = \frac{F[(l/\pi d_m) + \mu]}{1 - (\mu l/\pi d_m)} \tag{e}$$

$$P = \frac{F[\mu - (l/\pi d_m)]}{1 + (\mu l/\pi d_m)} \tag{f}$$

Finally, noting that the torque is the product of the force P and the mean radius $d_m/2$, for raising the load we can write

$$T = \frac{F d_m}{2} \left(\frac{l + \pi \mu d_m}{\pi d_m - \mu l} \right) \tag{8-1}$$

where T is the torque required for two purposes: to overcome thread friction and to raise the load.

The torque required to lower the load, from Eq. (f), is found to be

$$T = \frac{F d_m}{2} \left(\frac{\pi \mu d_m - l}{\pi d_m + \mu l} \right) \tag{8-2}$$

This is the torque required to overcome a part of the friction in lowering the load. It may turn out, in specific instances where the lead is large or the friction is low, that the load will lower itself by causing the screw to spin without any external effort. In such cases, the torque T from Eq. (8-2) will be negative or zero. When a positive torque is obtained from this equation, the screw is said to be *self-locking*. Thus the condition for self-locking is

$$\pi \mu d_m > l$$

Now divide both sides of this inequality by πd_m. Recognizing that $l/\pi d_m = \tan \lambda$, we get

$$\mu > \tan \lambda \tag{8-3}$$

This relation states that self-locking is obtained whenever the coefficient of thread friction is equal to or greater than the tangent of the thread lead angle.

An expression for efficiency is also useful in the evaluation of power screws. If we let $\mu = 0$ in Eq. (8-1), we obtain

$$T_0 = \frac{Fl}{2\pi} \qquad\qquad (h)$$

which, since thread friction has been eliminated, is the torque required only to raise the load. The efficiency is therefore

$$e = \frac{T_0}{T} = \frac{Fl}{2\pi T} \qquad\qquad (8\text{-}4)$$

The preceding equations have been developed for square threads where the normal thread loads are parallel to the axis of the screw. In the case of Acme or Unified threads, the normal thread load is inclined to the axis because of the thread angle 2α and the lead angle λ. Since lead angles are small, this inclination can be neglected and only the effect of the thread angle (Fig. 8-6a) considered. The effect of the angle α is to increase the frictional force by the wedging action of the threads. Therefore the frictional terms in Eq. (8-1) must be divided by $\cos \alpha$. For raising the load, or for tightening a screw or bolt, this yields

$$T = \frac{Fd_m}{2}\left(\frac{l + \pi\mu d_m \sec \alpha}{\pi d_m - \mu l \sec \alpha}\right) \qquad\qquad (8\text{-}5)$$

FIGURE 8-6 (a) Normal thread force is increased because of angle α; (b) thrust collar has frictional diameter d_c.

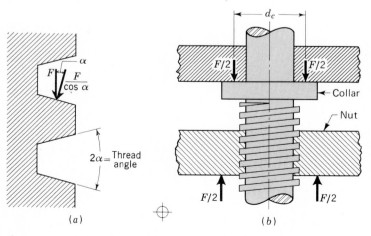

In using Eq. (8-5), remember that it is an approximation because the effect of the lead angle has been neglected.

For power screws the Acme thread is not so efficient as the square thread because of the additional friction due to the wedging action, but it is often preferred because it is easier to machine and permits the use of a split nut, which can be adjusted to take up for wear.

Usually a third component of torque must be applied in power-screw applications. When the screw is loaded axially, a thrust or collar bearing must be employed between the rotating and stationary members in order to take out the axial component. Figure 8-6b shows a typical thrust collar in which the load is assumed to be concentrated at the mean collar diameter d_c. If μ_c is the coefficient of collar friction, the torque required is

$$T_c = \frac{F\mu_c d_c}{2} \tag{8-6}$$

For large collars the torque should probably be computed in a manner similar to that employed for disk clutches.*

Example 8-1 A power screw has 6 square threads per inch, double threads, and a major diameter of 1 in and is to be used in an application similar to that of Fig. 8-3. The given data include $\mu = \mu_c = 0.08$, $d_c = 1.25$ in, and $F = 1500$ lb per screw.

    (a) Find the pitch, thread depth, thread width, mean diameter, minor diameter, and lead.
    (b) Find the torque required to rotate the screw "against" the load.
    (c) Find the torque required to rotate the screw "with" the load.
    (d) Find the overall efficiency.

Solution (a) Since $N = 6$, $p = \frac{1}{6}$ in. From Fig. 8-2b the thread depth and width are the same and equal to half the pitch, or $\frac{1}{12}$ in. Also,

$$d_m = d - \frac{p}{2} = 1 - \tfrac{1}{12} = 0.9167 \text{ in} \qquad\qquad\qquad Ans.$$

$$d_r = d - p = 1 - \tfrac{1}{6} = 0.8333 \text{ in} \qquad\qquad\qquad Ans.$$

$$l = np = (2)(\tfrac{1}{6}) = 0.333 \text{ in} \qquad\qquad\qquad Ans.$$

---

* See Sec. 16-5.

(b) Using Eqs. (8-1) and (8-6), the torque required to turn the screw against the load is

$$T = \frac{Fd_m}{2}\left(\frac{l + \pi\mu d_m}{\pi d_m - \mu l}\right) + \frac{F\mu_c d_c}{2}$$

$$= \frac{(1500)(0.9167)}{2}\left[\frac{0.333 + \pi(0.08)(0.9167)}{\pi(0.9167) - (0.08)(0.333)}\right] + \frac{(1500)(0.08)(1.25)}{2}$$

$$= 136 + 75 = 211 \text{ lb} \cdot \text{in} \qquad\qquad Ans.$$

(c) The torque required to lower the load, that is, to rotate the screw with the load, is obtained using Eqs. (8-2) and (8-6). Thus

$$T = \frac{Fd_m}{2}\left(\frac{\pi\mu d_m - l}{\pi d_m + \mu l}\right) + \frac{F\mu_c d_c}{2}$$

$$= \frac{(1500)(0.9167)}{2}\left[\frac{\pi(0.08)(0.9167) - 0.333}{\pi(0.9167) + (0.08)(0.333)}\right] + \frac{(1500)(0.08)(1.25)}{2}$$

$$= -24.4 + 75 \approx 50 \text{ lb} \cdot \text{in} \qquad\qquad Ans.$$

The minus sign in the first term indicates that the screw alone is not self-locking and would rotate due to the action of the load except for the fact that collar friction is present and must be overcome too. Thus the torque required to rotate the screw "with" the load is less than is necessary to overcome collar friction alone.

(d) The overall efficiency is

$$e = \frac{Fl}{2\pi T} = \frac{(1500)(0.333)}{(2\pi)(211)} = 0.377 \qquad\qquad Ans.$$

$$////$$

8-3 THREAD STRESSES

In Fig. 8-6b a force F is transmitted through a square-threaded screw into a nut. We are interested in finding the stresses in the nut threads and in the screw threads which might cause these threads to fail, say, by yielding.

If we assume that the load is uniformly distributed over the nut height h and that the screw threads would fail by shearing off on the minor diameter, then the average screw-thread shear stress is

$$\tau = \frac{2F}{\pi d_r h} \qquad\qquad (8-7)$$

The threads on the nut would shear off on the major diameter, and so the average nut-thread shear stress is

$$\tau = \frac{2F}{\pi d h} \qquad (8\text{-}8)$$

We note very particularly that these are *average stresses* because we have assumed that the threads share the load equally. There are many cases, as we shall discover later, where this assumption is grossly in error. In view of this, rather large factors of safety, $n > 2$, should be used when Eqs. (8-7) and (8-8) are used for design purposes.

The bearing stress in the threads is

$$\sigma = \frac{-4pF}{\pi h (d^2 - d_r^2)} \qquad (8\text{-}9)$$

and this is an average stress too, because the force is assumed to be uniformly distributed over the face of the threads. Actually, there may be some bending of the thread, and so a high factor of safety should be employed in this case too.

Similar thread-stress formulas can be developed for other thread forms. In the case of threaded fasteners, ANSI has given several methods for the calculation of thread shear stresses. The simplest of these assumes that the bolt and nut materials are of similar tensile strengths. If thread stripping occurs in this case, it will occur simultaneously in both the internal and external threads at or near the pitch diameter. The shear area A_s can be computed as

$$A_s = \frac{\pi d_m L_e}{2} \qquad (8\text{-}10)$$

where L_e is the engaged length of the thread. Thread stresses calculated from such areas are highly approximate. Furthermore the nut material is often significantly different from that of the bolt. So at the best these equations only provide educated guesses as to the stresses. Remember, a good engineer never went wrong confirming his or her analysis with lots of good laboratory tests.

8-4 THREADED FASTENERS

Fasteners are named according to how they are intended to be used rather than how they are actually employed in specific applications. If this basic fact is remembered, it will not be difficult to distinguish between a *screw* and a *bolt*.

If a product is designed so that its primary purpose is assembly into a tapped hole, it is a *screw*. Thus, a screw is tightened by exerting torque on the *head*.

If the product is designed so that it is intended to be used with a nut, it is a *bolt*. A bolt is tightened by exerting torque on the *nut*.

A *stud* resembles a threaded rod; one end assembles into a tapped hole, the other end receives a nut.

It is the intent, rather than the actual use, which determines the name of a product. Thus, it may be desirable on various occasions to drill through two sheets of steel, say, and join them using a machine screw and a nut.

Space does not permit a complete tabulation of the dimensions of the large variety of threaded products, but Tables A-28 to A-31 show some of the sizes of bolts, screws, and nuts.

Cap screws (finished screws with hexagonal or other type head) are employed in sizes from $\frac{1}{4}$ in up to and including 3 in, as shown in Table A-29. Some of the head configurations, such as socket heads and flat heads, require no wrench clearance for installation. Metric cap screws are made in sizes from 5 mm up to 100 mm in diameter. They are available in fine- as well as coarse-pitch series.

There is no difference between a hexagon-head cap screw and a *finished* hexagon-head bolt. Of course, bolts are also made in *semifinished* and in *unfinished*, forms which may be dimensionally different. The following characteristics should be noted in the figure in Table A-30: the chamfer on the head and on the body, the length of thread of $\left(2d + \frac{1}{4}\right)$ in, the washer face under the head, and the fillet radius.

8-5   BOLTED JOINTS IN TENSION

When a connection is desired which can be disassembled without destructive methods and which is strong enough to resist both external tensile loads and shear loads, or a combination of these, then the simple bolted joint using hardened washers is a good solution. However, when the major loads are of the shear type, it is recommended that rivets be considered since rivets fill their holes, thus helping to assure a uniform distribution of loads among the supporting rivets. Bolted joints have clearance between the bolt and the hole. Manufacturing tolerances will allow certain bolts to carry an unpredictable share of the load.

A portion of a bolted joint is illustrated in Fig. 8-7. Notice the clearance

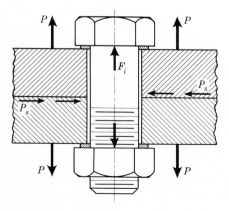

FIGURE   8-7   A   bolted connection.

space between the bolt and the hole. The bolt in this application has been preloaded to an initial tensile load F_i, then the external load P and the external shear load P_s are applied. The effect of the preload is to place the bolted member components in compression for better resistance to the external tensile load and to create a friction force between the parts to resist the shear load. The shear load does not affect the final bolt tension, and we shall neglect this load for the time being in order to study the effect of the external tensile load on the compression of the parts and the resultant bolt tension.

The *spring constant*, or *stiffness constant*, of an elastic member such as a bolt, as we learned in Chap. 3, is the ratio of the force applied to the member to the deflection produced by that force. The deflection of a bar in simple tension or compression was found to be

$$\delta = \frac{Fl}{AE} \tag{a}$$

where δ = deflection
$\quad F$ = force
$\quad A$ = area
$\quad E$ = modulus of elasticity

Therefore the stiffness constant is

$$k = \frac{F}{\delta} = \frac{AE}{l} \tag{b}$$

In finding the stiffness of a bolt, A is the area based on the nominal or major diameter because the effect of the threads is neglected. The grip l is the total thickness of the parts which have been fastened together. Note that this is somewhat less than the length of the bolt.

Let us now visualize a tension-loaded bolted connection. We use the following nomenclature:

$\quad P$ = total external load on bolted assembly
$\quad F_i$ = preload on bolt due to tightening and in existence before P is applied
$\quad P_b$ = portion of P taken by bolt
$\quad P_m$ = portion of P taken by members
$\quad F_b$ = resultant bolt load
$\quad F_m$ = resultant load on members

When the external load P is applied to the preloaded assembly, there is a change in the deformation of the bolt and also in the deformation of the connected members. The bolt, initially in tension, gets longer. This *increase* in deformation of the bolt is

$$\Delta\delta_b = \frac{P_b}{k_b} \tag{c}$$

The connected members have initial compression due to the preload. When the external load is applied, this compression will *decrease*. The decrease in deformation of the members is

$$\Delta\delta_m = \frac{P_m}{k_m} \tag{d}$$

On the assumption that the members have not separated, the increase in deformation of the bolt must equal the decrease in deformation of the members, and consequently

$$\frac{P_b}{k_b} = \frac{P_m}{k_m} \tag{e}$$

Since $P = P_b + P_m$, we have

$$P_b = \frac{k_b P}{k_b + k_m} \tag{f}$$

Therefore the resultant load on the bolt is

$$F_b = P_b + F_i = \frac{k_b P}{k_b + k_m} + F_i \tag{8-11}$$

In the same manner, the resultant load in the connected members is found to be

$$F_m = \frac{k_m P}{k_b + k_m} - F_i \tag{8-12}$$

Equations (8-11) and (8-12) hold only as long as some of the initial compression remains in the members. If the external force is large enough to remove this compression completely, the members will separate and the entire load will be carried by the bolt.

Figure 8-8 is a plot of the force-deflection characteristics and shows what is happening. The line k_m is the stiffness of the members; any force, such as the preload F_i, will cause a compressive deformation δ_m in the members. The same force will cause a tensile deformation δ_b in the bolt. When an external load is applied, δ_m is reduced by the amount $\Delta\delta_m$ and δ_b is increased by the same amount

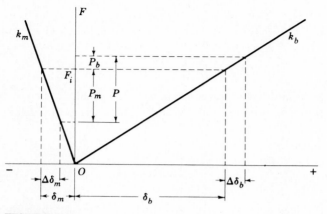

FIGURE 8-8

$\Delta\delta_b = \Delta\delta_m$. Thus the load on the bolt increases and the load in the members decreases.

The following example is used to illustrate the meaning of Eqs. (8-11) and (8-12). In Fig. 8-9a the fish scale with the 150-lb weight is analogous to a bolt tightened to a tensile preload of 150 lb. Then in Fig. 8-9b a block is forced in position as shown and the 150-lb weight removed and replaced with a 20-lb weight. Figure 8-9b now represents a bolted assembly having a bolt preload of 150 lb and an external load of 20 lb. Adding the 20-lb weight does not increase the tension in shank A, which represents the bolt. If the tension were greater than

FIGURE 8-9

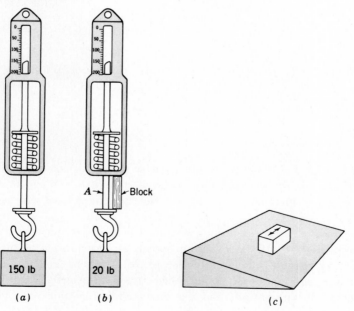

150 lb, the scale would read more than 150 lb and the block would fall out. This is an extreme example, since the stiffness constant of the block k_m is a great deal more than that of the scale k_b, but it does illustrate the advantages to be gained by proper preloading. The following example is more realistic.

Example 8-2 In Fig. 8-7, let $k_m = 8k_b$. If the preload is $F_i = 1000$ lb and the external load is $P = 1100$ lb, what is the resultant tension in the bolt and the compression in the members?

Solution   From Eq. (8-11) the resultant bolt tension is

$$F_b = \frac{k_b P}{k_b + k_m} + F_i = \frac{k_b(1100)}{k_b + 8k_b} + 1000 = 1122 \text{ lb} \qquad \textit{Ans.}$$

The resultant compression of the members, from Eq. (8-12), is

$$F_m = \frac{k_m P}{k_b + k_m} - F_i = \frac{8k_b(1100)}{k_b + 8k_b} - 1000 = -22 \text{ lb} \qquad \textit{Ans.}$$

This example shows that the proportion (11 percent) of the load taken by the bolt is small and that it depends on the relative stiffness. The members are still in compression, and hence there is no separation of the parts even though the external load is greater than the preload.   ////

The importance of preloading of bolts cannot be overestimated. A high preload improves both the fatigue resistance of a bolted connection and the locking effect. To see why this is true, imagine an external tensile load which varies from 0 to P. If the bolts are preloaded, only about 10 percent of this load will cause a fluctuating bolt stress. Thus we will be operating with a very small σ_a/σ_m slope on the modified Goodman fatigue diagram.

To see why preloading of bolts improves the locking effect, visualize an inclined plane representing a screw thread (Fig. 8-9c), and on this plane we place a block, which is analogous to the nut. Now if the block is wiggled back and forth, it will eventually work its way down the plane. This is analogous to the loosening of a nut. To retain a tight nut, the resultant bolt tension should vary as little as possible; in other words, σ_a should be very small compared with σ_m.

8-6   COMPRESSION OF BOLTED MEMBERS

In the preceding section we learned of the importance of designing a bolted connection such that the stiffness of the members is large compared to the stiffness of the bolt. In this section we are concerned with the determination of the stiffness of the members.

There may be more than two members included in the grip of the bolt. These act like compressive springs in series, and hence the total spring rate of the *members* is

$$\frac{1}{k_m} = \frac{1}{k_1} + \frac{1}{k_2} + \frac{1}{k_3} + \cdots + \frac{1}{k_i} \qquad (a)$$

If one of the members is a soft gasket, its stiffness relative to the other members is usually so small that for all practical purposes the others can be neglected and only the gasket stiffness used.

If there is no gasket, the stiffness of the members is rather difficult to obtain, except by experimentation, because the compression spreads out between the bolt head and nut and hence the area is not uniform. In many cases the geometry is such that this area can be determined.

When the area under compression cannot be determined Osman[*] suggests that a hollow cylinder be used of height equal to the grip length and of annular area equal to 1.25A, where A is the nominal area of the bolt. This method does not account for spreading of the pressure as it proceeds down into the material. Ito[†] has used ultrasonic techniques to determine the pressure distribution at the member interface. The results show that the pressure stays high out to about 1.5 bolt radii, which corresponds to Osman's annular area of 1.25A. The pressure, however, falls off farther away from the bolt. Thus Ito suggests the use of the Rotsher's pressure-cone method for stiffness calculations with a variable cone angle. This method is quite complicated, and so Mischke[‡] has developed a simpler approach using a fixed-cone angle.

Figure 8-10 illustrates the cone geometry. The top surface has a diameter d_w equal to the diameter of the washer face of the bolt. As shown in Fig. 8-10b, the surface of the cone makes a fixed cone angle of 45° to the centerline of the bolt. An inverted duplicate of this cone contacts the washer face of the nut. Thus the height of each cone is half the grip or $l/2$ as shown.

The elongation of an element of the cone of thickness dx subjected to a tensile force P is, from Eq. (3-3),

$$d\delta = \frac{P\,dx}{EA} \qquad (b)$$

---

[*] M. O. M. Osman, W. M. Mansour, and R. V. Dukkipati, "On the Design of Bolted Connections with Gaskets Subjected to Fatigue Loading," ASME Paper no. 76-DET-57, 1976.

[†] Y. Ito, J. Toyoda, and S. Nagata, "Interface Pressure Distribution in a Bolt-Flange Assembly," ASME Paper no. 77-WA/DE-11, 1977.

[‡] Charles R. Mischke, class notes, Iowa State University, 1978.

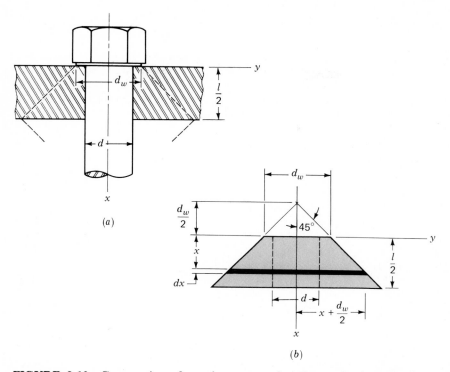

FIGURE 8-10 Compression of members assumed to be confined to the frustum of a hollow cone.

The area of the element is

$$A = \pi(r_0^2 - r_i^2) = \pi\left[\left(x + \frac{d_w}{2}\right)^2 - \left(\frac{d}{2}\right)^2\right]$$

$$= \pi\left[\left(x + \frac{d_w + d}{2}\right)\left(x + \frac{d_w - d}{2}\right)\right] \qquad (c)$$

Substituting this into Eq. (b) and integrating the left side gives the elongation as

$$\delta = \frac{P}{\pi E} \int_0^{l/2} \frac{dx}{\left(x + \dfrac{d_w + d}{2}\right)\left(x + \dfrac{d_w - d}{2}\right)} \qquad (d)$$

Using a table of integrals we find

$$\int \frac{dx}{(ax + b)(cx + d)} = \frac{1}{bc - ad} \ln \frac{cx + d}{ax + b} \qquad bc - ad \neq 0$$

After evaluating the integral, Eq. (d) becomes

$$\delta = \frac{P}{\pi Ed} \, ln \left[\frac{(l + d_w - d)(d_w + d)}{(l + d_w + d)(d_w - d)} \right] \qquad (e)$$

The diameter of the washer face is very nearly equal to the width across flats of the hexagon which is usually 50 percent greater than the bolt diameter. (See Table A-30.) By substituting $d_w = 1.5d$, Eq. (e) simplifies to

$$\delta = \frac{P}{\pi Ed} \, ln \left[5 \left(\frac{l + 0.5d}{l + 2.5d} \right) \right] \qquad (f)$$

Thus the spring rate of a single frustum is

$$k = \frac{P}{\delta} = \frac{\pi Ed}{ln \left[5 \left(\dfrac{l + 0.5d}{l + 2.5d} \right) \right]} \qquad (g)$$

But the two cones act like springs in series and so, from Eq. (a), we learn that $k_m = k/2$. Thus the spring rate or stiffness of the members is

$$k_m = \frac{\pi Ed}{2 \, ln \left[5 \left(\dfrac{l + 0.5d}{l + 2.5d} \right) \right]} \qquad (8\text{-}13)$$

Figure 8-11 shows how the spring rate varies as the grip-length-to-diameter ratio increases. Note the tendency to level off for values of l/d greater than 1.

FIGURE 8-11   Plot of Eq. (8-13) showing how the stiffness softens with increasing grip length. Note that both ratios are dimensionless.

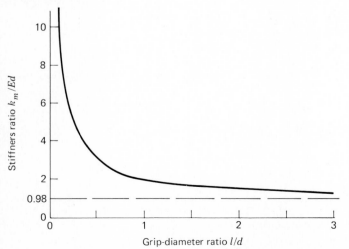

8-7 TORQUE REQUIREMENTS

Having learned that a high preload is very desirable in important bolted connections, we must next consider means of assuring that the preload is actually developed when the parts are assembled.

If the overall length of the bolt can actually be measured with a micrometer when it is assembled, the bolt elongation due to the preload F_i can be computed using the formula $\delta = F_i l/AE$. Then simply tighten the nut until the bolt elongates through the distance δ. This assures that the desired preload has been attained.

The elongation of a screw cannot usually be measured because the threaded end is often in a blind hole. It is also impractical in many cases to measure bolt elongation. In such cases the wrench torque required to develop the specified preload must be estimated. Then torque wrenching, pneumatic-impact wrenching, or the turn-of-the-nut method may be used.

The torque wrench has a built-in dial which indicates the proper torque.

With impact wrenching the air pressure is adjusted so that the wrench stalls when the proper torque is obtained, or in some wrenches, the air automatically shuts off at the desired torque.

The turn-of-the-nut method requires that we first define the meaning of snug-tight. The *snug-tight* condition is the tightness attained by a few impacts of an impact wrench, or the full effort of a person using an ordinary wrench. When the snug-tight condition is attained, all additional turning develops useful tension in the bolt. The turn-of-the-nut method requires that one compute the fractional number of turns necessary to develop the required preload from the snug-tight condition. For example, for heavy hexagon structural bolts the turn-of-the-nut specification states that the nut should be turned a minimum of 180° from the snug-tight condition under optimum conditions. Note that this is also about the correct rotation for the wheel nuts of a passenger car.

Although the coefficients of friction may vary widely, we can obtain a good estimate of the torque required to produce a given preload by combining Eqs. (8-5) and (8-6).

$$T = \frac{F_i d_m}{2}\left(\frac{1 + \pi\mu d_m \sec\alpha}{\pi d_m - \mu l \sec\alpha}\right) + \frac{F_i \mu_c d_c}{2} \tag{a}$$

Since $\tan\lambda = l/\pi d_m$, we divide the numerator and denominator of the first term by πd_m and get

$$T = \frac{F_i d_m}{2}\left(\frac{\tan\lambda + \mu\sec\alpha}{1 - \mu\tan\lambda\sec\alpha}\right) + \frac{F_i \mu_c d_c}{2} \tag{b}$$

Examination of Table A-30 shows that the diameter of the washer face of a hex nut is the same as the width across flats and equal to $1\frac{1}{2}$ times the nominal size. Therefore the mean collar diameter is $d_c = (d + 1.5d)/2 = 1.25d$. Equation (b) can now be arranged to give

$$T = \left[\left(\frac{d_m}{2d} \right) \left(\frac{\tan \lambda + \mu \sec \alpha}{1 - \mu \tan \lambda \sec \alpha} \right) + 0.625 \mu_c \right] F_i d \qquad (c)$$

We now define a *torque coefficient* K as the term in brackets, and so

$$K = \left(\frac{d_m}{2d} \right) \left(\frac{\tan \lambda + \mu \sec \alpha}{1 - \mu \tan \lambda \sec \alpha} \right) + 0.625 \mu_c \qquad (8\text{-}14)$$

Equation (c) can now be written

$$T = KF_i d \qquad (8\text{-}15)$$

Orthwein* found the range of frictional coefficients for both collar and thread friction in power screws to be

Metal-to-metal surfaces	0.40 to 0.80
Lubricated surfaces	0.005 to 0.20

The actual value selected depends upon the surface smoothness, accuracy, and degree of lubrication. On the average, both μ and μ_c are about 0.15. The interesting thing about Eq. (8-14) is that $K \approx 0.20$ for $\mu = \mu_c = 0.15$, no matter what size bolts are employed and no matter whether the threads are coarse or fine.† Thus Eq. (8-15) is more convenient as

$$T = 0.20 F_i d \qquad (8\text{-}16)$$

In this form it is very simple to compute the wrench torque T needed to create a desired preload F_i when the size d of the fastener is known.

Blake and Kurtz have published results of numerous tests of the torquing of bolts.‡ By subjecting their data to a statistical analysis we can learn something about the distribution of the torque coefficients and the resulting preload. Blake and Kurtz determined the preload in quantities of unlubricated and lubricated bolts of size $\frac{1}{2}$ in-20 UNF when torqued to 800 lb·in. The statistical analyses of these two groups of bolts are displayed in Tables 8-3 and 8-4.

* William C. Orthwein, "Simplified Power Screw Design," *Machine Design*, vol. 53, no. 19, Aug. 20, 1981, pp. 79–81.

† See Table 7-3, p. 246, in the original (1963) edition of this book for a complete listing of these values.

‡ J. C. Blake and H. J. Kurtz, "The Uncertainties of Measuring Fastener Preload," *Machine Design*, vol. 37, Sept. 30, 1965, pp. 128–131.

Table 8-3 THE DISTRIBUTION OF PRELOAD FOR 20 TESTS OF UNLUBRICATED BOLTS, SIZE $\frac{1}{2}$–20 UNF, TORQUED TO 800 lb·in

F_i, kip
5.3, 6.2, 6.3, 6.6, 6.8, 6.9, 7.4, 7.6, 7.6, 7.6,
7.8, 8.0, 8.0, 8.4, 8.5, 8.5, 8.8, 9.0, 9.1, 9.6

Mean value, $\bar{F_i}$ = 7.7 kip   Standard deviation, $\hat{\sigma}$ = 1.107 kip

We first note that both groups have about the same mean preload, 7700 lb. The unlubricated bolts have a standard deviation of 1100 lb, which is about 15 percent of the mean. The lubricated bolts have a standard deviation of 680 lb, or about 9 percent of the mean, a substantial reduction. These deviations are quite large, though, and emphasize the necessity for quality-control procedures throughout the entire manufacturing and assembly process to assure uniformity.

The means obtained from the two samples are nearly identical, 7700 lb; using Eq. (8-15), we find, for both samples, $K = 0.208$, which is close to the recommended value.

8-8 STRENGTH SPECIFICATIONS

Table 8-5 lists the grades and the specifications of most of the threaded fasteners used today. SAE grades 1 and 2 should only be used for unimportant or non-loaded connections. Their carbon content is too low and the ductility is too high for heavily loaded connections. This same table shows how bolt grades are identified by markings on the tops or sides of the heads.

The terms *proof load* and *proof strength* appear frequently in the literature. The *proof load* of a bolt is the maximum load (force) that a bolt can withstand without acquiring a permanent set. The *proof strength* is the limiting value of the stress determined using the proof load and the tensile-stress area. Although proof strength and yield strength have something in common, the yield strength is usually the higher of the two because it is based on a 0.2 percent permanent deformation.

Table 8-4 THE DISTRIBUTION OF PRELOAD FOR 10 TESTS OF LUBRICATED BOLTS, SIZE $\frac{1}{2}$–20 UNF, TORQUED TO 800 lb·in

F_i, kip
6.8, 7.3, 7.3, 7.4, 7.4, 7.6, 7.7, 7.8, 8.4, 9.1

Mean value, $\bar{F_i}$ = 7.68 kip   Standard deviation $\hat{\sigma}$ = 0.681 kip

Table 8-5 SPECIFICATIONS AND IDENTIFICATION MARKINGS FOR BOLTS, SCREWS, STUDS, SEMS[a]AND U BOLTS[b]
(Multiply the strengths in kpsi by 6.89 to get the strength in MPa.)

SAE grade	ASTM grade	Metric[c] grade	Nominal diameter in	Proof strength kpsi	Tensile strength kpsi	Yield[d] strength kpsi	Core hardness Rockwell min/max	Grade identification marking	Products[e]	Material
1	A307	4.6	¼ thru 1½	33	60	36	B70/B100	None	B, Sc, St	Low- or medium-carbon steel
2	...	5.8	¼ thru ¾	55	74	57	B80/B100	None	B, Sc, St	Low- or medium-carbon steel
		4.6	Over ¾ thru 1½	33	60	36	B70/B100	None	B, Sc, St	Low- or medium-carbon steel
4	...	8.9	¼ thru 1½	65[f]	115	100	C22/C32	None	St	Medium-carbon, cold-drawn steel
5	A449 or A325 Type 1	8.8	¼ thru 1	85	120	92	C25/C34	(marking)	B, Sc, St	Medium-carbon steel, Q&T
		7.8	Over 1 thru 1½	74	105	81	C19/C30	(marking)	B, Sc, St	Medium-carbon steel, Q&T
		8.6	Over 1½ to 3	55	90	58	...	(marking)	B, Sc, St	Medium-carbon steel, Q&T
5.1		8.8	No. 6 thru ⅝	85	120	...	C25/C40	 (marking)	Se	Low- or medium-carbon, Q&T
		8.8	No. 6 thru ½	85	120	...	C25/C40	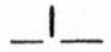 (marking)	B, Sc, St	Low- or medium-carbon, Q&T
5.2	A325 Type 2	8.8	¼ thru 1	85	120	92	C26/C36	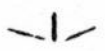 (marking)	B, Sc	Low-carbon martensite steel, fully killed, fine-grained, Q&T
7[g]	... A354	10.9	¼ thru 1½	105	133	115	C28/C34	(marking)	B, Sc	Medium-carbon alloy steel, Q&T
8	Grade BD	10.9	¼ thru 1½	120	150	130	C33/C39	 (marking)	B, Sc, St	Medium-carbon alloy steel, Q&T
8.1	...	10.9	¼ thru 1½	120	150	130	C32/C38	None	St	Elevated temperature drawn steel-medium carbon alloy or G15410
8.2	...	10.9	¼ thru 1	120	150	130	C35/C42	 (marking)	B, Sc	Low-carbon martensite steel, fully killed, fine-grained, Q&T
...	A574	12.9	0 thru ½	140	180	160	C39/C45	12.9	SHCS	Alloy steel, Q&T
		12.9	⅝ thru 1½	135	170	160	C37/C45	12.9	SHCS	Alloy steel, Q&T

[a] Sems = screw and washer assemblies.
[b] Compiled from ANSI/SAE J429j; ANSI B18.3.1-1978; and ASTM A307, A325, A354, A449, and A574.
[c] Metric grade is xx.x where xx is approximately $0.01S_{ut}$ in MPa and .x is the ratio of the minimum S_y to S_{ut}.
[d] Yield strength is stress at which a permanent set of 0.2% of gauge length occurs.
[e] B = bolt, Sc = screws, St = studs, Se = Sems, and SHCS = socket head cap screws.
[f] Entry appears to be in error but conforms to the standard ANSI/SAE J429j.
[g] Grade 7 bolts and screws are roll threaded after heat treatment.

8-9 BOLT PRELOAD: STATIC LOADING

In Sec. 8-5 we learned that the portion of the external load P taken by the bolt, in a tension-loaded joint, is

$$F_b = \frac{k_b P}{k_b + k_m} + F_i \qquad (a)$$

By designating

$$C = \frac{k_b}{k_b + k_m} \qquad (8\text{-}16)$$

Eq. (a) can be written

$$F_b = CP + F_i \qquad (b)$$

And so the condition for joint separation is the bolt carrying all the load. If P is renamed P_0 for this condition, then, at opening $F_b = P_0$ and

$$P_0 = CP_0 + F_i$$

Thus the preload F_i must always be greater than $(1 - C)P$. Applying a factor of safety n to the external load P, we find the safe value of F_i to be

$$F_i = nP(1 - C) \qquad (8\text{-}17)$$

Of course F_i must always be less than $A_t S_P$.

If a factor of safety $n = 1.5$ is chosen, this will account for about three standard deviations of the preload, as indicated in Table 8-3. Thus, in general, choose a factor of safety $n \geq 1.5$, depending upon the design considerations.

It is important to note that Eq. (8-17) and the recommendations for n apply to *static loads only*. The recommendations for preload, when fatigue is present, are so important that we shall discuss this problem separately in another section.

The Mohr's circle diagram of Fig. 8-12 shows what happens during and after tightening. The results shown can only occur with high-quality bolt materials, in which the stress-strain diagram rises all the way to fracture. With such bolt materials preloads greater than the proof strength can be used, though with considerable caution. With such materials, the torsion that occurs during tightening relaxes in time, resulting in a lower final bolt tension. For this reason, it is sometimes said that if a bolt subjected only to static loading does not fail during tightening, there is a very good chance that it never will fail.

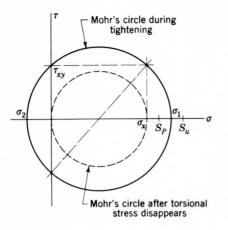

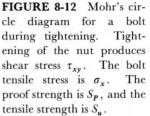

FIGURE 8-12 Mohr's circle diagram for a bolt during tightening. Tightening of the nut produces shear stress τ_{xy}. The bolt tensile stress is σ_x. The proof strength is S_P, and the tensile strength is S_u.

It is necessary to emphasize the importance of maintaining a high preload. Any relaxation of the preload may cause the entire external load to be carried by the bolt and result in failure of the joint. See Fig. 6-12 where failure was caused by lack of preload. When a bolted joint is placed into service, vibration and racking produce small deformations causing flattening of the high spots, dirt, and paint, and reducing the original preload. So the use of a high preload is a way of creating a margin of safety to guard against such occurrences. For this reason, it is suggested that the preload be within the range

$$0.6F_P \le F_i \le 0.9F_P \tag{8-18}$$

where F_P is the proof load, obtained from the equation

$$F_P = A_t S_P \tag{8-19}$$

Here, S_P is the proof strength obtained from Table 8-5. For bolts made from materials other than those of Table 8-5, an approximate value for the proof strength is $S_P = 0.85S_y$.

8-10 SELECTION OF THE NUT

Imagine three annular rings, analogous to square threads on a screw, cut on a male member and three corresponding grooves with clearances, as shown in Fig. 8-13a, on a female member or nut. Now apply a tension load P to the screw and let the nut react to this load as shown in the figure. If the load is assumed to be uniformly divided among the three threads, stress elements at A, B, and C on the screw will have tensile loads of $F_A = P$, $F_B = 2P/3$, and $F_C = P/3$. Corresponding stress elements in the nut, not shown, will have compressive loads $F_A = -P$, $F_B = -2P/3$, and $F_C = -P/3$. Now, with the screw in tension, the screw gets

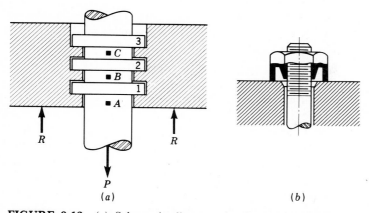

FIGURE 8-13 (*a*) Schematic diagram showing how the first nut thread takes the entire load. (*b*) Typical method of distributing the thread load.

longer, and so threads 1, 2, and 3 will tend to move apart. However, the nut is in compression, and so the nut threads will tend to move closer together. But these actions prevent sharing of the loads as assumed in the beginning. We therefore conclude that the load will not be shared at all and that, instead, the first thread takes the entire force.

This tendency may be partially corrected by proportioning the nut so as to cause more deformation to exist at the bottom. Figure 8-13*b* shows a nut design in which material has been removed from the lower portion of the nut in order to equalize the stress distribution.

In practice, conditions are not quite as severe as pictured, since yielding of the threads in the nut will permit the other threads to transfer some of the load. However, since such a tendency is present, it must be guarded against and knowledge of it used in selecting the nut.

Another factor which acts to reduce the tendency of the bottom thread to take the entire load is that the wedging action of the threads tends to spread, or dilate, the nut.

These conditions point to the fact that, when preloading is desired, careful attention should be given to the nut material. Selecting a soft nut ensures plastic yielding, which will enable the nut threads to divide the load more evenly.

Nuts are tested by determining their stripping strength. The test is made by threading a nut on a hardened-steel mandrel and pulling it through the nut. The strength is the load divided by the mean thread area. Common nuts have a stripping strength of approximately 90 kpsi.

It is interesting to know that three full threads are all that are required to develop the full bolt strength.

Another factor which must be considered in the design of bolted joints is the maintenance of the initial preload. This load may be relaxed by yielding of the clamped material, by extrusion of paint or plating from the contact surfaces, or by a compression of rough places. Extra contact area may be provided by hardened washers. This is especially necessary if the bolted parts are relatively soft and the bolt head or nut does not provide sufficient bearing area.

8-11 BOLT PRELOAD: FATIGUE LOADING

Most of the time the type of fatigue loading encountered in the analysis of bolted joints is one in which the externally applied load fluctuates between zero and some maximum force P. This would be the situation in a pressure cylinder, for example, where a pressure either exists or does not exist. In order to determine the mean and alternating bolt stresses for such a situation, let us employ the notation of Sec. 8-5. Then $F_{max} = F_b$ and $F_{min} = F_i$. Therefore, the alternating component of bolt stress is, from Eqs. (8-11) and (8-16),

$$\sigma_a = \frac{F_b - F_i}{2A_t} = \frac{k_b}{k_b + k_m} \frac{P}{2A_t} = \frac{CP}{2A_t} \tag{8-20}$$

Then, since the mean stress is equal to the alternating component plus the minimum stress, we have

$$\sigma_m = \sigma_a + \frac{F_i}{A_t} = \frac{CP}{2A_t} + \frac{F_i}{A_t} \tag{8-21}$$

Equation (7-28) gives the formula for the modified Goodman criterion of failure as

$$\frac{\sigma_a}{S_e} + \frac{\sigma_m}{S_{ut}} = 1 \tag{8-22}$$

We can substitute Eqs. (8-20) and (8-21) into (8-22) and solve for F_i. This yields

$$F_i = A_t S_{ut} - \frac{CP}{2}\left(\frac{S_{ut}}{S_e} + 1\right) \tag{8-23}$$

which is the limiting value of F_i. A safe value can be obtained by multiplying the external load P by a factor of safety n. Thus the proper preload when fatigue loading is present is given by

$$F_i = A_t S_{ut} - \frac{CnP}{2}\left(\frac{S_{ut}}{S_e} + 1\right) \tag{8-24}$$

Note that S_{ut} and S_e of this equation can be replaced by S_y to yield

$$F_i = A_t S_y - CnP \tag{8-25}$$

Table 8-6 FATIGUE-STRENGTH REDUCTION FACTORS K_f FOR THREADED ELEMENTS

SAE grade	Metric grade	Rolled threads	Cut threads	Fillet
0 to 2	3.6 to 5.8	2.2	2.8	2.1
4 to 8	6.6 to 10.9	3.0	3.8	2.3

which is identical with Eq. (8-17). In order to avoid static failure as well as fatigue failure, use the value of F_i, that is, the least from these two equations.*

8-12 FATIGUE LOADING

Tension-loaded bolted joints subjected to fatigue action can be analyzed directly by the methods of Chap. 7. Table 8-6 lists average fatigue-strength reduction factors for the fillet under the bolt head and also at the beginning of the threads on the bolt shank. These are already corrected for notch sensitivity and for surface finish. Designers should be aware that situations may arise in which it would be advisable to investigate these factors more closely, since they are only average values. In fact, Peterson† observes that the distribution of typical bolt failures is about 15 percent under the head, 20 percent at the end of the thread, and 65 percent in the thread at the nut face.

In using Table 8-6, it is usually safe to assume that the fasteners have rolled threads, unless specific information is available. Also, in computing endurance limit, use a machined finish for the body of the bolt if nothing is stated.

Example 8-3 Based on 50 percent reliability, find the endurance limit of the bolt shown in Fig. 8-14.

Solution From Table 8-5 we find $S_u = 115$ kpsi. Therefore, from Eq. (7-17),

$$S'_e = 19.2 + 0.314 S_{uc} = 19.2 + 0.314(115) = 55.31 \text{ kpsi}$$

Note that we have assumed $S_{uc} = S_{ut}$. The value of S'_e just found is corrected for size, but not for stress concentration. From Table 8-6 we select $K_f = 3.0$ for rolled threads, because it is larger than K_f at the fillet. Then $k_e = 1/K_f =$

---

* For another analysis of this problem, see G. A. Fazekas, "On the Optimal Bolt Preload," ASME Paper no. 75-WA/DE 14, 1975.

† R. E. Peterson, *Stress Concentration Factors*, Wiley, New York, 1974, p. 253.

$1/3.0 = 0.333$. The endurance limit of the bolt for axial loading is now found to be

$$S_e = k_e S'_e = 0.333(55.31) = 18.4 \text{ kpsi} \qquad\qquad Ans.$$

$////$

Example 8-4   The section of Fig. 8-14 is from a pressure cylinder. A total of N bolts is to be used to resist a separating force that varies from 0 to 36 kip. Use a factor of safety of 3.0 and find the proper bolt preload and the minimum number of bolts required. Use the strengths found in Example 8-3.

Solution   The stiffness constant of the bolt is

$$k_b = \frac{AE}{l} = \frac{\pi d^2 E}{4l} = \frac{\pi(0.625)^2(30)(10)^6}{4(1.5)} = 6.13(10)^6 \text{ lb/in}$$

where the grip is $l = 1.5$ in. The modulus of elasticity of No. 25 cast iron is $E = 12$ Mpsi. Thus the stiffness of the members, from Eq. (8-13), is

$$k_m = \frac{\pi E d}{2 \ln \left[\dfrac{5(l + 0.5d)}{l + 2.5d} \right]} = \frac{\pi(12)(0.625)(10)^6}{2 \ln \left\{ \dfrac{5[1.5 + 0.5(0.625)]}{1.5 + 2.5(0.625)} \right\}} = 10.86(10)^6 \text{ lb/in}$$

The constant C is now found to be

$$C = \frac{k_b}{k_b + k_m} = \frac{6.13}{6.13 + 10.86} = 0.361$$

From Table 8-2 we find $A_t = 0.226$ in² for $\frac{5}{8}$-in bolts. By substituting $36/N$ for P

$\frac{5}{8}''$-11 UNC x $2\frac{1}{4}''$ grade 4
finished hexagonal head bolt

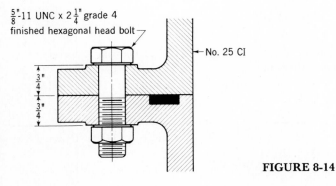

← No. 25 CI

$\frac{3}{4}''$

$\frac{3}{4}''$

FIGURE 8-14

in Eq. (8-24), we get

$$F_i = A_t S_{ut} - \frac{CnP}{2}\left(\frac{S_{ut}}{S_e} + 1\right)$$

$$= 0.226(115) - \frac{0.361(3.0)(36)}{2N}\left(\frac{115}{18.4} + 1\right) = 25.99 - \frac{141.3}{N}\ \text{kip}$$

By substituting various values for N, we get the following results:

N	6	8	9	10	12
F_i, kpsi	2.44	8.32	10.29	11.86	14.21

From Table 8-5 find $S_P = 65$ kpsi. Thus the proof load is

$$F_P = A_t S_P = 0.226(65) = 14.69\ \text{kpsi}$$

Using Eq. (8-18) we find the upper and lower bounds on F_i to be

$$F_i(\text{min}) = 0.6F_P = 0.6(14.69) = 8.81\ \text{kpsi}$$

$$F_i(\text{max}) = 0.9F_P = 0.9(14.69) = 13.22\ \text{kpsi}$$

The table above shows that good choices are either $N = 9$ or $N = 10$. If we choose 10 bolts, then $F_i = 11.86$ kip and $P = \frac{36}{10} = 3.6$ kip per bolt.

To check on the possibility of a static failure, we rearrange Eq. (8-25) to the form

$$n = \frac{A_t S_y - F_i}{CP}$$

From Table 8-5 we find $S_y = 100$ kpsi. So the factor of safety guarding against a static failure is

$$n = \frac{0.226(100) - 11.86}{0.361(3.6)} = 8.26$$

Thus the use of 10 bolts is a satisfactory solution.

It is also worth noting that the preload equations can be solved using multiple factors of safety, say separate n's for each strength as well as for the applied load. ////

8-13   GASKETED JOINTS

Figure 8-15 shows three basic gasket configurations. Joints having confined gaskets have the members in metal-to-metal contact, and so the methods already discussed apply to these situations.

The gasket in an unconfined joint is subject to the full compressive load between the members, its stiffness predominates, and so the characteristics of the gasket rule the design of the connection. Table 8-7 provides the modulus of elasticity E needed to solve for the gasket stiffness for some types of gaskets and materials. Note that the values of E are quite small in comparison to those for metals, except for copper-asbestos gaskets. This means that the stiffness of the metal parts within the bolt grip can be considered infinite and only the gasket stiffness need be used for k_m.

The modulus of elasticity of copper-asbestos is quite large and so Eq. (a) of Sec. 8-6 will have to be used to combine the stiffnesses of the several parts in the bolt grip, including that of the gasket.

A gasketed joint must satisfy Eqs. (8-24) and (8-25) for preload. In addition, the preload must be large enough to achieve the minimum sealing pressure required for the gasket material. Thus, the preload must satisfy the relation

$$F_i \geq A_g p_o \tag{8-26}$$

where A_g = gasket area
p_o = minimum gasket seal pressure

In Sec. 8-5 we learned that the resultant clamping load in a bolted connection is given by the equation

$$F_m = \frac{k_m P}{k_b + k_m} - F_i \tag{a}$$

FIGURE 8-15   Some types of gaskets used to prevent leakage. (a) Unconfined gasket; (b) O-ring or confined gasket in which the seal is achieved due to the pressure p; (c) confined gasket in which the seal is achieved by compressing the gasket.

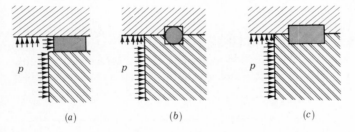

$\qquad\qquad(a)\qquad\qquad\qquad\qquad(b)\qquad\qquad\qquad\qquad(c)$

Table 8-7 MODULUS OF ELASTICITY FOR GASKET MATERIALS

Material	Modulus of elasticity, E	
	kpsi	MPa
Cork	12.5	86
Compressed asbestos	70.0	480
Copper-asbestos	$13.5(10)^3$	$93(10)^3$
Plain rubber	10.0	69
Spiral wound	41.0	280
Teflon	35.0	240
Vegetable fiber	17.0	120

Source: Compiled from John W. Axelson and F. B. Pintard, "Pressure Components (Seals)," in Harold A. Rothbart (ed.), *Mechanical Design and Systems Handbook*, McGraw-Hill, New York, 1964, p. 24–11; also see M. O. M. Osman, W. M. Mansour, and R. V. Dukkipati, "On the Design of Bolted Connections with Gaskets Subjected to Fatigue Loading," ASME Paper no. 76-DET-57, 1976.

We can simplify this relation by using the stiffness ratio of Eq. (8-16). Equation (*a*) can then be written

$$F_m = P(1 - C) - F_i \tag{8-27}$$

where P is the external tensile load per bolt. In a gasketed joint this value of the clamping load must satisfy the relation

$$F_m \geq A_g mp \tag{8-28}$$

where m is called the *gasket factor*; values of this factor, which is analogous to a factor of safety, vary from about 2 to 4 for most materials. The term p is the pressure which tends to separate the halves of the joint; don't confuse it with the seal pressure p_o.

Example 8-5 The pressure vessel of Fig. 8-16 is to be sealed using an asbestos gasket having dimensions as shown, and with a minimum seal pressure of 11 MPa. The head is to be fastened down using socket-head cap screws having a nominal diameter of 14 mm. Since these thread into the cylinder, only half of the depth of the threaded hole should be included in the bolt grip. The cylinder is to be rated for an internal pressure of 2000 kPa and a factor of safety of 1.5 is to be used.

(*a*) How many ASTM A325 type 2 cap screws should be used to prevent a fatigue failure based on 90 percent reliability?

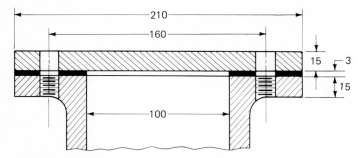

FIGURE 8-16   Dimensions in millimeters.

(*b*)   Does the result found in (*a*) satisfy both of the gasket-sealing requirements?

Solution   (*a*) The effective grip of the bolt is $l = 15 + 3 + 7.5 = 25.5$ mm.   Therefore the stiffness of the bolt is

$$k_b = \frac{\pi d^2 E}{4l} = \frac{\pi(14)^2(207)}{4(25.5)} = 1250 \text{ MN/m}$$

To get the spring rate of the gasket note that $l = 3$ mm and $E = 480$ MPa from
Table 8-7.   Using Eq. (8-13), we find

$$k_m = \frac{\pi E d}{2 \ln \left[\dfrac{5(l + 0.5d)}{l + 2.5d} \right]} = \frac{\pi(480)(14)(10)^{-3}}{2 \ln \left\{ \dfrac{5[3 + 0.5(14)]}{3 + 2.5(14)} \right\}} = 38.5 \text{ MN/m}$$

Note that in a gasketed joint the bolt is many times stiffer than the gasket.   The
stiffness ratio is, from Eq. (8-16),

$$C = \frac{k_b}{k_b + k_m} = \frac{1250}{1250 + 38.5} = 0.970$$

The total separating force P is

$$P = Ap = \frac{\pi(100)^2(2000)}{4(10)^6} = 15.7 \text{ kN}$$

From Table 8-1 we find $A_t = 115$ mm^2.   From Table 8-5 we find $S_{ut} = 120$ kpsi.
Thus $S_{ut} = 120(6.89) = 827$ MPa.   Also, from Eq. (7-17),

$$S_e' = 19.2 + 0.314 S_{uc} = 19.2 + 0.314(120) = 56.88 \text{ kpsi}$$

So $S'_e = 56.88(6.89) = 392$ MPa. Using $k_c = 0.897$ for 90 percent reliability and $k_e = 0.333$ for rolled threads, we have

$$S_e = 0.897(0.333)(392) = 117 \text{ MPa}$$

We can now solve Eq. (8-24). Using P/N for the load per bolt, we have

$$F_i = A_t S_{ut} - \frac{CnP}{2N}\left(\frac{S_{ut}}{S_e} + 1\right) = 115(0.827) - \frac{0.970(1.5)(15.7)}{2N}\left(\frac{0.827}{0.117} + 1\right)$$

$$= 95.1 - \frac{92.2}{N} \text{ kN}$$

Solving this equation for F_i per bolt for various numbers N gives:

N	8	6	4	3	2
F_i, kN	83.6	79.7	72.1	64.4	49.0

From Table 8-5 we find $S_P = 85(6.89) = 586$ MPa. Therefore the proof load is $F_P = A_t S_P = 115(0.586) = 67.4$ kN. Using Eq. (8-18), the range of acceptable preloads is found to be

$$40.4 \leq F_i \leq 60.7 \text{ kN}$$

Thus the table above shows that the use of two bolts will satisfy the fatigue requirement. But this is an absurd solution.

To obtain a fairly uniform gasket pressure the bolts should not be spaced more than 10 bolt diameters apart. The circumference of the bolt circle is $160\pi = 503$ mm. Six bolts would give a spacing of $503/(6)(14) = 5.99$ bolt diameters. This is quite satisfactory and so we select six bolts for this cylinder and a preload of 50 kN per bolt, which is in the center of the range shown above.

(b) The first gasket requirement is that the preload must equal or exceed the product of the gasket area and the minimum seal pressure [Eq. (8-26)]. The total gasket area is

$$A_g = \frac{\pi}{4}\left[(210)^2 - (100)^2\right] - \frac{6\pi}{4}(14)^2 = 25.9(10)^3 \text{ mm}^2$$

Since the minimum seal pressure is given as $p_o = 11$ MPa, we have

$$A_g p_o = 25.9(10)^3(0.011) = 284.9 \text{ kN}$$

The total preload is $NF_i = 6(50) = 300$ kN. Thus the seal pressure requirement is satisfied.

The second requirement is that the clamping load must satisfy Eq. (8-28). From Eq. (8-27) we find the total clamping force to be

$$F_m = P(1 - C) - F_i = 15.7(1 - 0.970) - 300 = -299.5 \text{ kN}$$

Selecting $m = 2$, the right-hand side of Eq. (8-28) is

$$A_g m p = 25.9(10)^3(2)(2000)(10)^{-6} = 103.6 \text{ kN}$$

Thus the second requirement is satisfied too.   ////

8-14 BOLTED AND RIVETED JOINTS LOADED IN SHEAR*

Riveted and bolted joints loaded in shear are treated exactly alike in design and analysis.

In Fig. 8-17a is shown a riveted connection loaded in shear. Let us now study the various means by which this connection might fail.

Figure 8-17b shows a failure by bending of the rivet or of the riveted members. The bending moment is approximately $M = Ft/2$, where F is the shearing force and t is the grip of the rivet, that is, the total thickness of the connected parts. The bending stress in the members or in the rivet is, neglecting stress concentration,

$$\sigma = \frac{M}{I/c} \tag{8-29}$$

where I/c is the section modulus for the weakest member or for the rivet or rivets, depending upon which stress is to be found. The calculation of the bending stress in this manner is an assumption, because we do not know exactly how the load is distributed to the rivet nor the relative deformations of the rivet and the members. Although this equation can be used to determine the bending stress, it is seldom used in design; instead its effect is compensated for by an increase in the factor of safety.

In Fig. 8-17c failure of the rivet by pure shear is shown; the stress in the rivet is

$$\tau = \frac{F}{A} \tag{8-30}$$

---

* The design of bolted and riveted connections for boilers, bridges, buildings, and other structures in which danger to human life is involved is strictly governed by various construction codes. When designing these structures the engineer should refer to the *American Institute of Steel Construction Handbook*, the American Railway Engineering Association specifications, or the Boiler Construction Code of the American Society of Mechanical Engineers.

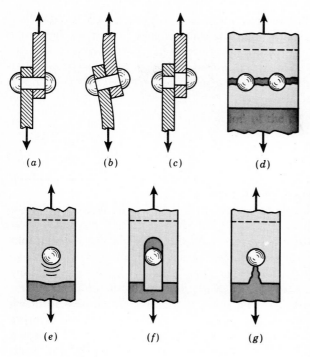

(a)   (b)   (c)   (d)

(e)   (f)   (g)

FIGURE 8-17 Modes of failure in shear loading of a bolted or riveted connection. (*a*) Shear loading; (*b*) bending of rivet; (*c*) shear of rivet; (*d*) tensile failure of members; (*e*) bearing of rivet on members or bearing of members on rivet; (*f*) shear tearout; (*g*) shear tearout.

where A is the cross-sectional area of all the rivets in the group. It may be noted that it is standard practice in structural design to use the nominal diameter of the rivet rather than the diameter of the hole, even though a hot-driven rivet expands and nearly fills up the hole.

Rupture of one of the connected members or plates by pure tension is illustrated in Fig. 8-17*d*. The tensile stress is

$$\sigma = \frac{F}{A} \tag{8-31}$$

where A is the net area of the plate, that is, reduced by an amount equal to the area of all the rivet holes. For brittle materials and static loads and for either ductile or brittle materials loaded in fatigue, the stress-concentration effects must be included. It is true that the use of a bolt with an initial preload and, sometimes, a rivet will place the area around the hole in compression and the thus tend to nullify the effects of stress concentration, but unless definite steps are taken to assure that the preload does not relax, it is on the conservative side to design as if the full stress-concentration effect were present. The stress-concentration effects are not considered in structural design because the loads are static and the materials ductile.

In calculating the area for Eq. (8-31) the designer should, of course, use the combination of rivet or bolt holes which gives the smallest area.

Figure 8-17e illustrates a failure by crushing of the rivet or plate. Calculation of this stress, which is usually called a *bearing stress*, is complicated by the distribution of the load on the cylindrical surface of the rivet. The exact values of the forces acting upon the rivet are unknown, and so it is customary to assume that the components of these forces are uniformly distributed over the projected contact area of the rivet. This gives for the stress

$$\sigma = \frac{F}{A} \tag{8-32}$$

where the projected area for a single rivet is $A = td$. Here, t is the thickness of the thinnest plate and d is the rivet or bolt diameter.

Shearing, or tearing, of the margin is shown in Fig. 8-17f and g. In structural practice this failure is avoided by spacing the rivet at least $1\frac{1}{2}$ diameters away from the margin. Bolted connections usually are spaced an even greater distance than this for satisfactory appearance, and hence this type of failure may usually be neglected.

In structural design it is customary to select in advance the number of rivets and their diameters and spacing. The strength is then determined for each method of failure. If the calculated strength is not satisfactory, a change is made in the diameter, spacing, or number of rivets used, to bring the strength in line with expected loading conditions. It is not usual, in structural practice, to consider the combined effects of the various methods of failure.

8-15   CENTROIDS OF BOLT GROUPS

In Fig. 8-18 let A_1 through A_5 be the respective cross-sectional areas of a group of five bolts. These bolts need not be of the same diameter. In order to determine the shear forces which act upon each bolt, it is necessary to know the location of the centroid of the bolt group. Using statics, we learn that the centroid G is located by the coordinates

$$\bar{x} = \frac{A_1 x_1 + A_2 x_2 + A_3 x_3 + A_4 x_4 + A_5 x_5}{A_1 + A_2 + A_3 + A_4 + A_5} = \frac{\sum_{1}^{n} A_i x_i}{\sum_{1}^{n} A_i}$$

$$\text{8-33}$$

$$\bar{y} = \frac{A_1 y_1 + A_2 y_2 + A_3 y_3 + A_4 y_4 + A_5 y_5}{A_1 + A_2 + A_3 + A_4 + A_5} = \frac{\sum_{1}^{n} A_i y_i}{\sum_{1}^{n} A_i}$$

where x_i and y_i are the distances to the respective bolt centers. In many instances these centroidal distances can be located by symmetry. Note that the $\sum$ keys on your calculator are especially useful for finding these centroidal distances.

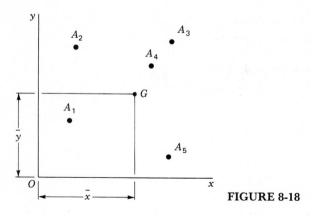

FIGURE 8-18

8-16   SHEAR OF BOLTS AND RIVETS DUE TO ECCENTRIC LOADING

An example of eccentric loading of fasteners is shown in Fig. 8-19.   This is a portion of a machine frame containing a beam A subjected to the action of a bending load.   In this case, the beam is fastened to vertical members at the ends

FIGURE 8-19   (*a*) Beam bolted at both ends with distributed load; (*b*) free-body diagram of beam; (*c*) enlarged view of bolt group showing primary and secondary shear forces.

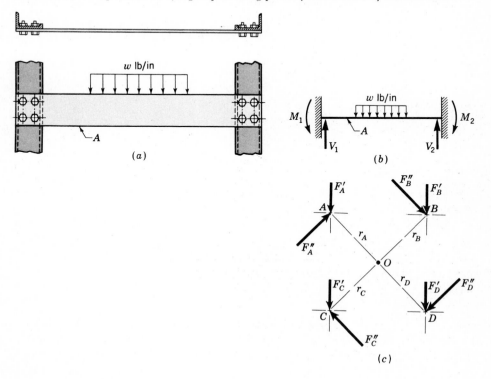

with bolts. You will recognize the schematic representation in Fig. 8-19b as an indeterminate beam with both ends fixed and with the moment reaction M and the shear reaction V at the ends.

For convenience, the centers of the bolts at one end of the beam are drawn to a larger scale in Fig. 8-19c. Point O represents the centroid of the group, and it is assumed in this example that all the bolts are of the same diameter. The total load taken by each bolt will be calculated in three steps. In the first step the shear V is divided equally among the bolts so that each bolt takes $F' = V/n$, where n refers to the number of bolts in the group, and the force F' is called the *direct load*, or *primary shear*.

It is noted that an equal distribution of the direct load to the bolts assumes an absolutely rigid member. The arrangement of the bolts or the shape and size of the members sometimes justify the use of another assumption as to the division of the load. The direct loads F' are shown as vectors on the loading diagram (Fig. 8-19c).

The *moment load*, or *secondary shear*, is the additional load on each bolt due to the moment M. If r_A, r_B, r_C, etc., are the radial distances from the centroid to the center of each bolt, the moment and moment load are related as follows:

$$M = F_A'' r_A + F_B'' r_B + F_C'' r_C + \cdots \qquad (a)$$

where F'' is the moment load. The force taken by each bolt depends upon its radius; that is, the bolt farthest from the center of gravity takes the greatest load while the nearest bolt takes the smallest. We can therefore write

$$\frac{F_A''}{r_A} = \frac{F_B''}{r_B} = \frac{F_C''}{r_C} \qquad (b)$$

Solving Eqs. (a) and (b) simultaneously, we obtain

$$F_n'' = \frac{M r_n}{r_A^2 + r_B^2 + r_C^2 + \cdots} \qquad (8\text{-}34)$$

where the subscript n refers to the particular bolt whose load is to be found. These moment loads are also shown as vectors on the loading diagram.

In the third step the direct and moment loads are added vectorially to obtain the resultant load on each bolt. Since all the bolts or rivets are usually the same size, only that bolt having the maximum load need to be considered. When the maximum load is found, the strength may be determined, using the various methods already described.

Example 8-6 Shown in Fig. 8-20 is a 15- by 200-mm rectangular steel bar cantilevered to a 250-mm steel channel using four bolts. Based on the external

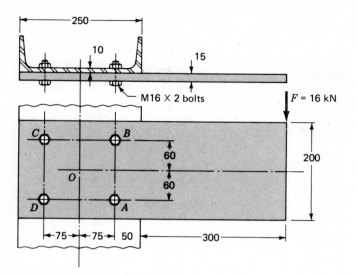

FIGURE 8-20   Dimensions in millimeters.

load of 16 kN, find:

(a)   The resultant load on each bolt
(b)   The maximum bolt shear stress
(c)   The maximum bearing stress
(d)   The critical bending stress in the bar

Solution   (a) Point O, the centroid of the bolt group in Fig. 8-20, is found by symmetry. If a free-body diagram of the beam were constructed, the shear reaction V would pass through O and the moment reaction M would be about O. These reactions are

$$V = 16 \text{ kN} \qquad M = 16(425) = 6800 \text{ N} \cdot \text{m}$$

In Fig. 8-21, the bolt group has been drawn to a larger scale and the reactions are shown. The distance from the centroid to the center of each bolt is

$$r = \sqrt{(60)^2 + (75)^2} = 96.0 \text{ mm}$$

The primary shear load per bolt is

$$F' = \frac{V}{n} = \frac{16}{4} = 4 \text{ kN}$$

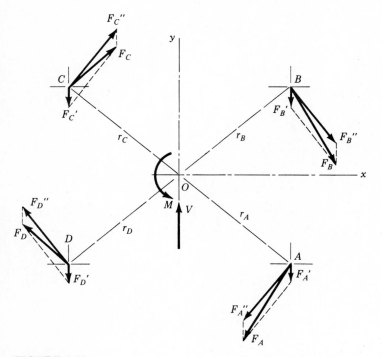

FIGURE 8-21

Since the secondary shear forces are equal, Eq. (8-34) becomes

$$F'' = \frac{Mr}{4r^2} = \frac{M}{4r} = \frac{6800}{4(96.0)} = 17.7 \text{ kN}$$

The primary and secondary shear forces are plotted to scale in Fig. 8-21 and the resultants obtained by using the parallelogram rule. The magnitudes are found by measurement (or analysis) to be

$$F_A = F_B = 21.0 \text{ kN} \qquad\qquad\qquad\qquad\qquad Ans.$$

$$F_C = F_D = 13.8 \text{ kN} \qquad\qquad\qquad\qquad\qquad Ans.$$

(b) Bolts A and B are critical because they carry the largest shear load. The bolt will tend to shear across its major diameter. Therefore the shear-stress area is $A_s = \pi d^2/4 = \pi(16)^2/4 = 201 \text{ mm}^2$. So the shear stress is

$$\tau = \frac{F}{A_s} = \frac{21.0(10)^3}{201} = 104 \text{ MPa} \qquad\qquad\qquad Ans.$$

(c) The channel is thinner than the bar so the largest bearing stress is due to the bolt pressing against the channel web. The bearing area is $A_b = td = 10(16)) = 160$ mm^2. So the bearing stress is

$$\sigma = \frac{F}{A_b} = -\frac{21.0(10)^3}{160} = -131 \text{ MPa} \qquad Ans.$$

(d) The critical bending stress in the bar is assumed to occur in a section parallel to the y axis and through bolts A and B. At this section the bending moment is

$$M = 16(300 + 50) = 5600 \text{ N} \cdot \text{m}$$

The moment of inertial through this section is obtained by the use of the transfer formula, as follows:

$$I = I_{bar} - 2(I_{holes} + d^2 A)$$

$$= \frac{15(200)^3}{12} - 2\left[\frac{15(16)^3}{12} + (60)^2(15)(16)\right] = 8.26(10)^6 \text{ mm}^4$$

Then

$$\sigma = \frac{Mc}{I} = \frac{5600(100)}{8.26(10)^6} (10)^3 = 67.8 \text{ MPa} \qquad Ans.$$

////

8-17 KEYS, PINS, AND RETAINERS

Keys, pins, and retainers are normally used to secure elements such as gears or pulleys to shafts so that torque can be transferred between them. A pin may serve the double purpose of transferring torque as well as preventing relative axial motion between the mating parts. Relative axial motion can also be prevented by using a press or shrink fit, setscrews, or retainers or cotters. Figures 8-22 and 8-23 illustrate a variety of these devices and methods of employing them. Notice that the gib-head key is tapered so that, when firmly driven, it also acts to prevent relative axial motion. The head makes removal possible without access to the other end, but this projection is hazardous with rotating parts. The Woodruff key is of general usefulness, especially when a wheel is to be positioned against a shaft shoulder, since the key slot need not be cut into the shoulder stress-concentration region. Standard sizes and numbering systems for keys and pins may be found in any of the mechanical engineering or machinery handbooks.

The usual practice is to choose a key whose size is one-fourth the shaft diameter. The length of the key is then adjusted according to the hub length and

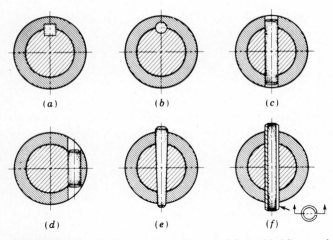

FIGURE 8-22 (*a*) Square key; (*b*) round key; (*c*)-(*d*) round pins; (*e*) taper pin; (*f*) split tubular spring pin. The pins in (*e*) and (*f*) are shown longer than necessary to illustrate the chamfer on the ends; but their lengths should not exceed the hub diameters to avoid injuries which might be caused by projections on rotating parts.

the strength required. It is sometimes necessary to use two keys to obtain the required strength.

In determining the strength of a key, the assumption may be made that the forces are uniformly distributed throughout the key length. This assumption is probably not true, since the torsional stiffness of the shaft will usually be less than that of the hub, causing large forces at one end of the key and small forces at the

FIGURE 8-23 (*a*) Gib-head key; (*b*) Woodruff key.

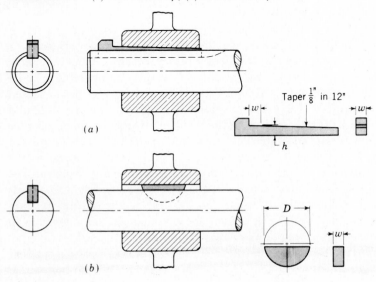

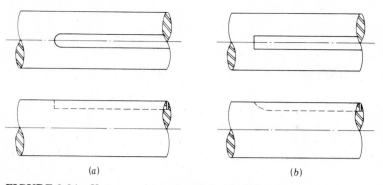

(a) (b)

FIGURE 8-24 Keyways: (a) end-milled; (b) sled-runner.

other end. This distribution may be still more complicated by the stiffening effect of the arms or web at the middle of the hub.

Geometric stress-concentration factors for keyways, when the shaft is in bending, are given by Peterson* as 1.79 for an end-milled keyway (Fig. 8-24) and 1.38 for the sled-runner keyway; $K_t = 3$ should be used when the shafts are subject to combined bending and torsion. Charts A-26-10, A-26-11, and A-26-13 to A-26-15 should be used for shafts containing grooves or holes.

A force distribution having been assumed, it is customary to base the strength of the key on failure by crushing or by shearing. This is illustrated in the following example.

Example 8-7 A UNS G 10350 steel shaft, heat-treated to a yield strength of 75 kpsi, has a diameter of $1\frac{7}{16}$ in. The shaft rotates at 600 rpm and transmits 40 hp through a gear. Select an appropriate key for the gear.

Solution A $\frac{3}{8}$-in square key is selected, UNS G10200 cold-drawn steel being used. The design will be based on a yield strength of 65 kpsi. A factor of safety of 2.80 will be employed in the absence of exact information about the nature of the load.

The torque is obtained from the horsepower equation

$$T = \frac{63\ 000\ H}{n} = \frac{(63\ 000)(40)}{600} = 4200\ \text{lb} \cdot \text{in}$$

Referring to Fig. 8-25, the force F at the surface of the shaft is

$$F = \frac{T}{r} = \frac{4200}{0.719} = 5850\ \text{lb}$$

---

* Op. cit., pp. 245–247.

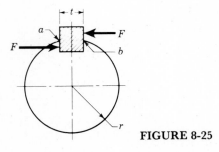

FIGURE 8-25

By the distortion-energy theory, the shear strength is

$$S_{sy} = 0.577S_y = (0.577)(65) = 37.5 \text{ kpsi}$$

Failure by shear across the area ab will create a stress of $\tau = F/tl$. Substituting the strength divided by the factor of safety for τ gives

$$\frac{S_{sy}}{n} = \frac{F}{tl} \quad \text{or} \quad \frac{37.5(10)^3}{2.80} = \frac{5850}{0.375l}$$

or $l = 1.16$ in. To resist crushing, the area of one-half the face of the key is used:

$$\frac{S_y}{n} = \frac{F}{tl/2} \quad \text{or} \quad \frac{65(10)^3}{2.80} = \frac{5850}{0.375l/2}$$

and $l = 1.34$ in. The hub length of a gear is usually greater than shaft diameter,

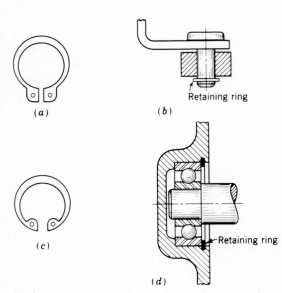

(a)

(b)

Retaining ring

(c)

Retaining ring

(d)

FIGURE 8-26 Typical uses for retaining rings. (a) External ring; (c) internal ring.

for stability. If the key in this example is made equal in length to the hub, it would therefore have ample strength, since it would probably be $1\frac{7}{16}$ in or longer.

$/\!/\!/\!/$

A retaining ring is frequently used instead of a shaft shoulder to axially position a component on a shaft or in a housing bore. As shown in Fig. 8-26, a groove is cut in the shaft or bore to receive the spring retainer. The tapered design of both the internal and external rings assures uniform pressure against the bottom of the groove. For sizes, dimensions, and ratings, the manufacturers' catalogs should be consulted.

PROBLEMS*

Section 8-1

8-1   Verify the tensile-stress area of the M2 × 0.4, M20 × 1.5, and M90 × 2 bolts given in Table 8-1.

8-2   Verify the tensile-stress area of the 10-24 UNC, 10-32 UNF, and $\frac{7}{8}$-9 UNC bolts given in Table 8-2.

8-3*   Determine the designation for a Unified cap screw based upon the following conditions: fatigue is present, $A_t > 0.19$ in^2, the screw is required to provide precision positioning of structural components.

8-4*   Find an adequate metric screw for the service described in Problem 8-3.

Section 8-2

8-5   Show that for zero collar friction the efficiency of a square-thread screw is given by the equation

$$e = \tan \lambda \, \frac{1 - \mu \tan \lambda}{\tan \lambda + \mu}$$

Plot a curve of the efficiency for lead angles up to 45°. Use $\mu = 0.08$.

8-6   A 20-mm power screw has single square threads with a pitch of 4 mm and is used in a power-operated press. Each screw is subjected to a load of 5 kN. The coefficients of friction are 0.075 for the threads and 0.095 for the collar friction. The frictional diameter of the collar is 30 mm.

(a)   Find the thread depth, the thread width, the mean and root diameters, and the lead.

(b)   Find the torque required to "lower" and to "raise" the load.

(c)   Find the overall efficiency.

8-7*   The C clamp shown in the figure uses a $\frac{3}{8}$-in screw with 12 square threads per inch. The frictional coefficients are 0.15 for the threads and for the collar. The collar,

* Problems with asterisks are design-type problems. Answers, if given, are typical.

which in this case is the anvil striker's swivel joint, has a friction diameter of $\frac{5}{8}$ in. The handle is cold-drawn UNS G10100 steel. The capacity of the clamp is to be 150 lb.

(a)   What torque is required to tighten the clamp to full capacity?

(b)   Specify the length and the diameter of the handle such that it will bend with a permanent set when the rated capacity of the clamp is exceeded. Use 3 lb as the handle force.

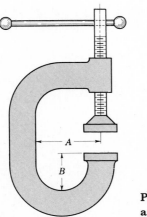

PROBLEMS   8-7,   8-10, and 8-11

8-8   Find the horsepower required to drive a $1\frac{1}{2}$-in power screw having four square threads per inch. The threads are double and the load is 2.40 kip. The nut is to move at a velocity of 8 fpm. The frictional coefficients are 0.10 for the threads and collar. The frictional diameter of the collar is 3 in.

8-9   A single square-thread power screw is to raise a load of 70 kN. The screw has a major diameter of 36 mm and a pitch of 6 mm. The frictional coefficients are 0.13 for the threads and 0.10 for the collar. If the collar frictional diameter is 90 mm and the screw turns at a speed of $1\ s^{-1}$, find:

(a)   The power input to the screw

(b)   The combined efficiency of the screw and collar

Sections 8-3 and 8-4

8-10   The C clamp of Prob. 8-7 is to have the screw machined from a cold-rolled UNS G10100 steel. Assume that the user can overload the handle by a factor of 3 before realizing that overload has occurred. The design load is 150 lb and the nut height is $1\frac{1}{8}$ in. Will the screw thread withstand this overload?

8-11   The C clamp in Prob. 8-10 is made of ASTM No. 20 cast iron. Given the same conditions as in Prob. 8-10, what is the safety factor on the nut?

8-12   A large wing nut adjuster has a metric bolt cast into it as shown in the figure. The nut threads on the bolt. It is prevented from turning upon tightening of the wing nut by a cast-in hexagonal recess in member 2. The two clamped members cause a braking material to be pressed on a rotating member. This is the drag adjustment on the mechanism. This device has experienced repeated thread failures. Experience has shown that the operators apply heavy loads to the wing nut while adjusting with two hands. We learn from the *Handbook of Engineering Psychology*† that the maximum force that we can expect to be comfortably applied per hand is 50 lbs (222 N). Use the relation $F = 5T/d$ to determine the axial bolt load [See Eq. (8-15)]. When the braking materials become glazed and ineffective, the applied torque is expected to double. Under these conditions are the threads strong enough to carry the load?

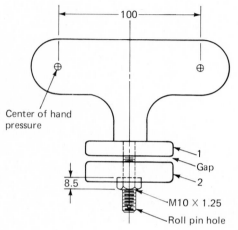

Center of hand pressure

PROBLEM 8-12 Dimensions in millimeters.

Sections 8-5 to 8-9

8-13 The figure illustrates the connection of a cylinder head to a pressure vessel using 10 bolts and a confined gasket. The dimensions in millimeters are: $A = 100$, $B = 200$, $C = 300$, $D = 20$, and $E = 25$. The cylinder is used to store gas at a static pressure of 6 MPa. Metric grade 6.8 bolts are to be used with a factor of safety of at least 3. What size bolts should be used for this application?

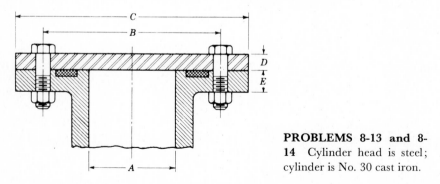

PROBLEMS 8-13 and 8-14 Cylinder head is steel; cylinder is No. 30 cast iron.

8-14 The figure illustrates the connection of a cylinder head which is to be bolted using 24 coarse-thread SAE grade 4 bolts. The dimensions are: $A = 20$ in, $B = 24$ in,

---

† Edwin H. Hilborn, *Handbook of Engineering Psychology*, TAD, Cambridge, Mass., 1965, p. 11–12.

$C = 26.5$ in, $D = \frac{3}{4}$ in, and $E = 1$ in. The cylinder will be used to store liquid at a static pressure of 100 psi. Use a factor of safety of 3 and determine the size of bolts to be used.

8-15 The upside-down A frame shown in the figure is to be bolted to steel beams on the ceiling of a machine room using metric grade 8.8 bolts. This frame is to support the 40-kN radial load as illustrated. The total bolt grip is 48 mm, which includes the thickness of the steel beam, the A-frame feet, and the steel washers used.

(*a*) What tightening torque should be used?

(*b*) What portion of the external load is taken by the bolts? By the members?

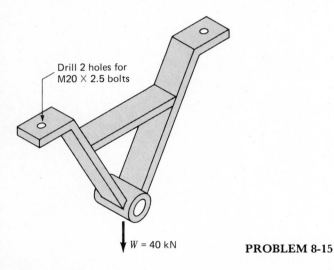

Drill 2 holes for
M20 × 2.5 bolts

$W = 40$ kN   **PROBLEM 8-15**

8-16 The figure shows a welded steel bracket which is to support a force F as shown. The bracket is to be bolted to a smooth vertical face, not shown, by means of four SAE grade 5 $\frac{3}{8}''$-16 UNC bolts, two on centerline (abbreviated CL) A and the other two on CL B. One way of analyzing such a connection would be to assume that the bolts on CL A carry the entire moment load and those on CL B carry the entire shear load.

(*a*) Compute the external shear load carried by the bolts at B.

PROBLEM 8-16   The bolt grip is 1 in.

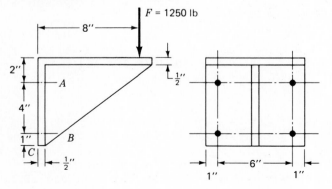

$F = 1250$ lb

(*b*) Compute the external tensile load carried by the bolts at *A*.

(*c*) What is the factor of safety of the connection based on the *A* bolts?

(*d*) What is the factor of safety for the connection based on shear of the *B* bolts?

8-17* The brick or stonework above a fireplace opening is supported by a steel lintel which is usually a 4″ × 4″ or 4″ × 6″ steel angle with legs $\frac{1}{2}$ in thick. The figure illustrates an ASTM No. 25 cast-iron heat exchanger bolted to such a lintel using $\frac{1}{4}$-in cap screws which thread into the lintel. The heat exchanger weighs 200 lb and this exerts a tension load on the cap screws. There have been a number of complaints that the bolts "look" too small. In analyzing a bolted connection using cap screws, use only half of the depth of the threaded hole to determine the grip. The cap screws are SAE grade 5.

(*a*) Find the spring rate of the screw and of the members.

(*b*) What values of initial tension and screw torque should be used for installation?

(*c*) What are the loads in the screws and in the members after installation?

(*d*) Are the complaints justified?

Section of lintel angle

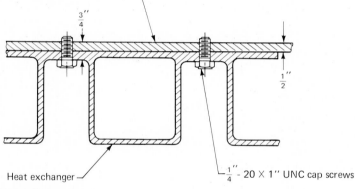

Heat exchanger

$\frac{1}{4}$″- 20 X 1″ UNC cap screws

PROBLEM 8-17

Sections 8-10 to 8-12

8-18 In the figure for Prob. 8-13, let *A* = 0.9 m, *B* = 1 m, *C* = 1.10 m, *D* = 20 mm, and *E* = 25 mm. The cylinder is made of ASTM No. 40 cast iron (*E* = 120 GPa), and the head of low-carbon steel. There are 36 M10 × 1.5 bolts of metric grade 10.9 steel. These bolts are to be tightened so that the preload is 60 percent of the proof

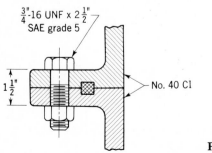

$\frac{3}{4}$″-16 UNF x 2$\frac{1}{2}$″
SAE grade 5

No. 40 CI

PROBLEM 8-19

load. During use, the cylinder pressure will vary between 0 and 550 kPa. Find the factor of safety that guards against a fatigue failure based on 90 percent reliability.

8-19 The section of the gasketed joint shown in the figure is loaded by a tensile force P which fluctuates between 0 and 6 kip. The bolts have been carefully preloaded to $F_i = 25$ kip per bolt. The members have $E = 16$ Mpsi.

(*a*)   Find the endurance limit of the bolts based on 90 percent reliability.

(*b*)   Find the stiffness constants k_m and k_b.

(*c*)   Find the factor of safety of the connection based on the possibility of a fatigue failure.

8-20 The figure shows a fluid-pressure linear actuator, or hydraulic cylinder, in which $D = 100$ mm, $t = 10$ mm, $L = 300$ mm, and $w = 20$ mm. Both brackets as well as the cylinder are of steel. The actuator has been designed for a working pressure of 4 MPa. Five M12 × 1.75 metric grade 6.8 bolts are used, tightened to 50 percent of the proof load.

(*a*)   Find the stiffnesses of the bolts and members, assuming that the entire cylinder is compressed uniformly and that the end brackets are perfectly rigid.

(*b*)   Find the mean and alternating stresses in the bolts.

(*c*)   Find the endurance limit of the bolts based on 50 percent reliability.

(*d*)   What factor of safety guards against a fatigue failure? A static failure?

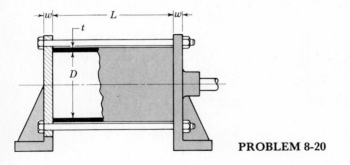

PROBLEM 8-20

Section 8-13

8-21 The figure shows a gasketed joint to be used in a pressure vessel. The dimensions are $A = 4$ in, $B = 6$ in, $C = \frac{3}{4}$ in, $D = \frac{1}{8}$ in, and $E = 8$ in. The gasket manufacturers recommend a pressure of at least 2200 psi to obtain a proper leakproof seal.

(*a*)   Determine the number of $\frac{5}{8}$"-18 UNF grade 5 cap screws necessary to obtain a satisfactory seal.

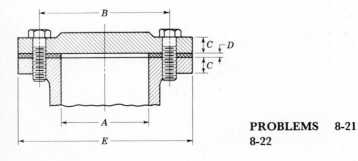

PROBLEMS   8-21   and 8-22

(b) Screws should be spaced about 5 or more diameters for good wrench clearance. What spacing results from your result in (a)?

8-22 The dimensions of the gasketed pressure-vessel joint shown in the figure are: $A = 120$ mm, $B = 200$ mm, $C = 24$ mm, $D = 4$ mm, and $E = 250$ mm. A leakproof seal can be obtained if the average gasket pressure is at least 15 MPa.

(a) Determine the number of 12 mm, 16 mm, and 20 mm grade 8.8 metric cap screws necessary to develop the required gasket pressure.

(b) If the screw spacing is less than five screw diameters, the wrench clearance will be tight. And if the spacing is over ten diameters, the seal pressure may not be uniform from screw to screw. Based on these criteria, which result of (a) is the optimum?

Sections 8-14 to 8-16

8-23 The connections shown are made of UNS G10180 cold-drawn steel and SAE grade 5 bolts. In each case find the static load F that would cause failure.

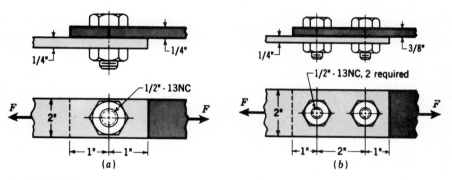

PROBLEM 8-23

8-24 The figure shows two cold-drawn bars of UNS G10180 steel bolted together to form a lap joint. Determine the factor of safety of the connection.

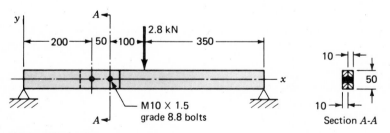

PROBLEM 8-24 Dimensions in millimeters.

8-25 A cold-drawn UNS G41300 steel bar is to be fastened using three M12 × 1.75 grade 8.8 bolts to the 150-mm channel shown in the figure. What maximum force F can be applied to this cantilever if the factor of safety is to be at least 2.8? Do not consider the channel.

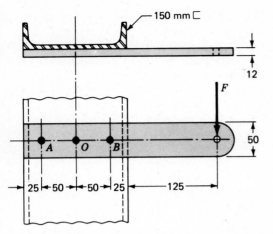

PROBLEM 8-25 Dimensions in millimeters.

8-26 Find the total shear load on each of the three bolts for the connection shown in the figure and compute the significant bolt shear stress and bearing stress. Compute the moment of inertia of the 8-mm plate on a section through the three bolt holes and find the maximum bending stress in the plate.

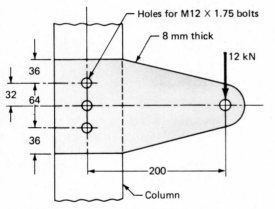

PROBLEM 8-26 Dimensions in millimeters.

8-27 A $\frac{3}{8} \times 2''$ UNS G10180 cold-drawn steel cantilever bar supports a static load of 300 lb as shown in the figure. The bar is secured to the support using two $\frac{1}{2}''$-13 UNC

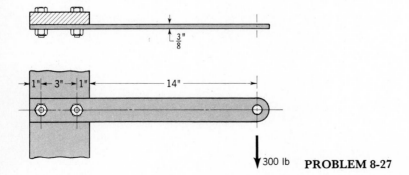

PROBLEM 8-27

grade 5 bolts as shown. Find the factor of safety for the following modes of failure: shear of bolt, bearing on bolt, bearing on member, and strength of member.

8-28 The figure shows a welded fitting which has been tentatively designed to be bolted to a channel so as to transfer the 2500-lb load into the channel. The channel is made of hot-rolled low-carbon steel with $S_y = 46$ kpsi; the two fitting plates are of hot-rolled stock, $S_y = 45.5$ kpsi. The fitting is to be bolted using six standard SAE grade 2 bolts. Check the strength of the design by calculating the factor of safety for all possible modes of failure.

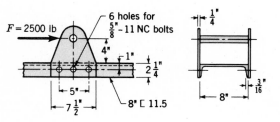

PROBLEM 8-28

CHAPTER
9

WELDED, BRAZED, AND BONDED JOINTS

Processes such as welding, brazing, soldering, cementing, and gluing are used extensively in manufacturing today. Whenever parts have to be assembled or fabricated it is probable that one of these processes should be considered in preliminary design work. Especially when the sections to be joined are thin, one of these fastening methods may lead to significant savings. The elimination of individual fasteners and the adaptability of the method to rapid machine assembly are among the advantages.

One of the difficulties encountered by the design engineer in dealing with the subject of joint design is that it has not benefited by the rigorous treatment that so many other processes, materials, and mechanical elements have had. It is not clear why this should be so. It may be that the geometry does not lend itself well

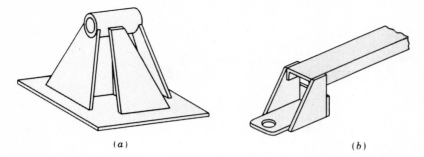

(a)   (b)

FIGURE 9-1 Examples of weldments. (a) A bearing base; can be made of relatively thin parts, yet provides a good stiffness in both directions. (b) Another base detail; made from a hot-rolled channel section and an assortment of cut plates. (*From Procedure Handbook of Arc Welding, 11th ed., Lincoln Electric Company, Cleveland, 1957, pp. 5-219, 5-226; reproduced by permission of the publishers.*)

to mathematical treatment. Of course this means that an added element of uncertainty has been introduced and that this must be compensated for by the use of larger factors of safety in design. The fact that so many safe and reliable structures and devices utilizing these processes are employed today attests to the fact that engineers have been successful in overcoming these handicaps.

9-1 WELDING

Weldments are usually fabricated by clamping, jigging, or holding a collection of hot-rolled low- or medium-carbon steel shapes, cut to particular configurations, while the several parts are welded together. Two examples are shown in Fig. 9-1. An ingenious designer, familiar with various hot-rolled shapes and methods of cutting them, should be able to design strong and lightweight weldments that can be easily and quickly welded with simple holding fixtures.

Figures 9-2 to 9-5* illustrate the type of welds used most frequently by designers. For general machine elements most welds are fillet welds, though butt welds are used a great deal in designing pressure vessels. Of course, the parts to be joined must be arranged so that there is sufficient clearance for the welding operation. If unusual joints are required because of insufficient clearance or because of the section shape, the design may be a poor one and the designer should begin again and endeavor to synthesize another solution.

Since heat is used in the welding operation, there is a possibility of metallurgical changes in the parent metal in the vicinity of the weld. Also, residual stresses

---

* From *Designers Guide for Welded Construction*, The Lincoln Electric Co., Cleveland, by permission. This is a six-page folder containing a summary of all welding symbols and other useful information available to designers and students on request.

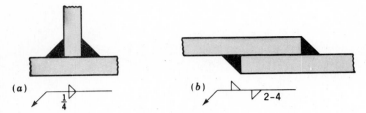

FIGURE 9-2 Fillet welds. (*a*) The fraction indicates the leg size; the arrow should point to only one weld when both sides are the same. (*b*) The symbol indicates that the welds are intermittent and staggered 2 in on 4-in centers.

may be introduced because of clamping or holding or, sometimes, because of the order of welding. Usually these residual stresses are not severe enough to cause concern; in some cases a light heat treatment after welding has been found helpful in relieving them. When the parts to be welded are thick, a preheating will also be of benefit. If the reliability of the component is to be quite high, a testing program should be established to learn what changes or additions to the operations are necessary to assure the best quality.

9-2 BUTT AND FILLET WELDS

Figure 9-6 shows a single V-groove weld loaded by the tensile force F. For either tension or compression loading, the average normal stress is

$$\sigma = \frac{F}{hl} \tag{9-1}$$

where h is the weld throat and l is the length of the weld, as shown in the figure. Note that the value of h does not include the reinforcement. The reinforcement is

FIGURE 9-3 The circle on the weld symbol indicates that the welding is to go all around.

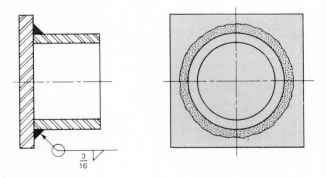

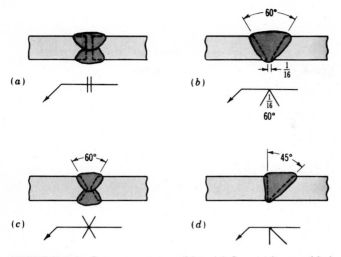

FIGURE 9-4 Butt or groove welds. (a) Square butt-welded on both sides; (b) single V with 60° bevel and root opening of 1/16 in; (c) double V; (d) single bevel.

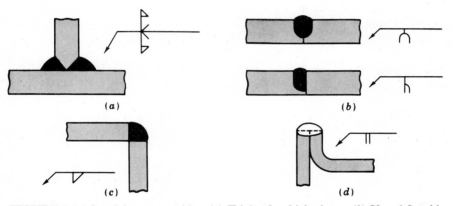

FIGURE 9-5 Special groove welds. (a) T joint for thick plates; (b) U and J welds for thick plates; (c) corner weld may also have a bead weld on inside for greater strength but should not be used for heavy loads; (d) edge weld for sheet metal and light loads.

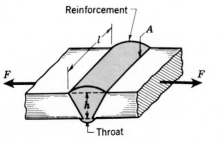

FIGURE 9-6 A typical butt joint.

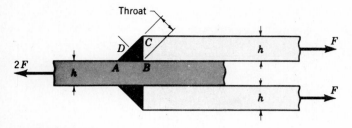

FIGURE 9-7   A transverse fillet weld.

desirable in order to compensate for flaws, but it varies somewhat and does produce stress concentration at point A in the figure.   If fatigue loads exist, it is good practice to grind or machine off the reinforcement.

The average stress in a butt weld due to shear loading is

$$\tau = \frac{F}{hl} \tag{9-2}$$

Figure 9-7 illustrates a typical transverse fillet weld.   Attempts to solve for the stress distribution in such welds, using the methods of theory of elasticity, have not been very successful.   Conventional practice in welding engineering design has always been to base the size of the weld upon the magnitude of the stress on the throat area DB.

In Fig. 9-8a a portion of the weld has been selected from Fig. 9-7 so as to

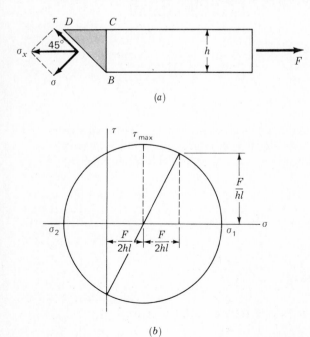

(a)

(b) **FIGURE 9-8**

treat the weld throat as a problem in free-body analysis. The throat area is $A = hl \cos 45° = 0.707hl$, where l is the length of the weld. Thus the stress σ_x is

$$\sigma_x = \frac{F}{A} = \frac{F}{0.707hl} \qquad (a)$$

This stress can be divided into two components, a shear stress τ and a normal stress σ. These are

$$\tau = \sigma_x \cos 45° = \frac{F}{hl} \qquad \sigma = \sigma_x \cos 45° = \frac{F}{hl} \qquad (b)$$

In Fig. 9-8b these are entered into a Mohr's circle diagram. The largest principal stress is seen to be

$$\sigma_1 = \frac{F}{2hl} + \sqrt{\left(\frac{F}{2hl}\right)^2 + \left(\frac{F}{hl}\right)^2} = 1.618\frac{F}{hl} \qquad (c)$$

The maximum shear stress is

$$\tau_{\text{max}} = \sqrt{\left(\frac{F}{2hl}\right)^2 + \left(\frac{F}{hl}\right)^2} = 1.118\frac{F}{hl} \qquad (d)$$

However, for design purposes it is customary to base the shear stress on the throat area and to neglect the normal stress altogether. Thus the equation for *average stress* is

$$\tau = \frac{F}{0.707hl} \qquad (9\text{-}3)$$

and is normally used in designing joints having fillet welds. Note that this gives a shear stress $(1/0.707)(1.118) = 1.27$ times greater than that given by Eq. (d).

There are some experimental and analytical results that are helpful in evaluating Eq. (9-3). A model of the transverse fillet weld of Fig. 9-7 is easily constructed for photoelastic purposes and has the advantage of a balanced loading condition. Norris constructed such a model and reported the stress distribution along the sides AB and BC of the weld.* An approximate graph of the results he obtained is shown as Fig. 9-9a. Note that stress concentration exists at A and B on the horizontal leg and at B on the vertical leg. Norris states that he could not determine the stresses at A and B with any certainty.

---

* C. H. Norris, "Photoelastic Investigation of Stress Distribution in Transverse Fillet Welds," *Welding J.*, vol. 24, 1945, p. 557s.

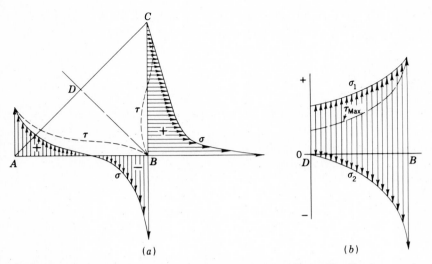

FIGURE 9-9 Stress distribution in fillet welds. (*a*) Stress distribution on the legs as reported by Norris; (*b*) distribution of principal stresses and maximum shear stress as reported by Salakian.

Salakian* presents data for the stress distribution across the throat of a fillet weld (Fig. 9-9*b*). This graph is of particular interest because we have just learned that it is the throat stresses that are used in design. Again, the figure shows stress concentration at point *B*. Note that Fig. 9-9*a* applies either to the weld metal or to the parent metal, and that Fig. 9-9*b* applies only to the weld metal.

More recently Bagci† has made an important contribution to the analysis using the finite-element approach. Figure 9-10 shows the joint analyzed by Bagci, and Fig. 9-11 the results he obtained. Do not compare the results of Fig. 9-11 with those of Fig. 9-9 because the joint geometry is different. But it is interesting to compare the results of Fig. 9-11*b* with a result from Eq. (9-3). Using the geometry of Fig. 9-10 and Eq. (9-3) we find

$$\tau = \frac{F}{0.707hl} = \frac{1000}{4(0.707)(0.125)(1)} = 2830 \text{ psi}$$

since there are four welds. But the maximum shear stress from Fig. 9-11*b* is seen to be

$$\tau_{\text{max}} = 3800 \text{ psi}$$

---

* A. G. Salakian and G. E. Claussen, "Stress Distribution in Fillet Welds; A Review of the Literature," *Welding J.*, vol. 16, May 1937, pp. 1–24.

† Cemil Bagci, "Finite Stress Elements and Applications in Machine Design," tutorial paper presented at the Sixth OSU Applied Mechanisms Conference, Denver, Oct. 1979.

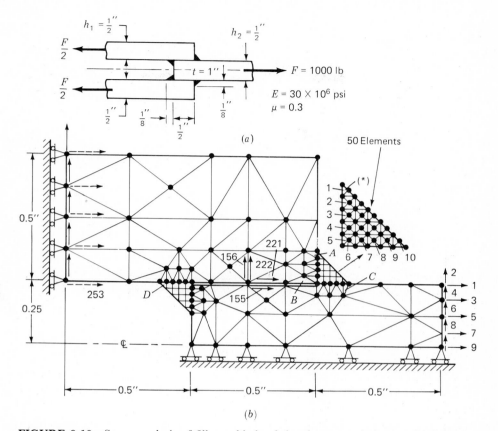

FIGURE 9-10 Stress analysis of fillet welds by finite-element techniques. (*a*) Specimen used showing dimensions and loading; (*b*) mathematical model showing the triangular elements. (*Published with the permission of the author, Cemil Bagci, Tennessee Technological University.*)

Thus a generous factor of safety should always be used when Eq. (9-3) is used in design. When the equation is used with the maximum stresses permitted by various construction codes, it turns out that the resulting welds are perfectly safe. However a great many failures are attributable to defective welds. So you must be assured that the welds are truly made in accordance with the specifications.

In the case of parallel fillet welds (Fig. 9-12), the assumption of a shear stress along the throat is more realistic. Since there are two welds, the throat area for both is $A = 2(0.707hl) = 1.414hl$. The average shear stress is therefore

$$\tau = \frac{F}{1.414hl} \tag{e}$$

It is quite probable that the stress distribution along the length of the welds is *not* uniform.

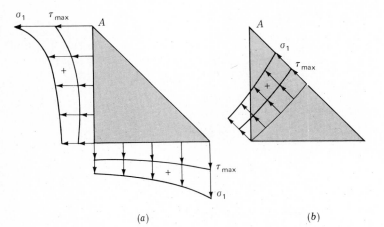

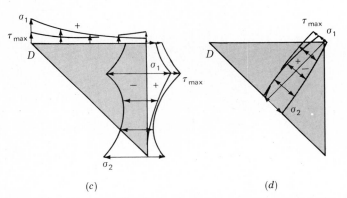

FIGURE 9-11 (*a*) Stress distribution on legs of upper weld; when quite small the σ_2 stresses are not shown. Vertical leg, $\sigma_{1,max} = 12.91$ kpsi, $\tau_{max,max} = 6.29$ kpsi. Horizontal leg, $\sigma_{1,max} = 9.47$ kpsi, $\tau_{max,max} = 4.62$ kpsi. (*b*) Stresses in throat of upper weld; $\sigma_{1,max} = 7.57$ kpsi, $\tau_{max,max} = 3.80$ kpsi. (*c*) Stresses on legs of lower weld. Horizontal leg, $\sigma_{1,max} = 4.26$ kpsi, $\tau_{max,max} = 2.06$ kpsi. Vertical leg, $\sigma_{1,max} = 3.96$ kpsi, $\sigma_{2,max} = -7.60$ kpsi, $\tau_{max,max} = 5.53$ kpsi. (*d*) Stresses in throat of lower weld, $\sigma_{1,max} = 1.80$ kpsi, $\sigma_{2,max} = -3.66$ kpsi, $\tau_{max,max} = 2.48$ kpsi.

FIGURE 9-12 A double-filleted lap joint.

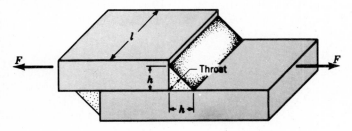

9-3 TORSION IN WELDED JOINTS

Figure 9-13 illustrates a cantilever of length l welded to a column by two fillet welds. The reaction at the support of a cantilever always consists of a shear force V and a moment M. The shear force produces a *primary shear* in the welds of magnitude

$$\tau' = \frac{V}{A} \tag{9-4}$$

where A is the throat area of all the welds.

The moment at the support produces *secondary shear* or *torsion* of the welds, and this stress is

$$\tau'' = \frac{Mr}{J} \tag{9-5}$$

where r is the distance from the centroid of the weld group to the point in the weld of interest, and J is the polar moment of inertia of the weld group about the centroid of the group. When the sizes of the welds are known, these equations can be solved and the results combined to obtain the maximum shear stress. Note that r is usually the farthest distance from the weld centroid.

Figure 9-14 shows two welds in a group. The rectangles represent the throat areas of the welds. Weld 1 has a throat width $b_1 = 0.707h_1l$; and weld 2 has a throat width $d_2 = 0.707h_2$. Note that h_1 and h_2 are the respective weld sizes. The throat area of both welds is

$$A = A_1 + A_2 = b_1d_1 + b_2d_2 \tag{a}$$

This is the area that is to be used in Eq. (9-4).

FIGURE 9-13   A moment connection.

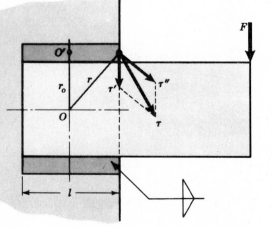

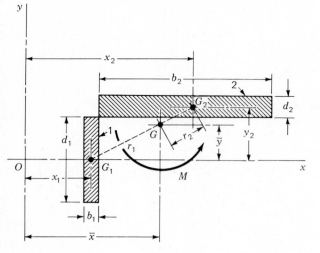

FIGURE 9-14

The x axis in Fig. 9-14 passes through the center gravity G_1 of weld 1. The moment of inertia about this axis is

$$I_x = \frac{b_1 d_1^3}{12}$$

Similarly, the moment of inertia about an axis through G_1 parallel to the y axis is

$$I_y = \frac{d_1 b_1^3}{12}$$

Thus the polar moment of inertia of weld 1 about its own center of gravity is

$$J_{G_1} = I_x + I_y = \frac{b_1 d_1^3}{12} + \frac{d_1 b_1^3}{12} \tag{b}$$

In a similar manner, the polar moment of inertia of weld 2 about its center of gravity is

$$J_{G_2} = I_x + I_y = \frac{b_2 d_2^3}{12} + \frac{d_2 b_2^3}{12} \tag{c}$$

We must next locate the center of gravity G of the weld group. Thus,

$$\bar{x} = \frac{A_1 x_1 + A_2 x_2}{A} \qquad \bar{y} = \frac{A_1 y_1 + A_2 y_2}{A}$$

Using Fig. 9-14 again, we see that the distance r_1 from G_1 to G is

$$r_1 = \sqrt{(\bar{x} - x_1)^2 + \bar{y}^2}$$

Similarly, the distance r_2 from G_2 to G is

$$r_2 = \sqrt{(y_2 - \bar{y})^2 + (x_2 - \bar{x})^2}$$

Now, using the parallel-axis theorem we find the polar moment of inertia of the weld group to be

$$J = (J_{G_1} + A_1 r_1^2) + (J_{G_2} + A_2 r_2^2) \qquad (d)$$

This is the quantity to be used in Eq. (9-5). The distance r must be measured from G and the moment M computed about G.

The reverse procedure is that in which the allowable shear stress is given and we wish to find the weld size. The usual procedure would be to estimate the weld size, compute J and A, and then find and combine τ' and τ''. If the resulting maximum stress were too large, a larger weld size would be estimated and the procedure repeated. After a few such trials, a satisfactory result would be obtained.

Observe in Eq. (b) that the second term contains the quantity b_1^3 which is the cube of the weld width; and the quantity d_2^3 in the first term of Eq. (c) is also the cube of the weld width. Both of these quantities are quite small and can be neglected. This leads to the idea of treating each fillet weld as a line. The resulting polar moment of inertia is then equivalent to a *unit polar moment of inertia*. The advantage of treating the weld as a line is that the unit polar moment of inertia is the same, regardless of the weld size. Since the throat width of a fillet weld is $0.707h$, the relationship between the unit polar moment of inertia and the polar moment of inertia of a fillet weld is

$$J = 0.707hJ_u \qquad (9\text{-}6)$$

in which J_u is found by conventional methods for an area having unit width. The transfer formula for unit polar moment of inertia must be employed when the welds occur in groups, as in Fig. 9-13. Table 9-1 lists the throat areas and unit polar moments of inertia for the most common fillet welds encountered. The example that follows is illustrative of the calculations normally made.

Example 9-1 A 50-kN load is transferred from a welded fitting into a 200-mm steel channel as illustrated in Fig. 9-15. Compute the maximum stress in the weld.

Table 9-1 TORSIONAL PROPERTIES OF FILLET WELDS*

Weld	Throat area	Location of G	Unit polar moment of inertia
	$A = 0.707hd$	$\bar{x} = 0$ $\bar{y} = d/2$	$J_u = d^3/12$
	$A = 1.414hd$	$\bar{x} = b/2$ $\bar{y} = d/2$	$J_u = \dfrac{d(3b^2 + d^2)}{6}$
	$A = 0.707h(b + d)$	$\bar{x} = \dfrac{b^2}{2(b + d)}$ $\bar{y} = \dfrac{2bd + d^2}{2(b + d)}$	$J_u = \dfrac{(b + d)^4 - 6b^2d^2}{12(b + d)}$
	$A = 0.707h(2b + d)$	$\bar{x} = \dfrac{b^2}{2b + d}$ $\bar{y} = d/2$	$J_u = \dfrac{8b^3 + 6bd^2 + d^3}{12} - \dfrac{b^4}{2b + d}$
	$A = 1.414h(b + d)$	$\bar{x} = b/2$ $\bar{y} = d/2$	$J_u = \dfrac{(b + d)^3}{6}$
	$A = 1.414\pi hr$		$J_u = 2\pi r^3$

* G is centroid of weld group; h is weld size; plane of torque couple is in the plane of the paper; all welds are of the same size.

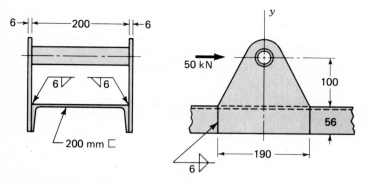

FIGURE 9-15 Dimensions in millimeters.

Solution As shown by the figure, each plate is welded to the channel using three 6-mm fillet welds. We shall divide the load in half and consider only a single plate in the analysis to follow. Two of the three welds are 56 mm long, and the length of the third is 190 mm. Using Table 9-1, we first locate the centroid of the weld group.

$$\bar{y} = \frac{b^2}{2b + d} = \frac{(56)^2}{2(56) + 190} = 10.4 \text{ mm}$$

$$\bar{x} = d/2 = 190/2 = 95 \text{ mm}$$

These dimensions are shown on the free-body diagram of Fig. 9-16*a*. Note that these dimensions also locate the origin O of the xy reference system. The reaction moment, a torque, is

$$M = 25(110.4) = 2760 \text{ N} \cdot \text{m}$$

per plate.

FIGURE 9-16 Dimensions in millimeters.

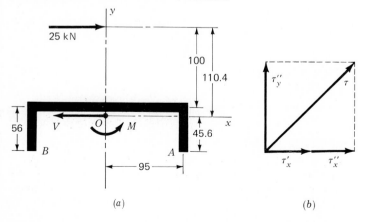

(*a*)   (*b*)

Next, we refer to Table 9-1 and find the unit polar moment of inertia to be

$$
\begin{aligned}
J_u &= \frac{8b^3 + 6bd^2 + d^3}{12} - \frac{b^4}{2b + d} \\
&= \frac{8(56)^3 + 6(56)(190)^2 + (190)^3}{12} - \frac{(56)^4}{2(56) + 190} \\
&= 1.67(10)^6 \text{ mm}^3
\end{aligned}
$$

Then, from Eq. (9-6),

$$
J = 0.707hJ_u = 0.707(6)(1.67)(10)^6 = 7.07(10)^6 \text{ mm}^4
$$

Using Table 9-1 again, we find the throat area for the welds on one plate to be

$$
A = 0.707h(2b + d) = 0.707(6)[2(56) + 190] = 1280 \text{ mm}^2
$$

The primary shear stress is

$$
\tau'_x = \frac{V}{A} = \frac{25(10)^3}{1280} = 19.5 \text{ MPa}
$$

We can find the secondary shear stress in components parallel to x and y. The y component is

$$
\tau''_y = \frac{Mr_x}{J} = \frac{2760(10)^3(95)}{7.07(10)^6} = 37.1 \text{ MPa}
$$

$$
\tau''_x = \frac{Mr_y}{J} = \frac{2760(10)^3(45.6)}{7.07(10)^3} = 17.8 \text{ MPa}
$$

These stress components should be combined to yield the maximum stresses, which occur at ends A and B. Thus, from Fig. 9-16b, we have

$$
\tau = \sqrt{\tau_y^2 + \tau_x^2} = \sqrt{(37.1)^2 + (19.5 + 17.8)^2} = 52.6 \text{ MPa} \qquad Ans.
$$

////

9-4 BENDING IN WELDED JOINTS

Figure 9-17a shows a cantilever welded to a support by fillet welds at top and bottom. A free-body diagram of the beam would show a shear-force reaction V and a moment reaction M. The shear force produces a primary shear in the

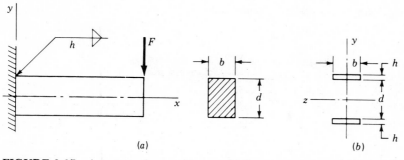

FIGURE 9-17 A rectangular cantilever welded to a support at the top and bottom edges.

welds of magnitude

$$\tau' = V/A \tag{a}$$

where A is the total throat area.

The moment M produces a normal bending stress σ in the welds. Though not rigorous, it is customary in the stress analysis of welds to assume that this stress acts normal to the throat area. By treating the two welds in Fig. 9-17b as lines, we find the unit moment of inertia to be

$$I_u = \frac{bd^2}{2} \tag{b}$$

Then the moment of inertia based on the weld throat is

$$I = 0.707h\,\frac{bd^2}{2} \tag{c}$$

The normal stress is now found to be

$$\sigma = \frac{Mc}{I} = \frac{M(d/2)}{0.707bd^2h/2} = \frac{1.414M}{bdh} \tag{d}$$

The moment of inertia in Eq. (d) is based on the distance d between the two welds. If the moment of inertia is found by treating the two welds as rectangles instead, the distance between the weld centroids would be $(d + h)$. This would produce a slightly larger moment of inertia and result in a smaller value of the stress σ. Thus the method of treating welds as lines produces more conservative results. Perhaps the added safety is appropriate in view of the stress distributions of Figs. 9-9 and 9-11.

Once the stress components σ and τ have been found for welds subjected to bending, they may be combined by using a Mohr's circle diagram to find the principal stresses or the maximum shear stress. Then an appropriate failure theory is applied to determine the likelihood of failure or safety. Because of the greater uncertainties in the analysis of weld stresses, the more conservative maximum-shear-stress theory is generally preferred.

Table 9-2 lists the bending properties most likely to be encountered in the analysis of welded beams.

Table 9-2   BENDING PROPERTIES OF FILLET WELDS*

Weld	Throat area	Location of G	Unit moment of inertia
	$A = 0.707hd$	$\bar{x} = 0$ $\bar{y} = d/2$	$I_u = \dfrac{d^3}{12}$
	$A = 1.414hd$	$\bar{x} = b/2$ $\bar{y} = d/2$	$I_u = \dfrac{d^3}{6}$
	$A = 1.414hb$	$\bar{x} = b/2$ $\bar{y} = d/2$	$I_u = \dfrac{bd^2}{2}$
	$A = 0.707h(2b + d)$	$\bar{x} = \dfrac{b^2}{2b + d}$ $\bar{y} = d/2$	$I_u = \dfrac{d^2}{12}(6b + d)$
	$A = 0.707h(b + 2d)$	$\bar{x} = b/2$ $\bar{y} = \dfrac{d^2}{b + 2d}$	$I_u = \dfrac{2d^3}{3} - 2d^2\bar{y} + (b + 2d)\bar{y}^2$

Table 9-2 (*Continued*)

Weld	Throat area	Location of G	Unit moment of inertia
	$A = 1.414h(b + d)$	$\bar{x} = b/2$ $\bar{y} = d/2$	$I_u = \dfrac{d^2}{6}(3b + d)$
	$A = 0.707h(b + 2d)$	$\bar{x} = b/2$ $\bar{y} = \dfrac{d^2}{b + 2d}$	$I_u = \dfrac{2d^3}{3} - 2d^2\bar{y} + (b + 2d)\bar{y}^2$
	$A = 1.414h(b + d)$	$\bar{x} = b/2$ $\bar{y} = d/2$	$I_u = \dfrac{d^2}{6}(3d + b)$
	$A = 1.414\pi hr$		$I_u = \pi r^3$

* I_u, unit moment of inertia, is taken about a horizontal axis through G, the centroid of the weld group; h is weld size; the plane of the bending couple is normal to the plane of the paper; all welds are of the same size.

9-5 THE STRENGTH OF WELDED JOINTS

The matching of the electrode properties with those of the parent metal is usually not so important as speed, operator appeal, and the appearance of the completed joint. The properties of electrodes vary considerably, but Table 9-3 lists the minimum properties for some electrode classes.

It is preferable, in designing welded components, to select a steel that will result in a fast, economical weld even though this may require a sacrifice of other

Table 9-3 MINIMUM WELD-METAL PROPERTIES

AWS electrode number*	Tensile strength, kpsi	Yield strength, kpsi	Percent elongation
E60xx	62	50	17–25
E70xx	70	57	22
E80xx	80	67	19
E90xx	90	77	14–17
E100xx	100	87	13–16
E120xx	120	107	14

* The American Welding Society (AWS) specification code numbering system for electrodes. This system uses an E prefix to a four- or five-digit numbering system in which the first two or three digits designate the approximate tensile strength. The last digit indicates variables in the welding technique, such as current supply. The next to the last digit indicates the welding position, as, for example, flat, or vertical, or overhead. The complete set of specifications may be obtained from the AWS upon request.

qualities such as machinability. Under the proper conditions all steels can be welded, but best results will be obtained if steels having a UNS specification between G10140 and G10230 are chosen. All these steels have a tensile strength in the hot-rolled condition in the range of 60 to 70 kpsi.

The designer can choose factors of safety or permissible working stresses with more confidence if he or she is aware of the values of those used by others. One of the best standards to use is the American Institute of Steel Construction (AISC) code for building construction.* The permissible stresses are now based on the yield strength of the material instead of the ultimate strength, and the code permits the use of a variety of ASTM structural steels having yield strengths varying from 33 to 50 kpsi. Provided the loading is the same, the code permits the same stress in the weld metal as in the parent metal. For these ASTM steels, $S_y = 0.5S_u$. Table 9-4 lists the formulas specified by the code for calculating these permissible stresses for various loading conditions. The factors of safety implied by this code are easily calculated. For tension, $n = 1/0.60 = 1.67$. For shear, $n = 0.577/0.40 = 1.44$, if we accept the distortion-energy theory as the criterion of failure.

It is important to observe that the electrode material is often the strongest material present. If a bar of AISI 1010 steel is welded to one of 1018 steel, the weld metal is actually a mixture of the electrode material and the 1010 and 1018 steels. Furthermore, a welded cold-drawn bar has its cold-drawn properties replaced with the hot-rolled properties in the vicinity of the weld. Finally, remembering that the weld metal is usually the strongest, do check the stresses in the parent metals.

The AISC code, as well as the AWS code, for bridges includes permissible stresses when fatigue loading is present. The designer will have no difficulty in

* For a copy write the AISC, New York.

Table 9-4 STRESSES PERMITTED BY THE AISC CODE FOR WELD METAL

Type of loading	Type of weld	Permissible stress	n^*
Tension	Butt	$0.60S_y$	1.67
Bearing	Butt	$0.90S_y$	1.11
Bending	Butt	$0.60-0.66S_y$	1.52–1.67
Simple compression	Butt	$0.60S_y$	1.67
Shear	Butt or fillet	$0.40S_y$	1.44

* The factor of safety n has been computed using the distortion-energy theory.

using these codes, but their empirical nature tends to obscure the fact that they have been established by means of the same knowledge of fatigue failure already discussed in Chap. 7. Of course, for structures covered by these codes the actual stresses *cannot* exceed the permissible stresses; otherwise the designer is legally liable. But in general, codes tend to conceal the actual margin of safety involved.

The fatigue-strength reduction factors listed in Table 9-5, as proposed by Jennings,* are suggested for use. These factors should be used for the parent metal as well as for the weld metal.

Table 9-5 FATIGUE-STRENGTH REDUCTION FACTORS

Type of weld	K_f
Reinforced butt weld	1.2
Toe of transverse fillet weld	1.5
End of parallel fillet weld	2.7
T-butt joint with sharp corners	2.0

Example 9-2 Brackets, such as the one of Fig. 9-18, are used in mooring small watercraft. Failure of such brackets is usually caused by the bearing pressure of the mooring-line clip against the side of the hole and the fatigue failure of either the weld metal or the parent metal. To get an idea of the margins of safety involved we use a bracket $\frac{1}{4}$ in thick made of hot-rolled AISI 1018 steel. We then assume that wave action on the boat will create no greater force F than 500 lb. A reliability of 95 percent is desired. Determine all appropriate factors of safety.

Solution Figure 9-19 shows a free-body diagram of the bracket with the external force F acting through the center of the hole. The center of gravity of the weld

---

* C. H. Jennings, "Welding Design," *Trans. ASME*, vol. 58, 1936, pp. 497–509.

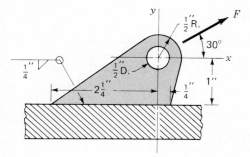

FIGURE   9-18 Welded
mooring bracket.

group and of the bottom of the bracket is G. The force F_G is the force of the weld
group acting on the bracket. Since F_G has a different line of action than F, we
also have a moment M. The dimensions shown are obtained from the trigonometry of the diagram.

Note that the forces and moments of the bracket acting on the welds are
equal and opposite to those shown. Thus

- The moment M produces a bending stress in the welds with tension at A
and compression at C.
- The force component F_y produces tension throughout the weld.
- The force component F_x produces shear throughout the weld.

These effects are

$$M = 500(0.632) = 316 \text{ lb} \cdot \text{in} \qquad F_x = 500 \cos 30° = 433 \text{ lb}$$

$$F_y = 500 \sin 30° = 250 \text{ lb}$$

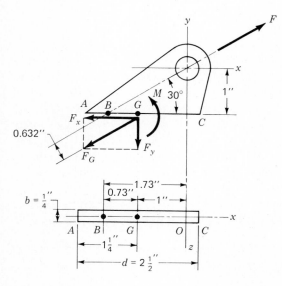

FIGURE 9-19

From Table 9-2 we find

$$A = 1.414h(b + d) = 1.414(0.25)(0.25 + 2.5) = 0.972 \text{ in}^2$$

$$I_u = \frac{d^2}{6}(3b + d) = \frac{(2.5)^2}{6}[3(0.25) + 2.5] = 3.39 \text{ in}^4$$

Then the moment of inertia about an axis through G parallel to z is

$$I = 0.707hI_u = 0.707(0.25)(3.39) = 0.599 \text{ in}^4$$

At end A the bending stress and the tensile stress due to F_y add. For the weld metal the total normal stress is

$$\sigma_y = \frac{F_y}{A} + \frac{Mc}{I} = \frac{250}{0.972} + \frac{316(1.25)}{0.599} = 916 \text{ psi}$$

The shear stress is

$$\tau_{yx} = \frac{F_x}{A} = \frac{433}{0.972} = 445 \text{ psi}$$

From Table A-17 we find the mechanical properties for the bracket material to be $S_u = 58$ kpsi, $S_y = 32$ kpsi, and $H_B = 116$ Bhn. We shall use these same properties for the weld metal. Now use Eq. (6-15) to find the von Mises stress in the weld metal. The result is

$$\sigma' = \sqrt{\sigma_y^2 + 3\tau_{yx}^2} = \sqrt{(916)^2 + 3(445)^2} = 1200 \text{ psi}$$

Thus the factor of safety guarding against a static failure in the weld metal is

$$n = \frac{S_y}{\sigma'} = \frac{32(10)^3}{1200} = 26.7 \qquad\qquad Ans.$$

Next, we proceed to calculate the stresses in the parent metal. The area subject to shear is

$$A = bd = 0.25(2.5) = 0.625 \text{ in}^2$$

Thus the shear stress in the parent metal is

$$\tau_{yx} = \frac{F_x}{A} = \frac{433}{0.625} = 693 \text{ psi}$$

The section modulus of the bracket at the weld interface is

$$\frac{I}{c} = \frac{bd^2}{6} = \frac{0.25(2.5)^2}{6} = 0.260 \text{ in}^3$$

Thus the tensile stress at A in the parent metal is

$$\sigma_y = \frac{F_y}{A} + \frac{M}{I/c} = \frac{250}{0.625} + \frac{316}{0.260} = 1615 \text{ psi}$$

Using Eq. (6-15) again we find

$$\sigma' = \sqrt{\sigma_y^2 + 3\tau_{yx}^2} = \sqrt{(1615)^2 + 3(693)^2} = 2010 \text{ psi}$$

Then the factor of safety guarding against a failure in the parent metal is

$$n = \frac{S_y}{\sigma'} = \frac{32(10)^3}{2010} = 15.9 \qquad\qquad Ans.$$

The clip on the mooring line bears against the side of the $\frac{1}{2}$-in hole. If we assume the clip fills the hole, then the average bearing stress is

$$\sigma = \frac{F}{td} = -\frac{500}{0.25(0.50)} = -4000 \text{ psi}$$

Thus the factor of safety is

$$n = \frac{S_y}{|\sigma|} = \frac{32(10)^3}{4000} = 8 \qquad\qquad Ans.$$

The next problem is to investigate the probability of a fatigue failure. The endurance limit of the rotating-beam specimen is

$$S_e' = 0.50S_u = 0.50(58) = 29 \text{ kpsi}$$

The surface factor k_a for welds and for the parent material in the vicinity of the weld should always be based on an as-forged surface. Thus, from Fig. 7-10 we find $k_a = 0.53$.

To obtain the size factor we use Fig. 7-15b and Eq. (7-16). Thus

$$d_{eq} = \sqrt{\frac{0.05bd}{0.0766}} = \sqrt{\frac{0.05(0.25)(2.5)}{0.0766}} = 0.639 \text{ in}$$

$$k_b = 0.869(d_{eq})^{-0.097} = 0.869(0.639)^{-0.097} = 0.908$$

This is for the parent metal and we shall use this same factor for the weld metal.

For 95 percent reliability we find $k_c = 0.868$ from Table 7-7.

From Table 9-5 we find $K_f = 2.7$ at the end of a parallel fillet weld. Therefore $k_e = 1/K_f = 1/2.7 = 0.370$.

We now find the corrected endurance limit to be

$$S_e = k_a k_b k_c k_e S_e' = (0.53)(0.908)(0.868)(0.370)(29) = 4.48 \text{ kpsi}$$

for both the weld metal and the parent metal.

Since the load is applied through a mooring line, the tensile force F is either zero or it is not zero; this is a repeated load. Therefore the alternating and mean von Mises stresses in the weld metal are

$$\sigma_a' = \sigma_m' = \frac{\sigma'}{2} = \frac{1200}{2} = 600 \text{ psi}$$

We now wish to compute the factor of safety guarding against a fatigue failure using the modified Goodman relations. Substituting $S_a = n\sigma_a'$ and $S_m = n\sigma_m'$ into Eq. (7-34) yields

$$\frac{S_a}{S_e} + \frac{S_m}{S_{ut}} = 1 \quad \text{or} \quad \frac{600n}{4480} + \frac{600n}{58(10)^3} = 1$$

Solving for the factor of safety gives $n = 6.93$ *Ans.*

For the parent metal $\sigma_a' = \sigma_m' = \sigma'/2 = 2010/2 = 1005$ psi. Using Eq. (7-34) again gives

$$\frac{1005n}{4480} + \frac{1005n}{58(10)^3} = 1$$

The solution to this equation yields $n = 4.14$ *Ans.*

                                                      ////

9-6 RESISTANCE WELDING

The heating and consequent welding which occur when an electrical current is passed through several parts which are pressed together is called resistance welding. *Spot welding* and *seam welding* are the forms of resistance welding most often used. The advantages of resistance welding over other forms are the speed, the accurate regulation of time and heat, the uniformity of the weld and the mechanical properties which result, the elimination of filler rods or fluxes, and the fact that the process is easy to automate.

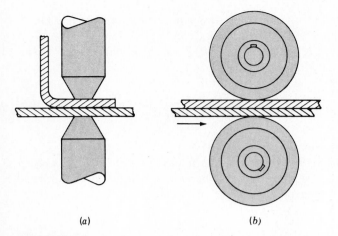

(a)   (b)

FIGURE 9-20   (a) Spot welding; (b) seam welding.

The spot- and seam-welding processes are illustrated schematically in Fig. 9-20. Seam welding is actually a series of overlapping spot welds, since the current is applied in pulses as the work moves between the rotating electrodes.

Failure of a resistance weld is either by shearing of the weld or by tearing of the metal around the weld. Because of the tearing, it is good practice to avoid loading a resistance-welded joint in tension. Thus, for the most part, design so that the spot or seam is loaded in pure shear. The shear stress is then simply the load divided by the area of the spot. Because of the fact that the thinner sheet of the pair being welded may tear, the strength of spot welds is often specified by stating the load per spot based on the thickness of the thinnest sheet. Such strengths are best obtained by experiment.

Somewhat larger factors of safety should be used when parts are fastened by spot welding, rather than by bolts or rivets, to account for the metallurgical changes in the materials due to the welding.

9-7 BONDED JOINTS

When two parts or materials are connected together by a third material unlike the base materials, the process is called *bonding*. Thus *brazing, soldering,* and *cementing,* or the use of adhesives, are all means of bonding parts together. Figure 9-21 shows some examples of bonded joints that represent good joint design.

Brazing Parts that are joined by heating them to more than 800°F to allow the filler metal to flow into the clearance space by capillary action are said to be brazed. Parts, like that in Fig. 9-21*b*, which are self-jigging, can be furnace-brazed. Another method of brazing is by torch, and this may be either automatic

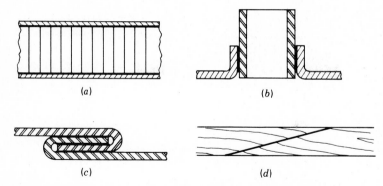

FIGURE 9-21 (*a*) Airplane wing section fabricated by bonding aluminum honeycomb to the skins, using resin bonding under heat and pressure; (*b*) tubing joined to sheet-metal section by brazing metal; (*c*) sheet-metal parts joined by soldering; (*d*) wood parts joined by gluing.

or manual. Some of the advantages offered by brazing are:

1 The ability to join materials of different thicknesses
2 The lack of radical change in the mechanical properties of the base materials after they are joined
3 The ability to join both cast and rolled materials as well as unlike materials
4 The fact that the completed joint requires no additional finishing procedures

A well-designed brazed joint is one that permits the base materials rather than the filler material to carry the load. Thus if a stress analysis appears to be needed, it is good practice to review the joint design to see if a more effective geometry can be obtained.

Soldering A soldered joint is much like a brazed joint, except that the filler metal is softer and the process is carried out at temperatures less than 800°F. Soldering is used to join sheet metals, such as tin cans and duct work, and for numerous electrical applications. Connections to be joined by soldering should always be designed so that the base materials carry the entire external load.

Cementing In many instances parts that must be connected together can be joined using adhesives, creating a significant cost advantage over the use of screws, rivets, or other mechanical fasteners. The stresses in an adhesive joint are also much more uniform than in, say, a riveted joint, where the load is shared by each rivet. Adhesive joints should be carefully designed so as to carry only compression or shear. Figure 9-22 shows that even a shear-loaded joint will have stress con-

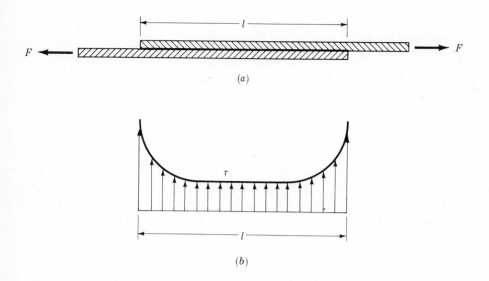

FIGURE 9-22 (*a*) Simple lap joint, adhesively bonded, and loaded in shear. (*b*) Approximate stress distribution showing stress concentration at the ends.

centration at the ends. The eccentricity in such lap joints may cause some slight bending and eventual failure by peeling.

Some of the advantages of adhesive joints are:

1   They involve low cost.
2   They result in smooth surfaces.
3   Base materials can be lightweight.
4   Unlike materials can be joined.
5   Joints are sealed against pressure to avoid leakage.
6   They provide some vibration damping.

PROBLEMS*

9-1 to 9-3   The permissible shear stress for the welds shown is 20 kpsi in English units and 140 MPa in SI units. For each case, find the load F that would cause such a stress.

---

* The asterisk indicates a design-type problem or one that may have no unique result.

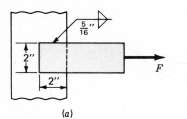

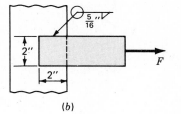

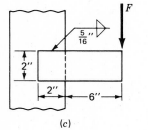

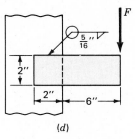

PROBLEM 9-1.   All bars 3/8 in thick.

PROBLEM 9-2   Dimensions in millimeters; all bars 10 mm thick.

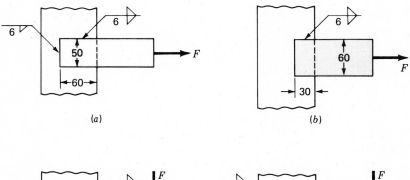

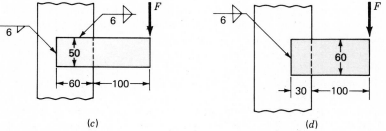

440

DESIGN OF MECHANICAL ELEMENTS

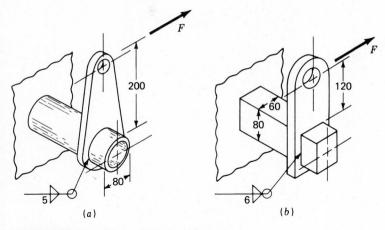

PROBLEM 9-3 Dimensions in millimeters.

9-4 For each weldment shown, find the torque T that can be applied if the permissible weld shear stress is 20 kpsi.

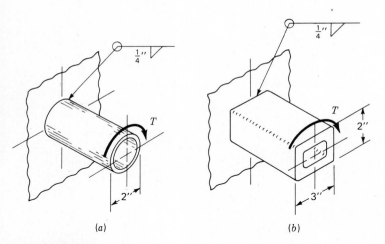

PROBLEM 9-4

9-5 to 9-7 The beams shown in the figures are welded to fixed supports or to plates as shown. In each case find the maximum combined shear stress in the weld metal.

9-8* A fuel tank 4 ft in diameter holds 750 gal of fuel oil. It is made of hot-rolled low-carbon steel $\frac{3}{16}$ in thick. We wish to design a pair of connectors to be welded to the tank at each end so that a hand-operated chain hoist can be used to lower the tank into the excavation. Design the connectors, make a detailed drawing of one of them, and specify the welds in detail.

9-9* A forged-steel wrecking ball is 8 in in diameter. Design a suitable connector so that the crane cable can be clipped to it. Show all details of bolts or welds used.

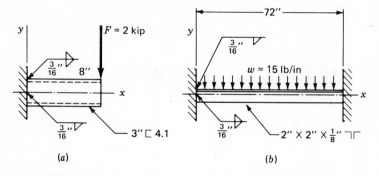

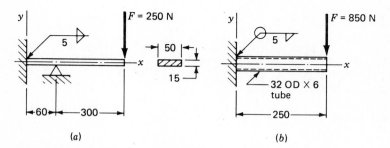

PROBLEM 9-5 (*a*) Beam is a 4.1-lb structural-steel channel; (*b*) beam is two structural angles, back-to-back, with same welds at each end.

PROBLEM 9-6   Dimensions in millimeters.

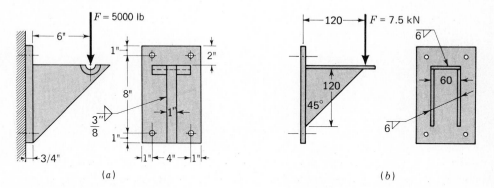

PROBLEM 9-7   (*a*) Dimensions in inches; (*b*) dimensions in millimeters.

CHAPTER

10

MECHANICAL SPRINGS

Mechanical springs are used in machines to exert force, to provide flexibility, and to store or absorb energy. In general, springs may be classified as either wire springs, flat springs, or special-shaped springs, and there are variations within these divisions. Wire springs include helical springs of round or square wire and are made to resist tensile, compressive, or torsional loads. Under flat springs are included the cantilever and elliptical types, the wound motor- or clock-type power springs, and the flat spring washers, usually called Belleville springs.

10-1 STRESSES IN HELICAL SPRINGS

Figure 10-1a shows a round-wire helical compression spring loaded by the axial force F. We designate D as the *mean spring diameter* and d as the *wire diameter*.

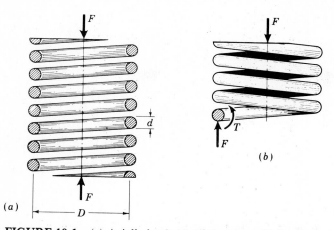

FIGURE 10-1 (*a*) Axially loaded helical spring; (*b*) free-body diagram showing that the wire is subjected to a direct shear and a torsional shear.

Now imagine that the spring is cut at some point (Fig. 10-1*b*), a portion of it removed, and the effect of the removed portion replaced by the internal forces. Then, as shown in the figure, the cut portion would exert a direct shear force *F* and a torsion *T* on the remaining part of the spring.

To visualize the torsion, picture a coiled garden hose. Now pull one end of the hose in a straight line perpendicular to the plane of the coil. As each turn of hose is pulled off the coil, the hose twists or turns about its own axis. The flexing of a helical spring creates a torsion in the wire in a similar manner.

Using superposition, the maximum stress in the wire may be computed using the equation

$$\tau_{\max} = \pm \frac{Tr}{J} + \frac{F}{A} \tag{a}$$

where the term Tr/J is the torsion formula of Chap. 2. Replacing the terms by $T = FD/2$, $r = d/2$, $J = \pi d^4/32$, and $A = \pi d^2/4$ gives

$$\tau = \frac{8FD}{\pi d^3} + \frac{4F}{\pi d^2} \tag{b}$$

In this equation the subscript indicating maximum shear stress has been omitted as unnecessary. The positive signs of Eq. (*a*) have been retained, and hence Eq. (*b*) gives the shear stress at the inside fiber of the spring.

Now define *spring index*

$$C = \frac{D}{d} \tag{10-1}$$

as a measure of coil curvature. With this relation, Eq. (b) can be arranged to give

$$\tau = \frac{8FD}{\pi d^3}\left(1 + \frac{0.5}{C}\right) \tag{c}$$

Or designating

$$K_s = 1 + \frac{0.5}{C} \tag{10-2}$$

then

$$\tau = K_s \frac{8FD}{\pi d^3} \tag{10-3}$$

where K_s is called a *shear-stress multiplication factor*. For most springs, C will range from about 6 to 12. Equation (10-3) is quite general and applies for both static and dynamic loads. It gives the maximum shear stress in the wire, and this stress occurs at the inner fiber of the spring.

Many writers present the stress equation as

$$\tau = K \frac{8FD}{\pi d^3} \tag{10-4}$$

where K is called the *Wahl correction factor*.* This factor includes the direct shear, together with another effect due to curvature. As shown in Fig. 10-2, curvature of the wire increases the stress on the inside of the spring but decreases it only slightly on the outside. The value of K may be obtained from the equation

$$K = \frac{4C - 1}{4C - 4} + \frac{0.615}{C} \tag{10-5}$$

By defining $K = K_c K_s$, where K_c is the effect of curvature alone, we have

$$K_c = \frac{K}{K_s} \tag{10-6}$$

Investigation reveals that curvature shear stress is highly localized on the inside of the spring. Springs subjected only to static loads will yield at the inside fiber and relieve this stress. Thus, for static loads, the curvature stress can be neglected and

---

* See A. M. Wahl, *Mechanical Springs*, 2d ed., McGraw-Hill, New York, 1963. This book is the standard reference on springs.

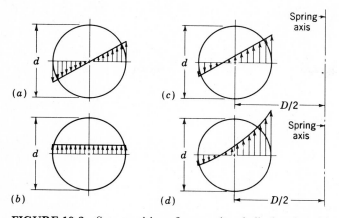

FIGURE 10-2 Superposition of stresses in a helical spring. (*a*)
Pure torsional stress; (*b*) direct-shear stress; (*c*) resultant of
direct- and torsional-shear stresses; (*d*) resultant of direct-,
torsional-, and curvature-shear stresses.

Eq. (10-3) used. For fatigue loads, K_c is used as a *fatigue-strength reduction factor*;
therefore Eq. (10-3) gives the correct stress when fatigue is a factor too. Thus we
shall not generally make use of Eq. (10-4) in this book.

The use of square or rectangular wire is not recommended for springs unless
space limitations make it necessary. Special-wire shapes are not made in large
quantities, as are those of round wire; they have not had the benefit of refining
development and hence may not be as strong as springs made from round wire.
When space is severely limited, the use of nested round-wire springs should always
be considered. They may have an economical advantage over the special-section
springs, as well as a strength advantage.

10-2 DEFLECTION OF HELICAL SPRINGS

To obtain the equation for the deflection of a helical spring, we shall consider an
element of wire formed by two adjacent cross sections. Figure 10-3 shows such an
element, of length *dx*, cut from wire of diameter *d*. Let us consider a line *ab* on the
surface of the wire which is parallel to the spring axis. After deformation it will
rotate through the angle γ and occupy the new position *ac*. From Eq. (2-16),

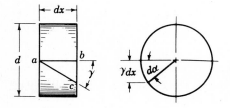

FIGURE   **10-3** Cross-
sectional element of a helical spring.

which is the expression of Hooke's law for torsion, we have

$$\gamma = \frac{\tau}{G} = \frac{8FD}{\pi d^3 G} \tag{a}$$

where the value of τ is obtained from Eq. (10-4), unity being used for the value of the Wahl correction factor.* The distance bc is $\gamma\,dx$, and the angle $d\alpha$, through which one section rotates with respect to the other, is

$$d\alpha = \frac{\gamma\,dx}{d/2} = \frac{2\gamma\,dx}{d} \tag{b}$$

If the number of active coils is denoted by N, the total length of the wire is πDN. Upon substituting γ from Eq. (a) into Eq. (b) and integrating, the angular deflection of one end of the wire with respect to the other is

$$\alpha = \int_0^{\pi DN} \frac{2\gamma}{d}\,dx = \int_0^{\pi DN} \frac{16FD}{\pi d^4 G}\,dx = \frac{16FD^2 N}{d^4 G} \tag{c}$$

The load F has a moment arm of $D/2$, and so the deflection is

$$y = \alpha\,\frac{D}{2} = \frac{8FD^3 N}{d^4 G} \tag{10-7}$$

The deflection can also be obtained by using strain-energy methods. From Eq. (3-26) the strain energy for torsion is

$$U = \frac{T^2 l}{2GJ} \tag{d}$$

Substituting $T = FD/2$, $l = \pi DN$, and $J = \pi d^4/32$ gives

$$U = \frac{4F^2 D^3 N}{d^4 G} \tag{e}$$

and so the deflection is

$$y = \frac{\partial U}{\partial F} = \frac{8FD^3 N}{d^4 G} \tag{f}$$

---

* Wahl quotes the results of tests to show that K can be made unity for calculating deflections with very accurate results, Ibid., p. 29.

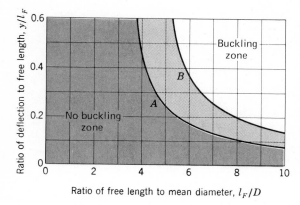

FIGURE 10-4 Curves show when buckling of compression coil springs may occur. Both curves are for springs having squared and ground ends. For curve *A* one end of the spring is compressed against a flat surface, the other against a rounded surface. For curve *B* both ends of the spring are compressed against flat and parallel surfaces.

To find the spring constant, use Eq. (3-2), and substitute the value of y from Eq. (10-7). This gives

$$k = \frac{d^4 G}{8 D^3 N} \tag{10-8}$$

The equations presented in this section are valid for both compression and extension springs. Long coil springs having a free length more than four times the mean diameter may fail by buckling. This condition may be corrected by mounting the spring over a round bar or in a tube. Figure 10-4 will be helpful in deciding whether a compression spring is likely to buckle.

10-3 EXTENSION SPRINGS

Extension springs necessarily must have some means of transferring the load from the support to the body of the spring. Although this can be done with a threaded plug or a swivel hook, both of these add to the cost of the finished product, and so one of the methods shown in Fig. 10-5 is usually employed. In designing a spring with a hook end, the stress-concentration effect must be considered.

FIGURE 10-5 Types of ends used on extension springs. (*Courtesy of Associated Spring Corporation.*)

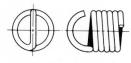

Machine half loop – open

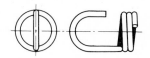

Raised hook

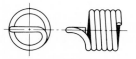

Short twisted loop

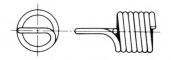

Full twisted loop

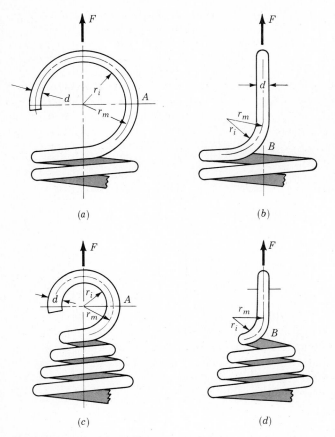

FIGURE 10-6 Ends for extension springs. (*a*) Usual design; stress at *A* is due to combined axial and bending forces. (*b*) Side view of (*a*); stress is mostly torsion at *B*. (*c*) Improved design; stress at *A* is due to combined axial and bending forces. (*d*) Side view of (*c*); stress at *B* is mostly torsion.

In Fig. 10-6*a* and *b* is shown a much-used method of designing the end. Stress concentration due to the sharp bend makes it impossible to design the hook as strong as the body. Tests show that the stress-concentration factor is given approximately by

$$K = \frac{r_m}{r_i} \tag{10-9}$$

which holds for bending stress and occurs when the hook is offset, and for torsional stress. Figure 10-6*c* and *d* shows an improved design due to a reduced coil diameter, not to elimination of stress concentration. The reduced coil diameter results in a lower stress because of the shorter moment arm. No stress-concentration factor is needed for the axial component of the load.

Initial Tension

When extension springs are made with coils in contact with one another, they are said to be *close-wound*. Spring manufacturers prefer some initial tension in close-wound springs in order to hold the free length more accurately. The initial tension is created in the winding process by twisting the wire as it is wound onto the mandrel. When the spring is completed and removed from the mandrel, the initial tension is locked in because the spring cannot get any shorter.

The direction of the stresses can be visualized by reference to Fig. 10-7. In (*a*) block *A* simulates the effect of the stacked coils, and the free length of the spring is the length l_F with no external force applied. In Fig. 10-7*b* an external force *F* has been applied, causing the spring to elongate through the distance *y*. Note very particularly that the stresses in the spring are in the *same direction* in Fig. 10-7*a* and (*b*).

Figure 10-7*c* shows the relation between the external force and the spring elongation. Here we see that *F* must exceed the initial tension F_i before a deflection *y* is experienced.

10-4 COMPRESSION SPRINGS

The four types of ends commonly used for compression springs are illustrated in Fig. 10-8. A spring with *plain ends* has a noninterrupted helicoid; the ends are the

FIGURE 10-7 Simulation of an extension spring with initial tension. (*a*) No external force; spring compresses block *A* with initial force F_i. The free length is l_F. (*b*) Spring extended a distance *y* by external force *F*. (*c*) Force-deflection relation.

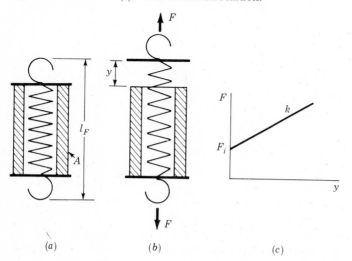

(*a*)   (*b*)   (*c*)

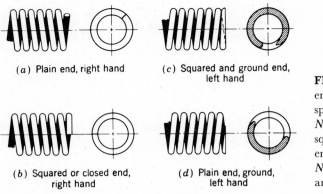

(a) Plain end, right hand   (c) Squared and ground end,
                                        left hand

(b) Squared or closed end,   (d) Plain end, ground,
    right hand   left hand

FIGURE 10-8 Types of ends for compression springs; (a) both ends plain, $N_D = \frac{1}{2}$; (b) both ends squared, $N_D = 1$; (c) both ends squared and ground, $N_D = 2$; (d) both ends plain and ground, $N_D = 1$.

same as if a long spring had been cut into sections. A spring with plain ends that are *squared* or *closed* is obtained by deforming the ends to a zero-degree helix angle. A better transfer of the load into the spring is obtained by grinding the ends, as can be seen in Fig. 10-8.

The type of end used results in *dead* or *inactive* turns at each end of the spring, and these must be subtracted from the total number of turns to obtain the number of active turns. The formula is

$$N = N_T - N_D \tag{10-10}$$

where N = number of active coils
N_T = total number of coils
N_D = number of inactive coils

The caption of Fig. 10-8 contains an estimate of the number of dead coils to be expected.

It is customary in the design of springs to neglect the effects of eccentricity of loading due to the end turns. It is also customary to neglect the effect of residual stresses caused by heat treatment or overstressing. Instead, these two effects are usually accounted for by an increase in the factor of safety. It is the usual practice in the manufacture of compression springs to close them to their solid height; this induces a residual stress opposite in direction to the working stress and has the effect of improving the strength of the spring.

10-5 SPRING MATERIALS

Springs are manufactured either by hot- or cold-working processes, depending upon the size of the material, the spring index, and the properties desired. In general, prehardened wire should not be used if $D/d < 4$ or if $d > \frac{1}{4}$ in. Winding of the spring induces residual stresses through bending, but these are normal to the direction of the torsional working stresses in a coil spring. Quite frequently in spring manufacture, they are relieved, after winding, by a mild thermal treatment.

A great variety of spring materials is available to the designer, including plain carbon steels, alloy steels, and corrosion-resisting steels, as well as nonferrous materials such as phosphor bronze, spring brass, beryllium copper, and various nickel alloys. Descriptions of the most commonly used steels will be found in Table 10-1. The UNS steels listed in the Appendix should be used in designing hot-worked, heavy-coil springs, as well as flat springs, leaf springs, and torsion bars.

Table 10-1 HIGH-CARBON AND ALLOY SPRING STEELS

Name of material	Similar specifications	Description
Music wire, 0.80–0.95C	UNS G10850 AISI 1085 ASTM A228-51	This is the best, toughest, and most widely used of all spring materials for small springs. It has the highest tensile strength and can withstand higher stresses under repeated loading than any other spring material. Available in diameters 0.12 to 3 mm (0.005 to 0.125 in). Do not use above 120°C (250°F) or at subzero temperatures.
Oil-tempered wire, 0.60–0.70C	UNS G10650 AISI 1065 ASTM 229-41	This general-purpose spring steel is used for many types of coil springs where the cost of music wire is prohibitive and in sizes larger than available in music wire. Not for shock or impact loading. Available in diameters 3 to 12 mm (0.125 to 0.5000 in), but larger and smaller sizes may be obtained. Not for use above 180°C (350°F) or at subzero temperatures.
Hard-drawn wire, 0.60–0.70C	UNS G10660 AISI 1066 ASTM A227-47	This is the cheapest general-purpose spring steel and should be used only where life, accuracy, and deflection are not too important. Available in diameters 0.8 to 12 mm (0.031 to 0.500 in). Not for use above 120°C (250°F) or at subzero temperatures.
Chrome vanadium	UNS G61500 AISI 6150 ASTM 231-41	This is the most popular alloy spring steel for conditions involving higher stresses than can be used with the high-carbon steels and for use where fatigue and long endurance are needed. Also good for shock and impact loads. Widely used for aircraft-engine valve springs and for temperatures to 220°C (425°F). Available in annealed or pretempered sizes 0.8 to 12 mm (0.031 to 0.500 in) in diameter.
Chrome silicon	UNS G92540 AISI 9254	This alloy is an excellent material for highly stressed springs requiring long life and subjected to shock loading. Rockwell hardnesses of C50 to C53 are quite common, and the material may be used up to 250°C (475°F). Available from 0.8 to 12 mm (0.031 to 0.500 in) in diameter.

Source: By permission from Harold C. R. Carlson, "Selection and Application of Spring Materials," *Mech. Eng.*, vol. 78, 1956, pp. 331–334.

Table 10-2   CONSTANTS FOR USE IN EQ. (10-11) TO ESTIMATE THE TENSILE
STRENGTH OF SELECTED SPRING STEELS

Material	Size range, in	Size range, mm	Exponent, m	Constant, A kpsi	Constant, A MPa
Music wire[a]	0.004–0.250	0.10–6.5	0.146	196	2170
Oil-tempered wire[b]	0.020–0.500	0.50–12	0.186	149	1880
Hard-drawn wire[c]	0.028–0.500	0.70–12	0.192	136	1750
Chrome vanadium[d]	0.032–0.437	0.80–12	0.167	169	2000
Chrome silicon[e]	0.063–0.375	1.6–10	0.112	202	2000

[a] Surface is smooth, free from defects, and has a bright, lustrous finish.
[b] Has a slight heat-treating scale which must be removed before plating.
[c] Surface is smooth and bright, with no visible marks.
[d] Aircraft-quality tempered wire; can also be obtained annealed.
[e] Tempered to Rockwell C49 but may also be obtained untempered.

Spring materials may be compared by an examination of their tensile strengths; these vary tremendously with wire size and, to a lesser extent, with the material and processing. In the past it has been customary to tabulate these strengths for various wire sizes and materials.* But the availability of the scientific electronic calculator now makes such a tabulation unnecessary. The reason for this is that a log-log plot of the tensile strengths versus wire diameters is a straight line. The equation of this line can be written in terms of the ordinary logarithms of the strengths and wire diameters. This equation can then be solved to give

$$S_{ut} = \frac{A}{d^m} \tag{10-11}$$

where A is a constant related to a strength intercept and m is the slope of the line on the log-log plot. Of course such an equation is only valid for a limited range of wire sizes. Table 10-2 gives values of m and the constant A for both English and SI units for the materials listed in Table 10-1.

Although the torsional yield strength is needed to design springs, surprisingly little information on this property is available. Using an approximate relationship between yield strength and ultimate strength in tension,

$$S_y = 0.75 S_{ut} \tag{10-12}$$

and then applying the distortion-energy theory gives

$$S_{sy} = 0.577 S_y \tag{10-13}$$

* See, for example, Joseph E. Shigley, *Mechanical Engineering Design*, 2d ed., McGraw-Hill, New York, 1972, p. 362.

and provides us with a means of estimating the torsional yield strength S_{sy}. But this method should not be used if experimental data are available; if used, a generous factor of safety should be employed, especially for extension springs, because of the uncertainty involved.

Variations in the wire diameter and in the coil diameter of the spring have an effect on the stress as well as on the spring scale. Large tolerances will result in more economical springs, and so the defining of tolerances is an important phase of spring design. The commercial tolerance on wire diameter is usually not more than plus or minus 1.5 percent of the diameter. The tolerance on coil diameters varies from about 5 percent for springs having an index $D/d = 4$ up to more than 25 percent for D/d values of 16 or more. These tolerances correspond roughly to three standard deviations.

Example 10-1 A helical compression spring is made of No. 16 gauge (0.037 in) music wire. The outside diameter of the spring is $\frac{7}{16}$ in. The ends are squared and there are $12\frac{1}{2}$ total turns.

(a) Estimate the torsional yield strength of the wire.
(b) Find the maximum static load corresponding to the yield strength.
(c) What is the scale of the spring?
(d) What deflection would be caused by the load in (b)?
(e) Compute the solid height of the spring.
(f) What should be the length of the spring so that when it is compressed solid and then released there will be no permanent change in the free length?
(g) Corresponding to the length found in (f), is buckling a possibility?

Solution (a) Using Eq. (10-11) and Table 10-2 we find $A = 196$ kpsi and $m = 0.146$. Therefore

$$S_{ut} = \frac{A}{d^m} = \frac{196}{(0.037)^{0.146}} = 317 \text{ kpsi}$$

Then, from Eqs. (10-12) and (10-13) we find

$$S_y = 0.75 S_{ut} = 0.75(317) = 238 \text{ kpsi}$$

$$S_{sy} = 0.577 S_y = 0.577(238) = 137 \text{ kpsi} \qquad\qquad Ans.$$

(b) The mean spring diameter is $D = \left(\frac{7}{16}\right) - 0.037 = 0.400$ in. So the spring index is $C = 0.400/0.037 = 10.8$. Then, from Eq. (10-2) we find

$$K_s = 1 + \frac{0.5}{C} = 1 + \frac{0.5}{10.8} = 1.05$$

Now rearrange Eq. (10-3) and solve for F_{max} by substituting the yield strength for the shear stress. This gives

$$F_{max} = \frac{S_{sy} \pi d^3}{8K_s D} = \frac{137(10)^3(\pi)(0.037)^3}{8(1.05)(0.400)} = 6.49 \text{ lb} \qquad\qquad Ans.$$

(c) From Fig. 10-8 and Eq. (10-10) we find the number of active turns to be

$$N = 12.5 - 1 = 11.5$$

Using $G = 11.5$ Mpsi, the scale of the spring is found using Eq. (10-8). Thus

$$k = \frac{d^4 G}{8D^3 N} = \frac{(0.037)^4(11.5)(10)^6}{8(0.400)^3(11.5)} = 3.66 \text{ lb/in} \qquad\qquad Ans.$$

(d) $y = \dfrac{F}{k} = \dfrac{6.49}{3.66} = 1.77$ in $\qquad\qquad Ans.$

(e) The solid height is the total number of coils times the wire diameter, or

$$h = 12.5(0.037) = 0.4625 \text{ in} \qquad\qquad Ans.$$

(f) To avoid yielding, the spring can be no longer than the solid height plus the deflection caused by a load just short of the amount required to initiate yielding. Summing the results of (d) and (e) gives this length as

$$l_F = 1.77 + 0.4625 = 2.23 \text{ in} \qquad\qquad Ans.$$

(g) Figure 10-4 doesn't quite apply because the ends are not ground. Nevertheless, we find the two parameters to be

$$\frac{y}{l_F} = \frac{1.77}{2.23} = 0.794 \qquad \frac{l_F}{D} = \frac{2.23}{0.400} = 5.57$$

Entering Fig. 10-4 with these results indicates quite clearly that buckling would occur even if both ends were ground and were seated against a flat and parallel surface. ////

Example 10-2 The following specifications were taken from a tension spring removed from a junked washing machine: $D = 10$ mm, $d = 1.80$ mm, $N = 122$ active coils; raised hook ends 244 mm between hooks (Fig. 10-5); $r_m = 5$ mm in Fig. 10-6a; $r_m = 2.5$ mm in Fig. 10-6b; preload, 25 N; material appears to be painted, hard-drawn wire.

(a) Estimate the tensile and torsional yield strengths of the wire.

(b) Compute the initial torsional stress in the wire.

(c) What is the spring rate?

(d) What force is required to cause the body of the spring to be stressed to the yield strength?

(e) What force is required to cause the torsional stress in the hook ends to reach the yield strength?

(f) What force is required to cause the normal stress in the hook ends to reach the tensile yield strength?

(g) What is the distance between the hook ends if the smallest of the three forces found in (d), (e), and (f) is applied?

Solution (a) From Eq. (10-11) and Table 10-2 we have $A = 1750$ MPa and $m = 0.192$. So

$$S_{ut} = \frac{A}{d^m} = \frac{1750}{(1.80)^{0.192}} = 1560 \text{ MPa}$$

Then the yield strengths are estimated from Eqs. (10-12) and (10-13) to be

$$S_y = 0.75(1560) = 1170 \text{ MPa} \qquad S_{sy} = 0.577(1170) = 675 \text{ MPa}$$

(b) The spring index is $C = D/d = 10/1.8 = 5.56$; and so the shear-stress multiplication factor is

$$K_s = 1 + \frac{0.5}{C} = 1 + \frac{0.5}{5.56} = 1.09$$

The initial torsional stress is found from Eq. (10-3) using the preload $F = 25$ N. Thus

$$\tau = K_s \frac{8FD}{\pi d^3} = \frac{1.09(8)(25)(10)}{\pi(1.8)^3} = 119 \text{ MPa} \qquad\qquad Ans.$$

where Table A-4 has been used to obtain preferred units.

(c) Using $G = 79.3$ GPa and Eq. (10-8) we find the spring rate to be

$$k = \frac{d^4 G}{8D^3 N} = \frac{(1.8)^4(79.3)(10)^6}{8(10)^3(122)} = 853 \text{ N/m} \qquad\qquad Ans.$$

(d) Rearranging Eq. (10-3) and substituting the torsional yield strength for the shear stress gives

$$F_{max} = \frac{\pi d^3 S_{sy}}{8K_s D} = \frac{\pi(1.8)^3(675)}{8(1.09)(10)} = 141.8 \text{ N} \qquad\qquad Ans.$$

(e) Referring to Fig. 10-6b we note that $r_m = 2.5$ mm and so $r_i = 2.5 - (1.8/2) = 1.6$ mm. Thus, from Eq. (10-9) we have

$$K = \frac{r_m}{r_i} = \frac{2.5}{1.6} = 1.56$$

This stress-concentration factor should be used instead of K_s in Eq. (10-3). The maximum force is then found to be

$$F_{max} = \frac{\pi d^3 S_{sy}}{8KD} = \frac{\pi(1.8)^3(675)}{8(1.56)(10)} = 99.1 \text{ N} \qquad\qquad Ans.$$

Since we are dealing with very hard materials, it is perfectly appropriate to use stress-concentration factors as stress-increasing factors when the loads are static.

(f) The normal stress in the hook results from an axial component and a bending component. Thus the stress is

$$\sigma = \frac{M}{I/c} + \frac{F}{A} = K \frac{32 F r_m}{\pi d^3} + \frac{4F}{\pi d^2} \qquad\qquad (1)$$

where K is obtained from Fig. 10-6a. Thus $r_i = r_m - (d/2) = 5 - (1.8/2) = 4.1$ mm. Therefore

$$K = \frac{r_m}{r_i} = \frac{5}{4.1} = 1.22$$

Substituting $\sigma = S_y$ and the known values in Eq. (1) yields

$$1170 = \frac{1.22(32)(F_{max})(5)}{(1.8)^3} + \frac{4F_{max}}{(1.8)^2} \qquad\qquad (2)$$

when this equation is solved, we get $F_{max} = 106$ N   Ans.

(g) The smallest of the three forces is $F_{max} = 99.1$ N. The spring preload of 25 N must be overcome before the spring will elongate. Thus the elongation is

$$y = \frac{F}{k} = \frac{99.1 - 25}{853}(10)^3 = 86.9 \text{ mm}$$

The distance between the hook ends will then be

$$l = 86.9 + 244 = 331 \text{ mm} \qquad\qquad Ans.$$

////

10-6 DESIGN OF HELICAL SPRINGS

The design of a new spring involves the following considerations:

- Space into which the spring must fit and operate
- Values of working forces and deflections
- Accuracy and reliability needed
- Tolerances and permissible variations in specifications
- Environmental conditions such as temperature and a corrosive atmosphere
- Cost and quantities needed

The designer uses these factors to select a material and specify suitable values for the wire size, the number of turns, the diameter and free length, the type of ends, and the spring rate needed to satisfy the working force-deflection requirements. Because of the large number of interdependent variables, the problem is not as simple as inserting numbers into a procedure that will produce a complete set of results. Because of this fact, there are a rather large number of researchers who have attempted to simplify the spring-design problem by the use of charts and nomographs.*

Nomographs for use with the two fundamental design equations [Eqs. (10-4) and (10-8)] are given by Chironis† and also by Camm.‡ These equations are solved by separate nomographs in which vertical lines are used for the various design parameters.

Tsai§ superposes nomographs to solve Eqs. (10-3) and (10-8) on the same graph. This is done by using both horizontal and vertical lines for the various design parameters.

The widespread availability of home computers and programmable calculators now makes a simple iteration procedure perfectly feasible for designs. There is usually only a limited, finite number of wire sizes that can be used. Furthermore, the limits on the diameter, the number of turns, and the choice of material are such that not too many iterations are needed to get a satisfactory design. If the quantity needed is not large, then it will be more economical to choose a spring from a catalog of stock sizes. This also reduces the number of iterations needed.

There are almost as may ways to create a spring-design program as there are

---

* See M. Massoud and L. Hubert, "Brief Survey of Spring Design Nomographs," ASME Paper no. 76-DET/77, 1977. See also My Dao-Thien and M. Massoud, "Design Nomographs of Compression Helical Springs for Predetermined Reliability Levels," ASME Paper no. 80-C2/DET 85, 1980.

† N. P. Chironis, *Spring Design and Application*, McGraw-Hill, New York, 1961.

‡ F. J. Camm, *Newnes Engineer's Reference Book*, 8th ed., Newnes, 1958.

§ T. K. Tsai, "Speedy Design of Helical Compression Springs by Nomography Method," *Journal of Engineering for Industry*, Feb. 1975, pp. 373–374.

programmers; and there is nothing unusual about the program presented here. It works. The program, which is for the design of compression springs, can be used as a starting point for the creation of other programs. It consists of seven separate subroutines, all of which utilize the same memory locations. The subroutines, which should be used in the order in which they are presented, are:

1 Enter and display (or print) the outside diameter.
2 Enter and display the total number of coils. Enter and display the number of dead coils. Compute and display the number of active coils.
3 Select a material and enter and display the exponent and coefficient.
4 Enter and display the wire diameter. Compute and display the torsional yield strength.
5 Enter and display the maximum torsional stress desired when the spring is closed solid. Compute and display the solid height, the free length, and the force required to compress the spring solid.
6 Compute and display the spring constant.
7 Enter and display any desired operating force F. Compute and display the corresponding values of the torsional stress and the spring deflection.

Separate subroutines are used in this program in order to avoid reentering all the data when only a single parameter is to be changed. In this way it is easy to see the effect of the single change. For example, having run through the program once it may be desirable to try a different wire size. This can be done by entering the new wire size in subroutine 4 and proceeding from that point.

Example 10-3 Indexing is used in machine operations when a circular part being manufactured must be divided into a certain number of segments. Figure 10-9

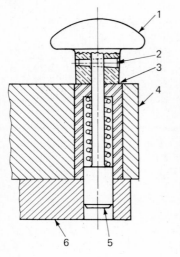

FIGURE 10-9 Part 1, pull knob; part 2, tapered retaining pin; part 3, hardened bushing with press fit; part 4, body of fixture; part 5, indexing pin; part 6, work-piece holder. Space for spring is $\frac{5}{8}$ in OD by $\frac{1}{4}$ in ID and $1\frac{3}{8}$ in long, with the pin down as shown. The pull knob must be raised $\frac{3}{4}$ in to permit indexing.

shows a portion of an indexing fixture used to successively position a part for the operation. When the knob is pulled up, part 6, which holds the work piece, is rotated to the next position and locked in place by releasing the index pin. In this example we wish to design the spring to exert a force of about 3 lb and to fit in the space defined in the figure caption.

Solution Since only a limited number of springs is needed, a stock spring will be selected. These are available in music wire. In one catalog there are 76 stock springs available having an outside diameter of 0.480 in and designed to work in a $\frac{1}{2}$-in hole. These are made in seven different wire sizes, ranging from 0.038 in up to 0.063 in, and in free lengths from $\frac{1}{2}$ in to $2\frac{1}{2}$ in, depending upon the wire size.

Since the pull knob must be raised $\frac{3}{4}$ in for indexing, and the space for the spring is $1\frac{3}{8}$ in long when the pin is down, the solid length cannot be more than $\frac{5}{8}$ in.

Let us begin by selecting a spring having an outside diameter of 0.480 in, a wire size of 0.051 in, a free length of $1\frac{3}{4}$ in, $11\frac{1}{2}$ total turns, and plain ends. Then $m = 0.146$ and $A = 196$ kpsi for music wire. This gives $S_{sy} = 131$ kpsi, $k = 11.2$ lb/in, and a solid height $h = 0.586$ in. The spring force, when the pin is down, is

$$F_{min} = ky = 11.2(1.75 - 1.375) = 4.20 \text{ lb}$$

When the pin is pulled up, the spring force is

$$F_{max} = k(l_F - h) = 11.2(1.75 - 0.586) = 13.0 \text{ lb}$$

The stress at the solid height is $\tau_{max} = 114$ kpsi. But this spring is too strong and so we must go to a smaller wire size.

For a second trial we select a spring having an outside diameter of 0.480 in, a wire size $d = 0.042$ in, a free length of $1\frac{3}{4}$ in, $10\frac{1}{2}$ total turns, and plain ends. Running this selection through the same routine gives $S_{sy} = 135$ kpsi, $k = 5.32$ lb/in, $F_{min} = 2.00$ lb, $F_{max} = 6.96$ lb, and $\tau_{max} = 110$ kpsi. This is getting closer.

For a third trial use the same wire size as before, but use $l_F = 2$ in and $N_T = 11.5$ turns. The results are now found to be $S_{sy} = 135$ kpsi as before, $k = 4.84$ lb/in, $F_{min} = 3.02$ lb, $F_{max} = 7.34$ lb, and $\tau_{max} = 116$ kpsi. All of this seems quite satisfactory and so we specify the following spring:

Material, music wire
Ends, plain
Outside diameter, 0.480 in
Wire diameter, 0.042 in
Total number of coils, 11.5
Free length, 2 in

////

10-7   CRITICAL FREQUENCY OF HELICAL SPRINGS

If a wave is created by a disturbance at one end of a swimming pool, this wave will travel down the length of the pool, be reflected back at the far end, and continue this back-and-forth motion until it is finally damped out. The same effect happens to helical springs, and it is called *spring surge*. If one end of a compression spring is held against a flat surface and the other end is disturbed, a compression wave is created that travels back and forth from one end to the other exactly like the swimming-pool wave.

Spring manufacturers have taken slow-motion movies of automotive valve-spring surge. These pictures show a very violent surging, with the spring actually jumping out of contact with the end plates. Figure 10-10 is a photograph of such a failure.

When helical springs are used in applications requiring a rapid reciprocating motion, the designer must be certain that the physical dimensions of the spring are not such as to create a natural vibratory frequency close to the frequency of the applied force; otherwise resonance may occur resulting in damaging stresses, since the internal damping of spring materials is quite low.

Mischke* and also Wolford and Smith† have shown that the governing equation for a spring placed between two flat and parallel plates is the wave equation and is

$$\frac{\partial^2 u}{\partial y^2} = \frac{W}{kgl^2} \frac{\partial^2 u}{\partial t^2} \tag{10-14}$$

where k = spring rate
$\quad\quad g$ = acceleration due to gravity
$\quad\quad l$ = length of spring between plates
$\quad\; W$ = weight of spring
$\quad\quad y$ = coordinate along length of spring
$\quad\quad u$ = motion of any particle at distance y

Partial differential equations like Eq. (10-14) are sometimes solved by assuming the solution is a product-type function. For example

$$u = ye^t \quad\quad \text{and} \quad\quad u = \sin y \log t$$

are product-type functions. If u should just happen to be this kind of a solution, we could write

$$u(y, t) = T(t) Y(y) \tag{a}$$

---

* Charles R. Mischke, *Elements of Mechanical Analysis*, Addison-Wesley, Reading, Mass., 1963, pp. 324–327.

† J. C. Wolford and G. M. Smith, "Surge of Helical Springs," *Mechanical Engineering News*, vol. 13, no. 1, Feb. 1976, pp. 4–9.

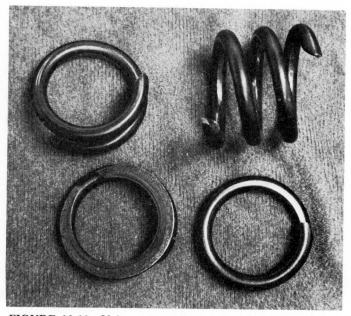

FIGURE 10-10 Valve-spring failure caused by spring surge in an overrevved engine. The fractures exhibit the classic 45° shear failure.

solve for Y and T independently, and then put them back together to get the solution. For this example it happens that y is indeed this kind of solution; so let us proceed.

First, take the second derivative of Eq. (a) with respect to y.

$$\frac{\partial^2 u}{\partial y^2} = T\frac{d^2 Y}{dy^2} \tag{b}$$

Next, take the second derivative of Eq. (a) with respect to t.

$$\frac{\partial^2 u}{\partial t^2} = Y\frac{d^2 T}{dt^2} \tag{c}$$

Now substitute Eqs. (b) and (c) into Eq. (10-14).

$$T\frac{d^2 Y}{dy^2} = \frac{W}{kgl^2}\, Y\frac{d^2 T}{dt^2} \tag{d}$$

Now separate the variables by dividing both sides by YT; the result is

$$\frac{1}{Y}\frac{d^2 Y}{dy^2} = \frac{W}{kgl^2}\frac{1}{T}\frac{d^2 T}{dt^2} \tag{e}$$

But Y and T were assumed to be independent functions. That being so, both sides must equal a constant. Therefore

$$\frac{kgl^2}{W}\frac{1}{Y}\frac{d^2Y}{dy^2} \equiv \frac{1}{T}\frac{d^2T}{dt^2} \equiv -p^2$$

Thus we get the two ordinary differential equations

$$\frac{d^2T}{dt^2} + p^2 T = 0$$

$$\frac{d^2Y}{dy^2} + p^2\frac{W}{kgl^2}Y = 0 \tag{f}$$

The solutions to these two equations are

$$Y = C_1 \sin\frac{p}{l}\sqrt{\frac{W}{kg}}y + C_2 \cos\frac{p}{l}\sqrt{\frac{W}{kg}}y$$

$$T = C_3 \sin pt + C_4 \cos pt \tag{g}$$

Putting these back together in accordance with Eq. (a) yields

$$u = \left(C_1 \sin\frac{p}{l}\sqrt{\frac{W}{kg}}y + C_2 \cos\frac{p}{l}\sqrt{\frac{W}{kg}}y\right)(C_3 \sin pt + C_4 \cos pt) \tag{h}$$

The constants are evaluated by specifying the boundary or end conditions. We will not require all of them, but if the spring operates between two flat and parallel plates, then

at $y = 0$, $u = 0$ and   at $y = l$, $u = 0$

The first condition yields

$$u = [C_1(0) + C_2(1)](C_3 \sin pt + C_4 \cos pt) = 0$$

and hence $C_2 = 0$. The second condition yields

$$u = \left(C_1 \sin p\sqrt{\frac{W}{kg}}\right)(C_3 \sin pt + C_4 \cos pt) = 0$$

The solution with $C_1 = 0$ means that nothing happens; this is of no interest, and so it must be true that

$$\sin p \sqrt{\frac{W}{kg}} = 0$$

This requires that

$$p \sqrt{\frac{W}{kg}} = m\pi \qquad \text{where } m = 1, 2, 3, \ldots$$

Thus

$$p = m\pi \sqrt{\frac{kg}{W}} \qquad\qquad\qquad (i)$$

and the solution, thus far, is

$$u = \left(C_3 \sin m\pi \sqrt{\frac{kg}{W}}\, t + C_4 \cos m\pi \sqrt{\frac{kg}{W}}\, t \right)(C_1 \sin m\pi y) \qquad (10\text{-}15)$$

It is not necessary to evaluate the remaining constants since Eq. (10-15) contains all the information required to guard against spring surge failures. The natural frequencies in radians per second are

$$\omega = m\pi \sqrt{\frac{kg}{W}}$$

where the fundamental frequency is found for $m = 1$, the second harmonic for $m = 2$ and so on. We are usually interested in the frequency in cycles per second; since $\omega = 2\pi f$, we have, for the fundamental frequency

$$f = \frac{1}{2} \sqrt{\frac{kg}{W}} \qquad\qquad\qquad (10\text{-}16)$$

Wolford and Smith show that the frequency is

$$f = \frac{1}{4} \sqrt{\frac{kg}{W}} \qquad\qquad\qquad (10\text{-}17)$$

where the spring has one end against a flat plate and the other end free. They also point out that Eq. (10-16) applies when one end is against a flat plate and the other end is driven with a sine-wave motion.

The weight of a helical spring is

$$W = AL\rho = \frac{\pi d^2}{4}(\pi DN)(\rho) = \frac{\pi^2 d^2 DN\rho}{4}$$
(10-18)

where ρ is the weight (not mass) per unit volume.

The fundamental critical frequency should be from 15 to 20 times the frequency of the force or motion of the spring in order to avoid resonance with the harmonics. If the frequency is not high enough, the spring should be redesigned to increase k or decrease W.

10-8   OPTIMIZATION

In Sec. 10-6 we learned that the normal or basic procedure for designing springs is an iterative process because of the large number of requirements and relations that are to be satisfied. Indeed it is for these very reasons that the optimal design of springs has interested so many engineers. Specific methods of optimization will not be presented here because of space limitations. However, you should be aware of these approaches, what they can do for you, and where to go for additional information.

Published optimization procedures are found in two categories. In the first category the objective is to *minimize* (1) weight, (2) volume, (3) wire diameter, (4) length, and (5) spring rate. In the second cateogry the objective is to *maximize* (1) work done by the spring, (2) deflection, (3) factor of safety, (4) reliability, and (5) fatigue strength.

Hinkle and Morse[*] use both graphs and equations to achieve a spring index C that will result in an optimum design. Using computer search techniques Bachtler and Rommel[†] and Bachtler[‡] present methods of minimizing weight. Using geometric programming Agrawal[§] also interested himself in the minimum-weight problem. Kothari[¶] uses a series of graphs and equations to obtain a direct solution to the problem of minimizing wire diameter and spring rate when a specific fatigue life is desired.

[*] R. T. Hinkle and I. E. Morse, "Design of Helical Springs for Minimum Weight, Volume, and Length," *ASME Jour. of Eng. for Ind.*, vol. 81, no. 1, 1959, pp. 37–42.

[†] C. S. Bachtler and J. B. Rommel, "Optimum Design of a High Duty Helical Spring," ASME Paper no. 73-DET-39.

[‡] C. S. Bachtler, "Discrete Synthesis of a Minimum Weight Helical Spring," ASME Paper no. 75-DET-134.

[§] G. K. Agrawal, "Optimal Design of Helical Springs for Minimum Weight by Geometric Programming," ASME Paper no. 78-WA/DE-1.

[¶] Harsked Kothari, "Optimum Design of Helical Springs," *Machine Design*, vol. 52, no. 25, Nov. 6, 1980, pp. 69–73.

Dilpare* uses an iteration technique to maximize the work delivered by a helical spring. Ray C. Johnson† of Worcester Polytechnic Institute presents a large variety of optimization solutions including both categories of spring optimization.

10-9 FATIGUE LOADING

Springs are made to be used, and consequently they are almost always subject to fatigue loading. In many instances the number of cycles of required life may be small, say, several thousand for a padlock spring or a toggle-switch spring. But the valve spring of an automotive engine must sustain millions of cycles of operation without failure; so it must be designed for infinite life.

In the case of shafts and many other machine members, fatigue loading in the form of completely reversed stresses is quite ordinary. Helical springs, on the other hand, are never used as both compression and extension springs. In fact, they are usually assembled with a preload so that the working load is additional. Thus the stress-time diagram of Fig. 7-22d expresses the usual condition for helical springs. The worst condition, then, would occur when there is no preload, that is, when $\tau_{min} = 0$.

In analyzing springs for the cause of a fatigue failure or in designing springs to resist fatigue, it is proper to apply the shear-stress multiplication factor K_s both to the mean stress τ_m and to the stress amplitude τ_a. The reason for this is that K_s is not really a stress-concentration factor at all, as indicated in Sec. 10-1, but merely a convenient means of calculating the shear stress at the inside of the coil. Now, we define

$$F_a = \frac{F_{max} - F_{min}}{2} \tag{10-19}$$

and

$$F_m = \frac{F_{max} + F_{min}}{2} \tag{10-20}$$

where the subscripts have the same meaning as those of Fig. 7-22d when applied to the axial spring force F. Then the stress components are

$$\tau_a = K_s \frac{8F_a D}{\pi d^3} \tag{10-21}$$

---

* Armand L. Dilpare, "Maximum-Work Springs," *Machine Design*, vol. 40, no. 16, July 4, 1968, pp. 111, 112.

† Ray C. Johnson, *Mechanical Design Synthesis*, Van Nostrand Reinhold, New York, 1971, pp. 194–206, 234–239, 261–263, 290–291.

$$\tau_m = K_s \frac{8F_m D}{\pi d^3} \tag{10-22}$$

As indicated in Sec. 7-10, recent investigation indicates that notch sensitivities are higher than they were formerly thought to be. Furthermore, most of the knowledge gained refers to reversed bending rather than to alternating torsion. This, coupled with the fact that the sensitivity of high-hardness steels in Fig. 7-18 approaches unity, is a strong argument in favor of making the fatigue-strength reduction factor for steels equal to the full value of the Wahl curvature-correction factor. In the past some authorities have used a reduced value, but this does not now seem proper. We have already seen (Sec. 7-16) that a torsional failure will occur whenever

$$\tau_a = S_{se} \tag{10-23}$$

or whenever

$$\tau_{\max} = \tau_a + \tau_m = S_{sy} \tag{10-24}$$

Consequently, these two equations will be the basis of our design to resist fatigue failure.

The best data on the torsional endurance limits of spring steels are those reported by Zimmerli.* He discovered the surprising fact that size, material, and tensile strength have no effect on the endurance limits (infinite life only) of spring steels in sizes under $\frac{3}{8}$ in (10 mm). We have already observed that endurance limits tend to level out at high tensile strengths (Fig. 7-4), but the reason for this is not clear. Zimmerli suggests that it may be because the original surfaces are alike or because plastic flow during testing makes them the same.

Interpreted in terms of the nomenclature of this book, Zimmerli's results are

$$S'_{se} = 45.0 \text{ kpsi } (310 \text{ MPa}) \qquad \text{for unpeened springs}$$
$$S'_{se} = 67.5 \text{ kpsi } (465 \text{ MPa}) \qquad \text{for peened springs}$$

These results are valid for all the spring steels in Table 10-2. They are corrected for surface finish and size, but not for reliability, temperature, or stress concentration.

It is necessary to approximate the S-N diagram when springs are to be designed for finite life. One point on this diagram is provided by the endurance limits quoted above. The other point is more difficult to obtain, because no published data exist for the value of the modulus of rupture S_{su} (ultimate torsional

---

* F. P. Zimmerli, "Human Failures in Spring Applications," *The Mainspring*, no. 17, Associated Spring Corporation, Bristol, Conn., August–September 1957.

strength) for spring steels. Until published data becomes available, it is suggested that the following relation be used:

$$S_{su} = 0.60S_u \qquad\qquad (10\text{-}25)$$

When analyzing or designing springs to resist fatigue it is always important to check the critical frequency to be sure spring surge will not be a problem.

Example 10-4 A No. 13 W & M gauge (0.091-in) music-wire compression spring has an outside diameter of $\frac{9}{16}$ in, a free length of $4\frac{1}{8}$ in, 21 active coils, and squared and ground ends. The spring is to be assembled with preload of 10 lb and will operate to a maximum load of 50 lb during use. Determine the factor of safety guarding against a fatigue failure based on a life of $50(10)^3$ cycles and 99 percent reliability.

Solution The mean diameter is $D = 0.5625 - 0.091 = 0.4715$ in. Then $C = D/d = 0.4715/0.091 = 5.19$, and, from Eq. (10-2)

$$K_s = 1 + \frac{0.5}{C} = 1 + \frac{0.5}{5.19} = 1.096$$

Using Eqs. (10-19) and (10-20), find

$$F_a = \frac{F_{max} - F_{min}}{2} = \frac{50 - 10}{2} = 20 \text{ lb}$$

$$F_m = \frac{F_{max} + F_{min}}{2} = \frac{50 + 10}{2} = 30 \text{ lb}$$

Then, employing Eqs. (10-21) and (10-22), we have for the stresses

$$\tau_a = K_s \frac{8F_a D}{\pi d^3} = \frac{1.096(8)(20)(0.4715)}{\pi(0.091)^3} = 34.9(10)^3 \text{ psi}$$

$$\tau_m = K_s \frac{8F_m D}{\pi d^3} = \frac{1.096(8)(30)(0.4715)}{\pi(0.091)^3} = 52.4(10)^3 \text{ psi}$$

The endurance limit is $S'_{se} = 45$ kpsi, but this must be corrected for reliability and stress concentration and then for a finite life. Using Table 7-7, find $k_c = 0.814$. Next, from Eq. (10-5)

$$K = \frac{4C - 1}{4C - 4} + \frac{0.615}{C} = \frac{4(5.19) - 1}{4(5.19) - 4} + \frac{0.615}{5.19} = 1.30$$

Thus, from Eq. (10-6), the curvature factor is

$$K_c = \frac{K}{K_s} = \frac{1.30}{1.096} = 1.186$$

The notch sensitivity of spring steels is very close to unity because they have such a high tensile strength; so $K_f = K_c$. Therefore the modifying factor for stress concentration is $k_e = 1/K_f = 1/1.186 = 0.843$. Therefore

$$S_{se} = k_c \, k_e \, S'_{se} = 0.814(0.843)(45) = 30.9 \text{ kpsi}$$

From Table 10-2 we find $A = 196$ and $m = 0.146$. Therefore, from Eq. (10-11)

$$S_{ut} = \frac{A}{d^m} = \frac{196}{(0.091)^{0.146}} = 278 \text{ kpsi}$$

Thus $S_{su} = 0.60 S_u = 0.60(278) = 167$ kpsi. Next, since $0.8 S_{su} = 0.8(167) = 134$ kpsi, we have from Eqs. (7-11) and (7-12) that

$$b = -\frac{1}{3} \log \frac{0.8 S_{su}}{S_{se}} = -\frac{1}{3} \log \frac{134}{30.9} = -0.212$$

$$C = \log \frac{(0.8 S_{su})}{S_{se}} = \log \frac{(134)^2}{30.9} = 2.764$$

Then, using Eq. (7-13) we get the strength at $50(10)^3$ cycles as

$$S_{sf} = (10)^{2.764} [50(10)^3]^{-0.212} = 58.6 \text{ kpsi}$$

Therefore the factor of safety guarding against a fatigue failure is

$$n = \frac{S_{sf}}{\tau_a} = \frac{58.6}{34.9} = 1.68 \qquad\qquad Ans.$$

Using Eq. (10-8) we find the spring rate to be

$$k = \frac{d^4 G}{8 D^3 N} = \frac{(0.091)^4 (11.5)(10)^6}{8(0.5625)^3 (21)} = 26.4 \text{ lb/in}$$

Using $\rho = 0.282 \text{ lb/in}^3$ and Eq. (10-18) we find the weight of the spring to be

$$W = \frac{\pi^2 d^2 D N \rho}{4} = \frac{\pi^2 (0.091)^2 (0.5625)(21)(0.282)}{4} = 0.0681 \text{ lb}$$

From Eq. (10-16) we find the critical frequency as

$$f = \frac{1}{2}\sqrt{\frac{kg}{W}} = \frac{1}{2}\sqrt{\frac{26.4(386)}{0.0681}} = 193 \text{ cy/s}$$

Thus if the operational frequency is more than $193/20 \cong 10$ cy/s the spring may need to be redesigned according to the recommendations of Sec. 10-7. ////

10-10 HELICAL TORSION SPRINGS

The torsion springs illustrated in Fig. 10-11 are used in door hinges and automobile starters and, in fact, for any application where torque is required. They are wound in the same manner as extension or compression springs but have the ends shaped to transmit torque.

A torsion spring is subjected to the action of a bending moment $M = Fr$, producing a normal stress in the wire. Note that this is in contrast to a compression or extension helical spring, in which the load produces a torsional stess in the wire. This means that the residual stresses built in during winding are in the same direction but of opposite sign of the working stresses which occur during use. These residual stresses are useful in making the spring stronger by opposing the working stress, *provided* the load is always applied so as to cause the spring to wind up. Because the residual stress opposes the working stress, torsional springs can be designed to operate at stress levels which equal or even exceed the yield strength of the wire.

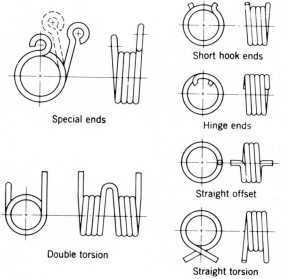

Special ends

Double torsion

Short hook ends

Hinge ends

Straight offset

Straight torsion

FIGURE 10-11 Torsion springs. (*Courtesy of Associated Spring Corporation.*)

The bending stress can be obtained by using curved-beam theory as explained in Sec. 2-20. It is convenient to write the expression in the form

$$\sigma = K \frac{Mc}{I} \tag{a}$$

where K is a stress-concentration factor and, in this case, is treated as such, rather than as a strength-reduction factor. The value of K depends upon the shape of the wire and upon whether or not the stress is desired on the inner fiber of the coil or on the outer fiber. Wahl has analytically determined the following values for K for round wire:

$$K_i = \frac{4C^2 - C - 1}{4C(C - 1)} \qquad K_o = \frac{4C^2 + C - 1}{4C(C + 1)} \tag{10-26}$$

where C is the spring index and the subscripts i and o refer to the inner and outer fibers, respectively. When the bending moment $M = Fr$ and the section modulus $I/c = \pi d^3/32$ are substituted in Eq. (a), we obtain

$$\sigma = K \frac{32Fr}{\pi d^3} \tag{10-27}$$

which gives the bending stress for a round-wire torsion spring.

Deflection

The strain energy in bending is, from Eq. (3-28),

$$U = \int \frac{M^2 \, dx}{2EI} \tag{b}$$

For the torsion spring, $M = Fr$, and integration must be accomplished over the length of the wire. The force F will deflect through the distance $r\theta$, where θ is the total angular deflection of the spring. Applying Castigliano's theorem,

$$r\theta = \frac{\partial U}{\partial F} = \int_0^{\pi DN} \frac{\partial}{\partial F}\left(\frac{F^2 r^2 \, dx}{2EI}\right) = \int_0^{\pi DN} \frac{Fr^2 \, dx}{EI} \tag{c}$$

Substituting $I = \pi d^4/64$ for round wire and solving Eq. (c) for θ gives

$$\theta = \frac{64FrDN}{d^4 E} \tag{10-28}$$

where θ is the angular deflection of the spring in radians. The spring rate is therefore

$$k = \frac{Fr}{\theta} = \frac{d^4 E}{64DN} \tag{10-29}$$

The spring rate may also be expressed as the torque required to wind up the spring one turn. This is obtained by multiplying Eq. (10-29) by 2π. Thus

$$k' = \frac{d^4 E}{10.2DN} \tag{10-30}$$

These deflection equations have been developed without taking into account the curvature of the wire. Actual tests show that the constant 10.2 should be increased slightly. Thus the equation

$$k' = \frac{d^4 E}{10.8DN} \tag{10-31}$$

will give better results. Corresponding corrections may be made to Eqs. (10-28) and (10-29) if desired.

Torsion springs are frequently used over a round bar or pin. And when a load is applied to a torsion spring the spring winds up, causing a decrease in the inside diameter. It is necessary to assure that the inside diameter of the spring never becomes equal to the diameter of the bar or pin, otherwise a spring failure will occur. The inside diameter of a loaded torsion spring can be found from the equation

$$D'_i = \frac{N}{N'} D_i \tag{10-32}$$

where N = number of coils at no load
$\quad\quad D_i$ = inside diameter at no load
$\quad\quad N'$ = number of coils when loaded
$\quad\quad D'_i$ = inside diameter when loaded

Example 10-5 A stock torsion spring is shown in Fig. 10-12. It is made of 0.070-in music wire and has $4\frac{1}{4}$ total turns.

(a) Find the maximum operating torque and the angular rotation.
(b) Compute the inside diameter corresponding to the result found in (a).
(c) Find the maximum operating torque and the angular rotation for an infinite number of cycles of operation.

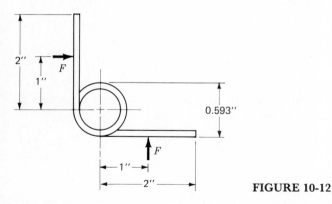

FIGURE 10-12

Solution   (*a*) For music wire we find from Table 10-2 that $m = 0.146$ and $A = 196$ kpsi. Therefore, from Eq. (10-11),

$$S_{ut} = \frac{A}{d^m} = \frac{196}{(0.070)^{0.146}} = 289 \text{ kpsi}$$

Next, using Eq. (10-12), the yield strength is estimated to be

$$S_y = 0.75 S_{ut} = 0.75(289) = 217 \text{ kpsi}$$

The mean diameter is $D = 0.593 - 0.070 = 0.523$ in. Thus the spring index is $C = D/d = 0.523/0.070 = 7.47$. Equations (10-26) give the inside and outside stress-concentration factors as

$$K_i = \frac{4C^2 - C - 1}{4C(C - 1)} = \frac{4(7.47)^2 - 7.47 - 1}{4(7.47)(7.47 - 1)} = 1.111$$

$$K_o = \frac{4C^2 + C - 1}{4C(C + 1)} = \frac{4(7.47)^2 + 7.47 - 1}{4(7.47)(7.47 + 1)} = 0.908$$

Thus the critical stress occurs on the inside of the coils. Now we rearrange Eq. (10-27), substitute S_y for σ, and solve for the maximum rated torque *Fr*. This gives

$$Fr = \frac{\pi d^3 \sigma}{32 K_i} = \frac{\pi (0.070)^3 (217)(10)^3}{32(1.111)} = 6.577 \text{ lb} \cdot \text{in} \qquad\qquad Ans.$$

Note that no factor of safety has been used. This is considered safe because the assistance provided by the built-in winding stresses is significant. This is also evidenced by the fact that the manufacturer gives the spring a rating of 7.50 lb · in.

Next, from Eq. (10-31) we find the spring rate to be

$$k' = \frac{d^4 E}{10.8 DN} = \frac{(0.070)^4 (30)(10)^6}{10.8(0.523)(4.25)} = 30 \text{ lb} \cdot \text{in/turn}$$

Thus a torque of 6.577 lb · in will wind the spring

$$n = \frac{Fr}{k'} = \frac{6.577}{30} = 0.219 \text{ turns} \qquad\qquad Ans.$$

The corresponding angular deflection is

$$\theta = 0.219(360°) = 78.9° \qquad\qquad Ans.$$

(b) With no load, the inside diameter of the spring is $D_i = 0.593 - 2(0.070) = 0.453$ in. At full torque Eq. (10-32) is used to find the new inside diameter.

$$D_i' = \frac{N}{N'} D_i = \frac{4.25}{4.25 + 0.219} (0.453) = 0.431 \text{ in} \qquad\qquad Ans.$$

(c) Since $S_{ut} > 200$ kpsi, $S_e' = 100$ kpsi. While the surface-finish factor for music wire should be somewhat greater than values listed for cold-drawn bars, we shall use the same factor. Thus, from Fig. 7-10, we find $k_a = 0.63$. Equation (7-16) gives $k_b = 1$. Also, assuming 50 percent reliability, $k_c = 1$. And $k_e = 1$ because for torsion springs the stress-concentration factor is used to increase the stress. Therefore $S_e = 0.63(100) = 63$ kpsi.

If we assume $(Fr)_{min} = 0$, then

$$(Fr)_a = (Fr)_m = \frac{(Fr)_{max}}{2}$$

where the subscripts a and m refer to the alternating and mean components. Thus

$$\sigma_a = \sigma_m = S_a = S_m = K \frac{32(Fr)_{max}}{2\pi d^3} = \frac{1.111(32)(Fr)_{max}}{2\pi(0.070)^3} = 16.5(10)^3 (Fr)_{max} \text{ psi}$$

We now choose the Gerber relation [Eq. (7-30)] for the fatigue analysis. It is

$$\frac{S_a}{S_e} + \left(\frac{S_m}{S_{ut}}\right)^2 = 1$$

Then, in kpsi, we have

$$\frac{16.5(Fr)_{max}}{63} + \left[\frac{16.5(Fr)_{max}}{289}\right]^2 = 1$$

Performing the indicated operations, simplifying, and rearranging gives the quadratic

$$(Fr)^2_{\max} + 80.58(Fr)_{\max} - 307.7 = 0$$

This equation has one positive and one negative root. The positive root is

$$(Fr)_{\max} = 3.65 \text{ lb} \cdot \text{in} \hspace{4cm} Ans.$$

Corresponding to this torque

$$n = \frac{3.65}{30} = 0.1217 \text{ turns} \hspace{4cm} Ans.$$

and

$$\theta = 0.1217(360°) = 43.8° \hspace{4cm} Ans.$$

$$////$$

10-11 BELLEVILLE SPRINGS

The inset of Fig. 10-13 shows a coned-disk spring, commonly called a *Belleville spring*. Although the mathematical treatment is beyond the purposes of this book, you should at least be familiar with the remarkable characteristics of these springs.

FIGURE 10-13 Load-deflection curves for Belleville springs. (*Courtesy of Associated Spring Corporation.*)

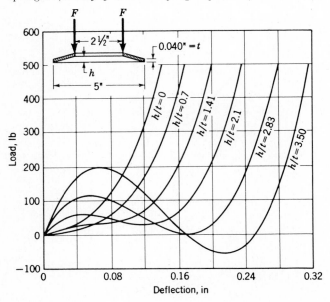

Aside from the obvious advantage of occupying a small space, a variation of the h/t ratio will produce a wide variety of load-deflection curve shapes, as illustrated in Fig. 10-13. For example, using an h/t ratio of 2.83 or larger gives an S curve which might be useful for snap-acting mechanisms. A reduction of the ratio to a value between 1.41 and 2.1 causes the central portion of the curve to become horizontal, which means that the load is constant over a considerable deflection range.

A higher load for a given deflection may be obtained by nesting, that is, by stacking the springs in parallel. On the other hand, stacking in series provides a larger deflection for the same load, but in this case there is danger of instability.

10-12 MISCELLANEOUS SPRINGS

The extension spring shown in Fig. 10-14 is made of slightly curved strip steel, not flat, so that the force required to uncoil it remains constant; thus it is called a *constant-force spring*. This is equivalent to a zero spring rate. Such springs can also be manufactured having either a positive or a negative spring rate.

A *volute spring* is a wide, thin strip, or "flat," of steel wound on the flat so that the coils fit inside one another. Since the coils do not stack, the solid height of the spring is the width of the strip. A variable-spring scale, in a compression volute spring, is obtained by permitting the coils to contact the support. Thus, as the deflection increases, the number of active coils decreases. The volute spring, shown in Fig. 10-15a, has another important advantage which cannot be obtained with round-wire springs: if the coils are wound so as to contact or slide on one another during action, the sliding friction will serve to damp out vibrations or other unwanted transient disturbances.

A *conical spring*, as the name implies, is a coil spring wound in the shape of a cone. Most conical springs are compression springs and are wound with round wire. But a volute spring is a conical spring too. Probably the principal advantage of this type of spring is that it can be wound so that the solid height is only a single wire diameter.

Flat stock is used for a great variety of springs, such as clock springs, power

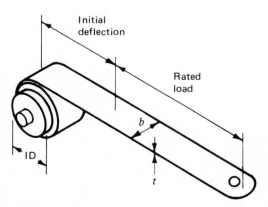

FIGURE 10-14 Constant-force spring. (*Courtesy of Vulcan Spring & Mfg. C., Huntingdon Valley, Pa.*)

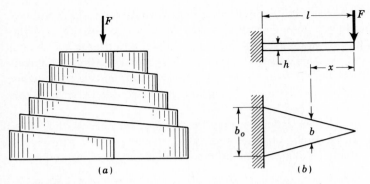

FIGURE 10-15   (*a*) A volute spring; (*b*) a flat triangular spring.

springs, torsion springs, cantilever springs, and hair springs; frequently it is specially shaped to create certain spring actions for fuse clips, relay springs, spring
washers, snap rings, and retainers.

In designing many springs of flat stock or strip material it is often economical
and of value to proportion the material so as to obtain a constant stress throughout the spring material.   A uniform-section cantilever spring has a stress

$$\sigma = \frac{M}{I/c} = \frac{Fx}{I/c} \qquad\qquad (a)$$

which is proportional to the distance x if I/c is a constant.   But there is no reason
why I/c need be a constant.   For example, one might design such a spring as that
shown in Fig. 10-15b, in which the thickness h is constant but the width b is
permitted to vary.   Since, for a rectangular section, $I/c = bh^2/6$, we have, from Eq.
(a),

$$\frac{bh^2}{6} = \frac{Fx}{\sigma}$$

or

$$b = \frac{6Fx}{h^2\sigma} \qquad\qquad (b)$$

Since b is linearly related to x, the width b_σ at the base of the spring is

$$b_\sigma = \frac{6Fl}{h^2\sigma} \qquad\qquad (10\text{-}33)$$

But the deflection of this triangular flat spring is more difficult to obtain, because
the moment of inertia is now a variable.   Probably the quickest solution could be
obtained by using singularity functions, or the method of graphical integration.

The methods of stress and deflection analysis illustrated in previous sections of this chapter have served to illustrate that springs may be analyzed and designed by using the fundamentals discussed in the earlier chapters of this book. This is also true for most of the miscellaneous springs mentioned in this section, and you should now experience no difficulty in reading and understanding the literature of such springs.

10-13 ENERGY-STORAGE CAPACITY

Quite frequently, in the selection and design of springs, the capacity of a spring to store energy is of major importance. Sometimes the designer is interested in absorbing shock and impact loads; at other times he or she is simply interested in storing the maximum energy in the smallest space. Equations (3-29) for strain energy can be particularly useful to the designer in choosing a particular form of spring. These equations are, or may be written,

$$u = \frac{\sigma^2}{2E} \qquad u = \frac{\tau^2}{2G} \tag{10-34}$$

where u is the strain energy per unit volume. Of course, the particular equation to be used depends upon whether the spring is stressed axially, that is, in tension or compression, or whether it is stressed in shear. Maier[*] prefers to divide springs into two classes, which he calls E springs or G springs, depending upon which formula is applicable. Since the stress is usually not uniform, a form coefficient C_F is defined as follows:

$$u = C_F \frac{\sigma^2}{2E} \qquad u = C_F \frac{\tau^2}{2G} \tag{10-35}$$

where $C_F = 1$, a maximum value, if the stress is uniformly distributed, meaning that the material is used most efficiently. For most springs the stress is not uniformly distributed, and so C_F will be less than unity. Thus the value of the form coefficient is a measure of the spring's capacity to store energy.

To calculate the form coefficient for a helical extension or compression spring, we write

$$u = \frac{U}{v} = \frac{Fy}{2v} \tag{a}$$

[*] Karl W. Maier, "Springs That Store Energy Best," *Prod. Eng.*, vol. 29, no. 45, Nov. 10, 1958, p. 71.

where F = force

$\qquad y$ = deflection

$\qquad v$ = volume of active wire

Since $y = 8FD^3N/d^4G$, $\tau = 8FDK/\pi d^3$, and $v = lA = (\pi DN)(\pi d^2/4)$, we have, from Eq. (a),

$$u = \frac{1}{2K^2}\left(\frac{\tau^2}{2G}\right) \tag{b}$$

And so $C_F = \frac{1}{2}K^2$. Note that, for $K = 1.20$, $C_F = 0.35$.

For a torsion bar, we use the relation

$$u = \frac{U}{v} = \frac{T\theta}{2v} \tag{c}$$

where θ is the angle of twist. Here $\tau = 16T/\pi d^3$, $\theta = 32Tl/\pi d^4G$, and $v = \pi d^2l/4$. The strain energy per unit volume is

$$u = \frac{1}{2}\frac{\tau^2}{2G} \tag{d}$$

and so $C_F = 0.50$.

Table 10-3 contains a list of form coefficients computed by Maier, which should be useful in selecting springs for energy-storage purposes.

Table 10-3   FORM COEFFICIENTS—
A MEASURE OF THE CA-
PACITY OF SPRINGS TO
STORE ENERGY

Name of spring	Type	C_F
Tension bar	E	1.0
Clock spring	E	0.33
Torsion spring	E	0.25
Belleville washers	E	0.05–0.20
Cantilever beam	E	0.11
Torsion tube	G	About 0.90
Torsion bar	G	0.50
Compression spring	G	About 0.35

PROBLEMS*

Sections 10-1 to 10-6

10-1 A helical compression spring is made of No. 18 (0.047-in) wire having a torsional yield strength of 108 kpsi. It has an outside diameter of $\frac{1}{2}$ in and has 14 active coils.
   (a) Find the maximum static load corresponding to the yield point of the material.
   (b) What deflection would be caused by the load in (a)?
   (c) Calculate the scale of the spring.
   (d) If the spring has one dead turn at each end, what is the solid height?
   (e) What should be the length of the spring so that when it is compressed solid the stress will not exceed the yield point?

10-2 The same as Prob. 10-1 except that the wire is No. 13 (0.092-in), the outside spring diameter is $\frac{3}{4}$ in, and the torsional strength is 90 kpsi.

10-3 A helical tension spring is made of 1.2 mm wire having yield strengths of 1280 MPa and 740 MPa in tension and torsion, respectively. The spring has an OD of 12 mm, 36 active coils, and hook ends as in Fig. 10-6a and b. The mean radii at the ends is 5.4 mm for bending and 3 mm for torsion (Fig. 10-6b), respectively. The spring is prestressed to 75 MPa during winding, which keeps it closed solid until an external load of sufficient magnitude is applied. When wound, the distance between hook ends is 70 mm.
   (a) What is the spring preload?
   (b) What load would cause yielding?
   (c) What is the spring rate?
   (d) What is the distance between the hook ends if the spring is extended until the stress just reaches the yield strength?

10-4 The compression helical spring shown in the figure is made of spring steel wire having a torsional yield strength of 640 MPa.

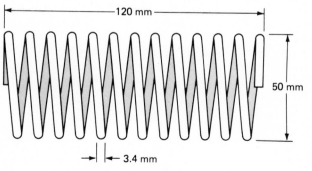

PROBLEM 10-4

   (a) Compute the spring rate.
   (b) What force is required to close the spring to its solid height?
   (c) After the spring has been closed to its solid height once, and the compressive force removed, will it spring back to its original free length?

---

* The asterisk indicates a design-type problem or one that may have no unique results.

10-5 Two steel compression coil springs are to be nested. The outer spring has an inside diameter of $1\frac{1}{2}$ in, a wire diameter of 0.120 in, and 10 active coils. The inner spring has an outside diameter of 1.25 in, a wire diameter of 0.091 in, and 13 active coils.

(*a*) Compute the spring rate of each spring.

(*b*) What force is required to deflect the nested spring assembly a distance of 1 in? (Both have the same free length.)

(*c*) Which spring will be stressed the most? Calculate this stress using the result of (*b*).

10-6 A compression coil spring of 3.4-mm music wire, having an outside diameter of 22 mm, has 8 active coils. Determine the stress and deflection caused by a static load of 270 N.

10-7 A compression coil spring has 18 active coils, an outside diameter of $\frac{9}{32}$ in, and plain ends; it is made of No. 15 gauge (0.035-in) music wire.

(*a*) What should be the free length of the spring such that no permanent deformation will occur when it is compressed solid?

(*b*) What force is necessary to compress the spring to its solid length?

10-8 A helical compression spring uses No. 14 gauge (0.080-in) oil-tempered wire and has six active coils and squared ends. If the spring has an outside diameter of $\frac{1}{2}$ in, what should be the free length of the spring in order that, when it is compressed solid, the stress will not exceed 90 percent of the yield strength?

10-9* Design a compression coil spring of music wire having squared ends. The spring is to be assembled with a preload of 10 N and exert a force of 50 N when it is compressed an additional 140 mm. Find the wire diameter to the nearest 0.2 mm and the coil diameter to the nearest millimeter using $C = 12$, but do not use a larger wire diameter than is necessary. What is the free length and the solid height? The force corresponding to the solid height should be more than 50 N, say about 60 N.

10-10* Design a compression coil spring of hard-drawn wire having plain ends. When the spring is compressed a distance of 2.25 in, it is to exert a force of 18 lb, but the force corresponding to the solid height should be a little more, say 24 lb, for safety. The spring index should be about 10 to avoid a large coil diameter. Use the smallest even-numbered W & M gauge wire, as tabulated in Table A-25, consistent with safety. Specify the number of coils, the free length, and the OD of the spring.

10-11 Solve part (*f*) of Example 10-2 using curved-beam theory.

10-12 The extension spring shown in the figure has full twisted loop ends as shown. The material is UNS G10650 oil-tempered wire. The spring has 84 coils and is close-wound with a preload stress equal to 15 percent of the torsional yield strength.

(*a*) Find the overall closed length of the spring.

(*b*) Compute the preload force.

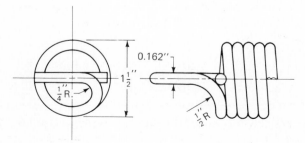

PROBLEM 10-12

(c)   Compute the spring rate.

(d)   What load would cause yielding?

(e)   What is the spring elongation corresponding to the force found in (d)?

10-13* Design a compression spring for the clamping fixture shown in the figure. The spring should exert a force of about 10 lb corresponding to a length of $1\frac{1}{2}$ in. It is to have squared and ground ends and fit over the $\frac{1}{2}$-in-diameter clamp screw.

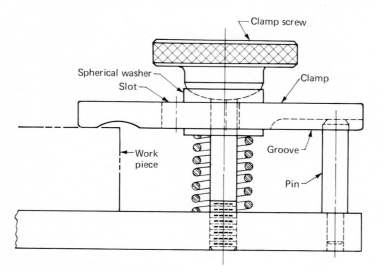

PROBLEM 10-13   Clamping fixture.

Sections 10-7 to 10-9

10-14   A compression spring has 18 active coils, a free length of $1\frac{1}{4}$ in, and an outside diameter of $\frac{9}{32}$ in; it is made of No. 15 gauge (0.035-in) music wire and has plain ends.

(a)   Compute the spring rate, the solid height, and the stress in the spring when it is compressed to the solid height.

(b)   The spring operates with a minimum force of 2 lb and a maximum force of 4.5 lb. Compute the factor of safety guarding against a fatigue failure based on 50 percent reliability.

(c)   Find the critical frequency.

10-15   A helical compression spring is made of $\frac{1}{4}$-in-diameter steel wire and has an outside diameter of $2\frac{1}{4}$ in with squared and ground ends and 12 total coils. The length of the spring is such that, when it is compressed solid, the torsional stress is 120 kpsi.

(a)   Determine the spring rate and critical frequency.

(b)   Determine the free length of the spring.

(c)   This is a shot-peened spring; based on 90 percent reliability and infinite life, determine whether a fatigue failure can be expected if the spring is cycled between the loads $F_{\min} = 50$ lb and $F_{\max} = 250$ lb. Show computations to verify your decision.

10-16   An extension spring is made of 0.60-mm music wire and has an outside diameter of 4.8 mm. The spring is wound with a pretension of 1.10 N and the load fluctuates between this value and 6.8 N. Since the spring might fail statically or in fatigue, find the factor of safety for both types of failure.

10-17 A compression coil spring is made of 2-mm music wire and has an outside diameter of 12.5 mm. The maximum and minimum values of the fatigue load to which the spring is subjected are 90 and 45 N, respectively. For infinite life and 95 percent reliability, find the factor of safety.

10-18 A compression coil spring is wound from $\frac{1}{2}$-in-diameter bar stock over a mandrel which is 5 in in diameter. Assume that springback results in a spring having an ID 10 percent larger than the mandrel diameter. The spring is wound with 12 total coils and has squared and ground ends. After heat treatment the Brinell hardness is found to be $H_B = 380$. The free length of the spring is 20 in. The spring is assembled into a machine by compressing it to a length of 18 in. When the machine runs, the spring is compressed an additional 10 in so that the maximum load corresponds to a spring length of 8 in and the minimum load to a length of 18 in.

(*a*) Would this spring develop a permanent set if compressed solid? Why?

(*b*) What is the spring rate?

(*c*) Is the spring likely to buckle?

(*d*) Based on 50 percent reliability and infinite life, will the spring fail by fatigue? If not, what is the factor of safety guarding against a fatigue failure?

(*e*) What is the critical frequency?

Section 10-10

10-19 The rat trap shown in the figure uses two opposite-image torsion springs. The wire size is about 0.080 in in diameter and the outside diameter of the springs is $\frac{1}{2}$ in in

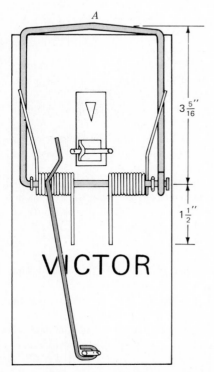

PROBLEM 10-19

the position shown. Each spring has 11 turns. Use of a fish scale revealed a force of about 8 lb needed to set the trap.

(a) Find the probable configuration of the spring prior to assembly.

(b) Find the maximum stress in the spring when the trap is set.

10-20 A stock torsion spring is made of 0.054-in music wire, has 6 coils, and straight ends 2 in long and 180° apart. The outside diameter is 0.654 in.

(a) What value of torque would cause a maximum stress equal to the yield strength?

(b) If the torque found in (a) is used as the maximum working torque, what is the smallest value of the inside diameter?

(c) Compute the angle of rotation corresponding to the torque found in (a).

10-21 The spring of Prob. 10-20 is to be used in an application subject to fatigue loading. The reliability of the application should be 95 percent. If the minimum torque is to be 25 percent of the maximum torque, what value of maximum torque can be used for an infinite life?

10-22 The figure shows a finger exerciser used by law-enforcement officials and handgun enthusiasts to strengthen their grip. It is formed by winding cold-drawn steel wire around a mandrel so as to obtain $2\frac{1}{2}$ turns when the grip is in the closed position. After winding, the wire is cut so as to leave the two legs as handles. The plastic handles are then molded on, the grip squeezed together, and a wire clip placed around the legs to obtain initial "tension" and to space the handles for the best gripping position. The clip is formed like a figure eight to prevent it from coming off. The wire material is hard-drawn 0.60 carbon having a yield strength of 150 kpsi. The stress in the wire when the grip is closed should not, of course, exceed this figure. Based on the stress not exceeding 150 kpsi, find the configuration of the exerciser before the clip is assembled and the hand force necessary to close the grip.

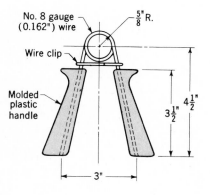

PROBLEM 10-22

CHAPTER
11

ROLLING
CONTACT
BEARINGS

The terms *rolling-contact bearing, antifriction bearing,* and *rolling bearing* are all used to describe that class of bearing in which the main load is transferred through elements in rolling contact rather than in sliding contact. In a rolling bearing the starting friction is about twice the running friction, but still it is negligible in comparison to the starting friction of a sleeve bearing. Load, speed, and the operating viscosity of the lubricant do affect the frictional characteristics of a rolling bearing. It is probably a mistake to describe a rolling bearing as "antifriction," but the term is used generally throughout the industry.

From the mechanical designer's standpoint, the study of antifriction bearings differs in several respects when compared with the study of other topics. The specialist in antifriction-bearing design is confronted with the problem of designing a group of elements which compose a rolling bearing; these elements must be designed to fit into a space whose dimensions are specified; they must be designed to receive a load having certain characteristics; and, finally, these elements must be designed to have a satisfactory life when operated under the specified conditions. Bearing specialists must therefore consider such matters as fatigue loading, friction, heat, corrosion resistance, kinematic problems, material properties, lubrication, machining tolerances, assembly, use, and cost. From a consideration of all these factors, bearing specialists arrive at a compromise which, in their judgment, is a good solution to the problem as stated.

11-1 BEARING TYPES

Bearings are manufactured to take pure radial loads, pure thrust loads, or a combination of these two. The nomenclature of a ball bearing is illustrated in Fig. 11-1, which also shows the four essential parts of a bearing. These are the outer ring, the inner ring, the balls or rolling elements, and the separator. In low-priced bearings the separator is sometimes omitted, but it has the important function of separating the elements so that rubbing contact will not occur.

Some of the various types of standardized bearings which are manufactured are shown in Fig. 11-2. The single-row, deep-groove bearing will take radial load as well as some thrust load. The balls are inserted into the grooves by moving the inner ring to an eccentric position. The balls are separated after loading, and the separator is then assembled.

The use of a filling notch (Fig. 11-2b) in the inner and outer rings enables a

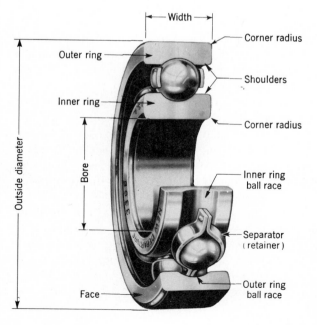

FIGURE 11-1 Nomenclature of a ball bearing. (*Courtesy of New-Departure-Hyatt Division, General Motors Corporation.*)

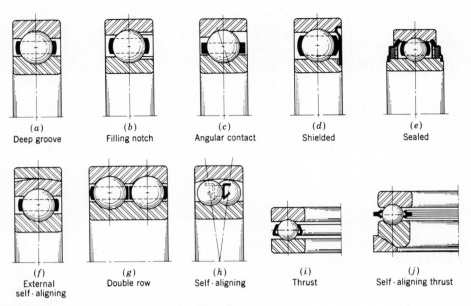

FIGURE 11-2   Various types of ball bearings.

greater number of balls to be inserted, thus increasing the load capacity. The thrust capacity is decreased, however, because of the bumping of the balls against the edge of the notch when thrust loads are present.

The angular-contact bearing (Fig. 11-2c) provides a greater thrust capacity. All these bearings may be obtained with shields on one or both sides. The shields are not a complete closure but do offer a measure of protection against dirt. A variety of bearings are manufactured with seals on one or both sides. When the seals are on both sides, the bearings are lubricated at the factory. Although a sealed bearing is supposed to be lubricated for life, a method of relubrication is sometimes provided.

Single-row bearings will withstand a small amount of shaft misalignment or deflection, but where this is severe, self-aligning bearings may be used. Double-row bearings are made in a variety of types and sizes to carry heavier radial and thrust loads. Sometimes two single-row bearings are used together for the same reason, although a double-row bearing will generally require fewer parts and occupy less space. The one-way ball thrust bearings (Fig. 11-2i) are made in many types and sizes.

Some of the large variety of standard roller bearings available are illustrated in Fig. 11-3. Straight roller bearings (Fig. 11-3a) will carry a greater load than ball bearings of the same size because of the greater contact area. However, they have the disadvantage of requiring almost perfect geometry of the raceways and rollers. A slight misalignment will cause the rollers to skew and get out of line. For this reason, the retainer must be heavy. Straight roller bearings will not, of course, take thrust loads.

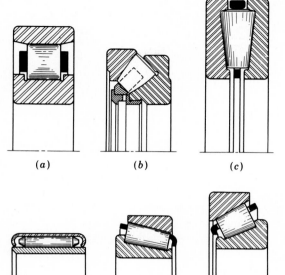

(a)   (b)   (c)

(d)   (e)   (f)

FIGURE 11-3 Types of roller bearings: (a) straight roller; (b) spherical roller thrust; (c) tapered roller thrust; (d) needle; (e) tapered roller; (f) steep-angle tapered roller. (*Courtesy of The Timken Company.*)

Helical rollers are made by winding rectangular material into rollers, after which they are hardened and ground. Because of the inherent flexibility, they will take considerable misalignment. If necessary, the shaft and housing can be used for raceways instead of separate inner and outer races. This is especially important if radial space is limited.

The spherical-roller thrust bearing (Fig. 11-3b) is useful where heavy loads and misalignment occur. The spherical elements have the advantage of increasing their contact area as the load is increased.

Needle bearings (Fig. 11-3d) are very useful where radial space is limited. They have a high load capacity when separators are used, but may be obtained without separators. They are furnished both with and without races.

Tapered roller bearings (Fig. 11-3e, f) combine the advantages of ball and straight roller bearings, since they can take either radial or thrust loads of any combination of the two, and in addition, they have the high load-carrying capacity of straight roller bearings. The tapered roller bearing is designed so that all elements in the roller surface and the raceways intersect at a common point on the bearing axis.

The bearings described here represent only a small portion of the many available for selection. Many special-purpose bearings are manufactured, and bearings are also made for particular classes of machinery. Typical of these are:

• Instrument bearings, which are high-precision and are available in stainless steel and high-temperature materials

- Nonprecision bearings, usually made with no separator and sometimes having split or stamped sheet-metal races
  - Ball bushings, which permit either rotation or sliding motion or both
  - Bearings with flexible rollers

11-2 BEARING LIFE*

When the ball or roller of an antifriction bearing rolls into the loading zone, Hertzian stresses occur on the inner ring, the rolling element, and the outer ring. Because the curvature of the contacting elements is different in the axial direction than it is in the radial direction, the formulas for these stresses are much more complicated than the Hertzian equations presented in Sec. 2-21.† If a bearing is clean and properly lubricated, is mounted and sealed against the entrance of dust or dirt, is maintained in this condition, and is operated at reasonable temperatures, then metal fatigue will be the only cause of failure. Since this implies many millions of stress applications, the term *bearing life* is in very general use.

The *life* of an *individual bearing* is defined as the total number of revolutions, or the number of hours at a given constant speed, of bearing operation required for the failure criteria to develop. Under ideal conditions the fatigue failure will consist of a spalling of the load-carrying surfaces. The Anti-Friction Bearing Manufacturers Association (AFBMA) standard states that the failure criterion is the first evidence of fatigue. It is noted, however, that the *useful life* is often used as the definition of fatigue life. The failure criterion used by the Timken Company laboratories‡ is the spalling or pitting of an area of 0.01 in². But Timken observes that the useful life may extend considerably beyond this point.

Rating life is a term sanctioned by the AFBMA and used by most bearing manufacturers. The rating life of a group of apparently identical ball or roller bearings is defined as the number of revolutions, or hours at some given constant speed, that 90 percent of a group of bearings will complete or exceed before the failure criterion develops. The terms *minimum life* and L_{10} *life* are also used to denote rating life.

The terms *average life* and *median life* are both used quite generally in discussing the longevity of bearings. Both terms are intended to have the same import. When groups consisting of large numbers of bearings are tested to failure, the median lives of the groups are averaged. Thus, these terms are really intended to denote the *average median life*. In this book we shall use the term "median life" to signify the average of these medians.

In testing groups of bearings, the objective is to determine the median life and

---

* For additional information see *AFBMA Standards*, Anti-Friction Bearing Manufacturers Association, New York, 1972.

† These equations are not required here. For a complete presentation, see Hudson T. Morton, *Anti-Friction Bearings*, 2d ed., Hudson T. Morton, Ann Arbor, Mich., 1965, pp. 223–236.

‡ *Timken Engineering Journal*, vol. 1, Timken Company, 1972, p. 22.

the L_{10}, or rated life. When many groups of bearings are tested, it is found that the median life is somewhere between four and five times the L_{10} life. The graph of Fig. 11-4 shows approximately how the failures are distributed. This curve is only approximate; it must not be used for analytical or prediction purposes.

The importance of knowing the probable survival of a group of bearings can be examined. Assume that the probability of any single bearing failure is independent of the others in the same machine. If the machine is assembled with a total of N bearings, each having the same reliability R, then the reliability of the group must be

$$R_N = (R)^N$$

Suppose we have a gear-reduction unit consisting of six bearings, all loaded so that the L_{10} lives are equal. Since the reliability of each bearing is 90 percent, the reliability of all the bearings in the assembly is

$$R_6 = (0.90)^6 = 0.531$$

This points up the need to select bearings having reliabilities greater than 90 percent.

The distribution of bearing failures can be approximated by the Weibull distribution. By substituting $R = 1 - F(x)$ and $x = L/L_{10}$, Eq. (5-24) becomes

$$R = \exp\left[-\left(\frac{\dfrac{L}{L_{10}} - x_0}{\theta - x_0} \right)^b \right] \tag{11-1}$$

where b, θ, and x_0 are the three Weibull parameters. Examination of Fig. 11-4 shows that x_0 is quite small. For a three-parameter evaluation of Eq. (11-1) we shall select a life $L = 0.02L_{10}$, so that $x_0 = 0.02$. To evaluate b and θ requires experimental data.

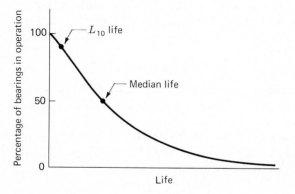

FIGURE 11-4 Typical curve of bearing life expectancy.

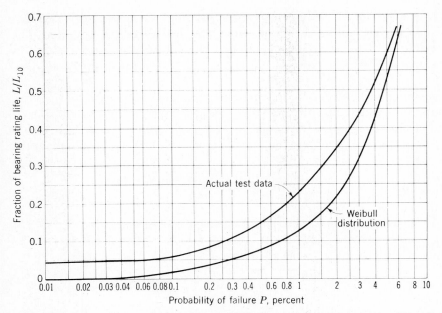

FIGURE 11-5 Reduction in life for reliabilities greater than 90 percent. Note that the abscissa is the probability of failure $P = 100 - R$ in percent. (*By permission from Tedric A. Harris, "Predicting Bearing Reliability," Machine Design, vol. 35, no. 1, Jan. 3, 1963, pp. 129-132.*)

The actual test data shown in Fig. 11-5 were obtained from over 2500 bearings tested. In order to evaluate b and θ, we extract from Fig. 11-5 the data points shown in Table 11-1. Figure 11-6 shows these points plotted on Weibull paper and the Weibull line drawn through them. The slope is found to be $b = 1.40$. To get the other parameters, we note the point $(L/L_{10}) - 0.02 = 0.5$ and $P = 4$ percent lies directly on the Weibull line of Fig. 11-6. The corresponding decimal reliability is $R = 0.96$. Substituting these two values and $b = 1.40$

Table 11-1 POINTS OBTAINED FROM CURVE OF ACTUAL TEST DATA IN FIG. 11-5

P Percent	P Decimal	L/L_{10}	$(L/L_{10}) - 0.02$
6	0.06	0.67	0.65
4	0.04	0.52	0.50
2	0.02	0.35	0.33
1	0.01	0.23	0.21
0.4	0.004	0.13	0.11
0.1	0.001	0.055	0.035
0.04	0.0004	0.05	0.03

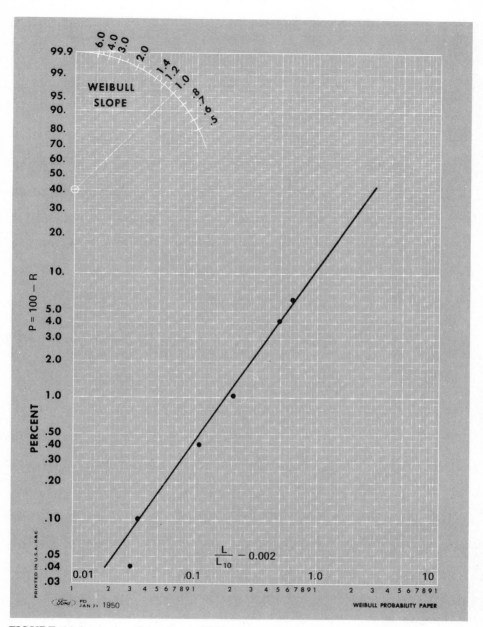

FIGURE 11-6 A plot of the data in Table 11-1 on Weibull probability paper. The slope of the line is $b = 1.40$.

into Eq. (11-1) gives

$$0.96 = \exp\left[-\left(\frac{0.5}{\theta - 0.02}\right)^{1.40}\right]$$

Solving for the unknown parameter gives $\theta - 0.02 = 4.91$. Thus, the three-parameter equation for bearings is

$$R = \exp\left[-\left(\frac{\dfrac{L}{L_{10}} - 0.02}{4.91} \right)^{1.40} \right] \qquad (11\text{-}2)$$

The two-parameter equation is obtained by making $x_0 = 0$ in Eq. (11-1). A result obtained by Mischke* in a manner quite similar to the procedure illustrated above is $b = 1.17$ and $\theta = 6.84$.† With these substitutions, Eq. (11-1) becomes

$$R = \exp\left[-\left(\frac{L}{6.84 L_{10}} \right)^{1.17} \right] \qquad (11\text{-}3)$$

Example 11-1 A certain application requires a bearing to last for 1800 h with a reliability of 99 percent. What should be the rated life of the bearing selected for this application?

Solution Substitute into Eq. (11-3) as follows:

$$0.99 = \exp\left[-\left(\frac{1800}{6.84 L_{10}} \right)^{1.17} \right]$$

Now take the natural logarithm of both sides and simplify the result. This gives

$$-0.010\ 050 = -\frac{(1800)^{1.17}}{(6.84)^{1.17}(L_{10})^{1.17}} = -\frac{678.7}{(L_{10})^{1.17}}$$

$$L_{10} = \left(\frac{678.7}{0.010\ 050} \right)^{1/1.17} = 13.4(10)^3 \text{ h} \qquad\qquad Ans.$$

////

11-3 BEARING LOAD

Experiments show that two groups of identical bearings tested under different loads F_1 and F_2 will have respective lives L_1 and L_2 according to the relation

---

* Charles Mischke, "Bearing Reliability and Capacity," *Machine Design*, vol. 37, no. 22, Sept. 30, 1965, pp. 139–140.

† There seems to be no agreement on the exponent b. Harris (see Fig. 11-5) uses 1.125; Mischke uses 1.17; others use 1.34, but feel unsure about it.

$$\frac{L_1}{L_2} = \left(\frac{F_2}{F_1}\right)^a \tag{11-4}$$

where $a = 3$ for ball bearings
$\qquad a = \frac{10}{3}$ for roller bearings

The AFBMA has established a standard load rating for bearings in which speed is not a consideration. This rating is called the basic load rating. The *basic load rating C* is defined as *the constant radial load which a group of apparently identical bearings can endure for a rating life of one million revolutions of the inner ring* (stationary load and stationary outer ring). The rating life of one million revolutions is a base value selected for ease of computation. The corresponding load rating is so high that plastic deformation of the contacting surfaces would occur were it actually applied. Consequently the basic load rating is purely a reference figure; such a large load would probably never be applied.

Other names in common use for the basic load rating are *dynamic load rating*, *basic dynamic capacity*, and *specific dynamic capacity*.

Using Eq. (11-4), the life of a bearing subjected to any other load F will be

$$L = \left(\frac{C}{F}\right)^a \tag{11-5}$$

where L is in millions of revolutions. The equation is more useful in the form

$$C = FL^{1/a} \tag{11-6}$$

For example, if we desire a life of 27 million revolutions for a roller bearing, then the basic load rating must be

$$C = F(27)^{3/10} = 2.69F$$

or 2.69 times the actual radial load.

It is customary practice with bearing manufacturers to specify the rated radial bearing load corresponding to a certain speed in rpm and a certain L_{10} life in hours. For example, the *Timken Engineering Journal* tabulates the load ratings at 3000 h of L_{10} life at 500 rpm. By adopting the subscripts D to refer to design or required values, and R as the catalog or rated values, then Eq. (11-6) can be rewritten as

$$C_R = F\left[\left(\frac{L_D}{L_R}\right)\left(\frac{n_D}{n_R}\right)\right]^{1/a} \tag{11-7}$$

where C_R is the basic load rating corresponding to L_R hours of L_{10} life at the speed n_R rpm. The force F is the actual radial bearing load; it is to be carried for L_D hours of L_{10} life at a speed of n_D rpm.

Example 11-2 A roller bearing is to be selected to withstand a radial load of 4 kN and have an L_{10} life of 1200 h at a speed of 600 rpm. What load rating would you look for in searching the *Timken Engineering Journal?*

Solution The quantities for use in Eq. (11-7) are $F = 4$ kN, $L_D = 1200$ h, $L_R = 3000$ h, $n_D = 600$ rpm, $n_R = 500$ rpm, and $a = \frac{10}{3}$. Therefore use

$$C_R = 4\left[\left(\frac{1200}{3000}\right)\left(\frac{600}{500}\right)\right]^{3/10} = 3.21 \text{ kN}$$

The Timken ratings are tabulated in English units and in dekanewtons (see Table A-1). Therefore the basic load rating used to enter the catalog is 321 daN.   ////

It is also possible to develop a relation to find the catalog rating corresponding to any desired reliability. To find this relation, note that the reciprocal of Eq. (11-3) is

$$\frac{1}{R} = \exp\left[\frac{L}{6.84L_{10}}\right]^{1.17} \tag{a}$$

where L is the desired life corresponding to the reliability R. Taking the natural logarithm of both sides gives

$$\ln\frac{1}{R} = \left(\frac{L}{6.84}\right)^{1.17}\frac{1}{(L_{10})^{1.17}}$$

Now, solve this expression for L_{10}. The result is

$$L_{10} = \frac{L}{6.84}\frac{1}{[\ln (1/R)]^{1/1.17}} \tag{11-8}$$

Equation (11-8) gives the rating life corresponding to any desired life L at the reliability R. Incorporating this into Eq. (11-7) yields

$$C_R = F\left[\left(\frac{L_D}{L_R}\right)\left(\frac{n_D}{n_R}\right)\left(\frac{1}{6.84}\right)\right]^{1/a}\frac{1}{[\ln (1/R)]^{1/1.17a}} \tag{11-9}$$

This useful expression can be used for any bearing, provided the product $L_R n_R$ is always taken as 10^6 and $L_D n_D$ as the corresponding design number.

The three-parameter Weibull solution can be obtained from Eq. (11-2) in a similar manner. The result is

$$C_R = F\left[\frac{(L_D n_D/L_R n_R)}{0.02 + 4.91[\ln (1/R)]^{1/1.40}}\right]^{1/a} \tag{11-10}$$

Example 11-3 What load rating would be used if the application in Example 11-2 is to have a reliability of 99 percent?

Solution The terms are identical with those of Example 11-2 and in addition $R = 0.99$. Equation (11-9) yields

$$C_R = 4\left[\left(\frac{1200}{3000}\right)\left(\frac{600}{500}\right)\left(\frac{1}{6.84}\right)\right]^{3/10} \frac{1}{[\ln{(1/0.99)}]^{1/(1.17)(10/3)}} = 5.86 \text{ kN}$$

Therefore, we enter the catalog with $C_R = 586$ daN.   ////

11-4 SELECTION OF BALL AND STRAIGHT ROLLER BEARINGS

Except for pure thrust bearings, as in Fig. 11-2*i*, ball bearings are usually operated with some combination of radial and thrust load. Since catalog ratings are based only on radial load, it is convenient to define an *equivalent radial load F_e* that will have the same effect on bearing life as do the applied loads. The AFBMA equation for equivalent radial load for ball bearings is the maximum of the two values

$$F_e = VF_r \tag{11-11}$$

$$F_e = XVF_r + YF_a \tag{11-12}$$

where F_e = equivalent radial load
  F_r = applied radial load
  F_a = applied thrust load
  V = a rotation factor
  X = a radial factor
  Y = a thrust factor

In using these equations the rotation factor V is to correct for the various rotating-ring conditions. For a rotating inner ring, $V = 1$. For a rotating outer ring, $V = 1.2$. The factor of 1.2 for outer-ring rotation is simply an acknowledgment that the fatigue life is reduced under these conditions. Self-aligning bearings are an exception; they have $V = 1$ for rotation of either ring.

The X and Y factors in Eq. (11-12) depend upon the geometry of the bearing, including the number of balls and ball diameter. When a theoretical derivation of the X and Y factors is made, it is found that the resulting curves can be approximated by pairs of straight lines. Thus there are two values of X and Y listed in Table 11-2. The set of values giving the largest equivalent load should always be used.

The AFBMA has established standard boundary dimensions for bearings

Table 11-2   EQUIVALENT RADIAL-LOAD FACTORS

Bearing type	X_1	Y_1	X_2	Y_2
Radial-contact ball bearings	1	0	0.5	1.4
Angular-contact ball bearings with shallow angle	1	1.25	0.45	1.2
Angular-contact ball bearings with steep angle	1	0.75	0.4	0.75
Double-row and duplex ball bearings (type DB or DF)	1	0.75	0.63	1.25

which define the bearing bore, the outside diameter, the width, and the fillet sizes on the shaft and housing shoulders. The basic plan covers all ball and straight roller bearings in the metric sizes. The plan is quite flexible in that, for a given bore, there is an assortment of widths and outside diameters. Furthermore, the outside diameters selected are such that, for a particular outside diameter, one can usually find a variety of bearings having different bores and widths.

This basic AFBMA plan is illustrated in Fig. 11-7. The bearings are identified by a two-digit number called the *dimension-series code*. The first number in the code is from the *width series* 0, 1, 2, 3, 4, 5, and 6. The second number is from the *diameter series* (outside) 8, 9, 0, 1, 2, 3, and 4. Figure 11-7 shows the variety of bearings which may be obtained with a particular bore. Since the dimension-series code does not reveal the dimensions directly, it is necessary to resort to tabulations. The 02- and 03-series bearings are the most widely used, and the dimensions of some of these are tabulated in Tables 11-3 and 11-4. Shaft and housing shoulder diameters listed in the tables should be used whenever possible to secure adequate support for the bearing and to resist the maximum thrust loads (Fig. 11-8). Table 11-5 lists the dimensions and load ratings of some straight roller bearings.

To assist the designer in the selection of bearings, most of the manufacturers' handbooks contain data on bearing life for many classes of machinery, as well as

FIGURE 11-7   The basic AFBMA plan for boundary dimensions. These apply to ball bearings, straight roller bearings, and spherical roller bearings, but not to tapered roller bearings or to inch-series ball bearings. The contour of the corner is not specified; it may be rounded or chamfered, but it must be small enough to clear the fillet radius specified in the standards.

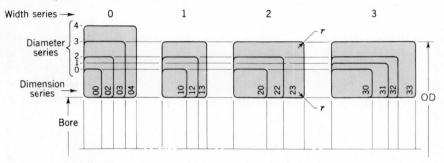

Table 11-3 DIMENSIONS AND BASIC LOAD RATINGS FOR THE 02-SERIES BALL BEARINGS

Bore, mm	OD, mm	Width, mm	Fillet radius, mm	Shoulder diameter, mm		Load rating, kN
				d_S	d_H	
10	30	9	0.6	12.5	27	3.58
12	32	10	0.6	14.5	28	5.21
15	35	11	0.6	17.5	31	5.87
17	40	12	0.6	19.5	34	7.34
20	47	14	1.0	25	41	9.43
25	52	15	1.0	30	47	10.8
30	62	16	1.0	35	55	14.9
35	72	17	1.0	41	65	19.8
40	80	18	1.0	46	72	22.5
45	85	19	1.0	52	77	25.1
50	90	20	1.0	56	82	26.9
55	100	21	1.5	63	90	33.2
60	110	22	1.5	70	99	40.3
65	120	23	1.5	74	109	44.1
70	125	24	1.5	79	114	47.6
75	130	25	1.5	86	119	50.7
80	140	26	2.0	93	127	55.6
85	150	28	2.0	99	136	64.1
90	160	30	2.0	104	146	73.9
95	170	32	2.0	110	156	83.7

information on load-application factors. Such information has been accumulated the hard way, that is, by experience, and the beginner designer should utilize this information until he or she gains enough experience to know when deviations are possible. Table 11-6 contains recommendations on bearing life for some classes of machinery. The load-application factors in Table 11-7 serve the same purpose as factors of safety; use them to increase the equivalent load before selecting a bearing.

11-5 SELECTION OF TAPERED ROLLER BEARINGS

The nomenclature for a tapered roller bearing differs in some respects from that of ball and straight roller bearings. The inner ring is called the cone, and the outer ring is called the cup, as shown in Fig. 11-9. It can also be seen that a tapered roller bearing is separable in that the cup can be removed from the cone-and-roller assembly.

A tapered roller bearing can carry both radial and thrust (axial) loads or any combination of the two. However, even when an external thrust load is not present, the radial load will induce a thrust reaction within the bearing because of

Table 11-4 DIMENSIONS AND BASIC LOAD RATINGS FOR THE
                 03-SERIES BALL BEARINGS

Bore, mm	OD, mm	Width, mm	Fillet radius, mm	Shoulder diameter, mm		Load rating, kN
				d_S	d_H	
10	35	11	0.6	12.5	31	6.23
12	37	12	1.0	16	32	7.48
15	42	13	1.0	19	37	8.72
17	47	14	1.0	21	41	10.37
20	52	15	1.0	25	45	12.24
25	62	17	1.0	31	55	16.2
30	72	19	1.0	37	65	21.6
35	80	21	1.5	43	70	25.6
40	90	23	1.5	49	80	31.4
45	100	25	1.5	54	89	40.5
50	110	27	2.0	62	97	47.6
55	120	29	2.0	70	106	55.2
60	130	31	2.0	75	116	62.7
65	140	33	2.0	81	125	71.2
70	150	35	2.0	87	134	80.1
75	160	37	2.0	93	144	87.2
80	170	39	2.0	99	153	94.8
85	180	41	2.5	106	161	101.9
90	190	43	2.5	111	170	110.8
95	200	45	2.5	117	179	117.9

the taper. To avoid separation of the races and rollers, this thrust must be resisted by an equal and opposite force. One way of generating this force is to always use at least two tapered roller bearings on a shaft. These can be mounted with the backs facing each other, called *indirect mounting*, or with the fronts facing each other, called *direct mounting*.

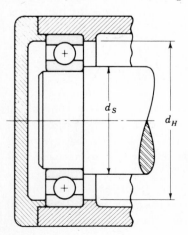

FIGURE 11-8 Shaft and housing shoulder diameters d_S and d_H should be adequate to assure good bearing support.

Table 11-5 DIMENSIONS AND BASIC LOAD RATINGS FOR STRAIGHT ROLLER BEARINGS

Bore, mm	02-series			03-series		
	OD, mm	Width, mm	Load, kN	OD, mm	Width, mm	Load, kN
25	52	15	10.9	62	17	23.1
30	62	16	18.0	72	19	30.3
35	72	17	26.0	80	21	39.2
40	80	18	34.0	90	23	46.3
45	85	19	35.6	100	25	63.6
50	90	20	36.9	110	27	75.7
55	100	21	45.4	120	29	92.6
60	110	22	55.6	130	31	103.0
65	120	23	65.0	140	33	116.0
70	125	24	65.8	150	35	136.0
75	130	25	80.1	160	37	162.0
80	140	26	87.2	170	39	163.0
85	150	28	99.7	180	41	196.0
90	160	30	126.0	190	43	211.0
95	170	32	140.0	200	45	240.0
100	180	34	154.0	215	47	274.0
110	200	38	205.0	240	50	352.0
120	215	40	220.0	260	55	416.0
130	230	40	239.0	280	58	489.0
140	250	42	280.0	300	62	538.0

The thrust component F_a produced by a pure radial load F_r is specified by Timken as

$$F_a = \frac{0.47F_r}{K} \tag{11-13}$$

Table 11-6 BEARING-LIFE RECOMMENDATIONS FOR VARIOUS CLASSES OF MACHINERY

Type of application	Life, kh
Instruments and apparatus for infrequent use	Up to 0.5
Aircraft engines	0.5–2
Machines for short or intermittent operation where service interruption is of minor importance	4–8
Machines for intermittent service where reliable operation is of great importance	8–14
Machines for 8-h service which are not always fully utilized	14–20
Machines for 8-h service which are fully utilized	20–30
Machines for continuous 24-h service	50–60
Machines for continuous 24-h service where reliability is of extreme importance	100–200

Table 11-7 LOAD-APPLICATION FACTORS

Type of application	Load factor
Precision gearing	1.0–1.1
Commercial gearing	1.1–1.3
Applications with poor bearing seals	1.2
Machinery with no impact	1.0–1.2
Machinery with light impact	1.2–1.5
Machinery with moderate impact	1.5–3.0

where K is the ratio of the radial rating of the bearing to the thrust rating. The constant 0.47 is derived from a summation of the thrust components from the individual rollers supporting the load. The value of K is approximately 1.5 for radial bearings and 0.75 for steep-angle bearings. These values may be used for a preliminary bearing selection, after which the exact values may be obtained from the *Timken Engineering Journal* in order to verify the selection.

Figure 11-10 shows a typical bearing mounting subjected to an external thrust load T_e. The radial reactions F_{rA} and F_{rB} are computed by taking moments about the effective load centers G. The distance a (Fig. 11-9) is obtained from the catalog rating sheets (*Timken Engineering Journal*). The equivalent radial loads are computed using an equation similar to Eq. (11-12), except that a

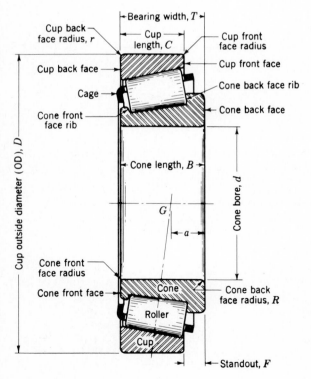

FIGURE 11-9 Nomenclature of a tapered roller bearing. Point G is the effective load center; use this point to calculate the radial bearing load. (*Courtesy of The Timken Company.*)

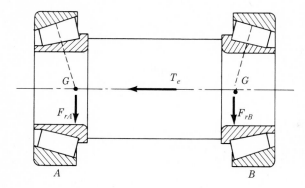

FIGURE 11-10 Schematic drawing showing a pair of tapered roller bearings assembled on a single shaft with indirect mounting. The radial bearing forces are F_{rA} and F_{rB}. T_e is the external thrust.

rotation factor is not used with tapered roller bearings. We shall use subscripts A and B to designate each of the two bearings in Fig. 11-10. The equivalent radial load on bearing A is

$$F_{eA} = 0.4F_{rA} + K_A\left(\frac{0.47F_{rB}}{K_B} + T_e\right) \tag{11-14}$$

For bearing B, we have

$$F_{eB} = 0.4F_{rB} + K_B\left(\frac{0.47F_{rA}}{K_A} - T_e\right) \tag{11-15}$$

If the actual radial load on either bearing should happen to be larger than the corresponding value of F_e, then use the actual radial load instead of F_e for that bearing.

Figure 11-11 is a reproduction of a portion of a typical catalog page from the *Timken Engineering Journal.*

Example 11-4 The gear-reduction unit shown in Fig. 11-12 is arranged to rotate the cup, while the cone is stationary. Bearing A takes the thrust load of 250 lb and, in addition, has a radial load of 875 lb. Bearing B is subjected to a pure radial load of 625 lb. The speed is 150 rpm. The desired L_{10} life is 90 kh. The desired shaft diameters are $1\frac{3}{8}$ in at A and $1\frac{1}{4}$ in at B. Select suitable tapered roller bearings using an application factor of unity.

Solution Since B carries only radial load, the thrust on A is augmented by the induced thrust due to B. Equation (11-14) applies. Using a trial value of 1.5 for K, we obtain

$$F_{eA} = 0.4F_{rA} + K_A\left(\frac{0.47F_{rB}}{K_B} + T_e\right) = 0.4(875) + 1.5\left[\frac{0.47(625)}{1.5} + 250\right]$$

$$= 1020 \text{ lb}$$

SINGLE-ROW STRAIGHT BORE—TS

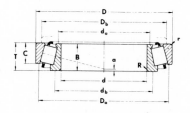

bore	outside diameter	width	rating at 500 RPM for 3000 hours L10		fac-tor	eff. load center	part numbers		cone max. shaft fillet radius	width	backing shoulder diameters		cup max. hous-ing fillet radius	width	backing shoulder	
			one row radial	thrust												
							cone	cup								
d	D	T	lb	lb	K	a			R	B	d_b	d_a	r	C	D_b	D_a
1.2500	2.3125	0.5781	1280	1040	1.23	−0.05	08125	08231	0.04	0.5937	1.48	1.42	0.04	0.4219	2.05	2.17
1.2500	2.3280	0.6250	1580	1110	1.42	−0.12	▲LM67048	LM67010	Spec.	0.6600	1.67	1.42	0.05	0.4650	2.05	2.20
1.2500	2.4404	0.6250	1580	1110	1.42	−0.12	▲LM67049A	LM67014	0.03	0.6600	1.46	1.42	0.05	0.4650	2.13	2.24
1.2500	2.4409	0.7150	1990	1190	1.67	−0.19	15123	15245	Spec.	0.7500	1.67	1.44	0.05	0.5625	2.17	2.28
1.2500	2.4409	0.7500	1990	1190	1.67	−0.23	15125	15245	0.14	0.8125	1.67	1.44	0.05	0.5625	2.17	2.28
1.2500	2.4409	0.7500	1990	1190	1.67	−0.23	15126	15245	0.03	0.8125	1.46	1.44	0.05	0.5625	2.17	2.28
1.3125	3.0000	1.1563					HM89443	HM89410		1.1250	1.83	1.75	0.13	0.9063	2.44	2.87
1.3125	3.0000	1.1563	3880	3630	1.07	−0.22	HM89444	HM89411	0.15	1.1250	2.09	1.75	0.03	0.9063	2.56	2.87
1.3125	3.4843	1.0000	3180	4250	0.75	0.09	44131	44348	0.08	0.9330	2.01	1.89	0.06	0.6875	2.95	3.31
1.3750	2.5625	0.7100	2140	1380	1.55	−0.15	▲LM48548	LM48510	Spec.	0.7200	1.81	1.57	0.05	0.5500	2.28	2.40
1.3750	2.5625	0.8300	2140	1380	1.55	−0.15	▲LM48548A	LM48511A	0.03	0.7200	1.59	1.66	0.06	0.6700	2.28	2.40
1.3750	2.6250	0.8125	2520	1520	1.66	−0.22	M38549	M38510	0.14	0.8125	1.83	1.57	0.09	0.6563	2.28	2.44
1.3750	2.6875	0.8125	2330	1410	1.66	−0.23	14585	14525	0.14	0.8125	1.81	1.57	0.09	0.6250	2.32	2.48
1.3750	2.7148	0.7813	2180	1420	1.53	−0.17	14137A	14274A	0.06	0.7710	1.65	1.57	0.13	0.6250	2.32	2.48
1.3750	2.7148	0.7813	2180	1420	1.53	−0.17	14138A	14274A	0.14	0.7710	1.81	1.57	0.13	0.6250	2.32	2.48
1.3750	2.8438	1.0000	3190	2980	1.07	−0.18	HM88649	HM88610	0.09	1.0000	1.91	1.69	0.09	0.7812	2.36	2.72
1.3750	2.8750	0.8750	2620	2030	1.29	−0.15	02877	02820	0.14	0.8750	1.91	1.65	0.13	0.6875	2.44	2.68
1.3750	2.8750	0.8750	2620	2030	1.29	−0.15	02878	02820	0.03	0.8750	1.67	1.65	0.13	0.6875	2.44	2.68

FIGURE 11-11 A portion of the TS bearing tables from the *Timken Engineering Journal*, sec. 1. The original page contains SI equivalents in red print below the U. S. customary values.

Thus $F_{eA} > F_{rA}$ and so we use 1020 lb as the equivalent radial load to select bearing A. We next use Eq. (11-7) to obtain the L_{10} rating. Using $L_R = 3$ kh and $n_R = 500$ rpm, we get

$$C_R = F\left[\left(\frac{L_D}{L_R}\right)\left(\frac{n_D}{n_R}\right)\right]^{1/a} = 1020\left[\left(\frac{90}{3}\right)\left(\frac{150}{500}\right)\right]^{3/10} = 1970 \text{ lb}$$

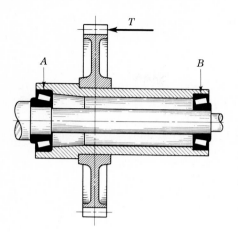

FIGURE 11-12 Tapered roller bearings applied to a gear-reduction unit. (*Courtesy of The Timken Company.*)

Using this figure and a bore of $1\frac{3}{8}$ in we enter the catalog sheets (Fig. 11-11 is typical) and select an LM48548 cone and an LM48510 cup. This bearing has an L_{10} rating of 2140 lb and $K = 1.55$. Since we assumed 1.5 for K, the difference is small and we need not recalculate F_{eA}.

For bearing B, Eq. (11-15) applies. Thus

$$F_{eB} = 0.4F_{rB} + K_B\left(\frac{0.47F_{rA}}{K_A} - T_e\right) = 0.4(625) + 1.5\left[\frac{0.47(875)}{1.55} - 250\right]$$

$$= 273 \text{ lb}$$

Note that the actual value of K_A was used, but K_B was assumed to be 1.5 as before. Since $F_{eB} < F_{rB}$, we use F_{rB}. Using Eq. (11-7) again, we find the L_{10} desired rating as

$$C_R = F\left[\left(\frac{L_D}{L_R}\right)\left(\frac{n_D}{n_R}\right)\right]^{1/a} = 625\left[\left(\frac{90}{3}\right)\left(\frac{150}{500}\right)\right]^{3/10} = 1210 \text{ lb}$$

This bearing is to have a bore of $1\frac{1}{4}$ in. Therefore, from Fig. 11-11 we select an 08125 cone and 08231 cup. The L_{10} rating is 1280 lb with $K = 1.23$. However, the actual load was used instead of the smaller equivalent load and so we need not recalculate. ////

11-6 LUBRICATION

The contacting surfaces in rolling bearings have a relative motion that is both rolling and sliding, and so it is difficult to understand exactly what happens. If the relative velocity of the sliding surfaces is high enough, then the lubricant action is hydrodynamic (see Chap. 12). *Elastohydrodynamic lubrication* (EHD) is the phenomenon that occurs when a lubricant is introduced between surfaces that are in pure rolling contact. The contact of gear teeth, rolling bearings, and cam-and-follower surfaces are typical examples. When a lubricant is trapped between two surfaces in rolling contact, a tremendous increase in the pressure within the lubricant film occurs. But viscosity is exponentially related to pressure and so a very large increase in viscosity occurs in the lubricant that is trapped between the surfaces. Leibensperger* observes that the change in viscosity in and out of contact pressure is equivalent to the difference between cold asphalt and light sewing machine oil.

The purposes of an antifriction-bearing lubricant may be summarized as follows:

1 To provide a film of lubricant between the sliding and rolling surfaces
2 To help distribute and dissipate heat
3 To prevent corrosion of the bearing surfaces
4 To protect the parts from the entrance of foreign matter

Either oil or grease may be employed as a lubricant. The following rules may help in deciding between them.

Use grease when	Use oil when
1. The temperature is not over 200°F.	1. Speeds are high.
2. The speed is low.	2. Temperatures are high.
3. Unusual protection is required from the entrance of foreign matter.	3. Oiltight seals are readily employed.
4. Simple bearing enclosures are desired.	4. Bearing type is not suitable for grease lubrication.
5. Operation for long periods without attention is desired.	5. The bearing is lubricated from a central supply which is also used for other machine parts.

11-7 MOUNTING AND ENCLOSURE

There are so many methods of mounting antifriction bearings that each new design is a real challenge to the ingenuity of the designer. The housing bore and shaft outside diameter must be held to very close limits, which of course is expensive. There are usually one or more counterboring operations, several facing operations, and drilling, tapping, and threading operations, all of which must be performed on the shaft, housing, or cover plate. Each of these operations contributes to the cost of production, so that the designer, in ferreting out a trouble-free and low-cost mounting, is faced with a difficult and important problem. The various bearing manufacturers' handbooks give many mounting details in almost every design area. In a text of this nature, however, it is possible to give only the barest details.

---

* R. L. Leibensperger, "When Selecting a Bearing," *Machine Design*, vol. 47, no. 8, April 3, 1975, pp. 142–147.

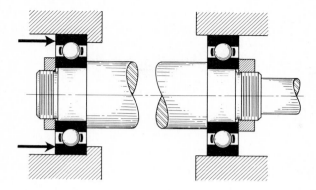

FIGURE 11-13 A common bearing mounting.

The most frequently encountered mounting problem is that which requires one bearing at each end of a shaft. Such a design might use one ball bearing at each end, one tapered roller bearing at each end, or a ball bearing at one end and a straight roller bearing at the other. One of the bearings usually has the added function of positioning or axially locating the shaft. Figure 11-13 shows a very common solution to this problem. The inner rings are backed up against the shaft shoulders and are held in position by round nuts threaded onto the shaft. The outer ring of the left-hand bearing is backed up against a housing shoulder and is held in position by a device which is not shown. The outer ring of the right-hand bearing floats in the housing.

There are many variations possible to the method shown in Fig. 11-13. For example, the function of the shaft shoulder may be performed by retaining rings, by the hub of a gear or pulley, or by spacing tubes or rings. The round nuts may be replaced by retaining rings or by washers locked in position by screws, cotters, or taper pins. The housing shoulder may be replaced by a retaining ring; the outer ring of the bearing may be grooved for a retaining ring, or a flanged outer ring may be used. The force against the outer ring of the left-hand bearing is usually applied by the cover plate, but if no thrust is present, the ring may be held in place by retaining rings.

Figure 11-14 shows an alternative method of mounting in which the inner

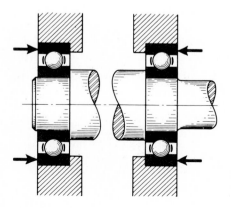

FIGURE 11-14 An alternative bearing mounting.

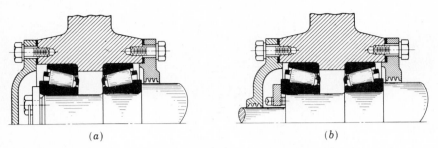

(a)   (b)

FIGURE 11-15   Two-bearing mountings.   (*Courtesy of The Timken Company.*)

races are backed up against the shaft shoulders as before but no retaining devices
are required.   With this method the outer races are completely retained.   This
eliminates the grooves or threads, which cause stress concentration on the overhanging end, but it requires accurate dimensions in an axial direction or the
employment of adjusting means.   This method has the disadvantage that, if the
distance between the bearings is great, the temperature rise during operation may
expand the shaft enough to wreck the bearings.

It is frequently necessary to use two or more bearings at one end of a shaft.
For example, two bearings could be used to obtain additional rigidity or increased
load capacity or to cantilever a shaft.   Several two-bearing mountings are shown
in Fig. 11-15.   These may be used with tapered roller bearings, as shown, or with
ball bearings.   In either case it should be noted that the effect of the mounting is
to preload the bearings in an axial direction.

Figure 11-16 shows another two-bearing mounting.   Note the use of washers
against the cone backs.

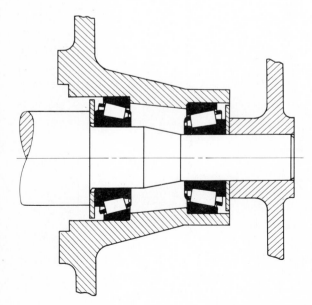

FIGURE 11-16   Mounting
for a washing-machine spindle.   (*Courtesy of The Timken
Company.*)

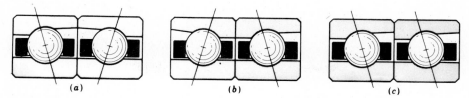

FIGURE 11-17 Duplex arrangements of angular-contact bearings. (*a*) DF mounting; (*b*) DB mounting; (*c*) DT mounting. (*Courtesy of Miniature Precision Bearings, Inc.*)

When maximum stiffness and resistance to shaft misalignment is desired, pairs of angular-contact ball bearings (Fig. 11-2) are often used in an arrangement called *duplexing*. Bearings manufactured for duplex mounting have their rings ground with an offset, so that when a pair of bearings is tightly clamped together, a preload is automatically established. As shown in Fig. 11-17, three mounting arrangements are used. The face-to-face mounting, called DF, will take heavy radial loads and thrust loads from either direction. The DB mounting (back to back) has the greatest aligning stiffness and is also good for heavy radial loads and thrust loads from either direction. The tandem arrangement, called the DT mounting, is used where the thrust is always in the same direction; since the two bearings have their thrust functions in the same direction, a preload, if required, must be obtained in some other manner.

Bearings are usually mounted with the rotating ring a press fit, whether it be the inner or outer ring. The stationary ring is then mounted with a push fit. This permits the stationary ring to creep in its mounting slightly, bringing new portions of the ring into the load-bearing zone to equalize wear.

Preloading

The object of preloading is to remove the internal clearance usually found in bearings, to increase the fatigue life, and to decrease the shaft slope at the bearing. Figure 11-18 shows a typical bearing in which the clearance is exaggerated for clarity.

Preloading of straight roller bearings may be obtained by:

1 Mounting the bearing on a tapered shaft or sleeve to expand the inner ring

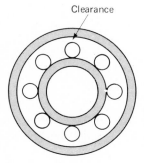

Clearance

Figure 11-18 Clearance in an off-the-shelf bearing; exaggerated for clarity.

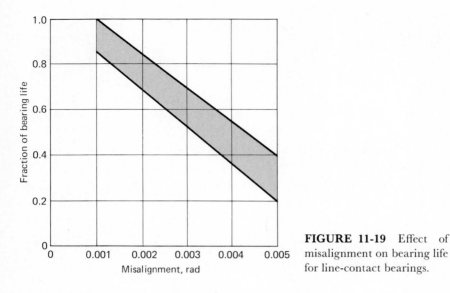

FIGURE 11-19 Effect of misalignment on bearing life for line-contact bearings.

2 Using an interference fit for the outer ring
3 Purchasing a bearing with the outer ring preshrunk over the rollers

Ball bearings are usually preloaded by the axial load built in during assembly. However, the bearings of Fig. 11-17a and b are preloaded in assembly because of the differences in widths of the inner and outer rings.

It is always good practice to follow manufacturer's recommendations in determining preload since too much will lead to early failure.

Alignment*

Based on the general experience with rolling bearings as expressed in manufacturers' catalogs, the permissible misalignment in cylindrical and tapered roller bearings is limited to 0.001 rad. For spherical ball bearings the misalignment should not exceed 0.0087 rad. But for deep-groove ball bearings, the allowable range of misalignment is 0.0035 to 0.0047 rad.

The life of the bearing decreases significantly when the misalignment exceeds the allowable limits. Figure 11-19 shows that there is about a 20 percent loss in life for every 0.001 rad of neutral-axis slope beyond 0.001 rad.

Additional protection against misalignment is obtained by providing the full shoulders (see Fig. 11-8) recommended by the manufacturer. Also, if there is any misalignment at all, it is good practice to provide a safety factor of around 2 to account for possible increases during assembly.

---

* We are grateful to Professor Charles R. Mischke of Iowa State University for this topic. J.E.S., L.D.M.

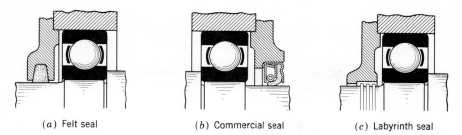

(a) Felt seal   (b) Commercial seal   (c) Labyrinth seal

FIGURE 11-20 Typical sealing methods. (*Courtesy of New-Departure-Hyatt Division, General Motors Corporation.*)

Enclosures

To exclude dirt and foreign matter and to retain the lubricant, the bearing mountings must include a seal. The three principal methods of sealings are the felt seal, the commercial seal, and the labyrinth seal (Fig. 11-20).

Felt seals may be used with grease lubrication when the speeds are low. The rubbing surfaces should have a high polish. Felt seals should be protected from dirt by placing them in machined grooves or by using metal stampings as shields.

The *commercial seal* is an assembly consisting of the rubbing element and, generally, a spring backing, which are retained in a sheet-metal jacket. These seals are usually made by press-fitting them into a counterbored hole in the bearing cover. Since they obtain the sealing action by rubbing, they should not be used for high speeds.

The *labyrinth seal* is especially effective for high-speed installations and may be used with either oil or grease. It is sometimes used with flingers. At least three grooves should be used, and they may be cut on either the bore or the outside diameter. The clearance may vary from 0.010 to 0.040 in, depending upon the speed and temperature.

PROBLEMS*

11-1   The geared printing roll shown in the figure is driven at 300 rpm by a force $F = 200$ lb acting as shown. A uniform force distribution $w = 20$ lb/in acts against the bottom surface of roll 3 in the positive y direction. Radial-contact 02-series ball bearings are to be selected for this application and mounted at O and A. Use an application factor of 1.2 and an L_{10} life of 30 kh and find the size of bearings to be used. Both bearings are to be the same size.

11-2   The figure shows a geared countershaft with an overhanging pinion at C. Select a plain radial-contact ball bearing for mounting at O and a straight roller bearing for mounting at B. The force on gear A is $F_A = 600$ lb and the shaft is to run at a speed of 480 rpm. Solution of the statics problem gives the force of the bearings

---

* The asterisk indicates a design-type problem or one which may have no unique result.

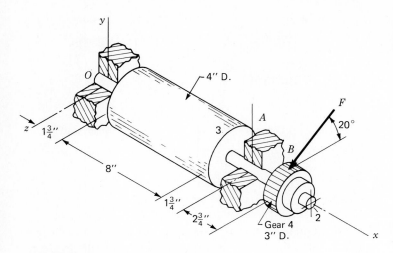

PROBLEM 11-1

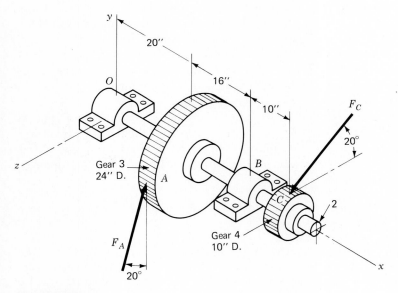

PROBLEM 11-2

against the shaft at O as $\mathbf{R}_O = -388\mathbf{j} + 471\mathbf{k}$ lb and at B as $\mathbf{R}_B = 317\mathbf{j} - 1620\mathbf{k}$ lb. Find the size of 02-series bearings required using an application factor of 1.4 and an L_{10} life of 50 kh.

11-3 The figure is a schematic drawing of a countershaft that supports two V-belt pulleys. The countershaft runs at 1100 rpm and the bearings are to have a life of 12 kh at 99 percent reliability using an application factor of unity. The belt tension on the loose side of pulley A is 15 percent of the tension on the tight side. An analysis of this problem gave shaft forces at B and D of $\mathbf{F}_B = -253\mathbf{j} - 253\mathbf{k}$ N and $\mathbf{F}_D = -320\mathbf{k}$ N, with the belt pulls assumed to be parallel. Based on bending deflection, a shaft diameter of 35 mm has been selected. Determine the size of ball bearings to be used if both are to be of the same size.

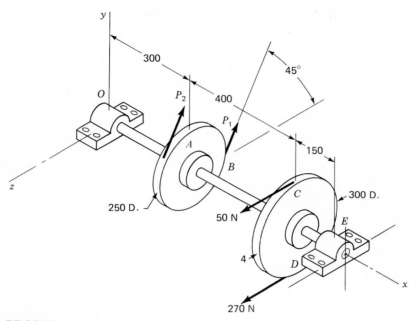

PROBLEM 11-3 Dimensions in millimeters.

11-4 Radial-contact 02-series ball bearings are to be selected and located at O and B to support the overhung countershaft shown in the figure. The belt tensions shown are parallel to each other. The tension on the loose side on pulley A is 20 percent of the tension on the tight side. Shaft speed is 720 rpm. The bearings are to have

PROBLEM 11-4 Dimensions in millimeters.

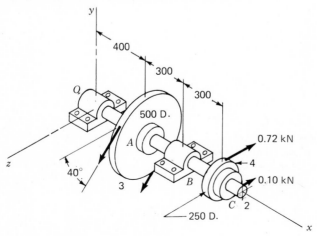

a reliability of 90 percent corresponding to a life of 24 kh. Use an application
factor of unity and the same size bearings at each location. Find an appropriate
bearing size.

11-5   The shaft in the figure has parallel belt pulls with the tension on the loose side of
pulley 4 being 20 percent of the tension on the tight side. The shaft rotates at 840
rpm and the radial ball bearings, to be selected for locations at O and B, are to
have a reliability of 99 percent corresponding to a life of 24 kh. Use an application
factor of unity, same-size bearings, and specify the size of 02-series bearings needed.

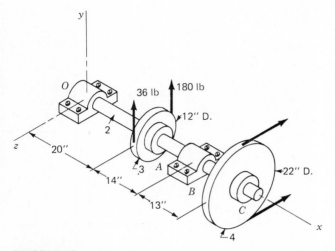

PROBLEM 11-5

11-6   An analysis of the shaft forces for the part shown in the figure gave radial bearing
forces of 112 lb at the left-hand bearing, and 299 lb at the right-hand bearing. The
shaft rotates at 400 rpm. Tapered roller bearings are to be used, having an L_{10} life
of 40 kh, unity for the application factor, and 1.5 for K. Determine the required
radial rating.

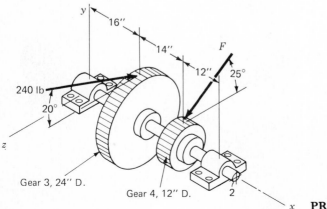

PROBLEM 11-6

11-7   When analyzed, the gear-transmission shaft shown in the figure has a radial bearing force at D of 6.06 kN, and at C, 11.4 kN. Tapered roller bearings are to be used with a life of 24 kh corresponding to 99.9 percent reliability. The shaft speed is 360 rpm. Use 1.2 for the application factor, 1.5 for K, and determine the required radial ratings in dekanewtons.

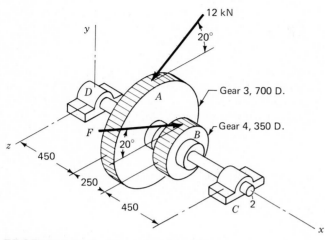

PROBLEM 11-7   Dimensions in millimeters.

11-8*   The figure shows a shaft contained in a helical-gear reduction unit in which a force $\mathbf{F} = -1700\mathbf{i} + 6400\mathbf{j} - 2300\mathbf{k}$ lb is applied to gear B as shown. The forces $\mathbf{F}_A$ and

PROBLEM 11-8

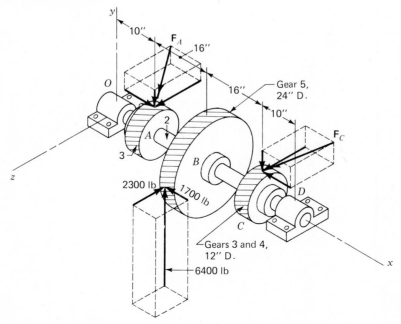

$\mathbf{F}_C$, of equal magnitudes, resist the applied force. The directions of these two forces can be indicated by the unit vectors $\hat{\mathbf{F}}_A = 0.470\mathbf{i} - 0.342\mathbf{j} + 0.814\mathbf{k}$ and $\hat{\mathbf{F}}_C = -0.470\mathbf{i} - 0.342\mathbf{j} + 0.814\mathbf{k}$. The notation $\hat{\mathbf{F}}$ means $\mathbf{F}/|\mathbf{F}|$. In this problem we wish to determine the required radial ratings of tapered roller bearings to be mounted in the housings at O and D. The shaft dimensions shown in the figure locate the effective load centers of the bearings and gears. The bearings are to have an L_{10} life of 60 kh. Use unity for the application factor and 1.5 for K. The shaft speed is 1200 rpm, and the bearing bore is about $3\frac{1}{2}$ in.

11-9* The figure shows a portion of a transmission containing an ordinary helical gear and an overhung bevel gear. Tapered roller bearings are to be mounted in housings at O and B, with the bearing at O intended to take out the major thrust component. The dimensions are to the effective load centers of the gears and bearings. The vector bevel-gear force $\mathbf{F}_D$ can be expressed in the general form $\mathbf{F}_D = F_x\mathbf{i} + F_y\mathbf{j} + F_z\mathbf{k}$ or, for this particular bevel gear, in the form

$$\mathbf{F}_D = -0.242F_D\mathbf{i} - 0.242F_D\mathbf{j} + 0.940F_D\mathbf{k}$$

The bearings are to have an L_{10} life of 36 kh corresponding to a shaft speed of 900 rpm. Use 1.5 for K, unity for the application factor, and find the required radial rating of each bearing. Bearing bores are 90 mm at O and 60 mm at B.

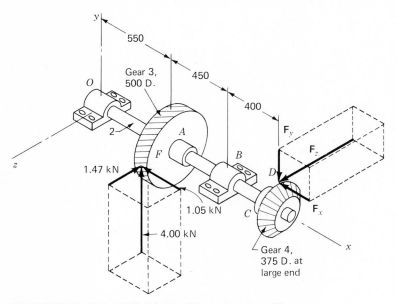

PROBLEMS 11-9 AND 11-10 Dimensions in millimeters.

11-10* An angular-contact ball bearing with shallow angle is to be housed at O in the figure to take both radial and thrust loads. The bearing at B is to be a straight roller bearing. Determine the required radial ratings of each bearing based on an L_{10} life of 36 kh at a shaft speed of 900 rpm.

CHAPTER

12

LUBRICATION AND JOURNAL BEARINGS

The object of lubrication is to reduce friction, wear, and heating of machine parts which move relative to each other. A lubricant is any substance which, when inserted between the moving surfaces, accomplishes these purposes. In a sleeve bearing, a shaft, or *journal*, rotates or oscillates within a sleeve, or *bearing*, and the relative motion is sliding. In an antifriction bearing, the main relative motion is rolling. A follower may either roll or slide on the cam. Gear teeth mate with each other by a combination of rolling and sliding. Pistons slide within their cylinders. All these applications require lubrication to reduce friction, wear, and heating.

The field of application for journal bearings is immense. The crankshaft and connecting-rod bearings of an automotive engine must operate for thousands of miles at high temperatures and under varying load conditions. The journal bearings used in the steam turbines of power-generating stations are said to have reliabilities approaching 100 percent. At the other extreme there are thousands of applications in which the loads are light and the service relatively unimportant. A simple, easily installed bearing is required, using little or no lubrication. In such cases an antifriction bearing might be a poor answer because of the cost, the elaborate enclosures, the close tolerances, the radial space required, the high speeds, or the increased inertial effects. Instead, a nylon bearing requiring no lubrication, a powder-metallurgy bearing with the lubrication "built in," or a bronze bearing with ring-oiled, wick-feed, solid-lubricant film or grease lubrication might be a very satisfactory solution. Recent metallurgy developments in bearing materials, combined with increased knowledge of the lubrication process, now make it possible to design journal bearings with satisfactory lives and very good reliabilities.

Much of the material we have studied thus far in this book has been based on fundamental engineering studies, such as statics, dynamics, the mechanics of solids, metal processing, mathematics, and metallurgy. In the study of lubrication and journal bearings, additional fundamental studies, such as chemistry, fluid mechanics, thermodynamics, and heat transfer, must be utilized in developing the material. While we shall not utilize all of them in the material to be included here, you can now begin to appreciate better how the study of mechanical engineering design is really an integration of most of your previous studies and a directing of this total background toward the resolution of a single objective.

12-1 TYPES OF LUBRICATION

Five distinct forms of lubrication may be identified:

1 Hydrodynamic
2 Hydrostatic
3 Elastohydrodynamic
4 Boundary
5 Solid-film

Hydrodynamic lubrication means that the load-carrying surfaces of the bearing are separated by a relatively thick film of lubricant, so as to prevent metal-to-metal contact, and that the stability thus obtained can be explained by the laws of fluid mechanics. Hydrodynamic lubrication does not depend upon the introduction of the lubricant under pressure, though that may occur; but it does require the existence of an adequate supply at all times. The film pressure is created by the moving surface itself pulling the lubricant into a wedge-shaped zone at a velocity sufficiently high to create the pressure necessary to separate the surfaces against the load on the bearing. Hydrodynamic lubrication is also called *full-film*, or *fluid*, *lubrication*.

Hydrostatic lubrication is obtained by introducing the lubricant, which is sometimes air or water, into the load-bearing area at a pressure high enough to separate the surfaces with a relatively thick film of lubricant. So, unlike hydrodynamic lubrication, motion of one surface relative to another is not required. We shall not deal with hydrostatic lubrication in this book,* but the subject should be considered in designing bearings where the velocities are small or zero and where the frictional resistance is to be an absolute minimum.

Elastohydrodynamic lubrication is the phenomenon that occurs when a lubricant is introduced between surfaces which are in rolling contact, such as mating gears or rolling bearings. The mathematical explanation requires the Hertzian theory of contact stress and fluid mechanics.†

Insufficient surface area, a drop in the velocity of the moving surface, a lessening in the quantity of lubricant delivered to a bearing, an increase in the bearing load, or an increase in lubricant temperature resulting in a decrease in viscosity—any one of these—may prevent the buildup of a film thick enough for full-film lubrication. When this happens, the highest asperities may be separated by lubricant films only several molecular dimensions in thickness. This is called *boundary lubrication*. The change from hydrodynamic to boundary lubrication is not at all a sudden or abrupt one. It is probable that a mixed hydrodynamic- and boundary-type lubrication occurs first, and as the surfaces move closer together, the boundary-type lubrication becomes predominant. The viscosity of the lubricant is not of as much importance with boundary lubrication as is the chemical composition.

When bearings must be operated at extreme temperatures, a *solid-film lubricant* such as graphite or molybdenum disulfide must be used because the ordinary mineral oils are not satisfactory. Much research is currently being carried out in an effort, too, to find composite bearing materials with low wear rates as well as small frictional coefficients.

12-2 VISCOSITY‡

In Fig. 12-1 let a plate A be moving with a velocity U on a film of lubricant of thickness h. We imagine the film as composed of a series of horizontal layers and the force F causing these layers to deform or slide on one another just like a deck of cards. The layers in contact with the moving plate are assumed to have a

---

* See Oscar Pinkus and Beno Sternlicht, *Theory of Hydrodynamic Lubrication*, McGraw-Hill, New York, 1961, chap. 6. See also Dudley D. Fuller, *Theory and Practice of Lubrication for Engineers*, Wiley, New York, 1956, chaps. 3 and 4.

† See A. Cameron, *Principles of Lubrication*, Wiley, New York, 1966, chaps. 7–9.

‡ For a complete discussion see any fluid-mechanics text, for example, W. M. Swanson, *Fluid Mechanics*, Holt, New York, 1970, pp. 17–30 and 740.

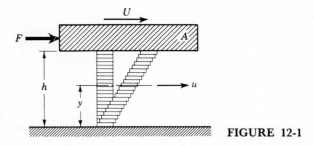

FIGURE 12-1

velocity U; those in contact with the stationary surface are assumed to have a zero velocity. Intermediate layers have velocities which depend upon their distances y from the stationary surface. Newton's law of viscous flow states that the shear stress in the fluid is proportional to the rate of change of velocity with respect to y. Thus

$$\tau = \frac{F}{A} = \mu \frac{du}{dy} \qquad (12\text{-}1)$$

where μ is the constant of proportionality and defines *absolute viscosity*. The derivative du/dy is the rate of change of velocity with distance and may be called the rate of shear, or the velocity gradient. The viscosity μ is thus a measure of the internal frictional resistance of the fluid. If the assumption is made that the rate of shear is a constant, then $du/dy = U/h$, and from Eq. (12-1),

$$\tau = \frac{F}{A} = \mu \frac{U}{h} \qquad (12\text{-}2)$$

The unit of viscosity in the IPS system is seen to be the pound-force-second per square inch; this is the same as stress or pressure multiplied by time. The IPS unit is called the *reyn* in honor of Sir Osborne Reynolds.

The absolute viscosity, also called *dynamic viscosity*, is measured by the pascal-second (Pa · s) in SI; this is the same as a newton-second per square meter. The conversion from IPS units to SI is the same as for stress. For example, multiply the absolute viscosity in reyns by 6890 to convert to units of Pa · s.

The American Society of Mechanical Engineers (ASME) has published a list of cgs units which are not to be used in ASME documents.* This list results from a recommendation by the International Committee of Weights and Measures (CIPM) that the use of cgs units with special names be discouraged. Included in this list is a unit of force called the *dyne* (dyn), a unit of dynamic viscosity called the *poise* (P), and a unit of kinematic viscosity called the *stoke* (St). All of these units have been, and still are, used extensively in lubrication studies.

---

* *ASME Orientation and Guide for Use of Metric Units*, 2d ed., American Society of Mechanical Engineers, 1972, p. 13.

The poise is the cgs unit of dynamic or absolute viscosity, and its unit is the dyne-second per square centimeter ($\mathrm{dyn \cdot s/cm^2}$). It has been customary to use the centipoise (cP) in analysis because its value is more convenient. When the viscosity is expressed in centipoises, it is designated by Z. The conversion from cgs units to SI and IPS units is as follows:

$$\mu(\mathrm{Pa \cdot s}) = (10)^{-3} Z(\mathrm{cP})$$

$$\mu(\mathrm{reyn}) = \frac{Z(\mathrm{cP})}{6.89(10)^6}$$

The ASTM standard method for determining viscosity uses an instrument called the Saybolt Universal Viscosimeter. The method consists of measuring the time in seconds for 60 ml of lubricant at a specified temperature to run through a tube 17.6 mm in diameter and 12.25 mm long. The result is called the *kinematic viscosity* and in the past the unit of square centimeter per second has been used. One square centimeter per second is defined as a stoke. By the use of the *Hagen-Poiseuille law** the kinematic viscosity based upon seconds Saybolt, also called *Saybolt Universal viscosity* (SUV) in seconds, is

$$Z_k = \left(0.22t - \frac{180}{t} \right) \tag{12-3}$$

where Z_k is in centistokes (cSt) and t is the number of seconds Saybolt.

In SI the kinematic viscosity ν has the unit of square meter per second ($\mathrm{m^2/s}$) and the conversion is

$$\nu(\mathrm{m^2/s}) = 10^{-6} Z_k(\mathrm{cSt})$$

Thus, Eq. (12-3) becomes

$$\nu = \left(0.22t - \frac{180}{t} \right) (10^{-6}) \tag{12-4}$$

To convert to dynamic viscosity we multiply ν by the density in SI units. Designating the density as ρ with the unit of kilogram per cubic meter, we have

$$\mu = \rho \left(0.22t - \frac{180}{t} \right) (10^{-6}) \tag{12-5}$$

where μ is in pascal-seconds.

---

* See any text on fluid mechanics for example, Chia-Shun Yih, *Fluid Mechanics*, McGraw-Hill, New York, 1969, p. 314.

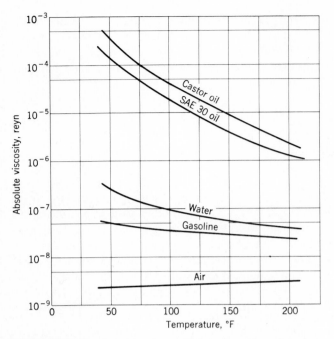

FIGURE 12-2 A comparison of the viscosities of various fluids.

Figure 12-2 shows the absolute viscosity in the IPS system of a number of fluids often used for lubrication purposes and their variation with temperature.

12-3 PETROFF'S LAW

The phenomenon of bearing friction was first explained by Petroff using the assumption that the shaft is concentric. Though we shall seldom make use of Petroff's method of analysis in the material to follow, it is important because it defines groups of dimensionless parameters and because the coefficient of friction predicted by this law turns out to be quite good even when the shaft is not concentric.

Let us now consider a vertical shaft rotating in a guide bearing. It is assumed that the bearing carries a very small load, that the clearance space c is completely filled with oil, and that leakage is negligible (Fig. 12-3). We denote the radius of the shaft by r, the radial clearance by c, and the length of the bearing by l, all dimensions being in inches. If the shaft rotates at N rps, then its surface velocity is $U = 2\pi rN$ in/s. Since the shearing stress in the lubricant is equal to the velocity gradient times the viscosity, from Eq. (12-2) we have

$$\tau = \mu\,\frac{U}{h} = \frac{2\pi r\mu N}{c} \qquad\qquad (a)$$

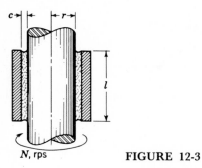

FIGURE 12-3

where the radial clearance c has been substituted for the distance h. The force required to shear the film is the stress times the area. The torque is the force times the lever arm. Thus

$$T = (\tau A)(r) = \left(\frac{2\pi r \mu N}{c}\right)(2\pi r l)(r) = \frac{4\pi^2 r^3 l \mu N}{c} \qquad (b)$$

If we now designate a small force on the bearing by W, in pounds-force, then the pressure P, in pounds-force per square inch of projected area, is $P = W/2rl$. The frictional force is fW, where f is the coefficient of friction, and so the frictional torque is

$$T = fWr = (f)(2rlP)(r) = 2r^2 flP \qquad (c)$$

Substituting the value of the torque from Eq. (c) into Eq. (b) and solving for the coefficient of friction, we find

$$f = 2\pi^2 \frac{\mu N}{P} \frac{r}{c} \qquad (12\text{-}6)$$

Equation (12-6) is called *Petroff's law* and was first published in 1883. The two quantities $\mu N/P$ and r/c are very important parameters in lubrication. Substitution of the appropriate dimensions in each parameter will show that they are dimensionless.

12-4 STABLE LUBRICATION

The difference between boundary and hydrodynamic lubrication can be explained by reference to Fig. 12-4. This plot of the change in the coefficient of friction versus the bearing characteristic $\mu N/P$ was obtained by the McKee brothers in an actual test of friction.* The plot is important because it defines stability of

---

* S. A. McKee and T. R. McKee, "Journal Bearing Friction in the Region of Thin Film Lubrication," *SAE Journal*, vol. 31, 1932, pp. (T)371–377.

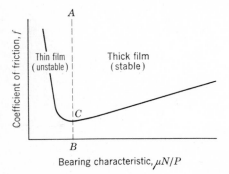

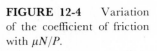

FIGURE 12-4   Variation
of the coefficient of friction
with $\mu N/P$.

lubrication and helps us to understand hydrodynamic and boundary, or thin-film, lubrication.

Suppose we are operating to the right of ordinate BA and something happens, say, an increase in lubricant temperature. This results in a lower viscosity and hence a smaller value of $\mu N/P$. The coefficient of friction decreases, not as much heat is generated in shearing the lubricant, and consequently the lubricant temperature drops. Thus the region to the right of ordinate BA defines *stable lubrication* because variations are self-correcting.

To the left of ordinate BA a decrease in viscosity would increase the friction. A temperature rise would ensue, and the viscosity would be reduced still more. The result would be compounded. Thus the region to the left of ordinate BA represents *unstable lubrication*.

It is also helpful to see that a small viscosity, and hence a small $\mu N/P$, means that the lubricant film is very thin and that there will be a greater possibility of some metal-to-metal contact, and hence of more friction. Thus, point C represents what is probably the beginning of metal-to-metal contact as $\mu N/P$ becomes smaller.

12-5   THICK-FILM LUBRICATION

Let us now examine the formation of a lubricant film in a journal bearing. Figure 12-5a shows a journal which is just beginning to rotate in a clockwise direction. Under starting conditions the bearing will be dry, or at least partly dry, and hence the journal will climb or roll up the right side of the bearing as shown in Fig. 12-5a. Under the conditions of a dry bearing, equilibrium will be obtained when the friction force is balanced by the tangential component of the bearing load.

Now suppose a lubricant is introduced into the top of the bearing as shown in Fig. 12-5b. The action of the rotating journal is to pump the lubricant around the bearing in a clockwise direction. The lubricant is pumped into a wedge-shaped space and forces the journal over to the other side. A *minimum film thickness* h_0 occurs, not at the bottom of the journal, but displaced clockwise from the bottom as in Fig. 12-5b. This is explained by the fact that a film pressure in the

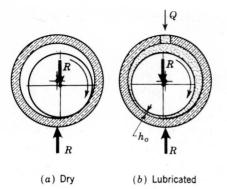

(a) Dry   (b) Lubricated

FIGURE 12-5   Formation of a film.

converging half of the film reaches a maximum somewhere to the left of the bearing center.

Figure 12-5 shows how to decide whether the journal, under hydrodynamic lubrication, is eccentrically located on the right or on the left side of the bearing. Visualize the journal beginning to rotate. Find the side of the bearing upon which the journal tends to roll. Then, if the lubrication is hydrodynamic, place the journal on the opposite side.

The nomenclature of a journal bearing is shown in Fig. 12-6. The dimension c is the *radial clearance* and is the difference in the radii of the bearing and journal. In Fig. 12-6 the center of the journal is at O and the center of the bearing at O'. The distance between these centers is the *eccentricity* and is denoted by e. The *minimum film thickness* is designated by h_0, and it occurs at the line of centers. The

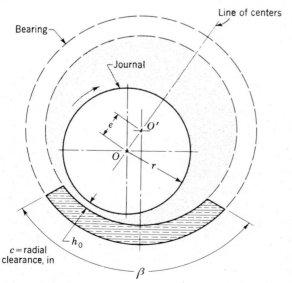

FIGURE 12-6   Nomenclature of a journal bearing.

film thickness at any other point is designated by h. We also define an *eccentricity ratio* ϵ as

$$\epsilon = \frac{e}{c}$$

The bearing shown in the figure is known as a *partial bearing*. If the radius of the bearing is the same as the radius of the journal, it is known as a *fitted bearing*. If the bearing encloses the journal, as indicated by the dashed lines, it becomes a *full bearing*. The angle β describes the angular length of a partial bearing. For example, a 120° partial bearing has the angle β equal to 120°.

12-6 HYDRODYNAMIC THEORY

The present theory of hydrodynamic lubrication originated in the laboratory of Beauchamp Tower in the early 1880s in England. Tower had been employed to study the friction in railroad journal bearings and learn the best methods of lubricating them. It was an accident or error, during the course of this investigation, that prompted Tower to look at the problem in more detail and that resulted in a discovery that eventually led to the development of the theory.

Figure 12-7 is a schematic drawing of the journal bearing which Tower investigated. It is a partial bearing, 4 in in diameter by 6 in long, with a bearing arc of 157°, and having bath-type lubrication, as shown. The coefficients of friction obtained by Tower in his investigations on this bearing were quite low, which is not now surprising. After testing this bearing, Tower later drilled a $\frac{1}{2}$-in-diameter lubricator hole through the top. But when the apparatus was set in motion, oil flowed out of this hole. In an effort to prevent this, a cork stopper was used, but this popped out, and so it was necessary to drive a wooden plug into the hole. When the wooden plug was pushed out too, Tower, at this point, undoubtedly realized that he was on the verge of discovery. A pressure gauge connected to the hole indicated a pressure of more than twice the unit-bearing load. Finally,

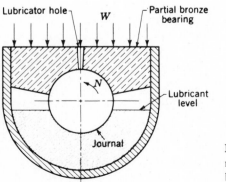

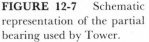

FIGURE 12-7 Schematic representation of the partial bearing used by Tower.

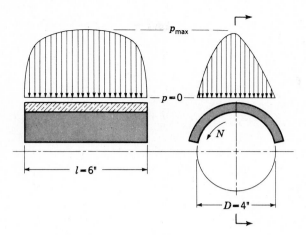

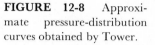

FIGURE 12-8 Approximate pressure-distribution curves obtained by Tower.

he investigated the bearing film pressures in detail throughout the bearing width and length and reported a distribution similar to that of Fig. 12-8.*

The results obtained by Tower had such regularity that Osborne Reynolds concluded that there must be a definite law relating the friction, the pressure, and the velocity. The present mathematical theory of lubrication is based upon Reynolds' work following the experiments by Tower.† The original differential equation, developed by Reynolds, was used by him to explain Tower's results. The solution is a challenging problem which has interested many investigators ever since then, and it is still the starting point for lubrication studies.

Reynolds pictured the lubricant as adhering to both surfaces and being pulled by the moving surface into a narrowing, wedge-shaped space so as to create a fluid or film pressure of sufficient intensity to support the bearing load. One of the important simplifying assumptions resulted from Reynolds' realization that the fluid films were so thin in comparison with the bearing radius that the curvature could be neglected. This enabled him to replace the curved partial bearing with a flat bearing, called a *plane slider bearing*. Other assumptions made are:

1 The lubricant obeys Newton's law of viscous flow.
2 The forces due to the inertia of the lubricant are neglected.
3 The lubricant is assumed to be incompressible.
4 The viscosity is assumed to be constant throughout the film.
5 The pressure does not vary in the axial direction.

Figure 12-9a shows a journal rotating in the clockwise direction supported by a film of lubricant of variable thickness h on a partial bearing which is fixed. We

---

* Beauchamp Tower, "First Report on Friction Experiments," *Proc. Inst. Mech. Eng.*, November 1883, pp. 632–666; "Second Report," ibid., 1885, pp. 58–70; "Third Report," ibid., 1888, pp. 173–205; "Fourth Report," ibid., 1891, pp. 111–140.

† Osborne Reynolds, "Theory of Lubrication, Part I," *Phil. Trans. Roy. Soc. London*, 1886.

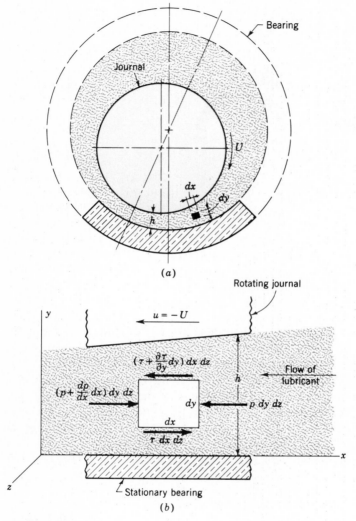

FIGURE 12-9

specify that the journal has a constant surface velocity U. Using Reynolds' assumption that curvature can be neglected, we fix a right-handed xyz reference system to the stationary bearing. We now make the following additional assumptions:

6   The bearing and journal extend infinitely in the z direction; this means there can be no lubricant flow in the z direction.

7   The film pressure is constant in the y direction. Thus the pressure depends only on the coordinate x.

8 The velocity of any particle of lubricant in the film depends only on the coordinates x and y.

We now select an element of lubricant in the film (Fig. 12-9a) of dimensions dx, dy, and dz, and compute the forces which act on the sides of this element. As shown in Fig. 12-9b, normal forces, due to the pressure, act upon the right and left sides of the element, and shear forces, due to the viscosity and to the velocity, act upon the top and bottom sides. Summing the forces gives

$$\sum F = \left(p + \frac{dp}{dx}\,dx\right) dy\,dz + \tau\,dx\,dz - \left(\tau + \frac{\partial \tau}{\partial y}\,dy\right) dx\,dz - p\,dy\,dz = 0 \qquad (a)$$

This reduces to

$$\frac{dp}{dx} = \frac{\partial \tau}{\partial y} \qquad (b)$$

From Eq. (12-1) we have

$$\tau = \mu\,\frac{\partial u}{\partial y} \qquad (c)$$

where the partial derivative is used because the velocity u depends upon both x and y. Substituting Eq. (c) into (b), we obtain

$$\frac{dp}{dx} = \mu\,\frac{\partial^2 u}{\partial y^2} \qquad (d)$$

Holding x constant, we now integrate this expression twice with respect to y. This gives

$$\frac{\partial u}{\partial y} = \frac{1}{\mu}\frac{dp}{dx}\,y + C_1$$

$$\qquad (e)$$

$$u = \frac{1}{2\mu}\frac{dp}{dx}\,y^2 + C_1 y + C_2$$

Note that the act of holding x constant means that C_1 and C_2 can be functions of x. We now assume that there is no slip between the lubricant and the boundary surfaces. This gives two sets of boundary conditions for evaluating the constants C_1 and C_2 :

$$
\begin{aligned}
y &= 0 & y &= h \\
u &= 0 & u &= -U
\end{aligned}
\qquad (f)
$$

Notice, in the second condition, that h is a function of x. Substituting these conditions in Eq. (e) and solving for the constants gives

$$C_1 = -\frac{U}{h} - \frac{h}{2\mu}\frac{dp}{dx} \qquad C_2 = 0$$

or

$$u = \frac{1}{2\mu}\frac{dp}{dx}(y^2 - hy) - \frac{U}{h}y \qquad\qquad (12\text{-}7)$$

This equation gives the velocity distribution of the lubricant in the film as a function of the coordinate y and the pressure gradient dp/dx. The equation shows that the velocity distribution across the film (from $y = 0$ to $y = h$) is obtained by superposing a parabolic distribution (the first term) onto a linear distribution (the second term). Figure 12-10 shows the superposition of these two terms to obtain the velocity for particular values of x and dp/dx. In general, the parabolic term may be additive or subtractive to the linear term, depending upon the sign of the pressure gradient. When the pressure is maximum, $dp/dx = 0$ and the velocity is

$$u = -\frac{U}{h}y \qquad\qquad (g)$$

a linear relation.

We next define Q as the volume of lubricant flowing in the x direction per unit time. By using a width of unity in the z direction, the volume may be obtained by the expression

$$Q = \int_0^h u\,dy \qquad\qquad (h)$$

FIGURE 12-10   Velocity of the lubricant.

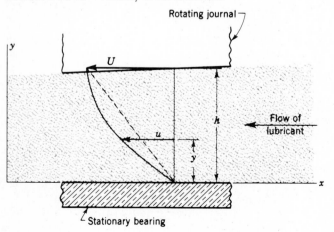

Substituting the value of u from Eq. (12-7) and integrating gives

$$Q = -\frac{Uh}{2} - \frac{h^3}{12\mu}\frac{dp}{dx} \tag{i}$$

The next step uses the assumption of an incompressible lubricant and states that the flow is the same for any cross section. Thus

$$\frac{dQ}{dx} = 0$$

From Eq. (i),

$$\frac{dQ}{dx} = -\frac{U}{2}\frac{dh}{dx} - \frac{d}{dx}\left(\frac{h^3}{12\mu}\frac{dp}{dx}\right) = 0$$

or

$$\frac{d}{dx}\left(\frac{h^3}{\mu}\frac{dp}{dx}\right) = -6U\frac{dh}{dx} \tag{12-8}$$

which is the classical Reynolds equation for one-dimensional flow. It neglects side leakage, that is, flow in the z direction. A similar development is used when side leakage is not neglected. The resulting equation is

$$\frac{\partial}{\partial x}\left(\frac{h^3}{\mu}\frac{\partial p}{\partial x}\right) - \frac{\partial}{\partial z}\left(\frac{h^3}{\mu}\frac{\partial p}{\partial z}\right) = -6U\frac{\partial h}{\partial x} \tag{12-9}$$

There is no general solution to Eq. (12-9); approximate solutions have been obtained by using electrical analogies, mathematical summations, relaxation methods, and numerical and graphical methods. One of the important solutions is due to Sommerfeld* and may be expressed in the form

$$\frac{r}{c}f = \phi\left[\left(\frac{r}{c}\right)^2\frac{\mu N}{P}\right] \tag{12-10}$$

where ϕ indicates a functional relationship. Sommerfeld found the functions for half-bearings and full bearings by using the assumption of no side leakage.

---

* A. Sommerfeld, "Zur Hydrodynamischen Theorie der Schmiermittel-Reibung," ("On the Hydrodynamic Theory of Lubrication"), Z. Math. Physik, vol. 50, 1904, pp. 97–155.

12-7 DESIGN CONSIDERATIONS

We may distinguish between two groups of variables in the design of sliding bearings. In the first group are those whose values are either given or are under the control of the designer. These are:

1 The viscosity μ
2 The load per unit of projected bearing area, P
3 The speed N
4 The bearing dimensions r, c, β, and l

Of these four variables, the designer usually has no control over the speed, because it is specified by the overall design of the machine. Sometimes the viscosity is specified in advance, as, for example, when the oil is stored in a sump and is used for lubricating and cooling a variety of bearings. The remaining variables, and sometimes the viscosity, may be controlled by the designer and are therefore the *decisions* he or she makes. In other words, when these four variables are defined, the design is complete.

In the second group are the dependent variables. The designer cannot control these except indirectly by changing one or more of the first group. These are:

1 The coefficient of friction f
2 The temperature rise ΔT
3 The flow of oil Q
4 The minimum film thickness h_0

This group of variables tells us how well the bearing is performing and hence we may regard them as *performance factors*. Certain limitations on their values must be imposed by the designer to assure satisfactory performance. These limitations are specified by the characteristics of the bearing materials and of the lubricant. The fundamental problem in bearing design, therefore, is to define satisfactory limits for the second group of variables and then to decide upon values for the first group such that these limitations are not exceeded.

Significant Angular Velocity*

Especial care must be used in computing the bearing speed N when both the bearing and the journal rotate. This situation occurs, for example, in epicyclic bearings. Let

ω_j = absolute angular velocity of the journal
ω_b = absolute angular velocity of the bearing
ω_f = absolute angular velocity of the load vector

---

* We are especially grateful to Dr. Charles R. Mischke of Iowa State University for these thoughts. J.E.S., L.D.M.

Then the velocity of the journal relative to the load vector is

$$\omega_{jf} = \omega_j - \omega_f \qquad\qquad (a)$$

The velocity of the bearing relative to the load vector is

$$\omega_{bf} = \omega_b - \omega_f \qquad\qquad (b)$$

The significant angular velocity is simply the sum of Eqs. (a) and (b) and is

$$\omega^* = \omega_{jf} + \omega_{bf} = \omega_j + \omega_b - 2\omega_f \qquad\qquad (12\text{-}11)$$

This is the angular velocity that should be used to compute the speed N as used in this chapter.

12-8 THE RELATION OF THE VARIABLES

Before proceeding to the problem of design, it is necessary to establish the relationships between the variables. A. A. Raimondi and John Boyd, of Westinghouse Research Laboratories, used an iteration technique to solve Reynolds' equation on the digital computer*. This is the first time such extensive data have been available for use by designers, and consequently we shall employ them in this book.†

The Raimondi and Boyd papers were published in three parts and contain 45 detailed charts and 6 tables of numerical information. In all three parts, charts are used to define the variables for length-diameter (l/d) ratios of $1:4$, $1:2$, and 1 and for beta angles of 60 to $360°$. Under certain conditions the solution to the Reynolds equation gives negative pressures in the diverging portion of the oil film. Since a lubricant cannot usually support a tensile stress, Part III of the Raimondi-Boyd papers assumes that the oil film is ruptured when the film pressure becomes zero. Part III also contains data for the infinitely long bearing; since it has no ends, this means that there is no side leakage. The charts appearing in this book are from Part III of the papers, and are for full journal bearings $(\beta = 360°)$ only. Space does not permit the inclusion of charts for partial bearings. This means that you must refer to the charts in the original papers when beta angles of less than $360°$ are desired. The notation is very nearly the same as in this book, and so no problems should arise.

---

* A. A. Raimondi and John Boyd, "A Solution for the Finite Journal Bearing and Its Application to Analysis and Design, Parts I, II, and III," *Trans. ASLE*, vol. 1, no. 1, in *Lubrication Science and Technology*, Pergamon, New York, 1958, pp. 159–209.

† A number of other sources of data are available; for a description of these see Fuller, op. cit., pp. 150, 157, 175, 177, 195, and 201. See also the earlier companion paper, John Boyd and Albert A. Raimondi, "Applying Bearing Theory to the Analysis and Design of Journal Bearings, Parts I and II," *J. Appl. Mechanics*, vol. 73, 1951, pp. 298–316.

The *bearing-characteristic number*, or the *Sommerfeld number*, is defined by the equation

$$S = \left(\frac{r}{c}\right)^2 \frac{\mu N}{P} \tag{12-12}$$

where S = bearing-characteristic number
$\quad r$ = journal radius, in
$\quad c$ = radial clearance, in
$\quad \mu$ = absolute viscosity, reyn
$\quad N$ = significant speed in revolutions per second (rps), based on Eq. (12-11)
$\quad P$ = load per unit of projected bearing area, psi

The Sommerfeld number contains all the variables usually specified by the designer; it is dimensionless, and so it has been used as the abscissa in all of the charts except, of course, the viscosity charts.

Viscosity Charts (Figs. 12-11 to 12-13)

One of the important assumptions made in the Raimondi-Boyd analysis is that *viscosity of the lubricant is constant as it passes through the bearing.* But since work is done on the lubricant during this flow, the temperature of the oil is higher when it leaves the loading zone than it was on entry. And the viscosity charts clearly indicate that the viscosity drops off significantly with a rise in temperature. Since the analysis is based on a constant viscosity, our problem now is to determine the value of viscosity to be used in the analysis.

Some of the lubricant that enters the bearing emerges as a side flow, which carries away some of the heat. The balance of the lubricant flows through the load-bearing zone and carries away the balance of the heat generated. In determining the viscosity to be used we shall employ a temperature that is the average of the inlet and outlet temperatures, or

$$T_{av} = T_1 + \frac{\Delta T}{2} \tag{12-13}$$

where T_1 is the inlet temperature and ΔT is the temperature rise of the lubricant from inlet to outlet. Of course the viscosity used in the analysis must correspond to T_{av}.

One of the objectives of lubrication analysis is to determine the oil outlet temperature when the oil and its inlet temperature are specified. This is a trial-and-error type of problem.

To illustrate, suppose we have decided to use SAE 30 oil in an application in

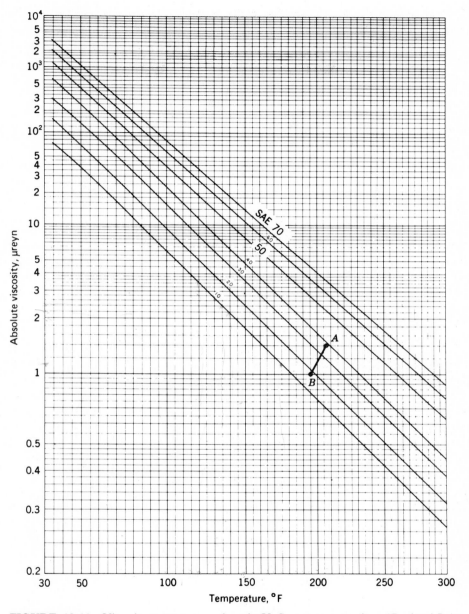

FIGURE 12-11 Viscosity-temperature chart in U. S. customary units. (*Boyd and Raimondi.*)

which the oil inlet temperature is $T_1 = 180°F$. We begin by estimating that the temperature rise will be $\Delta T = 30°F$. Then, from Eq. (12-13),

$$T_{av} = T_1 + \frac{\Delta T}{2} = 180 + \frac{30}{2} = 195°F$$

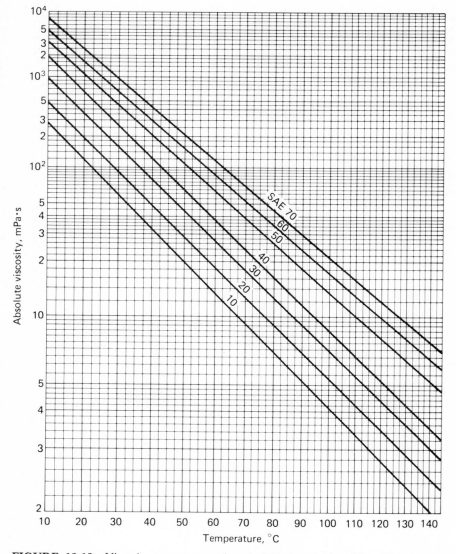

FIGURE 12-12   Viscosity-temperature chart in SI units.   (*Adapted from Fig. 12-11.*)

From Fig. 12-11 we follow the SAE 30 line and find that $\mu = 1.40$ μreyn at 195°F. So we use this viscosity in an analysis to be explained in detail in due time and find that the temperature rise is actually $\Delta T = 54°$F.   Thus Eq. (12-13) gives

$$T_{av} = 180 + \frac{54}{2} = 207°F$$

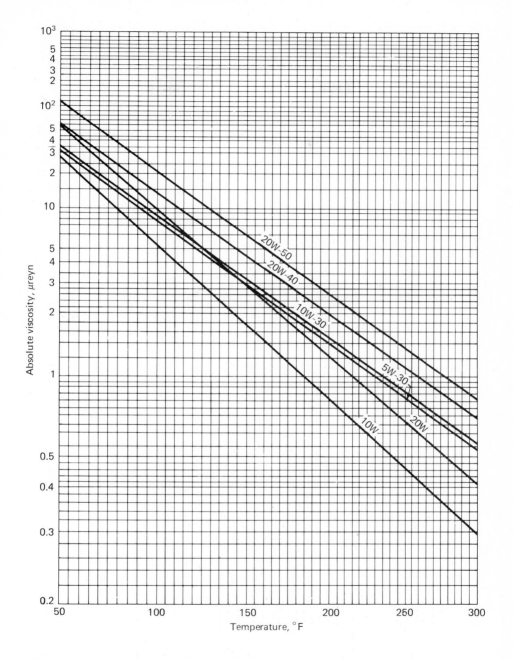

FIGURE 12-13 Chart for multiviscosity lubricants. This chart was derived from known viscosities at two points, 100 and 210°F, and the results are believed to be correct for other temperatures.

This corresponds to point A on Fig. 12-11 which is above the SAE 30 line and indicates that the viscosity used in the analysis was too high.

For a second guess, try $\mu = 1.00$ μreyn. Again we run through an analysis and this time find that $\Delta T = 30°$F. This gives an average temperature of

$$T_{av} = 180 + \frac{30}{2} = 195°F$$

and locates point B on Fig. 12-11.

If points A and B are fairly close to each other and on opposite sides of the SAE 30 line, a straight line can be drawn between them with the intersection locating the correct values of viscosity and average temperature to be used in the analysis. For this illustration we see from the viscosity chart that they are $T_{av} = 203°$F and $\mu = 1.26$ μreyn.

Minimum Film Thickness (Figs. 12-14 and 12-15)

Let us specify the following quantities for a full journal bearing:

$$\mu = 4 \ \mu\text{reyn}$$
$$N = 30 \text{ rps}$$
$$W = 500 \text{ lb (bearing load)}$$
$$r = 0.75 \text{ in}$$
$$c = 0.0015 \text{ in}$$
$$l = 1.50 \text{ in}$$

The unit load is

$$P = \frac{W}{2rl} = \frac{500}{2(0.75)(1.50)} = 222 \text{ psi}$$

The bearing characteristic number, from Eq. (12-12), is

$$S = \left(\frac{r}{c}\right)^2\left(\frac{\mu N}{P}\right) = \left(\frac{0.75}{0.0015}\right)^2\left[\frac{4(10)^{-6}(30)}{222}\right] = 0.135$$

Also, $l/d = 1.50/(2)(0.75) = 1$.

Entering Fig. 12-14 with $S = 0.135$ and $l/d = 1$ gives

$$\frac{h_0}{c} = 0.42 \qquad \epsilon = 0.58$$

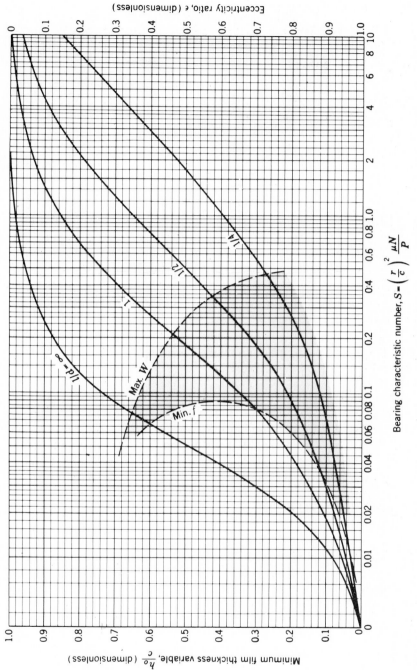

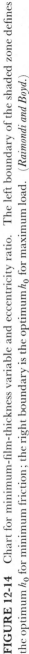

FIGURE 12-14 Chart for minimum-film-thickness variable and eccentricity ratio. The left boundary of the shaded zone defines the optimum h_0 for minimum friction; the right boundary is the optimum h_0 for maximum load. (*Raimondi and Boyd.*)

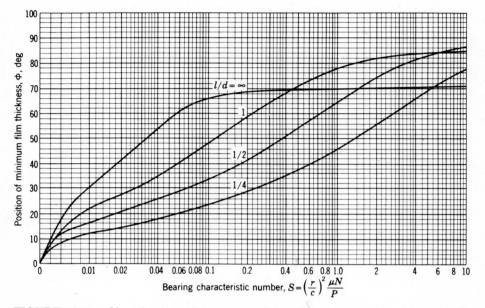

FIGURE 12-15   Chart for determining the position of the minimum film thickness h_0. For location of origin see Fig. 12-16.   (*Raimondi and Boyd.*)

The quantity h_0/c is called the *minimum-film-thickness variable*.   Since $c = 0.0015$ in, the minimum film thickness is

$$h_0 = 0.42(0.0015) = 0.000\ 63 \text{ in}$$

This is shown in Fig. 12-16.   We can find the angular location ϕ of the minimum film thickness from the chart of Fig. 12-15.   Entering with $S = 0.135$ and $l/d = 1$ gives $\phi = 53°$.

The eccentricity ratio is $\epsilon = e/c = 0.58$.   This means that the eccentricity is

$$e = 0.58(0.0015) = 0.000\ 87 \text{ in}$$

This is also shown in Fig. 12-16.   Note that if the bearing is centered, $e = 0$ and $h_0 = c$; this corresponds to a very light or zero load and the eccentricity ratio is zero.   As the load is increased, the journal is forced downward and the limiting position is reached when $h_0 = 0$ and $e = c$; that is, the journal is touching the bearing.   For this condition the eccentricity ratio is unity.

Since

$$h_0 = c - e \tag{12-14}$$

we have, by dividing both sides by c,

$$\frac{h_0}{c} = 1 - \epsilon \tag{12-15}$$

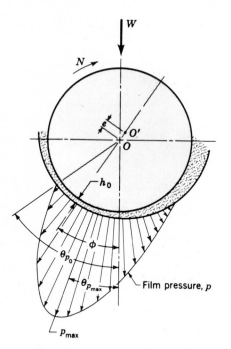

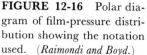

FIGURE 12-16 Polar diagram of film-pressure distribution showing the notation used. (*Raimondi and Boyd.*)

Design optimums frequently used are *maximum load*, which is a load-carrying characteristic, and *minimum power loss*, which is a function of the departure from thick-film relationships. Dashed lines for both these conditions have been constructed in Fig. 12-14 so that the optimum values of h_0 or ϵ can readily be found. The shaded zone between the boundaries defined by these two optimums may therefore be considered as a recommended operating zone.

From this discussion and by examination of Fig. 12-14 you should have concluded that lightly loaded bearings operate with a large Sommerfeld number, while heavily loaded bearings will operate at a small number.

Coefficient of Friction (Fig. 12-17)

The friction chart has the *friction variable* $(r/c)f$ plotted against S for various values of the l/d ratio. Using the same data as before, we enter Fig. 12-17 with $S = 0.135$ and $l/d = 1$. We then find the friction variable to be

$$\frac{r}{c}f = 3.50$$

Therefore the coefficient of friction is

$$f = 3.50\,\frac{c}{r} = 3.50\left(\frac{0.0015}{0.75}\right) = 0.007$$

With this known, other things can be learned about the bearing performance.

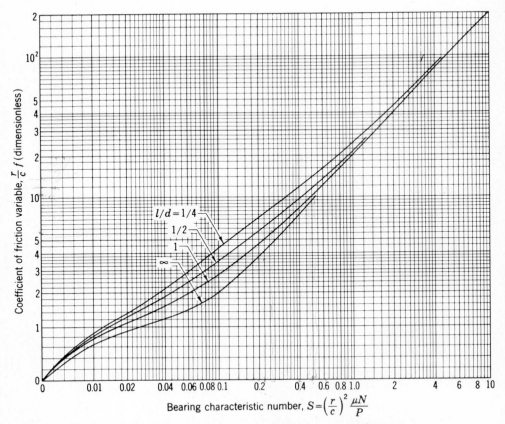

FIGURE 12-17   Chart for coefficient-of-friction variable.   (*Raimondi and Boyd.*)

For example, the torque required to overcome friction is

$$T = fWr = 0.007(500)(0.75) = 2.62 \text{ lb} \cdot \text{in}$$

The power lost in the bearing, in horsepower, is

$$H = \frac{TN}{1050} = \frac{2.62(30)}{1050} = 0.0748 \text{ hp}$$

or, expressed in Btu, we have

$$H = \frac{2\pi TN}{778(12)} = \frac{2\pi(2.62)(30)}{778(12)} = 0.0529 \text{ Btu/s}$$

Lubricant Flow (Figs. 12-18 and 12-19)

The *flow variable* $Q/rcNl$, found from the chart of Fig. 12-18, is used to find the volume of lubricant Q which is pumped into the converging space by the rotating

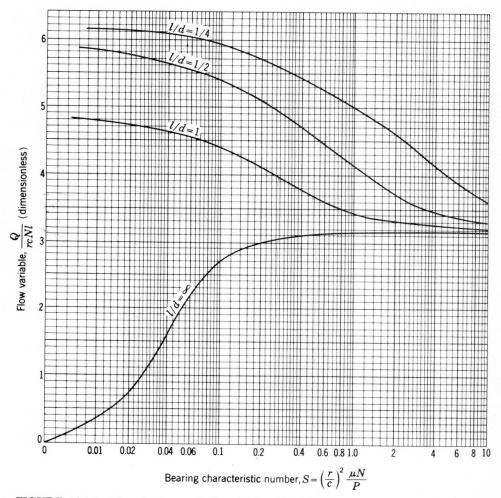

FIGURE 12-18 Chart for flow variable. (*Raimondi and Boyd.*)

journal. This chart is based on the *assumption of atmospheric pressure and the absence of oil grooves or holes in the bearing.* The amount of oil supplied to the bearing must, at least, be equal to Q if the bearing is to perform according to the charts.

Of the amount of oil Q pumped by the rotating journal, an amount Q_s flows out the ends, and hence is called the *side leakage.* This side leakage can be computed from the *flow ratio* Q_s/Q of Fig. 12-19.

Using the same data as before, we enter Fig. 12-18 with $S = 0.135$ and $l/d = 1$. From the chart we find

$$\frac{Q}{rcNl} = 4.28$$

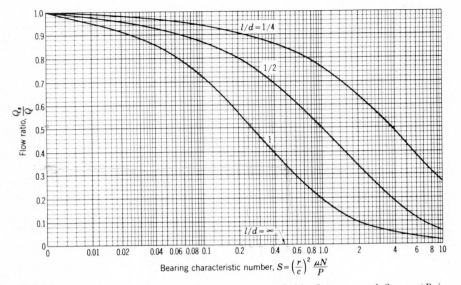

FIGURE 12-19   Chart for determining the ratio of side flow to total flow.   (*Raimondi and Boyd.*)

Therefore, the total flow is

$$Q = 4.28rcNl = 4.28(0.75)(0.0015)(30)(1.5) = 0.216 \text{ in}^3/s$$

From Fig. 12-19 we find the flow ratio to be

$$\frac{Q_s}{Q} = 0.655$$

Therefore the side leakage is

$$Q_s = 0.655Q = 0.655(0.216) = 0.142 \text{ in}^3/s$$

Film Pressure (Figs. 12-20 and 12-21)

The maximum film pressure developed in the film can be obtained by finding the *pressure ratio* P/p_{max} from the chart of Fig. 12-20.   Using the same data as before, we enter this chart with $S = 0.135$ and $l/d = 1$.   The maximum-film-pressure ratio is found to be

$$\frac{P}{p_{max}} = 0.42$$

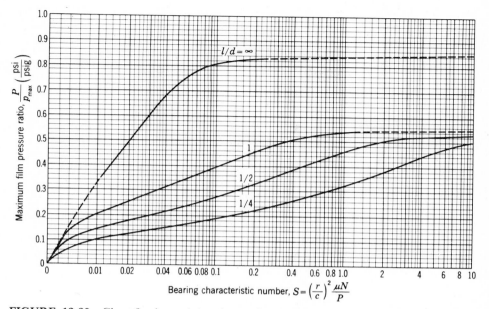

Maximum film pressure ratio, $\frac{P}{p_{\text{max}}}\left(\frac{\text{psi}}{\text{psig}}\right)$

$l/d = \infty$

1

1/2

1/4

Bearing characteristic number, $S = \left(\frac{r}{c}\right)^2 \frac{\mu N}{P}$

FIGURE 12-20 Chart for determining the maximum film pressure. (*Raimondi and Boyd.*)

Since $P = 222$ psi, the maximum pressure is found to be

$$p_{\text{max}} = \frac{P}{0.42} = \frac{222}{0.42} = 505 \text{ psi}$$

Figure 12-16 shows that the angular location of this point of maximum pressure is given by the angle $\theta_{p_{\text{max}}}$. Entering Fig. 12-21 with $S = 0.135$ and $l/d = 1$ gives $\theta_{p_{\text{max}}} = 18.5°$.

The terminating position of the oil film is θ_{p_o} according to Fig. 12-16. Entering Fig. 12-21 again we find this angle to be $\theta_{p_o} = 75°$.

Temperature Rise

Since the journal does work on the lubricant, heat is produced, as we have seen. This heat must be dissipated by conduction, convection, and radiation and carried away by the flow of oil. It is very difficult to calculate the rate of heat flow by each method with any accuracy. Later we shall examine this problem in more detail; but for the present we make the assumption that the oil flow carries away all the heat generated. Then, as far as oil temperature is concerned, we shall be on the conservative side.

The Raimondi-Boyd papers contain temperature-rise charts based on assumptions similar to these. Instead of presenting these charts in this book, we present an analytical approach based on information already obtained.

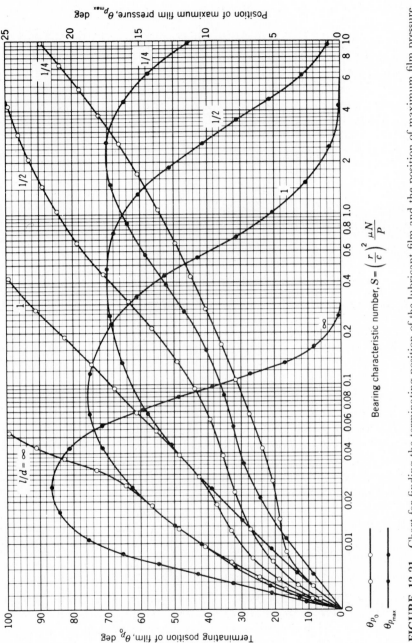

FIGURE 12-21 Chart for finding the terminating position of the lubricant film and the position of maximum film pressure. (*Raimondi and Boyd.*)

Let us use the following additional notation:

J = mechanical equivalent of heat, 9336 lbf-in per Btu

C_H = specific heat of lubricant, 0.42 Btu per lbf per °F being an average value for use

γ = weight per unit volume of the lubricant, at an average specific gravity of 0.86, $\gamma = (0.86)(62.4)/1728 = 0.0311$ lbf per in^3

ΔT_F = temperature rise, °F

$X = (r/c)f$ = friction variable

$Y = Q/rcNl$ = flow variable

The heat generated is

$$H = \frac{2\pi TN}{J} = \frac{2\pi f WrN}{J} \tag{a}$$

Substituting $(c/r)X$ for f gives

$$H = \left(\frac{2\pi WNc}{J}\right)X \tag{b}$$

We now assume that an oil flow Q is to carry away all the heat. Then the temperature rise of the oil will be

$$\Delta T_F = \frac{H}{\gamma C_H Q} \tag{c}$$

If for Q we substitute $(rcNl)Y$, then

$$\Delta T_F = \frac{H}{(\gamma C_H \, rcNl)Y} \tag{d}$$

We now multiply the numerator and denominator of Eq. (d) by the unit pressure P, noting that $P = W/2rl$, and substitute the value of H from Eq. (b). After cancelling terms, this gives

$$\Delta T_F = \frac{4\pi P}{J\gamma C_H} \frac{X}{Y} \tag{e}$$

If then we assume average lubrication conditions and substitute the values of J, γ, and C_H, we finally obtain

$$\Delta T_F = 0.103P \frac{(r/c)f}{Q/rcNl} \tag{12-16}$$

where ΔT is in degrees Fahrenheit. This equation is valid when *all* the oil flow carries away all the heat generated. But some of the oil flows out the side of the bearing before the hydrodynamic film is terminated. If we assume that the temperature of the side flow is the mean of the inlet and the outlet temperatures, the temperature rise of the side flow is $\Delta T_F/2$. This means that the heat generated raises the temperature of the flow $Q - Q_s$ an amount ΔT_F, and the flow Q_s an amount $\Delta T_F/2$. Consequently,

$$\gamma C_H(Q - Q_s)\Delta T_F + \frac{\gamma C_H Q_s \Delta T_F}{2} = H \qquad\qquad (f)$$

and so

$$\Delta T_F = \frac{H}{\gamma C_H Q[(1 - \frac{1}{2}(Q_s/Q)]} \qquad\qquad (g)$$

Equation (12-16) therefore becomes

$$\Delta T_F = \frac{0.103P}{[1 - \frac{1}{2}(Q_s/Q)]} \frac{(r/c)f}{Q/rcNl} \qquad\qquad (12\text{-}17)$$

In this equation the pressure P is in IPS units and ΔT_F in degrees Fahrenheit. The corresponding equation using SI is

$$\Delta T_C = \frac{8.30P}{[1 - \frac{1}{2}(Q_s/Q)]} \frac{(r/c)f}{Q/rcNl} \qquad\qquad (12\text{-}18)$$

where P is in MPa and ΔT_C in degrees Celsius.

Using the same data as before, Eq. (12-17) gives a temperature rise of

$$\Delta T_F = \frac{(0.103)(222)}{1 - (0.5)(0.655)} \frac{3.50}{4.28} = 26.6°\text{F}$$

Interpolation

According to Raimondi and Boyd, interpolation of the chart data for other l/d ratios can be done by using the equation

$$y = \frac{1}{(l/d)^3}\left[-\frac{1}{8}\left(1 - \frac{l}{d}\right)\left(1 - 2\frac{l}{d}\right)\left(1 - 4\frac{l}{d}\right)y_\infty + \frac{1}{3}\left(1 - 2\frac{l}{d}\right)\left(1 - 4\frac{l}{d}\right)y_1 \right.$$
$$\left. -\frac{1}{4}\left(1 - \frac{l}{d}\right)\left(1 - 4\frac{l}{d}\right)y_{1/2} + \frac{1}{24}\left(1 - \frac{l}{d}\right)\left(1 - 2\frac{l}{d}\right)y_{1/4} \right] \qquad (12\text{-}19)$$

where y is the desired variable within the interval $\infty > l/d > \frac{1}{4}$, and $y_\infty, y_1, y_{1/2}$, and $y_{1/4}$ are the variables corresponding to l/d ratios of ∞, 1, $\frac{1}{2}$, and $\frac{1}{4}$, respectively.

12-9 TEMPERATURE AND VISCOSITY CONSIDERATIONS

In a *self-contained bearing* there is no method of circulating or cooling the lubricant; it passes through the bearing, heats up, and is stored in a sump. Heat is removed by convection, conduction, and radiation, and eventually the system reaches an equilibrium temperature.

In a *force-feed lubricating system* cool, clean lubricant is fed to the bearing from an external source.

For most problems it is possible to specify the inlet temperature, but, since the viscosity used in the analysis should correspond to the mean of the inlet and outlet temperatures, this does not yield a value of viscosity to use in the analysis. As we have learned, a solution to this problem, when the lubricant grade is specified, is to assume two trial values of viscosity. To repeat, one of these should be somewhat lower than expected, the other higher. By using each of these viscosities, the temperature rise is computed and the average temperature determined from Eq. (12-13). When these pairs of results are plotted on Fig. 12-11, a straight line, such as AB, can be drawn between them, and the intersection of this line with the SAE grade of oil gives the correct viscosity to use in the analysis. It should be noted that a series of trial viscosities will yield a curved line instead of a straight line if their values differ considerably; therefore the viscosities chosen should not be too different from one another. The following example illustrates this procedure.

Example 12-1 If an SAE 20 oil at an inlet temperature of 100°F is the lubricant for the example of the preceding section, what viscosity should be used in the analysis?

Solution We have already determined that a viscosity of 4 μreyn gives a temperature rise of 26.6°F. The mean temperature is

$$T_{av} = T_1 + \frac{\Delta T_F}{2} = 100 + \frac{26.6}{2} = 113.3°F$$

This yields one point on the viscosity-temperature chart, and when plotted, it is found to be below the SAE 20 line. Therefore we choose $\mu = 6$ μreyn for the second trial value. Computing the S number gives

$$S = \left(\frac{r}{c}\right)^2 \frac{\mu N}{P} = \left(\frac{0.75}{0.0015}\right)^2 \frac{(6)(10)^{-6}(30)}{222} = 0.202$$

Then, using Figs. 12-17, 12-18, and 12-19 gives $(r/c)f = 4.7$, $Q/rcNl = 4.1$, and $Q_s/Q = 0.56$. Equation (12-17) gives

$$\Delta T_F = \frac{0.103P}{1 - \frac{1}{2}(Q_s/Q)} \frac{(r/c)f}{Q/rcNl} = \frac{(0.103)(222)}{1 - (0.5)(0.56)} \frac{4.7}{4.1} = 36.4°F$$

Therefore the average temperature is

$$T_{av} = 100 + \frac{36.4}{2} = 118.2°F$$

When both pairs of points are plotted on Fig. 12-11 and joined, they cross the SAE 20 line at $\mu = 5.5$ μreyn and $T_{av} = 117°F$. Therefore this is the correct viscosity to use in completing the analysis. The temperature rise is twice 17, or 34°F.   ////

12-10   OPTIMIZATION TECHNIQUES

In designing a journal bearing for thick-film lubrication, the engineer must select the grade of oil to be used, together with suitable values for P, N, r, c, and l. A poor selection of these or inadequate control of them during manufacture or in use may result in a film that is too thin, so that the oil flow is insufficient, causing the bearing to overheat and, eventually, fail. Furthermore, the radial clearance c is difficult to hold accurate during manufacture, and it may increase because of wear. What is the effect of an entire range of radial clearances, expected in manufacture, and what will happen to the bearing performance if c increases because of wear? Most of these questions can be answered and the design optimized by plotting curves of the performance as functions of the quantities over which the designer has control.

Figure 12-22 shows the results obtained when the performance of a particular bearing is calculated for a whole range of radial clearances and is plotted with clearance as the independent variable. The bearing used for this graph is the one of Example 12-1, with SAE 20 oil at an inlet temperature of 100°F. The graph shows that if the clearance is too tight, the temperature will be too high and the minimum film thickness too low. High temperatures may cause the bearing to fail by fatigue. If the oil film is too thin, dirt particles may not pass without scoring or may embed themselves in the bearing. In either event, there will be excessive wear and friction, resulting in high temperatures and possible seizing.

It would seem that a large clearance will permit the dirt to pass through and also will permit a large flow of oil. This lowers the temperature and increases the life of the bearing. However, if the clearance becomes too large, the bearing becomes noisy and the minimum film thickness begins to decrease again.

When both the production tolerance and the future wear on the bearing are

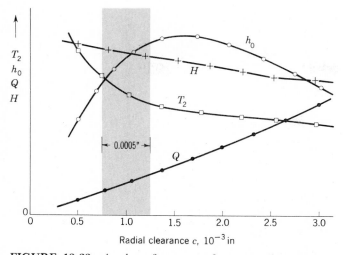

Radial clearance c, 10^{-3} in

FIGURE 12-22 A plot of some performance characteristics of the bearing of Example 12-1 for radial clearances of 0.0005 to 0.003 in. The bearing outlet temperature is designated T_2. New bearings should be designed for the shaded zone because wear will move the operating point to the right.

considered, it is seen, from Fig. 12-22, that the best compromise is a clearance range slightly to the left of the top of the minimum-film-thickness curve. In this way, future wear will move the operating point to the right and increase the film thickness and decrease the operating temperature.

Of course, if the tightest acceptable clearance or the maximum allowable temperature rise is specified, the statistical methods introduced in Chap. 5 may be used to determine the percentage of unacceptable bearings to be expected in a given lot or assembly.

12-11 PRESSURE-FED BEARINGS

When so much heat is generated by hydrodynamic action that the normal lubricant flow is insufficient to carry it away, an additional supply of lubricant must be furnished under pressure. To force a maximum flow through the bearing and thus obtain the greatest cooling effect, a common practice is to use a circumferential groove at the center of the bearing, with an oil-supply hole located opposite the load-bearing zone. Such a bearing is shown in Fig. 12-23. The effect of the groove is to create two half-bearings, each having a smaller l/d ratio than the original. The groove divides the pressure-distribution curve into two lobes and reduces the minimum film thickness, but it has wide acceptance among lubrication engineers and carries more load without overheating.

To set up a method of solution for oil flow, we shall assume a groove ample enough so that the pressure drop in the groove itself is small. Initially we will

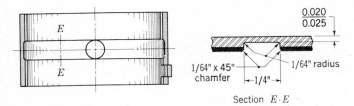

FIGURE 12-23 Centrally located, full annular groove. (*Courtesy of the Cleveland Graphite Bronze Company, Division of Clevite Corporation.*)

neglect eccentricity and then apply a correction factor for this condition. The oil flow, then, is the amount which flows out of the two halves of the bearing in the direction of the concentric shaft. If we neglect the rotation of the shaft, we obtain the force situation shown in Fig. 12-24. Here we designate the supply pressure by p_s and the pressure at any point by p. Laminar flow is assumed, and we are interested in the static equilibrium of an element of width dx, thickness $2y$, and unit depth. Note particularly that the origin of the reference system has been chosen at the midpoint of the clearance space.* The pressure is $p + dp$ on the left face and p on the right face, and the upper and lower surfaces are acted upon by the shear stresses τ. The equilibrium equation is

$$2y(p + dp) - 2yp - 2\tau\,dx = 0 \qquad (a)$$

Expanding and canceling terms, we find that

$$\tau = y\,\frac{dp}{dx} \qquad (b)$$

Newton's law for viscous flow [Eq. (12-1)] is

$$\tau = \mu\,\frac{du}{dy}$$

However, in this case we have taken τ in a negative direction. Also, du/dy is negative because u decreases as y increases. We therefore write Newton's law in the form

$$-\tau = \mu\left(-\frac{du}{dy}\right) \qquad (c)$$

---

* We are grateful to Professors Arthur W. Sear of California State University, Los Angeles, and Charles R. Mischke of Iowa State University, Ames, for suggestions concerning this analysis. J.E.S., L.D.M.

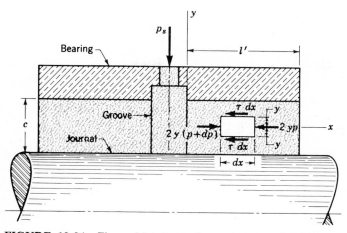

FIGURE 12-24 Flow of lubricant from a pressure-fed bearing having a central groove.

Now eliminating τ from Eqs. (b) and (c) gives

$$\frac{du}{dy} = \frac{1}{\mu} \frac{dp}{dx} y \qquad\qquad (d)$$

Treating dp/dx as a constant and integrating with respect to y gives

$$u = \frac{1}{2\mu} \frac{dp}{dx} y^2 + C_1 \qquad\qquad (e)$$

At the boundaries, where $y = \pm c/2$, the velocity u is zero. Using one of these conditions in Eq. (e) gives

$$0 = \frac{1}{2\mu} \frac{dp}{dx} \left(\frac{c}{2}\right)^2 + C_1$$

or

$$C_1 = -\frac{c^2}{8\mu} \frac{dp}{dx}$$

Substituting this constant into Eq. (e) yields

$$u = \frac{1}{8\mu} \frac{dp}{dx} (4y^2 - c^2) \qquad\qquad (f)$$

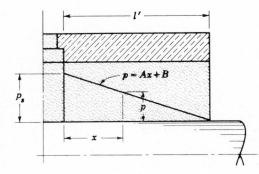

FIGURE 12-25 A linear oil-pressure distribution is assumed.

Next let us assume that the oil pressure varies linearly from the center to the end of the bearing as shown in Fig. 12-25. Since the equation of a straight line may be written

$$p = Ax + B$$

with $p = p_s$ at $x = 0$ and $p = 0$ at $x = l'$, then substituting these end conditions gives

$$A = -\frac{p_s}{l'} \qquad B = p_s$$

or

$$p = -\frac{p_s}{l'} x + p_s \tag{g}$$

And therefore

$$\frac{dp}{dx} = -\frac{p_s}{l'} \tag{h}$$

We can now substitute Eq. (h) into Eq. (f) to get the relationship between the oil velocity and the coordinate y:

$$u = \frac{p_s}{8\mu l'} (c^2 - 4y^2) \tag{12-20}$$

Figure 12-26 shows a graph of this relation fitted into the clearance space c so that you can see how the velocity of the lubricant varies from the journal surface to the bearing surface. The distribution is parabolic, as shown, with the maximum

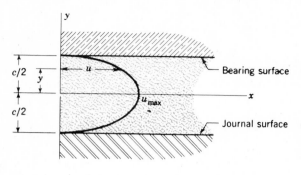

FIGURE 12-26 Parabolic distribution of the lubricant velocity.

velocity occurring at the center, where $y = 0$. The magnitude is, from Eq. (12-20),

$$u_{max} = \frac{p_s c^2}{8\mu l'} \qquad (i)$$

The average ordinate of a parabola is two-thirds of the maximum; if we also generalize by substituting h for c in Eq. (i), then the average velocity at any angular position θ (Fig. 12-27) is

$$u_{av} = \frac{2}{3} \frac{p_s h^2}{8\mu l'} = \frac{p_s}{12\mu l'} (c - e \cos \theta)^2 \qquad (j)$$

We still have a little further to go in this analysis; so be patient. Now that we have an expression for the lubricant velocity, we can compute the amount of lubricant that flows out both ends; so the elemental side flow at any position θ

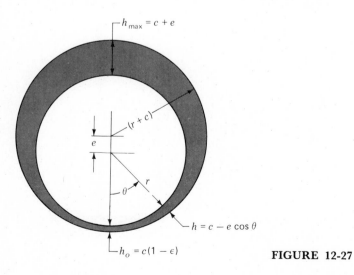

FIGURE 12-27

(Fig. 12-27) is

$$dQ_s = 2u_{av} \, dA = 2u_{av}(rh \, d\theta) \tag{k}$$

where dA is the elemental area. Substituting u_{av} from Eq. (j) and h from Fig. 12-27 gives

$$dQ_s = \frac{p_s r}{6\mu l'} (c - e \cos \theta)^3 \, d\theta \tag{l}$$

Integrating around the bearing gives the total side flow as

$$Q_s = \int dQ_s = \frac{p_s r}{6\mu l'} \int_0^{2\pi} (c - e \cos \theta)^3 \, d\theta$$

$$= \frac{\pi p_s r c^3}{3\mu l'} (1 + 1.5\epsilon^2) \tag{12-21}$$

In analyzing the performance of pressure-fed bearings, the bearing length should be taken as l', as defined in Fig. 12-24. Thus the unit load is

$$P = \frac{W/2}{2rl'} = \frac{W}{4rl'} \tag{12-22}$$

because each half of the bearing carries half of the load.

The charts of Figs. 12-18 and 12-19 for flow variable and flow ratio, of course, do not apply to pressure-fed bearings. Also, to the maximum film pressure, given by Fig. 12-20, must be added the supply pressure p_s, to obtain the total film pressure.

Since the oil flow has been increased by forced-feed, Eq. (12-17) will give a temperature rise that is too high. From Eq. (c) of Sec. 12-8, we have

$$\Delta T_F = \frac{H}{\gamma C_H Q_s} \tag{m}$$

and the heat loss is

$$H = \frac{2\pi f W r N}{J} \tag{n}$$

Substituting Eqs. (12-21) and Eq. (n) into Eq. (m) and canceling terms gives

$$\Delta T_F = \frac{6\mu l' f W N}{(1 + 1.5\epsilon^2) J \gamma C_H p_s c^3} \tag{12-23}$$

For average lubrication conditons, $\gamma = 0.0311$ lbf per in^3, $C_H = 0.42$ Btu per lbf per °F; also $J = 9336$ lbf·in per Btu, and so Eq. (12-23) can be written

$$\Delta T_F = \frac{0.0492\mu l' f WN}{(1 + 1.5\epsilon^2)p_s c^3} \tag{o}$$

Now multiply Eq. (o) by the Sommerfeld number S and divide by

$$S = \left(\frac{r}{c}\right)^2 \frac{\mu N}{P} = \left(\frac{r}{c}\right)^2 \frac{4rl'\mu N}{W} \tag{p}$$

Upon rearranging the terms, we find

$$\Delta T_F = \frac{0.0123}{1 + 1.5\epsilon^2} \frac{[(r/c) f]SW^2}{p_s r^4} \tag{12-24}$$

which is easier to solve than Eq. (o) because the number S must be computed anyway. Equation (12-24) is, of course, in the customary IPS units. The corresponding equation in SI is

$$\Delta T_C = \frac{978(10)^6}{1 + 1.5\epsilon^2} \frac{[(r/c) f]SW^2}{p_s r^4} \tag{12-25}$$

where ΔT_C = temperature rise, °C
$\quad\quad W$ = bearing load, kN
$\quad\quad p_s$ = supply pressure, kPa
$\quad\quad r$ = radius, mm

12-12 HEAT BALANCE

The case in which the lubricant carries away all the generated heat has already been discussed. We shall now investigate self-contained bearings, in which the lubricant is stored in the bearing housing itself. These bearings find many applications in industrial machinery; are often described as pedestal, or pillow-block, bearings; and are used on fans, blowers, pumps, motors, and the like. The problem is to balance the heat-dissipation capacity of the housing with the heat generated in the bearing itself.

The heat given up by the bearing housing may be approximated by the equation

$$H = CA(T_H - T_A) \tag{12-26}$$

where H = heat dissipated, Btu per h

C = combined coefficient of radiation and convection, Btu per h per ft^2 per °F

A = surface area of housing, ft^2

T_H = surface temperature of housing, °F

T_A = temperature of surrounding air, °F

The coefficient C depends upon the material, color, geometry, and roughness of the housing, the temperature difference between the housing and the surrounding objects, and the temperature and velocity of the air. Equation (12-26) should be used only when "ball-park" answers are sufficient. Exact results can be obtained by experimentation under actual, not simulated, operating conditions and environment. With these limitations, assume C is a constant having the values

$$C = \begin{cases} 2 \text{ Btu/(h)(ft}^2)(°F) & \text{for still air} \\ 2.7 \text{ Btu/(h)(ft}^2)(°F) & \text{for average design practice} \\ 5.9 \text{ Btu/(h)(ft}^2)(°F) & \text{for air moving at 500 fpm} \end{cases}$$

An expression quite similar to Eq. (12-26) can be written for the temperature difference $T_L - T_H$ between the lubricant film and the housing. Because the type of lubricating system and the quality of the lubricant circulation affect this relationship, the resulting expression is even more approximate than that of Eq. (12-26). An *oil-bath lubrication system* in which a part of the journal is actually immersed in the lubricant would provide good circulation. A *ring-oiled bearing* in which oil rings ride on top of the journal, dip into the oil sump, and hence carry a moderate amount of lubricant into the load-bearing zone would provide satisfactory circulation for many purposes. On the other hand, if the lubricant is supplied by *wick-feeding* methods, the circulation is so inadequate that it is doubtful if any heat at all can be carried away by the lubricant. No matter what type of self-contained lubrication system is used, a great deal of engineering judgment is necessary in computing the heat balance. Based on these limitations, the equation

$$T_L - T_H = B(T_H - T_A) \tag{a}$$

where T_L is the *average* film temperature and B is a constant which depends upon the lubrication system, may be used to get a rough estimate of the bearing temperature. Table 12-1 provides some guidance in deciding on a suitable value for B.

Since T_L and T_A are usually known, Eqs. (12-26) and (a) can be combined to give

$$H = \frac{CA}{B+1}(T_L - T_A) \tag{12-27}$$

In beginning a heat-balance computation, the film temperature, and hence

Table 12-1

Lubrication system	Conditions	Range of B
Oil ring	Moving air	1–2
	Still air	$\frac{1}{2}$–1
Oil bath	Moving air	$\frac{1}{2}$–1
	Still air	$\frac{1}{5}$–$\frac{2}{5}$

the viscosity of the lubricant, in a self-contained bearing is unknown. Thus, finding the equilibrium temperatures is an iterative procedure which starts with an estimate of the film temperature and ends with verification or nonverification of this estimate. Since the computations are lengthy, a computer should be used to make them if at all possible.

12-13 BEARING DESIGN

A typical sleeve-bearing design situation is depicted in Fig. 12-28. Here a rotating shaft is to be supported by bearings at A and B. It is evident that some of the decisions have been fixed by other considerations, such as the shaft dimensions, heat treatment, speed, and general geometry. It is also evident that the problem is not completely revealed. What is the purpose of the shaft? What is it that causes the external forces? Is the shaft completely enclosed by a housing or is it in the open? Are the bearings to be self-contained or does the lubricant come from a sump, and is it also used for other purposes? Once the answers to these questions are known, the design process can begin.

The diameter and length of the bearing depend upon the magnitude of the unit load. While the experienced bearing designer will have a rather good idea of a satisfactory range, the beginner needs a starting point. Table 12-2 is presented to indicate the range of unit loads in current use. These values will have to be modified upward or downward, depending upon the severity of the operating conditions, but they can be used to obtain a first-trial value of P. Having settled

FIGURE 12-28

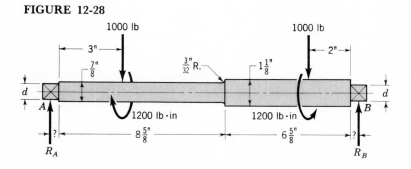

Table 12-2 RANGE OF UNIT LOADS
                IN CURRENT USE FOR
                SLEEVE BEARINGS

Application	Unit load	
	psi	MPa
Diesel engines:		
Main bearings	900–1700	6–12
Crankpin	1150–2300	8–15
Wristpin	2000–2300	14–15
Electric motors	120–250	0.8–1.5
Steam turbines	120–250	0.8–1.5
Gear reducers	120–250	0.8–1.5
Automotive engines:		
Main bearings	600–750	4–5
Crankpin	1700–2300	10–15
Air compressors:		
Main bearings	140–280	1–2
Crankpin	280–500	2–4
Centrifugal pumps	100–180	0.6–1.2

on a value of the unit load, suitable values for the bearing diameter d and the length l can be selected.

The next decision is what to use for the radial clearance. This depends somewhat on the bearing material as well as on the finish used and the relative velocity. For preliminary design use the figures given in Table 12-3. As an additional guide, the Cast Bronze Bearing Institute (CBBI)* has published a list of recommended radial clearances for full bronze bearings with various degrees of finish. These recommendations, which permit up to 20 percent deviation, are summarized by the graph of Fig. 12-29.

The length-diameter ratio l/d of a bearing depends upon whether it is expected to run under thin-film-lubrication conditions. A long bearing (large l/d ratio) reduces the coefficient of friction and the side flow of oil and therefore is desirable where thin-film or boundary-value lubrication is present. On the other hand, where forced-feed or positive lubrication is present, the l/d ratio should be relatively small. The short bearing results in a greater flow of oil out of the ends, and thus keeping the bearing cooler. Current practice is to use an l/d ratio of about unity, in general, and then to increase this ratio if thin-film lubrication is likely to occur and to decrease it for thick-film lubrication or high temperatures. If shaft deflection is likely to be severe, a short bearing should be used to prevent metal-to-metal contact at the ends of the bearings.

You should always consider the use of a partial bearing if high temperatures are a problem, because relieving the nonload-bearing area of a bearing can very substantially reduce the heat generated.

---

* Harry C. Rippel, *Cast Bronze Bearing Design Manual*, 2d ed., International Copper Research Assoc., New York, 1965, pp. 12 and 13.

Table 12-3   SOME CHARACTERISTICS OF BEARING ALLOYS

Alloy name	Thickness, in	SAE number	Clearance ratio r/c	Load capacity	Corrosion resistance
Tin-base babbit	0.022	12	600–1000	1.0	Excellent
Lead-base babbit	0.022	15	600–1000	1.2	Very good
Tin-base babbit	0.004	12	600–1000	1.5	Excellent
Lead-base babbit	0.004	15	600–1000	1.5	Very good
Leaded bronze	Solid	792	500–1000	3.3	Very good
Copper-lead	0.022	480	500–1000	1.9	Good
Aluminum alloy	Solid		400–500	3.0	Excellent
Silver plus overlay	0.013	17P	600–1000	4.1	Excellent
Cadmium (1.5% Ni)	0.022	18	400–500	1.3	Good
Trimetal 88*				4.1	Excellent
Trimetal 77†				4.1	Very good

* This is a 0.008-in layer of copper-lead on a steel back plus 0.001 in of tin-base babbit.

† This is a 0.013-in layer of copper-lead on a steel back plus 0.001 in of lead-base babbit.

FIGURE 12-29   Recommended radial clearances for cast-bronze bearings. The curves are identified as follows:

A—precision spindles made of hardened ground steel, running on lapped cast-bronze bearings (8- to 16-μin rms finish), with a surface velocity less than 10 fps

B—precision spindles made of hardened ground steel, running on lapped cast-bronze bearings (8- to 16-μin rms finish), with a surface velocity more than 10 fps

C—electric motors, generators, and similar types of machinery using ground journals in broached or reamed cast-bronze bearings (16- to 32-μin rms finish)

D—general machinery which continuously rotates or reciprocates and uses turned or cold-rolled steel journals in bored and reamed cast-bronze bearings (32- to 63-μin rms finish)

E—rough-service machinery having turned or cold-rolled steel journals operating on cast-bronze bearings (65- to 125-μin rms finish)

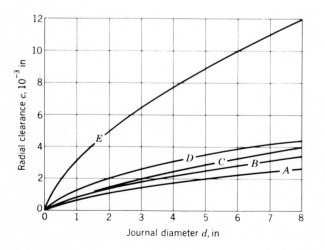

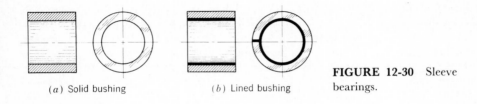

(a) Solid bushing   (b) Lined bushing   **FIGURE 12-30**   Sleeve
                                                       bearings.

With all these tentative decisions made, a lubricant can be selected and the
hydrodynamic analysis made as already presented.   The values of the various
performance parameters, if plotted as in Fig. 12-22, for example, will then indicate
whether a satisfactory design has been achieved or additional iterations are
necessary.

12-14   BEARING TYPES

A bearing may be as simple as a hole machined into a cast-iron machine member.
It may still be simple yet require detailed design procedures, as, for example, the
two-piece, grooved, pressure-fed, connecting-rod bearing in an automotive engine.
Or it may be as elaborate as the large, water-cooled, ring-oiled bearings with
built-in oil reservoirs used on heavy machinery.

Figure 12-30 shows two types of bearings which are often called bushings.
The solid bushing is made by casting, by drawing and machining, or by using a
powder-metallurgy process.   The lined bushing is usually a split type.   In one
method of manufacture the molten lining material is cast continuously on thin
strip steel.   The babbitted strip is then processed through presses, shavers, and
broaches, resulting in a lined bushing.   Any type of grooving may be cut into the
bushings.   Bushings are assembled as a press fit and finished by boring, reaming,
or burnishing.

Flanged and straight two-piece bearings are shown in Fig. 12-31.   These are
available in many sizes in both thick- and thin-wall types, with or without lining
material.   A locking lug positions the bearing and effectively prevents axial or
rotational movement of the bearing in the housing.

Some typical groove patterns are shown in Fig. 12-32.   In general, the lubricant may be brought in from the end of the bushing, through the shaft, or through

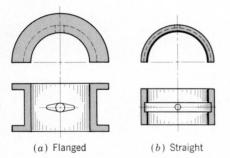

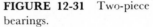

(a) Flanged   (b) Straight   **FIGURE 12-31**   Two-piece
                                            bearings.

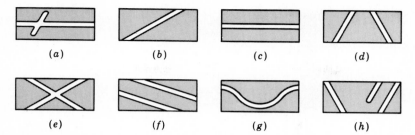

(a)   *(b)*   *(c)*   *(d)*

(e)   *(f)*   *(g)*   *(h)*

FIGURE 12-32 Developed views of typical groove patterns. (*Courtesy of the Cleveland Graphite Bronze Company, Division of Clevite Corporation.*)

the bushing. The flow may be intermittent or continuous. The preferred practice is to bring the oil in at the center of the bushing so that it will flow out both ends, thus increasing the flow and cooling action.

12-15 THRUST BEARINGS

This chapter is devoted to the study of the mechanics of lubrication and its application to the design and analysis of journal bearings. The design and analysis of thrust bearings is an important application of lubrication theory, too. A detailed study of thrust bearings is not included here because it would not contribute anything significantly different and because of space limitations. Having studied this chapter, you should experience no difficulty in reading the literature on thrust bearings and applying that knowledge to actual design situations.*

Figure 12-33 shows a fixed-pad thrust bearing consisting essentially of a runner sliding over a fixed pad. The lubricant is brought into the radial grooves

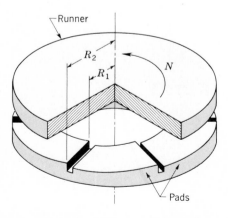

FIGURE 12-33 Fixed-pad thrust bearing. (*Courtesy of the Westinghouse Corporation Research Laboratories.*)

---

* Harry C. Rippel, "Cast Bronze Thrust Bearing Design Manual," International Copper Research Association, Inc., 825 Third Ave., New York, NY 10022, 1967. CBBI, 14600 Detroit Ave., Cleveland, OH, 44107, 1967.

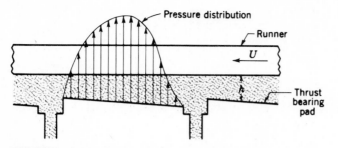

FIGURE 12-34   Pressure distribution of lubricant in a thrust bearing.   (*Courtesy of International Copper Research Corporation.*)

and pumped into the wedge-shaped space by the motion of the runner.   Full-film, or hydrodynamic, lubrication is obtained if the speed of the runner is continuous and sufficiently high, if the lubricant has the correct viscosity, and if it is supplied in sufficient quantity.   Figure 12-34 provides a picture of the pressure distribution under conditions of full-film lubrication.

We should note that bearings are frequently made with a flange, as shown in Fig. 12-35.   The flange positions the bearing in the housing and also takes a thrust load.   Even when it is grooved, however, and has adequate lubrication, such an arrangement is not a hydrodynamically lubricated thrust bearing.   The reason for this is that the clearance space is not wedge-shaped but has a uniform thickness. Similar reasoning would apply to various designs of thrust washers.

12-16   BEARING MATERIALS

When we refer to a lubricated assembly, it is necessary to distinguish between the following elements:

1   The shaft, journal, or moving member
2   The lubricant, if it exists
3   The bearing, or fixed member
4   The housing in which the bearing is assembled

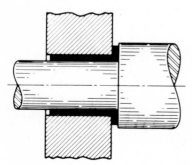

FIGURE   12-35   Flanged sleeve bearing takes both radial and thrust loads.

The two conflicting requirements of a good bearing material are that it must have a satisfactory compressive and fatigue strength to resist the externally applied loads and that it must be soft and have a low melting point and a low modulus of elasticity. The second set of requirements is necessary to permit the material to wear or break in, since the material can conform to slight irregularities and absorb and release foreign particles. The resistance to wear and the coefficient of friction are also important because all bearings must operate, at least for part of the time, with thin-film lubrication.

Additional considerations in the selection of a good bearing material are its ability to resist corrosion and, of course, the cost of producing the bearing. Some of the commonly used materials are listed in Table 12-3, together with their composition and characteristics.

Bearing life can be increased very substantially by depositing a layer of babbitt, or other white metal, in thicknesses from 0.001 to 0.014 in over steel backup material. In fact, a copper-lead on steel to provide strength, with a babbitt overlay to provide surface and corrosion characteristics, makes an excellent bearing.

Small bushings and thrust collars are often expected to run with thin-film lubrication. When this is the case, improvements over a solid bearing material can be made to add significantly to the life. A powder-metallurgy bushing is porous and permits the oil to penetrate into the bushing material. Sometimes such a bushing may be enclosed by oil-soaked material to provide additional storage space. Bearings are frequently ball-indented to provide small basins for the storage of lubricant while the journal is at rest. This supplies some lubrication during starting. Another method of reducing friction is to indent the bearing wall and to fill these indentations with graphite.

12-17 BOUNDARY-LUBRICATED BEARINGS

When two surfaces slide relative to each other with only a partial lubricant film between them, *boundary lubrication* is said to exist. Boundary- or thin-film lubrication occurs in hydrodynamically lubricated bearings when they are starting or stopping, when the load increases, when the supply of lubricant decreases, or whenever other operating changes happen to occur. There are, of course, a very large number of cases in design in which boundary-lubricated bearings must be used because of the type of application or the competitive situation.

The coefficient of friction for boundary-lubricated surfaces may be greatly decreased by the use of animal or vegetable oils mixed with the mineral oil or grease. Fatty acids, such as stearic acid, palmitic acid, or oleic acid, or several of these, which occur in animal and vegetable fats, are called *oiliness agents*. These acids appear to reduce friction, either because of their strong affinity for certain metallic surfaces or because they form a soap film which binds itself to the metallic surfaces by a chemical reaction. Thus the fatty-acid molecules bind themselves to the journal and bearing surfaces with such great strength that the metallic asperities of the rubbing metals do not weld or shear.

Fatty acids will break down at temperatures of 250°F or more, causing increased friction and wear in thin-film-lubricated bearings. In such cases the *extreme-pressure*, or EP, lubricants may be mixed with the fatty-acid lubricant. These are composed of chemicals such as chlorinated esters or tricresyl phosphate, which form an organic film between the rubbing surfaces. Though the EP lubricants make it possible to operate at higher temperatures, there is the added possibility of excessive chemical corrosion of the sliding surfaces.

When a bearing operates partly under hydrodynamic conditions and partly under dry or thin-film conditions, a *mixed-film lubrication* exists. If the lubricant is supplied by hand oiling, by drop or mechanical feed, or by wick feed, for example, the bearing is operating under mixed-film conditions. In addition to occurring with a scarcity of lubricant, mixed-film conditions may be present when

- The viscosity is too low.
- The bearing speed is too low.
- The bearing is overloaded.
- The clearance is too tight.
- Journal and bearing are not properly aligned.

The range of operating conditions encountered is so great that it is virtually impossible to devise reliable design procedures. The best approach is that of establishing design guidelines or targets and then designing to meet or exceed these standards.

One method of design is based on the ability of the bearing to dissipate heat; after all, a cool bearing is likely to have a long life. In this approach a *PV* value is computed using the equation

$$PV = \frac{k(T_B - T_A)}{f_M} \tag{12-28}$$

where P = load per unit of projected bearing area, psi
V = surface velocity of journal relative to bearing surface, fpm
T_A = ambient air temperature, °F
T_B = bearing bore temperature, °F
f_M = coefficient of mixed-film friction

Table 12-4 shows some of the materials commonly used when dry or mixed-film conditions are present. Note that all quantities listed are maximum values. However, they cannot all be maximum at the same time.

The coefficient of friction used depends upon whether there is any lubrication at all. Figure 12-36 is a graph of suggested frictional coefficients plotted against the percentage of mixed lubrication. Except for the graph at $f_B = 0.20$, these are values recommended by Rippel.*

---

* Ibid

Table 12-4 SOME MATERIALS FOR BOUNDARY-LUBRICATED BEAR-INGS AND THEIR OPERATING LIMITS

Material	Maximum load, psi	Maximum temperature, °F	Maximum speed, fpm	Maximum PV value*
Cast bronze	4 500	325	1 500	50 000
Porous bronze	4 500	150	1 500	50 000
Porous iron	8 000	150	800	50 000
Phenolics	6 000	200	2 500	15 000
Nylon	1 000	200	1 000	3 000
Teflon	500	500	100	1 000
Reinforced Teflon	2 500	500	1 000	10 000
Teflon fabric	60 000	500	50	25 000
Delrin	1 000	180	1 000	3 000
Carbon-graphite	600	750	2 500	15 000
Rubber	50	150	4 000	
Wood	2 000	150	2 000	15 000

* P = load, psi; V = speed, fpm.

The constant k in Eq. (12-28) depends upon the ability of the bearing to dissipate heat. The best way to evaluate k is to analyze the bearing performance in known situations or to obtain the value from previous designs known to be satisfactory. Here we can estimate a value of k by using minimum f_M and maximum PV and $(T_B - T_A)$. Using values for cast bronze from Table 12-4,

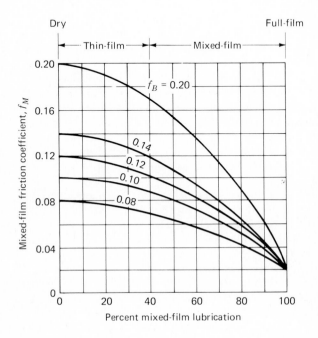

FIGURE 12-36 Coefficient of friction corresponding to various percentages of mixed-film lubrication. Coefficient of dry friction is f_B. Coefficient of mixed-film friction is f_M. The lower 40 percent is the approximate range of thin-film lubrication.

$f_M = 0.2$ from Fig. 12-36, and $T_A = 75°F$ gives

$$k = \frac{f_M (PV)_{max}}{(T_B - T_A)_{max}} = \frac{0.02(50\ 000)}{325 - 75} = 4.0$$

Thus the equation

$$PV = \frac{4(T_B - T_A)}{f_M} \tag{12-29}$$

can be used as a guideline in designing cast bronze bearings for mixed- or dry-film conditions

PROBLEMS

Section 12-8

12-1 A full journal bearing is 2 in long and 2 in in diameter. The bearing load is 700 lb, and the journal runs at 1200 rpm. Using a clearance of 0.001 in and an average viscosity of 20 μreyn, find the friction horsepower.

12-2 Purchase a quart of your favorite engine oil, say a multiviscosity grade, and determine the viscosity in your lubrication laboratory according to ASTM standards. Plot the absolute viscosity as a function of temperature on the chart of Fig. 12-11 for later use.

12-3 An 8-in-diameter bearing is 4 in long, has a load of 7500 lb, and runs at 900 rpm. Using a radial clearance of 0.004 in, find the friction horsepower for the following lubricants: SAE 10, 20, 30, and 40. Use an operating temperature of 160°F.

12-4 Repeat Prob. 12-3, but use an SAE 40 lubricant and the following clearances: 0.002, 0.003, 0.004, 0.005, and 0.006 in. Plot a curve showing the relation between the coefficient of friction and the clearance.

12-5 A 3-in-diameter bearing has a journal speed of 400 rpm, is 3 in long, and is subjected to a radial load of 600 lb. The bearing is lubricated with SAE 30 oil, which flows into the bearing at a temperature of 160°F. The radial clearance is 0.0014 in. Calculate the heat loss, the side flow, the total flow, the minimum film thickness, and the temperature rise.

12-6 A $1\frac{1}{4} \times 1\frac{1}{4}$-in sleeve bearing supports a load of 700 lb and has a journal speed of 3600 rpm. Using an SAE 10 oil at 160°F operating temperature, specify the radial clearance for an h_o/c value of 0.662.

12-7 A sleeve bearing has a diameter of 75 mm, a journal speed of 7 Hz, and a length of 75 mm. The oil supply is SAE 30 at an inlet temperature of 70°C. The bearing carries a radial load of 2.7 kN and has a radial clearance of 35 μm. Calculate the heat loss, the side flow, the total flow, the minimum film thickness, and the temperature rise.

12-8 A sleeve bearing is 32 mm in diameter and 32 mm long and has a journal speed of 60 Hz. The bearing supports a radial load of 3 kN. SAE 10 oil at an average operating temperature of 60°C is used. Determine the radial clearance for an h_o/c value of 0.50.

12-9 A sleeve bearing is $\frac{3}{4}$ in long and $1\frac{1}{2}$ in diameter. It has a clearance of 0.0015 in and uses SAE 20 lubricant at an operating temperature of 140°F. The bearing supports a load of 350 lb. Calculate the heat generated for speeds of 1000, 2000, 3000, and 4000 rpm, and construct a graph of the results.

Section 12-9

12-10 A bearing $1\frac{1}{2}$ in long and $1\frac{1}{2}$ in in diameter has an r/c ratio of 1000. The journal speed is 1200 rpm, the load is 600 lb, and the lubricant is SAE 40 at an inlet temperature of 100°F.
    (*a*)   Find the minimum film thickness and the oil outlet temperature.
    (*b*)   Determine the magnitude and location of the maximum film pressure.

12-11 An SAE 60 oil at 80°F inlet temperature is used to lubricate a sleeve bearing 6 in long and 2 in in diameter ($1/d = \infty$). The bearing load is 1800 lb and the journal speed is 160 rpm. Using a clearance ratio $r/c = 600$ find the temperature rise, the maximum film pressure and the minimum film thickness.

12-12 A sleeve bearing is $\frac{3}{8}$ in in diameter and $\frac{3}{8}$ in long. SAE 10 oil at an inlet temperature of 120°F is used to lubricate the bearing. The radial clearance is 0.0003 in. If the journal speed is 3600 rpm and the radial load on the bearing is 15 lb, find the temperature rise of the lubricant and the minimum film thickness.

12-13 A sleeve bearing is $1\frac{1}{4}$ in in diameter and $1\frac{1}{4}$ in long. The shaft rotates at 1750 rpm and subjects the bearing to a radial load of 250 lb. The clearance is 0.000 75 in. Using SAE 30 oil at an initial temperature of 120°F, find the temperature rise and the minimum film thickness.

12-14 Repeat Prob. 12-13 for SAE 10, 20, and 40 oils, and compare the results. Which lubricant is the best to use?

12-15 A sleeve bearing is 38 mm in diameter and has an l/d ratio of unity. Other data include a clearance ratio of 1000, a radial load of 2.5 kN, and a journal speed of 20 s^{-1}. The bearing is supplied with SAE 40 lubricant at an inlet temperature of 35°C.
    (*a*)   Find the average oil temperature.
    (*b*)   What is the minimum film thickness?
    (*c*)   Find the maximum oil-film pressure.

12-16 A sleeve bearing 60 mm in diameter and 60 mm long is lubricated using SAE 30 oil at an inlet temperature of 40°C. The bearing supports a 4-kN radial load and has a journal speed of 1120 rpm. The radial clearance is 45 μm.
    (*a*)   Find both the temperature rise and the average temperature of the lubricant.
    (*b*)   Find the coefficient of friction.
    (*c*)   Find the magnitude and location of the minimum oil-film thickness.
    (*d*)   Find the side flow and the total flow.
    (*e*)   Determine the maximum oil-film pressure and its angular location.
    (*f*)   Find the terminating position of the oil film.

12-17 An SAE 20 oil is used to lubricate a sleeve bearing 3 in long and 3 in in diameter. The oil enters the bearing at a temperature of 100°F. The journal rotates at 1200 rpm, and the bearing supports a radial load of 1500 lb. The radial clearance is 0.0015 in.
    (*a*)   Find the magnitude and location of the minimum oil-film thickness.
    (*b*)   Find the coefficient of friction.
    (*c*)   Compute the side flow and the total oil flow.

    (*d*)   Find the maximum oil-film pressure and its location.

    (*e*)   Find the terminating position of the oil film.

    (*f*)   Find the average temperature of the oil flowing out the sides of the bearing and the temperature of the oil at terminating position of the film.

12-18 A journal bearing has a diameter of $2\frac{1}{2}$ in and a length of $1\frac{1}{4}$ in. The journal is to operate at a speed of 1800 rpm and carry a load of 750 lb. If SAE 20 oil at an inlet temperature of 110°F is to be used, what should be the radial clearance for optimum load-carrying capacity?

Section 12-11

12-19 A $1\frac{3}{4}$-in-diameter bearing is 2 in long and has a central annular oil groove $\frac{1}{4}$ in wide which is fed by SAE 10 oil at 120°F and 30 psi supply pressure. The radial clearance is 0.0015 in. The journal rotates at 3000 rpm and the average load is 600 psi of projected area. Find the temperature rise, the minimum film thickness, and the maximum film pressure.

12-20 An eight-cylinder diesel engine has a front main bearing $3\frac{1}{2}$ in in diameter and 2 in long. The bearing has a central annular oil groove 0.250 in wide. It is pressure-lubricated with SAE 30 oil at an inlet temperature of 180°F and at a supply pressure of 50 psi. Corresponding to a radial clearance of 0.0025 in, a speed of 2800 rpm, and a radial load of 4600 lb, find the temperature rise and the minimum oil-film thickness.

12-21 A 50-mm-diameter bearing is 55 mm long and has a central annular oil groove 5 mm wide which is fed by SAE 30 oil at 55°C and 200-kPa supply pressure. The radial clearance is 42 μm. The journal speed is $48\,\text{s}^{-1}$ corresponding to a bearing load of 10 kN. Find the temperature rise of the lubricant, the total oil flow, and the minimum film thickness.

CHAPTER
13

SPUR GEARS

We study gears because the transmission of rotary motion from one shaft to another occurs in nearly every machine one can imagine. Gears constitute one of the best of the various means available for transmitting this motion.

When you realize that the gears in, say, your automotive differential can be made to run 100 000 miles or more before they need to be replaced, and when you count the actual number of meshes or revolutions, you begin to appreciate the fact that the design and manufacture of these gears is really a remarkable accomplishment. People do not generally realize how highly developed the design, engineering, and manufacture of gearing has become because gears are such ordinary machine elements. There are a great many lessons to be learned about engineering and design in general through the study of gears, because both the science and the art of engineering are employed. This is another reason for studying the design and analysis of gears. Maybe the lessons learned can be applied elsewhere.

You will find that this chapter consists essentially of four parts:

1 The kinematics of gear teeth and gear trains. In this part we shall learn something about the shape of the gear tooth itself, together with the problems caused by this shape and what to do about them. We shall also learn about the speed ratio of various kinds of gear trains. Students who have taken courses in mechanisms or kinematics of machines should use this part of the chapter as a review and a reintroduction to the nomenclature of gearing and pass on to the other parts.

2 The force analysis of gears and gear trains.

3 The design, that is, determination of the size, of gears based on the strength of the material used.

4 The design of gears based on wear considerations.

13-1 NOMENCLATURE

Spur gears are used to transmit rotary motion between parallel shafts; they are usually cylindrical in shape, and the teeth are straight and parallel to the axis of rotation.

The terminology of gear teeth is illustrated in Fig. 13-1. The *pitch circle* is a theoretical circle upon which all calculations are usually based. The pitch circles of a pair of mating gears are tangent to each other. A *pinion* is the smaller of two mating gears. The larger is often called the *gear*.

The *circular pitch p* is the distance, measured on the pitch circle, from a point

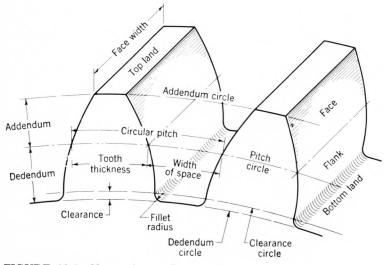

FIGURE 13-1 Nomenclature of gear teeth.

on one tooth to a corresponding point on an adjacent tooth. Thus the circular pitch is equal to the sum of the *tooth thickness* and the *width of space*.

The *module m* is the ratio of the pitch diameter to the number of teeth. The customary unit of length used is the millimeter. The module is the index of tooth size in SI.

The *diametral pitch P* is the ratio of the number of teeth on the gear to the pitch diameter. Thus, it is the reciprocal of the module. Since diametral pitch is only used with English units, it is expressed as teeth per inch.

The *addendum a* is the radial distance between the *top land* and the pitch circle. The *dedendum b* is the radial distance from the *bottom land* to the pitch circle. The *whole depth h_t* is the sum of the addendum and dedendum.

The *clearance circle* is a circle that is tangent to the addendum circle of the mating gear. The *clearance c* is the amount by which the dedendum in a given gear exceeds the addendum of its mating gear. The *backlash* is the amount by which the width of a tooth space exceeds the thickness of the engaging tooth measured on the pitch circles.

You should prove for yourself the validity of the following useful relations:

$$P = \frac{N}{d} \tag{13-1}$$

where P = diametral pitch, teeth per inch
$\quad\quad N$ = number of teeth
$\quad\quad d$ = pitch diameter

$$m = \frac{d}{N} \tag{13-2}$$

where m = module, mm
$\quad\quad d$ = pitch diameter, mm

$$p = \frac{\pi d}{N} = \pi m \tag{13-3}$$

where p = circular pitch

$$pP = \pi \tag{13-4}$$

13-2 CONJUGATE ACTION

The following discussion assumes the teeth to be perfectly formed, perfectly smooth, and absolutely rigid. Such an assumption is, of course, unrealistic, because the application of forces will cause deflections.

Mating gear teeth acting against each other to produce rotary motion are similar to cams. When the tooth profiles, or cams, are designed so as to produce a

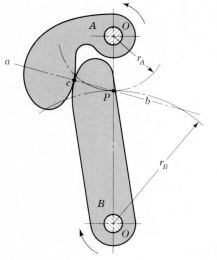

FIGURE 13-2

constant angular-velocity ratio during meshing, they are said to have *conjugate action*. In theory, at least, it is possible arbitrarily to select any profile for one tooth and then to find a profile for the meshing tooth which will give conjugate action. One of these solutions is the *involute profile*, which, with few exceptions, is in universal use for gear teeth and is the only one with which we shall be concerned.

When one curved surface pushes against another (Fig. 13-2), the point of contact occurs where the two surfaces are tangent to each other (point *c*), and the forces at any instant are directed along the common normal *ab* to the two curves. The line *ab*, representing the direction of action of the forces, is called the *line of action*. The line of action will intersect the line of centers *O-O* at some point *P*. The angular-velocity ratio between the two arms is inversely proportional to their radii to the point *P*. Circles drawn through point *P* from each center are called *pitch circles*, and the radius of each circle is called the *pitch radius*. Point *P* is called the *pitch point*.

To transmit motion at a constant angular-velocity ratio, the pitch point must remain fixed; that is, all the lines of action for every instantaneous point of contact must pass through the same point *P*. In the case of the involute profile it will be shown that all points of contact occur on the same straight line *ab*, that all normals to the tooth profiles at the point of contact coincide with the line *ab*, and thus, that these profiles transmit uniform rotary motion.

13-3 INVOLUTE PROPERTIES

An involute curve may be generated as shown in Fig. 13-3*a*. A partial flange *B* is attached to the cylinder *A*, around which is wrapped a cord *def* which is held tightly. Point *b* on the cord represents the tracing point, and as the cord is

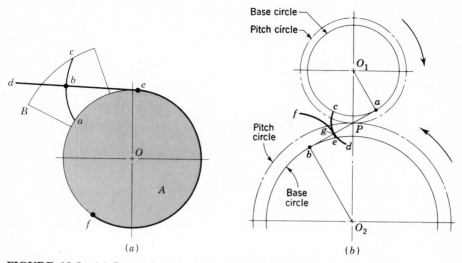

FIGURE 13-3 (*a*) Generation of an involute; (*b*) involute action.

wrapped and unwrapped about the cylinder, point *b* will trace out the involute curve *ac*. The radius of curvature of the involute varies continuously, being zero at point *a* and a maximum at point *c*. At point *b* the radius is equal to the distance *be*, since point *b* is instantaneously rotating about point *e*. Thus the generating line *de* is normal to the involute at all points of intersection and, at the same time, is always tangent to the cylinder *A*. The circle on which the involute is generated is called the *base circle*.

Let us now examine the involute profile to see how it satisfies the requirement for the transmission of uniform motion. In Fig. 13-3*b* two gear blanks with fixed centers at O_1 and O_2 are shown having base circles whose respective radii are O_1a and O_2b. We now imagine that a cord is wound clockwise around the base circle of gear 1, pulled tightly between points *a* and *b*, and wound counterclockwise around the base circle of gear 2. If, now, the base circles are rotated in different directions so as to keep the cord tight, a point *g* on the cord will trace out the involutes *cd* on gear 1 and *ef* on gear 2. The involutes are thus generated simultaneously by the tracing point. The tracing point, therefore, represents the point of contact, while the portion of the cord *ab* is the generating line. The point of contact moves along the generating line, the generating line does not change position because it is always tangent to the base circles, and since the generating line is always normal to the involutes at the point of contact, the requirement for uniform motion is satisfied.

13-4 FUNDAMENTALS

Among other things, it is necessary that you actually be able to draw the teeth on a pair of meshing gears. You should understand, however, that it is not necessary to draw the gear teeth for manufacturing or shop purposes. Rather, we make

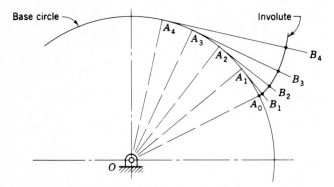

FIGURE 13-4 Construction of an involute curve.

drawings of gear teeth to obtain an understanding of the problems involved in the meshing of the mating teeth.

First, it is necessary to learn how to construct an involute curve. As shown in Fig. 13-4, divide the base circle into a number of equal parts and construct radial lines OA_0, OA_1, OA_2, etc. Beginning at A_1, construct perpendiculars A_1B_1, A_2B_2, A_3B_3, etc. Then along A_1B_1 lay off the distance A_1A_0, along A_2B_2 lay off twice the distance A_1A_0, etc., producing points through which the involute curve can be constructed.

To investigate the fundamentals of tooth action let us proceed step by step through the process of constructing the teeth on a pair of gears.

When two gears are in mesh, their pitch circles roll on one another without slipping. Designate the pitch radii as r_1 and r_2 and the angular velocities as ω_1 and ω_2, respectively. Then the pitch-line velocity is

$$V = r_1\omega_1 = r_2\omega_2$$

Thus the relation between the radii and the angular velocities is

$$\frac{\omega_1}{\omega_2} = \frac{r_2}{r_1} \tag{13-5}$$

Suppose now we wish to design a speed reducer such that the input speed is 1800 rpm and the output speed is 1200 rpm. This is a ratio of 3 : 2; the pitch diameters would be in the same ratio, for example, a 4-in pinion driving a 6-in gear. The various dimensions found in gearing are always based on the pitch circles.

We next specify that an 18-tooth pinion is to mesh with a 30-tooth gear and that the diametral pitch of the gearset is to be 2 teeth per inch. Then, from Eq. (13-1) the pitch diameters of the pinion and gear are, respectively,

$$d_1 = \frac{N_1}{P} = \frac{18}{2} = 9 \text{ in} \qquad d_2 = \frac{N_2}{P} = \frac{30}{2} = 15 \text{ in}$$

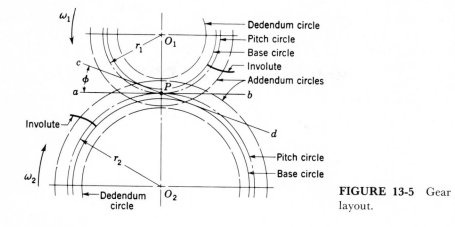

FIGURE 13-5   Gear layout.

The first step in drawing teeth on a pair of mating gears is shown in Fig. 13-5. The center distance is the sum of the pitch radii, in this case 12 in. So locate the pinion and gear centers O_1 and O_2, 12 in apart. Then construct the pitch circles of radii r_1 and r_2. These are tangent at P, the *pitch point*. Next draw line ab, the common tangent, through the pitch point. We now designate gear 1 as the driver, and since it is rotating counterclockwise, we draw a line cd through point P at an angle ϕ to the common tangent ab. The line cd has three names, all of which are in general use. It is called the *pressure line*, the *generating line*, and the *line of action*. It represents the direction in which the resultant force acts between the gears. The angle ϕ is called the *pressure angle*, and it usually has values of 20 or 25°, though $14\frac{1}{2}°$ was once used.

Next, on each gear draw a circle tangent to the pressure line. These circles are the *base circles*. Since they are tangent to the pressure line, the pressure angle determines their size. As shown in Fig. 13-6, the radius of the base circle is

$$r_b = r \cos \phi \qquad (13\text{-}6)$$

where r is the pitch radius.

Now generate an involute on each base circle as previously described and as

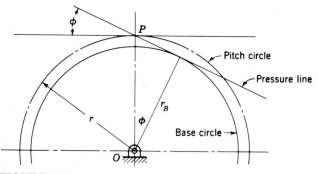

FIGURE 13-6

shown in Fig. 13-5. This involute is to be used for one side of a gear tooth. It is not necessary to draw another curve in the reverse direction for the other side of the tooth because we are going to use a template which can be turned over to obtain the other side.

The addendum and dedendum distances for standard interchangeable teeth are, as we shall learn later, $1/P$ and $1.25/P$, respectively. Therefore, for the pair of gears we are constructing,

$$a = \frac{1}{P} = \frac{1}{2} = 0.500 \text{ in} \qquad b = \frac{1.25}{P} = \frac{1.25}{2} = 0.625 \text{ in}$$

Using these distances, draw the addendum and dedendum circles on the pinion and on the gear as shown in Fig. 13-5.

Next, using heavy drawing paper, or preferably, a sheet of 0.015- to 0.020-in clear plastic, cut a template for each involute, being careful to locate the gear centers properly with respect to each involute. Figure 13-7 is a reproduction of the template used to create some of the illustrations for this book. Note that only one side of the tooth profile is formed on the template. To get the other side, turn the template over. For some problems you might wish to construct a template for the entire tooth.

To draw a tooth we must know the tooth thickness. From Eq. (13-4) the circular pitch is

$$p = \frac{\pi}{P} = \frac{\pi}{2} = 1.57 \text{ in}$$

Therefore the tooth thickness is

$$t = \frac{p}{2} = \frac{1.57}{2} = 0.785 \text{ in}$$

measured on the pitch circle. Using this distance for the tooth thickness as well as the tooth space, draw as many teeth as are desired, using the template, after the points have been marked on the pitch circle. In Fig. 13-8 only one tooth has been drawn on each gear. You may run into trouble in drawing these teeth if one of the base circles happens to be larger than the dedendum circle. The reason for this is that the involute begins at the base circle and is undefined below this

FIGURE 13-7 Template for drawing gear teeth.

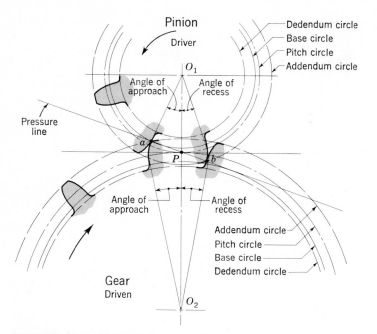

FIGURE 13-8   Tooth action.

circle. So, in drawing gear teeth, we usually draw a radial line for the profile below the base circle. The actual shape, however, will depend upon the kind of machine tool used to form the teeth in manufacture, that is, how the profile is generated.

The portion of the tooth between the clearance circle and the dedendum circle is the fillet. In this instance the clearance is

$$c = b - a = 0.625 - 0.500 = 0.125 \text{ in}$$

The construction is finished when these fillets have been drawn.

Referring again to Fig. 13-8, the pinion with center at O_1 is the driver and turns counterclockwise. The pressure, or generating, line is the same as the cord used in Fig. 13-3a to generate the involute, and contact occurs along this line. The initial contact will take place when the flank of the driver comes into contact with the tip of the driven tooth. This occurs at point a in Fig. 13-8, where the addendum circle of the driven gear crosses the pressure line. If we now construct tooth profiles through point a and draw radial lines from the intersections of these profiles with the pitch circles to the gear centers, we obtain the *angles of approach.*

As the teeth go into mesh, the point of contact will slide up the side of the driving tooth so that the tip of the driver will be in contact just before contact ends. The final point of contact will therefore be where the addendum circle of the driver crosses the pressure line. This is point b in Fig. 13-8. By drawing another set of tooth profiles through b, we obtain the *angles of recess* for each gear in a manner similar to that of finding the angles of approach. The sum of the angle of approach and the angle of recess for either gear is called the *angle of action*. The line ab is called the *line of action*.

We may imagine a *rack* as a spur gear having an infinitely large pitch diameter. Therefore the rack has an infinite number of teeth and a base circle which is an infinite distance from the pitch point. The sides of involute teeth on a rack are straight lines making an angle to the line of centers equal to the pressure angle. Figure 13-9 shows an involute rack in mesh with a pinion.

Corresponding sides of involute teeth are parallel curves; the *base pitch* is the constant and fundamental distance between them along a common normal as shown in Fig. 13-9. The base pitch is related to the circular pitch by the equation

$$p_b = p_c \cos \phi \tag{13-7}$$

where p_b is the base pitch.

Figure 13-10 shows a pinion in mesh with an *internal*, or *annular*, *gear*. Note that both of the gears now have their centers of rotation on the same side of the pitch point. Thus the positions of the addendum and dedendum circles with respect to the pitch circle are reversed; the addendum circle of the internal gear lies *inside* the pitch circle. Note, too, from Fig. 13-10, that the base circle of the internal gear lies inside the pitch circle near the addendum circle.

Another interesting observation concerns the fact that the operating diameters of the pitch circles of a pair of meshing gears need not be the same as the respective design pitch diameters of the gears, though this is the way they have been constructed in Fig. 13-8. If we increase the center distance we create two new operating pitch circles having larger diameters because they must be tangent to each other at the pitch point. Thus the pitch circles of gears really do not come into existence until a pair of gears is brought into mesh.

Changing the center distance has no effect on the base circles because these

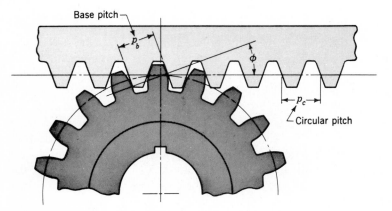

FIGURE 13-9 Involute pinion and rack.

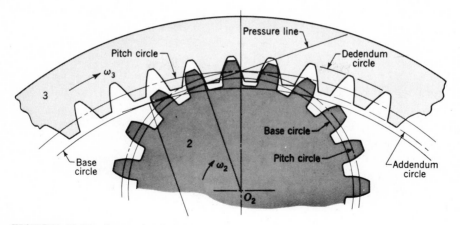

FIGURE 13-10 Internal gear and pinion.

were used to generate the tooth profiles. Thus the base circle is basic to a gear. Increasing the center distance increases the pressure angle and decreases the length of the line of action but the teeth are still conjugate, the requirement for uniform motion transmission is still satisfied, and the angular-velocity ratio has not changed.

Example 13-1 A gearset consists of a 16-tooth pinion driving a 40-tooth gear. The diametral pitch is 2, and the addendum and dedendum are $1/P$ and $1.25/P$, respectively. The gears are cut using a pressure angle of $20°$.

(*a*) Compute the circular pitch, the center distance, and the radii of the base circles.

(*b*) In mounting these gears, the center distance was incorrectly made $\frac{1}{4}$ in larger. Compute the new values of the pressure angle and the pitch-circle diameters.

Solution

(*a*) $$p = \frac{\pi}{P} = \frac{\pi}{2} = 1.57 \text{ in} \qquad\qquad Ans.$$

The pitch diameters of the pinion and gear are, respectively,

$$d_P = \tfrac{16}{2} = 8 \text{ in} \qquad d_G = \tfrac{40}{2} = 20 \text{ in}$$

Therefore the center distance is

$$\frac{d_P + d_G}{2} = \frac{8 + 20}{2} = 14 \text{ in} \qquad\qquad Ans.$$

Since the teeth were cut on the $20°$ pressure angle, the base-circle radii are found to be, using $r_b = r \cos \phi$,

$$r_b(\text{pinion}) = \tfrac{8}{2} \cos 20° = 3.76 \text{ in} \qquad\qquad Ans.$$

$$r_b(\text{gear}) = \tfrac{20}{2} \cos 20° = 9.40 \text{ in} \qquad\qquad Ans.$$

(b) Designating d'_P and d'_G as the new pitch-circle diameters, the increase of $\tfrac{1}{4}$ in in the center distance requires that

$$\frac{d'_P + d'_G}{2} = 14.250 \tag{1}$$

Also, the velocity ratio does not change, and hence

$$\frac{d'_P}{d'_G} = \frac{16}{40} \tag{2}$$

Solving Eqs. (1) and (2) simultaneously yields

$$d'_P = 8.143 \text{ in} \qquad d'_G = 20.357 \text{ in} \qquad\qquad Ans.$$

Since $r_b = r \cos \phi$, the new pressure angle is

$$\phi' = \cos^{-1} \frac{r_b(\text{pinion})}{d'_P/2} = \cos^{-1} \frac{3.76}{8.143/2} = 22.56° \qquad\qquad Ans.$$

////

13-5   CONTACT RATIO

The zone of action of meshing gear teeth is shown in Fig. 13-11. We recall that tooth contact begins and ends at the intersections of the two addendum circles with the pressure line. In Fig. 13-11 initial contact occurs at a and final contact at b. Tooth profiles drawn through these points intersect the pitch circle at A and B, respectively. As shown, the distance AP is called the *arc of approach* q_a, and the distance PB, the *arc of recess* q_r. The sum of these is the *arc of action* q_t.

Now, consider a situation in which the arc of action is exactly equal to the circular pitch, that is, $q_t = p$. This means that one tooth and its space will occupy the entire arc AB. In other words, when a tooth is just beginning contact at a, the previous tooth is simultaneously ending its contact at b. Therefore, during the tooth action from a to b, there will be exactly one pair of teeth in contact.

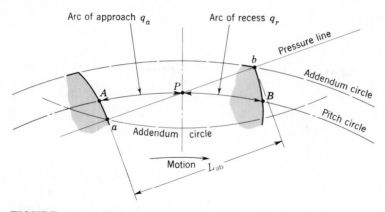

FIGURE 13-11 Definition of contact ratio.

Next, consider a situation in which the arc of action is greater than the circular pitch, but not very much greater, say, $q_t \approx 1.2p$. This means that when one pair of teeth is just entering contact at a, another pair, already in contact, will not yet have reached b. Thus, for a short period of time, there will be two pairs of teeth in contact, one near the vicinity of A and another near B. As the meshing proceeds, the pair near B must cease contact, leaving only a single pair of contacting teeth, until the procedure repeats itself.

Because of the nature of this tooth action, either one or two pairs of teeth in contact, it is convenient to define the term *contact ratio* m_c as

$$m_c = \frac{q_t}{p} \tag{13-8}$$

a number which indicates the average number of pairs of teeth in contact. Note that this ratio is also equal to the length of the path of contact divided by the base pitch. Gears should not generally be designed having contact ratios less than about 1.20 because inaccuracies in mounting might reduce the contact ratio even more, increasing the possibility of impact between the teeth as well as an increase in the noise level.

An easier way to obtain the contact ratio is to measure the line of action ab instead of the arc distance AB. Since ab in Fig. 13-11 is tangent to the base circle when extended, the base pitch p_b must be used to calculate m_c instead of the circular pitch as in Eq. (13-8). Designating the length of the line of action as L_{ab}, the contact ratio is

$$m_c = \frac{L_{ab}}{p \cos \phi} \tag{13-9}$$

in which Eq. (13-7) was used for the base pitch.

13-6   INTERFERENCE

The contact of portions of tooth profiles which are not conjugate is called *inter-ference*. Consider Fig. 13-12. Illustrated are two 16-tooth gears which have been cut using the now obsolete $14\frac{1}{2}°$ pressure angle. The driver, gear 2, turns clock-wise. The initial and final points of contact are designated A and B, respectively, and are located on the pressure line. Now notice that the points of tangency of the pressure line with the base circles C and D are located *inside* of points A and B. Interference is present.

The interference is explained as follows. Contact begins when the tip of the driven tooth contacts the flank of the driving tooth. In this case the flank of the driving tooth first makes contact with the driven tooth at point A, and this occurs *before* the involute portion of the driving tooth comes within range. In other words, contact is occurring below the base circle of gear 2 on the *noninvolute*

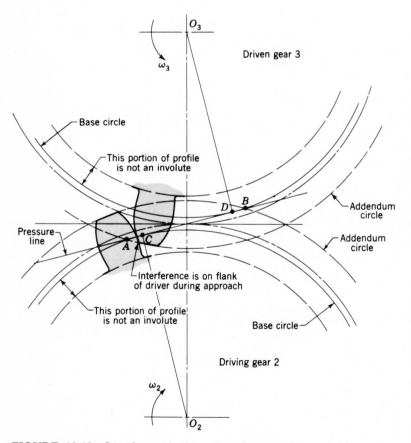

FIGURE 13-12   Interference in the action of gear teeth.

portion of the flank. The actual effect is that the involute tip or face of the driven gear tends to dig out the noninvolute flank of the driver.

In this example the same effect occurs again as the teeth leave contact. Contact should end at point D or before. Since it does not end until point B, the effect is for the tip of the driving tooth to dig out, or interfere with, the flank of the driven tooth.

When gear teeth are produced by a generation process, interference is automatically eliminated because the cutting tool removes the interfering portion of the flank. This effect is called *undercutting*; if undercutting is at all pronounced, the undercut tooth is considerably weakened. Thus the effect of eliminating interference by a generation process is merely to substitute another problem for the original one.

The importance of the problem of teeth which have been weakened by undercutting cannot be overemphasized. Of course, interference can be eliminated by using more teeth on the gears. However, if the gears are to transmit a given amount of power, more teeth can be used only by increasing the pitch diameter. This makes the gears larger, which is seldom desirable, and it also increases the pitch-line velocity. This increased pitch-line velocity makes the gears noisier and reduces the power transmission somewhat, although not in direct ratio. In general, however, the use of more teeth to eliminate interference or undercutting is seldom an acceptable solution.

Interference can also be reduced by using a larger pressure angle. This results in a smaller base circle, so that more of the tooth profile becomes involute. The demand for smaller pinions with fewer teeth thus favors the use of a 25° pressure angle even though the frictional forces and bearing loads are increased and the contact ratio decreased.

13-7 THE FORMING OF GEAR TEETH

There are a large number of ways of forming the teeth of gears, such as *sand casting, shell molding, investment casting, permanent-mold casting, die casting,* and *centrifugal casting.* Teeth can be formed by using the *powder-metallurgy process*; or, by using *extrusion,* a single bar of aluminum may be formed and then sliced into gears. Gears which carry large loads in comparison with their size are usually made of steel and are cut with either *form cutters* or *generating cutters.* In form cutting, the tooth space takes the exact shape of the cutter. In generating, a tool having a shape different from the tooth profile is moved relative to the gear blank so as to obtain the proper tooth shape. One of the newest and most promising of the methods of forming teeth is called *cold forming,* or *cold rolling,* in which dies are rolled against steel blanks to form the teeth. The mechanical properties of the metal are greatly improved by the rolling process, and a high-quality generated profile is obtained at the same time.

Gear teeth may be shaped by milling, shaping, or hobbing. They may be finished by shaving, burnishing, grinding, or lapping.

Milling

Gear teeth may be cut with a form milling cutter shaped to conform to the tooth space. With this method it is theoretically necessary to use a different cutter for each gear, because a gear having 25 teeth, for example, will have a different-shaped tooth space than one having, say, 24 teeth. Actually, the change in space is not too great, and it has been found that eight cutters may be used to cut with reasonable accuracy any gear in the range of 12 teeth to a rack. A separate set of cutters is, of course, required for each pitch.

Shaping

Teeth may be generated with either a pinion cutter or a rack cutter. The pinion cutter (Fig. 13-13) reciprocates along the vertical axis and is slowly fed into the gear blank to the required depth. When the pitch circles are tangent, both the cutter and blank rotate slightly after each cutting stroke. Since each tooth of the cutter is a cutting tool, the teeth are all cut after the blank has completed one revolution.

The sides of an involute rack tooth are straight. For this reason, a rack-

FIGURE 13-13 Generating a spur gear with a pinion cutter. (*Courtesy of Boston Gear Works, Inc.*)

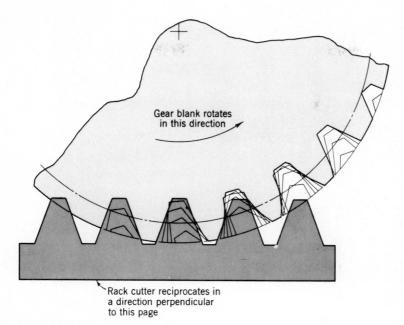

Gear blank rotates
in this direction

Rack cutter reciprocates in
a direction perpendicular
to this page

FIGURE 13-14 Shaping teeth with a rack cutter.

generating tool provides an accurate method of cutting gear teeth. This is also a shaping operation and is illustrated by the drawing of Fig. 13-14. In operation, the cutter reciprocates and is first fed into the gear blank until the pitch circles are tangent. Then, after each cutting stroke, the gear blank and cutter roll slightly on their pitch circles. When the blank and cutter have rolled a distance equal to the circular pitch, the cutter is returned to the starting point, and the process is continued until all the teeth have been cut.

Hobbing

The hobbing process is illustrated in Fig. 13-15. The hob is simply a cutting tool which is shaped like a worm. The teeth have straight sides, as in a rack, but the hob axis must be turned through the lead angle in order to cut spur-gear teeth. For this reason, the teeth generated by a hob have a slightly different shape than those generated by a rack cutter. Both the hob and the blank must be rotated at the proper angular-velocity ratio. The hob is then fed slowly across the face of the blank until all the teeth have been cut.

Finishing

Gears which run at high speeds and transmit large forces may be subjected to additional dynamic forces due to errors in tooth profiles. These errors may be diminished somewhat by finishing the tooth profiles. The teeth may be finished, after cutting, either by shaving or burnishing. Several shaving machines are

FIGURE 13-15   Hobbing a worm gear.   (*Courtesy of Boston Gear Works, Inc.*)

available which cut off a minute amount of metal, bringing the accuracy of the tooth profile within the limits of 250 μin.

Burnishing, like shaving, is used with gears which have been cut but not heat-treated.   In burnishing, hardened gears with slightly oversize teeth are run in mesh with the gear until the surfaces become smooth.

Grinding and lapping are used for hardened gear teeth after heat treatment. The grinding operation employs the generating principle and produces very accurate teeth.   In lapping, the teeth of the gear and lap slide axially so that the whole surface of the teeth is abraded equally.

13-8   TOOTH SYSTEMS

A *tooth system* is a standard* which specifies the relationships involving addendum, dedendum, working depth, tooth thickness, and pressure angle, for the purpose of attaining interchangeability of gears of all tooth numbers but of the same pressure

---

* Standardized by the American Gear Manufacturers Association (AGMA) and the American National Standards Institute (ANSI).   The AGMA standards may be quoted or extracted in their entirety, provided an appropriate credit line is included, for example, "Extracted from AGMA Information Sheet—Strength of Spur, Helical, Herringbone, and Bevel Gear Teeth (AGMA 225.01), with permission of the publisher, the American Gear Manufacturers Association, 1901 North Fort Myer Drive, Suite 1000, Arlington, Va. 22209." These standards have been used extensively in this chapter and in the chapter

angle and pitch. You should be aware of the advantages and disadvantages of the various systems so that you can choose the optimum tooth for a given design and have a basis of comparison when departing from a standard tooth profile.

Table 13-1 lists the tooth proportions for completely interchangeable English-system gears and for operation on standard center distances. No standard has been established in this country for tooth systems based wholly upon the use of SI units. In fact, it is probable that a number of years will elapse before agreement can be reached; the problems to be solved are complex as well as expensive. Even in England, which has been ahead of this country in the metric changeover, the inch system is still predominantly used for gearing. Merritt* states that among the reasons is that new standards had been approved and adopted shortly before metrication began.

The addenda listed in Table 13-1 are for gears having tooth numbers equal to or greater than the minimum numbers listed, and for these numbers there will be no undercutting. For fewer numbers of teeth a modification called the *long and short addendum system* should be used.† In this system the addendum of the gear is decreased just sufficiently so that contact does not begin before the interference point (point *a* in Fig. 13-11). The addendum of the pinion is then increased a corresponding amount. In this modification, there is no change in the pressure angle or in the pitch circles, so the center distance remains the same. The intent is to increase the recess action and decrease the approach action.

The 0.002-in additional dedendum shown in Table 13-1 for fine-pitch gears provides space for the accumulation of dirt at the roots of the teeth.

The working depths shown in Table 13-1 are for, and define, full-depth teeth; for stub teeth, use $1.60/P$.

It should be particularly noted that the standards shown in Table 13-1 are *not* intended to restrict the freedom of the designer. Standard tooth proportions lead to interchangeability and standard cutters which are economical to purchase, but the need for high-performance gears may well dictate considerable deviation from these systems.

Some of the tooth systems which are now obsolete are the two AGMA $14\frac{1}{2}°$ systems, the Fellows 20° stub-tooth system, and the Brown and Sharpe system. The obsolete standards should not be used for new designs, but you may need to refer to them when redesigning existing machinery which utilizes these older systems.

The diametral pitches listed in Table 13-2 should be used whenever possible in order to keep to a minimum the inventory of gear-cutting tools.

---

which follows. In each case the information sheet number is given. Table 13-1 is from AGMA Publication 201.02 and 201.02A, but see also 207.04. Write the AGMA for a complete list of standards because changes and additions are made from time to time.

* H. E. Merritt, *Gear Engineering*, Wiley, New York, 1971.

† For an explanation of these modifications and others, see Joseph E. Shigley and John J. Uicker, Jr., *Theory of Machines and Mechanisms*, McGraw-Hill, New York, 1980, pp. 267–277.

Table 13-1   STANDARD AGMA AND ANSI TOOTH SYSTEMS FOR
SPUR GEARS

Quantity	Coarse pitch (up to 20P)* full depth		Fine pitch (20P and up) full depth
Pressure angle ϕ	20°	25°	20°
Addendum a	$\dfrac{1.000}{P}$	$\dfrac{1.000}{P}$	$\dfrac{1.000}{P}$
Dedendum b	$\dfrac{1.250}{P}$	$\dfrac{1.250}{P}$	$\dfrac{1.200}{P} + 0.002$ in
Working depth h_k	$\dfrac{2.000}{P}$	$\dfrac{2.000}{P}$	$\dfrac{2.000}{P}$
Whole depth h_t (min.)	$\dfrac{2.25}{P}$	$\dfrac{2.25}{P}$	$\dfrac{2.200}{P} + 0.002$ in
Circular tooth thickness t	$\dfrac{\pi}{2P}$	$\dfrac{\pi}{2P}$	$\dfrac{1.5708}{P}$
Fillet radius of basic rack r_f	$\dfrac{0.300}{P}$	$\dfrac{0.300}{P}$	Not standardized
Basic clearance c (min.)	$\dfrac{0.250}{P}$	$\dfrac{0.250}{P}$	$\dfrac{0.200}{P} + 0.002$ in
Clearance c (shaved or ground teeth)	$\dfrac{0.350}{P}$	$\dfrac{0.350}{P}$	$\dfrac{0.3500}{P} + 0.002$ in
Minimum number of pinion teeth	18	12	18
Minimum number of teeth per pair	36	24	
Minimum width of top land t_0	$\dfrac{0.25}{P}$	$\dfrac{0.25}{P}$	Not standardized

* But not including 20P.

13-9 GEAR TRAINS

Consider a pinion 2 driving a gear 3. The speed of the driven gear is

$$n_3 = \frac{N_2}{N_3}\, n_2 = \frac{d_2}{d_3}\, n_2 \tag{13-10}$$

where n = rpm or number of turns
N = number of teeth
d = pitch diameter

Table 13-2 DIAMETRAL PITCHES IN GENERAL USE

Coarse pitch	2, $2\frac{1}{4}$, $2\frac{1}{2}$, 3, 4, 6, 8, 10, 12, 16
Fine pitch	20, 24, 32, 40, 48, 64, 80, 96, 120, 150, 200

For spur gears the directions correspond to the right-hand rule and are positive or negative corresponding to counterclockwise or clockwise. The gear train shown in Fig. 13-16 is made up of five gears. The speed of gear 6 is

$$n_6 = \frac{N_2}{N_3}\frac{N_3}{N_4}\frac{N_5}{N_6} n_2 \tag{a}$$

Here we notice that gear 3 is an idler, that its tooth numbers cancel in Eq. (a), and hence that it affects only the direction of rotation of 6. We notice, furthermore, that gears 2, 3, and 5 are drivers, while 3, 4, and 6 are driven members. We define *train value e* as

$$e = \frac{\text{Product of driving tooth numbers}}{\text{Product of driven tooth numbers}} \tag{13-11}$$

Note that pitch diameters can be used in Eq. (13-11) as well. When Eq. (13-11) is used for spur gears, e is positive if the last gear rotates in the same sense as the first, and negative if the last rotates in the opposite sense.

Now we can write

$$n_L = en_F \tag{13-12}$$

where n_L is the speed of the last gear in the train and n_F is the speed of the first.

Unusual effects can be obtained in a gear train by permitting some of the gear axes to rotate about others. Such trains are called *planetary*, or *epicyclic, gear trains*. Planetary trains always include a *sun gear*, a *planet carrier or arm*, and one or more *planet gears*, as shown in Fig. 13-17. Planetary gear trains are unusual mechanisms because they have two degrees of freedom; that is, for constrained motion a planetary train must have two inputs. For example, in Fig. 13-17 these two inputs could be the motion of any two of the elements of the train. We might, in Fig. 13-17, say, specify that the sun gear rotates at 100 rpm clockwise and that the ring gear rotates at 50 rpm counterclockwise; these are the inputs.

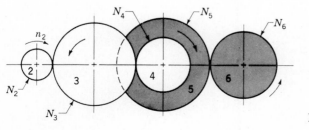

FIGURE 13-16

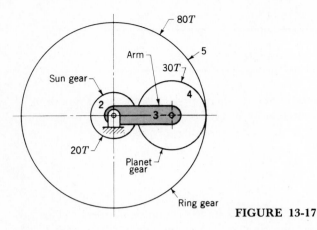

FIGURE 13-17

The output would be the motion of the arm. In most planetary trains one of the elements is attached to the frame and has a zero input motion.

Figure 13-18 shows a planetary train composed of a sun gear 2, an arm or carrier 3, and planet gears 4 and 5. The angular velocity of gear 2 relative to the arm in rpm is

$$n_{23} = n_2 - n_3 \tag{b}$$

Also, the velocity of gear 5 relative to the arm is

$$n_{53} = n_5 - n_3 \tag{c}$$

Dividing Eq. (c) by (b) gives

$$\frac{n_{53}}{n_{23}} = \frac{n_5 - n_3}{n_2 - n_3} \tag{d}$$

Equation (d) expresses the ratio of the relative velocity of gear 5 to that of gear 2, and both velocities are taken relative to the arm. Now this ratio is the same and is proportional to the tooth numbers, whether the arm is rotating or not. It is the train value. Therefore we may write

$$e = \frac{n_5 - n_3}{n_2 - n_3} \tag{e}$$

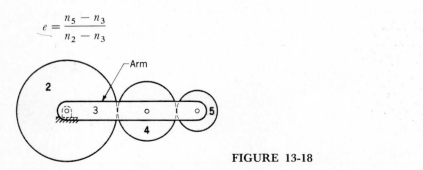

FIGURE 13-18

This equation can be used to solve for the output motion of any planetary train. It is more conveniently written in the form

$$e = \frac{n_L - n_A}{n_F - n_A} \tag{13-13}$$

where n_F = rpm of first gear in planetary train
n_L = rpm of last gear in planetary train
n_A = rpm of arm

Example 13-2 In Fig. 13-17 the sun gear is the input, and it is driven clockwise at 100 rpm. The ring gear is held stationary by being fastened to the frame. Find the rpm and direction of rotation of the arm.

Solution Designate $n_F = n_2 = -100$ rpm, and $n_L = n_5 = 0$. Unlocking gear 5 and holding the arm stationary, in our imagination, we find

$$e = -(\tfrac{20}{30})(\tfrac{30}{80}) = -0.25$$

Substituting in Eq. (13-13),

$$-0.25 = \frac{0 - n_A}{(-100) - n_A}$$

or

$$n_A = -20 \text{ rpm} \qquad\qquad Ans.$$

To obtain the rpm of gear 4, we follow the procedure outlined by Eqs. (*b*), (*c*), and (*d*). Thus

$$n_{43} = n_4 - n_3 \qquad n_{23} = n_2 - n_3$$

And so

$$\frac{n_{43}}{n_{23}} = \frac{n_4 - n_3}{n_2 - n_3} \tag{1}$$

But

$$\frac{n_{43}}{n_{23}} = -\frac{20}{30} = -\frac{2}{3} \tag{2}$$

Substituting the known values in Eq. (1) gives

$$-\frac{2}{3} = \frac{n_4 - (-20)}{(-100) - (-20)}$$

Solving gives

$$n_4 = 33\frac{1}{3} \text{ rpm}$$ *Ans.*

////

13-10   FORCE ANALYSIS

Before beginning the force analysis of gear trains, let us agree on the notation to be used. Beginning with the numeral 1 for the frame of the machine, we shall designate the input gear as gear 2, and then number the gears successively 3, 4, etc., until we arrive at the last gear in the train. Next, there may be several shafts involved, and usually one or two gears are mounted on each shaft as well as other elements. We shall designate the shafts, using lowercase letters of the alphabet, a, b, c, etc.

With this notation we can now speak of the force exerted by gear 2 against gear 3 as F_{23}. The force of gear 2 against shaft a is F_{2a}. We can also write F_{a2} to mean the force of shaft a against gear 2. Unfortunately, it is also necessary to use superscripts to indicate directions. The coordinate directions will usually be indicated by the x, y, and z coordinates, and the radial and tangential directions by superscripts r and t. With this notation F_{43}^t is the tangential component of the force of gear 4 acting against gear 3.

Figure 13-19a shows a pinion mounted on shaft a rotating clockwise at n_2 rpm

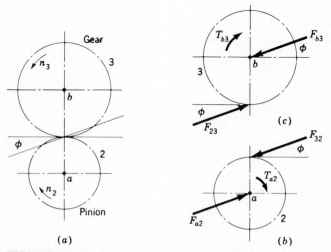

FIGURE 13-19   Free-body diagrams of the forces acting upon two gears in a simple gear train.

and driving a gear on shaft b at n_3 rpm. The reactions between the mating teeth occur along the pressure line. In Fig. 13-19b the pinion has been separated from the gear and from the shaft, and their effects replaced by forces. F_{a2} and T_{a2} are the force and torque, respectively, exerted by shaft a against pinion 2. F_{32} is the force exerted by gear 3 against the pinion. Using a similar approach, we obtain the free-body diagram of the gear shown in Fig. 13-19c.

In Fig. 13-20 the free-body diagram of the pinion has been redrawn and the forces resolved into tangential and radial components. We now define

$$W_t = F_{32}^t \qquad (a)$$

as the *transmitted load*. The transmitted load is really the useful component because the radial component F_{32}^r serves no useful purpose. It does not transmit power. The applied torque and the transmitted load are seen to be related by the equation

$$T = \frac{d}{2} W_t \qquad (13\text{-}14)$$

where we have used $T = T_{a2}$ and $d = d_2$ to obtain a general relation.

If next we designate the pitch-line velocity by V, where $V = \pi dn/12$ and is in feet per minute, the tangential load may be obtained from the equation

$$H = \frac{W_t V}{33\ 000} \qquad (13\text{-}15)$$

The corresponding equation in SI is

$$W_t = \frac{60(10)^3 H}{\pi dn} \qquad (13\text{-}16)$$

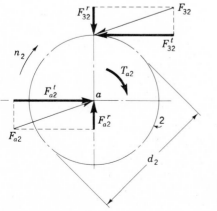

FIGURE 13-20 Resolution of gear forces.

where W_t = transmitted load
    H = power
    d = gear diameter, mm
    n = speed, rpm

Example 13-3 Pinion 2 in Fig. 13-21a runs at 1750 rpm and transmits 2.5 kW to idler gear 3. The teeth are cut on the 20° full-depth system and have a module of $m = 2.5$ mm. Draw a free-body diagram of gear 3 and show all the forces which act upon it.

Solution The pitch diameters of gears 2 and 3 are

$$d_2 = N_2 m = 20(2.5) = 50 \text{ mm}$$

$$d_3 = N_3 m = 50(2.5) = 125 \text{ mm}$$

From Eq. (13-16) we find the transmitted load to be

$$W_t = \frac{60(10)^3 H}{\pi d_2 n} = \frac{60(10)^3 (2.5)}{\pi(50)(1750)} = 0.546 \text{ kN}$$

Thus, the tangential force of gear 2 on gear 3 is $F_{23}^t = 0.546$ kN as shown in Fig. 13-21b. Therefore

$$F_{23}^r = F_{23}^t \tan 20° = (0.546) \tan 20° = 0.199 \text{ kN}$$

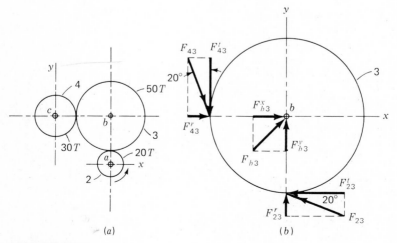

(a)

(b)

FIGURE 13-21

And so

$$F_{23} = \frac{F^t_{23}}{\cos 20°} = \frac{0.546}{\cos 20°} = 0.581 \text{ kN}$$

Since gear 3 is an idler, it transmits no power (torque) to its shaft, and so the tangential reaction of gear 4 on gear 3 is also equal to W_t. Therefore

$$F^t_{43} = 0.546 \text{ kN} \qquad F^r_{43} = 0.199 \text{ kN} \qquad F_{43} = 0.581 \text{ kN}$$

and the directions are as shown in Fig. 13-21*b*.

The shaft reactions in the *x* and *y* directions are

$$F^x_{b3} = -(F^t_{23} + F^r_{43}) = -(-0.546 + 0.199) = 0.347 \text{ kN}$$

$$F^y_{b3} = -(F^r_{23} + F^t_{43}) = -(0.199 - 0.546) = 0.347 \text{ kN}$$

The resultant shaft reaction is

$$F_{b3} = \sqrt{(0.347)^2 + (0.347)^2} = 0.491 \text{ kN}$$

These are shown on the figure. ////

13-11 TOOTH STRESSES*

The following considerations must be treated as important limiting design factors in specifying the capacity of any gear drive:

- The heat generated during operation
- Failure of the teeth by breakage (see Fig. 6-14)
- Fatigue failure of the tooth surfaces
- Abrasive wear of the tooth surfaces
- Noise as a result of high speeds, heavy loads, or mounting inaccuracies

In this book we shall study the strength of gear teeth based on three kinds of possible failures. These are static failure due to bending stress, fatigue failure due to bending stress, and surface fatigue failure due to contact or Hertzian stress.

---

* One of the most important papers published in many years on gear-tooth stresses has recently become available, and many of the results of this investigation are reviewed here; all of the tabular material in this section is from this paper. See R. G. Mitchiner and H. H. Mabie, "The Determination of the Lewis Form Factor and the AGMA Geometry Factor *J* for External Spur Gear Teeth," *ASME Journal of Mechanical Design*, vol. 104, no. 1, Jan. 1982, pp. 148–158.

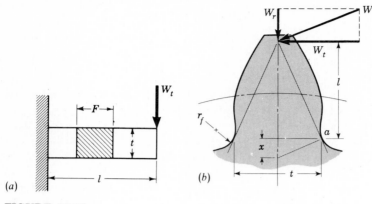

FIGURE 13-22

The particular purpose of this section is to obtain relations for the bending stresses in a loaded tooth. Wilfred Lewis first presented a formula for computing the bending stress in gear teeth in which the tooth form entered into the equation. The formula was announced in 1892 and it still remains the basis for most gear design today.

To derive the basic Lewis equation, refer to Fig. 13-22a which shows a cantilever of cross-sectional dimensions F and t, having a length l and a load W_t uniformly distributed across the distance F. The section modulus is $I/c = Ft^2/6$, and therefore the bending stress is

$$\sigma = \frac{M}{I/c} = \frac{6W_t l}{Ft^2} \tag{a}$$

Referring now to Fig. 13-22b, we assume that the maximum stress in a gear tooth occurs at point a. By similar triangles, you can write

$$\frac{t/2}{x} = \frac{l}{t/2} \qquad \text{or} \qquad x = \frac{t^2}{4l} \tag{b}$$

By rearranging Eq. (a),

$$\sigma = \frac{6W_t l}{Ft^2} = \frac{W_t}{F}\frac{1}{t^2/6l} = \frac{W_t}{F}\frac{1}{t^2/4l}\frac{1}{\frac{4}{6}} \tag{c}$$

If we now substitute the value of x from Eq. (b) into Eq. (c) and multiply the numerator and denominator by the circular pitch p, we find

$$\sigma = \frac{W_t p}{F(\frac{2}{3})xp} \tag{d}$$

Letting $y = 2x/3p$, we have

$$\sigma = \frac{W_t}{Fpy} \tag{13-17}$$

This completes the development of the original Lewis equation. The factor y is called the *Lewis form factor*, and it may be obtained by a graphical layout of the gear tooth or by digital computation.

In using this equation, most engineers prefer to employ the diametral pitch in determining the stresses. This is done by substituting $P = \pi/p$ and $Y = \pi y$ in Eq. (13-17). This gives

$$\sigma = \frac{W_t P}{FY} \tag{13-18}$$

where

$$Y = \frac{2xP}{3} \tag{13-19}$$

The use of this equation for Y means that only the bending of the tooth is considered and that the compression due to the radial component of the force is neglected. Values of Y obtained from this equation are tabulated in Table 13-3.

The use of Eq. (13-19) also implies that the teeth do not share the load and that the greatest force is exerted at the tip of the tooth. But we have already learned that the contact ratio should be somewhat greater than unity, say about 1.5, to achieve a quality gearset. If, in fact, the gears are cut with sufficient accuracy, the tip-load condition is not the worst because another pair of teeth will be in contact when this conditions occurs. Examination of run-in teeth will show that the heaviest loads occur near the middle of the tooth. Therefore the maximum stress probably occurs while a single pair of teeth is carrying the full load, at a point where another pair of teeth is just on the verge of coming into contact.

The AGMA equation for the Lewis form factor overcomes both of these objections.* The equation is

$$Y = \frac{1}{\dfrac{\cos \phi_L}{\cos \phi} \left(\dfrac{1.5}{x} - \dfrac{\tan \phi_L}{t} \right)} \tag{13-20}$$

where ϕ_L is the angle between the total load vector $\mathbf{W}$ and a perpendicular to the centerline of the tooth at the highest point of single-tooth contact. The determi-

---

* AGMA Publication 225.01.

Table 13-3 VALUES OF THE AGMA LEWIS FORM FACTOR Y*

Number of teeth	$\phi = 20°$ $a = 0.800$ $b = 1.000$	$\phi = 20°$ $a = 1.000$ $b = 1.250$	$\phi = 25°$ $a = 1.000$ $b = 1.250$	$\phi = 25°$ $a = 1.000$ $b = 1.350$
12	0.335 12	0.229 60	0.276 77	0.254 73
13	0.348 27	0.243 17	0.292 81	0.271 77
14	0.359 85	0.255 30	0.307 17	0.287 11
15	0.370 13	0.266 22	0.320 09	0.301 00
16	0.379 31	0.276 10	0.331 78	0.313 63
17	0.387 57	0.285 08	0.342 40	0.325 17
18	0.395 02	0.293 27	0.352 10	0.335 74
19	0.401 79	0.300 78	0.360 99	0.345 46
20	0.407 97	0.307 69	0.369 16	0.354 44
21	0.413 63	0.314 06	0.376 71	0.362 76
22	0.418 83	0.319 97	0.383 70	0.370 48
24	0.428 06	0.330 56	0.396 24	0.384 39
26	0.436 01	0.339 79	0.407 17	0.396 57
28	0.442 94	0.347 90	0.416 78	0.407 33
30	0.449 02	0.355 10	0.425 30	0.416 91
34	0.459 20	0.367 31	0.439 76	0.433 23
38	0.467 40	0.377 27	0.451 56	0.446 63
45	0.478 46	0.390 93	0.467 74	0.465 11
50	0.484 58	0.398 60	0.476 81	0.475 55
60	0.493 91	0.410 47	0.490 86	0.491 77
75	0.503 45	0.422 83	0.505 46	0.508 77
100	0.513 21	0.435 74	0.520 71	0.526 65
150	0.523 21	0.449 30	0.536 68	0.545 56
300	0.533 48	0.463 64	0.553 51	0.565 70
Rack	0.544 06	0.478 97	0.571 39	0.587 39

Source: R. G. Mitchiner and H. H. Mabie, "The Determination of the Lewis Form Factor and the AGMA Geometry Factor J for External Spur Gear Teeth, *ASME Journal of Mechanical Design*, vol. 104, no. 1, Jan. 1982, pp. 148–158.

* All dimensions in inches; values for a diametral pitch of unity.

nation of the distances x and t are shown in Figs. 13-23 and 13-24. Note that the load angle ϕ_L differs from the pressure angle ϕ because the centerline of the tooth does not coincide with the centerline of the gears when the tooth is at the particular position corresponding to the highest point of single-tooth contact.

Stress Concentration

When Wilfred Lewis first proposed the formula for bending stress, stress-concentration factors were not in use. But we know now (see Chaps. 6 and 7) that there are a great many situations in which they must be used. Bagci* has developed techniques recently for the determination of stresses in the fillet of a

---

* See Fig. 6-27.

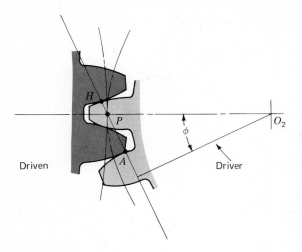

FIGURE 13-23 Driver turns clockwise; point *A* is initial point of contact. Point *H* is highest point of single-tooth contact. Line $O_2 P$ is usually *not* coincident with the centerline of the tooth.

gear tooth, but there is a great deal more work to be done before the results can be useful for analysis and design.

A photoelastic investigation (see Fig. 6-25) by Dolan and Broghamer conducted over 40 years ago still constitutes the primary source of information on stress concentration.† Mitchiner and Mabie interpret the results in terms of the

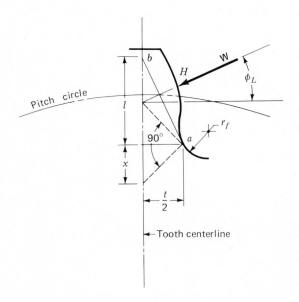

FIGURE 13-24 Graphical layout to obtain *x* and *t* when load **W** is exerted at highest point of single-tooth contact.

† T. J. Dolan and E. I. Broghamer, *A Photoelastic Study of the Stresses in Gear Tooth Fillets*, Bulletin 335, Univ. Ill. Expt. Sta., March, 1942. But see also R. E. Peterson, *Stress Concentration Factors*, Wiley, 1974, pp. 250, 251.

fatigue stress-concentration factor K_f as

$$K_f = H + \left(\frac{t}{r}\right)^L + \left(\frac{t}{l}\right)^M \tag{13-21}$$

where $H = 0.34 - 0.458\ 366\ 2\phi$ (a)

$L = 0.316 - 0.458\ 366\ 2\phi$ (b)

$M = 0.290 + 0.458\ 366\ 2\phi$ (c)

$$r = \frac{r_f + (b - r_f)^2}{(d/2) + b - r_f} \tag{d}$$

In these equations l and t are found from the layout of Fig. 13-22, ϕ is the pressure angle, r_f is the fillet radius, b is the dedendum, and d is the pitch diameter.

Geometry Factor

The AGMA has established a factor J, called the *geometry factor*, which uses the modified form factor Y of Eq. (13-20), the fatigue stress-concentration factor K_f of Eq. (13-21), and a *load-sharing ratio* m_N; the load-sharing ratio is based on the proportion of the total load carried by the most heavily loaded tooth. The AGMA equation is

$$J = \frac{Y}{K_f m_N} \tag{13-22}$$

Since the value of Y in Eq. (13-22) is based on the highest point of single-tooth contact, $m_N = 1$, and, for spur gears, Eq. (13-22) is written

$$J = \frac{Y}{K_f} \tag{13-23}$$

Here we emphasize that Y in Eq. (13-23) is the value given by Eq. (13-20) which is *not* the same as the values in Table 13-3.

With this definition of geometry factor, we can now write Eq. (13-18) in the form

$$\sigma = \frac{W_t P}{FJ} \tag{13-24}$$

which gives the normal stress corresponding to the total load W acting at the highest point of single-tooth contact and including the effects of stress concentration. Values of the geometry factor J are given in Tables 13-4 to 13-7 for some of the 20° and 25° standards. All of these are from the Mitchiner and Mabie paper; this paper includes data for other tooth geometries too.

13-12 DYNAMIC EFFECTS

When a pair of gears is driven at moderate or high speeds and noise is generated, it is certain that dynamic effects are present. One of the earliest efforts to account for an increase in dynamic load due to velocity employed a number of gears of the same size, material, and strength. Several of these gears were tested to destruction by meshing and loading them at zero velocity. The remaining gears were tested to destruction at various pitch-line velocities. Then, for example, if a pair of gears failed at 500 lb at zero velocity, and at 250 lb at a velocity V_1, then a *velocity factor*, designated as K_v, of 0.5 was specified for the gears at velocity V_1. Thus, another identical pair of gears running at a pitch-line velocity V_1 could be assumed to have a dynamic load equal to twice the transmitted load.

Table 13-4 AGMA GEOMETRY FACTOR J FOR TEETH HAVING $\phi = 20°$, $a = 0.800$ in, $b = 1.000$ in, AND $r_f = 0.304$ in

Number of teeth	Number of teeth in mating gear							
	1	17	25	35	50	85	300	1000
14	0.275 94	0.321 54	0.328 92	0.334 52	0.339 48	0.345 07	0.349 70	0.354 05
15	0.279 92	0.328 49	0.336 24	0.342 14	0.347 40	0.353 36	0.358 31	0.362 98
16	0.283 36	0.334 67	0.342 76	0.348 94	0.354 47	0.360 75	0.365 99	0.370 96
17	0.286 38	0.340 21	0.348 60	0.355 04	0.360 81	0.367 39	0.372 91	0.378 15
18	0.289 04	0.345 20	0.353 87	0.360 54	0.366 54	0.373 40	0.379 16	0.384 66
19	0.291 39	0.349 72	0.358 64	0.365 53	0.371 74	0.378 85	0.384 85	0.390 58
20	0.293 50	0.353 83	0.362 99	0.370 07	0.376 47	0.383 82	0.390 03	0.395 99
21	0.295 39	0.357 59	0.366 96	0.374 23	0.380 81	0.388 38	0.394 79	0.400 95
22	0.297 69	0.361 04	0.370 61	0.378 05	0.384 79	0.392 56	0.399 16	0.405 51
24	0.300 05	0.367 15	0.377 08	0.384 82	0.391 86	0.400 00	0.406 94	0.413 63
26	0.302 52	0.372 40	0.382 64	0.390 64	0.397 94	0.406 41	0.413 64	0.420 64
28	0.304 61	0.376 90	0.387 47	0.395 71	0.403 24	0.411 99	0.419 48	0.426 76
30	0.306 41	0.380 96	0.391 71	0.400 15	0.407 88	0.416 89	0.424 62	0.432 13
34	0.309 34	0.387 64	0.398 80	0.407 59	0.415 66	0.425 10	0.433 23	0.441 15
38	0.311 61	0.393 01	0.404 49	0.413 56	0.421 91	0.431 71	0.440 16	0.448 43
45	0.314 57	0.400 23	0.412 15	0.421 61	0.430 34	0.440 62	0.449 52	0.458 26
50	0.316 16	0.404 21	0.416 38	0.426 05	0.435 00	0.445 55	0.454 70	0.463 70
60	0.318 51	0.410 28	0.422 82	0.432 81	0.442 09	0.453 06	0.462 60	0.472 01
75	0.320 63	0.416 45	0.429 38	0.439 71	0.449 32	0.460 72	0.470 67	0.480 51
100	0.323 12	0.422 74	0.436 06	0.446 74	0.456 70	0.468 54	0.478 91	0.489 19
150	0.325 35	0.429 16	0.442 87	0.453 91	0.464 22	0.476 53	0.487 33	0.498 07
300	0.327 64	0.435 71	0.449 83	0.461 23	0.471 91	0.484 69	0.495 94	0.507 16
Rack	0.325 89	0.442 40	0.456 94	0.468 70	0.479 77	0.493 03	0.504 76	0.516 47

Table 13-5   AGMA GEOMETRY FACTOR J FOR TEETH HAVING $\phi = 20°$, $a = 1.000$ in, $b = 1.250$ in, AND $r_f = 0.300$ in

Number of teeth	Number of teeth in mating gear							
	1	17	25	35	50	85	300	1000
18	0.244 86	0.324 04	0.332 14	0.338 40	0.344 04	0.350 50	0.355 94	0.361 12
19	0.247 34	0.330 29	0.338 78	0.345 37	0.351 34	0.358 22	0.364 05	0.369 63
20	0.250 72	0.336 00	0.344 85	0.351 76	0.358 04	0.365 32	0.371 51	0.377 49
21	0.253 23	0.341 24	0.350 44	0.357 64	0.364 22	0.371 86	0.378 41	0.384 75
22	0.255 52	0.346 07	0.355 59	0.363 06	0.369 92	0.377 92	0.384 79	0.391 48
24	0.259 51	0.354 68	0.364 77	0.372 75	0.380 12	0.388 77	0.396 26	0.403 60
26	0.262 89	0.362 11	0.372 72	0.381 15	0.388 97	0.398 21	0.406 25	0.414 18
28	0.265 80	0.368 60	0.379 67	0.388 51	0.396 73	0.406 50	0.415 04	0.423 51
30	0.268 31	0.374 62	0.385 80	0.395 00	0.403 59	0.413 83	0.422 83	0.431 79
34	0.272 47	0.383 94	0.396 11	0.405 94	0.415 17	0.426 24	0.436 04	0.445 86
38	0.275 75	0.391 70	0.404 46	0.414 80	0.424 56	0.436 33	0.446 80	0.457 35
45	0.280 13	0.402 23	0.415 79	0.426 85	0.437 35	0.450 10	0.461 52	0.473 10
50	0.282 52	0.408 08	0.422 08	0.433 55	0.444 48	0.457 78	0.469 75	0.481 93
60	0.286 13	0.417 02	0.431 73	0.443 83	0.455 42	0.469 60	0.482 43	0.495 57
75	0.289 79	0.426 20	0.441 63	0.454 40	0.466 68	0.481 79	0.495 54	0.509 70
100	0.293 53	0.435 61	0.451 80	0.465 27	0.478 27	0.494 37	0.509 09	0.524 35
150	0.297 38	0.445 30	0.462 26	0.476 45	0.490 23	0.507 36	0.523 12	0.539 54
300	0.301 41	0.455 26	0.473 04	0.487 98	0.502 56	0.520 78	0.537 65	0.555 33
Rack	0.305 71	0.465 54	0.484 15	0.499 88	0.515 29	0.534 67	0.552 72	0.571 73

Carl G. Barth in the nineteenth century first expressed the velocity factor, also called *dynamic factor*, by the equation

$$K_v = \frac{600}{600 + V} \tag{13-25}$$

where V is the pitch-line velocity in feet per minute (fpm). This equation is called the Barth equation and it is known to be based on tests of cast-iron gears with cast teeth.   It is also highly probable that the tests were made on teeth having a cycloidal profile, instead of an involute; cycloidal teeth were in quite general use in those days because they were easier to cast than involute teeth.

The Barth equation is often modified to

$$K_v = \frac{1200}{1200 + V} \tag{13-26}$$

which is then used for cut or milled teeth or for gears not carefully generated.

The AGMA dynamic factors* are intended to account for

---

* *AGMA Information Sheet for Strength of Spur, Helical, Herringbone, and Bevel Gear Teeth*, AGMA 225.01, American Gear Manufacturers Association, Arlington, Va.

Table 13-6 AGMA GEOMETRY FACTOR J FOR TEETH HAVING $\phi = 25°$, $a = 1.000$ in, $b = 1.250$ in, AND $r_f = 0.300$ in

Number of teeth	Number of teeth in mating gear							
	1	17	25	35	50	85	300	1000
13	0.286 65	0.346 84	0.352 92	0.357 44	0.361 38	0.365 72	0.369 25	0.372 51
14	0.293 64	0.359 24	0.365 87	0.370 81	0.375 14	0.379 94	0.383 86	0.387 49
15	0.300 09	0.370 27	0.377 40	0.382 75	0.387 44	0.392 67	0.396 94	0.400 92
16	0.305 58	0.380 16	0.387 75	0.393 46	0.398 49	0.404 11	0.408 73	0.413 03
17	0.310 43	0.389 07	0.397 09	0.403 14	0.408 49	0.414 48	0.419 41	0.424 02
18	0.314 75	0.397 14	0.405 56	0.411 93	0.417 56	0.423 90	0.429 13	0.434 03
19	0.318 62	0.404 49	0.413 28	0.419 94	0.425 85	0.432 50	0.438 01	0.443 18
20	0.322 11	0.411 21	0.420 34	0.427 27	0.433 44	0.440 39	0.446 16	0.451 59
21	0.325 28	0.417 38	0.426 82	0.434 01	0.440 42	0.447 65	0.453 67	0.459 33
22	0.328 16	0.423 06	0.432 80	0.440 23	0.446 86	0.454 36	0.460 60	0.466 50
24	0.333 22	0.433 18	0.443 46	0.451 32	0.458 36	0.466 35	0.473 01	0.479 32
26	0.337 52	0.441 93	0.452 68	0.460 93	0.468 33	0.476 74	0.483 78	0.490 46
28	0.341 22	0.449 57	0.460 75	0.469 33	0.477 05	0.485 85	0.493 23	0.500 23
30	0.344 43	0.456 31	0.467 85	0.476 75	0.484 75	0.493 89	0.501 57	0.508 68
34	0.349 76	0.467 63	0.479 81	0.489 23	0.497 72	0.507 46	0.515 66	0.523 49
38	0.354 00	0.476 78	0.489 48	0.499 33	0.508 24	0.518 47	0.527 10	0.535 36
45	0.359 67	0.489 19	0.502 61	0.513 05	0.522 52	0.533 44	0.542 68	0.551 54
50	0.362 78	0.496 08	0.509 91	0.520 68	0.530 47	0.541 77	0.551 36	0.560 56
60	0.367 50	0.506 83	0.521 09	0.532 38	0.542 67	0.554 57	0.564 69	0.574 44
75	0.372 32	0.517 47	0.532 57	0.544 40	0.555 20	0.567 73	0.578 42	0.588 73
100	0.377 26	0.528 60	0.544 36	0.556 76	0.568 10	0.581 29	0.592 57	0.603 48
150	0.382 37	0.540 05	0.556 51	0.569 51	0.581 38	0.595 26	0.607 16	0.618 69
300	0.387 72	0.551 85	0.569 02	0.582 59	0.595 07	0.609 67	0.622 22	0.634 42
Rack	0.393 42	0.564 05	0.581 94	0.596 13	0.609 21	0.624 56	0.637 78	0.650 68

- The effect of tooth spacing and profile errors
- The effect of pitch-line velocity and rpm
- The inertia and stiffness of all rotating elements
- The transmitted load per inch of face width
- The tooth stiffness

For spur gears whose teeth are finished by hobbing or shaping, AGMA recommends the formula

$$K_v = \frac{50}{50 + \sqrt{V}} \tag{13-27}$$

If the gears have high-precision shaved or ground teeth and if an appreciable dynamic load is developed, then the AGMA dynamic factor is

$$K_v = \sqrt{\frac{78}{78 + \sqrt{V}}} \tag{13-28}$$

Table 13-7 AGMA GEOMETRY FACTOR J FOR TEETH HAVING $\phi = 25°$, $a = 1.000$ in, $b = 1.350$ in, AND $r_f = 0.245$ in

Number of teeth	Number of teeth in mating gear							
	1	17	25	35	50	85	300	1000
14	0.278 49	0.333 63	0.339 13	0.343 18	0.346 71	0.350 62	0.353 79	0.356 73
15	0.285 63	0.345 19	0.351 08	0.355 47	0.359 32	0.363 59	0.367 07	0.370 30
16	0.291 93	0.355 52	0.361 82	0.366 53	0.370 67	0.375 27	0.379 04	0.382 55
17	0.297 51	0.364 84	0.371 51	0.376 52	0.380 93	0.385 85	0.389 89	0.393 66
18	0.302 49	0.373 29	0.380 31	0.385 60	0.390 26	0.395 48	0.399 77	0.403 78
19	0.306 58	0.380 99	0.388 33	0.393 88	0.398 78	0.404 27	0.408 80	0.413 04
20	0.311 03	0.388 03	0.395 68	0.401 46	0.406 59	0.412 34	0.417 09	0.421 55
21	0.314 71	0.394 50	0.402 43	0.408 44	0.413 77	0.419 17	0.424 73	0.429 39
22	0.318 07	0.400 47	0.408 66	0.414 88	0.420 40	0.426 63	0.431 79	0.436 64
24	0.323 98	0.411 11	0.419 78	0.426 38	0.432 25	0.438 90	0.444 42	0.449 63
26	0.329 02	0.420 31	0.429 40	0.436 34	0.442 53	0.449 55	0.455 39	0.460 91
28	0.333 36	0.428 35	0.437 82	0.445 06	0.451 53	0.458 88	0.465 01	0.470 81
30	0.337 15	0.435 44	0.445 24	0.452 75	0.459 48	0.467 12	0.473 51	0.479 57
34	0.343 45	0.447 38	0.457 75	0.465 72	0.472 87	0.481 03	0.487 87	0.494 37
38	0.348 49	0.457 04	0.467 88	0.476 22	0.483 74	0.492 32	0.499 54	0.506 40
45	0.355 24	0.470 16	0.481 63	0.490 51	0.498 51	0.507 69	0.515 42	0.522 80
50	0.358 96	0.477 45	0.489 29	0.498 45	0.506 74	0.516 25	0.524 28	0.531 95
60	0.364 64	0.488 63	0.501 03	0.510 65	0.519 37	0.529 40	0.537 89	0.546 02
75	0.370 45	0.500 13	0.513 10	0.523 20	0.532 37	0.542 94	0.551 91	0.560 52
100	0.376 45	0.511 97	0.525 53	0.536 12	0.545 75	0.556 90	0.566 37	0.575 49
150	0.382 71	0.524 18	0.538 35	0.549 44	0.559 56	0.571 30	0.581 29	0.590 94
300	0.389 33	0.536 61	0.551 60	0.563 21	0.573 83	0.586 18	0.596 72	0.606 92
Rack	0.396 50	0.549 90	0.565 32	0.577 47	0.588 60	0.601 58	0.612 69	0.623 46

In both of these equations V is the pitch-line velocity in feet per minute.

If the gears have high-precision shaved or ground teeth and there is no appreciable dynamic load, then the AGMA recommends the dynamic factor $K_v = 1$. Thus, if the design involves high-accuracy gears, then you must decide whether there is an appreciable dynamic load or not. To do this, examine the driving and driven machinery. If the gears are between a motor and a fan, it is doubtful if much of a dynamic load will be developed. On the other hand, one would expect a considerable dynamic load if the gears were between, say, a one-cylinder engine and the blade of a chain saw.

Introducing the dynamic factor into Eqs. (13-18) and (13-24) gives

$$\sigma = \frac{W_t P}{K_v F Y} \tag{13-29}$$

$$\sigma = \frac{W_t P}{K_v F J} \tag{13-30}$$

Now you will need to know when to use each of these equations. Equation (13-29) is generally used when fatigue failure of the teeth is not a problem or when a quick estimate of gear size is needed for later use in a more detailed analysis. Equation (13-30) should be used for important applications and where the possibility of fatigue failure must be considered. It might be helpful in differentiating the two to call Eq. (13-29) the *Lewis equation* for bending stress, and Eq. (13-30) the *AGMA equation* for bending stress. Note that the form factor Y from Table 13-3 is to be used in Eq. (13-29).

13-13 ESTIMATING GEAR SIZE

In order to analyze a gearset to determine the reliability corresponding to a specified life, or to determine the factor of safety guarding against the several kinds of failures, it is necessary to know the size of the gears and the materials of which they are made. In this section we are concerned mostly with getting a preliminary estimate of the size of gears required to carry a given load. The method can also be used to design gearsets in which life and reliability are not very important design considerations.

The design approach presented here is based on the choice of a face width in the range $3p \geq F \geq 5p$. Gearsets having face widths greater than five times the circular pitch are quite likely to have a nonuniform distribution of the load across the face of the tooth because of the torsional deflection of the gear and shaft, because of the machining inaccuracies, and because of the necessity of maintaining very accurate and rigid bearing mountings. Thus a face width of five times the circular pitch is about the maximum, unless special precautions are taken with respect to machining, mounting, and stiffness of the entire assembly.

When the face width is less than three times the circular pitch, a larger gear is needed to carry the larger load per inch of face width. Large gears require more space in the gear enclosure and make the finished machine bigger and more expensive. Large gears are more expensive to manufacture because they require larger machines to generate the teeth, and these machines usually have a slower production rate. For these reasons a face width of three times the circular pitch is a good lower limit on the face width. It should be noted, however, that many other considerations arise in design that may dictate a face width outside the recommended range.

The gear size is obtained using iteration because both the transmitted load and the velocity depend, directly or indirectly, on the pitch P. The given information is usually:

- The horsepower H
- The speed n in rpm of the gear to be sized
- The number of teeth N on the gear to be sized
- The Lewis form factor Y (Table 13-3) for the gear to be sized
- The permissible bending stress σ_p

When estimating gear sizes it is a good idea to use a factor of safety of 3 or more, depending upon the material and application.

The computation procedure is to select a trial value for the diametral pitch and then to make the following successive computations:

1   The pitch diameter d in inches from the equation

$$d = \frac{N}{P} \tag{a}$$

2   The pitch-line velocity V in feet per minute from the equation

$$V = \frac{\pi d n}{12} \tag{b}$$

3   The transmitted load W_t in pounds from Eq. (13-15)

$$W_t = \frac{33(10)^3 H}{V} \tag{c}$$

4   The velocity factor K_v from Eq. (13-26)

$$K_v = \frac{1200}{1200 + V} \tag{d}$$

5   The face width F in inches from Eq. (13-29)

$$F = \frac{W_t P}{K_v Y \sigma_p} \tag{e}$$

6   The minimum and maximum face widths $3p$ and $5p$, respectively

These six steps can be programmed if desired. The procedure is illustrated in the example that follows.

Example 13-4   A pair of $4 : 1$ reduction gears is desired for a 100-hp 1120-rpm motor. The gears are to be $20°$ full depth with a clearance of $0.250/P$ and made of UNS G10400 steel heat-treated and drawn to $1000°$F. The teeth are generated using a rack cutter. Make an estimate of the size of gears required, assuming that the motor-starting torque is no more than the full-load torque at rated speed.

Solution   From Table 13-1 we find the minimum number of teeth to avoid undercutting is 18. Thus we choose a 72-tooth gear to go with the 18-tooth pinion to give a $4 : 1$ reduction.

Table 13-8

Quantities		Results	
Pitch P	3	4	5
Diameter d, in	6	4.5	3.6
Velocity V, fpm	1759	1319	1056
Load W_t, lb	1876	2501	3126
Face width F, in	2.25	3.41	4.77
$F_{min} = 3p$, in	3.14	2.36	1.88
$F_{max} = 5p$, in	5.24	3.93	3.14

When the gear and pinion are made of the same material the pinion is always the weaker of the two because the teeth of the smallest gear have a more undercut shape. (Observe how the Y factors vary in Table 13-3).

From Table 13-3 we find $Y = 0.293\ 27$ for the 18-tooth pinion. Also, from Table A-17 we find $S_y = 84$ kpsi. Choosing a factor of safety of 4 yields a permissible bending stress $\sigma_p = 21$ kpsi. This data as well as the given data, are entered into Eqs. (a) to (e) for $P = 3$, 4, and 5, chosen arbitrarily. The results are displayed in Table 13-8.

Table 13-8 shows that a pitch $P = 4$ gives a solution with $F = 3.41$ in, which is in the recommended range. A face width of $3\frac{1}{2}$ in would likely be chosen. ////

13-14 FATIGUE STRENGTH

Endurance limits for gear materials can be obtained using the methods of Chap. 7. Certain simplifications are possible for gearing and so we repeat Eq. (7-15) here as a convenience.

$$S_e = k_a k_b k_c k_d k_e k_f S_e' \tag{13-31}$$

where S_e = endurance limit of the gear tooth
S_e' = endurance limit of rotating-beam specimen
k_a = surface factor
k_b = size factor
k_c = reliability factor
k_d = temperature factor
k_e = modifying factor for stress concentration
k_f = miscellaneous-effects factor

Surface finish The surface factor k_a should always correspond to a machined finish, even when the flank of the tooth is ground or shaved. The reason for this is that the bottom land is usually not ground, but left as the original machined finish. For convenience, a chart of the corresponding surface factors from Fig. 7-10 is included here as Fig. 13-25.

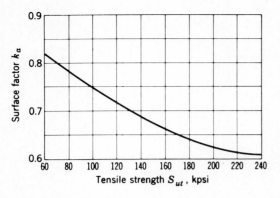

FIGURE 13-25 Surface-finish factors k_a for cut, shaved, and ground gear teeth.

Size The size factor, from Eq. (7-16) in inch units, is

$$k_b = \begin{cases} 0.869d^{-0.097} & 0.3 \text{ in} < d < 10 \text{ in} \\ 1 & 0.3 \text{ in} \geq d \end{cases} \qquad (a)$$

In this equation the dimension d is the diameter of a round specimen. A spur-gear tooth has a rectangular cross section and so the method of Fig. 7-15 must be used to get an equivalent value for d. For a rectangular cross section the formula for the equivalent diameter is

$$d_{\text{eq}} = \sqrt{\frac{0.05hb}{0.0766}} \qquad (b)$$

where h is the height of the section and b is the width. For a gear tooth h is the tooth thickness which is about half the circular pitch. And b is the face width F. Substituting $h = p/2$ and $F = 3p$ in Eq. (b) gives

$$d_{\text{eq}} = \sqrt{\frac{0.05(p/2)(3p)}{0.0766}} \cong p \qquad (c)$$

Thus a set of size factors can be worked out for various pitches using Eq. (a) and $p = \pi/P$. When this is done, we find

$$k_b = 1 \qquad P > 12 \text{ teeth/in} \qquad (13\text{-}32)$$

The corresponding size factors for gears having a diametral pitch less than 12 are listed in Table 13-9. Since many gears will have a face width greater than $3p$ the values in Table 13-9 are on the conservative side.

Reliability The reliability factors are applied exactly as in Chap. 7, and so a part of Table 7-7 is reproduced here as Table 13-10.

Table 13-9 SIZE FACTORS FOR SPUR-GEAR TEETH

Pitch P	Factor k_b	Pitch P	Factor k_b
2	0.832	6	0.925
$2\frac{1}{2}$	0.850	7	0.939
3	0.865	8	0.951
4	0.890	10	0.972
5	0.909	12	0.990

Temperature For the temperature factor use the methods of Sec. 7-9. For estimation purposes Eq. (7-22) can be used. Thus

$$k_d = \begin{cases} 1.0 & T \leq 450°C \ (840°F) \\ 1 - 5.8(10)^{-3}(T - 450) & 450°C < T \leq 550°C \\ 1 - 3.2(10)^{-3}(T - 840) & 840°F < T \leq 1020°F \end{cases} \tag{13-33}$$

But be sure to review the limitations of this equation in Chap. 7.

Stress concentration The fatigue stress-concentration factor K_f has been incorporated into the geometry factor J. Since it is fully accounted for, use $k_e = 1$ for gears.

Miscellaneous effects Gears that always rotate in the same direction and are not idlers are subjected to a tooth force that always acts on the same side of the tooth. Thus the fatigue load is repeated but not reversed and so the tooth is said to be subjected to one-way bending. For this situation the mean and alternating stress components are

$$\sigma_a = \sigma_m = \frac{\sigma}{2} \tag{d}$$

where σ is the tooth-bending stress as given by Eq. (13-30). This means that we can employ the miscellaneous-effects factor to increase the tooth endurance limit when it is subjected to only one-way bending. By substituting the values in Eq. (d) into Eq. (7-28) of the modified Goodman line we get

$$\sigma = \frac{2S_e S_{ut}}{S_{ut} + S_e} \tag{13-34}$$

Table 13-10 RELIABILITY FACTORS

Reliability R	0.50	0.90	0.95	0.99	0.999	0.9999
Factor k_c	1.000	0.897	0.868	0.814	0.753	0.702

Table 13-11 MISCELLANEOUS-EFFECTS FACTORS FOR ONE-
WAY BENDING

Tensile strength S_{ut}, kpsi	Up to 200	250	300	350	400
Factor k_f	1.33	1.43	1.50	1.56	1.60

In Chap. 7 we learned that $S'_e = 0.50S_{ut}$ when $S_{ut} \leq 200$ kpsi. By substituting $S_{ut} = S'_e/0.5$ in Eq. (13-34) we find that $\sigma = 1.33S'_e$. Thus the miscellaneous-effects factor k_f is 1.33 when $S_{ut} \leq 200$ kpsi. Equation (13-34) can be used to find other values of k_f when S_{ut} is greater than 200 kpsi. Some useful values have been calculated in this manner and are listed in Table 13-11.

Of course, for two-way bending, $k_f = 1.00$.

Cast iron For cast-iron gears use the endurance limits given by Eq. (7-2) or those in Table A-21. These values are fully corrected for surface finish but not for size, temperature, or miscellaneous effects. The low grades of cast irons should probably not be used if high reliabilities are desired. In any event the reliability factor k_c for cast irons should be evaluated in a laboratory testing program because the variance of the mechanical properties may be quite large.

13-15 FACTOR OF SAFETY

The formula

$$n_G = K_o K_m n \tag{13-35}$$

may be used to compute the factor of safety n_G for gears. In this formula K_o is the *overload factor*. Values recommended by AGMA are listed in Table 13-12. Factor K_m is an AGMA *load-distribution factor* which accounts for the possibility that the tooth force may not be uniformly distributed across the full face width. Use Table 13-13 for K_m. The factor n in Eq. (13-35) is the usual factor of safety as defined in Chap. 1. The AGMA practice is to use $n \geq 2$ to guard against fatigue failure.

Table 13-12 OVERLOAD CORRECTION FACTOR K_o

| Source of power | Driven machinery | | |
	Uniform	Moderate shock	Heavy shock
Uniform	1.00	1.25	1.75
Light shock	1.25	1.50	2.00
Medium shock	1.50	1.75	2.25

Source: Darle W. Dudley (ed.), *Gear Handbook*, McGraw-Hill, New York, 1962, p. 13–20.

Table 13-13 LOAD-DISTRIBUTION FACTOR K_m FOR SPUR GEARS

Characteristics of support	Face width, in			
	0 to 2	6	9	16 up
Accurate mountings, small bearing clearances, minimum deflection, precision gears	1.3	1.4	1.5	1.8
Less rigid mountings, less accurate gears, contact across full face	1.6	1.7	1.8	2.2
Accuracy and mounting such that less than full-face contact exists	Over 2.2			

* *Source*: Darle W. Dudley (ed.), *Gear Handbook*, McGraw-Hill, New York, 1962, p. 13–21.

Example 13-5   In Example 13-4 the size of a pair of 4 : 1 reduction gears for a 100-hp 1120-rpm motor was estimated as a face width of 3 1/2 in for a diametral pitch of 4, with 18 and 72 teeth, respectively, for the pinion and gear. The gears are 20° full depth with a clearance of 0.250/P and made of UNS G10400 steel heat-treated and drawn to 1000°F. Based on average mounting conditions, light shock in the driven machinery, and a reliability of 95 percent, find the factors of safety n_G and n guarding against a fatigue failure.

Solution   From Example 13-4 solution the pinion diameter is 4.5 in, the pitch-line velocity is 1319 fpm, and the transmitted load is 2501 lb. Using Eq. (13-27) we find the velocity factor to be

$$K_v = \frac{50}{50 + \sqrt{V}} = \frac{50}{50 + \sqrt{1319}} = 0.579$$

Next, we enter Table 13-5 with $N_2 = 18$ teeth and $N_3 = 72$ teeth. Interpolating between 50 and 85 gives $J = 0.348\ 10$. Substituting and solving Eq. (13-30) gives

$$\sigma = \frac{W_t P}{K_v FJ} = \frac{2501(4)(10)^3}{0.579(3.5)(0.348\ 10)} = 14.18 \text{ kpsi}$$

From Table A-17 we find $S_{ut} = 113$ kpsi. Thus Fig. 13-25 gives the surface factor as $k_a = 0.725$. The size factor is $k_b = 0.890$ from Table 13-9. Also, Table 13-10 gives $k_c = 0.868$ for a reliability of 95 percent.

We assume $k_d = 1$, and k_e is known to be unity. Also $k_f = 1.33$ from Table 13-11. Next, $S_e' = 0.5 S_{ut} = 0.5(113) = 56.5$ kpsi. The endurance limit is now found from Eq. (13-31) to be

$$S_e = k_a k_b k_c k_d k_e k_f S_e' = (0.725)(0.890)(0.868)(1)(1)(1.33)(56.5)$$

$$= 42.09 \text{ kpsi}$$

We now find $K_o = 1.25$ from Table 13-12 for moderate shock in the driven machinery and $K_m = 1.7$ for average mountings and gear accuracy. Thus, using Eq. (13-35), we get

$$n_G = K_o K_m n = (1.25)(1.70)n = 2.125n$$

The factor of safety n_G is

$$n_G = \frac{S_e}{\sigma} = \frac{42.09}{14.18} = 2.97 \qquad\qquad\qquad Ans.$$

Therefore

$$n = \frac{n_G}{2.125} = \frac{2.97}{2.125} = 1.40 \qquad\qquad\qquad Ans.$$

The value of Sec. 13-13 ("Estimating Gear Size") should now be quite evident. The reverse problem of beginning with the factor of safety and solving for the pitch and face width is somewhat more complicated when fatigue failure is a possibility to be considered. ////

13-16 SURFACE DURABILITY

The preceding sections have been concerned with the stress and strength of a gear tooth subjected to bending action and how to guard against the possibility of tooth breakage by static overloads or by fatigue action. In this section we are interested in the *failure of the surfaces* of gear teeth, generally called *wear*. *Pitting*, as explained in Sec. 7-19, is a surface fatigue failure due to many repetitions of high contact stresses. Other surface failures are *scoring*, which is a lubrication failure, or *abrasion*, which is wear due to the presence of foreign material.

To assure a satisfactory life, the gears must be designed so that the dynamic surface stresses are within the surface-endurance limit of the material. In many cases the first visible evidence of wear is seen near the pitch line; this seems reasonable because the maximum dynamic load occurs near this area.

To obtain an expression for the surface-contact stress, we shall employ the Hertz theory. In Eq. (2-90) it was shown that the contact stress between two cylinders may be computed from the equation

$$p_{max} = \frac{2F}{\pi b l} \qquad\qquad\qquad (13-36)$$

where p_{max} = surface pressure, psi
$\quad\quad\quad F$ = force pressing the two cylinders together, lb
$\quad\quad\quad l$ = length of cylinders, in

and b is obtained from the equation [Eq. (2-89)]

$$b = \sqrt{\frac{2F}{\pi l} \frac{[(1 - \mu_2^2)/E_1] + [(1 - \mu_2^2)/E_2]}{(1/d_1) + (1/d_2)}} \tag{13-37}$$

where μ_1, μ_2, E_1 and E_2 are the elastic constants and d_1 and d_2 are the diameters, respectively, of the two cylinders.

To adapt these relations to the notation used in gearing, we replace F by $W_t/\cos \phi$, d by $2r$, and l by the face width F. With these changes we can substitute the value of b as given by Eq. (13-37) into Eq. (13-36). Replacing p_{max} by σ_H, the *surface compressive stress (Hertzian stress)* is found to be

$$\sigma_H^2 = \frac{W_t}{\pi F \cos \phi} \frac{(1/r_1) + (1/r_2)}{[(1 - \mu_1^2)/E_1] + [(1 - \mu_2^2)/E_2]} \tag{13-38}$$

where r_1 and r_2 are the instantaneous values of the radii of curvature on the pinion- and gear-tooth profiles, respectively, at the point of contact. By accounting for load sharing in the value of W_t used, Eq. (13-38) can be solved for the Hertzian stress for any or all points from the beginning to the end of tooth contact. Of course, pure rolling exists only at the pitch point. Elsewhere, the motion is a mixture of rolling and sliding. Equation (13-38) does not account for any sliding action in the evaluation of stress.

As an example of the use of this equation let us find the contact stress when a pair of teeth is in contact at the pitch point. The radii of curvature r_1 and r_2 of the tooth profiles, when they are in contact at the pitch point, are

$$r_1 = \frac{d_P \sin \phi}{2} \quad\quad r_2 = \frac{d_G \sin \phi}{2} \tag{a}$$

where ϕ is the pressure angle. Then

$$\frac{1}{r_1} + \frac{1}{r_2} = \frac{2}{\sin \phi} \left(\frac{1}{d_P} + \frac{1}{d_G} \right) \tag{b}$$

Defining the *speed ratio* m_G as

$$m_G = \frac{N_G}{N_P} = \frac{d_G}{d_P} \tag{13-39}$$

we can write Eq. (b) as

$$\frac{1}{r_1} + \frac{1}{r_2} = \frac{2}{\sin\phi} \frac{m_G + 1}{m_G d_P} \tag{c}$$

After some rearranging and the use of Eq. (c), Eq. (13-38) becomes

$$\sigma_H = -\sqrt{\frac{W_t}{Fd_P} \frac{1}{\pi\left(\dfrac{1-\mu_P^2}{E_P} + \dfrac{1-\mu_G^2}{E_G}\right)} \frac{1}{\dfrac{\cos\phi\,\sin\phi}{2} \dfrac{m_G}{m_G+1}}} \tag{13-40}$$

The minus sign indicates that σ_H is a compressive stress. The subscripts P and G in Eq. (13-40) applied to μ and E refer to the pinion and gear, respectively.

The second term under the radical of Eq. (13-40) is called the *elastic coefficient* C_p. Thus, the formula for C_p is

$$C_p = \sqrt{\frac{1}{\pi\left(\dfrac{1-\mu_P^2}{E_P} + \dfrac{1-\mu_G^2}{E_G}\right)}} \tag{13-41}$$

Values of C_p have been worked out for various combinations of materials, and these are listed in Table 13-14.

The *geometry factor* I for spur gears is the denominator of the third term under the radical of Eq. (13-40). Thus

$$I = \frac{\cos\phi\,\sin\phi}{2} \frac{m_G}{m_G + 1} \tag{13-42}$$

Table 13-14 VALUES OF THE ELASTIC COEFFICIENT C_p FOR SPUR AND HELICAL GEARS WITH NONLOCALIZED CONTACT AND FOR $\mu = 0.30$.

	Gear			
Pinion*	Steel	Cast iron	Aluminum bronze	Tin bronze
Steel, $E = 30$	2300	2000	1950	1900
Cast iron, $E = 19$	2000	1800	1800	1750
Aluminum bronze, $E = 17.5$	1950	1800	1750	1700
Tin bronze, $E = 16$	1900	1750	1700	1650

Source: Darle W. Dudley (ed.), *Gear Handbook*, McGraw-Hill, New York, 1962, p. 13–22.

* In each case the modulus of elasticity is in Mpsi.

which is valid for external spur gears. For internal gears, the factor is

$$I = \frac{\cos \phi \sin \phi}{2} \frac{m_G}{m_G - 1} \qquad\qquad (13\text{-}43)$$

Now recall that a velocity factor K_v was used in the bending-stress equation to account for the fact that the force between the teeth is actually more than the transmitted load because of the dynamic effect. This factor must also be used in the equation for surface-compressive stress for exactly the same reasons. When used here, the velocity factor is designated as C_v, but it has the same values and so $C_v = K_v$; the same formulas are used.

With Eqs. (13-41) to (13-43) and the addition of the velocity factor, Eq. (13-40) can be written in the more useful form

$$\sigma_H = -C_P \sqrt{\frac{W_t}{C_v F d_P I}} \qquad\qquad (13\text{-}44)$$

13-17 SURFACE FATIGUE STRENGTH

The methods of Sec. 7-19 can be used to determine the surface strength of contacting gear teeth. Even though the contacting teeth are subjected to repeated compressive stresses, we learned in Sec. 7-19 that a critical stress element below the surface is quite likely to be subject to some stress reversal.

The surface fatigue strength for steels is given in Sec. 7-19 as

$$S_C = 0.4 H_B - 10 \qquad \text{kpsi} \qquad\qquad (13\text{-}45)$$

where H_B is the Brinell hardness of the softer of the two contacting surfaces. We also observe that the value given by Eq. (13-45) corresponds to a life of 10^8 stress applications.

The AGMA recommends that this contact fatigue strength be modified in a manner quite similar to that used for the bending endurance limit. The equation is

$$S_H = \frac{C_L C_H}{C_T C_R} S_C \qquad\qquad (13\text{-}46)$$

where S_H = corrected fatigue strength, or Hertzian strength
$\quad C_L$ = life factor
$\quad C_H$ = hardness-ratio factor; use 1.0 for spur gears
$\quad C_T$ = temperature factor; use 1.0 for temperatures less than 250°F
$\quad C_R$ = reliability factor

Table 13-15   LIFE AND RELIABILITY MODIFICATION FACTORS

Cycles of life	Life factor C_L	Reliability R	Reliability factor C_R
10^4	1.5	Up to 0.99	0.80
10^5	1.3	0.99 to 0.999	1.00
10^6	1.1	0.999 up	1.25 up
10^8 up	1.0		

The *life modification factor* C_L is used to increase the strength when the gear is to be used for short periods of time; use Table 13-15.   The *reliability modification factor* C_R, as presented by AGMA, is rather vague.   It is believed that the AGMA meant the values of C_R to be about as listed in Table 13-15.

The *hardness ratio factor* C_H was included by AGMA to account for differences in strength due to the fact that one of the mating gears might be softer than the other.   However, for spur gears, use $C_H = 1$.

The AGMA makes no recommendations on values to use for the *temperature factor* C_T when the temperature exceeds 250°F, except to imply that a value $C_T > 1.0$ should probably be used.   To a large extent this will depend upon the temperature limitations of the lubricant used, since the materials should withstand larger temperatures.   See Eq. (13-33), for example.

Factors of safety to guard against surface failures should be selected using the guidelines outlined in Sec. 13-15 and Eq. (13-35).   The AGMA uses C_o and C_m to designate the overload and load-distribution factors, but their values are the same as those for K_o and K_m.   These factors should be used in the numerator of Eq. (13-44) as load-multiplying factors.

Using Eq. (1-7) we now designate the permissible transmitted load $W_{t,p}$ as

$$W_{t,p} = n_G W_t \qquad\qquad (a)$$

where, from Eq. (13-35),

$$n_G = C_o C_m n \qquad\qquad (13\text{-}47)$$

Equation (13-44) can now be written as

$$S_H = C_p \sqrt{\frac{W_{t,p}}{C_v F d_p I}} \qquad\qquad (13\text{-}48)$$

Note that this step is necessary because σ_H and W_t in Eq. (13-44) are not linearly related.

As we have observed many times in this book, there is no satisfactory substitute for a thorough laboratory testing program to verify analytical results.   This is particularly true in the design of gearing for a long life.

Example 13-6 Determine the factors of safety n_G and n guarding against a surface fatigue failure for the gearset of Example 13-5.

Solution The material of both gears is UNS G10400 steel, heat-treated and drawn to 1000°F. Table A-17 gives the tensile strength as $S_{ut} = 113$ kpsi, which corresponds to a Brinell hardness of 235. Thus Eq. (13-45) gives the contact strength as

$$S_C = 0.4H_B - 10 = 0.4(235) - 10 = 84 \text{ kpsi}$$

Using Table 13-15 we select $C_L = 1.10$ for 10^6 cycles and $C_R = 0.80$ for a reliability of 95 percent, as specified in Example 13-5. We also choose $C_T = C_H = 1$. Thus the Hertzian strength is, from Eq. (13-46),

$$S_H = \frac{C_L C_H}{C_T C_R} S_C = \frac{(1.10)(1)}{(1)(0.80)} (84) = 115.5 \text{ kpsi}$$

Next, we use $C_o = K_o = 1.25$ and $C_m = K_m = 1.7$ from Example 13-5. Thus

$$n_G = (1.25)(1.7)n = 2.125n$$

The tooth numbers are 18 and 72, the pressure angle is 20°, and the pitch is 4 teeth/in. This gives pitch diameters of $d_P = 18/4 = 4.5$ in and $d_G = 72/4 = 18$ in. So the speed ratio is $m_G = d_G/d_P = 18/4.5 = 4$, which, of course, was given in the example.

Using Eq. (13-42) we find the geometry factor I to be

$$I = \frac{\cos \phi \sin \phi}{2} \frac{m_G}{m_G + 1} = \frac{\cos 20° \sin 20°}{2} \frac{4}{4 + 1} = 0.129$$

We use $C_v = K_v = 0.579$ from Example 13-5. From the same example $W_t = 2501$ lb. Table 13-14 gives $C_p = 2300$ for steel on steel. We now want to compute the permissible tangential or transmitted load. Substituting directly into Eq. (13-48) gives

$$115.5(10)^3 = 2300 \sqrt{\frac{W_{t,p}}{(0.579)(3.5)(4.5)(0.129)}}$$

Solving gives $W_{t,p} = 2967$ lb. Since $W_t = 2501$ lb from Example 13-5, the factor n_G is

$$n_G = \frac{W_{t,p}}{W_t} = \frac{2967}{2501} = 1.19 \qquad Ans.$$

Equation (13-47) then gives

$$n = \frac{n_G}{C_o C_m} = \frac{1.19}{2.125} = 0.560$$ *Ans.*

Thus, there is no safety against surface fatigue failure, and the gears can be expected to have a wear life somewhat less than 10^6 stress applications.   ////

13-18 HEAT DISSIPATION

The power loss at each tooth mesh for spur gears is usually less than 1 percent of the power transmitted. The magnitude of this loss depends upon the gear materials, the tooth system, the lubrication, the character of the tooth surface, and the pitch-line velocity. In addition, there is a power loss at the bearings which may reach 1 or 2 percent. When the gearset is mounted in a housing, it is suggested that the power loss of the gears be added to that for the bearings and that Eq. (12-26) be used.

It is sometimes necessary to direct a stream of cooling oil against the teeth in order to carry away the generated heat. A rule of thumb which is occasionally employed is to use 1 gpm of cooling oil for each 400 hp transmitted.

13-19 GEAR MATERIALS

Gears are commonly made of steel, cast iron, bronze, or phenolic resins. Recently nylon, Teflon, titanium, and sintered iron have been used successfully. The great variety of materials available provides the designer with the opportunity of obtaining the optimum material for any particular requirement, whether it be high strength, a long wear life, quietness of operation, or high reliability.

In many applications, steel is the only satisfactory material because it combines both high strength and low cost. Gears are made of both plain-carbon and alloy steels, and there is no one best material. In many cases the choice will depend upon the relative success of the heat-treating department with the various steels. When the gear is to be quenched and tempered, a steel with 40 to 60 points of carbon is used. If it is to be case-hardened, one with 20 points or less of carbon is used. The core as well as the surface properties must always be considered.

Cast iron is a very useful gear material because it has such good wear resistance. It is easy to cast and machine and transmits less noise than steel. The tensile strengths of AGMA grades of cast irons are the same as the ASTM grades listed in the Appendix.

Bronzes may be used for gears when corrosion is a problem, and they are quite useful for reducing friction and wear when the sliding velocity is high, as in worm-gear applications. The AGMA lists five tin bronzes containing small per-

centages of nickel, lead, or zinc which are suitable gear materials. Their hardness varies from 70 to 85 Bhn.

Nonmetallic gears are mated with steel or cast-iron gears to obtain the greatest load-carrying capacity. To secure good wear resistance, the metal gear should have a hardness of at least 300 Bhn. A nonmetallic gear will carry almost as much load as a good cast-iron or mild-steel gear, even though the strength is much lower, because of the low modulus of elasticity. This low modulus permits the nonmetallic gear to absorb the effects of tooth errors so that a dynamic load is not created. A nonmetallic gear also has the important advantage of operating well on marginal lubrication.

Thermosetting laminates are widely used for gears. They are made of sheet materials composed of a fibrous or woven material, together with a resin binder, or are cast. Both nylon and Teflon, as gear materials, have given excellent results in service.

13-20 GEAR-BLANK DESIGN

Gear blanks are produced by casting, forging, machining from a solid blank, and fabricating. Some typical fabrication methods are shown in Fig. 13-26. When

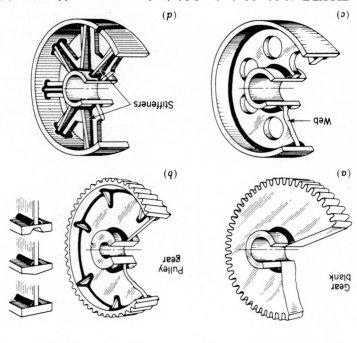

FIGURE 13-26 Methods of fabricating gears as weldments. (a) Solid gear blank to which hub is welded. (b) The gear has a solid web with stiffeners, giving additional support to the rim. (c) A satisfactory construction for small-diameter gears with short face widths. (d) A fabricated gear blank with spokes. (Courtesy of Lincoln Electric Company.)

the pinion is small, it is frequently made integral with the shaft, thus eliminating the key as well as an axial-locating device.

In designing a gear blank, rigidity is almost always a prime consideration. The hub must be thick enough to maintain a proper fit with the shaft and to provide sufficient metal for the key slot. This thickness must also be large enough so that the torque may be transmitted through the hub to the web or spokes without serious stress concentrations. The hub must have length so that the gear will rotate in a single plane without wobble. The arms or web and the rim must also have rigidity, but not too much, because of the effect upon the dynamic load.

There are no general rules for the design of hubs. If they are designed with sufficient rigidity, the stresses are usually quite low, especially when compared with the tooth stresses. The length of the hub should be at least equal to the face width, or greater if this does not give sufficient key length. Sometimes two keys are used. If the clearance between the bore and the shaft is large, the hub should have a length which is at least twice the bore diameter, because a slight inaccuracy here is magnified at the rim. Many designers prefer to make a scale drawing of the gear; the hub dimensions can then be adjusted by eye to obtain the necessary rigidity.

Figure 13-27 is a drawing of a portion of a cast-iron gear. The hub bead is

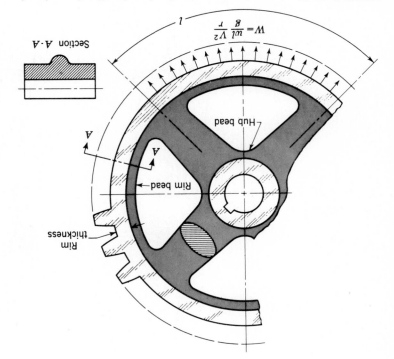

$$W = \frac{wl}{g}\frac{V^2}{r}$$

Section A-A

FIGURE 13-27 A cast-iron gear showing how bending is produced by the centrifugal force.

used to brace the arms and reduce the stress concentration caused by the torque transmitted from the hub to the arms. The arms are shown with an elliptical section, but they may also be designed with an H or I section, or any other shape, depending upon the stiffness and strength desired. The rim bead gives additional rigidity and strength to the rim.

If a gear rotates at a high pitch-line velocity, the weight of the rim and teeth may be sufficient to cause large bending stresses in the portion of the rim contained between any two arms. When the gear is made of steel, these stresses are usually not serious, but when cast iron is used, this stress should be checked. Although the problem is complicated, an approximation may be obtained by assuming that the rim is a uniformly loaded beam fixed at the ends by the spokes. The length of the beam would be the length of arc measured at the mean rim diameter between the spoke centerlines. Under these assumptions, the total bending load W is

$$W = \frac{wl}{g}\frac{V^2}{r} \qquad\qquad (13\text{-}49)$$

where w = unit weight of rim and teeth, lb per in
   V = pitch-line velocity, fps
   g = acceleration due to gravity, fps^2

The maximum bending moment occurs at the arms and is

$$M_{\text{max}} = \frac{Wl}{12} \qquad\qquad (13\text{-}50)$$

The stress may then be obtained by substituting the maximum moment and the section modulus into the equation for bending stress, $\sigma = Mc/I$. This solution neglects the curvature of the rim; the tensile, compressive, or bending forces in the rim due to the transfer of torque between the arms and the rim; and the stress-concentration effect where the arm joins the rim. In addition, we cannot be sure of the accuracy of the assumption that the ends are fixed.

The rim must also have rigidity in the direction parallel to the axis of the gear (Fig. 13-28a) to maintain a uniform load across the face. This means that the arm or web must be thick enough for adequate support.

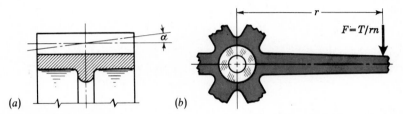

FIGURE 13-28 (a) Section of rim; the rim must have sufficient rigidity to avoid deflection through the angle α. (b) Bending force on a spoke.

The loading on the arms of a gear is complicated. The transmitted torque produces bending, the centrifugal force on the rim produces a combination of bending and tension, and the dynamic load acting between the teeth produces a vibrating bending force. An approximation can be obtained by neglecting all these except the bending produced by the transmitted torque. Then the bending force is (Fig. 13-28*b*)

$$F = \frac{T}{rn} \tag{13-51}$$

where T = transmitted torque, lb·in
r = length of spoke, in
n = number of spokes

The stress may then be determined by finding the maximum moment for a canti-lever beam and substituting this, together with the section modulus, in the bending-stress equation, $\sigma = Mc/I$. A high factor of safety should be used because this method is only a rough approximation and stress concentration is present.

The analytical methods investigated above are not used every time a gear is to be designed. In many cases the loads and velocities are not high, and the gear can be designed on the drawing board by using pleasing proportions. On the other hand, cases sometimes occur where the loads are extremely high, or where the weight of the gear is a very important consideration; in these situations it may be desirable to make a much more thorough investigation than is indicated here.

PROBLEMS*

Sections 13-1 to 13-8

13-1 A 21-tooth pinion has a diametral pitch of 7 teeth/in, runs at 1150 rpm, and drives a gear at 690 rpm. Find the number of teeth on the gear and the theoretical center distance.

13-2 A 19-tooth pinion has a module of 2.5 mm and runs at a speed of 1740 rpm. The driven gear is to operate at about 470 rpm. Find the circular pitch, the number of teeth on the gear, and the theoretical center distance.

13-3 A gearset has a circular pitch of $2\frac{1}{4}$ in and a velocity ratio of 3.00. The pinion has 18 teeth. Find the number of teeth on the driven gear, the diametral pitch, and the theoretical center distance.

13-4 A 21-tooth pinion mates with a 28-tooth gear. The diametral pitch is 3 teeth/in and the pressure angle is 20°. Make a drawing of the gears showing one tooth on each gear. Find and tabulate the following results: the addendum, dedendum,

---

* The asterisk indicates a design-type problem or one which may have no unique result.

clearance, circular pitch, tooth thickness, and base-circle diameters; the lengths of the arcs of approach, recess, and action; and the base pitch and contact ratio.

13-5 A 17-tooth pinion paired with a 50-tooth gear has a diametral pitch of $2\frac{1}{2}$ teeth/in and a 20° pressure angle. Make a drawing of the gears showing one tooth on each gear. Find the arcs of approach, recess, and action and the contact ratio.

13-6 Draw a 26-tooth pinion in mesh with a rack having a diametral pitch of 2 teeth/in and a pressure angle of 20°.

(a) Find the arcs of approach, recess, and action and the contact ratio.

(b) Draw a second rack in mesh with the same gear but offset $\frac{1}{8}$ in away from the pinion center. Determine the new contact ratio. Has the pressure angle changed?

13-7 A 15-tooth pinion having a 25° pressure angle and a diametral pitch of 3 teeth/in is to drive an 18-tooth gear. Without drawing the teeth, make a full-scale drawing and show the pitch circles, base circles, addendum circles, dedendum circles, and the pressure line. Locate both interference points and show the amount of interference, if it exists. Locate the initial and final points of contact and label them. Compute the base pitch and find the contact ratio.

13-8 We wish to establish a new gear-tooth system having teeth in which the addendum is still $1/P$ but such that a 12-tooth pinion will not be undercut when it is generated. We also wish to use the smallest possible pressure angle. Calculate the value of this angle if the mating gear is a rack.

Section 13-9

13-9 In part *a* of the figure, shaft *a* is attached to the planet carrier, shaft *b* is keyed to sun gear 5, and sun gear 2 is fixed to the housing. Planet gears 3 and 4 are attached to each other. Shaft *a* is driven at 800 rpm ccw; find the speed and direction of rotation of shaft *b*.

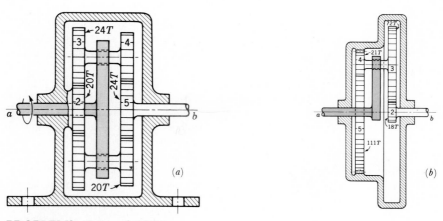

(a)   (b)

PROBLEMS 13-9 and 13-10

13-10 The epicyclic train shown in part *b* of the figure has the arm connected to shaft *a*, and sun gear 2 connected to shaft *b*. Gear 5, having 111 teeth, is a part of the frame, and planet gears 3 and 4 are both keyed to the same shaft. Note that gear 5 is an internal gear and that the teeth shown in the cross-sectional view are on the inside of the housing.

(*a*)   Find the speed and direction of rotation of shaft *a* if the frame is held stationary and shaft *b* is driven at 120 rpm cw.

(*b*)   Shaft *b* is held stationary, and the arm rotates at 8 rpm cw. Find the speed and direction of rotation of the housing.

13-11   Part *a* of the figure shows a reverted planetary train. Gear 2 is fastened to its shaft and is driven at 250 rpm in a clockwise direction. Gears 4 and 5 are planet gears which are joined to each other but are free to turn on the shaft carried by the arm. Gear 6 is stationary. Find the rpm and direction of rotation of the arm.

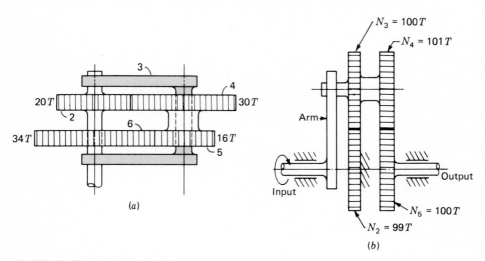

(*a*)

(*b*)

PROBLEMS 13-11 and 13-12

13-12   By using nonstandard gears, it is possible to mate a 99-T gear with a 100-T gear at the same distance between centers as would be required to mate a 100-T gear with a 101-T gear. The planetary train shown in part *b* of the figure is based on this idea.

(*a*)   Find the ratio of the speed of the output shaft to the speed of the input shaft.

(*b*)   The housing for this planetary train is cylindrical, with the axis of the cylinder coincident with the axis of the input and output shafts. If the pitch of gears 4 and 5 is 10 teeth per inch of pitch diameter, and if these gears have standard addenda, what should be the inside diameter of this housing? Allow $\frac{1}{2}$ in of radial clearance.

(*c*)   Suppose you designed an ordinary two-gear nonplanetary train having the same speed ratio. How many teeth would the gear have if the pinion had 20 teeth and was 10 diametral pitch? What size cylinder would be needed to enclose such a gear train? Use the same clearance as in part *b*.

13-13   The tooth numbers of the gears shown in the figure are: $N_2 = 20$, $N_3 = 18$, $N_5 = 20$, and $N_6 = 58$. Gear 6 is an internal stationary gear. Arm 4 is connected to output shaft *b*. Gear 2 is driven at 330 rpm cw by the input shaft *a*. Find the speed and direction of rotation of the output shaft.

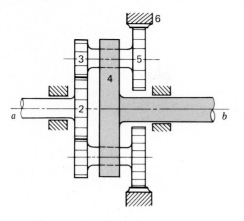

PROBLEM 13-13

13-14* Using not less than 15 teeth, design a gear train similar to that of Prob. 13-13 such that the output rpm is half of the input rpm.

13-15 The gear train shown in the figure has $N_2 = 24$ teeth, $N_3 = 18$ teeth, $N_4 = 30$ teeth, and $N_6 = 36$ teeth.

  (*a*) How many teeth must gear 7 have if all are of the same pitch?

  (*b*) Gear 7 is fixed and output shaft *b* is connected to the arm. If the speed of input shaft *a* is 250 rpm ccw, what is the speed and direction of rotation of shaft *b*?

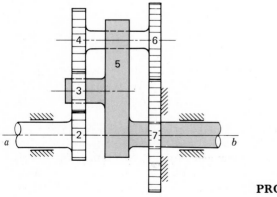

PROBLEM 13-15

13-16 Tooth numbers for the gear train shown in the figure are $N_2 = 18$, $N_3 = 24$, $N_4 = 18$, and $N_5 = 102$. Gear 5 is a stationary internal gear. Find the speed of the arm if input shaft *a* rotates at 500 rpm cw.

13-17 (*a*) In Prob. 13-16 gear 5 also rotates at 150 rpm cw. What is the speed of the arm?

  (*b*) How fast and in what direction must gear 5 of Prob. 13-16 rotate if the arm is not to rotate at all?

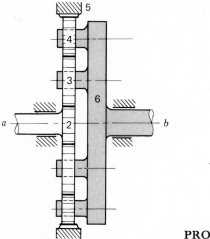

PROBLEM 13-16

Section 13-10

13-18 The gears shown in part *a* of the figure are 3 diametral pitch and 20° pressure angle, and are in the same plane or can be assumed to be. The pinion rotates counterclockwise at 600 rpm and transmits 25 hp through the idler to the 28-*T* gear on shaft *c*. Calculate the resulting shaft reaction on the 36-*T* idler.

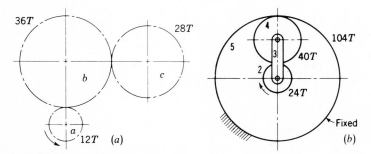

PROBLEMS 13-18 and 13-19

13-19 The 24-tooth 8-pitch 20° pinion shown in the figure for part *b* rotates clockwise at 900 rpm and transmits 3 hp to the planetary gear train. What torque can arm 3 deliver to its output shaft? Draw a free-body diagram of the arm and of each gear, and show all forces which act upon them.

13-20 Part *a* of the figure illustrates a double-reduction gear train. Shaft *a* is driven by a 1.25-kW source at 1720 rpm. The reduction between shafts *a* and *b* is 3.5 : 1, and between shafts *b* and *c* it is 4 : 1. The pinion on shaft *a* has 24 teeth, and the gear on shaft *c* has 160 teeth.

(*a*) Find the tooth numbers for the gears on shaft *b*.

(*b*) Find the speed of shafts *b* and *c*.

(*c*) If the power loss is 4 percent at each mesh, find the torque at each shaft.

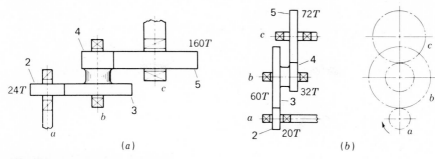

PROBLEMS 13-20 and 13-21

13-21 In part *b* of the figure, the gears connecting shafts *a* and *b* have a module of 2.5 mm and a 20° pressure angle. The mating gears on shafts *b* and *c* have a module of 3 mm with a 20° pressure angle. The pinion transmits 1.5 kW at a speed of $20\,\text{s}^{-1}$. Find the resulting shaft reactions by assuming that the forces acting on gears 3 and 4 are in the same plane.

13-22 In part *a* of the figure, gear 2 on shaft *a* is the driver and it transmits 10 hp at 600 rpm to gear 3 on shaft *b*. Gear 2 is 12-pitch and has 20 teeth and a 25° pressure angle. Gear 3 has 60 teeth. Gear 4 on shaft *b* has 24 teeth, is 6-pitch, and has a pressure angle of 20°. Gear 5 has 56 teeth.

(*a*) Find the shaft center distances.

(*b*) Find the force with which each gear pushes on its shaft.

(*c*) Assume gears 3 and 4 are in the same plane and find the resultant force on shaft *b*.

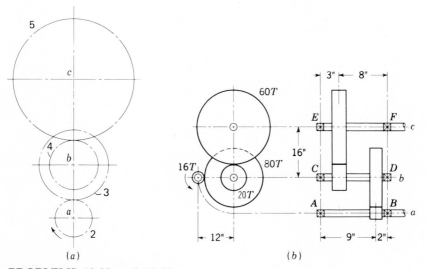

PROBLEMS 13-22 and 13-23

13-23 The 16-tooth pinion drives the double-reduction gear train shown in part *b* of the figure. All gears have 25° pressure angles. The pinion rotates counterclockwise at

1200 rpm and transmits 50 hp to the gear train. Calculate the magnitude and direction of the radial force exerted by each bearing on its shaft.

13-24 A 10-hp motor drives gear 2 in the figure at 1800 rpm. A flat belt on pulley 4 drives pulley 5. The shaft to which pulley 5 is attached drives a blower.

    (a) Find the speed and the torque input to the blower, assuming no frictional losses.

    (b) The belt tension on the loose side is 20 percent of the tension on the tight side. Assuming the belt tensile forces are vertical, find the tension in both sides.

    (c) Find the torque in countershaft b.

    (d) Compute the force exerted by pulley 4 on the countershaft.

    (e) Compute the y and z components of the force with which the bearings at A and D push against the shaft.

    (f) Locate and find the maximum bending moment on the countershaft.

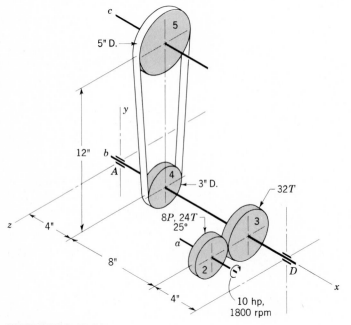

PROBLEM 13-24

Sections 13-11 and 13-12

13-25* Devise a program for a programmable calculator or desk-top computer to interpolate for the J factors of Tables 13-4 to 13-7. Note that interpolation may be necessary for the number of teeth in column 1, the number of teeth in mating gear in row 1, or both.

13-26 A gearset consists of a 16-tooth pinion that mates with a 32-tooth gear. The gears have a face width of $1\frac{1}{4}$ in and a diametral pitch of 10 teeth/in; they are cut by hobbing, using the 20° full-depth system, with $b = 1.25/P$. The material is UNS

G10350 cold-drawn steel. Estimate the horsepower capacity of this gearset based on the yield strength of the material, a factor of safety of 5, and a pinion speed of 1720 rpm.

13-27 A steel pinion has a pitch of 6 teeth/in, a 20° pressure angle, and 22 teeth. This pinion runs at 900 rpm and transmits 12.5 hp to a 60-tooth gear. Compute the bending stress on the pinion teeth using Eq. (13-29) based on a face width of $1\frac{1}{2}$ in.

13-28 Estimate the face width and pitch needed for an 8 : 1 reduction gearset to be connected between a 30-hp gas turbine running at 3600 rpm and a centrifugal pump. The gears are to be made of UNS G61500 steel heat-treated and drawn to 1000°F. Use 25° full-depth teeth with $b = 1.250/P$ and a factor of safety of 5.

13-29* A set of ASTM No. 25 cast-iron gears is to be designed to transmit 1.5 hp at a pinion speed of 400 rpm and a speed reduction of 1.5 : 1. Use a factor of safety of 4 and determine suitable values for the pitch, pitch diameters, tooth numbers, and face width. Use 20° full-depth teeth with $b = 1.250/P$.

Sections 13-14 and 13-15

13-30 A 15-tooth, 16-pitch pinion is paired with a 64-tooth gear. The gears have a face width of 1 in and are hobbed using a pressure angle of 25°, an addendum of $1/P$, and a dedendum of $1.25/P$. The pinion is cut from hot-rolled UNS G10350 steel and the gear is made of ASTM No. 30 cast iron. The application is such that light shock loads may be present in both the driving and driven machinery. Use a factor of safety $n = 2.5$, 50 percent reliability, average mounting conditions, a pitch-line velocity of 750 fpm, and find the safe horsepower capacity of this gearset.

13-31 A 14-tooth precision-made pinion is to drive a 21-tooth gear. A pitch of 8 teeth/in has been selected with a $2\frac{1}{4}$-in face width based upon a 25° pressure angle, a dedendum of $1.350/P$, a pinion speed of 1150 rpm and 25 hp to be transmitted under steady load conditions. The material selected is a forged UNS G10400 steel heat-treated to a hardness of 235 Bhn. Determine the factors of safety guarding against a fatigue failure for 99 percent reliability with better-than-average mountings and cutting accuracy.

13-32 A 15-tooth, 3-pitch pinion has a pressure angle of 25° and a dedendum of $1.250/P$. The teeth are hobbed and the material is forged UNS G10350 steel heat-treated to 220 Bhn. The pinion is to drive a 75-tooth gear made of ASTM No. 35 cast iron. The face widths are both 4 in. The pinion is driven at 575 rpm through V belts connected to an electric motor. The gear drives machinery subject to moderate shock loads. Use a reliability of 90 percent, a factor of safety of $n = 3$, $K_m = 1.60$, and find the safe horsepower capacity of the gearset.

Sections 13-16 and 13-17

13-33 A hobbed steel pinion has a pitch of 6 teeth/in, a 20° pressure angle with a dedendum of $1.250/P$, and 22 teeth. This pinion rotates at 575 rpm and transmits 10 hp to a 60-tooth steel gear. Determine the critical surface compressive stress on the teeth based on a face width of $1\frac{1}{2}$ in.

13-34 A 12-tooth, 10-pitch steel pinion hobbed using a 25° pressure angle with a dedendum of $1.350/P$ is mated with a 72-tooth cast-iron gear. Compute the surface contact stress that would be obtained if 2 hp is to be transmitted and the pinion speed is 1800 rpm. The face widths are $1\frac{1}{4}$ in.

13-35 A pair of mating 8-pitch gears have a 25° pressure angle, dedenda of $1.250/P$, and a face width of 2 in. Both gears are made of UNS G43400 steel, heat-treated to a Brinell hardness of 363, and have shaved teeth. The 14-tooth pinion rotates at 1150 rpm and drives a 21-tooth gear. The bearing mountings are quite rigid and accurate and the input and output loading is steady. A reliability of 99 percent is desired with a factor of safety $n = 2$. What horsepower can this gearset safely transmit?

13-36 A 12-tooth pinion has a pitch of 12 teeth/in and drives a 64-tooth gear. The gears have a face width of 1 in and are hobbed using a pressure angle of 25° and a dedendum of $1.250/P$. The pinion is cut from UNS G10150 hot-rolled steel and the gear is made of ASTM No. 20 cast iron. Both the driving and driven machinery are subject to slightly unsteady loading. Use a factor of safety $n = 1.5$, 50 percent reliability, below-average mounting conditions, a pitch-line velocity of 750 fpm, and find the horsepower capacity of this gearset based on the contact strength.

13-37 A 15-tooth, precision-made pinion is to drive a 25-tooth gear. A pitch of 6 teeth/in has been selected with a $2\frac{1}{4}$ in face width. Also, the teeth are to be 25° full depth, with a dedendum of $1.350/P$ for a pinion speed of 1150 rpm, and 10 hp is to be transmitted under steady load conditions. The material selected for both gears is a forged UNS G41400 steel heat-treated to a hardness of 302 Bhn. Determine the factors of safety guarding against a contact stress failure, using 99.9 percent reliability and superior mounting and cutting accuracy.

13-38* Solve Prob. 13-28 using contact strength as the primary design consideration.

13-39* Solve Prob. 13-29 based on the surface fatigue strength of the two gears. Use a life of 10^6 load applications, a reliability of 50 percent, normal temperatures, average mounting, hobbed teeth, and slightly unsteady loads in both the driving and the driven machinery.

General Problems

13-40* The pinion and gear shown in the figure have been tentatively designed to transmit 5 hp at 300 fpm pitch-line velocity. The tooth system is 20° full depth with $1.250/P$ for the dedendum. It is believed that the pinion can be hobbed from UNS G10180

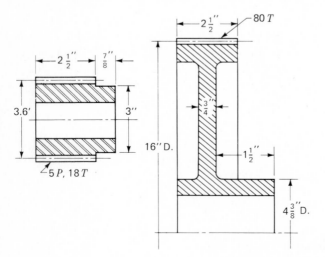

PROBLEM 13-40

cold-drawn steel and used with an ASTM No. 30 cast-iron gear. The gears are intended for mild service and average accuracy and mounting conditions. Given 90 percent reliability, find the factors of safety, if any, based on contact strength and bending endurance limit for a life not more than 10^6 load applications.

13-41* A gearset consists of an 8-pitch, 14-tooth pinion mating with an 18-tooth gear. The pinion is driven from a four-cycle gasoline engine at speeds varying from 1500 to 3200 rpm, depending to a large extent upon the whim of the operator. The gears are precision cut with a 25° pressure angle and a dedendum of $1.350/P$. The face width is $1\frac{1}{2}$ in. The material used is a UNS G10500 steel heat-treated to a hardness of 310 Bhn. Select bearing mountings are to be used in an effort to obtain very good life and reliability. Determine a safe value for the maximum tangential load to be transmitted.

13-42* Design the gears, including shaft sizes, for a double-reduction unit. The first pinion shaft is to be attached to a standard 50-hp squirrel-cage motor by means of a straight shaft coupling. The motor has a no-load speed of 1200 rpm. The output shaft of the unit is to rotate at 150 rpm and is to be connected to the driven machinery through a straight shaft coupling. The load is steady and continuous.

13-43* The same as Prob. 13-42, except the motor no-load speed is 1800 rpm.

CHAPTER
14

HELICAL, BEVEL, AND WORM GEARS

In the force analysis of spur gears, the forces are assumed to act in a single plane. In this chaper we shall study gears in which the forces have three dimensions. The reason for this, in the case of helical gears, is that the teeth are not parallel to the axis of rotation. And in the case of bevel gears, the rotational axes are not parallel to each other. There are also other reasons, as we shall learn.

In this chapter we shall rely heavily upon the fundamentals introduced in Chap. 13, especially the tables, charts, and graphs. And for each type of gearing the same general plan of presentation will be employed—kinematics, force analysis, bending strength, and surface strength, in that order.

14-1 PARALLEL HELICAL GEARS—KINEMATICS

Helical gears, used to transmit motion between parallel shafts, are shown in Fig. 14-1. The helix angle is the same on each gear, but one gear must have a right-hand helix and the other a left-hand helix. The shape of the tooth is an involute helicoid and is illustrated in Fig. 14-2. If a piece of paper cut in the shape of a parallelogram is wrapped around a cylinder, the angular edge of the paper becomes a helix. If we unwind this paper, each point on the angular edge generates an involute curve. This surface obtained when every point on the edge generates an involute is called an *involute helicoid*.

The initial contact of spur-gear teeth is a line extending all the way across the face of the tooth. The initial contact of helical-gear teeth is a point which changes into a line as the teeth come into more engagement. In spur gears the line of contact is parallel to the axis of rotation; in helical gears the line is diagonal across the face of the tooth. It is this gradual engagement of the teeth and the smooth transfer of load from one tooth to another which give helical gears the ability to transmit heavy loads at high speeds. Because of the nature of contact between helical gears, the contact ratio is of only minor importance, and it is the contact area, which is proportional to the face width of the gear, that becomes significant.

Helical gears subject the shaft bearings to both radial and thrust loads. When the thrust loads become high or are objectionable for other reasons, it may

FIGURE 14-1 A pair of helical gears. (*Courtesy of The Falk Corporation, Milwaukee, Wis.*)

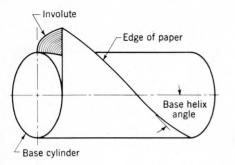

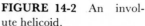

FIGURE 14-2 An involute helicoid.

be desirable to use double helical gears. A double helical gear (herringbone) is equivalent to two helical gears of opposite hand, mounted side by side on the same shaft. They develop opposite thrust reactions and thus cancel out the thrust load.

When two or more single helical gears are mounted on the same shaft, the hand of the gears should be selected so as to produce the minimum thrust load.

Figure 14-3 represents a portion of the top view of a helical rack. Lines *ab* and *cd* are the centerlines of two adjacent helical teeth taken on the pitch plane. The angle ψ is the *helix angle*. The distance *ac* is the *transverse circular pitch* p_t in the plane of rotation (usually called the *circular pitch*). The distance *ae* is the *normal circular pitch* p_n and is related to the transverse circular pitch as follows:

$$p_n = p_t \cos \psi \tag{14-1}$$

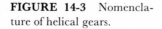

FIGURE 14-3 Nomenclature of helical gears.

The distance ad is called the *axial pitch* p_x and is related by the expression

$$p_x = \frac{p_t}{\tan \psi} \tag{14-2}$$

Since $p_n P_n = \pi$, the *normal diametral pitch* is

$$P_n = \frac{P_t}{\cos \psi} \tag{14-3}$$

The pressure angle ϕ_n in the normal direction is different from the pressure angle ϕ_t in the direction of rotation, because of the angularity of the teeth. These angles are related by the equation

$$\cos \psi = \frac{\tan \phi_n}{\tan \phi_t} \tag{14-4}$$

Figure 14-4 illustrates a cylinder cut by an oblique plane ab at an angle ψ to a right section. The oblique plane cuts out an arc having a radius of curvature of R. For the condition that $\psi = 0$, the radius of curvature is $R = D/2$. If we imagine the angle ψ to be slowly increased from zero to 90°, we see that R begins at a value of $D/2$ and increases until, when $\psi = 90°$, $R = \infty$. The radius R is the apparent pitch radius of a helical-gear tooth when viewed in the direction of the tooth elements. A gear of the same pitch and with the radius R will have a greater number of teeth, because of the increased radius. In helical-gear design this is called the *virtual number of teeth*. It can be shown by analytical geometry that the virtual number of teeth is related to the actual number by the equation

$$N' = \frac{N}{\cos^3 \psi} \tag{14-5}$$

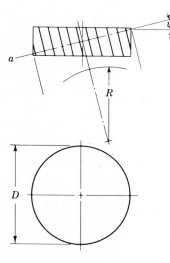

FIGURE 14-4 A cylinder cut by an oblique plane.

where N' is the virtual number of teeth and N is the actual number of teeth.   It is necessary to know the virtual number of teeth in applying the Lewis equation and also, sometimes, in cutting helical teeth.   This apparently larger radius of curvature means that fewer teeth may be used on helical gears, because there will be less undercutting.

14-2   HELICAL GEARS—TOOTH PROPORTIONS

Except for fine-pitch gears (20 diametral pitch and finer), there is no standard for the proportions of helical-gear teeth.   One reason for this is that it is cheaper to change the design slightly than it is to purchase special tooling.   Since helical gears are rarely used interchangeably anyway, and since many different designs will work well together, there is really little advantage in having them interchangeable.

As a general guide, tooth proportions should be based on a normal pressure angle of 20°.   Most of the proportions tabulated in Table 13-1 can then be used. The tooth dimensions should be calculated by using the normal diametral pitch. These proportions are suitable for helix angles from 0 to 30°, and all helix angles may be cut with the same hob.   Of course the normal diametral pitch of the hob and the gear must be the same.

An optional set of proportions can be based on a transverse pressure angle of 20° and the use of the transverse diametral pitch.   For these the helix angles are generally restricted to 15, 23, 30, or 45°. Angles greater than 45° are not recommended.   The normal diametral pitch must still be used to compute the tooth dimensions.   The proportions shown in Table 13-1 will usually be satisfactory.

Many authorities recommend that the face width of helical gears be at least two times the axial pitch $(F = 2p_x)$ to obtain helical-gear action.   Exceptions to this rule are automotive gears, which have a face width considerably less, and marine reduction gears, which often have a face width much greater.

14-3   HELICAL GEARS—FORCE ANALYSIS

Figure 14-5 is a three-dimensional view of the forces acting against a helical-gear tooth.   The point of application of the forces is in the pitch plane and in the center of the gear face.   From the geometry of the figure, the three components of the total (normal) tooth force W are

$$W_r = W \sin \phi_n$$

$$W_t = W \cos \phi_n \cos \psi \qquad\qquad\qquad (14\text{-}6)$$

$$W_a = W \cos \phi_n \sin \psi$$

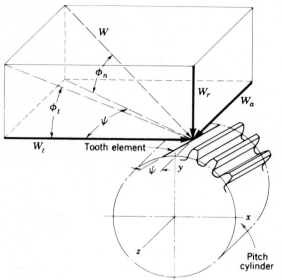

FIGURE 14-5 Tooth forces acting on a right-hand helical gear.

where W = total force
W_r = radial component
W_t = tangential component; also called transmitted load
W_a = axial component; also called thrust load

Usually W_t is given and the other forces are desired. In this case, it is not difficult to discover that

$$W_r = W_t \tan \phi_t$$

$$W_a = W_t \tan \psi \qquad\qquad (14\text{-}7)$$

$$W = \frac{W_t}{\cos \phi_n \cos \psi}$$

Example 14-1 In Fig. 14-6 a 1-hp electric motor runs at 1800 rpm in the clockwise direction, as viewed from the positive x axis. Keyed to the motor shaft is an 18-tooth helical pinion having a normal pressure angle of 20°, a helix angle of 30°, and a normal diametral pitch of 12 teeth/in. The hand of the helix is shown in the figure. Make a three-dimensional sketch of the motor shaft and pinion and show the forces acting on the pinion and the bearing reactions at A and B. The thrust should be taken out at A.

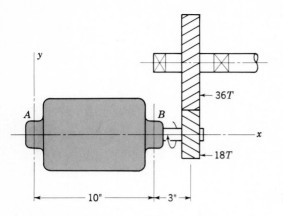

FIGURE 14-6

Solution   From Eq. (14-4) we find

$$\phi_t = \tan^{-1}\frac{\tan \phi_n}{\cos \psi} = \tan^{-1}\frac{\tan 20°}{\cos 30°} = 22.8°$$

Also $P_t = P_n \cos \psi = 12 \cos 30° = 10.4$ teeth/in.   Therefore the pitch diameter of the pinion is $d_p = 18/10.4 = 1.73$ in.   The pitch-line velocity is

$$V = \frac{\pi dn}{12} = \frac{\pi(1.73)(1800)}{12} = 815 \text{ fpm}$$

The transmitted load is

$$W_t = \frac{33000H}{V} = \frac{(33\ 000)(1)}{815} = 40.5 \text{ lb}$$

From Eq. (14-7) we find

$$W_r = W_t \tan \phi_t = (40.5)(0.422) = 17.1 \text{ lb}$$

$$W_a = W_t \tan \psi = (40.5)(0.577) = 23.4 \text{ lb}$$

$$W = \frac{W_t}{\cos \phi_n \cos \psi} = \frac{40.5}{(0.940)(0.866)} = 49.8 \text{ lb}$$

These three forces, W_r in the $-y$ direction, W_a in the $-x$ direction, and W_t in the $+z$ direction, are shown acting at point C in Fig. 14-7.   We assume bearing reactions at A and B as shown.   Then $F_a^x = W_a = 23.4$ lb.   Taking moments about the z axis,

$$-(17.1)(13) + (23.4)\left(\frac{1.73}{2}\right) + 10F_B^y = 0$$

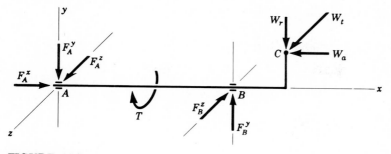

FIGURE 14-7

or $F_B^y = 20$ lb. Summing forces in the y direction then gives $F_A^y = 2.9$ lb. Taking moments about the y axis, next,

$$10F_B^z - (40.5)(13) = 0$$

or $F_B^z = 52.6$ lb. Summing forces in the z direction and solving gives $F_A^z = 12.1$ lb. Also, the torque is $T = W_t d_p/2 = (40.5)(1.73/2) = 35$ lb·in. ////

Example 14-2 Solve Example 14-1 using vectors.

Solution The force at C is

$$W = -23.4i - 17.1j + 40.5k$$

Position vectors to B and C from origin A are

$$R_B = 10i \qquad R_C = 13i + 0.865j$$

Taking moments about A, we have

$$R_B \times F_B + T + R_C \times W = 0$$

Using the directions assumed in Fig. 14-7 and substituting values gives

$$10i \times (F_B^y j - F_B^z k) - Ti + (13i + 0.865j) \times (-23.4i - 17.1j + 40.5k) = 0$$

When the cross products are formed, we get

$$(10F_B^y k + 10F_B^z j) - Ti + (35i - 526j - 200k) = 0$$

whence $T = 35$ lb·in, $F_B^y = 20$ lb, and $F_B^z = 52.6$ lb.
    Next, $F_A = -F_B - W$, and so $F_A = 23.4i - 2.9j + 12.1k$ lb. ////

14-4 HELICAL GEARS—STRENGTH ANALYSIS

We repeat here the equation for bending and surface stresses in spur gears because they also apply to helical gears.

$$\sigma = \frac{W_t P_t}{K_v FJ} \tag{14-8}$$

$$\sigma_H = -C_p \sqrt{\frac{W_t}{C_v F d_p I}} \tag{14-9}$$

where σ = bending stress, psi
$\quad \sigma_H$ = surface compressive stress, psi
$\quad W_t$ = transmitted load, lb
$\quad P_t$ = transverse diametral pitch, teeth/in
$\quad K_v = C_v$ = dynamic, or velocity, factor
$\quad d_p$ = pitch diameter of pinion, in
$\quad J$ = geometry factor (bending)
$\quad I$ = geometry factor (surface durability)

For helical gears the velocity factor is usually taken as

$$K_v = C_v = \sqrt{\frac{78}{78 + \sqrt{V}}} \tag{14-10}$$

where V is the pitch-line velocity in feet per minute (fpm).

Geometry factors for helical gears must account for the fact that contact takes place along a diagonal line across the tooth face and that we are usually dealing with the transverse pitch instead of the normal pitch. The worst loading occurs when the line of contact intersects the tip of the tooth, but the unloaded end strengthens the tooth.

The J factors for $\phi_n = 20°$ can be found in Fig. 14-8. The AGMA also publishes J factors for $\phi_n = 15°$ and $\phi_n = 22°$.

Geometry factors I for helical and herringbone gears are calculated from the equation*

$$I = \frac{\sin \phi_t \cos \phi_t}{2m_N} \frac{m_G}{m_G + 1} \tag{14-11}$$

for external gears. (Use a minus sign in the denominator of the second term for internal gears). In this equation ϕ_t is the transverse pressure angle and m_N is the load-sharing ratio and is found from the equation

$$m_N = \frac{p_N}{0.95Z} \tag{14-12}$$

---

* See *AGMA Information Sheet* 211.02, February 1969.

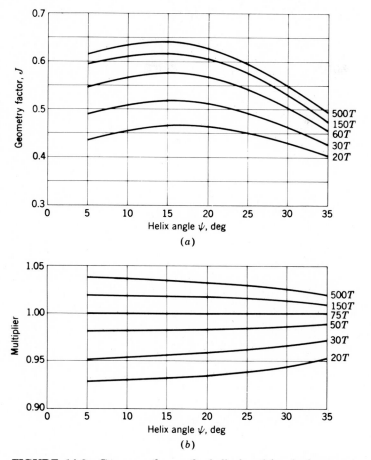

FIGURE 14-8 Geometry factors for helical and herringbone gears having a normal pressure angle of 20°. (a) Geometry factors for gears mating with a 75-tooth gear. (b) J-factor multipliers when tooth numbers other than 75 are used in the mating gear. (*AGMA Information Sheet 225.01.*)

Here p_N is the normal base pitch; it is related to the *normal circular pitch* p_n by the relation

$$p_N = p_n \cos \phi_n \qquad (14\text{-}13)$$

The quantity Z is the length of the line of action in the transverse plane. It is best obtained from a layout of the two gears, but may also be found from the equation*

$$Z = \sqrt{(r_p + a)^2 - r_{bP}^2} - \sqrt{(r_G + a)^2 - r_{bG}^2} - (r_P + r_G) \sin \phi_t \qquad (14\text{-}14)$$

---

* For a development, see Joseph E. Shigley and John J. Uicker, Jr., *Theory of Machines and Mechanisms*, McGraw-Hill, New York, 1980, p. 262.

Table 14-1 LOAD-DISTRIBUTION FACTORS C_m AND K_m FOR HELICAL GEARS

	Face width, in			
Characteristics of support	0–2	6	9	16 up
Accurate mountings, small bearing clearances, minimum deflection, precision gears	1.2	1.3	1.4	1.7
Less rigid mountings, less accurate gears, contact across full face	1.5	1.6	1.7	2.0
Accuracy and mounting such that less than full-face contact exists		Over 2.0		

Source: Darle W. Dudley (ed.), *Gear Handbook*, McGraw-Hill, New York, 1962, pp. 13–23.

where r_P and r_G are the pitch radii and r_{bP} and r_{bG} the base-circle radii, respectively, of the pinion and gear.   Certain precautions must be taken in using Eq. (14-14).   The tooth profiles are not conjugate below the base circle, and consequently, if either $\sqrt{(r_P + a)^2 - r_{bP}^2}$ or $\sqrt{(r_G + a)^2 - r_{bG}^2}$ is larger than $(r_P + r_G) \sin \phi_t$, that term should be replaced by $(r_P + r_G) \sin \phi_t$. In addition, the effective outside radius is sometimes less than $r + a$ owing to removal of burrs or rounding of the tips of the teeth.   When this is the case, always use the effective outside radius instead of $r + a$.

The modification and correction factors for helical gears are the same as for spur gears, except for the load-distribution factors K_m and C_m (Table 14-1) and the hardness-ratio factor C_H (Fig. 14-9).   With these changes Eq. (13-31) gives the endurance limit in bending, Eq. (13-45) the factor of safety, and Eqs. (13-45) and (13-46) the surface fatigue strength.

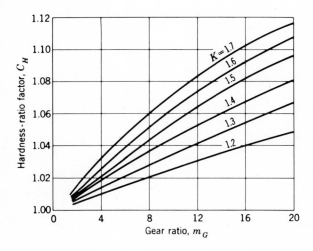

FIGURE 14-9 Hardness-ratio factor C_H for helical gears. The factor K is the Brinell hardness of the pinion divided by the Brinell hardness of the gear. Use $C_H = 1.00$ when $K < 1.2$. (*AGMA Information Sheet 215.01.*)

Example 14-3 A pair of 4 : 1 reduction gears for a 100-hp motor running at 1120 rpm has a face width of $3\frac{1}{2}$ in, a normal diametral pitch of 4 teeth/in, a normal pressure angle of 20°, and an 18-tooth pinion with a right-hand helix angle of 30°. The gears are made of UNS G10400 steel, heat-treated and drawn to 1000°F, and a reliability of 95 percent is desired. The mounting conditions are average and moderate shock may occur in the driven machinery.

(*a*)   Find the factors of safety n_G and n based on the bending endurance limit of the teeth.

(*b*)   Find the factors of safety n_G and n based on the surface fatigue strength of the teeth.

Solution   (*a*) Using Eq. (14-3)

$$P_t = P_n \cos \psi = 4 \cos 30° = 3.464 \text{ teeth/in}$$

So the pinion diameter is

$$d_P = \frac{N_P}{P_t} = \frac{18}{1.464} = 5.196 \text{ in}$$

based on the pitch circle. Therefore the pitch-line velocity is

$$V = \frac{\pi d_P n}{12} = \frac{\pi (5.196)(1120)}{12} = 1524 \text{ fpm}$$

Using Eq. (14-10) gives a velocity factor of

$$K_v = \sqrt{\frac{78}{78 + \sqrt{V}}} = \sqrt{\frac{78}{78 + \sqrt{1524}}} = 0.816$$

Also, the transmitted load is

$$W_t = \frac{33(10)^3 H}{V} = \frac{3(10)^3(100)}{1524} = 2165 \text{ lb}$$

Since the reduction ratio is 4, the gear has $N_G = 4(18) = 72$ teeth. Thus, from Fig. 14-8, we find $J = 0.43$ with a multiplying factor of about unity. Equation (14-8) now gives the bending stress as

$$\sigma = \frac{W_t P_t}{K_v F J} = \frac{2165(3.464)(10)^{-3}}{0.816(3.5)(0.43)} = 6.11 \text{ kpsi}$$

In preparing to obtain the tooth endurance limit we note from Table A-17 that $S_{ut} = 113$ kpsi. We find the surface factor to be $k_a = 0.725$ from Fig. 13-15. Basing the size factor on the normal pitch we find $k_b = 0.890$. And Table 13-10 gives $k_c = 0.868$ for a reliability of 95 percent. Also $k_f = 1.33$ from Table 13-11.

Since $S'_e = 0.5(113) = 56.5$ kpsi, the tooth endurance limit is

$$S_e = k_a k_b k_c k_d k_e k_f S'_e$$

$$= (0.725)(0.890)(0.868)(1)(1)(1.33)(56.5) = 42.09 \text{ kpsi}$$

From Table 14-1 we find $K_m = 1.5$ and from Table 13-12 $K_o = 1.25$. Thus

$$n_G = K_o K_m n = (1.25)(1.5)n = 1.875n$$

So the factor of safety n_G is

$$n_G = \frac{S_e}{\sigma} = \frac{42.09}{6.11} = 6.89 \qquad\qquad\qquad\qquad Ans.$$

and

$$n = \frac{6.89}{1.875} = 3.67 \qquad\qquad\qquad\qquad Ans.$$

(b) The contact fatigue strength is obtained from Eq. (13-45). Since $H_B = 235$ Bhn, we have

$$S_C = 0.4H_B - 10 = 0.4(235) - 10 = 84 \text{ kpsi}$$

Using Table 13-15, $C_L = 1$ and $C_R = 0.8$. From Fig. 14-9 $C_H = 1$. Also $C_T = 1$. Therefore, the Hertzian strength is

$$S_H = \frac{C_L C_H}{C_T C_R} S_C = \frac{(1)(1)}{(1)(0.8)} (84) = 105 \text{ kpsi}$$

Next, we find $C_P = 2300$ from Table 13-14, and $C_v = K_v = 0.816$.

We must next solve Eqs. (14-12) to (14-14) to obtain the load-sharing ratio m_N. The pitch radii of the pinion and gear are $r_P = d_{P/2} = 5.196/2 = 2.60$ in and $r_G = 4r_P = 10.4$ in. The addendum is $a = 1/P_n = 1/4 = 0.25$ in. The transverse pressure angle is

$$\phi_t = \tan^{-1} \frac{\tan \phi_n}{\cos \psi} = \tan^{-1} \frac{\tan 20°}{\cos 30°} = 22.8°$$

The base radii are

$$r_{bP} = r_P \cos \phi_t = 2.60 \cos 22.8° = 2.40 \text{ in}$$

$$r_{bG} = 10.4 \cos 22.8° = 9.59 \text{ in}$$

The length of the line of action is now found from Eq. (14-14). Substituting directly and in order gives

$$Z = \sqrt{(2.60 + 0.25)^2 - (2.40)^2} + \sqrt{(10.4 + 0.25)^2 - (9.59)^2}$$
$$- (2.60 + 10.4) \sin 22.8°$$
$$= 1.5370 + 4.6319 - 5.0377 = 1.13 \text{ in}$$

Since neither of the first two terms is larger than the third, the result is valid. Next, using Eq. (14-13) we find the normal base pitch to be

$$p_N = p_n \cos \phi_n = \frac{\pi}{4} \cos 20° = 0.738 \text{ in}$$

Thus the load-sharing ratio, from Eq. (14-12), is

$$m_N = \frac{p_N}{0.95Z} = \frac{0.738}{0.95(1.13)} = 0.688$$

In the development of geometry factors [Eq. (13-22)] we learned that the load-sharing ratio indicates the proportion of the total load carried by the most heavily loaded tooth. In analyzing spur gearing, a value $m_N = 1$ was used because the geometry factors were based on force transmission at the highest point of single-tooth contact. But there is always some load sharing in the action of helical gears and so m_N must be accounted for in the geometry factors. Thus, using Eq. (14-11), we get

$$I = \frac{\sin \phi_t \cos \phi_t}{2m_N} \frac{m_G}{m_G + 1} = \frac{\sin 22.8° \cos 22.8°}{2(0.688)} \frac{4}{4 + 1} = 0.208$$

We now rewrite Eq. (14-9) in the form

$$S_H = C_p \sqrt{\frac{W_{t,p}}{C_v F d_p I}}$$

where $W_{t,p}$ is the permissible transmitted load. Solving gives

$$W_{t,p} = \left(\frac{S_H}{C_p}\right)^2 (C_v F d_p I)$$
$$= \left[\frac{105(10)^3}{2300}\right]^2 (0.816)(3.5)(5.196)(0.208) = 6433 \text{ lb}$$

Since $W_{t,p} = n_G W_t$ we have

$$n_G = \frac{W_{t,p}}{W_t} = \frac{6433}{2165} = 2.97 \qquad\qquad Ans.$$

Also $n_G = 1.875n$ and so

$$n = \frac{2.97}{1.875} = 1.58 \qquad\qquad\qquad\qquad\qquad Ans.$$

These results indicate that the design appears to be quite satisfactory based on bending endurance limit as well as the contact strength.   ////

14-5   CROSSED HELICAL GEARS

Crossed helical, or spiral, gears are those in which the shaft centerlines are neither parallel nor intersecting. They are essentially nonenveloping worm gears, because the gear blanks have a cylindrical form. This class of gears is illustrated in Fig. 14-10.

The teeth of crossed helical gears have *point contact* with each other, which changes to *line contact* as the gears wear in. For this reason they will carry only very small loads. Crossed helical gears are for instrumental applications, and they are definitely not recommended for use in the transmission of power.

There is no difference between a crossed helical gear and a helical gear until they are mounted in mesh with each other. They are manufactured in the same way. A pair of meshed crosssed helical gears usually have the same hand; that is, a right-hand driver goes with a right-hand driven. The relation between thrust, hand, and rotation for crossed helical gears is shown in Fig. 14-11.

When specifying tooth sizes, the normal pitch should always be used. The reason for this is that, when different helix angles are used for the driver and driven, the transverse pitches are not the same. The relation between the shaft

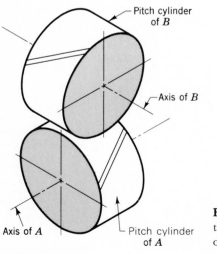

FIGURE 14-10 View of the pitch cylinders of a pair of crossed-helical gears.

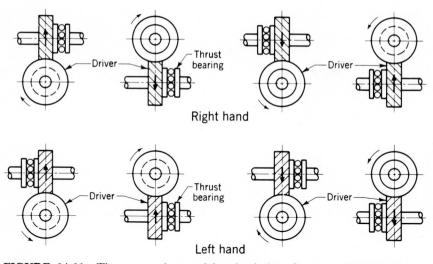

FIGURE 14-11 Thrust, rotation, and hand relations for crossed-helical gears. (*Courtesy of Boston Gear Works, Inc.*)

and helix angles is

$$\Sigma = \psi_1 \pm \psi_2 \tag{14-15}$$

where Σ is the shaft angle. The plus sign is used when both helix angles are of the same hand, and the minus sign when they are of opposite hand. Opposite-hand crossed helical gears are used when the shaft angle is small.

The pitch diameter is obtained from the equation

$$d = \frac{N}{P_n \cos \psi} \tag{14-16}$$

where N = number of teeth
P_n = normal diametral pitch
ψ = helix angle

Since the pitch diameters are not directly related to the tooth numbers, they cannot be used to obtain the angular-velocity ratio. This ratio must be obtained from the ratio of the tooth numbers.

In the design of crossed helical gears, the minimum sliding velocity is obtained when the helix angles are equal. However, when the helix angles are not equal, the gear with the larger helix angle should be used as the driver if both gears have the same hand.

There is no standard for crossed-helical-gear-tooth proportions. Many different proportions give good tooth action. Since the teeth are in point contact, an effort should be made to obtain a contact ratio of 2 or more. For this reason, crossed helical teeth are usually cut with a low pressure angle and a deep tooth.

14-6 WORM GEARING—KINEMATICS

Figure 14-12 shows a worm and a worm gear. Note that the shafts do not intersect and that the shaft angle is 90°; this is the usual shaft angle, though other angles can be used. The worm is the screwlike member in the figure, and you can see that it has, perhaps, five or six teeth (threads). A one-toothed worm would resemble an Acme screw thread very closely.

Worm gearsets are either single- or double-enveloping. A single-enveloping gearset is one in which the gear wraps around or partially encloses the worm, as in Fig. 14-12. A gearset in which each element partially encloses the other is, of course, a double-enveloping worm gearset. The important difference between the two is that *area contact* exists between the teeth of double-enveloping gears and only *line contact* between those of single-enveloping gears.

The nomenclature of a worm and worm gear is shown in Fig. 14-13. The worm and worm gear of a set have the same hand of helix as for crossed helical gears, but the helix angles are usually quite different. The helix angle on the worm is generally quite large, and that on the gear very small. Because of this, it is usual to specify the lead angle λ on the worm and helix angle ψ_G on the gear; the two angles are equal for a 90° shaft angle. The worm lead angle is the complement of the worm helix angle, as shown in Fig. 14-13.

FIGURE 14-12 A single-enveloping worm and worm gear. (*Courtesy of Horsburgh and Scot Company, Cleveland.*)

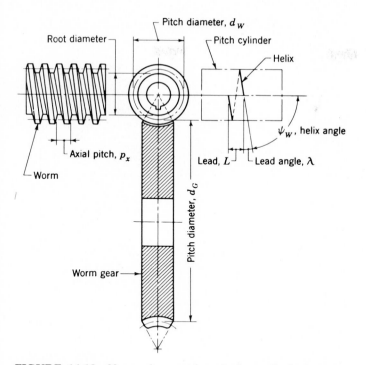

FIGURE 14-13 Nomenclature of a single-enveloping worm gearset.

In specifying the pitch of worm gearsets, it is customary to state the *axial pitch* p_x of the worm and the *transverse circular pitch* p_t, often simply called the circular pitch, of the mating gear. These are equal if the shaft angle is 90°. The pitch diameter of the gear is the diameter measured on a plane containing the worm axis, as shown in Fig. 14-13; it is the same as for spur gears and is

$$d_G = \frac{N_G p_t}{\pi} \qquad (14\text{-}17)$$

Since it is not related to the number of teeth, the worm may have any pitch diameter; this diameter should, however, be the same as the pitch diameter of the hob used to cut the worm-gear teeth. Generally, the pitch diameter of the worm should be selected so as to fall into the range

$$\frac{C^{0.875}}{3.0} \le d_W \le \frac{C^{0.875}}{1.7} \qquad (14\text{-}18)$$

where C is the center distance. These proportions appear to result in optimum horsepower capacity of the gearset.

Table 14-2   RECOMMENDED PRESSURE ANGLES AND TOOTH
DEPTHS FOR WORM GEARING

Lead angle λ, degrees	Pressure angle ϕ_n, degrees	Addendum a	Dedendum b_G
0–15	$14\frac{1}{2}$	$0.3683p_x$	$0.3683p_x$
15–30	20	$0.3683p_x$	$0.3683p_x$
30–35	25	$0.2865p_x$	$0.3314p_x$
35–40	25	$0.2546p_x$	$0.2947p_x$
40–45	30	$0.2228p_x$	$0.2578p_x$

The *lead L* and the *lead angle* λ of the worm have the following relations:

$$L = p_x N_W \tag{14-19}$$

$$\tan \lambda = \frac{L}{\pi d_W} \tag{14-20}$$

Tooth forms for worm gearing have not been highly standardized, perhaps because there has been less need for it. The pressure angles used depend upon the lead angles and must be large enough to avoid undercutting of the worm-gear tooth on the side at which contact ends. A satisfactory tooth depth, which remains in about the right proportion to the lead angle, may be obtained by making the depth a proportion of the axial circular pitch. Table 14-2 summarizes what may be regarded as good practice for pressure angle and tooth depth.

The *face width* F_G of the worm gear should be made equal to the length of a tangent to the worm pitch circle between its points of intersection with the addendum circle, as shown in Fig. 14-14.

14-7   WORM GEARING—FORCE ANALYSIS

If friction is neglected, then the only force exerted by the gear will be the force W, shown in Fig. 14-15, having the three orthogonal components W^x, W^y, and W^z.

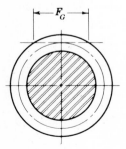

FIGURE 14-14

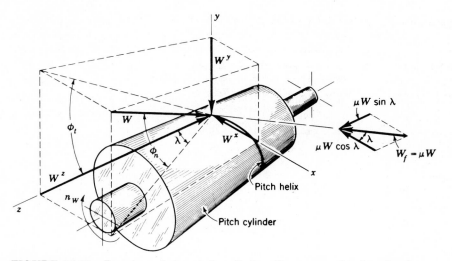

FIGURE 14-15 Drawing of the pitch cylinder of a worm, showing the forces exerted upon it by the worm gear.

From the geometry of the figure we see

$$W^x = W \cos \phi_n \sin \lambda$$

$$W^y = W \sin \phi_n \tag{14-21}$$

$$W^z = W \cos \phi_n \cos \lambda$$

We now use the subscripts W and G to indicate forces acting against the worm and gear, respectively. We note that W^y is the separating or radial, force for both the worm and gear. The tangential force on the worm is W^x and is W^z on the gear, assuming a 90° shaft angle. The axial force on the worm is W^z, and on the gear, W^x. Since the gear forces are opposite to the worm forces, we can summarize these relations by writing

$$W_{Wt} = -W_{Ga} = W^x$$

$$W_{Wr} = -W_{Gr} = W^y \tag{14-22}$$

$$W_{Wa} = -W_{Gt} = W^z$$

It is helpful in using Eq. (14-21) and also Eq. (14-22) to observe that *the gear axis is parallel to the x direction and the worm axis is parallel to the z direction* and that we are employing a right-handed coordinate system.

In our study of spur-gear teeth we have learned that the motion of one tooth relative to the mating tooth is primarily a rolling motion; in fact, when contact occurs at the pitch point, the motion is pure rolling. In contrast, the relative motion between worm and worm-gear teeth is pure sliding, and so we must expect that friction plays an important role in the performance of worm gearing. By introducing a coefficient of friction μ, we can develop another set of relations similar to those of Eq. (14-21). In Fig. 14-15 we see that the force W acting normal to the worm-tooth profile produces a frictional force $W_f = \mu W$, having a component $\mu W \cos \lambda$ in the negative x direction and another component $\mu W \sin \lambda$ in the positive z direction. Equation (14-21) therefore becomes

$$W^x = W(\cos \phi_n \sin \lambda + \mu \cos \lambda)$$

$$W^y = W \sin \phi_n \qquad\qquad\qquad\qquad\qquad\qquad\qquad (14\text{-}23)$$

$$W^z = W(\cos \phi_n \cos \lambda - \mu \sin \lambda)$$

Equation (14-22), of course, still applies.

If we substitute W^z into the third part of Eq. (14-22) and multiply both sides by μ, we find the frictional force to be

$$W_f = \mu W = \frac{\mu W_{Gt}}{\mu \sin \lambda - \cos \phi_n \cos \lambda} \qquad\qquad (14\text{-}24)$$

Another useful relation can be obtained by solving the first and third parts of Eq. (14-22) simultaneously to get a relation between the two tangential forces. The result is

$$W_{Wt} = W_{Gt} \frac{\cos \phi_n \sin \lambda + \mu \cos \lambda}{\mu \sin \lambda - \cos \phi_n \cos \lambda} \qquad\qquad (14\text{-}25)$$

Efficiency η can be defined by using the equation

$$\eta = \frac{W_{Wt}(\text{without friction})}{W_{Wt}(\text{with friction})} \qquad\qquad\qquad\qquad (a)$$

Substitute Eq. (14-25) with $\mu = 0$ in the numerator of Eq. (a) and the same equation in the denominator. After some rearranging you will find the efficiency to be

$$\eta = \frac{\cos \phi_n - \mu \tan \lambda}{\cos \phi_n + \mu \cot \lambda} \qquad\qquad\qquad\qquad (14\text{-}26)$$

Selecting a typical value of the coefficient of friction, say $\mu = 0.05$, and the pressure angles shown in Table 14-2, we can use Eq. (14-26) to get some useful design information. Solving this equation for helix angles from 1 to 30° gives the interesting results shown in Table 14-3.

Table 14-3 EFFICIENCY OF WORM GEARSETS FOR $\mu = 0.05$

Helix angle ψ, degrees	Efficiency η, percent
1.0	25.2
2.5	46.8
5.0	62.6
7.5	71.2
10.0	76.8
15.0	82.7
20.0	86.0
25.0	88.0
30.0	89.2

Many experiments have shown that the coefficient of friction is dependent on the relative or sliding velocity. In Fig. 14-16, V_G is the pitch-line velocity of the gear and V_W the pitch-line velocity of the worm. Vectorially, $\mathbf{V}_W = \mathbf{V}_G + \mathbf{V}_S$; consequently,

$$V_S = \frac{V_W}{\cos \lambda} \tag{14-27}$$

Published values of the coefficient of friction vary as much as 20 percent, undoubtedly because of the differences in surface finish, materials, and lubrication. The values on the chart of Fig. 14-17 are representative and indicate the general trend.

Example 14-4 A 2-tooth right-hand worm transmits 1 hp at 1200 rpm to a 30-tooth worm gear. The gear has a transverse diametral pitch of 6 teeth/in and

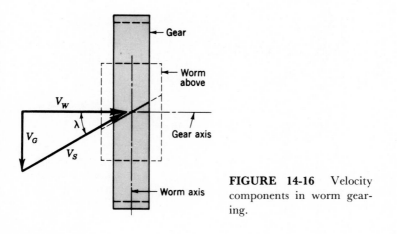

FIGURE 14-16 Velocity components in worm gearing.

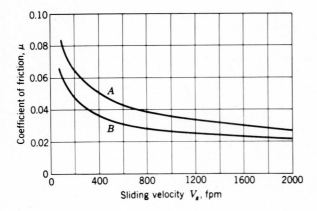

FIGURE 14-17 Representative values of the coefficient of friction for worm gearing. These values are based on good lubrication. Use curve *B* for high-quality materials, such as a case-hardened worm mating with a phosphor-bronze gear. Use curve *A* when more friction is expected, as for example a cast-iron worm and worm gear.

a face width of 1 in. The worm has a pitch diameter of 2 in and a face width of $2\frac{1}{2}$ in. The normal pressure angle is $14\frac{1}{2}°$. The materials and workmanship are such that curve *B* of Fig. 14-17 should be used to obtain the coefficient of friction.

(*a*) Find the axial pitch, the center distance, the lead, and the lead angle.

(*b*) Figure 14-18 is a drawing of the worm gear oriented with respect to the coordinate system described earlier in this section; the gear is supported by bearings *A* and *B*. Find the forces exerted by the bearings against the worm-gear shaft, and the output torque.

Solution (*a*) The axial pitch is the same as the transverse circular pitch of the gear, which is

$$p_t = \frac{\pi}{P} = \frac{\pi}{6} = 0.5236 \text{ in} \qquad\qquad Ans.$$

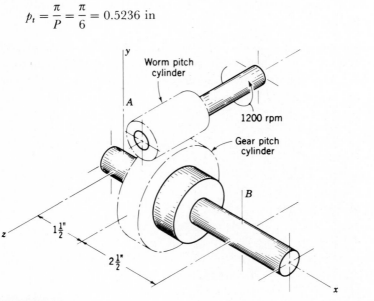

FIGURE 14-18

The pitch diameter of the gear is $d_G = N_G/P = 30/6 = 5$ in. Therefore the center distance is

$$C = \frac{d_W + d_G}{2} = \frac{2 + 5}{2} = 3.5 \text{ in} \qquad\qquad Ans.$$

From Eq. (14-19) the lead is

$$L = p_x N_W = (0.5236)(2) = 1.0472 \text{ in} \qquad\qquad Ans.$$

Also, using Eq. (14-20), find

$$\lambda = \tan^{-1} \frac{L}{\pi d_W} = \tan^{-1} \frac{1.0472}{\pi(2)} = 9.47° \qquad\qquad Ans.$$

(b) Using the right-hand rule for the rotation of the worm, you will see that your thumb points in the positive z direction. Now use the bolt-and-nut analogy (the worm is right-handed, as is the screw thread of a bolt), and turn the bolt clockwise with the right hand while preventing nut rotation with the left. The nut will move axially along the bolt toward your right hand. Therefore the surface of the gear (Fig. 14-18) in contact with the worm will move in the negative z direction. Thus the gear rotates clockwise about x, with your right thumb pointing in the negative x direction.

The pitch-line velocity of the worm is

$$V_W = \frac{\pi d_W n_W}{12} = \frac{\pi(2)(1200)}{12} = 628 \text{ fpm}$$

The speed of the gear is $n_G = (\frac{2}{30})(1200) = 80$ rpm. So the pitch-line velocity is

$$V_G = \frac{\pi d_G n_G}{12} = \frac{\pi(5)(80)}{12} = 105 \text{ fpm}$$

Then, using Eq. (14-27), the sliding velocity V_S is found to be

$$V_S = \frac{V_W}{\cos \lambda} = \frac{628}{\cos 9.47°} = 638 \text{ fpm}$$

Using Fig. 14-17, we find $\mu = 0.03$. We shall also require the normal pressure angle ϕ_n. Since the helix angle of the gear is the same as the lead angle of the worm, we can use Eq. (14-14). Thus

$$\phi_n = \tan^{-1} (\tan \phi_i \cos \psi) = \tan^{-1} (\tan 14.5° \cos 9.47°) = 14.3°$$

Getting to the forces now, we begin with the horsepower formula

$$W_{Wt} = \frac{33\ 000H}{V_W} = \frac{(33\ 000)(1)}{628} = 52.5 \text{ lb}$$

This force acts in the negative x direction, the same as in Fig. 14-15. Using the first part of Eq. (14-23), we next find

$$W = \frac{W^x}{\cos \phi_n \sin \lambda + \mu \cos \lambda}$$

$$= \frac{52.5}{\cos 14.3° \sin 9.47° + 0.03 \cos 9.47°} = 278 \text{ lb}$$

Also, from Eq. (14-23),

$$W^y = W \sin \phi_n = 278 \sin 14.3° = 68.6 \text{ lb}$$

$$W^z = W(\cos \phi_n \cos \lambda - \mu \sin \lambda)$$

$$= 278(\cos 14.3° \cos 9.47° - 0.03 \sin 9.47°) = 264 \text{ lb}$$

We now identify the components acting upon the gear as

$$W_{Ga} = -W^x = 52.5 \text{ lb}$$

$$W_{Gr} = -W^y = 68.6 \text{ lb}$$

$$W_{Gt} = -W^z = -264 \text{ lb}$$

At this point a three-dimensional line drawing should be made in order to simplify the work to follow. An isometric sketch, such as the one in Fig. 14-19, is easy to make and will help you to avoid errors. Note that the y axis is vertical, while the x and z axes make angles of 30° with the horizontal. The illusion of depth is enhanced by sketching lines parallel to each of the coordinate axes through every point of interest.

We shall make B a thrust bearing in order to place the gear shaft in compression. Thus, summing forces in the x direction gives

$$F_B^x = -52.5 \text{ lb} \qquad\qquad\qquad\qquad\qquad\qquad\qquad Ans.$$

Taking moments about the z axis,

$$-(52.5)(2.5) - (68.6)(1.5) + 4F_B^y = 0 \qquad F_B^y = 58.6 \text{ lb} \qquad\qquad Ans.$$

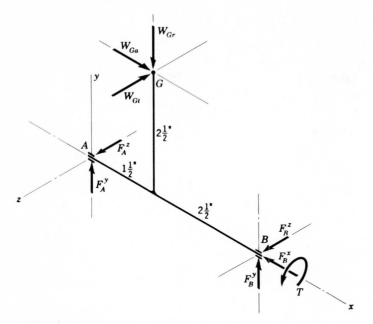

FIGURE 14-19

Taking moments about the y axis,

$$(264)(1.5) - 4F_B^z = 0 \qquad F_B^z = 99 \text{ lb} \qquad\qquad Ans.$$

These three components are now inserted on the sketch as shown at B in Fig. 14-19. Summing forces in the y direction,

$$-68.6 + 58.6 + F_A^y = 0 \qquad F_A^y = 10 \text{ lb} \qquad\qquad Ans.$$

Similarly, summing forces in the z direction,

$$-264 + 99 + F_A^z = 0 \qquad F_A^z = 165 \text{ lb} \qquad\qquad Ans.$$

These two components can now be placed at A on the sketch. We still have one more equation to write. Summing moments about x,

$$-(264)(2.5) + T = 0 \qquad T = 660 \text{ lb} \cdot \text{in} \qquad\qquad Ans.$$

It is because of the frictional loss that this output torque is less than the product of the gear ratio and the input torque. ////

Example 14-5   Solve (*b*) of Example 14-4 using vectors.

Solution   Using Fig. 14-19, write

$$\mathbf{W}_G = 52.5\mathbf{i} - 68.6\mathbf{j} - 264\mathbf{k}$$

Then define the position vectors

$$\mathbf{R}_G = 1.5\mathbf{i} + 2.5\mathbf{j} \qquad \mathbf{R}_B = 4\mathbf{i}$$

Writing the moment equation about *A* gives

$$\mathbf{R}_G \times \mathbf{W}_G + \mathbf{T} + \mathbf{R}_B \times \mathbf{F}_B = 0 \tag{1}$$

Substituting the known values,

$$(1.5\mathbf{i} + 2.5\mathbf{j}) \times (52.5\mathbf{i} - 68.6\mathbf{j} - 264\mathbf{k}) + T\mathbf{i} + (4\mathbf{i}) \times (F_B^x\mathbf{i} + F_B^y\mathbf{j} + F_B^z\mathbf{k}) = 0$$

When the cross products are formed, we have

$$(-660\mathbf{i} + 396\mathbf{j} - 234\mathbf{k}) + T\mathbf{i} + (-4F_B^z\mathbf{j} + 4F_B^y\mathbf{k}) = 0 \tag{2}$$

Thus

$$\mathbf{T} = 660\mathbf{i} \ \text{lb} \cdot \text{in} \qquad\qquad\qquad\qquad\qquad\qquad\qquad\qquad \textit{Ans.}$$

$$F_B^y = 58.6 \ \text{lb} \qquad F_B^z = 99 \ \text{lb}$$

Taking a summation of forces next yields

$$\mathbf{F}_A + \mathbf{F}_B + \mathbf{W}_G = 0 \tag{3}$$

or substituting known values,

$$(F_A^y\mathbf{j} + F_A^z\mathbf{k}) + (F_B^x\mathbf{i} + 58.6\mathbf{j} + 99\mathbf{k}) + (52.5\mathbf{i} - 68.6\mathbf{j} - 264\mathbf{k}) = 0 \tag{4}$$

whence

$$F_B^x = -52.5 \ \text{lb}$$

and so

$$\mathbf{F}_B = -52.5\mathbf{i} + 58.6\mathbf{j} + 99\mathbf{k} \ \text{lb} \qquad\qquad\qquad\qquad\qquad\qquad \textit{Ans.}$$

Also, from Eq. (4),

$$\mathbf{F}_A = 10\mathbf{j} + 165\mathbf{k} \text{ lb} \qquad\qquad\qquad\qquad Ans.$$

////

14-8 POWER RATING OF WORM GEARING

When worm gearsets are used intermittently or at slow gear speeds, the bending strength of the gear tooth may become a principal design factor. Since the worm teeth are inherently stronger than the gear teeth, they are usually not considered, though the methods of Chap. 8 can be used to compute worm-tooth stresses. The teeth of worm gears are thick and short at the two edges of the face and thin in the central plane, and this makes it difficult to determine the bending stress. Buckingham* adapts the Lewis equation as follows:

$$\sigma = \frac{W_{Gt}}{p_n F_G y} \qquad\qquad\qquad\qquad (14\text{-}28)$$

$$p_n = p_x \cos \lambda \qquad\qquad\qquad\qquad (14\text{-}29)$$

where σ = bending stress, psi
$\quad W_{Gt}$ = transmitted load, lb
$\quad p_n$ = normal circular pitch, in
$\quad p_x$ = axial circular pitch, in
$\quad F_G$ = face width of gear, in
$\quad y$ = Lewis form factor referred to the circular pitch
$\quad \lambda$ = lead angle

Since the equation is only a rough approximation, stress concentration is not considered. Also, for this reason, the form factors are not referred to the number of teeth, but only to the normal pressure angle. The values of y are listed in Table 14-4.

The AGMA equation for *input-horsepower rating* of worm gearing is

$$H = \frac{W_{Gt} d_G n_W}{126\,000 m_G} + \frac{V_s W_f}{33\,000} \qquad\qquad\qquad\qquad (14\text{-}30)$$

The first term on the right is the *output horsepower*, and the second is the *power loss*.

---

* Earle Buckingham, *Analytical Mechanics of Gears*, McGraw-Hill, New York, 1949, p. 495.

Table 14-4   VALUES OF _y_ FOR WORM GEARS

Normal pressure angle ϕ_n, degrees	Form factor _y_
$14\frac{1}{2}$	0.100
20	0.125
25	0.150
30	0.175

The *permissible* transmitted load W_{Gt} is computed from the equation

$$W_{Gt} = K_3 d_G^{0.8} F_e K_m K_v \qquad (14\text{-}31)$$

The notation of Eqs. (14-30) and (14-31) is as follows:

W_{Gt} = transmitted load, lb
d_G = pitch diameter of gear, in
n_W = speed of worm, rpm
m_G = gear ratio, N_G/N_W
V_S = sliding velocity at mean worm diameter, fpm
W_f = frictional force, lb
K_s = materials and size-correction factor
F_e = effective face width of gear; the effective face width is the face width of the gear or two-thirds of the worm pitch diameter, whichever is less
K_m = ratio-correction factor
K_v = velocity factor

Table 14-5   MATERIALS FACTOR K_s FOR CYLINDRICAL WORM GEARING*

Face width of gear F_G, in	Sand-cast bronze	Static chill-cast bronze	Centrifugal-cast bronze
Up to 3	700	800	1000
4	665	780	975
5	640	760	940
6	600	720	900
7	570	680	850
8	530	640	800
9	500	600	750

* For copper-tin and copper-tin-nickel bronze gears operating with steel worms case-hardened to Rockwell 58C minimum.

Source: Darle W. Dudley (ed.), *Gear Handbook*, McGraw-Hill, New York, 1962, pp. 13–38.

Table 14-6 RATIO-CORRECTION FACTOR K_m

Ratio m_G	K_m	Ratio m_G	K_m	Ratio m_G	K_m
3.0	0.500	8.0	0.724	30.0	0.825
3.5	0.554	9.0	0.744	40.0	0.815
4.0	0.593	10.0	0.760	50.0	0.785
4.5	0.620	12.0	0.783	60.0	0.745
5.0	0.645	14.0	0.799	70.0	0.687
6.0	0.679	16.0	0.809	80.0	0.622
7.0	0.706	20.0	0.820	100.0	0.490

Source: Darle W. Dudley (ed.), *Gear Handbook*, McGraw-Hill, New York, 1962, pp. 13–38.

Values of the materials factor for hardened steel worms mating with bronze gears are listed in Table 14-5. Note the effect of the size-correction factor as the face width increases.

Values of the ratio-correction factor K_m and the velocity factor K_v are tabulated in Tables 14-6 and 14-7, respectively.

Example 14-6 A gear catalog lists a 4-pitch, $14\frac{1}{2}°$ pressure angle, single-thread, hardened steel worm to mate with a 24-tooth gear. The gear has a $1\frac{1}{2}$-in face width. Specifications for the worm are: lead, 0.7854 in; lead angle, 4.767°; face width, $4\frac{1}{2}$ in; pitch diameter, 3 in. The material of the gear is sand-cast bronze.

Table 14-7 VELOCITY FACTOR K_v

Velocity V_S, fpm	K_v	Velocity V_S, fpm	K_v	Velocity V_S, fpm	K_v
1	0.649	300	0.472	1400	0.216
1.5	0.647	350	0.446	1600	0.200
10	0.644	400	0.421	1800	0.187
20	0.638	450	0.398	2000	0.175
30	0.631	500	0.378	2200	0.165
40	0.625	550	0.358	2400	0.156
60	0.613	600	0.340	2600	0.148
80	0.600	700	0.310	2800	0.140
100	0.588	800	0.289	3000	0.134
150	0.558	900	0.269	4000	0.106
200	0.528	1000	0.258	5000	0.089
250	0.500	1200	0.235	6000	0.079

Source: Darle W. Dudley (ed.), *Gear Handbook*, McGraw-Hill, New York, 1962, pp. 13–39.

The worm material is given as 40-point hardened carbon steel.

(a) Estimate the safe power rating of this gearset if the worm speed is 1800 rpm.

(b) What is the bending stress in the gear tooth at the above rating?

Solution (a) From Table 14-5 we find $K_s = 700$. The pitch diameter of the gear is

$$d_G = \frac{N_G}{P_t} = \frac{24}{4} = 6 \text{ in}$$

The pitch diameter of the worm is given as 3 in; two-thirds of this is 2 in. But since the face width of the gear is $1\frac{1}{2}$ in, $F_e = 1.5$ in. Next, using $m_G = N_G/N_W = 24/1 = 24$ and Table 14-6, we find $K_m = 0.823$ by interpolation. The pitch-line velocity of the worm is

$$V_W = \frac{\pi \, d_W \, n}{12} = \frac{\pi(3)(1800)}{12} = 1414 \text{ fpm}$$

Then, using Eq. (14-27), the sliding velocity is

$$V_S = \frac{V_W}{\cos \lambda} = \frac{1414}{\cos 4.767°} = 1419 \text{ fpm}$$

Therefore, from Table 14-7, $K_v = 0.215$. Using Eq. (14-31) we now find the transmitted load as

$$W_{Gt} = K_s d_G^{0.8} F_e K_m K_v = (700)(6)^{0.8}(1.5)(0.823)(0.215) = 779 \text{ lb}$$

Next, if we enter Fig. 14-17 with $V_S = 1419$ fpm and use curve B, we get a coefficient of friction $\mu = 0.023$. Then using Eq. (14-24) we find the friction load to be

$$W_f = \frac{\mu W_{Gt}}{\mu \sin \lambda - \cos \phi_n \cos \lambda} = \frac{0.023(779)}{(0.023) \sin 4.767° - \cos 14.5° \cos 4.767°}$$

$$= -18.6 \text{ lb}$$

Equation (14-30) gives the input horsepower to the worm as

$$H = \frac{W_{Gt} d_G n_W}{126\,000 m_G} + \frac{V_S W_f}{33\,000} = \frac{779(6)(1800)}{(126\,000)(24)} + \frac{1419(18.6)}{33\,000}$$

$$= 2.78 + 0.800 = 3.58 \text{ hp} \qquad\qquad Ans.$$

The output power of the gear is the first term and is 2.78 hp.

(*b*) From Table 14-4 we find the form factor to be $y = 0.100$. The normal diametral pitch is

$$P_n = \frac{P_t}{\cos \psi} = \frac{4}{\cos 4.767°} = 4.014$$

So $p_n = \pi/4.014 = 0.783$ in. Substituting this and $F_G = 1.5$ in in Eq. (14-28) gives the bending stress in the gear tooth as

$$\sigma = \frac{W_{Gt}}{p_n F_G y} = \frac{779}{0.783(1.5)(0.100)} = 6630 \text{ psi} \qquad\qquad Ans.$$

/////

14-9 STRAIGHT BEVEL GEARS—KINEMATICS

When gears are to be used to transmit motion between intersecting shafts, some form of bevel gear is required. A bevel gearset is shown in Fig. 14-20. Although bevel gears are usually made for a shaft angle of 90°, they may be produced for almost any angle. The teeth may be cast, milled, or generated. Only the generated teeth may be classed as accurate.

The terminology of bevel gears is illustrated in Fig. 14-21. The pitch of bevel gears is measured at the large end of the tooth, and both the circular pitch and the pitch diameter are calculated in the same manner as for spur gears. It should be noted that the clearance is uniform. The pitch angles are defined by the pitch

FIGURE 14-20 A straight bevel gear and pinion. (*Courtesy of Gleason Works, Rochester, N. Y.*)

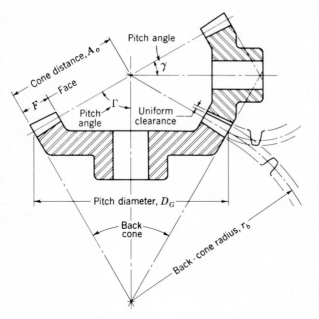

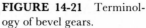

FIGURE 14-21 Terminology of bevel gears.

cones meeting at the apex, as shown in the figure. They are related to the tooth numbers as follows:

$$\tan \gamma = \frac{N_P}{N_G} \qquad \tan \Gamma = \frac{N_G}{N_P} \tag{14-32}$$

where the subscripts P and G refer to the pinion and gear, respectively, and where γ and Γ are, respectively, the pitch angles of the pinion and gear.

Figure 14-21 shows that the shape of the teeth, when projected on the back cone, is the same as in a spur gear having a radius equal to the back-cone distance r_b. This is called Tredgold's approximation. The number of teeth in this imaginary gear is

$$N' = \frac{2\pi r_b}{p} \tag{14-33}$$

where N' is the *virtual number of teeth* and p is the circular pitch measured at the large end of the teeth.

Standard straight-tooth bevel gears are cut by using a $20°$ pressure angle, unequal addenda and dedenda, and full-depth teeth. This increases the contact ratio, avoids undercut, and increases the strength of the pinion. Table 14-8 lists the standard tooth proportions at the large end of the teeth.

Table 14-8 TOOTH PROPORTIONS FOR 20° STRAIGHT BEVEL-GEAR TEETH

Item	Formula
Working depth	$h_k = 2.0/P$
Clearance	$c = (0.188/P) + 0.002$ in
Addendum of gear	$a_G = \dfrac{0.54}{P} + \dfrac{0.460}{P(m_{90})^2}$
Gear ratio	$m_G = N_G/N_P$
Equivalent 90° ratio	$m_{90} = m_G$ when $\Sigma = 90°$
	$m_{90} = \sqrt{m_G \dfrac{\cos \gamma}{\cos \Gamma}}$ when $\Sigma \neq 90°$
Face width	$F = \dfrac{A_o}{3}$ or $F = \dfrac{10}{P}$, whichever is smaller

Minimum number of teeth	Pinion	16	15	14	13
	Gear	16	17	20	30

14-10 BEVEL GEARS—FORCE ANALYSIS

In determining shaft and bearing loads for bevel-gear applications, the usual practice is to use the tangential or transmitted load which would occur if all the forces were concentrated at the midpoint of the tooth. While the actual resultant occurs somewhere between the midpoint and the large end of the tooth, there is only a small error in making this assumption. For the transmitted load, this gives

$$W_t = \frac{T}{r_{av}} \tag{14-34}$$

where T is the torque and r_{av} is the pitch radius of the gear under consideration at the midpoint of the tooth.

The forces acting at the center of the tooth are shown in Fig. 14-22. The resultant force W has three components, a tangential force W_t, a radial force W_r, and an axial force W_a. From the trigonometry of the figure,

$$W_r = W_t \tan \phi \cos \gamma \tag{14-35}$$
$$W_a = W_t \tan \phi \sin \gamma \tag{14-36}$$

The three forces W_t, W_r, and W_a are at right angles to each other and can be used to determine the bearing loads by using the methods of statics.

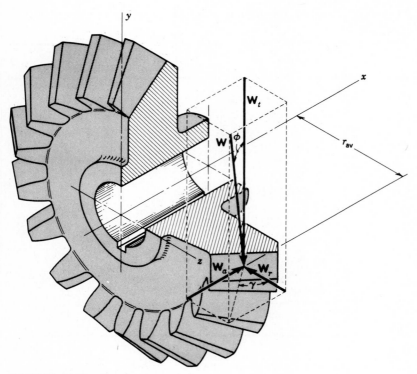

FIGURE 14-22   Bevel-gear tooth forces.

Example 14-7   The bevel pinion in Fig. 14-23 rotates at 600 rpm in the direction shown and transmits 5 hp to the gear. The mounting distances, the location of all bearings, and the average pitch radii of the pinion and gear are shown in the figure. For simplicity, the teeth have been replaced by the pitch cones. Bearings *A* and *C* should take the thrust loads. Find the bearing forces on the gearshaft.

Solution   The pitch angles are

$$\gamma = \tan^{-1} \left(\tfrac{3}{9}\right) = 18.4° \qquad \Gamma = \tan^{-1} \left(\tfrac{9}{3}\right) = 71.6°$$

The pitch-line velocity corresponding to the average pitch radius is

$$V = \frac{2\pi r_p n}{12} = \frac{2\pi(1.293)(600)}{12} = 406 \text{ fpm}$$

Therefore the transmitted load is

$$W_t = \frac{33\ 000H}{V} = \frac{(33\ 000)(5)}{406} = 406 \text{ lb}$$

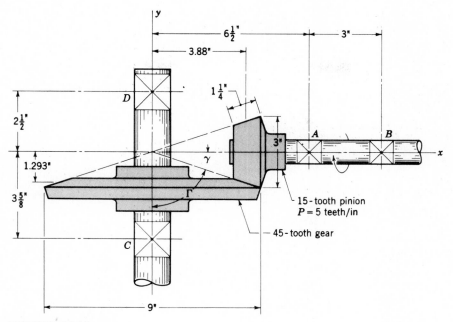

FIGURE 14-23

which acts in the positive z direction, as shown in Fig. 14-24. We next have

$$W_r = W_t \tan \phi \cos \Gamma = 406 \tan 20° \cos 71.6° = 46.6 \text{ lb}$$

$$W_a = W_t \tan \phi \sin \Gamma = 406 \tan 20° \sin 71.6° = 140 \text{ lb}$$

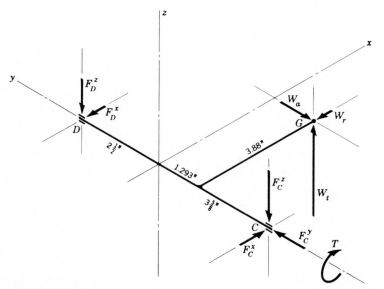

FIGURE 14-24

where W_r is in the $-x$ direction and W_a in the $-y$ direction, as illustrated in the isometric sketch of Fig. 14-24.

In preparing to take a sum of the moments about bearing D, define the position vector from D to G as

$$\mathbf{R}_G = 3.88\mathbf{i} - (2.5 + 1.293)\mathbf{j} = 3.88\mathbf{i} - 3.793\mathbf{j}$$

We shall also require a vector from D to C.

$$\mathbf{R}_C = -(2.5 + 3.625)\mathbf{j} = -6.125\mathbf{j}$$

Then, summing moments about D gives

$$\mathbf{R}_G \times \mathbf{W} + \mathbf{R}_C \times \mathbf{F}_C + \mathbf{T} = 0 \tag{1}$$

When we place the details in Eq. (1), we get

$$(3.88\mathbf{i} - 3.793\mathbf{j}) \times (-46.6\mathbf{i} - 140\mathbf{j} + 406\mathbf{k})$$
$$+ (6.125\mathbf{j}) \times (F_C^x\mathbf{i} + F_C^y\mathbf{j} + F_C^z\mathbf{k}) + T\mathbf{j} = 0 \tag{2}$$

After forming the two cross products, the equation becomes

$$(-1504\mathbf{i} - 1580\mathbf{j} - 721\mathbf{k}) + (-6.125F_C^z\mathbf{i} + 6.125F_C^x\mathbf{k}) + T\mathbf{j} = 0$$

from which

$$\mathbf{T} = 1580\mathbf{j} \text{ lb} \cdot \text{in} \qquad F_C^x = 118 \text{ lb} \qquad F_C^z = -246 \text{ lb} \tag{3}$$

Now sum the forces to zero. Thus

$$\mathbf{F}_D + \mathbf{F}_C + \mathbf{W} = 0 \tag{4}$$

When the details are inserted, Eq. (4) becomes

$$(F_D^x\mathbf{i} + F_D^z\mathbf{k}) + (118\mathbf{i} + F_C^y\mathbf{j} - 246\mathbf{k}) + (-46.6\mathbf{i} - 140\mathbf{j} + 406\mathbf{k}) = 0 \tag{5}$$

First we see that $F_C^y = 140$ lb, and so

$$\mathbf{F}_C = 118\mathbf{i} + 140\mathbf{j} - 246\mathbf{k} \text{ lb} \qquad\qquad\qquad Ans.$$

Then, from Eq. (5),

$$\mathbf{F}_D = -71\mathbf{i} - 160\mathbf{k} \text{ lb} \qquad\qquad\qquad Ans.$$

These are all shown in Fig. 14-24 in the proper directions. The analysis for the pinion shaft is quite similar. ////

14-11 BEVEL GEARING—BENDING STRESS AND STRENGTH

In a typical bevel-gear mounting, Fig. 14-23 for example, one of the gears is often mounted outboard of the bearings. This means that shaft deflections can be more pronounced and have a greater effect on the contact of the teeth. Another difficulty which occurs in predicting the stress in bevel-gear teeth is the fact that the teeth are tapered. Thus, to achieve perfect line contact passing through the cone center, the teeth ought to bend more at the large end than at the small end. To obtain this condition requires that the load be proportionately greater at the large end. Because of this varying load across the face of the tooth, it is desirable to have a fairly short face width.

The equation for bending stress in spur gears is used for bevel gears, too, and is repeated here for convenience.

$$\sigma = \frac{W_t P}{K_v F J} \tag{14-37}$$

where all relations are based on the large end of the teeth.

Caution: The transmitted load W_t must be computed using the pitch radius at the large end of the teeth in Eq. (14-37). Note that this is not the same transmitted load used in force analysis (Sec. 14-10), though the symbol is the same.

The geometry factor J is different for bevel gears because the long-and-short addendum system is used and because the teeth are tapered. Use Fig. 14-25.

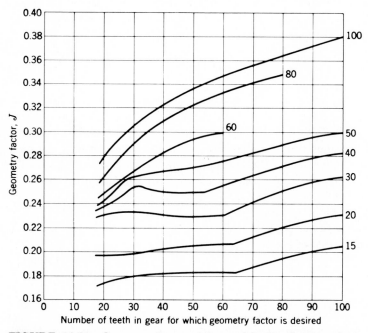

FIGURE 14-25 Geometry factors J for straight bevel gears; these are for a 90° shaft angle, 20° pressure angle, and a clearance of $c = 0.240/P$ in. (*AGMA Information Sheet 225.01.*)

Table 14-9 APPROXIMATE BEVEL-GEAR LOAD
DISTRIBUTION FACTORS K_m AND C_m

Application	Both gears inboard	One gear outboard	Both gears outboard
General industrial	1.00–1.10	1.10–1.25	1.25–1.40
Automotive	1.00–1.10	1.10–1.25	
Aircraft	1.00–1.25	1.10–1.40	1.25–1.50

Source: AGMA Information Sheet 225.01, 1967, table 4.

The modification and correction factors for bevel gears are the same as for spur gears except for the load-distribution factor K_m (Table 14-9).*

14-12 BEVEL GEARING—SURFACE DURABILITY

The Hertzian contact stress for bevel gears is given by the equation

$$\sigma_H = -C_p \sqrt{\frac{W_t}{C_v F d_p I}} \tag{14-38}$$

where, again, all values correspond to the large end of the teeth.

Since the contact of bevel-gear teeth tends to be localized, the elastic coefficient C_p must be based on a Hertzian analysis of contacting spheres rather than cylinders. This yields slightly different values. Thus, use Table 14-10.

Figure 14-26 is a chart of the geometry factor I for bevel gears. All other factors may be obtained using the methods of Chap. 13.

Example 14-8 A pair of miter gears listed in a catalog have a diametral pitch of 5, 25 teeth, 1.10 in face width, a 20° pressure angle, and are made of a 0.20 plain-carbon steel with the teeth case-hardened. In this example it is assumed that the case hardness is 500 Bhn. The gears are intended for general industrial use, and it is quite likely that applications will occur in which both gears must have outboard mountings.

(a) Specify a horsepower rating based on bending strength using a factor of safety of 1.8 and a speed of 600 rpm.
(b) The same as (a), based on surface durability and a factor of safety of 1.20.

---

* The AGMA uses a different size factor for bevel gears than for others. However, they compensate for this by recommending a different set of allowable stresses. See *AGMA Information Sheet* 225.01, 1967.

Table 14-10 VALUES OF THE ELASTIC COEFFICIENT C_p
FOR BEVEL GEARS AND OTHERS WITH
LOCALIZED CONTACT*

Pinion	Gear			
	Steel	Cast iron	Aluminum bronze	Tin bronze
Steel, $E = 30$	2800	2450	2400	2350
Cast iron, $E = 19$	2450	2250	2200	2150
Aluminum bronze, $E = 17.5$	2400	2200	2150	2100
Tin bronze, $E = 16$	2350	2150	2100	2050

Source: *AGMA Information Sheet* 212.02
* In each case the modulus of elasticity is in Mpsi.

Solution (*a*) The pitch diameter at the large end of the teeth is $d = \frac{25}{5} = 5$ in.
Since the face width is 1.10 in, and the pitch angle of a miter gear is 45°, the
average pitch diameter is

$$d_{av} = d - F \sin \gamma = 5 - 1.10 \sin 45° = 4.22 \text{ in}$$

The pitch-line velocity at the average diameter is

$$V = \frac{\pi d_{av} n}{12} = \frac{\pi (4.22)(600)}{12} = 663 \text{ fpm}$$

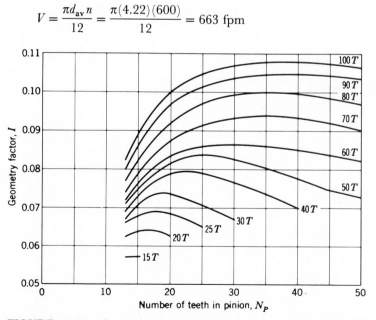

FIGURE 14-26 Geometry factors *I* for straight bevel gears of 20°
pressure angle mounted at a 90° shaft angle. (*AGMA Information Sheet*
212.02.)

We select Eq. (13-27) for the velocity factor. Thus

$$K_v = \frac{50}{50 + \sqrt{V}} = \frac{50}{50 + \sqrt{663}} = 0.660$$

Figure 14-25 gives $J = 0.21$ for the geometry factor. Equation (14-37) then gives

$$\sigma = \frac{W_t P}{K_v FJ} = \frac{5W_t}{(0.660)(1.10)(0.21)} = 32.8W_t \text{ psi} \tag{1}$$

The next step is to determine the corresponding strength. Based on $H_B = 500$, the tensile strength in the case is $S_{ut} = 500H_B = 500(500)(10)^{-3} = 250$ kpsi as indicated by Eq. (4-16). Then Fig. 13-25 gives the surface-finish factor as $k_a = 0.60$ approximately. Basing the size factor on the large end of the tooth, which is conservative, we find $k_b = 0.909$ from Table 13-9. Choosing a moderate reliability, say $R = 0.90$, gives $k_c = 0.897$ from Table 13-10. Both k_d and k_e are unity. Assuming one-way bending, we find $k_f = 1.43$ from Table 13-11. Then, since $S'_e = 100$ kpsi, we have

$$S_e = k_a k_b k_c k_d k_e k_f S'_e = (0.60)(0.909)(0.897)(1)(1)(1.43)(100) = 70.0 \text{ kpsi}$$

Next, we select $K_o = 1.50$ from Table 13-12 and $K_m = 1.40$ from Table 14-9. Using a factor of safety of 1.80, as given, we now have

$$n_G = K_o K_m n = (1.50)(1.40)(1.80) = 3.78$$

Thus, from Eq. (1), the permissible stress is found to be

$$\sigma_P = n_G \sigma = 32.8(3.78)W_t = 124W_t \text{ psi}$$

Equating the permissible stress to the strength, and solving for W_t, gives

$$W_t = \frac{70.0(10)^3}{124} = 565 \text{ lb}$$

The corresponding horsepower is

$$H = \frac{W_t V}{33(10)^3} = \frac{565(663)}{33(10)^3} = 11.4 \text{ hp} \qquad\qquad Ans.$$

(b) The contact strength is

$$S_C = 0.4H_B - 10 = 0.4(500) - 10 = 190 \text{ kpsi}$$

Using Table 13-15 we select $C_L = 1.0$ and $C_R = 0.80$. Also $C_H = C_T = 1$. Therefore, from Eq. (13-46) we get the Hertzian strength as

$$S_H = \frac{C_L\,C_H}{C_T\,C_R}\,S_C = \frac{(1)(1)}{(1)(0.8)}\,(190) = 237.5 \text{ kpsi}$$

Since the factor of safety is given as $n = 1.20$ for surface durability, we have $n_G = (1.50)(1.40)(1.20) = 2.52$. Also, from Table 14-10, $C_p = 2800$. And $C_v = K_v = 0.660$. Figure 14-26 gives $I = 0.065$.

We now write Eq. (14-38) in the form

$$S_H = C_p\sqrt{\frac{W_{t,\,p}}{C_v\,Fd_p\,I}}$$

where $W_{t,\,p} = n_G W_t$. Substituting gives

$$237.5(10)^3 = 2800\sqrt{\frac{2.52W_t}{(0.660)(1.10)(5)(0.065)}}$$

Solving this equation gives $W_t = 674$ lb. Therefore

$$H = \frac{W_t V}{33(10)^3} = \frac{674(663)}{33(10)^3} = 13.5 \text{ hp} \qquad\qquad Ans.$$

////

14-13   SPIRAL BEVEL GEARS

Straight bevel gears are easy to design and simple to manufacture and give very good results in service if they are mounted accurately and positively. As in the case of spur gears, however, they become noisy at the higher values of the pitch-line velocity. In these cases it is often good design practice to go to the spiral bevel gear, which is the bevel counterpart of the helical gear. Figure 14-27 shows a mating pair of spiral bevel gears, and it can be seen that the pitch surfaces and the nature of contact are the same as for straight bevel gears, except for the differences brought about by the spiral-shaped teeth.

The *hand* of the spiral is found using the right-hand rule, with the thumb pointing along the axis of rotation. In Fig. 14-27, a left-hand pinion is in mesh with a right-hand gear.

Spiral-bevel-gear teeth are conjugate to a basic crown rack $(2\Gamma = 180°)$, which is generated, as shown in Fig. 14-28, by using a circular cutter. The spiral angle ψ is measured at the mean radius of the gear. As in the case of helical gears, spiral bevel gears give a much smoother tooth action than straight bevel

FIGURE 14-27 Spiral bevel gears. (*Courtesy of Gleason Works, Rochester, N. Y.*)

gears, and hence are useful where high speeds are encountered. Antifriction bearings should be used to take the thrust loads, however, because these loads are larger than for straight bevel gears. The *face-contact ratio*, which is the face advance divided by the circular pitch (Fig. 14-28), should be at least 1.25 to obtain true spiral-tooth action.

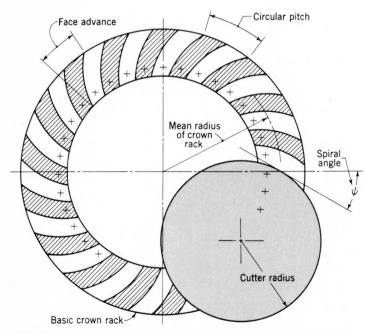

FIGURE 14-28 Cutting spiral-gear teeth on the basic crown rack.

Pressure angles used with spiral bevel gears are generally $14\frac{1}{2}$ to $20°$, while the spiral angle is usually $35°$. The hand of the spiral should be selected so as to cause the gears to separate from each other and not to force them together, which might cause jamming. For example, the left-hand pinion of Fig. 14-27 will be forced into the teeth of the gear if it is rotated in the direction of the fingers of your right hand when your thumb is pointing from left to right. In any event, the supporting bearings should always be designed so that there is no looseness or play in the axial direction.

Tooth proportions for spiral bevel gears having a $20°$ pressure angle, a $35°$ spiral angle, and a stub height are given in Table 14-11. For these the working depth is $1.700/P$ and the clearance $0.188/P$.

The total force W acting normal to the pinion tooth and assumed concentrated at the average radius of the pitch cone may be divided into three perpendicular components. These are the transmitted, or tangential, load W_t; the axial, or thrust, component W_a; and the separating, or radial, component W_r. The force W_t may, of course, be computed from the equation

$$W_t = \frac{T}{r_{av}} \tag{14-39}$$

where T is the input torque and r_{av} is the average radius of the pinion pitch cone. The forces W_a and W_r depend upon the hand of the spiral and the direction of rotation. Thus there are four possible cases to consider. For a *right-hand pinion spiral with clockwise pinion rotation* and for a *left-hand spiral with counterclockwise rotation*,

Table 14-11   GEAR ADDENDUM FOR 1-DIAMETRAL-PITCH SPIRAL
                   BEVEL GEARS

Ratios			Ratios			Ratios		
From	To	Addendum	From	To	Addendum	From	To	Addendum
1.00	1.00	0.850	1.23	1.26	0.710	1.82	1.90	0.570
1.00	1.02	0.840	1.26	1.28	0.700	1.90	1.99	0.560
1.02	1.03	0.830	1.28	1.31	0.690	1.99	2.10	0.550
1.03	1.05	0.820	1.31	1.34	0.680	2.10	2.23	0.540
1.05	1.06	0.810	1.34	1.37	0.670	2.23	2.38	0.530
1.06	1.08	0.800	1.37	1.41	0.660	2.38	2.58	0.520
1.08	1.09	0.790	1.41	1.44	0.650	2.58	2.82	0.510
1.09	1.11	0.780	1.44	1.48	0.640	2.82	3.17	0.500
1.11	1.13	0.770	1.48	1.52	0.630	3.17	3.67	0.490
1.13	1.15	0.760	1.52	1.57	0.620	3.67	4.56	0.480
1.15	1.17	0.750	1.57	1.63	0.610	4.56	7.00	0.470
1.17	1.19	0.740	1.63	1.68	0.600	7.00	∞	0.460
1.19	1.21	0.730	1.68	1.75	0.590			
1.21	1.23	0.720	1.75	1.82	0.580			

Source: From Gleason Works, Rochester, N. Y.

the equations are

$$W_a = \frac{W_t}{\cos \psi} (\tan \phi_n \sin \gamma - \sin \psi \cos \gamma)$$

$$\text{(14-40)}$$

$$W_r = \frac{W_t}{\cos \psi} (\tan \phi_n \cos \gamma + \sin \psi \sin \gamma)$$

The other two cases are a *left-hand spiral with clockwise rotation* and a *right-hand spiral with counterclockwise rotation*. For these two cases the equations are

$$W_a = \frac{W_t}{\cos \psi} (\tan \phi_n \sin \gamma + \sin \psi \cos \gamma)$$

$$\text{(14-41)}$$

$$W_r = \frac{W_t}{\cos \psi} (\tan \phi_n \cos \gamma - \sin \psi \sin \gamma)$$

where ψ = spiral angle
γ = pinion pitch angle
ϕ_n = normal pressure angle

and the rotation is observed from the input end of the pinion shaft. Equations

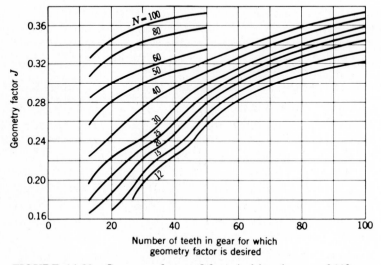

FIGURE 14-29 Geometry factors J for spiral bevel gears of 20° pressure angle and 35° spiral angle. (*By permission, from Gear Handbook, McGraw-Hill, New York, 1962, pp. 13–36.*)

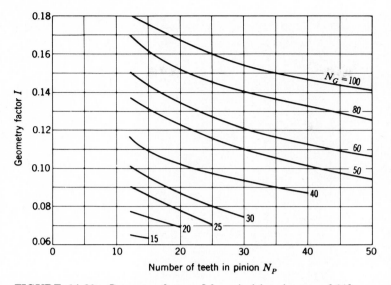

FIGURE 14-30 Geometry factors I for spiral bevel gears of 20° pressure angle and 35° spiral angle. (*By permission, from Gear Handbook, McGraw-Hill, New York, 1962, pp. 13–27.*)

(14-40) and (14-41) give the forces exerted by the gear on the pinion. A positive sign for either W_a or W_r indicates that it is directed away from the cone center.

The forces exerted by the pinion on the gear are equal and opposite. Of course, the opposite of an axial pinion load is a radial gear load, and the opposite of a radial pinion load is an axial gear load.

Except for the geometry factors I and J, the same stress and strength equations apply for both bending and wear as for straight bevel gears. Figures 14-29 and 14-30 are used to obtain the factors J and I, respectively.

The *Zerol bevel gear* is a patented gear having curved teeth but with a zero spiral angle. It thus can be generated using the same tool as for a regular spiral bevel gear. The curved teeth give somewhat better tooth action than can be obtained with straight-tooth bevel gears. In design it is probably best to proceed as for straight-tooth bevel gears and then simply substitute a Zerol bevel gear.

It is frequently desirable, as in the case of automotive differential applications, to have gearing similar to bevel gears but with the shafts offset. Such gears are called *hypoid gears* because their pitch surfaces are hyperboloids of revolution. The tooth action between such gears is a combination of rolling and sliding along a straight line and has much in common with that of worm gears. Figure 14-31 shows a pair of hypoid gears in mesh.

Figure 14-32 is included to assist in the classification of spiral-type bevel gearing. It is seen that the hypoid gear has a relatively small shaft offset. For larger offsets the pinion begins to resemble a tapered worm, and the set is then called *spiroid gearing*.

FIGURE 14-31   Hypoid gears.   (*Courtesy of Gleason Works, Rochester, N. Y.*)

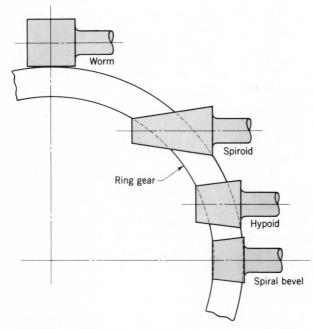

FIGURE 14-32   Comparison of intersecting- and offset-shaft bevel-type gearings.   (*By permission, from Gear Handbook, McGraw-Hill, New York, 1962, pp. 2–24.*)

Example 14-9 Let the pinion of Example 14-7 (Fig. 14-23) be cut with a left-hand 35° spiral. Using the remaining data of the example, find the forces exerted on the gear shaft by the bearings at C and D.

Solution Using $W_t = 406$ lb, we find, from Eq. (14-41),

$$W_a = \frac{406}{\cos 35°} (\tan 20° \sin 18.4° + \sin 35° \cos 18.4°) = 326 \text{ lb}$$

$$W_r = \frac{406}{\cos 35°} (\tan 20° \cos 18.4° - \sin 35° \sin 18.4°) = 81 \text{ lb}$$

These are the forces exerted by the gear on the pinion. Referring to Fig. 14-23, W_t is in the $-z$ direction, W_a in the $+x$ direction, and W_r in the $+y$ direction. The corresponding component forces on the gear are the same as in Fig. 14-24. Therefore, for the gear, we write

$$\mathbf{W} = -326\mathbf{i} - 81\mathbf{j} + 406\mathbf{k} \text{ lb}$$

The procedure is now the same as in Example 14-7. The results are

$$\mathbf{F}_C = 252\mathbf{i} + 81\mathbf{j} - 251\mathbf{k} \text{ lb} \qquad\qquad Ans.$$

$$\mathbf{F}_D = 74\mathbf{i} - 155\mathbf{k} \text{ lb} \qquad\qquad Ans.$$

$////$

PROBLEMS*

Sections 14-1 to 14-3

14-1 A parallel helical gearset consists of an 18-tooth pinion driving a 32-tooth gear. The pinion has a left-hand helix angle of 25°, a normal pressure angle of 20°, and a normal diametral pitch of 8 teeth/in.
   (*a*) Find the normal, transverse, axial, and normal base-circular pitches.
   (*b*) Find the transverse diametral pitch and the transverse pressure angle.
   (*c*) Find the addendum and dedendum and the pitch diameters of both gears.
14-2 The same as Prob. 14-1 except the normal diametral pitch is 6, the pinion has 16 teeth with a right-hand helix angle of 30°, and the gear has 48 teeth.
14-3 A pair of mating parallel helical gears has an angular-velocity ratio of 3.20. The

---

* The asterisk indicates a design-type problem or one which may have no unique result.

driver has 20 teeth, a transverse diametral pitch of 4, a 15° left-hand helix, and a transverse pressure angle of 20°.

(*a*)   Compute all the circular and diametral pitches and the normal pressure angle.

(*b*)   Find the addendum and dedendum and the pitch diameters of both gears.

14-4   A 14-tooth helical pinion runs at a speed of 2400 rpm and drives a gear on a parallel shaft at a speed of 600 rpm. The pinion has a normal diametral pitch of 12, and a 23° left-hand helix angle. The normal pressure angle is 25°.

(*a*)   Compute all the circular and diametral pitches and the transverse pressure angle.

(*b*)   Find the whole depth of the teeth based on shaved profiles using the proportions in Table 13-1.

(*c*)   What are the pitch diameters of both gears?

14-5   The gears shown in the figure have a normal diametral pitch of 6, a normal pressure angle of 20°, and a helix angle of 30°. For both gear trains the transmitted load is 400 lb. Find the force exerted by each gear on its shaft. The direction of pinion rotation is shown.

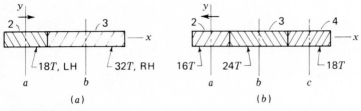

PROBLEM 14-5

14-6   A gear train is composed of four helical gears with the shaft axes in a single plane. The gears have a 20° transverse pressure angle and a 30° helix angle. In the figure, gear 2 drives gear 3 with a transmitted load of 500 lb. Find the radial and thrust forces exerted on each shaft. The gears on shaft *b* have 56 and 14 teeth and are each 7-diametral-pitch in the normal plane.

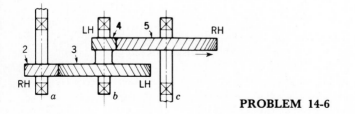

PROBLEM 14-6

14-7   Gear 2, in the figure has 16 teeth, a 20° transverse angle, a 15° helix angle, and a normal diametral pitch of 8 teeth/in. Gear 2 drives the idler on shaft *b*, which has 36 teeth. The driven gear on shaft *c* has 28 teeth. If the driver rotates at 1720 rpm and transmits $7\frac{1}{2}$ hp, find the radial and thrust load on each shaft.

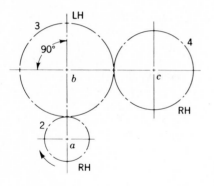

PROBLEM 14-7

14-8   The figure shows a double-reduction helical gearset. Pinion 2 is the driver, and it receives a torque of 1200 lb · in from its shaft in the direction shown. Pinion 2 has a normal diametral pitch of 8 teeth/in, 14 teeth, and a normal pressure angle of 20° and is cut right-handed with a helix angle of 30°. The mating gear 3 on shaft b has 36 teeth. Gear 4, which is the driver for the second pair of gears in the train, has a normal diametral pitch of 5 teeth/in, 15 teeth, and a normal pressure angle of 20° and is cut left-handed with a helix angle of 15°. Mating gear 5 has 45 teeth. Find the magnitude and direction of the force exerted by the bearings at C and D on shaft b if bearing C can take only radial load while bearing D is mounted to take both radial and thrust load.

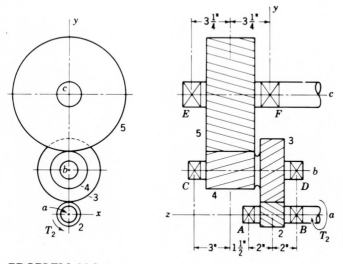

PROBLEM 14-8

14-9   The double-reduction helical gearset shown in the figure is driven through shaft a and receives $7\frac{1}{2}$ hp at a speed of 900 rpm. Gears 2 and 3 have a transverse diametral pitch of 10 teeth/in., a 30° helix angle, and a transverse pressure angle of

20°. Pinion 2 is cut with 14 teeth, left-handed; gear 3 has 54 teeth. Each of the second pair of gears in the train, 4 and 5, has a transverse diametral pitch of 6 teeth/in, a 23° helix angle, and a transverse pressure angle of 20°. Gear 4 is left-handed and has 16 teeth; gear 5 has 36 teeth. The gears are supported by bearings located as shown in the figure. Good design dictates that the thrust bearing should be located so as to load the shaft in compression. The thrust bearing then takes both radial and thrust loads, while the second bearing on the shaft is subject to pure radial load. In accordance with this convention, determine the magnitude and direction of the radial and thrust loads which bearings C and D exert on shaft b.

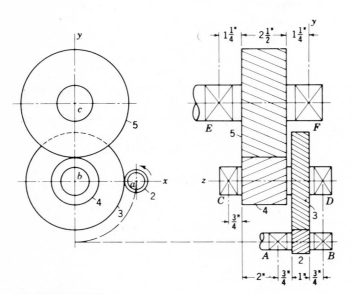

PROBLEM 14-9

Section 14-4

14-10   A parallel-shaft helical gearset is composed of a driving pinion with 20 teeth and a driven gear with 96 teeth. The gears have a transverse diametral pitch of 20 teeth/in, a 25° helix angle, and a pressure angle $\phi_n = 20°$ and are cut full depth. The face width is $\frac{3}{4}$ in, and the gears run at a pitch-line velocity of 1600 fpm. The driver is cut from cold-drawn UNS G10180 steel without heat treatment, and the driven of grade 30 cast iron. Basing computations on bending strength, a factor of safety $n_G = 4$, and unity for the life, temperature, and reliability factors, find the maximum safe capacity in horsepower for general industrial use.

14-11   A speed reducer is composed of a helical driver of 25 teeth and a mating gear of 64 teeth. The gears have a normal diametral pitch of 16, a 30° helix angle, and a 20° normal pressure angle. The face width is 1 in, and the maximum pinion speed is 2300 rpm. Both gears are made of nickel-alloy steel, case-hardened to 560 Bhn. Determine the maximum tangential load that these gears can safely transmit based on bending strength and a reliability of 99.9 percent or better. Use unity for overload and temperature factors and a factor of safety $n_G = 3.00$.

14-12 Find the horsepower capacity of the gears of Prob. 14-10 based upon surface durability.

14-13 Find the maximum tangential load for the gears of Prob. 14-11 based on surface durability.

14-14 The first pair of gears in a three-stage herringbone reducer consists of a 26-tooth driver and a 104-tooth driven. The gears have a normal diametral pitch of 14 teeth/in, a normal pressure angle of 20°, and a helix angle of 25°. The face width of each half is $1\frac{1}{4}$ in. The gears are forged of UNS G46500 steel and heat-treated to 410 Bhn. For a pinion speed of 3600 rpm and infinite life with 99 percent reliability, find the capacity in horsepower of the first pair of gears on the basis of bending strength. Use an overall factor of safety $n_G = 3.20$.

Section 14-5

14-15 Find the rpm and direction of rotation of gear 7 in the figure.

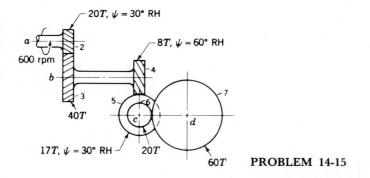

PROBLEM 14-15

Section 14-7

14-16 A 4-tooth right-hand worm transmits $\frac{1}{3}$ hp at 750 rpm to a 40-tooth gear having a transverse diametral pitch of 10 teeth/in. The worm has a pitch diameter of $1\frac{1}{4}$ in and a face width of $1\frac{3}{8}$ in. The normal pressure angle is $14\frac{1}{2}°$. The worm is made of low-carbon steel, without heat treatment, and the gear of cast iron. Find the input torque, the output torque, and the efficiency. Use $\mu = 0.07$.

14-17 A right-hand, single-tooth, hardened-steel worm (hardness not specified) has a catalog rating of 2.70 hp at 600 rpm when meshed with a 48-tooth cast-iron gear. The transverse diametral pitch is 3 teeth/in, the transverse pressure angle is $14\frac{1}{2}°$, the pitch diameter of the worm is 4 in, and the face widths of the worm and gear are, respectively, 4 and 2 in. The figure (see p. 684) shows bearings A and B on the worm shaft symmetrically located with respect to the worm and 8 in apart. Determine which should be the thrust bearing, and find the magnitudes and directions of the forces exerted by both bearings.

14-18 The hub diameter and projection for the gear of Prob. 14-17 are 4 and $1\frac{1}{2}$ in, respectively. The face width of the gear is 2 in. Locate bearings C and D on opposite sides, spacing them 1 in from the gear, and then find the magnitudes and directions of the forces exerted by the bearings on the gearshaft. Also, find the output torque.

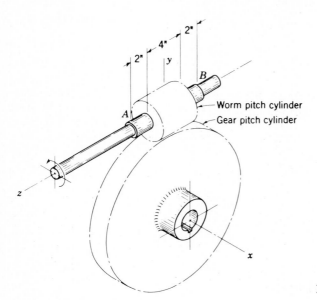

PROBLEM 14-17

Section 14-8

14-19 A 2-tooth, case-hardened worm of forged steel is to drive a 40-tooth sand-cast bronze worm gear. The diametral pitch is 10 teeth/in. The worm has a $14\frac{1}{2}°$ normal pressure angle, a lead angle of 9.083°, a pitch diameter of 1.25 in, and a face width of 2 in, and its teeth are ground and polished. The gear has a $\frac{5}{8}$-in face and a pitch diameter of 4.00 in. For a worm speed of 1720 rpm find the maximum safe horsepower output and the efficiency of this gearset.

14-20 A 5-diametral-pitch, 3-tooth steel worm is case-hardened and has ground and polished teeth. The worm has a 20° normal pressure angle, a 8.62° lead angle, a 2.30-in pitch diameter, and a 3-in face width. The worm rotates at 1740 rpm and drives a 50-tooth, $1\frac{3}{8}$-in face, sand-cast bronze worm gear. Find the maximum horsepower that can be transmitted by this gearset.

14-21 A 16-diametral-pitch, 4-tooth, case-hardened steel worm has a lead angle of 21.8° and a $\frac{5}{8}$-in pitch diameter. It drives a centrifugal-cast-bronze worm gear having a $\frac{5}{16}$-in face, 40 teeth, and a 20° normal pressure angle. The worm is right-handed and rotates at 600 rpm. Calculate the maximum rated horsepower and the efficiency of this gearset.

Section 14-9

14-22 The figure shows a gear train consisting of bevel gears, spur gears, and a worm and worm gear. The bevel pinion is mounted on a shaft driven by a V belt and pulleys. If pulley 2 rotates at 1200 rpm in the direction shown, find the speed and direction of rotation of gear 9.

14-23 The tooth numbers for the automotive differential shown in the figure are $N_2 = 17$, $N_3 = 54$, $N_4 = 11$, $N_5 = N_6 = 16$. The drive shaft turns at 1200 rpm. What is

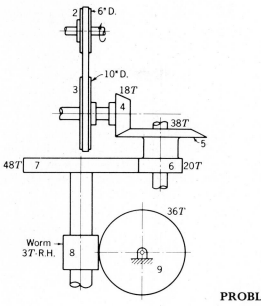

PROBLEM 14-22

the speed of the right wheel if it is jacked up and the left wheel is resting on the road surface?

14-24 A vehicle using the differential shown in the figure turns to the right at a speed of 30 mph on a curve of 80-ft radius. Use the same tooth numbers as in Prob. 14-23.

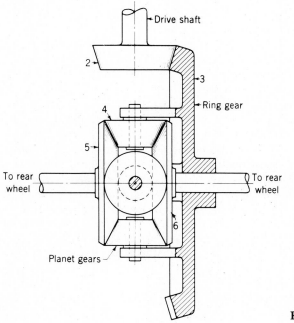

PROBLEMS 14-23 and 14-24

The tires are 15 in in diameter.   Use 60 in as the center-to-center distance between treads.

(*a*)   Calculate the speed of each rear wheel.

(*b*)   What is the speed of the ring gear?

Section 14-10

14-25   The figure shows a gear train composed of a pair of helical gears and a pair of straight bevel gears.   Shaft *c* is the output of the train, and it delivers 6 hp to the load at a speed of 370 rpm.   The bevel gears have a pressure angle of 20°.   If bearing *E* is to take both thrust and radial load while bearing *F* is to take only radial, determine the magnitude and direction of the forces which these bearings exert against shaft *c*.

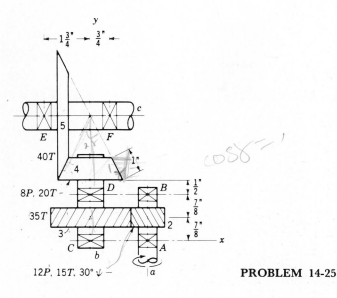

PROBLEM 14-25

14-26   Using the data of Prob. 14-25, find the forces exerted by bearing *C* and *D* on shaft *b*.   Which of these bearings should take the thrust load if the shaft is to be loaded in compression?   The helical gears have a 20° transverse pressure angle.

14-27   The gear train shown in the figure is composed of a right-hand helical pinion on shaft *a* which is cut with a $17\frac{1}{2}°$ normal pressure angle, the mating gear on shaft *b*, a straight-tooth miter gear on shaft *b*, and its mating miter gear on shaft *c*.   The miter gears operate at a 20° pressure angle.   The torque input to shaft *a* is 875 lb·in in the direction shown.   Find the forces exerted by bearings *C* and *D* on shaft *b*.   Specify which bearing is to take the thrust load.

14-28   Using the data of Prob. 14-27, find the forces exerted by bearings at *E* and *F* on shaft *c*.   Which of these two bearings should take the thrust component?

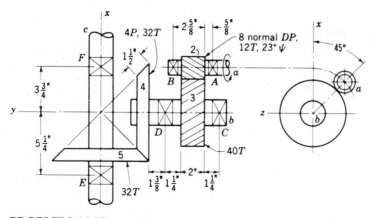

PROBLEM 14-27

Section 14-11

14-29 A straight-tooth bevel pinion has 15 teeth and a diametral pitch of 6 teeth/in, is made of cold-drawn UNS G10180 steel without heat treatment, and is formed by hobbing the teeth. The pinion drives a 60-tooth gear made of grade 30 cast iron. The shaft angle is 90°, the face widths $1\frac{1}{4}$ in, and the pressure angle 20°. The pinion rotates at 900 rpm and is mounted outboard of its bearings, while the gear is straddle-mounted. Based on bending strength, 50 percent reliability, and an overall factor of safety $n_G = 4$, specify the maximum safe capacity of the drive in horsepower for general industrial use.

14-30* The outboard-motor gears shown in Fig. 6-14 have a diametral pitch of 10 teeth/in and a face width of 0.55 in. There are 14 teeth on the pinion and 21 on the gear, as indicated on the figure caption. The material is unknown; assume it to be a through-hardened steel with $H_B = 550$ Bhn. Determine factors of safety based on bending stress and surface durability if 6 hp is transmitted to the propeller at a pinion speed of 3500 rpm.

14-31 A 16-tooth bevel pinion has a diametral pitch of 4 teeth/in, teeth cut by hobbing and then shaving, using the 20 system, and a face width of $1\frac{1}{2}$ in. The pinion drives a 32-tooth gear with a shaft angle of 90°. The pinion and gear are made of UNS G43400 forged steel, heat-treated to 350 Bhn. Based on bending strength, a reliability of 99 percent, and a pitch-line velocity of 1500 fpm, determine the horsepower capacity of this pair of gears. Use $n_G = 3.50$.

CHAPTER
15

SHAFTS

The design of shafts is a basic design problem. It utilizes most, if not all, of the fundamentals discussed in the first seven chapters of this book. You have by now gained some degree of maturity in applying these fundamentals to the analysis and design of various mechanical elements, and this, coupled with your knowledge of bearings and gears, makes it possible to reexamine shafting with somewhat more sophistication.

15-1 INTRODUCTION

A *shaft* is a rotating or stationary member, usually of circular cross section, having mounted upon it such elements as gears, pulleys, flywheels, cranks, sprockets, and other power-transmission elements. Shafts may be subjected to bending, tension, compression, or torsional loads, acting singly or in combination with one another. When they are combined, one may expect to find both static and fatigue strength to be important design considerations, since a single shaft may be subjected to static stresses, completely reversed stresses, and repeated stresses, all acting at the same time.

The word "shaft" covers numerous variations, such as axles and spindles. An *axle* is a shaft, either stationary or rotating, not subjected to a torsion load. A short rotating shaft is often called a *spindle*.

When either the lateral or the torsional deflection of a shaft must be held to close limits, the shaft should be sized on the basis of deflection before analyzing the stresses. The reason for this is that, if the shaft is made stiff enough so that the deflection is not too large, it is probable that the resulting stresses will be safe. But by no means should the designer *assume* that they are safe; it is almost always necessary to calculate them so as to *know* that they are within acceptable limits.

Whenever possible, the power-transmission elements, such as gears or pulleys, should be located close to the supporting bearings. This reduces the bending moment, and hence the deflection and bending stress.

In the following sections the design and analysis of shafting are discussed for the case when stress is the governing design consideration. When deflection is important, you should use the methods of Chap. 3 first, to size the members tentatively.

The design methods which follow differ in various respects from each other. Some are quite conservative while others are useful because they provide quick results. But you must not expect the methods to produce identical results.

15-2 DESIGN FOR STATIC LOADS

The stresses at the surface of a solid round shaft subjected to combined loading of bending and torsion are

$$\sigma_x = \frac{32M}{\pi d^3} \qquad \tau_{xy} = \frac{16T}{\pi d^3} \tag{15-1}$$

where σ_x = bending stress
τ_{xy} = torsional stress
d = shaft diameter
M = bending moment at critical section
T = torsional moment at critical section

By the use of a Mohr's circle it is found that the maximum shear stress is

$$\tau_{max} = \sqrt{\left(\frac{\sigma_x}{2}\right)^2 + \tau_{xy}^2} \tag{a}$$

Eliminating σ_x and τ_{xy} from Eq. (a) gives

$$\tau_{max} = \frac{16}{\pi d^3} \sqrt{M^2 + T^2} \tag{b}$$

The maximum-shear-stress theory of static failure states that $S_{sy} = S_y/2$. By employing a factor of safety n, we can now write Eq. (b) as

$$\frac{S_y}{2n} = \frac{16}{\pi d^3} \sqrt{M^2 + T^2} \tag{c}$$

or

$$d = \left[\left(\frac{32n}{\pi S_y}\right)(M^2 + T^2)^{1/2} \right]^{1/3} \tag{15-2}$$

A similar approach using the distortion-energy theory yields

$$d = \left[\frac{32n}{\pi S_y} \left(M^2 + \frac{3T^2}{4} \right)^{1/2} \right]^{1/3} \tag{d}$$

It is important to observe that these relations are only valid when the stresses are truly nonvariable.

15-3   A HISTORICAL APPROACH*

In 1927 the American Society of Mechanical Engineers established a code for the design of transmission shafting. This code has been obsolete for many years now, but it has considerable historical interest. The ASME code defines a permissible shear stress which is the smaller of the two following values:

$$\tau_p = 0.30S_{yt} \quad \text{or} \quad \tau_p = 0.18S_{ut} \tag{15-3}$$

The code states that these stresses should be reduced by 25 percent if stress concentration, possibly due to a shoulder fillet or a keyway, is present.

If we substitute τ_p for τ_{max} in Eq. (b) of the previous section, we have

$$\tau_p = \frac{16}{\pi d^3} \sqrt{M^2 + T^2} \tag{a}$$

---

* This section is included because of so many requests from users of previous editions of this book.

In the code, the bending moment M and the torsional moment T are multiplied by combined shock and fatigue factors C_m and C_t, respectively, depending on the conditions of the particular application. Thus

$$\tau_p = \frac{16}{\pi d^3} \sqrt{(C_m M)^2 + (C_t T)^2} \qquad (b)$$

Equation (b) can now be solved for the shaft diameter.

$$d = \left\{ \frac{5.1}{\tau_p} \left[(C_m M)^2 + (C_t T)^2 \right]^{1/2} \right\}^{1/3} \qquad (15\text{-}4)$$

Equation (15-4) is the ASME code formula, and as the development shows, it is based on the maximum-shear-stress theory of failure. Recommended values of C_m and C_t are listed in Table 15-1.

It is important to note that the design stress of Eq. (15-3) is the maximum permissible amount and that the designer is free to reduce it even more if circumstances justify such a reduction. For example, the code states that an additional 25 percent reduction should be taken if a shaft failure would cause serious consequences.

15-4 REVERSED BENDING AND STEADY TORSION

Any rotating shaft loaded by stationary bending and torsional moments will be stressed by completely reversed bending stress, because of shaft rotation, but the torsional stress will remain steady. This is a very common situation and probably occurs more often than any other loading. By using the subscript a for the alternating stress and the subscript m for mean stress, Eqs. (15-1) can be expressed as

$$\sigma_a = \frac{32M}{\pi d^3} \qquad \tau_m = \frac{16T}{\pi d^3} \qquad (a)$$

Table 15-1 VALUES OF BENDING-MOMENT
FACTOR C_m AND TORSIONAL
MOMENT FACTOR C_t

Type of loading	C_m	C_t
Stationary shaft:		
Load applied gradually	1.0	1.0
Load applied suddenly	1.5–2.0	1.5–2.0
Rotating shaft:		
Load applied gradually	1.5	1.0
Steady load	1.5	1.0
Load applied suddenly, minor shocks	1.5–2.0	1.0–1.5
Load applied suddenly, heavy shocks	2.0–3.0	1.5–3.0

Similar expressions can be written for shafts made of tubing.

Sines* states that experimental evidence shows that bending-fatigue strength is not affected by the existence of torsional mean stress until the torsional yield strength is exceeded by about 50 percent. This disclosure furnishes a very simple means of designing for the special case of combined reversed bending and steady torsion.

Designating S_e as the completely corrected endurance limit (see Sec. 7-5) and n as the factor of safety, the design equation is simply

$$\frac{S_e}{n} = \sigma_a \tag{b}$$

By substituting σ_a from Eq. (a) and solving the result for the diameter, we obtain

$$d = \left(\frac{32Mn}{\pi S_e}\right)^{1/3} \tag{15-5}$$

because the presence of τ_m does not affect the bending endurance limit according to Sines.

A word of caution with respect to the use of Eq. (15-5) is necessary. To be certain that a shaft whose diameter is obtained from Eq. (15-5) will not yield, compute the static factor of safety n using the methods of Sec. 15-2.

15-5   THE SODERBERG APPROACH†

In the simplest applications of a Soderberg diagram (Fig. 7-23), the diagram is used to determine the required dimensions of a machine part which must carry a steady stress and an alternating stress of the same kind. In the following discussion, an example will be given to show how a Soderberg diagram can be used to determine the required dimensions for a shaft subjected to a combination of steady torque and alternating bending, a common type of loading for shafts. The procedure will suggest how more complicated cases of loading might be treated. Both the example given below and the more general case where both the torque and the bending moment have steady and alternating components are treated in a paper by C. R. Soderberg.‡ In fact, the derivation given below follows exactly

---

* George Sines in George Sines, and J. L. Waisman (eds.), *Metal Fatigue*, McGraw-Hill, New York, 1959, p. 158.

† Except for slight changes in notation, this section is a faithful reproduction of the notes used at the University of California, Berkeley, and is reproduced with the permission of the engineering-design staff.

‡ C. R. Soderberg, "Working Stresses," *J. Appl. Mechanics*, vol. 57, 1935, p. A–106.

that published by Soderberg, except that some of Soderberg's equations are omitted and the purpose for which he uses these equations is accomplished graphically.
Also, the symbols used here differ from those used in Soderberg's paper.

Figure 15-1a shows a stress element on the surface of a solid round shaft
whose rotational speed is ω rad/s. Now suppose that a plane PQ is passed
through the upper right-hand corner of the element. Then, below plane PQ there
will be a wedge-shaped element, as shown in Fig. 15-1b. The angle α shown in
the figure is the angle between plane PQ and a horizontal plane. We shall
consider all possible values of α to see whether or not we can decide what it will be
for those planes on which failure occurs.

Since both bending and torsion are involved in this problem, it is necessary to
decide upon a theory of failure. Though failures due to variable stresses do not
occur on planes of maximum shear, we shall use the maximum-shear-stress theory
because both ratios S_{se}/S_e and S_{sy}/S_y are only slightly greater than 0.50.

Since we have decided to use the maximum-shear theory of failure, we are
interested in the value of the shear stress on the inclined face of the element. By
taking an equilibrium equation for all forces in the direction of τ_α, we get

$$\tau_\alpha + \sigma_x \sin \alpha \cos \alpha + \tau_{xy} \sin^2 \alpha - \tau_{xy} \cos^2 \alpha = 0$$

or

$$\tau_\alpha = \tau_{xy}(\cos^2 \alpha - \sin^2 \alpha) - \sigma_x \sin \alpha \cos \alpha \qquad (a)$$

Substituting the values of τ_{xy} and σ_x in Eq. (a) and using several trigonometric
identities yields

$$\tau_\alpha = \frac{16T}{\pi d^3} \cos 2\alpha - \frac{16M}{\pi d^3} \sin 2\alpha \cos \omega t \qquad (b)$$

In other words, for any plane making an angle α with the horizontal plane, the

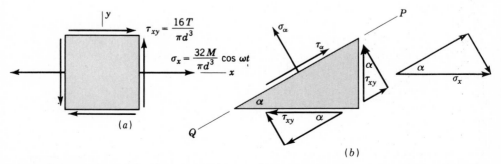

FIGURE 15-1 (a) Shaft stress element of unit depth has a steady shear stress τ_{xy} and an
alternating stress σ_x due to rotation; (b) element cut at angle α.

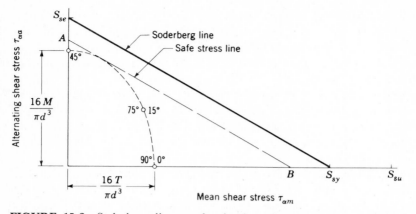

FIGURE 15-2 Soderberg diagram showing how the line of safe stress AB is drawn parallel to the Soderberg line and tangent to the ellipse.

shear stress has a mean value of

$$\tau_{am} = \frac{16T}{\pi d^3} \cos 2\alpha \qquad\qquad (c)$$

and an alternating component with amplitude

$$\tau_{aa} = \frac{16M}{\pi d^3} \sin 2\alpha \qquad\qquad (d)$$

Shown in Fig. 15-2 is the Soderberg diagram for shear strength. Alternating shear stresses are plotted on the vertical axis, while static or mean shear stresses are plotted on the horizontal axis. As shown, the Soderberg line is a straight line between the completely corrected shear endurance limit S_{se} and the yield strength in shear S_{sy}. Note, particularly, that the shear endurance limit is the endurance limit of the machine element after size, surface finish, reliability, life, stress concentration, and the like, have been accounted for by using Eq. (7-15).*

To determine whether or not failure will occur on certain planes making an angle α with the horizontal, we plot a point in Fig. 15-2 for each value of α. The coordinates of the point are (τ_{am}, τ_{aa}), as determined above. For example, for horizontal planes $(\alpha = 0)$, the coordinates of the point are

$$\left(\frac{16T}{\pi d^3}, 0\right)$$

---

* In the original paper, modification for many of these effects was made in the final equation.

For vertical planes, the coordinates are

$$\left(-\frac{16T}{\pi d^3}, 0\right)$$

but this is really the same as the point for $\alpha = 0$. For $\alpha = 45°$, the point is

$$\left(0, \frac{16M}{\pi d^3}\right)$$

It is recommended that additional points be plotted so as to verify the fact that all points fall on a quarter ellipse, as indicated in Fig. 15-2.

Consideration of Fig. 15-2 leads to the conclusion that the factor of safety should be taken to correspond to that point of the ellipse which is closest to the failure line. The problem is to draw a line parallel to the failure line and tangent to the ellipse. With such a line the factor of safety n can be determined graphically. Such a complete graphical solution would be in every way acceptable. However, by means of analytic geometry, it can be shown that the value of n thus obtained will be

$$n = \frac{\pi d^3}{16\sqrt{(T/S_{sy})^2 + (M/S_{se})^2}} \tag{15-6}$$

For use as a design formula, this equation can be rewritten

$$d = \left\{\frac{16n}{\pi}\left[\left(\frac{T}{S_{sy}}\right)^2 + \left(\frac{M}{S_{se}}\right)^2\right]^{1/2}\right\}^{1/3} \tag{15-7}$$

or more conveniently,

$$d = \left\{\frac{32n}{\pi}\left[\left(\frac{T}{S_y}\right)^2 + \left(\frac{M}{S_e}\right)^2\right]^{1/2}\right\}^{1/3} \tag{15-8}$$

since $S_{sy} = 0.5S_y$ and $S_{se} = 0.5S_e$ using the maximum-shear-stress theory. If the distortion-energy theory is used instead, then $S_{sy} = 0.577S_y$ and $S_{se} = 0.577S_e$. Substituting these in Eq. (15-7) gives

$$d = \left\{\frac{48n}{\pi}\left[\left(\frac{T}{S_y}\right)^2 + \left(\frac{M}{S_e}\right)^2\right]^{1/2}\right\}^{1/3} \tag{15-9}$$

based on the distortion-energy theory. The result is quite surprising, as it indicates that a diameter about 15 percent larger is obtained using the distortion-energy theory than is obtained with the maximum-shear-stress theory of failure.

For the most general case, where the bending and the torsion stress each contain both a steady component and a variable component, the equation corresponding to Eq. (15-8) is

$$d = \left\{ \frac{32n}{\pi} \left[\left(\frac{T_a}{S_e} + \frac{T_m}{S_y} \right)^2 + \left(\frac{M_a}{S_e} + \frac{M_m}{S_y} \right)^2 \right]^{1/2} \right\}^{1/3} \tag{15-10}$$

The use of the maximum-shear-stress theory is implicit in Eq. (15-10). Equation (15-10) is sometimes referred to (perhaps in a less complete form) as the *Westinghouse code formula*. Using the distortion-energy theory instead gives

$$d = \left\{ \frac{48n}{\pi} \left[\left(\frac{T_a}{S_e} + \frac{T_m}{S_y} \right)^2 + \left(\frac{M_a}{S_e} + \frac{M_m}{S_y} \right)^2 \right]^{1/2} \right\}^{1/3} \tag{15-11}$$

By noting that $I/c = (\frac{1}{2})(J/c) = \pi d^3/32$, Eq. (15-10) can be solved for the factor of safety in terms of the stresses and strengths. The result, which will be more useful for analysis, is

$$n = \frac{1}{\sqrt{\left(\frac{2\tau_a}{S_e} + \frac{2\tau_m}{S_y} \right)^2 + \left(\frac{\sigma_a}{S_e} + \frac{\sigma_m}{S_y} \right)^2}} \tag{15-12}$$

for the maximum-shear-stress theory. A similar analysis using the distortion-energy theory yields

$$n = \frac{2}{3\sqrt{\left(\frac{2\tau_a}{S_e} + \frac{2\tau_m}{S_y} \right)^2 + \left(\frac{\sigma_a}{S_e} + \frac{\sigma_m}{S_y} \right)^2}} \tag{15-13}$$

It is important to note that the analyses above do not account for the fact that torsional endurance limits may require different modification factors than those for the bending endurance limits. Stress-concentration factors, for example, are not the same for both bending and torsion. One possible solution for such problems is to apply the reciprocals of the differing modification factors to the stresses. This makes them act as stress-increasing factors instead of strength-reducing factors, but the result is the same.

Example 15-1 The integral pinion shaft shown in Fig. 15-3a is to be mounted in bearings at the locations shown and is to have a gear (not shown) mounted on the right-hand or overhanging end. The loading diagram (Fig. 15-3b) shows that the pinion force at A and the gear force at C are in the same xy plane. Equal and opposite torques T_A and T_C are assumed to be concentrated at A and C, as were the forces.

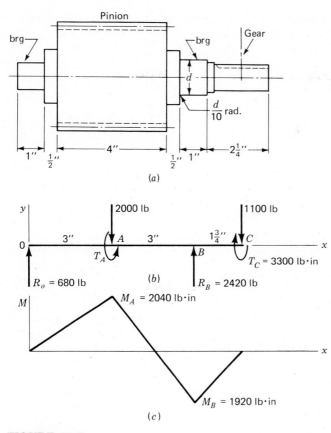

FIGURE 15-3

The bending-moment diagram of Fig. 15-3c shows a maximum moment at A, but the diameter of the pinion is quite large and so the stress due to this moment can be neglected. On the other hand, another moment M_B, almost as large, occurs at the center of the right-hand bearing. Find the diameter d of the shaft at the right-hand bearing based upon a material having a yield strength $S_y = 66$ kpsi, a fully corrected endurance limit $S_e = 20$ kpsi, and a factor of safety of 1.80.

Solution   Based on static loads only, Eq. (15-2) yields

$$d = \left[\frac{32n}{\pi S_y}(M^2 + T^2)^{1/2}\right]^{1/3} = \left\{\frac{(32)(1.80)}{\pi(66)(10)^3}[(1920)^2 + (3300)^2]^{1/2}\right\}^{1/3}$$

$$= 1.02 \text{ in} \qquad\qquad Ans.$$

But, if we consider fatigue, and use Eq. (15-5), we get

$$d = \left(\frac{32Mn}{\pi S_e}\right)^{1/3} = \left[\frac{(32)(1920)(1.80)}{\pi(20)(10)^3}\right]^{1/3} = 1.21 \text{ in} \qquad\qquad Ans.$$

This result is greater than the previous one, which means that a 1.21-in-diameter shaft is safe for fatigue loading and static loading.

The Soderberg approach is more conservative. Using Eq. (15-8), we get

$$d = \left\{ \frac{32n}{\pi} \left[\left(\frac{T}{S_y} \right)^2 + \left(\frac{M}{S_e} \right)^2 \right]^{1/2} \right\}^{1/3}$$

$$= \left\{ \frac{(32)(1.80)}{\pi} \left[\left(\frac{3300}{66\,000} \right)^2 + \left(\frac{1920}{20\,000} \right)^2 \right]^{1/2} \right\}^{1/3} = 1.26 \text{ in} \qquad Ans.$$

This result is based on the maximum-shear-stress theory. Using Eq. (15-9) in a similar fashion yields $d = 1.44$ in. This result, based on the distortion-energy theory is substantially more conservative.   ////

15-6   THE KIMMELMANN LOAD-LINE APPROACH

Whenever one or more of the shaft loads are subject to overloading, the method of Sec. 7-15 may be particularly useful. Though fully explained in that section, another example of its use in design is presented below.*

Example 15-2   The tensile strength of the shaft material of Example 15-1 is 92 kpsi. Determine the shaft diameter d (Fig. 15-3) using factors of safety as follows: $n_{es} = 1.2$ for the endurance limit, $n_{us} = 1.25$ for the ultimate tensile strength, and $n_1 = 1.15$ for the torsional mean stress. In this example the alternating stress σ_a is subject to overloading and for this we choose $n_2 = 1.5$. Using other data in Example 15-1, find a safe diameter d using the graphical load-line approach, combined with the modified Goodman criterion.

Solution   First, we apply factors of safety to the strengths. Thus

$$S_e(\text{min}) = \frac{S_e}{n_{es}} = \frac{20}{1.2} = 16.67 \text{ kpsi}$$

$$S_{ut}(\text{min}) = \frac{S_{ut}}{n_{us}} = \frac{92}{1.25} = 73.6 \text{ kpsi}$$

* For a more detailed presentation see L. D. Mitchell and D. T. Vaughan, "A General Method for the Fatigue-Resistant Design of Mechanical Components, Part I, Graphical," *ASME Journal of Engineering for Industry*, vol. 97, ser. B, no. 3, August 1975, pp. 965–969.

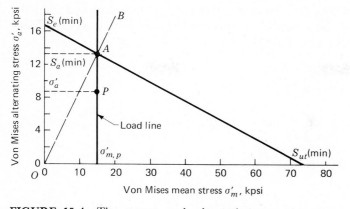

FIGURE 15-4 The mean- and alternating-stress axes are drawn to different scales.

These two points are shown plotted on Fig. 15-4. The straight line between them is the modified Goodman line.

Next, using Eq. (15-1), the torsional shear stress, which is a mean stress, is

$$\tau_m = \frac{16T}{\pi d^3} = \frac{16(3300)(10)^{-3}}{\pi d^3} = \frac{16.8}{d^3} \text{ kpsi}$$

Applying the factor of safety n_1 gives the permissible mean torsional stress as

$$\tau_{m,\,p} = n_1 \tau_m = 1.15\left(\frac{16.8}{d^3}\right) = \frac{19.32}{d^3} \text{ kpsi}$$

Now use Eq. (15-1) again to find the alternating stress component. The result is

$$\sigma_a = \frac{32M}{\pi d^3} = \frac{32(1920)(10)^{-3}}{\pi d^3} = \frac{19.6}{d^3} \text{ kpsi}$$

This is a combined stress state and so we must use Eq. (7-43) to obtain the von Mises stresses. The results are

$$\sigma'_{m,\,p} = \sqrt{3\tau_{m,\,p}^2} = \sqrt{3}\left(\frac{19.32}{d^3}\right) = \frac{33.46}{d^3} \text{ kpsi}$$

$$\sigma'_a = \sigma_a = \frac{19.6}{d^3} \text{ kpsi}$$

Overloading occurs on the alternating-stress component σ_a, while the mean stress remains constant. Thus the Kimmelmann or load line will be a vertical line through $\sigma'_{m,\,p}$ in Fig. 15-4. To find the intersection of this load line with the modified Goodman line, we imagine that σ'_a increases until failure occurs. Since $n_2 = 1.5$ this limiting value of σ'_a will be

$$\sigma'_{a,\,p} = n_2\,\sigma'_a = 1.5\left(\frac{19.6}{d^3}\right) = \frac{29.4}{d^3}\text{ kpsi}$$

Thus, the coordinates of the failure point A in Fig. 15-4 are $\sigma'_{a,\,p}$ and $\sigma'_{m,\,p}$. Taking the ratio of these two gives

$$\frac{\sigma'_{a,\,p}}{\sigma'_{m,\,p}} = \frac{29.4}{33.46} = 0.879$$

The line OB in Fig. 15-4, drawn to this slope, intersects the modified Goodman line at A. The load line is the vertical line through A. The alternating strength at A is $S_a(\min) = 13.2$ kpsi. Point P is the design point and overloading raises P in the direction of A. The stress at P is

$$\sigma'_a = \frac{S_a(\min)}{n_2} = \frac{13.2}{1.5} = 8.80\text{ kpsi}$$

Since $\sigma'_a = 19.6/d^3$, we have

$$d = \sqrt[3]{\frac{19.6}{8.80}} = 1.31\text{ in} \qquad\qquad\qquad\qquad\qquad\qquad\qquad Ans.$$

////

15-7 THE BASIC GRAPHICAL APPROACH

The advantages of the graphical method are that the theory is easily understood and that a visual indication of the margin of safety is obtained. This method employs the fatigue diagram first explained in Sec. 7-13 in combination with any of the theories of Sec. 7-14. The following example illustrates its use in design.

Example 15-3 Suppose, due to the dynamic action of shaft-mounted inertias, that the torque T_C, of Fig. 15-3, consists of a mean component of 2200 lb·in and an alternating component of 1100 lb·in. It is important to observe here that the fluctuations of the shaft inertias are in no way related to the rotation of the shaft. The variations of the bending stress, tension, compression, etc., are due to shaft rotation and are therefore in harmony with shaft speed. In this example we wish to obtain a safe shaft diameter d using an overall factor of safety of 2 based on the

Gerber criterion (Sec. 7-14). Except as noted, all of the data of Examples 15-1 and 15-2 are applicable.

Solution The worst condition will be obtained when the torque and bending-moment variations occur simultaneously, and so, for design purposes, we assume this condition. Thus the mean and alternating components of the torsional shear stress are

$$\tau_m = \frac{16T_m}{\pi d^3} = \frac{16(2200)(10)^{-3}}{\pi d^3} = \frac{11.2}{d^3} \text{ kpsi}$$

$$\tau_a = \frac{16T_a}{\pi d^3} = \frac{16(1100)(10)^{-3}}{\pi d^3} = \frac{5.60}{d^3} \text{ kpsi}$$

As in Example 15-2, the bending moment produces the alternating stress

$$\sigma_a = \frac{32M_a}{\pi d^3} = \frac{32(1920)(10)^{-3}}{\pi d^3} = \frac{19.6}{d^3} \text{ kpsi}$$

The von Mises stresses are found next, using Eq. (7-43). Thus

$$\sigma'_m = \sqrt{\sigma_{xm}^2 + 3\tau_{xym}^2} = \sqrt{3\left(\frac{11.2}{d^3}\right)^2} = \frac{19.4}{d^3} \text{ kpsi}$$

$$\sigma'_a = \sqrt{\sigma_{xa}^2 + 3\tau_{xya}^2} = \sqrt{\left(\frac{19.6}{d^3}\right)^2 + 3\left(\frac{5.60}{d^3}\right)^2} = \frac{21.9}{d^3} \text{ kpsi}$$

The strengths to be used in the design are $S_{ut} = 92$ kpsi and $S_e = 20$ kpsi. The Gerber criterion is expressed by Eq. (7-30), in the second form, as

$$S_a = S_e\left[1 - \left(\frac{S_m}{S_{ut}}\right)^2\right]$$

By substituting various values of S_m in this equation, it is easy to get enough points to plot the fatigue diagram of Fig. 15-5. A programmable calculator is helpful in doing this.

The mean and alternating components were assumed in this example to maintain the same ratio to each other. This ratio is

$$\frac{\sigma'_a}{\sigma'_m} = \frac{(21.9/d^3)}{(19.4/d^3)} = 1.129$$

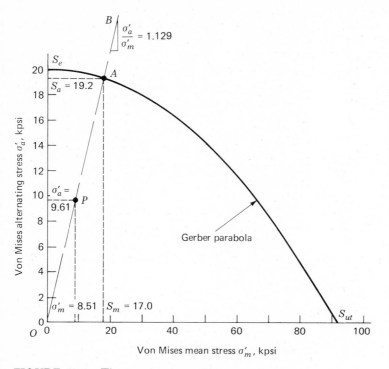

FIGURE 15-5   The two axes have different scales.

A line OB having this slope is constructed in Fig. 15-5. This line intersects the Gerber line at A and defines the two strengths $S_a = 19.2$ kpsi and $S_m = 17.0$ kpsi. Either value can be used to obtain d. Since the factor of safety is $n = 2$, we have

$$\frac{S_a}{n} = \sigma'_a \qquad \text{or} \qquad \frac{19.2}{2} = \frac{21.9}{d^3}$$

Solving for the diameter gives

$$d = \sqrt[3]{\frac{21.9(2)}{19.2}} = 1.316 \text{ in} \qquad\qquad\qquad Ans.$$

Corresponding to this diameter, the von Mises stresses are found to be

$$\sigma'_a = \frac{21.9}{(1.316)^3} = 9.61 \text{ kpsi} \qquad \sigma'_m = \frac{19.4}{(1.316)^3} = 8.51 \text{ kpsi}$$

These two points are plotted as P in Fig. 15-5. The distance PA represents the margin of safety.

In this example the Gerber theory was used. Any other theory, such as others in Sec. 7-14, can be employed in a similar manner. ////

15-8 A GENERAL APPROACH

Joseph Marin of Pennsylvania State University first proposed that strength relations be expressed in the first form of formulas like Eq. (7-30) for the $\sigma_a - \sigma_m$ plane.* For this reason, equations of this form are frequently called *Marin equations*. The most general form of the equation is

$$\left(\frac{S_a}{S_e}\right)^m + \left(\frac{KS_m}{S_{ut}}\right)^p = 1 \tag{15-14}$$

where K, m, and p depend upon the criterion used. Values of these constants for theories discussed in Chap. 7 are listed in Table 15-2.

Equation (15-14) can be used either for design purposes or for analysis. The following example is representative.

Example 15-4 The stresses in Example 15-3 were $\tau_m = 11.2/d^3$, $\tau_a = 5.60/d^3$, and $\sigma_a = 19.6/d^3$, all in kpsi. The strengths are $S_e = 20$ kpsi, $S_y = 66$ kpsi (from Example 15-1), and $S_{ut} = 92$ kpsi. Factors of safety selected to determine a safe diameter d are $n_{es} = 1.2$ for the endurance limit, $n_{ys} = 1.30$ for the yield strength, $n_{us} = 1.25$ for the ultimate tensile strength, $n_1 = 1.15$ for the torsional mean stress, $n_2 = 1.40$ for the alternating torsional stress, and $n_3 = 1.50$ for the alternating bending stress. Use the quadratic theory to find the diameter.

Solution Applying factors of safety to the strengths gives

$$S_e(\text{min}) = \frac{S_e}{n_{es}} = \frac{20}{1.2} = 16.67 \text{ kpsi}$$

$$S_y(\text{min}) = \frac{S_y}{n_{ys}} = \frac{66}{1.30} = 50.8 \text{ kpsi}$$

$$S_{ut}(\text{min}) = \frac{S_{ut}}{n_{us}} = \frac{92}{1.25} = 73.6 \text{ kpsi}$$

---

* Joseph Marin, "Design for Fatigue Loading, Part 3," *Machine Design*, vol. 29, no. 4, Feb. 21, 1957, pp. 128–131, and series of the same title, pts. 1–5, nos. 2–6, Jan.-April 1957.

Table 15-2 FAILURE CRITERIA AND CONSTANTS FOR USE IN EQ. (15-14)

Failure theory	K	m	p			
Soderberg	S_{ut}/S_y	1	1		0.8914	UNS G10180 $H_B = 130$
Modified Goodman	1	1	1			
Gerber parabolic	1	1	2	$a = $	0.9266	UNS G10380 $H_B = 164$
Quadratic (elliptic)	1	2	2		1.0176	UNS G41300 $H_B = 207$
Kececioglu	1	a	2			
Bagci	S_{ut}/S_y	1	4		0.9685	UNS G43400 $H_B = 233$

The permissible stresses are found to be

$$\tau_{m,\,p} = n_1 \tau_m = 1.15\left(\frac{11.2}{d^3}\right) = \frac{12.88}{d^3} \text{ kpsi}$$

$$\tau_{a,\,p} = n_2 \tau_a = 1.40\left(\frac{5.60}{d^3}\right) = \frac{7.84}{d^3} \text{ kpsi}$$

$$\sigma_{a,\,p} = n_3 \sigma_a = 1.50\left(\frac{19.6}{d^3}\right) = \frac{29.4}{d^3} \text{ kpsi}$$

Next we determine the von Mises mean and alternating stresses. These are

$$\sigma'_{m,\,p} = \sqrt{3\tau^2_{m,\,p}} = \sqrt{3}\left(\frac{12.88}{d^3}\right) = \frac{22.31}{d^3} \text{ kpsi}$$

$$\sigma'_{a,\,p} = \sqrt{\sigma^2_{a,\,p} + 3\tau^2_{a,\,p}} = \frac{1}{d^3}\sqrt{(29.4)^2 + 3(7.84)^2} = \frac{32.4}{d^3} \text{ kpsi}$$

The constants for the quadratic theory are $K = 1$, $m = 2$, and $p = 2$. Substituting these and the permissible stresses and strengths into Eq. (15-14) gives

$$\left[\frac{\sigma'_{a,\,p}}{S_e(\text{min})}\right]^2 + \left[\frac{\sigma'_{m,\,p}}{S_{ut}(\text{min})}\right]^2 = 1$$

in terms of the permissible or allowable values. Now we insert appropriate values and factor. Thus

$$\frac{1}{d^6}\left[\left(\frac{32.4}{16.67}\right)^2 + \left(\frac{22.31}{73.6}\right)^2\right] = 1$$

When this equation is solved we find the answer to be $d = 1.25$ in. In this case the equation could be solved directly for d because $m = p$. But some of the theories in Table 15-2 require iteration to obtain such a solution.   ////

15-9   THE SINES APPROACH

George Sines of the University of California at Los Angeles has proposed a solution which has been verified by many experiments.* Except for special cases, the Sines approach cannot be illustrated on the usual fatigue diagram.

To explain Sines' method, it is first necessary to introduce the concept of octahedral stresses. Visualize a principal stress element having the stresses σ_1, σ_2, and σ_3. Now cut the stress element by a plane that forms equal angles with each of the three principal stresses. The plane ABC in Fig. 15-6 satisfies this requirement. It is called an *octahedral plane*. Note that the solid cut from the stress element retains one of the original corners. Since there are eight of these corners in all, there is a total of eight of these planes.

Figure 15-6 may be regarded as a free-body diagram when each of the stress components shown is multiplied by the area over which it acts to obtain the corresponding force. It is possible, then, to sum these forces to zero in each of the three coordinate directions. When this is done, it is found that a force, called the *octahedral force*, exists on plane ABC. When this force is divided by the area of face ABC, it is found that the resulting stress has two components σ_{oct} and τ_{oct}, called, respectively, the *octahedral normal* and the *octahedral shear stresses*. The magnitudes of these stresses are†

$$\sigma_{oct} = \frac{\sigma_1 + \sigma_2 + \sigma_3}{3} \tag{15-15}$$

$$\tau_{oct} = \tfrac{1}{3}\sqrt{(\sigma_1 - \sigma_2)^2 + (\sigma_2 - \sigma_3)^2 + (\sigma_3 - \sigma_1)^2} \tag{15-16}$$

The Sines concept of fatigue failure is based on his observation that a linear relation must exist between the magnitude of the alternating octahedral shear stress components and the sum of the mean components of the principal stresses. This relation is expressed as

$$\tfrac{1}{3}\sqrt{(\sigma_{1a} - \sigma_{2a})^2 + (\sigma_{2a} - \sigma_{3a})^2 + (\sigma_{3a} - \sigma_{1a})^2}$$
$$= A - \alpha(\sigma_{1m} + \sigma_{2m} + \sigma_{3m}) \tag{15-17}$$

---

* George Sines, *Elasticity and Strength*, Allyn and Bacon, Boston, 1969, pp. 72–78.
† For a complete derivation see J. O. Smith and O. M. Sidebottom, *Elementary Mechanics of Deformable Bodies*, Macmillan, New York, 1969, p. 131.

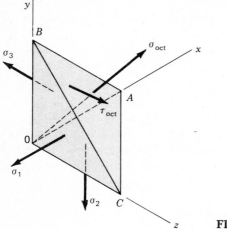

FIGURE 15-6

The constants A and α are to be determined from two S-N diagrams, or two fatigue-testing programs, in which the mean stresses are significantly different. For example, the rotating-beam fatigue test has $S_e = \sigma_{1a}$ and σ_{2a}, σ_{3a}, σ_{1m}, σ_{2m}, and σ_{3m} are all zero. Thus Eq. (15-17) gives

$$A = \frac{\sqrt{2}}{3} S_e \tag{15-18}$$

Another test suggested by Sines is the zero-tension fluctuating-stress cycle. In this cycle $\sigma_{1a} = \sigma_{1m}$ and σ_{2a}, σ_{3a}, σ_{2m}, and σ_{3m} are zero. Let us designate the strength as $S_A = \sigma_{1a} = \sigma_{1m}$. Substituting these in Eq. (15-17) gives

$$\frac{\sqrt{2}}{3} S_A = A - \alpha S_A \tag{a}$$

If we now substitute the value of A and solve for the constant α, we get

$$\alpha = \frac{\sqrt{2}}{3}\left(\frac{S_e}{S_A} - 1\right) \tag{15-19}$$

For biaxial stresses, σ_{3a} and σ_{3m} are zero and Eq. (15-17) simplifies to

$$\frac{\sqrt{2}}{3} \sigma_a' = A - \alpha(\sigma_{1m} + \sigma_{2m}) \tag{b}$$

where σ_a' is the von Mises alternating-stress component. Equation (b) gives the limiting value of the alternating-stress component σ_a' due to the mean stresses σ_{1m}

and σ_{2m}. Thus, it is convenient to replace σ'_a with S_a. If we also substitute the constants found in Eqs. (15-18) and (15-19) and simplify the result, we obtain

$$S_a = S_e - \left(\frac{S_e}{S_A} - 1\right)(\sigma_{1m} + \sigma_{2m}) \tag{15-20}$$

15-10 SUGGESTIONS FOR COMPUTER SOLUTION*

Shaft design can involve a considerable amount of iteration and for this reason almost any kind of machine computation facilities will be very helpful. The need for iteration is most frequently caused by:

- A size factor dependent on shaft diameter
- A stress-concentration factor dependent on shaft diameter as well as fillet radius
- The use of a nonlinear failure theory
- The design strategy selected

If the need for shaft design occurs frequently and over a long period of time, then a comprehensive computer program can be justified. Such a program might well consist of a number of callable subroutines. For example, stress-concentration charts can be reduced to a computer subprogram using curve-fitting techniques. Other subroutines can be used for the von Mises relations and the various σ_a-σ_m failure theories.

A series of independent short programs, which can be called directly, are of greater convenience when shaft-design problems arise infrequently. Convergence in most design situations is quite rapid, and so simple trial-and-error solutions can be quite satisfactory. For these problems a programmable calculator is often all that is needed to obtain a safe and reliable shaft design.

PROBLEMS

Sections 15-1 to 15-5

15-1 The small-geared industrial roll shown in the figure is driven at 300 rpm by a force $F = 200$ lb acting on the 3-in gear as shown. The material transported by the roll exerts a resultant uniform force distribution of $w = 20$ lb/in on the roll in the positive z direction. Analysis will show that these forces produce a maximum bending moment at A of 551 lb·in. A material selected has $S_{ut} = 64$ kpsi, $S_y = 54$ kpsi, and it is believed that an endurance limit of $S_e = 20$ kpsi can be obtained.

---

* See also L. D. Mitchell and J. H. Zinskie, *A Man-Machine Interactive Method for the Development of Fatigue Design Equations*, ASME Paper no. 79-DET-96, 1979.

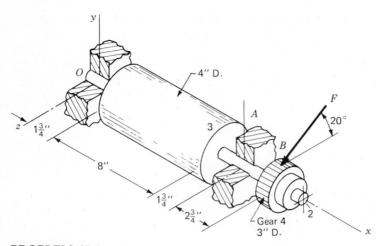

PROBLEM 15-1

Use an overall factor of safety of 3.50 and a reliability of 50 percent to obtain a tentative shaft diameter.

15-2 Based on the results of Prob. 15-1, the shaft has been sized as shown. Note that shoulders have been used to position the shaft between the bearings and to locate the gear. The roll is to be a force fit. Specify a suitable material for the shaft if all surfaces are machined.

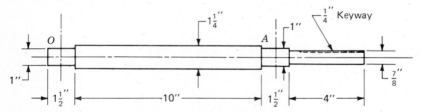

PROBLEM 15-2 All fillets are 1/16 in radius; sled-runner keyway is 3 1/2 in long.

15-3 In order to obtain the rigidity necessary for better roll performance, the shaft of Prob. 15-1 is cut from a UNS G10180 cold-drawn steel tubing having an OD of $1\frac{1}{4}$ in and a wall thickness of $\frac{3}{16}$ in. The roll is a force fit on the shaft and the roll ends are shaped to prevent end-play between the bearings. The gear is secured to the shaft using a taper pin through the hub. What factor of safety results from this design? Is it in any way superior to the solution of Prob. 15-2?

15-4 The resultant gear force $F_A = 600$ lb acts at an angle of 20° from the y axis of the overhanging countershaft shown in the figure. Though not in the same plane, critical resultant bending moments $M_A = 12.1(10)^3$ lb·in and $M_B = 14.4(10)^3$ lb·in are produced by this force and the force on the gear at C. The shaft is to be made of a solid round UNS G10400 cold-drawn steel shafting cut to length. The factor of safety is to be 2.60 corresponding to 99 percent reliability. Determine a safe shaft diameter to the nearest $\frac{1}{8}$ in.

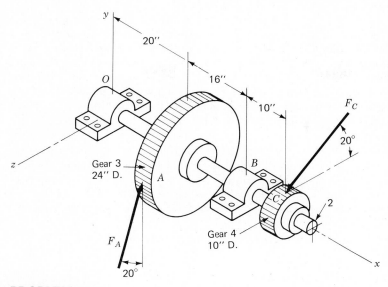

PROBLEM 15-4

15-5 A shaft having a diameter of $2\frac{3}{4}$ in has been selected for the counterhsaft of Prob. 15-4. Is shaft deflection likely to be a problem?

15-6 The figure is a schematic drawing of a countershaft that supports two V-belt pulleys. The loose belt tension in the pulley at A is 15 percent of the tension on the

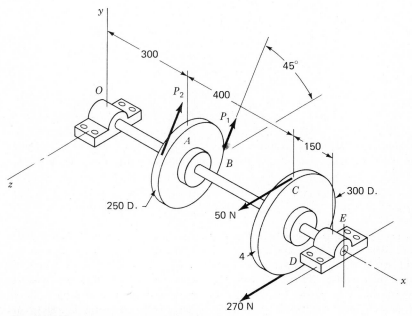

PROBLEM 15-6 Dimensions in millimeters.

tight side. For convenience, each pair of belt pulls is assumed to be parallel. An analysis of this problem gave shaft forces at B and D of $\mathbf{F_B} = -253\mathbf{j} - 253\mathbf{k}$ N and $\mathbf{F_D} = -320\mathbf{k}$ N. Resultant bending moments at B and D turned out to be $M_B = 58.7$ N·m and $M_D = 29.3$ N·m, but these were not in the same plane. Assuming shaft stiffness is the primary design factor, find a shaft diameter using cold-drawn steel and a bending deflection that is not to exceed 200 μm at each pulley.

15-7   A diameter of 35 mm has been tentatively chosen for the shaft of Prob. 15-6. Based on a reliability of 90 percent, will a UNS G10100 cold-drawn steel shafting be safe for this application?

15-8   The countershaft shown in the figure has two spur gears mounted on it with teeth cut with a 20° pressure angle. The gear forces shown in the figure cause resultant moments at A and B of 2720 N·m and 5150 N·m, respectively. A uniform-diameter, cold-drawn UNS G10150 steel is to be used for the shaft. Find a safe diameter for the shaft using a factor of safety of 2.5 and a reliability of 99 percent.

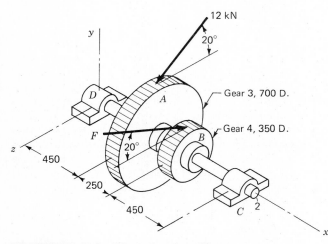

PROBLEM 15-8   Dimensions in millimeters.

15-9   An analysis of the forces acting on the shaft of Prob. 15-8 gave $\mathbf{F_A} = -4.1\mathbf{j} + 11.3\mathbf{k}$ kN at A and $\mathbf{F_B} = -8.23\mathbf{j} - 22.6\mathbf{k}$ kN at B. For proper meshing of the gears the shaft deflection should be within reasonable limits, say not over 250 μm. Determine the deflection at these two points if the diameter selected in Prob. 15-8 is 120 mm.

Sections 15-5 to 15-9

15-10   The speed-reducer shaft shown in the figure is designed to support a pulley and a worm. The shaft is made of plain carbon steel, heat-treated to $H_B = 226$, resulting in the strengths $S_{ut} = 113$ kpsi and $S_y = 86$ kpsi. All surfaces of the shaft are finished by grinding. The pulley subjects the shaft to a radial bending load of 618 lb and a torque of 3640 lb·in. The worm receives the torque and, in addition, subjects the shaft to a 3200-lb axial load, which is to be taken by the right-hand bearing, and a radial bending load of 335 lb. Note that this is a special case; in rotating machinery two radial bending loads are seldom in the same plane. This is

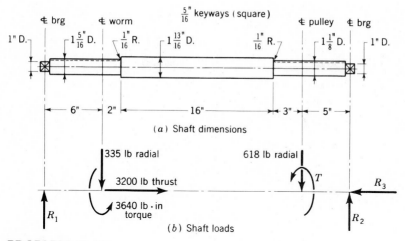

(a) Shaft dimensions

(b) Shaft loads

PROBLEM 15-10

evident from the previous problems. The shaft rotates at 60 rpm and is to have a life of 10 h corresponding to 99 percent reliability. Compute the factor of safety using the von Mises-Hencky-Goodman approach.

15-11 Part a of the figure (see p. 712) is a schematic drawing of a shaft, gear, and bearing subassembly that is part of a helical-gear reduction unit. A force $\mathbf{F}_B = -1700\mathbf{i} + 6400\mathbf{j} - 2300\mathbf{k}$ lb is applied to gear B as shown. The forces applied to gears A and C are then found to be $\mathbf{F}_A = 3690\mathbf{i} - 2690\mathbf{j} + 6400\mathbf{k}$ lb and $\mathbf{F}_C = -3690\mathbf{i} - 2690\mathbf{j} + 6400\mathbf{k}$ lb. Also, gears A and C each transmit torque in the amount of $38.4(10)^3$ lb·in to gear B. These forces produce resultant bending moments of $M_A = 51.5(10)^3$, $M_E = 36.4(10)^3$, $M_F = 51.1(10)^3$, and $M_C = 58.9(10)^3$, all in lb·in. Part b shows the dimensions tentatively selected for the shaft. The material is a UNS G43400 steel, heat-treated and tempered to 1000°F, with all important surfaces ground. Sled-runner keyways 8 in long are milled in the shaft at A, B, and C; they extend 4 in each way from these points. Analyze this shaft for safety, using a reliability of 99.9 percent and basing your analysis on the possibility of a fatigue failure initiated by the keyway at A or by the fillet on the shaft shoulder at E.

15-12 The same as Prob. 15-11, except find the factors of safety guarding against failure in the keyway at C and the shoulder fillet at F.

15-13 The figure (see p. 713) shows a countershaft subjected to alternating bending together with steady torsion and axial loads because of the action of the two gears. Part b shows all shaft dimensions. Sled-runner keyways are used at A and C, but these do not extend completely to the shaft shoulders. Note that the shaft design is such that only the bearing at O can take the axial load. The bevel-gear force is $\mathbf{F}_D = -0.242\mathbf{F}_D\mathbf{i} - 0.242\mathbf{F}_D\mathbf{j} + 0.940\mathbf{F}_D\mathbf{k}$. The shaft is made of UNS G10500 steel, heat-treated and drawn to 900°F, and all important surfaces are finished by grinding. The shaft is to be analyzed to determine how safe it is using a reliability of 99 percent and the von Mises-Hencky-Goodman approach.

(a) Determine the factor of safety at shoulder G.

(b) Determine the factor of safety at A where the keyway is located.

(c) Though the bending moment may be quite large at bearing B, there is no stress concentration here. Find the factor of safety at this point.

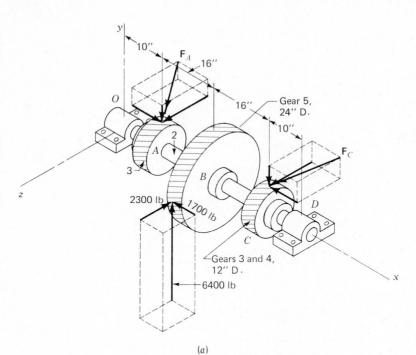

(a)

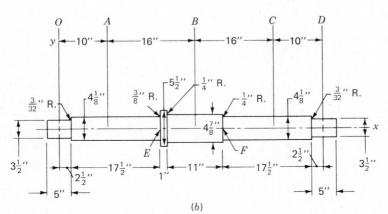

(b)

PROBLEM 15-11

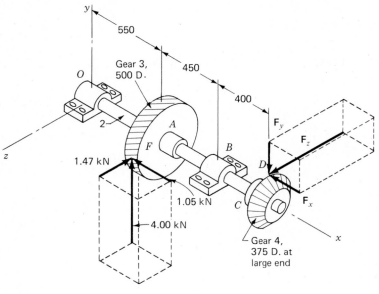

(a)

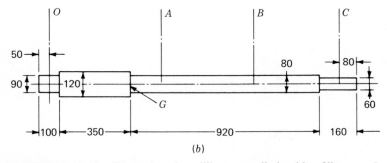

(b)

PROBLEM 15-13 Dimensions in millimeters; all shoulder fillets are 5 mm radius.

CHAPTER

16

CLUTCHES, BRAKES, COUPLINGS, AND FLYWHEELS

This chapter is concerned with a group of elements usually associated with rotation that have in common the function of storing and/or transferring rotating energy. Because of this similarity of function, clutches, brakes, couplings, and flywheels are treated together in this book.

A simplified dynamic representation of a friction clutch or brake is shown in Fig. 16-1a. Two inertias I_1 and I_2 traveling at the respective angular velocities ω_1 and ω_2, one of which may be zero in the case of brakes, are to be brought to the same speed by engaging the clutch or brake. Slippage occurs because the two elements are running at different speeds and energy is dissipated during actuation,

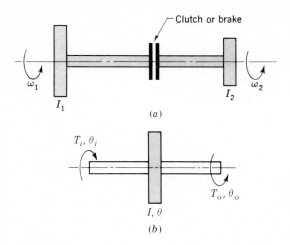

Clutch or brake

ω_1

I_1

(a)

I_2

ω_2

T_i, θ_i

T_o, θ_o

I, θ

(b)

FIGURE 16-1 (a) Dynamic representation of a clutch or brake; (b) mathematical representation of a flywheel.

resulting in a temperature rise. In analyzing the performance of these devices we shall be interested in:

1 The actuating force
2 The torque transmitted
3 The energy loss
4 The temperature rise

The torque transmitted is related to the actuating force, the coefficient of friction, and to the geometry of the clutch or brake. This is a problem in statics which will have to be studied separately for each geometric configuration. However, temperature rise is related to energy loss and can be studied without regard to the type of brake or clutch because the geometry of interest is the heat-dissipating surfaces.

The various types of devices to be studied may be classified as follows:

1 Rim types with internally expanding shoes
2 Rim types with externally contracting shoes
3 Band types
4 Disk or axial types
5 Cone types
6 Miscellaneous types

A flywheel is an inertial energy-storage device. It absorbs mechanical energy by increasing its angular velocity and delivers energy by decreasing its velocity. Fig. 16-1b is a mathematical representation of a flywheel. An input torque T_i, corresponding to a coordinate θ_i, will cause the flywheel speed to increase. And a load or output torque T_o, with coordinate θ_o, will absorb energy from the flywheel and cause it to slow down. We shall be interested in designing flywheels so as to obtain a specified amount of speed regulation.

16-1 STATICS

The analysis of all types of friction clutches and brakes uses the same general procedure. The following steps are necessary:

  1 Assume or determine the distribution of pressure on the frictional surfaces.
  2 Find a relation between the maximum pressure and the pressure at any point.
  3 Apply the conditions of static equilibrium to find (*a*) the actuating force, (*b*) the torque, and (*c*) the support reactions.

Let us now apply these steps to the theoretical problem shown in Fig. 16-2. The figure shows a short shoe hinged at A, having an actuating force F, a normal force N pushing the surfaces together, and a frictional force fN, f being the coefficient of friction. The body is moving to the right, and the shoe is stationary. We designate the pressure at any point by p and the maximum pressure by p_a. The area of the shoe is designated by A.

Step 1 Since the shoe is short, we assume the pressure is uniformly distributed over the frictional area.

Step 2 From step 1 it follows that

$$p = p_a \tag{a}$$

Step 3 Since the pressure is uniformly distributed, we may replace the unit normal forces by an equivalent normal force. Thus

$$N = p_a A \tag{b}$$

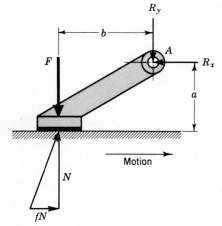

FIGURE 16-2 Forces acting upon a hinged friction shoe.

We now apply the conditions of static equilibrium by taking a summation of moments about the hinge pin. This gives

$$\sum M_A = Fb - Nb + fNa = 0 \qquad (c)$$

Substituting $p_a A$ for N and solving Eq. (c) for the actuating force,

$$F = \frac{p_a A(b - fa)}{b} \qquad (d)$$

Taking a summation of forces in the horizontal and vertical directions gives the hinge-pin reactions:

$$\sum F_x = 0 \qquad R_x = f p_a A \qquad (e)$$

$$\sum F_y = 0 \qquad R_y = p_a A - F \qquad (f)$$

This completes the analysis of the problem.

The preceding analysis is very useful when the dimensions of the clutch or brake are known and the characteristics of the friction material are specified. In design, however, we are interested more in synthesis than in analysis; that is, our purpose is to select a set of dimensions to obtain the best brake or clutch within the limitations of the frictional material we have specified.

In the preceding problem (Fig. 16-2) we make good use of the frictional material because the pressure is a maximum at all points of contact. Let us examine the dimensions. In Eq. (d), if we make $b = fa$, the numerator becomes zero, and no actuating force is required. This is the condition for *self-locking*. We are usually not interested in designing a brake to be self-locking, but on the other hand, we should take some advantage of the self-energizing effect. One way to do this is to obtain the dimensions a and b using a reduced value of the coefficient of friction. Thus, if we let $f' = 0.75 f$ to $0.85 f$, the equation

$$b = f'a$$

may be used to obtain the dimensions a and b, so that some self-energization is obtained.

16-2 INTERNAL-EXPANDING RIM CLUTCHES AND BRAKES

The internal-shoe rim clutch shown in Fig. 16-3 consists essentially of three elements: the mating frictional surfaces, the means of transmitting the torque to and from the surfaces, and the actuating mechanism. To analyze an internal-shoe device, refer to Fig. 16-4, which shows a shoe pivoted at point A, with the actuating force acting at the other end of the shoe. Since the shoe is long, we cannot

FIGURE 16-3 An internal-expanding centrifugal-acting rim clutch. (*Courtesy of the Hilliard Corporation.*)

make the assumption that the distribution of normal forces is uniform. The mechanical arrangement permits no pressure to be applied at the heel, and we will therefore assume the pressure at this point to be zero.

It is the usual practice to omit the friction material for a short distance away from the heel. This eliminates interference, and the material would contribute little to the performance anyway. In some designs the hinge pin is made movable to provide additional heel pressure. This gives the effect of a floating shoe. (Floating shoes will not be treated in this book, although their design follows the same general principles.)

Let us consider the unit pressure p acting upon an element of area of the

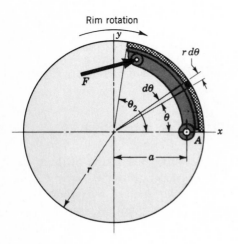

FIGURE 16-4 Internal friction shoe.

frictional material located at an angle θ from the hinge pin (Fig. 16-4). We designate the maximum pressure by p_a located at the angle θ_a from the hinge pin. We now make the assumption (step 1) that the pressure at any point is proportional to the vertical distance from the hinge pin. This vertical distance is proportional to $\sin \theta$, and (step 2) the relation between the pressures is

$$\frac{p}{\sin \theta} = \frac{p_a}{\sin \theta_a} \qquad (a)$$

Rearranging,

$$p = p_a \frac{\sin \theta}{\sin \theta_a} \qquad (16\text{-}1)$$

From Eq. (16-1) we see that p will be a maximum when $\theta = 90°$, or if the toe angle θ_2 is less than $90°$, then p will be a maximum at the toe.

When $\theta = 0$, Eq. (16-1) shows that the pressure is zero. The frictional material located at the heel therefore contributes very little to the braking action and might as well be omitted. A good design would concentrate as much frictional material as possible in the neighborhood of the point of maximum pressure. Such a design is shown in Fig. 16-5. In this figure the frictional material begins at an angle θ_1, measured from the hinge pin A, and ends at an angle θ_2. Any arrangement such as this will give a good distribution of the frictional material.

Proceeding now to step 3 (Fig. 16-5), the hinge-pin reactions are R_x and R_y.

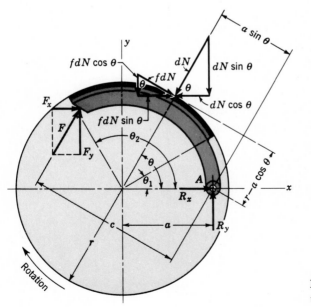

FIGURE 16-5 Forces on the shoe.

The actuating force F has components F_x and F_y and operates at distance c from the hinge pin. At any angle θ from the hinge pin there acts a differential normal force dN whose magnitude is

$$dN = pbr\, d\theta \qquad\qquad (b)$$

where b is the face width (perpendicular to the paper) of the friction material. Substituting the value of the pressure from Eq. (16-1), the normal force is

$$dN = \frac{p_a\, br \sin\theta\, d\theta}{\sin\theta_a} \qquad\qquad (c)$$

The normal force dN has horizontal and vertical components $dN\cos\theta$ and $dN\sin\theta$, as shown in the figure. The frictional force $f\, dN$ has horizontal and vertical components whose magnitudes are $f\, dN \sin\theta$ and $f\, dN \cos\theta$, respectively. By applying the conditions of static equilibrium, we may find the actuating force F, the torque T, and the pin reactions R_x and R_y.

We shall find the actuating force F, using the condition that the summation of the moments about the hinge pin is zero. The frictional forces have a moment arm about the pin of $r - a\cos\theta$. The moment M_f of these frictional forces is

$$M_f = \int f\, dN(r - a\cos\theta) = \frac{fp_a\, br}{\sin\theta_a} \int_{\theta_1}^{\theta_2} \sin\theta(r - a\cos\theta)\, d\theta \qquad\qquad (16\text{-}2)$$

which is obtained by substituting the value of dN from Eq. (c). It is convenient to integrate Eq. (16-2) for each problem, and we shall therefore retain it in this form. The moment arm of the normal force dN about the pin is $a\sin\theta$. Designating the moment of the normal forces by M_N and summing these about the hinge pin give

$$M_N = \int dN(a\sin\theta) = \frac{p_a\, bra}{\sin\theta_a} \int_{\theta_1}^{\theta_2} \sin^2\theta\, d\theta \qquad\qquad (16\text{-}3)$$

The actuating force F must balance these moments. Thus

$$F = \frac{M_N - M_f}{c} \qquad\qquad (16\text{-}4)$$

We see here that a condition for zero actuating force exists. In other words, if we make $M_N = M_f$, self-locking is obtained, and no actuating force is required. This furnishes us with a method for obtaining the dimensions for some self-energizing action. Thus, by using f' instead of f in Eq. (16-2), we may solve for a from the relation

$$M_N = M_f, \qquad\qquad (16\text{-}5)$$

where, as before, f' is made about 0.75 to 0.85 f.

The torque T, applied to the drum by the brake shoe, is the sum of the frictional forces $f\,dN$ times the radius of the drum:

$$T = \int f\,dN\,r = \frac{fp_a\,br^2}{\sin\theta_a}\int_{\theta_1}^{\theta_2}\sin\theta\,d\theta$$

$$= \frac{fp_a\,br^2(\cos\theta_1 - \cos\theta_2)}{\sin\theta_a} \tag{16-6}$$

The hinge-pin reactions are found by taking a summation of the horizontal and vertical forces. Thus, for R_x, we have

$$R_x = \int dN\,\cos\theta - \int f\,dN\,\sin\theta - F_x$$

$$= \frac{p_a\,br}{\sin\theta_a}\left(\int_{\theta_1}^{\theta_2}\sin\theta\cos\theta\,d\theta - f\int_{\theta_1}^{\theta_2}\sin^2\theta\,d\theta\right) - F_x \tag{d}$$

The vertical reaction is found in the same way:

$$R_y = \int dN\,\sin\theta + \int f\,dN\,\cos\theta - F_y$$

$$= \frac{p_a\,br}{\sin\theta_a}\left(\int_{\theta_1}^{\theta_2}\sin^2\theta\,d\theta + f\int_{\theta_1}^{\theta_2}\sin\theta\cos\theta\,d\theta\right) - F_y \tag{e}$$

The direction of the frictional forces is reversed if the rotation is reversed. Thus, for counterclockwise rotation the actuating force is

$$F = \frac{M_N + M_f}{c} \tag{16-7}$$

and since both moments have the same sense, the self-energizing effect is lost. Also, for counterclockwise rotation the signs of the frictional terms in the equations for the pin reactions change, and Eqs. (d) and (e) become

$$R_x = \frac{p_a\,br}{\sin\theta_a}\left(\int_{\theta_1}^{\theta_2}\sin\theta\cos\theta\,d\theta + f\int_{\theta_1}^{\theta_2}\sin^2\theta\,d\theta\right) - F_x \tag{f}$$

$$R_y = \frac{p_a\,br}{\sin\theta_a}\left(\int_{\theta_1}^{\theta_2}\sin^2\theta\,d\theta - f\int_{\theta_1}^{\theta_2}\sin\theta\cos\theta\,d\theta\right) - F_y \tag{g}$$

Equations (d), (e), (f), and (g) can be simplified to ease computations. Thus, let

$$A = \int_{\theta_1}^{\theta_2} \sin \theta \cos \theta \, d\theta = \left(\frac{1}{2} \sin^2 \theta \right)_{\theta_1}^{\theta_2}$$

$$B = \int_{\theta_1}^{\theta_2} \sin^2 \theta \, d\theta = \left(\frac{\theta}{2} - \frac{1}{4} \sin 2\theta \right)_{\theta_1}^{\theta_2}$$

(16-8)

Then, for clockwise rotation as shown in Fig. 16-5, the hinge-pin reactions are

$$R_x = \frac{p_a br}{\sin \theta_a} (A - fB) - F_x$$

$$R_y = \frac{p_a br}{\sin \theta_a} (B + fA) - F_y$$

(16-9)

For counterclockwise rotation Eqs. (f) and (g) become

$$R_x = \frac{p_a br}{\sin \theta_a} (A + fB) - F_x$$

$$R_y = \frac{p_a br}{\sin \theta_a} (B - fA) - F_y$$

(16-10)

In using these equations, the reference system always has its origin at the center of the drum. The positive x axis is taken through the hinge pin. The positive y axis is always in the direction of the shoe, even if this should result in a left-handed system.

The following assumptions are implied by the preceding analysis:

1 The pressure at any point on the shoe is assumed to be proportional to the distance from the hinge pin, being zero at the heel. This should be considered from the standpoint that pressures specified by manufacturers are averages rather than maximums.

2 The effect of centrifugal force has been neglected. In the case of brakes, the shoes are not rotating, and no centrifugal force exists. In clutch design, the effect of this force must be considered in writing the equations of static equilibrium.

3 The shoe is assumed to be rigid. Since this cannot be true, some deflection will occur, depending upon the load, pressure, and stiffness of the shoe. The resulting pressure distribution may be different from that which has been assumed.

4 The entire analysis has been based upon a coefficient of friction which does not vary with pressure. Actually, the coefficient may vary with a number of conditions, including temperature, wear, and environment.

Example 16-1 The brake shown in Fig. 16-6 is 300 mm in diameter and is actuated by a mechanism that exerts the same force F on each shoe. The shoes are identical and have a face width of 32 mm. The lining is a molded asbestos having a coefficient of friction of 0.32 and a pressure limitation of 1000 kPa.

  (*a*)   Determine the actuating force F.
  (*b*)   Find the braking capacity.
  (*c*)   Calculate the hinge-pin reactions.

Solution   (*a*) The right-hand shoe is self-energizing and so the force F is found on the basis that the maximum pressure will occur on this shoe. Here $\theta_1 = 0$, $\theta_2 = 126°$, $\theta_a = 90°$, and $\sin \theta_a = 1$. Also

$$a = \sqrt{(112)^2 + (50)^2} = 123 \text{ mm}$$

Integrating Eq. (16-2) from zero to θ_2 yields

$$M_f = \frac{f p_a\, br}{\sin \theta_a} \left\{ \left[-r \cos \theta \right]_0^{\theta_2} - a \left[\frac{1}{2} \sin^2 \theta \right]_0^{\theta_2} \right\}$$

$$= \frac{f p_a\, br}{\sin \theta_a} \left(r - r \cos \theta_2 - \frac{a}{2} \sin^2 \theta_2 \right)$$

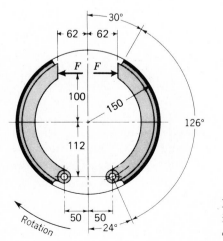

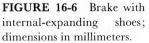

FIGURE 16-6   Brake with internal-expanding shoes; dimensions in millimeters.

Changing all lengths to meters, we have

$$M_f = (0.32)[1000(10)^3](0.032)(0.150)$$

$$\times \left[0.150 - 0.150 \cos 126° - \left(\frac{0.123}{2}\right) \sin^2 126° \right]$$

$$= 304 \text{ N} \cdot \text{m}$$

The moment of the normal forces is obtained from Eq. (16-3). Integrating from 0 to θ_2 gives

$$M_N = \frac{p_a \, bra}{\sin \theta_a} \left[\frac{\theta}{2} - \frac{1}{4} \sin 2\theta \right]_0^{\theta_2}$$

$$= \frac{p_a \, bra}{\sin \theta_a} \left(\frac{\theta_2}{2} - \frac{1}{4} \sin 2\theta_2 \right)$$

$$= [1000(10)^3](0.032)(0.150)(0.123) \left[\frac{\pi}{2} \frac{126}{180} - \frac{1}{4} \sin (2)(126°) \right]$$

$$= 790 \text{ N} \cdot \text{m}$$

From Eq. (16-4), the actuating force is

$$F = \frac{M_N - M_f}{c} = \frac{790 - 304}{100 + 112} = 2.29 \text{ kN} \qquad\qquad Ans.$$

(b) From Eq. (16-6), the torque applied by the right-hand shoe is

$$T_R = \frac{fp_a \, br^2 (\cos \theta_1 - \cos \theta_2)}{\sin \theta_a}$$

$$= \frac{0.32[1000(10)^3](0.032)(0.150)^2 (\cos 0° - \cos 126°)}{1} = 366 \text{ N} \cdot \text{m}$$

The torque contributed by the left-hand shoe cannot be obtained until we learn its maximum operating pressure. Equations (16-2) and (16-3) indicate that the frictional and normal moments are proportional to this pressure. Thus, for the left-hand shoe

$$M_N = \frac{790 p_a}{1000} \qquad M_f = \frac{304 p_a}{1000}$$

Then, from Eq. (16-7),

$$F = \frac{M_N + M_f}{c}$$

or

$$2.29 = \frac{(790/1000)p_a + (304/1000)p_a}{100 + 112}$$

Solving gives $p_a = 444$ kPa. Then, from Eq. (16-6), the torque on the left-hand shoe is

$$T_L = \frac{fp_a br^2(\cos \theta_1 - \cos \theta_2)}{\sin \theta_a}$$

Since $\sin \theta_a = 1$, we have

$$T_L = 0.32[444(10)^3](0.032)(0.150)^2(\cos 0° - \cos 126°) = 162 \text{ N} \cdot \text{m}$$

The braking capacity is the total torque:

$$T = T_R + T_L = 366 + 162 = 528 \text{ N} \cdot \text{m} \qquad\qquad Ans.$$

(c) In order to find the hinge-pin reactions we note $\sin \theta_a = 1$ and $\theta_1 = 0$. Then Eqs. (16-8) give

$$A = \frac{1}{2} \sin^2 \theta_2 = \frac{1}{2} \sin^2 126° = 0.3273$$

$$B = \frac{\theta_2}{2} - \frac{1}{4} \sin 2\theta_2 = \frac{\pi(126)}{2(180)} - \frac{1}{4} \sin (2)(126°) = 1.3373$$

Also, let

$$D = \frac{p_a br}{\sin \theta_a} = \frac{1000(0.032)(0.150)}{1} = 4.8 \text{ kN}$$

where $p_a = 1000$ kPa for the right-hand shoe. Then, using Eqs. (16-9), we have

$$R_x = D(A - fB) - F_x = 4.8[0.3273 - 0.32(1.3373)] - 2.29 \sin 24°$$
$$= -1.414 \text{ kN}$$

$$R_y = D(B + fA) - F_y = 4.8[1.3373 + 0.32(0.3273)] - 2.29 \cos 24°$$
$$= 4.830 \text{ kN}$$

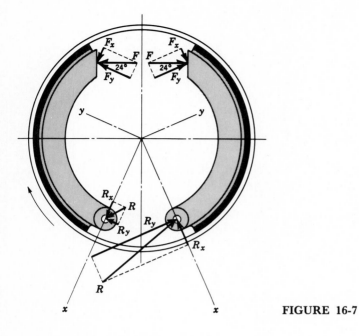

FIGURE 16-7

The resultant on this hinge pin is

$$R = \sqrt{(1.414)^2 + (4.830)^2} = 5.03 \text{ kN} \qquad \qquad \textit{Ans.}$$

The reactions at the hinge pin of the left-hand shoe are found using Eqs. (16-10) for a pressure of 444 kPa. They are found to be $R_x = 0.678$ kN and $R_y = 0.535$ kN. The resultant is

$$R = \sqrt{(0.678)^2 + (0.535)^2} = 0.864 \text{ kN} \qquad \qquad \textit{Ans.}$$

The reactions for both hinge pins, together with their directions, are shown in Fig. 16-7.

This example dramatically shows the benefit to be gained by arranging the shoes to be self-energizing. If the left-hand shoe were turned over so as to place the hinge pin at the top, it could apply the same torque as the right-hand shoe. This would make the capacity of the brake $(2)(366) = 732$ N·m instead of the present 528 N·m, a 30 percent improvement. In addition, some of the friction material at the heel could be eliminated without seriously affecting the capacity, because of the low pressure in this area. This change might actually improve the overall design because the additional rim exposure would improve the heat-dissipation capacity. ////

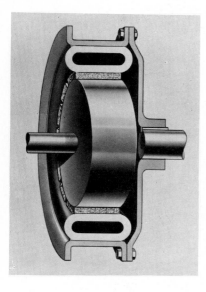

FIGURE 16-8 An external-contracting clutch-brake that is engaged by expanding the flexible tube with compressed air. (*Courtesy of Twin Disc Clutch Company.*)

16-3   EXTERNAL-CONTRACTING RIM CLUTCHES AND BRAKES

The patented clutch-brake of Fig. 16-8 has external-contracting friction elements, but the actuating mechanism is pneumatic.   Here we shall study only pivoted external shoe brakes and clutches, though the methods presented can easily be adapted to the clutch-brake of Fig. 16-8.

The notation for external-contracting shoes is shown in Fig. 16-9.   The moments of the frictional and normal forces about the hinge pin are the same as for the internal-expanding shoes.   Equations (16-2) and (16-3) apply and are repeated here for convenience:

$$M_f = \frac{fp_a br}{\sin \theta_a} \int_{\theta_1}^{\theta_2} \sin \theta (r - a \cos \theta) \, d\theta \qquad (16\text{-}2)$$

$$M_N = \frac{p_a bra}{\sin \theta_a} \int_{\theta_1}^{\theta_2} \sin^2 \theta \, d\theta \qquad (16\text{-}3)$$

Both these equations give positive values for clockwise moments (Fig. 16-9) when used for external-contracting shoes.   The actuating force must be large enough to balance both moments:

$$F = \frac{M_N + M_f}{c} \qquad (16\text{-}11)$$

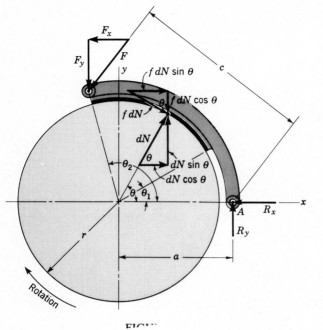

FIGURE 16-9 Notation for external-contracting shoe.

The horizontal and vertical reactions at the hinge pin are found in the same manner as for internal-expanding shoes. They are

$$R_x = \int dN \cos \theta + \int f \, dN \sin \theta - F_x \qquad\qquad (a)$$

$$R_y = \int f \, dN \cos \theta - \int dN \sin \theta + F_y \qquad\qquad (b)$$

By using Eqs. (16-8) we have

$$R_x = \frac{p_a b r}{\sin \theta_a} (A + fB) - F_x$$

$$\qquad\qquad\qquad\qquad\qquad (16\text{-}12)$$

$$R_y = \frac{p_a b r}{\sin \theta_a} (fA - B) + F_y$$

If the rotation is counterclockwise, the sign of the frictional term in each equation is reversed. Thus Eq. (16-11) for the actuating force becomes

$$F = \frac{M_N - M_f}{c} \qquad\qquad (16\text{-}13)$$

and self-energization exists for counterclockwise rotation. The horizontal and vertical reactions are found, in the same manner as before, to be

$$R_x = \frac{p_a br}{\sin \theta_a} (A - fB) - F_x$$

$$(16\text{-}14)$$

$$R_y = \frac{p_a br}{\sin \theta_a} (-fA - B) + F_y$$

It should be noted that, when external-contracting designs are used as clutches, the effect of centrifugal force is to decrease the normal force. Thus, as the speed increases, a larger value of the actuating force F is required.

A special case arises when the pivot is symmetrically located and also placed so that the moment of the friction forces about the pivot is zero. The geometry of such a brake will be similar to that of Fig. 16-10a. To get a pressure-distribution relation we assume that the lining wears so as always to retain its cylindrical shape. This means that the wear Δx in Fig. 16-10b is constant regardless of the angle θ. Thus the radial wear of the shoe is $\Delta r = \Delta x \cos \theta$. If, on any elementary area of the shoe, we assume that the energy or frictional loss is proportional to the radial pressure, and if we also assume that the wear is directly related to the frictional loss, then by direct analogy,

$$p = p_a \cos \theta \qquad (c)$$

and p is maximum at $\theta = 0$.

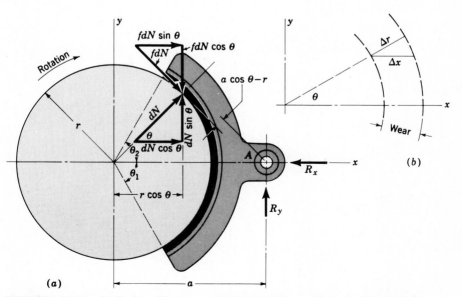

FIGURE 16-10   (a) Brake with symmetrical pivoted shoe; (b) wear of brake lining.

Proceeding to the force analysis, we observe that (Fig. 16-10a)

$$dN = pbr\ d\theta \tag{d}$$

or

$$dN = p_a\, br\ \cos\theta\ d\theta \tag{e}$$

The distance a to the pivot is to be chosen such that the moment of the frictional forces M_f is zero. Symmetry means that $\theta_1 = \theta_2$, and so

$$M_f = 2 \int_0^{\theta_2} (f\, dN)(a\cos\theta - r) = 0$$

Substituting Eq. (e) gives

$$2fp_a\, br \int_0^{\theta_2} (a\cos^2\theta - r\cos\theta)\ d\theta = 0$$

from which

$$a = \frac{4r\sin\theta_2}{2\theta_2 + \sin 2\theta_2} \tag{16-15}$$

With the pivot located according to this equation, the moment about the pin is zero, and the horizontal and vertical reactions are

$$R_x = 2 \int_0^{\theta_2} dN\cos\theta = \frac{p_a\, br}{2}(2\theta_2 + \sin 2\theta_2) \tag{16-16}$$

where, because of symmetry,

$$\int f\, dN \sin\theta = 0$$

Also

$$R_y = 2 \int_0^{\theta_2} f\, dN \cos\theta = \frac{p_a\, brf}{2}(2\theta_2 + \sin 2\theta_2) \tag{16-17}$$

where

$$\int dN \sin\theta = 0$$

also because of symmetry. Note, too, that $R_x = -N$ and $R_y = -fN$, as might be expected for the particular choice of the dimension a. Therefore the torque is

$$T = afN \qquad\qquad\qquad (16\text{-}18)$$

16-4 BAND-TYPE CLUTCHES AND BRAKES

Flexible clutch and brake bands are used in power excavators and in hoisting and other machinery. The analysis follows the notation of Fig. 16-11.

Because of friction and the rotation of the drum, the actuating force P_2 is less than the pin reaction P_1. Any element of the band, of angular length $d\theta$, will be in equilibrium under the action of the forces shown in the figure. Summing these forces in the vertical direction, we have

$$(P + dP)\,\sin\frac{d\theta}{2} + P\,\sin\frac{d\theta}{2} - dN = 0 \qquad\qquad (a)$$

$$dN = P\,d\theta \qquad\qquad\qquad (b)$$

since for small angles $\sin d\theta/2 = d\theta/2$. Summing the forces in the horizontal direction gives

$$(P + dP)\,\cos\frac{d\theta}{2} - P\,\cos\frac{d\theta}{2} - f\,dN = 0 \qquad\qquad (c)$$

$$dP - f\,dN = 0 \qquad\qquad\qquad (d)$$

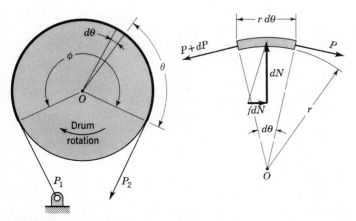

FIGURE 16-11   Forces on a brake band.

Substituting the value of dN from Eq. (b) into (d) and integrating,

$$\int_{P_1}^{P_1} \frac{dP}{P} = f \int_0^\phi d\theta \qquad \ln \frac{P_1}{P_2} = f\phi$$

and $\qquad \dfrac{P_1}{P_2} = e^{f\phi}$ $\hspace{5cm}$ (16-19)

The torque may be obtained from the equation

$$T = (P_1 - P_2)\frac{D}{2} \hspace{5cm} (16\text{-}20)$$

The normal force dN acting on an element of area of width b and length $r\, d\theta$ is

$$dN = pbr\, d\theta \hspace{6cm} (e)$$

where p is the pressure. Substitution of the value of dN from Eq. (b) gives

$$P\, d\theta = pbr\, d\theta$$

Therefore $\qquad p = \dfrac{P}{br} = \dfrac{2P}{bD}$ $\hspace{4.5cm}$ (16-21)

The pressure is therefore proportional to the tension in the band. The maximum pressure p_a will occur at the toe and has the value

$$p_a = \frac{2P_1}{bD} \hspace{6cm} (16\text{-}22)$$

16-5 FRICTIONAL-CONTACT AXIAL CLUTCHES

An axial clutch is one in which the mating frictional members are moved in a direction parallel to the shaft. One of the earliest of these is the cone clutch, which is simple in construction and quite powerful. However, except for relatively simple installations, it has been largely displaced by the disk clutch employing one or more disks as the operating members. Advantages of the disk clutch include the freedom from centrifugal effects, the large frictional area which can be installed in a small space, the more effective heat-dissipation surfaces, and the favorable pressure distribution. One very successful disk-clutch design is shown in Fig. 16-12. Let us now determine the capacity of such a clutch or brake in terms of the material and dimensions.

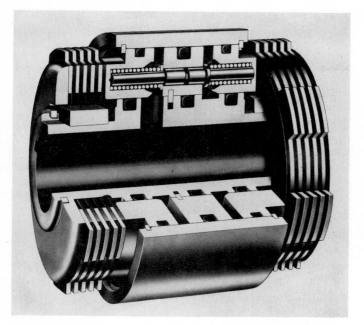

FIGURE 16-12 An oil-actuated multiple-disk clutch-brake for enclosed operation in an oil bath or spray. It is especially useful for rapid cycling. (*Courtesy of Twin Disc Clutch Company.*)

Figure 16-13 shows a friction disk having an outside diameter D and an inside diameter d. We are interested in obtaining the axial force F necessary to produce a certain torque T and pressure p. Two methods of solving the problem, depending upon the construction of the clutch, are in general use. If the disks are rigid, then, the greatest amount of wear will at first occur in the outer areas, since the work of friction is greater in those areas. After a certain amount of wear has taken place, the pressure distribution will change so as to permit the wear to be uniform. This is the basis of the first method of solution.

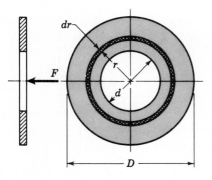

FIGURE 16-13 Disk friction member.

Another method of construction employs springs to obtain a uniform pressure over the area. It is this assumption of uniform pressure that is used in the second method of solution.

Uniform Wear

After initial wear has taken place and the disks have worn down to the point where uniform wear becomes possible, the greatest pressure must occur at $r = d/2$ in order for the wear to be uniform. Denoting the maximum pressure by p_a, we can then write

$$pr = p_a \frac{d}{2} \quad \text{or} \quad p = p_a \frac{d}{2r} \tag{a}$$

which is the condition for having the same amount of work done at radius r as is done at radius $d/2$. Referring to Fig. 16-13, we have an element of area of radius r and thickness dr. The area of this element is $2\pi r\, dr$, so that the normal force acting upon this element is $dF = 2\pi p r\, dr$. We can find the total normal force by letting r vary from $d/2$ to $D/2$ and integrating. Thus

$$F = \int_{d/2}^{D/2} 2\pi p r\, dr = \pi p_a d \int_{d/2}^{D/2} dr = \frac{\pi p_a d}{2}(D - d) \tag{16-23}$$

The torque is found by integrating the product of the frictional force and the radius:

$$T = \int_{d/2}^{D/2} 2\pi f p r^2\, dr = \pi f p_a d \int_{d/2}^{D/2} r\, dr = \frac{\pi f p_a d}{8}(D^2 - d^2) \tag{16-24}$$

By substituting the value of F from Eq. (16-23) we may obtain a more convenient expression for the torque. Thus

$$T = \frac{Ff}{4}(D + d) \tag{16-25}$$

In use, Eq. (16-23) gives the actuating force per friction surface pair for the selected maximum pressure p_a. Equation (16-25) is then used to obtain the torque capacity per friction surface.

Uniform Pressure

When uniform pressure can be assumed over the area of the disk, the actuating force F is simply the product of the pressure and the area. This gives

$$F = \frac{\pi p_a}{4}(D^2 - d^2) \tag{16-26}$$

As before, the torque is found by integrating the product of the frictional force and the radius:

$$T = 2\pi f p \int_{d/2}^{D/2} r^2 \, dr = \frac{2\pi f p}{24} (D^3 - d^3) \qquad (16\text{-}27)$$

Since $p = p_a$, we can rewrite Eq. (16-28) as

$$T = \frac{Ff}{3} \frac{D^3 - d^3}{D^2 - d^2} \qquad (16\text{-}28)$$

It should be noted for both equations that the torque is for a single pair of mating surfaces. This value must therefore be multiplied by the number of pairs of surfaces in contact.

16-6   CONE CLUTCHES AND BRAKES

The drawing of a *cone clutch* in Fig. 16-14 shows that it consists of a *cup* keyed or splined to one of the shafts, a *cone* which must slide axially on splines or keys on the mating shaft, and a helical *spring* to hold the clutch in engagement. The clutch is disengaged by means of a fork which fits into the shifting groove on the friction cone. The *cone angle* α and the diameter and face width of the cone are the important geometric design parameters. If the cone angle is too small, say less than about 8°, then the force required to disengage the clutch may be quite large. And the wedging effect lessens rapidly when larger cone angles are used. Depending upon the characteristics of the friction materials, a good compromise can usually be found using cone angles between 10 and 15°.

To find a relation between the operating force F and the torque transmitted, designate the dimensions of the friction cone as shown in Figure 16-15. As in the case of the axial clutch, we can obtain one set of relations for a uniform-wear and another set for a uniform-pressure assumption.

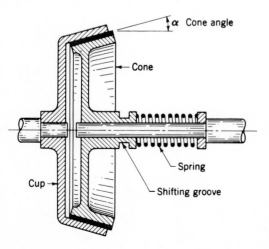

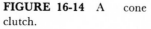

FIGURE 16-14 A cone clutch.

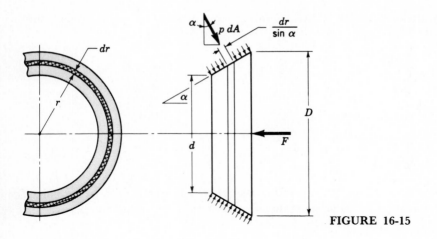

FIGURE 16-15

Uniform Wear

The pressure relation is the same as for the axial clutch:

$$p = p_a \frac{d}{2r} \qquad (a)$$

Next, referring to Fig. 16-15, we see that we have an element of area dA of radius r and width $dr/\sin \alpha$. Thus $dA = 2\pi r \, dr/\sin \alpha$. As shown in Fig. 16-15, the operating force will be the integral of the axial component of the differential force $p \, dA$. Thus

$$F = \int p \, dA \sin \alpha = \int_{d/2}^{D/2} \left(p_a \frac{d}{2r} \right)\left(\frac{2\pi r \, dr}{\sin \alpha} \right)(\sin \alpha)$$

$$= \pi p_a d \int_{d/2}^{D/2} dr = \frac{\pi p_a d}{2}(D - d) \qquad (16\text{-}29)$$

the same as Eq. (16-23).

The differential friction force is $fp \, dA$, and the torque is the integral of the product of this force with the radius. Thus

$$T = \int rfp \, dA = \int_{d/2}^{D/2} (rf)\left(pa \frac{d}{2r} \right)\left(\frac{2\pi r \, dr}{\sin \alpha} \right)$$

$$= \frac{\pi f p_a d}{\sin \alpha} \int_{d/2}^{D/2} r \, dr = \frac{\pi f p_a d}{8 \sin \alpha}(D^2 - d^2) \qquad (16\text{-}30)$$

Note that Eq. (16-24) is a special case of (16-30), with $\alpha = 90°$. Using Eq.

(16-29), we find the torque can also be written

$$T = \frac{Ff}{4 \sin \alpha}(D + d) \tag{16-31}$$

Uniform Pressure

Using $p = p_a$, the actuating force is found to be

$$F = \int p_a \, dA \sin \alpha = \int_{d/2}^{D/2} (p_a)\left(\frac{2\pi r \, dr}{\sin \alpha}\right)(\sin \alpha) = \frac{\pi p_a}{4}(D^2 - d^2) \tag{16-32}$$

The torque is

$$T = \int rfp_a \, dA = \int_{d/2}^{D/2} (rfp_a)\left(\frac{2\pi r \, dr}{\sin \alpha}\right) = \frac{\pi fp_a}{12 \sin \alpha}(D^3 - d^3) \tag{16-33}$$

Or using Eq. (16-32) in (16-33),

$$T = \frac{Ff}{3 \sin \alpha}\frac{D^3 - d^3}{D^2 - d^2} \tag{16-34}$$

16-7 ENERGY CONSIDERATIONS

When the rotating members of a machine are caused to stop by means of a brake, the kinetic energy of rotation must be absorbed by the brake. This energy appears in the brake in the form of heat. In the same way, when the members of a machine which are initially at rest are brought up to speed, slipping must occur in the clutch until the driven members have the same speed as the driver. Kinetic energy is absorbed during slippage of either a clutch or brake, and this energy appears as heat.

We have seen how the torque capacity of a clutch or brake depends upon the coefficient of friction of the material and upon a safe normal pressure. However, the character of the load may be such that, if this torque value is permitted, the clutch or brake may be destroyed by its own generated heat. The capacity of a clutch is therefore limited by two factors, the characteristics of the material and the ability of the clutch to dissipate heat. In this section we shall consider the amount of heat generated by a clutching or braking operation. If the heat is generated faster than it is dissipated, we have a temperature-rise problem; that is the subject of the next section.

To get a clear picture of what happens during a simple clutching or braking operation, refer to Fig. 16-1a, which is a mathematical model of a two-inertia system connected by a clutch. As shown, inertias I_1 and I_2 have initial angular velocities of ω_1 and ω_2, respectively. During the clutch operation both angular velocities change and eventually become equal. We assume that the two shafts are rigid and that the clutch torque is constant.

Writing the equation of motion for inertia 1 gives

$$I_1 \ddot{\theta}_1 = -T \qquad (a)$$

where $\ddot{\theta}_1$ is the angular acceleration of I_1 and T is the clutch torque. A similar equation for I_2 is

$$I_2 \ddot{\theta}_2 = T \qquad (b)$$

We can determine the instantaneous angular velocities $\dot{\theta}_1$ and $\dot{\theta}_2$ of I_1 and I_2 after any period of time t has elapsed by integrating Eqs. (a) and (b). The results are

$$\dot{\theta}_1 = -\frac{T}{I_1} t + \omega_1 \qquad (c)$$

$$\dot{\theta}_2 = \frac{T}{I_2} t + \omega_2 \qquad (d)$$

The difference in the velocities, sometimes called the relative velocity, is

$$\dot{\theta} = \dot{\theta}_1 - \dot{\theta}_2 = -\frac{T}{I_1} t + \omega_1 - \left(\frac{T}{I_2} t + \omega_2 \right)$$

$$= \omega_1 - \omega_2 - T \left(\frac{I_1 + I_2}{I_1 I_2} \right) t \qquad (16\text{-}35)$$

The clutching operation is completed at the instant in which the two angular velocities $\dot{\theta}_1$ and $\dot{\theta}_2$ become equal. Let the time required for the entire operation be t_1. Then $\dot{\theta} = 0$ when $\dot{\theta}_1 = \dot{\theta}_2$ and so Eq. (16-35) gives the time as

$$t_1 = \frac{I_1 I_2 (\omega_1 - \omega_2)}{T(I_1 + I_2)} \qquad (16\text{-}36)$$

This equation shows that the time required for the engagement operation is directly proportional to the velocity difference and inversely proportional to the torque.

We have assumed the clutch torque to be constant. Therefore, using Eq. (16-35), we find the rate of energy dissipation during the clutching operation to be

$$u = T\dot{\theta} = T \left[\omega_1 - \omega_2 - T \left(\frac{I_1 + I_2}{I_1 I_2} \right) t \right] \qquad (e)$$

This equation shows that the energy dissipation rate is greatest at the start when $t = 0$.

The total energy dissipated during the clutching operation or braking cycle is obtained by integrating Eq. (e) from $t = 0$ to $t = t_1$. The result is found to be

$$E = \int_0^{t_1} u \, dt = T \int_0^{t_1} \left[\omega_1 - \omega_2 - T\left(\frac{I_1 + I_2}{I_1 I_2}\right) t \right] dt$$

$$= \frac{I_1 I_2 (\omega_1 - \omega_2)^2}{2(I_1 + I_2)} \tag{16-37}$$

Note that the energy dissipated is proportional to the velocity difference and is independent of the clutch torque.

Note that E in Eq. (16-37) is the energy lost or dissipated; this is the energy that is absorbed by the clutch or brake. If the inertias are expressed in U.S. customary units (lbf·s^2/in), then the energy absorbed by the clutch assembly is in lbf·in. Using these units, the heat generated in Btu is

$$H = \frac{E}{9336} \tag{16-38}$$

In SI the inertias are expressed in kilogram meter units, and the energy dissipated is expressed in joules.

16-8 TEMPERATURE RISE

The temperature rise of the clutch or brake assembly can be approximated by the classic expression

$$\Delta T = \frac{H}{CW} \tag{16-39}$$

where ΔT = temperature rise, °F
C = specific heat; use 0.12 for steel or cast iron
W = weight of clutch or brake parts, 1bf

A similar equation can be written using SI units. It is

$$\Delta T = \frac{E}{Cm} \tag{16-40}$$

where ΔT = temperature rise, °C
C = specific heat; use 500 J/kg·°C for steel or cast iron
m = mass of clutch or brake parts, kg

The temperature-rise equations above can be used to explain what happens when a clutch or brake is operated. However, there are so many variables involved that it would be most unlikely that such an analysis would even approximate experimental results. For this reason such analyses are most useful, for repetitive cycling, in pinpointing those design parameters that have the greatest effect on performance.

An object heated to a temperature T_1 cools to an ambient temperature T_a according to the exponential relation

$$T_i - T_a = (T_1 - T_a)e^{-(AU/WC)t} \qquad (16\text{-}41)$$

where T_1 = instantaneous temperature at time t, °F
   A = heat transfer area, ft^2
   U = surface coefficient, Btu/(ft^2)(s)(°F)

A similar expression can be written in SI units.

Figure 16-16 shows an application of Eq. (16-41). At time t_A a clutching or braking operation causes the temperature to rise to T_1 at A. Though the rise occurs in a finite time interval, it is assumed to occur instantaneously. The temperature then drops along the decay line ABC unless interrupted by another clutching operation. If a second operation occurs at time t_B, the temperature will rise along the dashed line to T_2 and then begin an exponential drop as before.

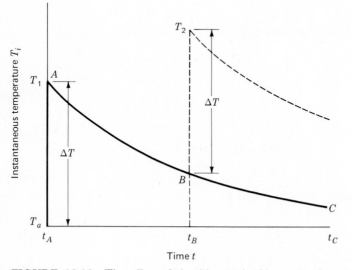

FIGURE 16-16 The effect of clutching or braking operations on temperature. T_a is the ambient temperature.

16-9 FRICTION MATERIALS

A brake or clutch friction material should have the following characteristics to a degree which is dependent upon the severity of the service:

1 A high and uniform coefficient of friction
2 Properties which are not affected by environmental conditions, such as moisture
3 The ability to withstand high temperatures, together with good heat conductivity
4 Good resiliency
5 High resistance to wear, scoring, and galling

Table 16-1 lists properties of typical brake linings. The linings may consist of a mixture of asbestos fibers to provide strength and ability to withstand high temperatures, various friction particles to obtain a degree of wear resistance as well as a higher coefficient of friction, and bonding materials.

Table 16-2 includes a wider variety of friction materials, together with some of their properties. Some of these materials may be run wet by allowing them to dip in oil or to be sprayed by oil. This reduces the coefficient of friction somewhat but carries away more heat and permits higher pressures to be used.

Table 16-1 SOME PROPERTIES OF BRAKE LININGS

	Woven lining	Molded lining	Rigid block
Compressive strength, kpsi	10–15	10–18	10–15
Compressive strength, MPa	70–100	70–125	70–100
Tensile strength, kpsi	2.5–3	4–5	3–4
Tensile strength, MPa	17–21	27–35	21–27
Max, temperature, °F	400–500	500	750
Max. temperature, °C	200–260	260	400
Max. speed, fpm	7500	5000	7500
Max. speed, m/s	38	25	38
Max. pressure, psi	50–100	100	150
Max. pressure, kPa	340–690	690	1000
Frictional coefficient mean	0.45	0.47	0.40–45

Source: From Z. J. Zania, "Friction Clutches and Brakes," in Harold A. Rothbart, (ed), *Mechanical Design and Systems Handbook*, McGraw-Hill, 1964, sec. 28, pp. 28–39. This material is included with the permission of the publishers.

Table 16-2 FRICTION MATERIALS FOR CLUTCHES

Material	Friction coefficient		Max. temperature		Max. pressure	
	Wet	**Dry**	**°F**	**°C**	**psi**	**kPa**
Cast iron on cast iron	0.05	0.15–0.20	600	320	150–250	1000–1750
Powdered metal* on cast iron	0.05–0.1	0.1–0.4	1000	540	150	1000
Powdered metal* on hard steel	0.05–0.1	0.1–0.3	1000	540	300	2100
Wood on steel or cast iron	0.16	0.2–0.35	300	150	60–90	400–620
Leather on steel or cast iron	0.12	0.3–0.5	200	100	10–40	70–280
Cork on steel or cast iron	0.15–0.25	0.3–0.5	200	100	8–14	50–100
Felt on steel or cast iron	0.18	0.22	280	140	5–10	35–70
Woven asbestos* on steel or cast iron	0.1–0.2	0.3–0.6	350–500	175–260	50–100	350–700
Molded asbestos* on steel or cast iron	0.08–0.12	0.2–0.5	500	260	50–150	350–1000
Impregnated asbestos* on steel or cast iron	0.12	0.32	500–750	260–400	150	1000
Carbon graphite on steel	0.05–0.1	0.25	700–1000	370–540	300	2100

* The friction coefficient can be maintained within ± 5 percent for specific materials in this group.

16-10 MISCELLANEOUS CLUTCHES AND COUPLINGS

The square-jaw clutch shown in Fig. 16-17*a* is one form of positive-contact clutch. These clutches have the following characteristics:

1 They do not slip.
2 No heat is generated.
3 They cannot be engaged at high speeds.

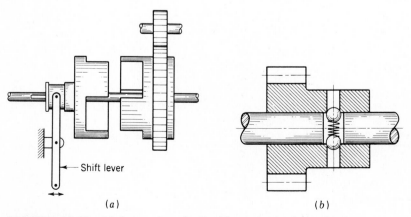

          (*a*)   (*b*)

FIGURE 16-17 (*a*) Square-jaw clutch; (*b*) overload-release clutch.

4 Sometimes they cannot be engaged when both shafts are at rest.

5 Engagement at any speed is accompanied by shock.

The greatest differences among the various types of positive clutches are concerned with the design of the jaws. To provide a longer period of time for shift action during engagement, the jaws may be ratchet-shaped, spiral-shaped, or gear-tooth-shaped. Sometimes a great many teeth or jaws are used, and they may be cut either circumferentially, so that they engage by cylindrical mating, or on the faces of the mating elements.

Although positive clutches are not used to the extent of the frictional-contact types, they do have important applications where synchronous operation is required, as, for example, in power presses or rolling-mill screw-downs.

Devices such as linear drives or motor-operated screw drivers must run to a definite limit and then come to a stop. An overload-release type of clutch is required for these applications. Figure 16-17b is a schematic drawing illustrating the principle of operation of such a clutch. These clutches are usually spring-loaded so as to release at a predetermined torque. The clicking sound which is heard when the overload point is reached is considered to be a desirable signal.

Both fatigue and shock loads must be considered in obtaining the stresses and deflections of the various portions of positive clutches. In addition, wear must generally be considered. The application of the fundamentals discussed in Part 1 is usually sufficient for the complete design of these devices.

An overrunning clutch or coupling permits the driven member of a machine to "freewheel" or "overrun" because the driver is stopped or because another source of power increases the speed of the driven mechanism. The construction uses rollers or balls mounted between an outer sleeve and an inner member having cam flats machined around the periphery. Driving action is obtained by wedging the rollers between the sleeve and the cam flats. This clutch is therefore equivalent to a pawl and ratchet with an infinite number of teeth.

There are many varieties of overrunning clutches available, and they are

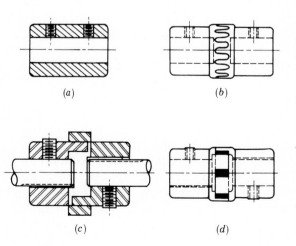

(a)

(b)

(c)

(d)

FIGURE 16-18 Shaft couplings. (a) Plain, (b) light-duty toothed coupling; (c) BOST-FLEX® through-bore design has elastomer insert to transmit torque by compression; insert permits 1° misalignment; (d) three-jaw coupling available with bronze, rubber, or polyurethane insert to minimize vibration. (*Reproduced by permission, Boston Gear Division, Incom International, Inc., Quincy, Mass.*)

built in capacities up to hundreds of horsepower. Since no slippage is involved, the only power loss is that due to bearing friction and windage.

The shaft couplings shown in Fig. 16-18 are representative of the selection available in catalogs.

16-11   FLYWHEELS

The equation of motion for the flywheel represented in Fig. 16-1b is

$$\sum M = T_i\,(\theta_i,\,\dot{\theta}_i) - T_o(\theta_o,\,\dot{\theta}_o) - I\ddot{\theta} = 0$$

or

$$I\ddot{\theta} = T_i\,(\theta_i,\,\omega_i) - T_o(\theta_o,\,\omega_o) \tag{a}$$

where T_i is considered positive and T_o negative, and where $\dot{\theta}$ and $\ddot{\theta}$ are the first and second time derivatives of θ, respectively. Note that both T_i and T_o may depend for their values on the angular displacements θ_i and θ_o as well as their angular velocities ω_i and ω_o. In many cases the torque characteristic depends upon only one of these. Thus, the torque delivered by an induction motor depends upon the speed of the motor. In fact, motor manufacturers publish charts detailing the torque-speed characteristics of their various motors.

When the input and output torque functions are given, Eq. (a) can be solved for the motion of the flywheel using well-known techniques for solving linear and nonlinear differential equations. We can dispense with this here by assuming a rigid shaft, giving $\theta_i = \theta = \theta_o$. Thus, Eq. ($a$) becomes

$$I\ddot{\theta} = T_i\,(\theta,\,\omega) - T_o(\theta,\,\omega) \tag{b}$$

When the two torque functions are known and the starting values of the displacement θ and velocity ω are given, Eq. (b) can be solved for ω and $\ddot{\theta}$ as functions of time. However, we are not really interested in the instantaneous values of these terms at all. Primarily we want to know the overall performance of the flywheel. What should its moment of inertia be? How do we match the power source to the load? And what are the resulting performance characteristics of the system that we have selected?

To gain insight into the problem, a hypothetical situation is diagrammed in Fig. 16-19. An input power source subjects a flywheel to a constant torque T_i while the shaft rotates from θ_1 to θ_2. This is a positive torque and is plotted upward. Equation (b) indicates that a positive acceleration $\ddot{\theta}$ will be the result, and so the shaft velocity increases from ω_1 to ω_2. As shown, the shaft now rotates from θ_2 to θ_3 with zero torque and hence, from Eq. (b), with zero acceleration. Therefore $\omega_3 = \omega_2$. From θ_3 to θ_4 a load, or output torque, of constant magnitude is applied, causing the shaft to slow down from ω_3 to ω_4. Note that the output torque is plotted in the negative direction in accordance with Eq. (b).

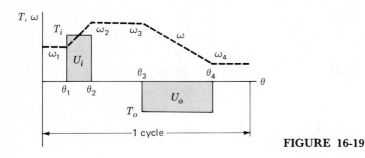

FIGURE 16-19

The work input to the flywheel is the area of the rectangle between θ_1 and θ_2, or

$$U_i = T_i(\theta_2 - \theta_1) \tag{c}$$

The work output of the flywheel is the area of the rectangle from θ_3 to θ_4, or

$$U_o = T_o(\theta_4 - \theta_3) \tag{d}$$

If U_o is greater than U_i, the load uses more energy than has been delivered to the flywheel and so ω_4 will be less than ω_1. If $U_o = U_i$, ω_4 will be equal to ω_1 because the gains and losses are equal; we are assuming no friction losses. And finally, ω_4 will be greater than ω_1 if $U_i > U_o$.

We can also write these relations in terms of kinetic energy. At $\theta = \theta_1$ the flywheel has a velocity of ω_1 rad/s, and so its kinetic energy is

$$E_1 = \tfrac{1}{2}I\omega_1^2 \tag{e}$$

At $\theta = \theta_2$ the velocity is ω_2, and so

$$E_2 = \tfrac{1}{2}I\omega_2^2 \tag{f}$$

Thus the change in kinetic energy is

$$E_2 - E_1 = \tfrac{1}{2}I(\omega_2^2 - \omega_1^2) \tag{16-42}$$

Many of the torque displacement functions encountered in practical engineering situations are so complicated that they must be integrated by approximate methods. Figure 16-20, for example, is a typical plot of the engine torque for one cycle of motion of a single-cylinder gas engine. Since a part of the torque curve is negative, the flywheel must return part of the energy back to the engine. Approximate integration of this curve for a cycle of 4π yields a mean torque T_m available to drive a load.

The simplest integration routine is Simpson's rule; this approximation can be

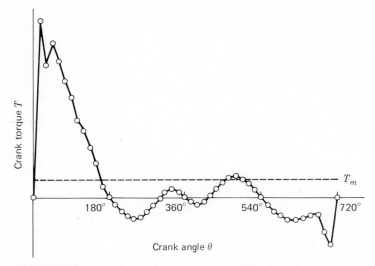

FIGURE 16-20   Relation between torque and crank angle for a one-cylinder four-cycle internal combustion engine.

handled on any computer and is short enough to use on the smallest programmable calculators.   In fact, this routine is usually found as part of the library for most calculators and minicomputers.   The equation used is

$$\int_{x_0}^{x_n} f(x) \; dx = \frac{h}{3} \, (f_0 + 4f_1 + 2f_2 + 4f_3 + 2f_4$$

$$+ \cdots + 2f_{n-2} + 4f_{n-1} + f_n) \tag{16-43}$$

where

$$h = \frac{x_n - x_0}{n} \qquad x_n > x_0$$

and n is the number of subintervals used, 2, 4, 6, ...   If memory is limited, solve Eq. (16-42) in two or more steps, say from 0 to $n/2$ and then from $n/2$ to n.

It is convenient to define a *coefficient of speed fluctuation* as

$$C_s = \frac{\omega_2 - \omega_1}{\omega} \tag{16-44}$$

where ω is the nominal angular velocity, given by

$$\omega = \frac{\omega_2 + \omega_1}{2} \tag{16-45}$$

Equation (16-42) can be factored to give

$$E_2 - E_1 = \frac{I}{2}\,(\omega_2 - \omega_1)(\omega_2 + \omega_1)$$

Since $\omega_2 - \omega_1 = C_s\omega$ and $\omega_2 + \omega_1 = 2\omega$, we have

$$E_2 - E_1 = C_s I\omega^2 \tag{16-46}$$

Equation (16-46) can be used to obtain an appropriate flywheel inertia corresponding to the energy change $E_2 - E_1$.

Example 16-2   Table 16-3 lists values of the torque used to plot Fig. 16-20. The nominal speed of the engine is to be 250 rad/s.

  (a)   Integrate the torque-displacement function for one cycle and find the energy that can be delivered to a load during the cycle.

  (b)   Determine the mean torque T_m (see Fig. 16-20).

  (c)   The greatest energy fluctuation is approximately between $\theta = 15°$ and $\theta = 150°$ on the torque diagram; see Fig. 16-20 and note that $T_o = -T_m$. Using a coefficient of speed fluctuation $C_s = 0.1$, find a suitable value for the flywheel inertia.

  (d)   Find ω_2 and ω_1.

Solution   (a) Using $n = 48$ and $h = 4\pi/48$, we enter the data of Table 16-3 into a computer program and get $E = 3490$ lb · in.   This is the energy that can be delivered to the load.

  (b) $T_m = \dfrac{3490}{4\pi} = 278$ lb · in   *Ans.*

Table 16-3

θ deg	T lb · in	θ deg	T lb · in	θ deg	T lb · in	θ deg	T lb · in
0	0	180	0	360	0	540	0
15	2800	195	−107	375	−85	555	−107
30	2090	210	−206	390	−125	570	−206
45	2430	225	−280	405	−89	585	−292
60	2160	240	−323	420	8	600	−355
75	1840	255	−310	435	126	615	−371
90	1590	270	−242	450	242	630	−362
105	1210	285	−126	465	310	645	−312
120	1066	300	−8	480	323	660	−272
135	803	315	89	495	280	675	−274
150	532	330	125	510	206	690	−548
165	184	345	85	525	107	705	−760

(c) The largest positive loop on the torque-displacement diagram occurs between $\theta = 0$ and $\theta = 180°$. We select this loop as yielding the largest speed change. Subtracting 278 lb·in from the values in Table 16-3 for this loop gives, respectively, -278, 2522, 1812, 2152, 1882, 1562, 1312, 932, 788, 525, 254, -94, and -278 lb·in. Entering Simpson's approximation again, using $n = 12$ and $h = 4\pi/48$, gives $E_2 - E_1 = 3660$ lb·in. We now solve Eq. (16-46) for I and substitute. This gives

$$I = \frac{E_2 - E_1}{C_s \omega^2} = \frac{3660}{0.1(250)^2} = 0.586 \text{ lb·s}^2\text{·in} \qquad\qquad Ans.$$

(d) Equations (16-44) and (16-45) can be solved simultaneously for ω_2 and ω_1. Substituting appropriate values in these two equations yields

$$\omega_2 = \frac{\omega}{2}(2 + C_s) = \frac{250}{2}(2 + 0.1) = 262.5 \text{ rad/s} \qquad\qquad Ans.$$

$$\omega_1 = 2\omega - \omega_2 = 2(250) - 262.5 = 237.5 \text{ rad/s} \qquad\qquad Ans.$$

These two speeds occur at $\theta = 180°$ and $\theta = 0°$, respectively.   ////

PROBLEMS

Section 16-2

16-1   An internal-expanding rim-type brake is shown in the figure. The brake drum has an inside diameter of 12 in and the radius to the hinge pins is $R = 5$ in. The shoes have a face width of $1\frac{1}{2}$ in, both of which are actuated by the same force F. The design coefficient of friction is 0.28, with a maximum pressure of 120 psi.

(a)   Find the actuating force F.

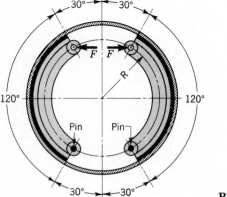

PROBLEM 16-1

(b) Calculate the torque capacity.

(c) Find the hinge-pin reactions.

16-2 If the actuating forces may differ from each other in Prob. 16-1, what values should they have in order that each shoe will be loaded by the maximum pressure of 120 psi?

16-3 In the figure for Prob. 16-1 $R = 100$ mm, the drum diameter is 250 mm, and the face width of the shoes is 28 mm. Calculate the actuating force F, the torque capacity, and the hinge-pin reactions if $p_a = 600$ kPa and $f = 0.32$.

16-4 The figure shows a 400-mm-diameter brake drum with four internally expanding shoes. Each of the hinge pins A and B supports a pair of shoes. The actuating mechanism is to be arranged to produce the same force F on each shoe. The face width of the shoes is 75 mm. The material used permits a coefficient of friction of 0.24 and a maximum pressure of 1.0 MPa.

(a) Determine the actuating force.

(b) Calculate the braking capacity.

(c) Noting that rotation may be in either direction, compute the hinge-pin reactions.

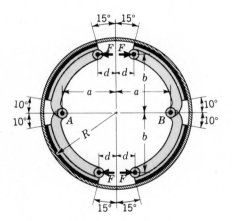

PROBLEM 16-4 The dimensions in millimeters are $a = 150$, $b = 165$, $R = 200$, $d = 50$.

Section 16-3

16-5 The block-type hand brake shown in the figure has a face width of 45 mm. The frictional material permits a maximum pressure of 550 kPa with a coefficient of friction of 0.24.

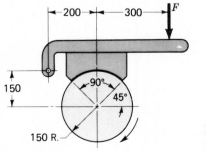

PROBLEM 16-5 Dimensions in millimeters.

(a) Determine the force F.

(b) What is the torque capacity?

(c) If the speed is 100 rpm and the brake is applied for 5 s at full capacity to bring the shaft to stop, how much heat is generated?

16-6 The brake whose dimensions are shown in the figure has a coefficient of friction of 0.30 and is to have a maximum pressure of 150 psi against the friction material.

(a) Using an actuating force of 400 lb, determine the width of face of the shoes. (Both shoes are to have the same width.)

(b) What torque will the brake absorb?

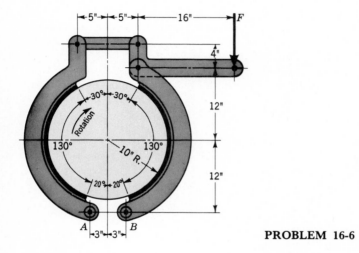

PROBLEM 16-6

16-7 Solve Prob. 16-6 for counterclockwise drum rotation.

Section 16-4

16-8 The band brake shown in the figure is to have a maximum lining pressure of 600 kPa. The drum is 350 mm in diameter, and the band width is 100 mm. The coefficient of friction is 0.25, and the angle of wrap 270°.

(a) Find the band tensions.

(b) Calculate the torque capacity.

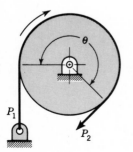

PROBLEM 16-8

16-9 The brake shown in the figure has a coefficient of friction of 0.30 and is to operate at a maximum band pressure of 1000 kPa. The width of the band is 50 mm.
  (*a*) What is the limiting value of the force *F*?
  (*b*) Determine the torque capacity of the brake.

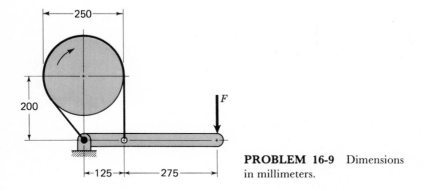

PROBLEM 16-9 Dimensions in millimeters.

16-10 Solve Prob. 16-9, for counterclockwise drum rotation.

16-11 The figure shows a 16-in differential band brake. The maximum pressure is to be 60 psi, with a coefficient of friction of 0.26 and a band width of 4 in. Determine the band tensions and the actuating force for clockwise rotation.

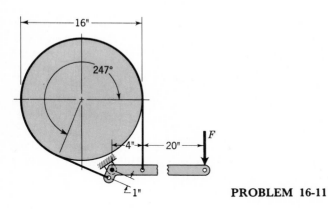

PROBLEM 16-11

16-12 Solve Prob. 16-11 for counterclockwise rotation.

Section 16-5

16-13 A plate clutch has a single pair of mating friction surfaces 300 mm OD by 225 mm ID. The coefficient of friction is 0.25 and the maximum pressure is 825 kPa. Find the torque capacity using the uniform-wear assumption.

16-14 Employing the uniform-pressure assumption, use the data of Prob. 16-13 to find the torque capacity of a similar clutch.

16-15 A plate clutch has a single pair of mating friction surfaces, is 200 mm OD by 100 mm ID, and has a coefficient of friction of 0.30. What is the maximum pressure corresponding to an actuating force of 15 kN? Use the uniform-wear method.

16-16 Solve Prob. 16-15 using the uniform-pressure method.

16-17 A disk clutch has four pairs of mating friction surfaces, is 5 in OD × 3 in ID, and has a coefficient of friction of 0.10.

(*a*)   What actuating force is required for a pressure of 120 psi?

(*b*)   What is the torque rating? (Use the uniform-pressure method.)

16-18 A plate clutch has two pairs of mating friction surfaces 14 in OD × 6 in ID. The coefficient of friction is 0.24, and the pressure is not to exceed 80 psi.

(*a*)   Determine the actuating force.

(*b*)   Calculate the torque capacity.

(Use the uniform-wear method.)

Section 16-6

16-19 A leather-faced cone clutch is to transmit 1200 lb·in of torque. The cone angle is 10°, the mean diameter of the friction surface is 12 in, and the face width is 2 in. Based on a coefficient of friction of 0.25, find the operating force and pressure.

(*a*)   Use the uniform-pressure assumption.

(*b*)   Use the uniform-wear assumption.

16-20 A cone clutch has a cone angle of 11.5°, a mean frictional diameter of 320 mm, and a face width of 60 mm. The clutch is to transmit a torque of 200 N·m. The coefficient of friction is 0.26. Find actuating force and pressure using the assumption of uniform pressure.

16-21 A one-cylinder, two-cycle engine develops its maximum torque at a speed of 3400 rpm when delivering 12 hp. The tentative design of a cone clutch to couple this engine to its load has $\alpha = 13°$, $D = 4$ in, and $p_a = 50$ psi, with $f = 0.20$, corresponding to a woven-asbestos friction facing. Determine the required face width and operating force. Use the uniform-wear assumption.

Sections 16-7 to 16-10

16-22 A two-jaw clutch has the dimensions shown in the figure and is made of ductile steel. The clutch has been designed to transmit 2 kW at 500 rpm. Find the bearing and shear stress in the key and in the jaws.

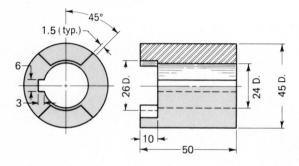

PROBLEM 16-22   Dimensions in millimeters.

16-23 A 0.500-m externally contracting brake drum has heat-dissipating surfaces whose mass is 20 kg. The brake has a rated torque capacity of 300 M · m. The brake is used to bring a rotating inertia to a complete stop. Determine the temperature rise if the inertia has an initial speed of 1800 rpm and is brought to rest in 8 s using the full torque capacity of the brake.

16-24 A flywheel is made from a steel disk 250 mm in diameter and 20 mm thick. It rotates at 500 rpm and is to be brought to rest in 0.40 s by a brake. Compute the total amount of energy to be absorbed and the required torque capacity of the brake.

16-25 A brake permits a mass of 250 kg to be lowered at a velocity of 3 m/s from a drum, as shown in the figure. The drum is 400 mm in diameter, weighs 1.40 kN, and has a radius of gyration of 180 mm.

(*a*) Compute the energy of the system.

(*b*) How much additional braking torque must be applied if the load is to be stopped in 0.50 s?

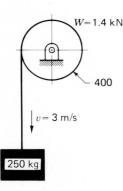

W = 1.4 kN

400

v = 3 m/s

250 kg

PROBLEM 16-25

Section 16-11

16-26 A flywheel has a mean rim diameter of 56 in and varies in speed from 260 to 240 rpm. If the energy fluctuation is 5000 lb · ft, find the coefficient of speed fluctuation and the weight of the rim.

16-27 A single-geared blanking press has a stroke of 8 in and a rated capacity of 35 tons. The crankshaft has a speed of 90 rpm and is geared to the flywheel shaft at a 6 : 1 ratio. Use a frictional allowance of 16 percent, and assume the full load is delivered during 15 percent of the stroke.

(*a*) Calculate the maximum energy fluctuation.

(*b*) Find the rim weight for a mean rim diameter of 48 in and a 10 percent slowdown.

16-28 The load torque required by a 200-ton punch press is displayed in Table 16-4 for one revolution of the flywheel. The flywheel is to have a nominal speed of 240 rpm and is to be designed for a coefficient of speed fluctuation of 0.075.

(*a*) Determine the mean motor torque required at the flywheel shaft and the

Table 16-4

θ deg	T lb·in	θ deg	T lb·in	θ deg	T lb·in	θ deg	T lb·in
0	857	90	7888	180	1801	270	857
10	857	100	8317	190	1629	280	857
20	857	110	8488	200	1458	290	857
30	857	120	8574	210	1372	300	857
40	857	130	8403	220	1115	310	857
50	1287	140	7717	230	1029	320	857
60	2572	150	3515	240	943	330	857
70	5144	160	2144	250	857	340	857
80	6859	170	1972	260	857	350	857

motor horsepower needed, assuming a constant torque-speed characteristic for the motor.

(b)   Find the moment of inertia needed for the flywheel.

16-29   Using the data of Table 16-3, find the mean output torque and flywheel inertia required for a three-cylinder, in-line engine corresponding to a nominal speed of 2400 rpm.   Use $C_s = 0.30$.

CHAPTER
17

FLEXIBLE MECHANICAL ELEMENTS

Flexible machine elements, such as belts, ropes, or chains, are used for the transmission of power over comparatively long distances. When these elements are employed, they usually replace a group of gears, shafts, and bearings or similar power-transmission devices. They thus greatly simplify a machine and consequently are a major cost-reducing element.

In addition, since these elements are elastic and usually long, they play an important part in absorbing shock loads and in damping out the effects of vibrating forces. Although this advantage is important as concerns the life of the driving machine, the cost-reduction element is generally the major factor in the selection of this means of power transmission.

17-1 BELTS

Ordinarily, belts are used to transmit power between two parallel shafts. The shafts must be separated a certain minimum distance, which is dependent upon the type of belt used, in order to work most efficiently. Belts have the following characteristics:

- They may be used for long center distances.
- Because of the slip and creep of belts, the angular-velocity ratio between the two shafts is neither constant nor exactly equal to the ratio of the pulley diameters.
- When using flat belts, clutch action may be obtained by shifting the belt from a loose to a tight pulley.
- When V belts are used, some variation in the angular-velocity ratio may be obtained by employing a small pulley with spring-loaded sides. The diameter of the pulley is then a function of the belt tension and may be varied by changing the center distance.
- Some adjustment of the center distance is usually necessary when belts are used.
- By employing step pulleys, an economical means of changing the velocity ratio may be obtained.

Flat belts are commonly made of oak-tanned leather or of a fabric which has been impregnated with rubber. The modern flat belt consists of a strong elastic core, such as steel or nylon cords, to take the tension and transmit the power, combined with a flexible envelope to provide friction between the belt and pulley. Flat belts are very efficient for high speeds, they are quiet, they can transmit large amounts of power over long center distances, they don't require large pulleys, and they can transmit power around corners or between pulleys at right angles to each other. Flat belts are particularly useful in group-drive installations because of the clutching action that can be obtained.

A *V belt* is made of fabric and cord, usually cotton, rayon, or nylon, and impregnated with rubber. In contrast to flat belts, V belts are used with smaller sheaves and at shorter center distances. V belts are slightly less efficient than flat belts, but a number of them can be used on a single sheave, thus making a multiple drive. They are endless, which eliminates the joint used in flat belts.

A *link V belt* is composed of a large number of rubberized-fabric links joined by suitable metal fasteners. This type of belt may be disassembled at any point and adjusted to length by removing some of the links. This eliminates the necessity for adjustable centers and simplifies the installation. It makes it possible to change the tension for maximum efficiency and also reduces the inventory of belt sizes which would usually be stocked. A typical application of the link V belt is shown in Fig. 17-1.

A *timing belt* is made of rubberized fabric and steel wire, having teeth which fit into grooves cut on the periphery of the pulleys (Fig. 17-2). The timing belt

FIGURE 17-1 Link V-belts successfully applied to a press. The use of endless belts would necessitate dismantling the press to replace belts. (*Courtesy of Mannheim Manufacturing and Belting Compnay.*)

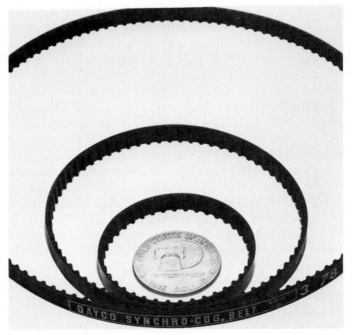

FIGURE 17-2 Timing belts; used for synchronized drives. (*Courtesy of Dayco Corporation.*)

does not stretch or slip and consequently transmits power at a constant angular-velocity ratio. The fact that the belt is toothed provides several advantages over ordinary belting. One of these is that no initial tension is necessary, so that fixed center drives may be used. Another is the elimination of the restriction on speeds; the teeth make it possible to run at nearly any speed, slow or fast. Disadvantages are the first cost of the belt, the necessity of grooving the pulleys, and the attendant dynamic fluctuations caused at the belt-tooth meshing frequency.

17-2 FLAT-BELT DRIVES

Modern flat belts consist of a strong elastic core surrounded by chrome leather or an elastomer; these belts have distinct advantages over a geared or V-belt drive. A flat-belt drive has an efficiency of about 98 percent, which is about the same as for a geared drive. But the efficiency of a V-belt drive is in the vicinity of 70 to 96 percent.* Flat-belt drives are quieter and absorb more torsional vibrations from the system than either gears or V belts.

Figure 17-3 illustrates open and crossed belts and gives equations for the angle of contact θ and the total belt length L for each case. When a horizontal open belt arrangement is used, the driver should rotate so that the slack side is on top. This makes for a larger angle of contact on both pulleys. When the drive is vertical or the center distance is short, a larger angle of contact may be obtained by using an idler tension pulley.

Firbank† explains the theory of flat-belt drives in the following way. A change in belt tension due to friction forces between the belt and pulley will cause the belt to elongate or contract and move relative to the surface of the pulley. This motion is caused by *elastic creep* and is associated with sliding friction as opposed to static friction. The action at the driving pulley, through that portion of the angle of contact that is actually transmitting power, is such that the belt moves more slowly than the surface speed of the pulley because of the elastic creep. The angle of contact is made up of the *effective arc*, through which power is transmitted, and the *idle arc*. For the driving pulley the belt first contacts the pulley with a *tight-side tension* F_1 and a velocity V_1, which is the same as the surface velocity of the pulley. The belt then passes through the idle arc with no change in F_1 or V_1. Then creep or sliding contact begins, and the belt tension changes in accordance with the friction forces. At the end of the effective arc the belt leaves the pulley with a *loose-side tension* F_2 and a reduced speed V_2.

Firbank has used this theory to express the mechanics of flat-belt drives in mathematical form and has verified the results by experiment. His observations include the fact that substantial amounts of power are transmitted by static fric-

---

* A. W. Wallin, "Efficiency of Synchronous Belts and V-Belts," *Proceedings of the National Conference on Power Transmission*, vol. 5, Illinois Institute of Technology, Chicago, Nov. 7–9, 1978, pp. 265–271.

† T. C. Firbank, *Mechanics of the Flat Belt Drive*, ASME Paper no. 72-PTG-21.

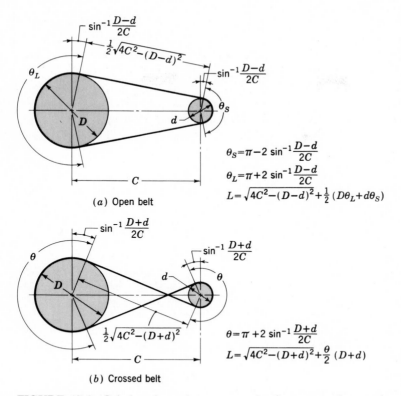

$$\theta_S = \pi - 2\sin^{-1}\frac{D-d}{2C}$$

$$\theta_L = \pi + 2\sin^{-1}\frac{D-d}{2C}$$

$$L = \sqrt{4C^2 - (D-d)^2} + \tfrac{1}{2}(D\theta_L + d\theta_S)$$

(a) Open belt

$$\theta = \pi + 2\sin^{-1}\frac{D+d}{2C}$$

$$L = \sqrt{4C^2 - (D+d)^2} + \frac{\theta}{2}(D+d)$$

(b) Crossed belt

FIGURE 17-3 Belt lengths and contact angles for open and crossed belts.

tion as opposed to sliding friction. He also found that the coefficient of friction for a belt having a nylon core and leather surface was typically 0.7, but that it could be raised to 0.9 by employing special surface finishes.

Because of the space required to present the Firbank analysis, we present here a simplified approach instead; it is a conventional analysis that has been used for many years. We assume that the friction force on the belt is uniform throughout the entire arc of contact and that the centrifugal forces on the belt can be neglected. Then the relation between the tight-side tension F_1 and the slack-side tension F_2 is the same as for band brakes, and is

$$\frac{F_1}{F_2} = e^{f\theta} \tag{17-1}$$

where f is the coefficient of friction and θ is the contact angle. The power transmitted is

$$P = (F_1 - F_2)V \tag{17-2}$$

In this equation the power P is in watts when the tension F is in newtons and the belt velocity V is in meters per second. The horsepower transmitted is

$$H = \frac{(F_1 - F_2)V}{33\ 000} \tag{17-3}$$

where the tensions F are in pounds and the velocity is in feet per minute.

The centrifugal force was neglected in writing Eq. (17-1). This force is given by the equation

$$F_c = mv^2 \tag{17-4}$$

where m is the mass of the belt per unit length and v is in units of length per second. When the centrifugal force is included, Eq. (17-1) becomes

$$\frac{F_1 - F_c}{F_2 - F_c} = e^{f\theta} \tag{17-5}$$

We note that the net tension ratio must be below $e^{f\theta}$, since this is the point of potential slippage of the belt-sheave interface. Now consider an added constraint. When the belt is installed, an initial tension F_i is set into the belt. Now think of each belt segment leaving the sheave as a spring under an initial tension F_i. As power is demanded, the sheave rotates, stretching the high-tension side and contracting the low-tension side. Thus

$$F_1 = F_i + \Delta F \tag{a}$$

$$F_2 = F_i - \Delta F \tag{b}$$

Solving for the initial tension gives

$$F_i = \frac{F_1 + F_2}{2} \tag{17-6}$$

The importance of Eq. (17-6) is that it really defines the maximum belt tension. Consider this: When no power is being transmitted the belt tensions on both sides are equal and so $F_1 = F_2 = F_i$. If now a slight load is added, some power is transmitted and F_1 increases by ΔF and F_2 decreases by the same amount. If the load is increased more and more, then eventually F_2 will become zero because the belt cannot support compression. At this point $F_1 = 2F_i$, which is the maximum belt tension. Thus, the only way to transmit more power is to increase the initial belt tension.

Based on the reasoning above, a flat-belt drive is designed by limiting the maximum tension F_1 according to the permissible tensile stress specified for the

Table 17-1 PROPERTIES OF SOME FLAT-BELT MATERIALS

Material	Joint	Size	Allowable tension* lb	Ultimate load* lb	Ultimate strength kpsi	Weight lb/in^3
Oak-tanned leather	Solid	$\frac{5}{32}$" ply		700	3–4	0.035–0.045
Oak-tanned leather	Riveted	$\frac{5}{32}$" ply		300–600	1–2	0.035–0.045
Oak-tanned leather	Laced	$\frac{5}{32}$" ply		300–600	1–2	0.035–0.045
Rubber cotton duck	Vulcanized	28 oz	15–25	280		0.041
Rubber cotton duck	Vulcanized	32 oz	15–25	300		0.047
	Vulcanized	35 oz	15–25	320		0.051
All cotton	Woven				5	0.045
All cotton	Sewn				7	0.044
Nylon	Core only				35	
Balatta duck	Vulcanized		22–25			0.040

* Pounds-force per inch of width per ply.

belt material. The values listed in Table 17-1 come from a variety of sources and can be used as a rough guide to determine belt sizes by limiting the tensile stress. But, when available, manufacturers' recommendations should always be used for these values.

If we solve Eq. (17-6) for F_2 and substitute it into Eqs. (17-2) and (17-3), we get

$$H = \frac{(F_1 - F_i)V}{16\ 500}$$

$$P = \frac{(F_1 - F_i)V}{2}$$

(17-7)

where, of course, consistent units must be used.

Having selected an initial tension and evaluated $F_1 = F_2$, our next problem is to check Eq. (17-1), or (17-5) if centrifugal force is included, to determine whether the belt will slip. To avoid slippage, the term $e^{f\theta}$ must be greater than the tension ratio. If slippage seems to be a possibility, we can increase the angle of contact θ, change the material to increase the friction, or use a larger belt section. For an all-leather belt, a conservative permissible stress is 1.75 MPa (250 psi). Coefficients of friction of leather, rubber, and cotton-canvas on steel are 0.40, 0.35, and 0.30, respectively. Others may vary from 0.30 to as high as 0.70 in exceptional cases.

17-3   V BELTS

In many places in this book we have learned that scientific theory does not always explain experimental results.   In some cases the correlation is close; in other cases the two sets of results differ so much from each other that purely empirical methods must be used.   Nowhere is this more evident than in the study of the performance of V belts.   While future investigations may resolve this problem, for now we must be content to use experimental results to design and analyze power-transmission systems using V belts.

The cross sections and lengths of V belts have been standardized by ANSI in U.S. customary units and in SI units.   Automotive belts are a special category, standardized by ANSI and SAE in both systems too.   The appropriate designations for these standards are shown in Table 17-2.

In this book we shall devote our attention only to *heavy-duty conventional V-belt drives* in order to save space.   The details of this standard for both English and SI series are shown in Table 17-3.   We note that manufacturers will also supply a *heavy-duty narrow belt*, called the V series and a *light-duty belt*, called the L series. Narrow V belts have many advantages in certain applications where space is limited.   They operate on small sheave diameters, permit higher speed, more efficient drive motors, and fewer belts, and cost less than the heavy-duty conventional V belt.   Erickson* gives an example in which five narrow belts were used to replace eight conventional belts.   This resulted in a 33 percent cost saving, if the costs of the smaller sheaves are included.

V belts are specified by combining the section identifications (see Table 17-2) with the belt lengths.   For example, a heavy-duty conventional D173 belt has a D section and an *inside length* of 173 in.   Standard lengths for this type are given in Table 17-7.   The pitch length of this belt is obtained by adding a conversion quantity, shown in Table 17-4, to the inside belt length.   Thus a D173 belt has a pitch length of $L_p = 173 + 3.3 = 176.3$ in.   Calculations of velocity ratios are based upon the pitch or effective sheave diameters.   For this reason, sheave diameters are understood to be pitch diameters unless otherwise specified.

The lengths of metric belts are standardized as the *pitch, or effective, length.* See Table 17-8 for the standard lengths.   A heavy-duty SI conventional belt designated as 16C1700 has a 16C section and an effective length of 1700 mm.

The automotive inch-type belts in Table 17-2 are available in lengths up to 80 in in 0.5-in increments, and in 1-in increments from 80 to 100 in.   The automotive SI belts are available to 2032 mm in 12.7-mm increments, and in 254-mm increments from 2032 to 2540 mm.

The groove angle of a sheave is made somewhat less than the belt-section angle.   This causes the belt to wedge itself into the groove, thus increasing fric-

---

* W. Erickson, "Cutting Drive Size and Cost with Narrow V Belts," *Machine Design, vol. 25,* January 1979, pp. 116–119.

Table 17-2 STANDARD DESIGNATIONS FOR VARIOUS V BELTS

Type	Section	Minimum sheave diameter*	Standard†
Heavy-duty conventional	A	3.0 in	ANSI/RMA-IP-20-1977
	B	5.4 in	
	C	9.0 in	
	D	13.0 in	
	E	21.0 in	
Heavy-duty SI conventional	13C	80 mm	ANSI/RMA-IP-20-1977
	16C	140 mm	
	22C	224 mm	
	32C	355 mm	
Heavy-duty narrow	3V	2.65 in	RMA-IP-22
	5V	7.1 in	
	8V	12.3 in	
Notched narrow	3VX	2.2 in	
	5VX	4.4 in	
Light-duty	2L	0.8 in	RMA-IP-23
	3L	1.5 in	
	4L	2.5 in	
	5L	3.5 in	
Automotive inch	0.25	2.25 in	ANSI/SAE J636C
	0.315	2.25 in	
	0.380	2.40 in	
	0.440	2.75 in	
	0.500	3.00 in	
	$\frac{11}{16}$	3.00 in	
	$\frac{3}{4}$	3.00 in	
	$\frac{7}{8}$	3.50 in	
	1.0	4.00 in	
Automotive SI	6A	57 mm	ANSI/SAE J636C
	8A	57 mm	
	10A	61 mm	
	11A	70 mm	
	13A	76 mm	
	15A	76 mm	
	17A	76 mm	
	20A	89 mm	
	23A	102 mm	

* Sheaves may be available smaller than this, but their use may shorten the belt life.

† Other standards cover double V belts, synchronous belts, variable-speed belts, and V-ribbed belts as RMA-IP-21, 24, 25, and 26, respectively.

Table 17-3 HEAVY-DUTY CONVENTIONAL V-BELT SECTIONS

Belt designation	Width, a in (mm)	Single belt thickness, b in (mm)	Joined multiple belt thickness, b' in (mm)	Power range per belt hp (kW)	Typical standard sheave sizes in (mm)
Inch series					
A	0.50	0.31	0.41	0.2–5	2.6 up by 0.2 increments
B	0.66	0.41	0.50	0.7–10	4.6 up by 0.2 increments
C	0.88	0.53	0.66	1–21	7.0 up by 0.5 increments
D	1.25	0.75	0.84	2–50	12.0 up by 0.5 increments
E	1.50	0.91	1.03	4–80	18.0 up by 1.0 increments
SI series					
13C	(13)	(8)	(10)	(0.1–3.6)	(65 up by 5 increments)
16C	(16)	(10)	(13)	(0.5–72)	(115 up by 5 increments)
22C	(22)	(13)	(17)	(0.7–15.0)	(180 up by 10 increments)
32C	(32)	(19)	(21)	(1.3–39.0)	(300 up by 20 increments)

Source: Partially compiled from ANSI/RMA-IP-20-1977.

Table 17-4 LENGTH CONVERSION QUAN-
TITIES FOR HEAVY-DUTY CON-
VENTIONAL INCH-SERIES BELTS

Belt section	Size range in	Conversion quantity* in
A	26 to 128	1.3
B	35 to 240	1.8
B	240 up	2.1
C	51 to 210	2.9
C	210 up	3.8
D	120 to 210	3.3
D	210 up	4.1
E	180 to 240	4.5
E	240 up	5.5

* The conversion quantities shown are in inches and are to be added to the inside circumference to get the pitch length.

Source: Compiled from ANSI/RMA-IP-20-1977.

tion. The exact value of this angle depends upon the belt section, the sheave diameter, and the angle of the contact. If it is made too much smaller than the belt angle, the force required to pull the belt out of the groove as the belt leaves the pulley will be excessive. Optimum values are given in commercial literature.

The design of a V-belt drive is similar to that of a flat-belt drive, both of them requiring the same initial information. Moreover, flat-to-V-belt drives can be designed. Information concerning belt-ratings corrections for V-to-flat-belt drives will be given later.

The minimum sheave diameters have been listed in Table 17-2. Table 17-3 gives the expected power capacity range per belt. This table will aid in your selection of an applicable cross section for your particular design.

For best results, a V belt should be run quite fast; 4000 fpm is a good speed. Trouble may be encountered if the belt runs much faster than 5000 fpm or much slower than 1000 fpm. Therefore, when possible, the sheaves should be sized for a belt speed in the neighborhood of 4000 fpm (20m/s).

The length of a V belt is obtained in a manner quite similar to that used for finding the length of a flat belt.

$$L_p = 2C + 1.57(D + d) + \frac{(D - d)^2}{4C}$$

$$(17\text{-}8)$$

Table 17-5 CONSTANTS FOR USE IN THE POWER-
RATING EQUATION

Belt section	C_1	C_2	C_3	C_4
A	0.8542	1.342	$2.436(10)^{-4}$	0.1703
B	1.506	3.520	$4.193(10)^{-4}$	0.2931
C	2.786	9.788	$7.460(10)^{-4}$	0.5214
D	5.922	34.72	$1.522(10)^{-3}$	1.064
E	8.642	66.32	$2.192(10)^{-3}$	1.532
13C	$3.316(10)^{-2}$	1.088	$1.161(10)^{-8}$	$5.238(10)^{-3}$
16C	$5.185(10)^{-2}$	2.273	$1.759(10)^{-8}$	$7.934(10)^{-3}$
22C	$1.002(10)^{-1}$	7.040	$3.326(10)^{-8}$	$1.500(10)^{-2}$
32C	$2.205(10)^{-1}$	26.62	$7.037(10)^{-8}$	$3.174(10)^{-2}$

where C = center distance
D = pitch diameter of large sheave
d = pitch diameter of small sheave
L_p = pitch length of belt (same as effective length)

The contact-angle calculations are the same as those used for flat belts (see Fig. 17-3). We have noted that flat belts have virtually no limit to their center distances. Long center distances, however, are not recommended for V belts because the excessive vibration of the slack side of the belt will shorten the belt life. In general, the center distance should not be greater than three times the sum of the sheave diameters nor less than the diameter of the larger sheave. Link-type V belts have less vibration; hence they may have larger center distances.

The selection of V belts is based upon obtaining a long and trouble-free life. Accordingly, the standards* provide a method for rating V belts for satisfactory performance under most conditions. The method can be summarized by the use of a *power-rating equation.* In both U.S. customary units and in SI units, this highly empirical equation is

$$H_r = \left[C_1 - \frac{C_2}{d} - C_3(rd)^2 - C_4 \log (rd) \right](rd) + C_2 r \left(1 - \frac{1}{K_A} \right) \qquad (17\text{-}9)$$

The constants C_1 through C_4 depend upon the belt section and are tabulated in Table 17-5 for both systems of units. The terms common to both systems are

r = rpm of high-speed shaft, divided by 1000
K_A = speed-ratio factor (see Table 17-6)

---

* ANSI/RMA-IP-1977 standard.

Table 17-6 SPEED-RATIO FACTORS FOR USE IN THE POWER-RATING EQUATION

D/d range	K_A
1.00 to 1.01	1.0000
1.02 to 1.04	1.0112
1.05 to 1.07	1.0226
1.08 to 1.10	1.0344
1.11 to 1.14	1.0463
1.15 to 1.20	1.0586
1.21 to 1.27	1.0711
1.28 to 1.39	1.0840
1.40 to 1.64	1.0972
over 1.64	1.1106

When U.S. customary units are used, the remaining terms in Eq. (17-9) are

H_r = rated horsepower
d = pitch diameter of small sheave, in

When SI units are used, the terms are

H_r = rated power, kW
d = pitch diameter of small sheave, mm

We next note that the power ratings obtained by Eq. (17-9) are for an arc of contact of 180° and an average belt length. The correction factor K_1 corrects the power rating for angle of contact and is obtained from Fig. 17-4. The length-

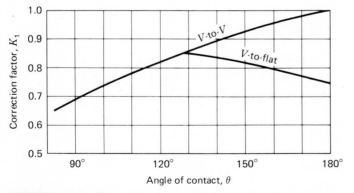

FIGURE 17-4 Correction factors K_1 for angles of contact up to 180°. The V-to-flat refers to a V-belt running onto a flat pulley surface. (*Compiled from ANSI/RMA-IP-20-1977 standard.*)

Table 17-7 STANDARD LENGTHS L_s AND LENGTH-CORRECTION FACTORS K_2 FOR HEAVY-DUTY CONVENTIONAL ENGLISH V BELTS*

L_s	A	B	C	D	L_s	B	C	D	E
26	0.78				144	1.10	1.00	0.91	
31	0.82				158	1.12	1.02	0.93	
35	0.85	0.80			173	1.14	1.04	0.94	
38	0.87	0.82			180	1.15	1.05	0.95	0.92
42	0.89	0.84			195	1.17	1.06	0.96	0.93
46	0.91	0.86			210	1.18	1.07	0.98	0.95
51	0.93	0.88	0.80		240	1.22	1.10	1.00	0.97
55	0.95	0.89			270	1.24	1.13	1.02	0.99
60	0.97	0.91	0.83		300	1.27	1.15	1.04	1.01
68	1.00	0.94	0.85		330		1.17	1.06	1.03
75	1.02	0.96	0.87		360		1.18	1.07	1.04
80	1.04				390		1.20	1.09	1.06
81		0.98	0.89		420		1.21	1.10	1.07
85	1.05	0.99	0.90		480			1'13	1.09
90	1.07	1.00	0.91		540			1.15	1.11
96	1.08		0.92		600			1.17	1.13
97		1.02			660			1.18	1.15
105	1.10	1.03	0.94						
112	1.12	1.05	0.95						
120	1.13	1.06	0.96	0.88					
128	1.15	1.08	0.98	0.89					

* Length designations correspond to the inside circumferences.

Source: From ANSI/RMA-IP-20-1977 standard.

correction factor is K_2 and is obtained from Tables 17-7 and 17-8. These two factors are used in the equation

$$H'_r = K_1 K_2 H_r \tag{17-10}$$

to obtain the corrected power rating H'_r.

Moreover, the characteristics of both the driving and the driven machinery must be considered in the selection of the belt or belts. If, for example, the load is started frequently by a source of power which develops 200 percent of full-load torque in starting, such as a squirrel-cage motor, then the power required of the belt drive must be multiplied by an overload service factor. The characteristics of the driven machinery must be considered in the same manner. The overload service factors listed in Table 17-9 may be used as a rough guide in selecting V belts. In general, the lowest factor in the suggested range should be used for intermittent service, the highest factor for continuous service, and the average for normal 8-to-10-hour-per-day service. Other factors can be obtained from the manufacturers' bulletins.

Table 17-8 STANDARD PITCH LENGTHS L_p AND
LENGTH-CORRECTION FACTORS K_2
FOR HEAVY-DUTY CONVENTIONAL
SI BELTS

13C		16C		22C		32C	
L_p	K_2	L_p	K_2	L_p	K_2	L_p	K_2
710	0.83	960	0.81	1 400	0.83	3 190	0.89
750	0.84	1 040	0.83	1 500	0.85	3 390	0.90
800	0.86	1 090	0.84	1 630	0.86	3 800	0.92
850	0.88	1 120	0.85	1 830	0.89	4 160	0.94
900	0.89	1 190	0.86	1 900	0.90	4 250	0.94
950	0.90	1 250	0.87	2 000	0.91	4 540	0.95
1 000	0.92	1 320	0.88	2 160	0.92	4 720	0.96
1 075	0.93	1 400	0.90	2 260	0.93	5 100	0.98
1 120	0.94	1 500	0.91	2 390	0.94	5 480	0.99
1 150	0.95	1 600	0.92	2 540	0.96	5 800	1.00
1 230	0.97	1 700	0.94	2 650	0.96	6 180	1.01
1 300	0.98	1 800	0.95	2 800	0.98	6 560	1.02
1 400	1.00	1 900	0.96	3 030	0.99	6 940	1.03
1 500	1.02	1 980	0.97	3 150	1.00	7 330	1.04
1 585	1.03	2 110	0.99	3 350	1.01	8 090	1.06
1 710	1.05	2 240	1.00	3 550	1.02	8 470	1.07
1 790	1.06	2 360	1.01	3 760	1.04	8 850	1.08
1 865	1.07	2 500	1.02	4 120	1.06	9 240	1.09
1 965	1.08	2 620	1.03	4 220	1.06	10 000	1.10
2 120	1.10	2 820	1.05	4 500	1.07	10 760	1.11
2 220	1.11	2 920	1.06	4 680	1.08	11 530	1.13
2 350	1.13	3 130	1.07	5 060	1.10	12 290	1.14
2 500	1.14	3 330	1.09	5 440	1.11		
2 600	1.15	3 530	1.10	5 770	1.13		
2 730	1.17	3 740	1.11	6 150	1.14		
2 910	1.18	4 090	1.13	6 540	1.15		
3 110	1.20	4 200	1.14	6 920	1.16		
3 310	1.21	4 480	1.15	7 300	1.17		
		4 650	1.16	7 680	1.18		
		5 040	1.18	8 060	1.19		
		5 300	1.19	8 440	1.20		
		5 760	1.21	8 820	1.21		
		6 140	1.23	9 200	1.22		
		6 520	1.24				
		6 910	1.25				
		7 290	1.26				
		7 670	1.27				

Source: From ANSI/RMA-IP-20-1977 standard.

Table 17-9   SUGGESTED SERVICE FACTORS K_S FOR
                 V-BELT DRIVES

| | Source of power | |
Driven machinery	Normal torque characteristic	High or nonuniform torque
Uniform	1.0 to 1.2	1.1 to 1.3
Light shock	1.1 to 1.3	1.2 to 1.4
Medium shock	1.2 to 1.4	1.4 to 1.6
Heavy shock	1.3 to 1.5	1.5 to 1.8

Source: Derived from ANSI/RMA-IP-20-1977 standard.

Example 17-1   A 10-hp split-phase motor running at 1750 rpm is to be used to belt-drive a centrifugal pump which operates 24 h per day.   The pump should run at approximately 1175 rpm.   The center distance should not exceed 44 in. Space limits the diameter of the driven sheave to 11.5 in.   Determine the sheave diameters, the belt size, and the number of belts to be used.

Solution   The following design decisions must be made first:

1   Using Table 17-9 we select a service factor of 1.2.
2   From Table 17-2 a B-section belt is selected so that an excessive number of belts can be avoided.
3   Since the driven sheave should not exceed 11.5 in, the smaller size, 11 in, will be tentatively selected.
4   A center distance of 42 in is tentatively selected, which is less than three times the sum of the large and small sheave diameters.

Multiplying the given horsepower by the service factor gives the design horsepower as $1.2(10) = 12$ hp.   The diameter of the small sheave is

$$d = D\left(\frac{n_1}{n_2}\right) = 11\left(\frac{1175}{1750}\right) = 7.4 \text{ in} \qquad\qquad Ans.$$

This is a standard pitch diameter of a B-section belt and it is over the minimum diameter listed in Table 17-2.

From Fig. 17-3a, the small angle of contact is

$$\theta_S = 2 \cos^{-1}\left(\frac{D - d}{2C}\right) = 2 \cos^{-1}\left[\frac{11 - 7.4}{2(42)}\right] = 3.056 \text{ rad } (175°)$$

Equation (17-8) gives the pitch length as

$$L_P = 2C + 1.57(D + d) + \frac{(D - d)^2}{4C}$$

$$= 2(42) + 1.57(11 + 7.4) + \frac{(11 - 7.4)^2}{4(42)} = 112.97 \text{ in}$$

The inside circumference is calculated using Table 17-4.

$$L = L_P - 1.8 = 112.97 - 1.8 = 111.2 \text{ in}$$

The nearest standard size, from Table 17-7, is a B112. Table 17-4 gives the length conversion quantity as 1.8 in; so the pitch length is $L_P = 112 + 1.8 = 113.8$ in. In order to use Eq. (17-9) we find C_1 through C_4 from Table 17-5. And $K_A = 1.0972$ from Table 17-6 with $D/d = 1.49$. Also $r = 1750/1000 = 1.75$ krpm. Equation (17-9) then gives

$$H_r = \left[C_1 - \frac{C_2}{d} - C_3(rd)^2 - C_4 \log (rd) \right](rd) + C_2 r \left(1 - \frac{1}{K_A} \right)$$

$$= \left\{ 1.506 - \frac{3.520}{7.4} - 4.193(10)^{-4}[1.75(7.4)]^2 \right.$$

$$\left. - 0.2931 \log [1.75(7.4)] \right\}(1.75)(7.4) + 3.520(1.75)\left(1 - \frac{1}{1.0972} \right)$$

$$= 8.76 \text{ hp}$$

This rating is based on an arc of contact of 180° and an average belt length; so it must be corrected using Eq. (17-10). Entering Fig. 17-4 with $\theta_S = 175°$ gives $K_1 = 0.99$; and $K_2 = 1.05$ from Table 17-7. Thus Eq. (17-10) gives

$$H_r' = K_1 K_2 H_r = (0.99)(1.05)(8.76) = 9.1 \text{ hp/belt}$$

Since the design horsepower is 12, the number of belts needed is $12/9.1 = 1.32$. We therefore select two B112 belts for this application. ////

Efficiency

The selection of a belt-drive element may sometimes hinge on its efficiency. In general, the efficiency of V belts ranges from 70 to 95 percent. Well-selected belts have efficiencies of 90 to 95 percent.*

* Wallin, op. cit.

The torque load has the largest effect on the efficiency. The lower the load, the lower the efficiency. Excessively high loads also lower efficiency. At about 20 percent of design torque, a particular V belt had an 80 percent efficiency, while at the design torque the efficiency was 91 percent. Next, efficiency drops with smaller pulley size. A drop of the sheave diameter from 5.6 in to 2.6 in on the V-belt system cited above caused an efficiency drop from 95 percent to 91 percent.

17-4 ROLLER CHAIN

Basic features of chain drives include a constant ratio, since no slippage or creep is involved; long life; and the ability to drive a number of shafts from a single source of power.

Roller chains have been standardized as to sizes by the ANSI. Figure 17-5 shows the nomenclature. The pitch is the linear distance between the centers of the rollers. The width is the space between the inner link plates. These chains are manufactured in single, double, triple, and quadruple strands. The dimensions of standard sizes are listed in Table 17-10.

In Fig. 17-6 is shown a sprocket driving a chain in a counterclockwise direction. Denoting the chain pitch by p, the pitch angle by γ, and the pitch diameter of the sprocket by D, from the trigonometry of the figure we see

$$\sin \frac{\gamma}{2} = \frac{p/2}{D/2} \qquad \text{or} \qquad D = \frac{p}{\sin (\gamma/2)} \tag{a}$$

Since $\gamma = 360/N$, where N is the number of sprocket teeth, Eq. (a) can be written

$$D = \frac{p}{\sin (180/N)} \tag{17-11}$$

The angle $\gamma/2$, through which the link swings as it enters contact, is called the *angle of articulation*. It can be seen that the magnitude of this angle is a function of

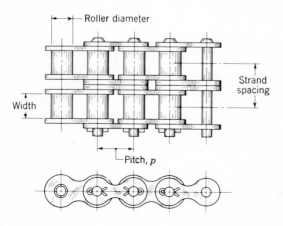

FIGURE 17-5 Portion of a double-strand roller chain.

Table 17-10 DIMENSIONS OF AMERICAN STANDARD ROLLER CHAIN-SINGLE STRAND

ANSI chain number	Pitch in (mm)	Width in (mm)	Minimum tensile strength lb (N)	Average weight lb/ft (N/m)	Roller diameter in (mm)	Multiple strand spacing in (mm)
25	0.250 (6.35)	0.125 (3.18)	780 (3 470)	0.09 (1.31)	0.130 (3.30)	0.252 (6.40)
35	0.375 (9.52)	0.188 (4.76)	1 760 (7 830)	0.21 (3.06)	0.200 (5.08)	0.399 (10.13)
41	0.500 (12.70)	0.25 (6.35)	1 500 (6 670)	0.25 (3.65)	0.306 (7.77)	··· ···
40	0.500 (12.70)	0.312 (7.94)	3 130 (13 920)	0.42 (6.13)	0.312 (7.92)	0.566 (14.38)
50	0.625 (15.88)	0.375 (9.52)	4 880 (21 700)	0.69 (10.1)	0.400 (10.16)	0.713 (18.11)
60	0.750 (19.05)	0.500 (12.7)	7 030 (31 300)	1.00 (14.6)	0.469 (11.91)	0.897 (22.78)
80	1.000 (25.40)	0.625 (15.88)	12 500 (55 600)	1.71 (25.0)	0.625 (15.87)	1.153 (29.29)
100	1.250 (31.75)	0.750 (19.05)	19 500 (86 700)	2.58 (37.7)	0.750 (19.05)	1.409 (35.76)
120	1.500 (38.10)	1.000 (25.40)	28 000 (124 500)	3.87 (56.5)	0.875 (22.22)	1.789 (45.44)
140	1.750 (44.45)	1.000 (25.40)	38 000 (169 000)	4.95 (72.2)	1.000 (25.40)	1.924 (48.87)
160	2.000 (50.80)	1.250 (31.75)	50 000 (222 000)	6.61 (96.5)	1.125 (28.57)	2.305 (58.55)
180	2.250 (57.15)	1.406 (35.71)	63 000 (280 000)	9.06 (132.2)	1.406 (35.71)	2.592 (65.84)
200	2.500 (63.50)	1.500 (38.10)	78 000 (347 000)	10.96 (159.9)	1.562 (39.67)	2.817 (71.55)
240	3.00 (76.70)	1.875 (47.63)	112 000 (498 000)	16.4 (239)	1.875 (47.62)	3.458 (87.83)

Source: Compiled from ANSI B29.1-1975.

the number of teeth. Rotation of the link through this angle causes impact between the rollers and the sprocket teeth and also wear in the chain joint. Since the life of a properly selected drive is a function of the wear and the surface fatigue strength of the rollers, it is important to reduce the angle of articulation as much as possible. The values of this angle have been plotted as a function of the number of teeth in Fig. 17-7a.

The number of sprocket teeth also affects the velocity ratio during the rota-

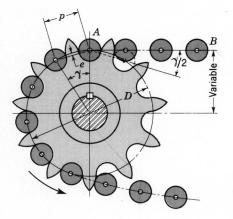

FIGURE 17-6 Engagement
of a chain and sprocket.

tion through the pitch angle γ. At the position shown in Fig. 17-6 the chain AB is
tangent to the pitch circle of the sprocket. However, when the sprocket has
turned an angle $\gamma/2$, the chain line AB moves closer to the center of rotation of the
sprocket. This means that the chain line AB is moving up and down, that the
lever arm varies, and that the chain velocity varies with rotation through the pitch
angle. Using the geometry of Fig. 17-6, a measure of this velocity change in
percent is

$$\frac{\Delta V}{V} = 100\left[\sec\left(\frac{180}{N}\right) - 1\right]\left(\cos\frac{180}{N}\right) \qquad\qquad (b)$$

This is called the *chordal speed variation* and is plotted in Fig. 17-7b. When chain
drives are used to synchronize precision components or processes, due consideration must be given to these variations. For example, if a chain drive synchronized the cutting of photographic film with the forward drive of the film, the
lengths of the cut sheets of film might vary too much because of this chordal speed
variation. Such variations can also cause vibrations within the system.

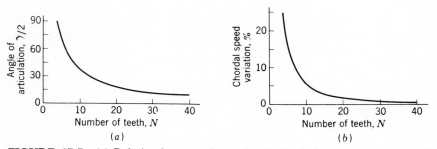

FIGURE 17-7 (a) Relation between the angle of articulation and the number of
sprocket teeth; (b) relation between the number of sprocket teeth and the chordal
speed variation.

The chain velocity is usually defined as the number of feet coming off the sprocket in unit time. Thus the velocity V in feet per minute is

$$V = \frac{\pi Dn}{12} = \frac{Npn}{12} \tag{17-12}$$

where N = number of teeth on sprocket
p = chain pitch
n = speed of sprocket, rpm

Although a large number of teeth is considered desirable for the driving sprocket, in the usual case it is advantageous to obtain as small a sprocket as possible, and this requires one with a small number of teeth. For smooth operation at moderate and high speeds it is considered good practice to use a driving sprocket with at least 17 teeth; 19 or 21 will, of course, give a better life expectancy with less chain noise. Where space limitations are severe or for very slow speeds, smaller tooth numbers may be used by sacrificing the life expectancy of the chain.

Driven sprockets are not made in standard sizes over 120 teeth because the pitch elongation will eventually cause the chain to "ride" high long before the chain is worn out. The most successful drives have velocity ratios up to 6 : 1, but higher ratios may be used at the sacrifice of chain life.

Roller chains seldom fail because they lack tensile strength; they more often fail because they have been subjected to a great many hours of service. Actual failure may be due either to wear of the rollers on the pins or to fatigue of the surfaces of the rollers. Roller-chain manufacturers have compiled tables which give the horsepower capacity corresponding to a life expectancy of 15 kh for various sprocket speeds. These capacities are tabulated in Table 17-11 for 17-tooth sprockets.

The characteristics of the load are important considerations in the selection of roller chain. In general, extra chain capacity is required for any of the following conditions:

- The small sprocket has less than 9 teeth for low-speed drives or less than 16 teeth for high-speed drives.
- The sprockets are unusually large.
- Shock loading occurs, or there are frequent load reversals.
- There are three or more sprockets in the drive.
- The lubrication is poor.
- The chain must operate under dirty or dusty conditions.

To account for these and other conditions of operation the ratings in Table 17-11 must be modified by two factors to get the corrected value for a single- or

Table 17-11 RATED HORSEPOWER CAPACITY OF SINGLE-STRAND SINGLE-PITCH ROLLER CHAIN FOR A 17-TOOTH SPROCKET

Sprocket speed rpm	ANSI chain number					
	25	35	40	41	50	60
50	0.05	0.16	0.37	0.20	0.72	1.24
100	0.09	0.29	0.69	0.38	1.34	2.31
150	0.13*	0.41*	0.99*	0.55*	1.92*	3.32
200	0.16*	0.54*	1.29	0.71	2.50	4.30
300	0.23	0.78	1.85	1.02	3.61	6.20
400	0.30*	1.01*	2.40	1.32	4.67	8.03
500	0.37	1.24	2.93	1.61	5.71	9.81
600	0.44*	1.46*	3.45*	1.90*	6.72*	11.6
700	0.50	1.68	3.97	2.18	7.73	13.3
800	0.56*	1.89*	4.48*	2.46*	8.71*	15.0
900	0.62	2.10	4.98	2.74	9.69	16.7
1000	0.68*	2.31*	5.48	3.01	10.7	18.3
1200	0.81	2.73	6.45	3.29	12.6	21.6
1400	0.93*	3.13*	7.41	2.61	14.4	18.1
1600	1.05*	3.53*	8.36	2.14	12.8	14.8
1800	1.16	3.93	8.96	1.79	10.7	12.4
2000	1.27*	4.32*	7.72*	1.52*	9.23*	10.6
2500	1.56	5.28	5.51*	1.10*	6.58*	7.57
3000	1.84	5.64	4.17	0.83	4.98	5.76

Type A   Type B   Type C

multiple-strand chain. These are

- The tooth correction factor K_1, which accounts for the fact that the driving sprocket may have more or less than 17 teeth (use Table 17-12)
- The multiple-strand factor K_2, which accounts for the fact that the rating is not linearly related to the number of strands (see Table 17-13)

The corrected horsepower is obtained by applying these two factors to the rated horsepower from Table 17-11 as follows:

$$H'_r = K_1 K_2 H_r \tag{17-13}$$

where H'_r is the fully corrected rating.

The service factor K_S in Table 17-14 is used to account for variations in the driving and driven sources. Multiply the computed or given horsepower by K_S to get the design horsepower.

The length of a chain should be determined in pitches. It is preferable to

Table 17-11 RATED HORSEPOWER CAPACITY OF SINGLESTRAND SINGLEPITCH ROLLER CHAIN FOR A 17-TOOTH SPROCKET. (CONCLUDED)

Sprocket speed rpm	ANSI chain number							
	80	100	120	140	160	180	200	240
50	2.88	5.52	9.33	14.4	20.9	28.9	38.4	61.8
100	5.38	10.3	17.4	26.9	39.1	54.0	71.6	115.
150	7.75	14.8	25.1	38.8	56.3	77.7	103.	166.
200	10.0	19.2	32.5	50.3	72.9	101.	134.	215.
300	14.5	27.7	46.8	72.4	105.	145.	193.	310.
400	18.7	35.9	60.6	93.8	136.	188.	249.	359.
500	22.9	43.9	74.1	115.	166.	204.	222.	0
600	27.0	51.7	87.3	127.	141.	155.	169.	
700	31.0	59.4	89.0	101.	112.	123.	0	
800	35.0	63.0	72.8	82.4	91.7	101.		
900	39.9	52.8	61.0	69.1	76.8	84.4		
1000	37.7	45.0	52.1	59.0	65.6	72.1		
1200	28.7	34.3	39.6	44.9	49.9	0		
1400	22.7	27.2	31.5	35.6	0			
1600	18.6	22.3	25.8	0				
1800	15.6	18.7	21.6					
2000	13.3	15.9	0					
2500	9.56	0.40						
3000	7.25	0						

Type B

Type C Type C′

* Estimated from ANSI tables by linear interpolation.

Note: Type A—Manual or drip lubrication; Type B—Bath or disc lubrication; Type C—Oil-stream lubrication; Type C′—Type C, but this is a galling region; submit design to manufacturer for evaluation.

Source: Compiled from ANSI B29.1-1975 information only section, and from B29.9-1958.

have an even number of pitches; otherwise an offset link is required. The approximate length may be obtained from the following equation:

$$\frac{L}{p} = \frac{2C}{p} + \frac{N_1 + N_2}{2} + \frac{(N_2 - N_1)^2}{4\pi^2(C/p)} \tag{17-14}$$

where L = chain length
p = chain pitch
C = center distance
N_1 = number of teeth on small sprocket
N_2 = number of teeth on large sprocket

Table 17-12 TOOTH CORRECTION FACTORS

Number of teeth on driving sprocket	Tooth correction factor K_1	Number of teeth on driving sprocket	Tooth correction factor K_1
11	0.53	22	1.29
12	0.62	23	1.35
13	0.70	24	1.41
14	0.78	25	1.46
15	0.85	30	1.73
16	0.92	35	1.95
17	1.00	40	2.15
18	1.05	45	2.37
19	1.11	50	2.51
20	1.18	55	2.66
21	1.26	60	2.80

Table 17-13 MULTIPLE-STRAND
FACTORS K_2

Number of strands	K_2
1	1.0
2	1.7
3	2.5
4	3.3

The length of chain for a multiple-sprocket drive is most easily obtained by making an accurate scale layout and determining the length by measurement.

Lubrication of roller chains is essential in order to obtain a long and trouble-free life. Either a drip feed or a shallow bath in the lubricant is satisfactory. A medium or light mineral oil, without additives, should be used. Except for unusual conditions, heavy oils and greases are not recommended because they are too viscous to enter the small clearances in the chain parts.

Example 17-2 A $7\frac{1}{2}$-hp speed reducer which runs at 300 rpm is to drive a conveyor at 200 rpm. The center distance is to be approximately 28 in. Select a suitable chain drive.

Solution Although an odd number of sprocket teeth is preferred, sprockets of 20 and 30 teeth are tentatively chosen in order to obtain the proper velocity ratio. A 20-tooth sprocket will have a longer life and generate less noise than a 16- or an 18-tooth sprocket. It is chosen because space does not seem to be at a premium.

From Table 17-14 a service factor of $K_S = 1.3$ is chosen for operation with moderate shock. Thus, the design horsepower is

$$H = 1.3(7.5) = 9.75 \text{ hp}$$

Table 17-14 LOAD SERVICE FACTORS K_S

Driven machinery	Driving source		
	Internal combustion engine with hydraulic drive	Electric motor or turbine	Internal combustion engine with mechanical drive
Smooth	1.00	1.00	1.2
Moderate shock	1.2	1.3	1.4
Heavy shock	1.4	1.5	1.7

Examination of Table 17-11 indicates that multiple strands of a No. 50 or 60 chain may be satisfactory. For a triple-strand No. 50, $H_r = 3.61$ hp. Tables 17-12 and 17-13 give $K_1 = 1.18$ and $K_2 = 2.5$. Thus, from Eq. (17-13)

$$H'_r = K_1 K_2 H_r = 1.18(2.5)(3.61) = 10.65 \text{ hp}$$

which is quite satisfactory. This is designated as a 50-3 chain.

For a 60-2 chain, we find $H_r = 6.20$ hp and $K_2 = 1.7$. Therefore

$$H'_r = 1.18(1.7)(6.20) = 12.44 \text{ hp}$$

which is also satisfactory.

The No. 60 chain would require larger sprockets, and hence it would run at a higher velocity, generate more noise, and have a shorter life. The No. 50 seems to be the better choice and is selected for this example. A comparison of the prices of the sprockets and chain for both cases might, however, make the No. 60 a better solution.

From Table 17-10, the pitch of No. 50 chain is $\frac{5}{8}$ in. Using a center distance of 28 in in Eq. (17-14), the required length of triple-strand chain in pitches is

$$
\begin{aligned}
\frac{L}{p} &= \frac{2C}{p} + \frac{N_1 + N_2}{2} + \frac{(N_2 - N_1)^2}{4\pi^2(C/p)} \\
&= \frac{(2)(28)}{0.625} + \frac{20 + 30}{2} + \frac{(30 - 20)^2}{4\pi^2(28/0.625)} \\
&= 114.7 \text{ pitches}
\end{aligned}
$$

The nearest even number of pitches is 114, and this will be used. A slight adjustment in the center distance is required. Substituting $L/p = 114$ into Eq. (17-14) and solving for C gives approximately $27\frac{3}{4}$ in as the new center distance.

In general, the center distance should not exceed 80 pitches; 30 to 50 pitches is a better value. In this problem the center distance is $27.75/0.625 = 44.4$ pitches, which is satisfactory. ////

17-5 ROPE DRIVES

Rope drives consisting of manila, cotton, or Dacron rope on multiple-grooved pulleys may often be a most economical form of drive over long distances and for large amounts of power. Accurate alignment of the pulleys is not necessary because they are grooved. The velocity of the rope should be quite high; 5000 fpm is a good speed for greatest economy.

17-6 WIRE ROPE

Wire rope is made with two types of winding, as shown in Fig. 17-8. The *regular lay*, which is the accepted standard, has the wire twisted in one direction to form the strands, and the strands twisted in the opposite direction to form the rope. In the completed rope the visible wires are approximately parallel to the axis of the rope. Regular-lay ropes do not kink or untwist and are easy to handle.

Lang-lay ropes have the wires in the strand and the strands in the rope twisted in the same direction, and hence the outer wires run diagonally across the axis of the rope. Lang-lay ropes are more resistant to abrasive wear and failure due to fatigue than are regular-lay ropes, but they are more likely to kink and untwist.

Standard ropes are made with a hemp core which supports and lubricates the strands. When the rope is subjected to heat, either a steel center or a wire-strand center must be used.

Wire rope is designated as, for example, a $1\frac{1}{8}$-in 6 × 7 haulage rope. The first figure is the diameter of the rope (Fig. 17-8c). The second and third figures are the number of strands and the number of wires in each strand, respectively. Table 17-15 lists some of the various ropes which are available, together with their characteristics and properties. The area of the metal in standard hoisting and haulage ropes is $A_m = 0.38d.^2$

When a wire rope passes around a sheave, there is a certain amount of readjustment of the elements. Each of the wires and strands must slide on one another, and presumably some individual bending takes place. It is probable that in this complex action there exists some stress concentration. The stress in one of the wires of a rope passing around a sheave may be calculated as follows. From solid mechanics we have

$$M = \frac{EI}{r} \qquad \text{and} \qquad M = \frac{\sigma I}{c} \qquad\qquad (a)$$

where the quantities have their usual meaning. Eliminating M and solving for the stress gives

$$\sigma = \frac{Ec}{r} \qquad\qquad (b)$$

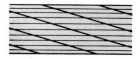

(a) Regular lay

(b) Lang lay

(c) Section of 6 x 7 rope

FIGURE 17-8 Types of wire rope; both lays are available in either right- or left-hand.

Table 17-15   WIRE-ROPE DATA

Rope	Weight per ft lb	Minimum sheave diameter in	Standard sizes d, in	Material	Size of outer wires	Modulus of elasticity* Mpsi	Strength† kpsi
6 × 7 haulage	$1.50d^2$	$42d$	$\frac{1}{4}$–$1\frac{1}{2}$	Monitor steel	$d/9$	14	100
				Plow steel	$d/9$	14	88
				Mild plow steel	$d/9$	14	76
6 × 19 standard hoisting	$1.60d^2$	$26d$–$34d$	$\frac{1}{4}$–$2\frac{3}{4}$	Monitor steel	$d/13$–$d/16$	12	106
				Plow steel	$d/13$–$d/16$	12	93
				Mild plow steel	$d/13$–$d/16$	12	80
6 × 37 special flexible	$1.55d^2$	$18d$	$\frac{1}{4}$–$3\frac{1}{2}$	Monitor steel	$d/22$	11	100
				Plow steel	$d/22$	11	88
8 × 19 extra flexible	$1.45d^2$	$21d$–$26d$	$\frac{1}{4}$–$1\frac{1}{2}$	Monitor steel	$d/15$–$d/19$	10	92
				Plow steel	$d/15$–$d/19$	10	80
7 × 7 aircraft	$1.70d^2$	$\cdots$	$\frac{1}{16}$ $\frac{3}{8}$	Corrosion-resistant steel	$\cdots$	$\cdots$	124
				Carbon steel	$\cdots$	$\cdots$	124
7 × 9 aircraft	$1.75d^2$	$\cdots$	$\frac{1}{8}$ $\frac{3}{8}$	Corrosion-resistant steel	$\cdots$	$\cdots$	135
				Carbon steel	$\cdots$	$\cdots$	143
19-wire aircraft	$2.15d^2$	$\cdots$	$\frac{1}{32}$ $\frac{5}{16}$	Corrosion-resistant steel	$\cdots$	$\cdots$	165
				Carbon steel	$\cdots$	$\cdots$	165

* The modulus of elasticity is only approximate; it is affected by the loads on the rope and, in general, increases with the life of the rope.

† The strength is based on the nominal area of the rope. The figures given are only approximate and are based on 1-in rope sizes and $\frac{1}{4}$-in aircraft-cable sizes.

Source: Compiled from *American Steel and Wire Company Handbook.*

For the radius of curvature r we can substitute the radius of the sheave $D/2$. Also $c = d_w/2$ where d_w is the diameter of the wire. This substitution gives

$$\sigma = E\,\frac{d_w}{D} \qquad\qquad (17\text{-}15)$$

Professor Charles R. Mischke of Iowa State University* has this to say about Eq. (17-15): "Now the individual wire makes a corkscrew figure in space and if you pull on it to determine E it will give more than its native E would suggest. Therefore E is still the modulus of elasticity of the *wire*, but in its peculiar configuration due to being part of the rope." A value for E equal to the modulus of elasticity of the *rope* gives an approximately correct value for the stress σ. For this reason we say that E in Eq. (17-15) is the modulus of elasticity of the rope, not the wire, recognizing that one can quibble over the name used.

Equation (17-15) gives the bending stress σ in the individual wires. The sheave diameter is represented by D. Equation (17-15) illustrates the importance

---

* By personal communication.

of using a large-diameter sheave. The suggested minimum sheave diameters in Table 17-15 are based on a D/d_w ratio of 400. If possible, the sheaves should be designed for a larger ratio. For elevators and mine hoists, D/d_w is usually taken from 800 to 1000. If the ratio is less than 200, heavy loads will often cause a permanent set in the rope.

A wire rope may fail because the static load exceeds the ultimate strength of the rope. Failure of this nature is generally not the fault of the designer, but rather that of the operator in permitting the rope to be subjected to loads for which it was not designed.

The first consideration in selecting a wire rope is to determine the static load. This load is composed of the following items:

- The known or dead weight
- Additional loads caused by sudden stops or starts
- Shock loads
- Sheave bearing friction

When these loads are summed, the total can be compared to the ultimate strength of the rope to find the factor of safety. However, the ultimate strength used in this determination must be reduced by the strength loss that occurs when the rope passes over a curved surface such as a stationary sheave or a pin; see Fig. 17-9.

For an average operation, use a factor of safety of 5. Factors of safety up to 8 or 9 are used if there is danger to human life and for very critical situations. Table 17-16 lists minimum factors of safety for a variety of design situations.

Having made a tentative selection of a rope based upon static strength, the next consideration is to assure that the wear life of the rope and the sheave or sheaves meets certain requirements. When a loaded rope is bent over a sheave,

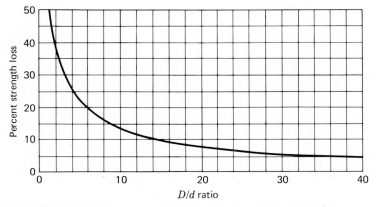

FIGURE 17-9 Percent strength loss due to different D/d ratios; derived from standard test data for 6 × 19 and 6 × 17 class ropes. (*From Wire Rope Users Manual; reproduced by permission from American Iron and Steel Institute.*)

Table 17-16   MINIMUM FACTORS OF SAFETY FOR WIRE
                  ROPE*

Track cables	3.2	Passenger elevators, fpm:	
Guys	3.5	50	7.60
Mine shafts, ft:		300	9.20
Up to 500	8.0	800	11.25
1000–2000	7.0	1200	11.80
2000–3000	6.0	1500	11.90
Over 3000	5.0	Freight elevators, fpm:	
Hoisting	5.0	50	6.65
Haulage	6.0	300	8.20
Cranes and derricks	6.0	800	10.00
Electric hoists	7.0	1200	10.50
Hand elevators	5.0	1500	10.55
Private elevators	7.5	Powered dumbwaiters, fpm:	
Hand dumbwaiter	4.5	50	4.8
Grain elevator	7.5	300	6.6
		500	8.0

* Use of these factors does not preclude a fatigue failure.

Source: Compiled from a variety of sources, including ANSI A17.1-1978.

the rope stretches like a spring, rubs against the sheave, and causes wear of both the rope and the sheave.   The amount of wear that occurs depends upon the pressure of the rope in the sheave groove.   This pressure is called *the bearing pressure*; a good estimate of its magnitude is given by

$$p = \frac{2F}{dD} \tag{17-16}$$

where F = tensile force on rope
      d = rope diameter
      D = sheave diameter

Note that if the rope makes a 180° bend, the area in Eq. (17-16) is only half the projected area.   The allowable pressures given in Table 17-17 are only to be used as a rough guide; they may not prevent a fatigue failure or severe wear.   They are presented here because they represent past practice and furnish a starting point in design.

A fatigue diagram not unlike an *S-N* diagram can be obtained for wire rope. Such a diagram is shown in Fig. 17-10.   Here the ordinate is the pressure-strength ratio p/S_u, and S_u is the ultimate tensile strength of the *wire*.   The abscissa is the number of bends which occur in the total life of the rope.   The curve implies that a wire rope has a fatigue limit; but this is not true at all.   A wire rope that is used over sheaves will eventually fail in fatigue or in wear, whichever comes first. However, the graph does show that the rope will have a long life if the ratio p/S_u is

Table 17-17   MAXIMUM ALLOWABLE BEARING PRESSURES OF
ROPES ON SHEAVES (in psi)

Rope	Wood*	Material Cast iron†	Cast steel‡	Chilled cast iron§	Manganese steel¶
Regular lay:					
6 × 7	150	300	550	650	1470
6 × 19	250	480	900	1100	2400
6 × 37	300	585	1075	1325	3000
8 × 19	350	680	1260	1550	3500
Lang lay:					
6 × 7	165	350	600	715	1650
6 × 19	275	550	1000	1210	2750
6 × 37	330	660	1180	1450	3300

* On end grain of beech, hickory, or gum.
† For H_B(min.) = 125.
‡ 30-40 carbon; H_B(min.) = 160.
§ Use only with uniform surface hardness.
¶ For high speeds with balanced sheaves having ground surfaces.
Source: Wire Rope Users Manual, AISI, 1979.

less than 0.001.   Substitution of this ratio in Eq. (17-16) gives

$$S_u = \frac{2F}{dD} \qquad\qquad (17\text{-}17)$$

where S_u is the ultimate strength of the *wire*, not the rope, and the units of S_u are
kpsi, *not* psi.   This interesting equation contains the wire strength, the load, the
rope diameter, and the sheave diameter—all four variables in a single equation!
Unfortunately the lack of data on the strength of the wire of which the rope is
composed makes the equation virtually useless for design purposes.   If the rope is

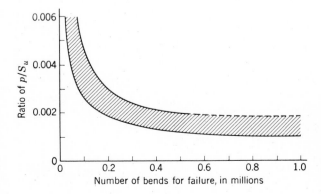

FIGURE 17-10 Experimentally determined relation between the fatigue life of wire rope and the sheave pressure; graph includes all rope sizes.

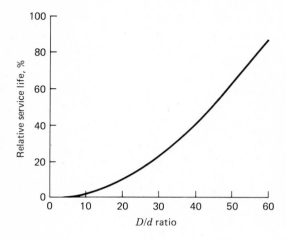

FIGURE 17-11 Service-life curve based on bending and tensile stresses only. This curve shows, for example, that the life corresponding to $D/d = 48$ is twice that of $D/d = 33$. (*From Wire Rope Users Manual; reproduced by permission from American Iron and Steel Institute.*)

made of plow steel, the wires are probably a hard-drawn AISI 1070 or 1080 carbon steel. Referring to Table 10-1 we see that this lies somewhere between hard-drawn spring wire and music wire. But the constants m and A needed to solve Eq. (10-11) for S_u are lacking.

Practicing engineers who desire to solve Eq. (17-17) should determine the wire strength S_u for the rope under consideration by unraveling enough wire to test for the Brinell hardness. Then S_u can be found using Eq. (4-16) or (4-17).

Figure 17-11 is another graph showing the gain in life to be obtained by using large D/d ratios. In view of the fact that the life of wire rope used over sheaves is only finite, it is extremely important that the designer specify and insist that periodic inspection, lubrication, and maintenance procedures be carried out during the life of the rope.

17-7 FLEXIBLE SHAFTS

One of the greatest limitations of the solid shaft is that it cannot transmit motion or power around corners. It is therefore necessary to resort to belts, chains, or gears, together with bearings and the supporting framework associated with them. The flexible shaft may often be an economical solution to the problem of transmitting motion around corners. In addition to the elimination of costly parts, its use may reduce noise considerably.

There are two main types of flexible shafts: the power-drive shaft for the transmission of power in a single direction, and the remote-control or manual-control shaft for the transmission of motion in either direction.

The construction of a flexible shaft is shown in Fig. 17-12. The cable is made by winding several layers of wire around a central core. For the power-drive shaft, rotation should be in a direction such that the outer layer is wound up. Remote-control cables have a different lay of the wires forming the cable, with more wires in each layer, so that the torsional deflection is approximately the same for either direction of rotation.

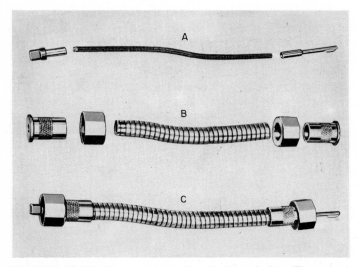

FIGURE 17-12 Construction of a flexible shaft. The wire-wound cable and end fittings are shown at *A*. These must be enclosed by the flexible casing and end fittings, shown at *B*. At *C* is shown the assembled flexible shaft. (*Courtesy of F. W. Stewart Corporation.*)

Flexible shafts are rated by specifying the torque corresponding to various radii of curvature of the casing. A 15-in radius of curvature, for example, will give from two to five times more torque capacity than a 7-in radius. When flexible shafts are used in a drive in which gears are also used, the gears should be placed so that the flexible shaft runs at as high a speed as possible. This permits the transmission of the maximum amount of horsepower.

PROBLEMS*

Section 17-2

17-1 A flat belt is 6 in wide, $\frac{9}{32}$ in thick, and transmits 15 hp. The pulley axes are parallel, in a horizontal plane, and 8 ft apart. The driving pulley is 6 in in diameter and rotates at 1750 rpm such that the loose side of the belt is on top. The driven pulley is 18 in in diameter. The weight of the belt material is 0.035 lb/in³.

    (*a*) Determine the tension in the tight and slack sides of the belt if the coefficient of friction is 0.30.

    (*b*) What belt tensions would result if adverse conditions caused the coefficient of friction to drop to 0.20? Would the belt slip?

    (*c*) Calculate the length of the belt.

---

* The asterisk indicates a design-type problem or one which may have no unique result.

17-2   A nylon-core flat belt has an elastomer envelope, is 200 mm wide, and transmits 60 kW at a belt speed of 25 m/s. The belt has a mass of 2 kg/m of belt length. The belt is used in a crossed configuration to connect a 300-mm-diameter driving pulley to a 900-mm-diameter driven pulley at a shaft spacing of 6 m.

   (*a*)   Calculate the belt length and the angles of wrap.

   (*b*)   Compute the belt tensions based on a coefficient of friction of 0.38.

17-3   A flat-belt drive consists of two 4-ft cast-iron pulleys spaced 16 ft apart. The tensile stress in the belt should not exceed 300 psi. Use a coefficient of friction of 0.33 and a unit belt weight of 0.032 lb/in^3. Find the width of a belt 0.36 in thick if 60 hp is to be transmitted at a pulley speed of 380 rpm.

17-4   A flat belt is 250 mm wide and 8 mm thick. The open belt connects a 400-mm-diameter cast-iron driving pulley to another cast-iron pulley 900 mm in diameter. The mass of the belt is 1.9 kg/m of belt length. The coefficient of friction of the drive is 0.28.

   (*a*)   Using a center distance of 5 m, a maximum tensile stress in the belt of 1400 kPa, and a belt velocity of 20 m/s, calculate the maximum power that can be transmitted.

   (*b*)   What are the shaft loads?

Section 17-3

17-5*   A single V belt is to be designed to deliver engine power to the wheel-drive transmission for a riding tractor. A 5-hp, single cylinder, internal-combustion engine is used. At most, 60 percent of this power is transmitted by the belt. The driving sheave is 6.2 in in diameter; the driven has a diameter of 12.0 in. The belt is to be as close to a 92-in pitch length as possible. The engine speed is governor-controlled to a maximum of 3100 rpm. An idler system is employed which causes the contact angles to be 180° and 135°.

   (*a*)   Select an adequate single belt.

   (*b*)   Are the sheaves large enough?

   (*c*)   What is the standard designation for the belt you selected?

17-6*   Rework Prob. 17-5 to take advantage of the fact that this tractor already uses a B90 belt elsewhere. A selection of a second B90 may result in overall cost savings when the inventory costs, parts supply problems, and other factors are considered.

17-7   Can a C-section belt be used for Prob. 17-5? Why, or why not?

17-8   The flat-belt drive of Prob. 17-3 may be too wide at 8 in. Suppose the driver is a multiple-cylindered, internal-combustion engine driving brick machinery under a schedule of two shifts per day (16 h/day). Select a V-belt arrangement that will handle this job. What approximate reduction in width will be obtained for this new design? Is the distance between shaft centers any problem?

17-9   Rework Prob. 17-8 using another cross-sectional size of V belt.

17-10   It has been discovered that the pulley system designed for Prob. 17-5 is too large. Suppose a belt system can be found to carry the load for any of the cross sections shown in Table 17-2. Which cross section should be investigated first for potential replacement of the existing design? Why did you choose this section?

17-11*   Select an SI heavy-duty V belt to transmit 50 kW at a speed of 1590 rpm for the small sheave. There is to be a speed reduction of 3 : 1. Select pulley sizes, centerline distance, belt type, and number. The driver is a shunt motor and the driven is a liquid agitator with 24-h service.

17-12*   You are to design an all-SI radial drill press for sale by a large catalog house. The drive motor is a 370-W split-phase. The drill press is expected to have intermittent use, as it is expected to be used mostly in the home in the pursuit of hobbies. The motor speed is 1725 rpm and the spindle speed is 3450 rpm. The drive system is to be as compact as possible, but the centerline distance is to be maintained at 500 mm for reasons of maintaining drilling throat depth on the drill press. Specify all details of the belt-drive system.

17-13 Discuss the design solution of Prob. 17-12. Is it an overdesign, underdesign, or satisfactory? What design alternatives do you suggest, if any?

Section 17-4

17-14 A double-strand No. 60 roller chain is used to transmit power between a 13-tooth driving sprocket rotating at 300 rpm and a 52-tooth driven sprocket.
(a) What is the rated horsepower of this drive?
(b) Determine the approximate center distance if the chain length is 82 pitches.

17-15 Calculate the torque and the bending force on the driving shaft produced by the chain of Prob. 17-14 if the actual horsepower transmitted is 30 percent less than the corrected rating.

17-16 A 4-strand No. 40 roller chain transmits power from a 21-tooth driving sprocket which turns at 1200 rpm. The velocity ratio is 4 : 1.
(a) Calculate the rated horsepower of this drive.
(b) Find the tension in the chain.
(c) What is the factor of safety for the chain based upon the minimum tensile strength?
(d) What should be the chain length if the center distance is to be about 20 in?
(e) Estimate the maximum Hertzian shear stress in a roller; assume a very large radius of curvature of the sprocket tooth at the point of contact and that one tooth takes the full load.

17-17* A roller chain is to transmit 90 hp from a 17-tooth sprocket to a 34-tooth sprocket at a speed of 300 rpm. The load characteristics are moderate shock with abnormal service conditions (poor lubrication, cold temperatures, and dirty surroundings). The equipment is to run 18 h/day. Specify the length and size of chain required for a center distance of about 25 pitches.

17-18* A 720-rpm, 25-hp squirrel-cage motor is to drive a two-cylinder reciprocating pump which is to be located out-of-doors under a shed. The pump is to run at full load for 24 h/day and freedom from breakdowns is especially desired. The pump speed is 144 rpm. Select suitable chain and sprocket sizes.

Section 17-6

17-19* A mine hoist uses a 2-in 6 × 19 monitor-steel wire rope. The rope is used to haul loads up to 4 tons from a shaft 480 ft deep. The drum has a diameter of 6 ft; the sheaves are of good quality cast steel and the smallest is 3 ft in diameter.
(a) Using a maximum hoisting speed of 1200 fpm and a maximum acceleration of 2 fps², determine the stresses in the rope.
(b) What are the factors of safety?

17-20* A temporary construction elevator is to be designed to carry workers and materials to a height of 90 ft. The maximum estimated load to be hoisted is 5000 lb at a velocity not exceeding 2 fps. Based on minimum sheave diameters, a minimum factor of safety, and an acceleration of 4 fps², find the number of ropes required. Use 1-in plow-steel 6 × 19 standard hoisting ropes.

ANSWERS TO SELECTED PROBLEMS

2-17 For $\phi = 30°$ cw, $\sigma_x = 9.5$ kpsi, $\sigma_y = 4.5$ kpsi, $\tau_{xy} = 4.33$ kpsi cw

2-19 **2-1:** 10, 0, -4, 7 kpsi; **2-3:** 0, 0, -10, 5 kpsi; **2-5:** 0, -0.95, -20, 10 kpsi; **2-7:** 10.2, 0, -6.2, 8.2 kpsi

2-21 **2-9:** 0, -11.5, -78.5, 39.3 MPa; **2-11:** 88.9, 1.1, 0, 44.4 MPa; **2-13:** 136, 0, -93.8. 115 MPa; **2-15:** 20, 0, -80, 50 MPa

2-23 -0.989 GPa, $-4.78(10)^{-3}$ m/m, -0.382 mm, 0.112 mm

2-25 (a) 15.3 kpsi, 0.0204 in; (b) 0.852 in; (c) 5.24 kpsi

2-27 $49.985(10)^{-3}$ in, $0.45(10)^{-3}$ in/in

2-29 475 lb

2-31 $X = 39.6$ in, 6.73 kpsi, 1.54 kpsi

2-33 $\epsilon_x = (\sigma_x/E)(1 - \mu)$

2-35 (a) $R_1 = -120$ lb, $R_2 = 200$ lb; (b) $R_1 = 14$ lb, $M_1 = 50$ lb·in ccw; (c) $R_1 = R_2 = 87.5$ lb; (d) $R_1 = R_2 = 120$ lb

2-37 Answer omitted

2-39 (a) $M = R_1\langle x\rangle^1 - F\langle x - a\rangle^1 + R_2\langle x - l\rangle^1$; (b) $M = -M_1\langle x\rangle^0 + R_1\langle x\rangle^1 - (w/2)\langle x\rangle^2 + (w/2)\langle x - l\rangle^2$

2-41 (a) $M = -R_1\langle x\rangle^1 + R_2\langle x - (l/2)\rangle^1 - F\langle x - l\rangle^1$; (b) $M = R_1\langle x\rangle^1 - (w/2)\langle x - a\rangle^2 + (w/2)\langle x - (l - a)\rangle^2 + R_2\langle x - l\rangle^1$

2-43 Use $1\frac{1}{2}'' \times \frac{1}{4}''$ tube

2-45 $M_{max} = 188$ N·m, $d = 23$ mm

2-47 At B, $\sigma_x = -16Fx/\pi d^3$, $\tau_{xy} = 8F/3\pi d^2$; at D, $\tau_x = 16Fx/\pi d^3$

2-49 1.105 in based on normal stress, 0.98 in based on shear, so use $d = 1\frac{1}{8}$ in

2-51 (a) $M_A = 420$ N·m, $M_B = 270$ N·m; (b) $\sigma_x = 131$ MPa, $\tau_{xz} = 93.3$ MPa; (c) $\phi = 27.4°$ from x to σ_1

2-53 (a) $\tau_{xz} = 71.3$ MPa, $\sigma_x = -47.5$ MPa, (b) $\sigma_1 = 51.4$ MPa, $\sigma_2 = -98.9$ MPa, $\phi = 54.2°$ from x to σ_1

2-55 At A: $\sigma_x = 9.016$ kpsi, $\tau_{xz} = 12.7$ kpsi; at B: $\sigma_x = 60.9$ kpsi, $\tau_{xy} = 12.7$ kpsi; at C: $\sigma_x = -9.344$ kpsi, $\tau_{xz} = 12.7$ kpsi

2-57 (b) $\sigma_z = 18.5$ kpsi at B, $\tau_{max} = 12.9$ kpsi at the middle of the $1\frac{1}{4}$-in side; (c) $\sigma_x = 43.4$ kpsi, $\tau_{xz} = 14.5$ kpsi

2-59 At $r = a$, $\sigma_r = -4$ kpsi, $\sigma_t = 18.2$ kpsi, $\epsilon_t = 30.1(10)^{-4}$, $\epsilon_r = -15.9(10)^{-4}$, $\epsilon_l = -7.65(10)^{-4}$; at $r = b$, $\sigma_r = 0$ kpsi, $\sigma_t = 14.2$ kpsi, $\epsilon_t = 21.8(10)^{-4}$, $\epsilon_l = \epsilon_r = -7.65(10)^{-4}$

2-61 $\sigma_x = \sigma_y = p_i D/4t, \sigma_r = 0$

2-63 At $r = a$, $\sigma_t = 250$ MPa; at $r = b$, $\sigma_t = 100$ MPa

2-65 ID of tire is 15.980 in

2-67 $p = 60$ kpsi; use, say, AISI 3130 steel drawn at 1000°F

2-69 $\sigma_i = 19.7$ kpsi, $\sigma_0 = -13.6$ kpsi

2-71 $\bar{y} = 18.74$ mm, $e = 2.45$ mm

3-1 $k = 16.6$ kip·in/rad

3-3 $\sigma = 707$ MPa, $\delta = 27.3$ mm, $k = 45.7$ kN/m

3-5 $k_A = (N_1/N_2)^2 k_1 k_2/[k_1 + (N_1/N_2)^2 k_2]$
$k_B = (N_2/N_1)^2 k_1 k_2/[(N_2/N_1)^2 k_1 + k_2]$

3-7 Use $d = 21$ mm, $\tau = 21.9$ MPa; the stress is rather small

3-11 0.215 in

3-13 Use a $2 \times 2 \times \frac{1}{8}$ in angle

3-15 $b = 21.7$ mm, $\sigma = 411$ MPa

3-17 $W = 84$ kip, $y_{max} = 0.0333$ in

3-19 $\theta_0 = -1.5(10)^{-3}$ rad, $y_{max} = 1.54(10)^{-2}$ in

3-21 2.12 in

3-23 $F = 10.3$ kN

3-25 $\theta = (Tb/GJ_1) + (Ta/3EI_2)$

3-27 $y_A = (F_1 l_1^3/3EI_1) + (F_1 l_1^2 l_2/3EI_2)$

3-29 5.72 mm

3-31 45.6 kN, 42.6 kN, 26.6 kN, 11.8 kN

3-33 53 kip

3-35 $\frac{5}{8}$ in, $\frac{3}{4}$ in, $1\frac{3}{8}$ in

3-37 $2\frac{1}{2}$ in, $1\frac{7}{8}$ in, $1\frac{3}{8}$ in

4-1 $E = 30$ Mpsi, $S_y = 45.5$ kpsi, $S_u = 85.5$ kpsi, $R = 45.8$ percent

4-3 $E = 207$ GPa, $S_y = 286$ MPa, $S_u = 372$ MPa, $R = 46.8$ percent

5-1 Yes; $S_{se} = 0.726 + 0.526 S_e$

5-3 (a) $\bar{k} = 1765$ N/m, $l_F = 80.00$ mm, $s_k = 8.97$ N/m, $s_l = 0.178$ mm; (b) $r = 0.948$; yes

5-5 $\bar{L} = 2019$ h, $s = 365$ h

5-7 (a) 1; (b) none; (c) 68 percent

5-9 $\mu = 0.375\ 148$ in, $\hat{\sigma} = 1629\ \mu$ in

5-11 2.5 kpsi

5-13 $c_{min} = 0.001$ in, 0.1 percent

6-1 (a) 4, 2.86, 3.20; (b) 3.51, 3.12, 3.29; (c) 4, 4, 4; (d) 3.92, 3.92, 4.53

6-3 (a) 5, 5, 5; (b) 5, 5, 5.78; (c) 5, 2.5, 2.89; (d) 5, 5, 5

6-5 416.4 lb

6-7 (a) $\sigma_1 = 37.5$ kpsi, $\sigma_2 = 0$, $\sigma_3 = -8.97$ kpsi; (b) 2.20; (c) 2.02

6-9 Use the equation $d = \{[32 n_s/(\pi S_y)][M^2 + (3T^2/4)]^{1/2}\}^{1/3}$

6-11 Using the distortion-energy theory a satisfactory size is $d = 25$ mm, $t = 6$ mm, and $d_i = 13$ mm

6-13 8.39 MPa

6-15 The critical von Mises stress is 81.4 kpsi at the hole

6-17 The outer member at $r = \frac{3}{8}$ in has a von Mises stress of 86.01 kpsi

6-19 (a) 2.34, 2.22, 2.34; (b) 6.80, 6.80, 6.80; (c) 3.25, 3.25, 3.25; (d) 3.84, 2.66, 3.22

6-21 $b = 3\frac{3}{4}$ in, $h = \frac{5}{8}$ in, $k = 107$ lb/in; maximum deflections are 2.55 in based on crack growth and 3.42 in based on yielding.

7-1 $S_f' = 31.9$ kpsi, $N = 8858$ cycles

7-3 Omitted

7-5 $S_e = 32.7$ kpsi

7-7 UNS G43400, $S_e = 12.59$ kpsi; UNS G10400, $S_e = 13.16$ kpsi

7-9 $F = 2.38$ kN

7-11 $n = 4.75$

7-13 $n = 3.51$

7-15 $F = 1572$ N

7-17 $n = 1.49$

7-21 995 000 cycles

7-23 $t = 28$ mm

7-25 $t = 14$ mm

7-29 $n = 5.90$
7-31 $d = 1\frac{3}{4}$ in
7-33 $n = 5.27$
7-35 $n(\text{fatigue}) = 2.84$, $n(\text{static}) = 2.66$
7-37 (*a*) $n(\text{fatigue}) = 1.06$, $n(\text{static}) = 1.66$
          (*b*) $n(\text{fatigue}) = 2.31$, $n(\text{static}) = 1.15$
          (*c*) $n(\text{fatigue}) = 1.18$, $n(\text{static}) = 1.34$
          (*d*) $N = 20.8(10)^3$ cycles, $n(\text{static}) = 1.15$
          (*e*) $n(\text{fatigue}) = 1.20$, $n(\text{static}) = 2$
7-39 $n = 1.51$
7-41 At $\theta = 15°$, $\sigma_1 = 11.7$ kpsi, $\sigma_2 = -3.1$ kpsi, $\phi = 27.1°$

8-3 $\frac{9}{16}''$—18 UNRF
8-7 (*a*) 12.83 lb·in; (*b*) $l = 4.46$ in, $d = 0.144$ in
8-9 3.36 kW, 12.5 percent
8-11 $n(\text{min}) = 34$
8-13 Use 6 mm bolts
8-15 $F_i = 0.75F_P = 108$ kN; $F_b = 111.7$ kN/bolt, $F_m = -91.7$ kN/bolt
8-17 (*a*) $k_b = 2.95(10)^6$ lb/in, $k_m = 4.42(10)^6$ lb/in; (*b*) $F_i = 2000$ lb, $T = 100$ lb·in;
          (*c*) $F_b = 2040$ lb, $F_m = -1940$ lb; (*d*) complaint not justified
8-19 $S_e = 17$ kpsi, $n = 1.93$
8-21 Use 5 bolts; $F_i = 15.9$ kip
8-23 (*a*) 6.75 kip based on bearing on member; (*b*) 13.5 kip based on bearing on
          member
8-25 $F = 5.85$ kN based on shear of bolt
8-27 Shear of bolt 5.35, bearing on bolt 8.85, bearing on member 5.63, strength of
          member 2.95

9-1 (*a*) 17.7 kip; (*b*) 35.4 kip; (*c*) 2.17 kip; (*d*) 4.33 kip
9-3 (*a*) 49.8 kN; (*b*) the weld has zero shear stress
9-5 (*a*) $\tau_{\text{max}} = 10.8$ kpsi; (*b*) $\tau_{\text{max}} = 11$ kpsi at bottom
9-7 (*a*) $\tau_{\text{max}} = 2910$ psi; (*b*) $\tau_{\text{max}} = 17.6$ MPa

10-1 (*a*) 9.25 lb; (*b*) 1.73 in; (*c*) 5.35 lb/in; (*d*) 0.752 in; (*e*) 2.48 in
10-3 (*a*) 4.46 N; (*b*) 34.9 N; (*c*) 453 N/m; (*d*) 137 mm
10-5 (*a*) $k_o = 7.01$ lb/in, $k_i = 4.87$ lb/in; (*b*) 11.9 lb; (*c*) $\tau_o = 17.4$ kpsi, $\tau_i = 19.8$ kpsi
10-7 1.74 in, 8.82 lb
10-9 $d = 1.6$ mm, $l_F = 263$ mm, $l_S = 52.8$ mm
10-11 111.6 N
10-15 (*a*) 70 lb/in, 88.1 cy/s; (*b*) 7.96 in; (*c*) safe, $n = 1.56$
10-17 $n(\text{fatigue}) = 2.76$, $n(\text{static}) = 2.58$
10-19 (*a*) About 210° end to end; (*b*) 313 kpsi
10-21 1.61 lb·in

11-1 Use a 30-mm 02 series or a 25-mm 03 series
11-3 Use 10-mm bore
11-5 Use 35-mm 02 series bearings
11-7 $C_R = 7.6$ daN
11-9 $C_R = 21.1$ kN at O, $C_R = 17.4$ kN at B

12-1   0.627 hp

12-3   1.37, 1.80, 2.31, and 2.79 hp

12-5   0.0238 Btu/s, 0.0798 in^3/s, 0.167 in^3/s, 0.000 86 in, 14.4°F

12-7 $H = 24.1$ J/s, $Q_s = 1350$ mm^3/s, $Q = 2740$ mm^3/s, $h_o = 21$ μm, $T_C = 7$°C

12-9   0.0109, 0.0351, 0.0723, 0.118 Btu/s

12-11 $\Delta T_F = 36$°F, $p_{max} = 181$ psi, $h_o = 157$ μin

12-13 54°F, 543 μin

12-15 59°C, 13.7 μm, 3360 kPa

12-17 (a) 0.0011 in at 71°; (b) 0.0095; (c) 0.200 in^3/s, 0.498 in^3/s; (d) 321 psi at 12.5°; (e) about 104°; (f) 128°F, 156°F

12-19 78°F, 1.8(10)$^{-4}$ in, 2820 psi

12-21 56°C, 3000 mm^3/s, 5.63 μm

13-1   35 teeth, 4 in

13-3   54 teeth, 1.396 teeth/in, 25.78 in

13-5 $q_a = 1.07$ in, $q_r = 0.99$ in, $p = 1.257$ in, $m_c = 1.64$

13-7 $p_b = 0.95$ in, $m_c = 1.34$

13-9   244.4 rpm ccw

13-11 114.3 rpm ccw

13-13 91.4 rpm cw

13-15 54 teeth, 86.96 rpm ccw

13-17 75 rpm ccw, 88.2 rpm cw

13-19 $F_{42} = F_{12} = F_{54} = F_{24} = 149$ lb, $F_{34} = F_{43} = F_{13} = 280$ lb, $T_3 = 1120$ lb $\cdot$ in

13-21 For shaft a, $F_{a2} = 0.508$ kN; for shaft b, $\mathbf{F}_{b3} = 0.508 \angle 200°$ kN, $\mathbf{F}_{b4} = 0.761 \angle 160°$ kN, $\mathbf{F}_b = 1.20 \angle 175.9°$ kN; for shaft c, $F_{c5} = 0.761$ kN

13-23 $\mathbf{F}_A = 112\mathbf{i} + 240\mathbf{j}$, $\mathbf{F}_B = 501\mathbf{i} + 1074\mathbf{j}$, $\mathbf{F}_C = 2274\mathbf{i} + 873\mathbf{j}$, $\mathbf{F}_D = 393\mathbf{i} - 657\mathbf{j}$, $\mathbf{F}_E = -2380\mathbf{i} - 1110\mathbf{j}$, $\mathbf{F}_F = -896\mathbf{i} - 418\mathbf{j}$, all answers in lb

13-27 10.3 kpsi

13-29 $P = 8$ teeth/in, $F = 1\frac{1}{4}$ in, $d_2 = 2\frac{1}{4}$ in, $d_3 = 3\frac{3}{8}$ in, $N_2 = 18$ teeth, $N_3 = 27$ teeth

13-31 $n_G = 2.23$, $n = 1.71$

13-33 -84.7 kpsi

13-35 11.7 hp

13-37 $n = 2.19$, $n_G = 3.07$

13-39 $P = 5$ teeth/in, $d_P = 3.6$ in, $d_G = 5.4$ in, $N_P = 18$, $N_G = 27$, $F = 2\frac{1}{2}$ in

14-1   (a) $p_n = 0.393$ in, $p_t = 0.433$ in, $p_x = 0.929$ in, $p_N = 0.369$ in; (b) $P_t = 7.25$, $\phi_t = 21.88°$; (c) $a = 0.125$ in, $b = 0.156$ in, $d_P = 2.48$ in, $d_G = 4.41$ in

14-3   (a) $P_n = 4.14$, $p_t = 0.785$ in, $p_n = 0.758$ in, $p_x = 2.93$ in, $\phi_n = 19.4°$; (b) $a = 0.242$ in, $b = 0.302$ in, $d_P = 5$ in, $d_G = 16$ in

14-5   (a) $\mathbf{F}_{2a} = -168\mathbf{i} - 231\mathbf{j} + 400\mathbf{k}$, $\mathbf{F}_{3b} = +168\mathbf{i} + 231\mathbf{j} - 400\mathbf{k}$; (b) $\mathbf{F}_{2a} = -168\mathbf{i} + 231\mathbf{j} - 400\mathbf{k}$, $\mathbf{F}_{3b} = -800\mathbf{k}$, $\mathbf{F}_{4c} = +168\mathbf{i} - 231\mathbf{j} - 400\mathbf{k}$, all answers in lb

14-7 $\mathbf{F}_{2a} = -266\mathbf{i} - 96.8\mathbf{j} - 71.3\mathbf{k}$ lb, $\mathbf{F}_{3b} = 169\mathbf{i} - 169\mathbf{j}$ lb, $F_{4c} = 96.8\mathbf{i} + 266\mathbf{j} + 71.3\mathbf{k}$ lb

14-9 $\mathbf{F}_C = 1230\mathbf{i} + 684\mathbf{j} - 1080\mathbf{k}$ lb, $\mathbf{F}_D = 654\mathbf{i} + 618\mathbf{j}$ lb

14-11 598 lb

14-13 420 lb

14-15 47.1 rpm cw

14-17 $\mu = 0.042$, $\mathbf{T} = 284\mathbf{k}$ lb·in, $\mathbf{F}_A = 71\mathbf{i} - 422\mathbf{j} - 1110\mathbf{k}$ lb, $\mathbf{F}_B = 71\mathbf{i} + 132\mathbf{j}$ lb
14-19 1.04 hp, 83 percent
14-21 0.343 hp, 84 percent
14-23 756 rpm
14-25 $\mathbf{F}_E = 163\mathbf{i} - 192\mathbf{j} + 355\mathbf{k}$ lb, $\mathbf{F}_F = 110\mathbf{j} + 145\mathbf{k}$ lb, $\mathbf{T} = -1030\mathbf{i}$ lb·in
14-27 $\mathbf{F}_C = 239\mathbf{i} + 87\mathbf{k}$ lb, $\mathbf{F}_D = 563\mathbf{i} - 239\mathbf{j} + 1250\mathbf{k}$ lb
14-29 1.43 hp
14-31 47.4 hp

15-1 $d = 1$ in
15-3 $n = 5.90$
15-5 $\delta = 0.0238$ in at overhanging end; this is too much
15-7 Yes, $n = 8.30$
15-9 $\delta_A = 210$ μm, $\delta_B = 232$ μm
15-11 At A, $n = 4.05$; at E, $n = 4.93$
15-13 (a) $n = 5.87$; (b) $n = 5.05$; (c) $n = 6.29$

16-1 (a) 538 lb; (b) 4130 lb·in; (c) 1130 lb on right shoe, 230 lb on left shoe
16-3 (a) 0.947 kN; (b) 183 N·m; (c) 2.16 kN on right shoe, 0.366 kN on left shoe
16-5 (a) 1.63 kN; (b) 151 N·m; (c) 3960 J
16-7 1.86 in, 24.9 kip·in
16-9 0.622 kN, 532 N·m
16-11 1920 lb, 627 lb, 24.6 lb
16-13 718 N·m
16-15 955 kPa
16-17 1510 lb, 1230 lb·in
16-19 (a) 140 lb, 10.7 psi; (b) 139 lb, 11 psi
16-21 $d = 3\ 1/2$ in, $b = 1.11$ in, $F = 137$ lb
16-23 $\Delta T = 22.7°C$
16-25 $E = 1650$ J, $T = 440$ N·m
16-27 $76.5(10)^3$ lb·in, 161 lb
16-29 $T_m = 833$ lb·in, $I = 1.013$ lb·s²·in

17-1 (a) 349 lb, 169 lb; (b) 321 lb, 196 lb, but the belt slips; (c) 230.1 in
17-3 Use an 8-in belt
17-5 One A90 belt using a service factor of 1.5; sheaves are properly sized
17-8 An E540 belt with a service factor of 1.4 provides a 61 percent reduction in width; center distance is no problem
17-11 A typical solution is a 32C6180 belt; service factor is 1.2; small sheave 355 mm; center distance, 2 m
17-15 1086 lb·in, 694 lb
17-17 A typical solution is a 140-3 chain 133 in long
17-19 $n = 5.03$ based on an estimate of the wire strength; $n = 26.0$ based on rope strength; $n = 2.75$ based on wear

APPENDIX

TABLES

Table A-1 STANDARD SI PREFIXES* †

Name	Symbol	Factor
exa	E	$1\ 000\ 000\ 000\ 000\ 000\ 000 = 10^{18}$
peta	P	$1\ 000\ 000\ 000\ 000\ 000 = 10^{15}$
tera	T	$1\ 000\ 000\ 000\ 000 = 10^{12}$
giga	G	$1\ 000\ 000\ 000 = 10^{9}$
mega	M	$1\ 000\ 000 = 10^{6}$
kilo	k	$1\ 000 = 10^{3}$
hecto‡	h	$100 = 10^{2}$
deka‡	da	$10 = 10^{1}$
deci‡	d	$0.1 = 10^{-1}$
centi‡	c	$0.01 = 10^{-2}$
milli	m	$0.001 = 10^{-3}$
micro	μ	$0.000\ 001 = 10^{-6}$
nano	n	$0.000\ 000\ 001 = 10^{-9}$
pico	p	$0.000\ 000\ 000\ 001 = 10^{-12}$
femto	f	$0.000\ 000\ 000\ 000\ 001 = 10^{-15}$
atto	a	$0.000\ 000\ 000\ 000\ 000\ 001 = 10^{-18}$

* If possible use multiple and submultiple prefixes in steps of 1000.

† Spaces are used in SI instead of commas to group numbers to avoid confusion with the practice in some European countries of using commas for decimal points.

‡ Not recommended but sometimes encountered.

Table A-2   CONVERSION OF U.S. CUSTOMARY UNITS TO SI UNITS

To convert from	To	Multiply by	
		Accurate	Common
foot(ft)	meter(m)	3.048 000*E − 01	0.305
foot-pound-force (ft · lbf)	joule (J)	1.355 818 E + 00	1.35
foot-pound-force/second (ft · lbf/s)	watt (W)	1.355 818 E + 00	1.35
inch	meter (m)	2.540 000*E − 02	0.0254
gallon (gal U.S.)	meter^3 (m^3)	3.785 412 E − 03	0.003 78
horsepower (hp)	kilowatt (kW)	7.456 999 E − 01	0.746
mile (mi U.S. Statute)	kilometer (km)	1.609 344*E + 00	1.610
pascal (Pa)	newton/meter^2 (N/m^2)	1.000 000*E + 00	1
pound-force (lbf avoirdupois)	newton (N)	4.448 222 E + 00	4.45
pound-mass (lbm avoirdupois)	kilogram (kg)	4.535 924 E − 01	0.454
pound-force/foot^2 (lbf/ft^2)	pascal (Pa)	4.788 026 E + 01	47.9
pound-force/inch^2 (psi)	pascal (Pa)	6.894 757 E + 03	6890
slug	kilogram (kg)	1.459 390 E + 01	14.6
ton (short 2000 lbm)	kilogram (kg)	9.071 847 E + 02	907

*Exact

Table A-3   CONVERSION OF SI UNITS TO U.S. CUSTOMARY UNITS

To convert from	To	Multiply by	
		Accurate	Common
joule (J)	foot-pound-force (ft · lb)	7.375 620 E − 01	0.737
kilogram (kg)	slug	6.852 178 E − 02	0.0685
kilogram (kg)	pound-mass (lbm avoirdupois)	2.204 622 E + 00	2.20
kilogram (kg)	ton (short 2000 lbm)	1.102 311 E − 03	0.001 10
kilometer (km)	mile (mi U.S. Statute)	6.213 712 E − 01	0.621
kilowatt (kW)	horsepower (hp)	1.341 022 E + 00	1.34
meter (m)	foot (ft)	3.280 840 E + 00	3.28
meter (m)	inch (in)	3.937 008 E + 02	39.4
meter^3 (m^3)	gallon (gal U.S.)	2.641 720 E + 02	264
newton (N)	pound-force (lb avoirdupois)	2.248 089 E − 01	0.225
pascal (Pa)	pound-force/foot^2 (lb/ft^2)	2.088 543 E − 02	0.0209
pascal (Pa)	pound-force/inch^2 (psi)	1.450 370 E − 04	0.000 145
watt (W)	foot-pound-force/second (ft · lb/s)	7.375 620 E − 01	0.737

Table A-4 PREFERRED SI UNITS FOR BENDING STRESS $\sigma = Mc/I$ AND TORSION STRESS $\tau = Tr/J$

M, T	I, J	c, r	σ, τ
$N \cdot m$	m^4	m	Pa
$N \cdot m$	cm^4	cm	MPa
$N \cdot m$	mm^4	mm	GPa
$kN \cdot m$	m^4	m	kPa
$kN \cdot m$	cm^4	cm	GPa
$kN \cdot m$	mm^4	mm	TPa
$N \cdot mm$	mm^4	mm	MPa
$kN \cdot mm$	mm^4	mm	GPa

Table A-5 PREFERRED SI UNITS FOR AXIAL STRESS $\sigma = F/A$ AND DIRECT SHEAR STRESS $\tau = F/A$

F	A	σ, τ
N	m^2	Pa
N	mm^2	MPa
kN	m^2	kPa
kN	mm^2	GPa

Table A-6 PREFERRED SI UNITS FOR DEFLECTION OF BEAMS $y = f\,(Fl^3/EI)$ OR $y = f\,(wl^4/EI)$

F, wl	l	I	E	y
N	m	m^4	MPa	μm
kN	m	m^4	GPa	μm
N	cm	mm^4	GPa	mm
N	mm	mm^4	GPa	μm

Table A-7 PHYSICAL CONSTANTS OF MATERIALS

Material	Modulus of elasticity, E		Modulus of rigidity, G		Poisson's ratio	Unit weight, w		
	Mpsi	GPa	Mpsi	GPa		lb/in³	lb/ft³	kN/m³
Aluminum (all alloys)	10.3	71.0	3.80	26.2	0.334	0.098	169	26.6
Beryllium copper	18.0	124.0	7.0	48.3	0.285	0.297	513	80.6
Brass	15.4	106.0	5.82	40.1	0.324	0.309	534	83.8
Carbon steel	30.0	207.0	11.5	79.3	0.292	0.282	487	76.5
Cast iron, gray	14.5	100.0	6.0	41.4	0.211	0.260	450	70.6
Copper	17.2	119.0	6.49	44.7	0.326	0.322	556	87.3
Douglas fir	1.6	11.0	0.6	4.1	0.33	0.016	28	4.3
Glass	6.7	46.2	2.7	18.6	0.245	0.094	162	25.4
Inconel	31.0	214.0	11.0	75.8	0.290	0.307	530	83.3
Lead	5.3	36.5	1.9	13.1	0.425	0.411	710	111.5
Magnesium	6.5	44.8	2.4	16.5	0.350	0.065	112	17.6
Molybdenum	48.0	331.0	17.0	117.0	0.307	0.368	636	100.0
Monel metal	26.0	179.0	9.5	65.5	0.320	0.319	551	86.6
Nickel silver	18.5	127.0	7.0	48.3	0.322	0.316	546	85.8
Nickel steel	30.0	207.0	11.5	79.3	0.291	0.280	484	76.0
Phosphor bronze	16.1	111.0	6.0	41.4	0.349	0.295	510	80.1
Stainless steel (18-8)	27.6	190.0	10.6	73.1	0.305	0.280	484	76.0

Table A-8 PROPERTIES OF STRUCTURAL SHAPES—EQUAL ANGLES (L)

w_a = weight per foot of aluminum sections, lb
w_s = weight per foot of steel sections, lb
A = area, in^2
I = moment of inertia, in^4
k = radius of gyration, in
y = centroidal distance, in
Z = section modulus, in^3

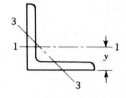

Size	w_a	w_s	A	I_{1-1}	k_{1-1}	Z_{1-1}	y	I_{3-3}	k_{3-3}
$1 \times 1 \times \frac{1}{8}$	0.28	0.80	0.23	0.02	0.30	0.03	0.30	0.008	0.19
$1 \times 1 \times \frac{1}{4}$	0.53	1.49	0.44	0.04	0.29	0.05	0.34	0.016	0.19
$1\frac{1}{2} \times 1\frac{1}{2} \times \frac{1}{8}$	0.44	1.23	0.36	0.07	0.45	0.07	0.41	0.031	0.29
$1\frac{1}{2} \times 1\frac{1}{2} \times \frac{1}{4}$	0.83	2.34	0.69	0.14	0.44	0.13	0.46	0.057	0.29
$2 \times 2 \times \frac{1}{8}$	0.59	1.65	0.49	0.18	0.61	0.13	0.53	0.08	0.40
$2 \times 2 \times \frac{1}{4}$	1.14	3.19	0.94	0.34	0.60	0.24	0.58	0.14	0.39
$2 \times 2 \times \frac{3}{8}$	1.65	4.70	1.37	0.47	0.59	0.35	0.63	0.20	0.39
$2\frac{1}{2} \times 2\frac{1}{2} \times \frac{1}{4}$	1.45	4.1	1.19	0.69	0.76	0.39	0.71	0.29	0.49
$2\frac{1}{2} \times 2\frac{1}{2} \times \frac{3}{8}$	2.11	5.9	1.74	0.98	0.75	0.56	0.76	0.41	0.48
$3 \times 3 \times \frac{1}{4}$	1.73	4.9	1.43	1.18	0.91	0.54	0.82	0.49	0.58
$3 \times 3 \times \frac{3}{8}$	2.55	7.2	2.10	1.70	0.90	0.80	0.87	0.70	0.58
$3 \times 3 \times \frac{1}{2}$	3.32	9.4	2.74	2.16	0.89	1.04	0.92	0.91	0.58
$3\frac{1}{2} \times 3\frac{1}{2} \times \frac{1}{4}$	2.05	4.9	1.69	1.93	1.07	0.76	0.94	0.80	0.69
$3\frac{1}{2} \times 3\frac{1}{2} \times \frac{3}{8}$	3.01	7.2	2.49	2.79	1.06	1.11	1.00	1.15	0.68
$3\frac{1}{2} \times 3\frac{1}{2} \times \frac{1}{2}$	3.94	11.1	3.25	3.56	1.05	1.45	1.05	1.49	0.68
$4 \times 4 \times \frac{1}{4}$	2.35	6.6	1.94	2.94	1.23	1.00	1.07	1.21	0.79
$4 \times 4 \times \frac{3}{8}$	3.46	9.8	2.86	4.26	1.22	1.48	1.12	1.75	0.78
$4 \times 4 \times \frac{1}{2}$	4.54	12.8	3.75	5.46	1.21	1.93	1.17	2.26	0.78
$4 \times 4 \times \frac{5}{8}$	5.58	15.7	4.61	6.56	1.19	2.36	1.22	2.76	0.77
$6 \times 6 \times \frac{3}{8}$	5.27	14.9	4.35	14.85	1.85	3.38	1.60	6.07	1.18
$6 \times 6 \times \frac{1}{2}$	6.95	19.6	5.74	19.38	1.84	4.46	1.66	7.92	1.17
$6 \times 6 \times \frac{5}{8}$	8.59	24.2	7.10	23.64	1.82	5.51	1.71	9.70	1.17
$6 \times 6 \times \frac{3}{4}$	10.20	28.7	8.43	27.64	1.81	6.52	1.76	11.43	1.16

Table A-9 PROPERTIES OF STRUCTURAL SHAPES—UNEQUAL ANGLES (L)

w_a = weight per foot of aluminum sections, lb
w_s = weight per foot of steel sections, lb
A = area, in^2
I = moment of inertia, in^4
k = radius of gyration, in
x and y = respective centroidal distances, in
Z = section modulus, in^3

Size	w_a	w_s	A	I_{1-1}	k_{1-1}	Z_{1-1}	y	I_{2-2}	k_{2-2}	Z_{2-2}	x	I_{3-3}	k_{3-3}
$2 \times 1\frac{1}{2} \times \frac{1}{8}$	0.51	1.44	0.42	0.17	0.63	0.12	0.60	0.08	0.44	0.07	0.36	0.04	0.32
$2 \times 1\frac{1}{2} \times \frac{1}{4}$	0.98	2.77	0.81	0.31	0.62	0.23	0.66	0.15	0.43	0.14	0.41	0.08	0.32
$3 \times 2 \times \frac{3}{16}$	1.10	3.07	0.91	0.82	0.95	0.40	0.94	0.29	0.56	0.19	0.46	0.17	0.43
$3 \times 2\frac{1}{2} \times \frac{1}{4}$	1.58	4.5	1.31	1.12	0.92	0.53	0.89	0.70	0.73	0.38	0.64	0.35	0.52
$3 \times 2\frac{1}{2} \times \frac{3}{8}$	2.32	6.6	1.92	1.60	0.91	0.78	0.94	1.00	0.72	0.55	0.69	0.51	0.51
$3 \times 2\frac{1}{2} \times \frac{1}{2}$	3.02	9.4	2.49	2.03	0.90	1.01	0.99	1.26	0.71	0.72	0.74	0.65	0.51
$4 \times 3 \times \frac{1}{4}$	2.05	5.8	1.69	2.68	1.26	0.96	1.21	1.29	0.87	0.56	0.72	0.70	0.64
$4 \times 3 \times \frac{1}{2}$	3.95	11.1	3.25	4.96	1.24	1.85	1.31	2.36	0.85	1.08	0.82	1.30	0.63
$6 \times 4 \times \frac{3}{8}$	4.36	12.3	3.60	13.02	1.90	3.17	1.90	4.63	1.13	1.50	0.91	2.67	0.86
$6 \times 4 \times \frac{1}{2}$	5.74	16.2	4.74	16.95	1.89	4.19	1.96	6.01	1.13	1.98	0.97	3.47	0.86

Table A-10 PROPERTIES OF ROUND TUBING

w_a = weight per foot of aluminum tubing, lb/ft
w_s = weight per foot of steel tubing, lb/ft
A = area, in^2
I = moment of inertia, in^4
k = radius of gyration, in
Z = section modulus, in^3

Size	w_a	w_s	A	I	k	Z
$1 \times \frac{1}{8}$	0.416	1.128	0.344	0.034	0.313	0.067
$1 \times \frac{1}{4}$	0.713	2.003	0.589	0.046	0.280	0.092
$1\frac{1}{2} \times \frac{1}{8}$	0.653	1.769	0.540	0.129	0.488	0.172
$1\frac{1}{2} \times \frac{1}{4}$	1.188	3.338	0.982	0.199	0.451	0.266
$2 \times \frac{1}{8}$	0.891	2.670	0.736	0.325	0.664	0.325
$2 \times \frac{1}{4}$	1.663	4.673	1.374	0.537	0.625	0.537
$2\frac{1}{2} \times \frac{1}{8}$	1.129	3.050	0.933	0.660	0.841	0.528
$2\frac{1}{2} \times \frac{1}{4}$	2.138	6.008	1.767	1.132	0.800	0.906
$3 \times \frac{1}{4}$	2.614	7.343	2.160	2.059	0.976	1.373
$3 \times \frac{3}{8}$	3.742	10.51	3.093	2.718	0.938	1.812
$4 \times \frac{3}{16}$	2.717	7.654	2.246	4.090	1.350	2.045
$4 \times \frac{3}{8}$	5.167	14.52	4.271	7.090	1.289	3.544

Table A-11 PROPERTIES OF STRUCTURAL SHAPES—CHANNELS (C)

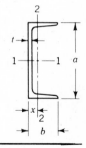

w_a = weight per foot of aluminum sections, lb
w_s = weight per foot of steel sections, lb
A = area, in^2
I = moment of inertia, in^4
k = radius of gyration, in
x = centroidal distance, in
Z = section modulus, in^3

a	b	t	A	w_a	w_s	I_{1-1}	k_{1-1}	Z_{1-1}	I_{2-2}	k_{2-2}	Z_{2-2}	x
3	1.410	0.170	1.21	1.46	4.1	1.66	1.17	1.10	0.20	0.40	0.20	0.44
3	1.498	0.258	1.47	1.78	5.0	1.85	1.12	1.24	0.25	0.41	0.23	0.44
3	1.596	0.356	1.76	2.13	6.0	2.07	1.08	1.38	0.31	0.42	0.27	0.46
4	1.580	0.180	1.57	1.90	5.4	3.83	1.56	1.92	0.32	0.45	0.28	0.46
4	1.720	0.320	2.13	2.58	7.25	4.58	1.47	2.29	0.43	0.45	0.34	0.46
5	1.750	0.190	1.97	2.38	6.7	7.49	1.95	3.00	0.48	0.49	0.38	0.48
5	1.885	0.325	2.64	3.20	9.0	8.90	1.83	3.56	0.63	0.49	0.45	0.48
6	1.920	0.200	2.40	2.91	8.2	13.12	2.34	4.37	0.69	0.54	0.49	0.51
6	2.034	0.314	3.09	3.73	10.5	15.18	2.22	5.06	0.87	0.53	0.56	0.50
6	2.157	0.437	3.82	4.63	13.0	17.39	2.13	5.80	1.05	0.52	0.64	0.51
7	2.090	0.210	2.87	3.47	9.8	21.27	2.72	6.08	0.97	0.58	0.63	0.54
7	2.194	0.314	3.60	4.36	12.25	24.24	2.60	6.93	1.17	0.57	0.70	0.52
7	2.299	0.419	4.33	5.24	14.75	27.24	2.51	7.78	1.38	0.56	0.78	0.53
8	2.260	0.220	3.36	4.10	11.5	32.30	3.10	8.10	1.30	0.63	0.79	0.58
8	2.343	0.303	4.04	4.89	13.75	36.11	2.99	9.03	1.53	0.61	0.85	0.55
8	2.527	0.487	5.51	6.67	18.75	43.96	2.82	10.99	1.98	0.60	1.01	0.57
9	2.430	0.230	3.91	4.74	13.4	47.68	3.49	10.60	1.75	0.67	0.96	0.60
9	2.485	0.285	4.41	5.34	15.0	51.02	3.40	11.34	1.93	0.66	1.01	0.59
9	2.648	0.448	5.88	7.11	20.0	60.92	3.22	13.54	2.42	0.64	1.17	0.58
10	2.600	0.240	4.49	5.43	15.3	67.37	3.87	13.47	2.28	0.71	1.16	0.63
10	2.739	0.379	5.88	7.11	20.0	78.95	3.66	15.79	2.81	0.69	1.32	0.61
10	2.886	0.526	7.35	8.89	25.0	91.20	3.52	18.24	3.36	0.68	1.48	0.62
10	3.033	0.673	8.82	10.67	30.0	103.45	3.43	20.69	3.95	0.67	1.66	0.65
12	3.047	0.387	7.35	8.89	25.0	144.37	4.43	24.06	4.47	0.78	1.89	0.67
12	3.170	0.510	8.82	10.67	30.0	162.08	4.29	27.01	5.14	0.76	2.06	0.67

Table A-12   SHEAR, MOMENT, AND DEFLECTION OF BEAMS

1   Cantilever—end load

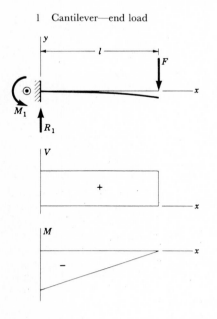

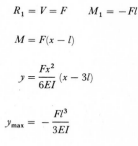

$$R_1 = V = F \qquad M_1 = -Fl$$

$$M = F(x - l)$$

$$y = \frac{Fx^2}{6EI}(x - 3l)$$

$$y_{max} = -\frac{Fl^3}{3EI}$$

2   Cantilever—intermediate load

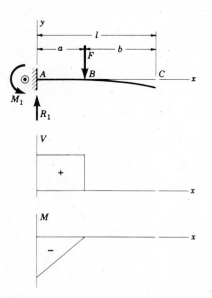

$$R_1 = V = F \qquad M_1 = -Fa$$

$$M_{AB} = F(x - a) \qquad M_{BC} = 0$$

$$y_{AB} = \frac{Fx^2}{6EI}(x - 3a)$$

$$y_{BC} = \frac{Fa^2}{6EI}(a - 3x)$$

$$y_{max} = \frac{Fa^2}{6EI}(a - 3l)$$

Table A-12 SHEAR, MOMENT, AND DEFLECTION OF BEAMS (*continued*)

3   Cantilever—uniform load

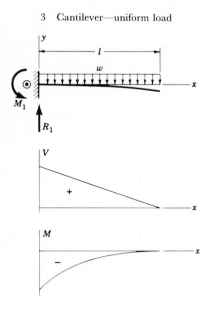

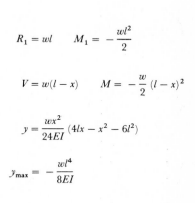

$$R_1 = wl \qquad M_1 = -\frac{wl^2}{2}$$

$$V = w(l - x) \qquad M = -\frac{w}{2}(l - x)^2$$

$$y = \frac{wx^2}{24EI}(4lx - x^2 - 6l^2)$$

$$y_{\text{max}} = -\frac{wl^4}{8EI}$$

4   Cantilever—moment load

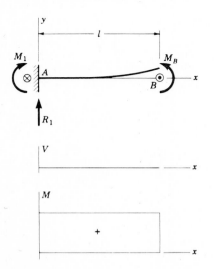

$$R_1 = 0 \qquad M_1 = M_B \qquad M = M_B$$

$$y = \frac{M_B x^2}{2EI} \qquad y_{\text{max}} = \frac{M_B l^2}{2EI}$$

Table A-12　SHEAR, MOMENT, AND DEFLECTION OF BEAMS *(continued)*

5　Simple supports—center load

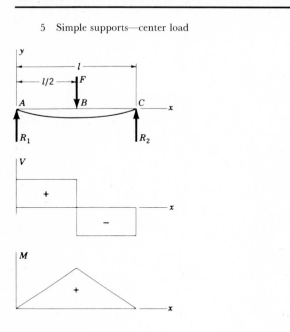

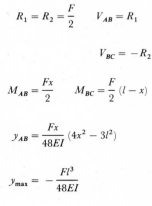

$$R_1 = R_2 = \frac{F}{2} \qquad V_{AB} = R_1$$

$$V_{BC} = -R_2$$

$$M_{AB} = \frac{Fx}{2} \qquad M_{BC} = \frac{F}{2}(l - x)$$

$$y_{AB} = \frac{Fx}{48EI}(4x^2 - 3l^2)$$

$$y_{max} = -\frac{Fl^3}{48EI}$$

6　Simple supports—intermediate load

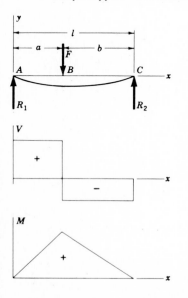

$$R_1 = \frac{Fb}{l} \qquad R_2 = \frac{Fa}{l} \qquad V_{AB} = R_1$$

$$V_{BC} = -R_2$$

$$M_{AB} = \frac{Fbx}{l} \qquad M_{BC} = \frac{Fa}{l}(l - x)$$

$$y_{AB} = \frac{Fbx}{6EIl}(x^2 + b^2 - l^2)$$

$$y_{BC} = \frac{Fa(l - x)}{6EIl}(x^2 + a^2 - 2lx)$$

Table A-12 SHEAR, MOMENT, AND DEFLECTION OF BEAMS (*continued*)

7 Simple supports—uniform load

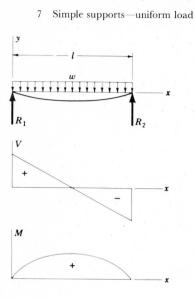

$$R_1 = R_2 = \frac{wl}{2} \qquad V = \frac{wl}{2} - wx$$

$$M = \frac{wx}{2}(l - x)$$

$$y = \frac{wx}{24EI}(2lx^2 - x^3 - l^3)$$

$$y_{max} = -\frac{5wl^4}{384EI}$$

8 Simple supports—moment load

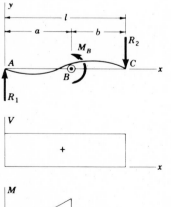

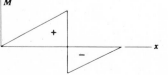

$$R_1 = -R_2 = \frac{M_B}{l} \qquad V = \frac{M_B}{l}$$

$$M_{AB} = \frac{M_B x}{l} \qquad M_{BC} = \frac{M_B}{l}(x - l)$$

$$y_{AB} = \frac{M_B x}{6EIl}(x^2 + 3a^2 - 6al + 2l^2)$$

$$y_{BC} = \frac{M_B}{6EIl}[x^3 - 3lx^2 + x(2l^2 + 3a^2) - 3a^2 l]$$

Table A-12 SHEAR, MOMENT, AND DEFLECTION OF BEAMS (*continued*)

9   Simple supports—twin loads

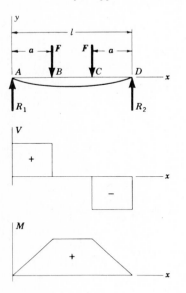

$$R_1 = R_2 = F \qquad V_{AB} = F \qquad V_{BC} = 0$$

$$V_{CD} = -F$$

$$M_{AB} = F_x \qquad M_{BC} = Fa$$

$$M_{CD} = F(l - x)$$

$$y_{AB} = \frac{Fx}{6EI} (x^2 + 3a^2 - 3la)$$

$$y_{BC} = \frac{Fa}{6EI} (3x^2 + a^2 - 3lx)$$

$$y_{max} = \frac{Fa}{24EI} (4a^2 - 3l^2)$$

10   Simple supports—overhanging load

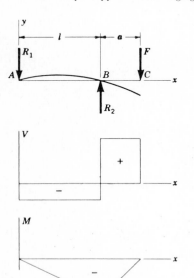

$$R_1 = -\frac{Fa}{l} \qquad R_2 = \frac{F}{l} (l + a)$$

$$V_{AB} = -\frac{Fa}{l}$$

$$V_{BC} = F \qquad M_{AB} = -\frac{Fax}{l}$$

$$M_{BC} = F(x - l - a)$$

$$y_{AB} = \frac{Fax}{6EIl} (l^2 - x^2)$$

$$y_{BC} = \frac{F(x - l)}{6EI} [(x - l)^2 - a(3x - l)]$$

$$y_C = -\frac{Fa^2}{3EI} (l + a)$$

Table A-12 SHEAR, MOMENT, AND DEFLECTION OF BEAMS (*continued*)

11 One fixed and one simple support
    —center load

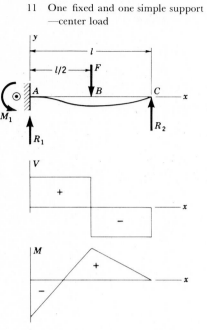

$$R_1 = \frac{11F}{16} \qquad R_2 = \frac{5F}{16} \qquad M_1 = -\frac{3Fl}{16}$$

$$V_{AB} = R_1 \qquad V_{BC} = -R_2$$

$$M_{AB} = \frac{F}{16}(11x - 3l) \qquad M_{BC} = \frac{5F}{16}(l - x)$$

$$y_{AB} = \frac{Fx^2}{96EI}(11x - 9l)$$

$$y_{BC} = \frac{F(l - x)}{96EI}(5x^2 + 2l^2 - 10lx)$$

12 One fixed and one simple support
    —intermediate load

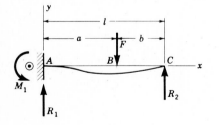

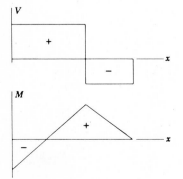

$$R_1 = \frac{Fb}{2l^3}(3l^2 - b^2) \qquad R_2 = \frac{Fa^2}{2l^3}(3l - a)$$

$$M_1 = \frac{Fb}{2l^2}(b^2 - l^2) \qquad V_{AB} = R_1$$

$$V_{BC} = -R_2$$

$$M_{AB} = \frac{Fb}{2l^3}\left[b^2 l - l^3 + x(3l^2 - b^2)\right]$$

$$M_{BC} = \frac{Fa^2}{2l^3}(3l^2 - 3lx - al + ax)$$

$$y_{AB} = \frac{Fbx^2}{12EIl^3}\left[3l(b^2 - l^2) + x(3l^2 - b^2)\right]$$

$$y_{BC} = y_{AB} - \frac{F(x - a)^3}{6EI}$$

Table A-12   SHEAR, MOMENT, AND DEFLECTION OF BEAMS (*continued*)

13   One fixed and one simple support
     —uniform load

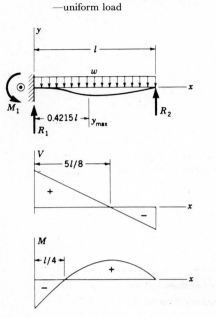

$$R_1 = \frac{5wl}{8} \qquad R_2 = \frac{3wl}{8} \qquad M_1 = -\frac{wl^2}{8}$$

$$V = \frac{5wl}{8} - wx$$

$$M = \frac{w}{8}(-4x^2 + 5lx - l^2)$$

$$y = \frac{wx^2}{48EI}(l-x)(2x-3l)$$

$$y_{max} = -\frac{wl^4}{185EI}$$

14   Fixed supports—center load

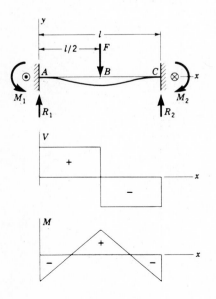

$$R_1 = R_2 = \frac{F}{2} \qquad M_1 = M_2 = -\frac{Fl}{8}$$

$$V_{AB} = -V_{BC} = \frac{F}{2}$$

$$M_{AB} = \frac{F}{8}(4x-l) \qquad M_{BC} = \frac{F}{8}(3l-4x)$$

$$y_{AB} = \frac{Fx^2}{48EI}(4x-3l)$$

$$y_{max} = -\frac{Fl^3}{192EI}$$

Table A-12 SHEAR, MOMENT, AND DEFLECTION OF BEAMS (*continued*)

15 Fixed supports—intermediate load

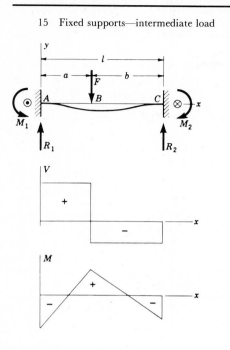

$$R_1 = \frac{Fb^2}{l^3}(3a+b) \qquad R_2 = \frac{Fa^2}{l^3}(3b+a)$$

$$M_1 = -\frac{Fab^2}{l^2} \qquad M_2 = -\frac{Fa^2b}{l^2}$$

$$V_{AB} = R_1$$

$$V_{BC} = -R_2$$

$$M_{AB} = \frac{Fb^2}{l^3}[x(3a+b)-la]$$

$$M_{BC} = M_{AB} - F(x-a)$$

$$y_{AB} = \frac{Fb^2x^2}{6EIl^3}[x(3a+b)-3al]$$

$$y_{BC} = \frac{Fa^2(l-x)^2}{6EIl^3}[(l-x)(3b+a)-3bl]$$

16 Fixed supports—uniform load

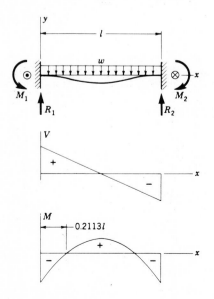

$$R_1 = R_2 = \frac{wl}{2} \qquad M_1 = M_2 = -\frac{wl^2}{12}$$

$$V = \frac{w}{2}(l-2x)$$

$$M = \frac{w}{12}(6lx - 6x^2 - l^2)$$

$$y = -\frac{wx^2}{24EI}(l-x)^2$$

$$y_{max} = -\frac{wl^4}{384EI}$$

Table A-13   PREFERRED SIZES

When a choice can be made, use one of these sizes. However not all parts or items are available in all of the sizes shown in the table.

Fraction of inches

$\frac{1}{64}$, $\frac{1}{32}$, $\frac{1}{16}$, $\frac{3}{32}$, $\frac{1}{8}$, $\frac{5}{32}$, $\frac{3}{16}$, $\frac{1}{4}$, $\frac{5}{16}$, $\frac{3}{8}$, $\frac{7}{16}$, $\frac{1}{2}$, $\frac{9}{16}$, $\frac{5}{8}$, $\frac{11}{16}$, $\frac{3}{4}$, $\frac{7}{8}$, 1, $1\frac{1}{4}$, $1\frac{1}{2}$, $1\frac{3}{4}$, 2, $2\frac{1}{4}$, $2\frac{1}{2}$, $2\frac{3}{4}$, 3, $3\frac{1}{4}$, $3\frac{1}{2}$, $3\frac{3}{4}$, 4, $4\frac{1}{4}$, $4\frac{1}{2}$, $4\frac{3}{4}$
5, $5\frac{1}{4}$, $5\frac{1}{2}$, $5\frac{3}{4}$, 6, $6\frac{1}{2}$, 7, $7\frac{1}{2}$, 8, $8\frac{1}{2}$, 9, $9\frac{1}{2}$, 10, $10\frac{1}{2}$, 11, $11\frac{1}{2}$, 12, $12\frac{1}{2}$, 13, $13\frac{1}{2}$, 14, $14\frac{1}{2}$, 15, $15\frac{1}{2}$, 16, $16\frac{1}{2}$, 17, $17\frac{1}{2}$, 18, $18\frac{1}{2}$, 19, $19\frac{1}{2}$, 20

Decimal inches

0.010, 0.012, 0.016, 0.020, 0.025, 0.032, 0.040, 0.05, 0.06, 0.08, 0.10, 0.12, 0.16, 0.20, 0.24, 0.30, 0.40, 0.50, 0.60, 0.80, 1.00, 1.20, 1.40, 1.60, 1.80, 2.0, 2.4, 2.6, 2.8, 3.0, 3.2, 3.4, 3.6, 3.8, 4.0, 4.2, 4.4, 4.6, 4.8, 5.0, 5.2, 5.4, 5.6, 5.8, 6.0, 7.0, 7.5, 8.0, 8.5, 9.0, 9.5, 10.0, 10.5, 11.0, 11.5, 12.0, 12.5, 13.0, 13.5, 14.0 ,14.5, 15.0, 15.5, 16.0, 16.5, 17.0, 17.5, 18.0, 18.5, 19.0, 19.5, 20

Millimeters

0.05, 0.06, 0.08, 0.10, 0.12, 0.16, 0.20, 0.25, 0.30, 0.40, 0.50, 0.60, 0.70, 0.80, 0.90, 1.0, 1.1, 1.2, 1.4, 1.5, 1.6, 1.8, 2.0, 2.2, 2.5, 2.8, 3.0, 3.5, 4.0, 4.5, 5.0, 5.5, 6.0, 6.5, 7.0, 8.0, 9.0, 10, 11, 12, 14, 16, 18, 20, 22, 25, 28, 30, 32, 35, 40, 45, 50, 60, 80, 100, 120, 140, 160, 180, 200, 250, 300

Table A-14   PROPERTIES OF SECTIONS

A = area
I = moment of inertia
J = polar moment of inertia
Z = section modulus
k = radius of gyration
$\bar{y}$ = centroidal distance

Rectangle

$$A = bh \qquad k = 0.289h$$

$$I = \frac{bh^3}{12} \qquad \bar{y} = \frac{h}{2}$$

$$Z = \frac{bh^2}{6}$$

Triangle

$$A = \frac{bh}{2} \qquad k = 0.236h$$

$$I = \frac{bh^3}{36} \qquad \bar{y} = \frac{h}{3}$$

$$Z = \frac{bh^2}{24}$$

Circle

$$A = \frac{\pi d^2}{4} \qquad J = \frac{\pi d^4}{32}$$

$$I = \frac{\pi d^4}{64} \qquad k = \frac{d}{4}$$

$$Z = \frac{\pi d^3}{32} \qquad \bar{y} = \frac{d}{2}$$

Hollow circle

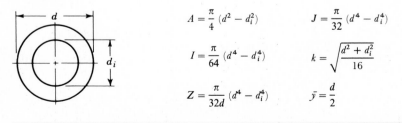

$$A = \frac{\pi}{4}\left(d^2 - d_i^2\right) \qquad J = \frac{\pi}{32}\left(d^4 - d_i^4\right)$$

$$I = \frac{\pi}{64}\left(d^4 - d_i^4\right) \qquad k = \sqrt{\frac{d^2 + d_i^2}{16}}$$

$$Z = \frac{\pi}{32d}\left(d^4 - d_i^4\right) \qquad \bar{y} = \frac{d}{2}$$

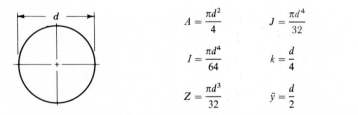

Table A-15 LIMITS AND FITS FOR CYLINDRICAL PARTS*

The limits shown in the accompanying tabulations are in thousandths of an inch. The size ranges include all sizes *over* the smallest size in the range, up to and *including* the largest size in the range. The letter symbols are defined as follows:

RC *Running and sliding fits* are intended to provide a similar running performance, with suitable lubrication allowance, throughout the range of sizes. The clearance for the first two classes, used chiefly as slide fits, increases more slowly with diameter than the other classes, so that accurate location is maintained even at the expense of free relative motion.

RC1 *Close sliding fits* are intended for the accurate location of parts which must assemble without perceptible play.

RC2 *Sliding fits* are intended for accurate location but with greater maximum clearance than class RC1. Parts made to this fit move and turn easily, but are not intended to run freely, and in the larger sizes may seize with small temperature changes.

RC3 *Precision running fits* are about the closest fits which can be expected to run freely and are intended for precision work at slow speeds and light journal pressures, but are not suitable where appreciable temperature differences are likely to be encountered.

RC4 *Close running fits* are intended chiefly for running fits on accurate machinery with moderate surface speeds and journal pressures, where accurate location and minimum play is desired.

RC5—RC6 *Medium running fits* are intended for higher running speeds or heavy journal pressures, or both.

RC7 *Free running fits* are intended for use where accuracy is not essential or where large temperature variations are likely to be encountered, or under both of these conditions.

RC8—RC9 *Loose running fits* are intended for use where wide commercial tolerances may be necessary, together with an allowance, on the external member.

L *Locational fits* are fits intended to determine only the location of the mating parts; they may provide rigid or accurate location, as with interference fits, or provide some freedom of location, as with clearance fits. Accordingly, they are divided into three groups: clearance fits, transition fits, and interference fits.

LC *Locational clearance fits* are intended for parts which are normally stationary but which can be freely assembled or disassembled. They run from snug fits for parts requiring accuracy of location, through the medium clearance fits for parts such as ball, race, and housing, to the looser fastener fits where freedom of assembly is of prime importance.

LT *Locational transition fits* are a compromise between clearance and interference fits, for application where accuracy of location is important, but either a small amount of clearance or interference is permissible.

LN *Locational interference fits* are used where accuracy of location is of prime importance and for parts requiring rigidity and alignment with no special requirements for bore pressure. Such fits are not intended for parts designed to transmit frictional loads from one part to another by virtue of the tightness of fit, since these conditions are covered by force fits.

FN *Force and shrink fits* constitute a special type of interference fit, normally characterized by maintenance of constant bore pressures throughout the range of sizes. The interference therefore varies almost directly with diameter, and the difference between its minimum and maximum value is small so as to maintain the resulting pressures within reasonable limits.

FN1 *Light drive fits* are those requiring light assembly pressures and producing more or less permanent assemblies. They are suitable for thin sections or long fits or in cast-iron external members.

FN2 *Medium drive fits* are suitable for ordinary steel parts or for shrink fits on light sections. They are about the tightest fits that can be used with high-grade cast-iron external members.

FN3 *Heavy drive fits* are suitable for heavier steel parts or for shrink fits in medium sections.

FN4—FN5 *Force fits* are suitable for parts which can be highly stressed or for shrink fits where the heavy pressing forces required are impractical.

---

* Extracted from American Standard Limits for Cylindrical Parts ANSI B4.1-1978, with the permission of the publishers, The American Society of Mechanical Engineers, United Engineering Center, 345 East 47th Street, New York 10017. Limit dimensions are tabulated in this standard for nominal sizes up to and including 200 in. An SI version is also available.

Table A-15-1 RUNNING AND SLIDING FITS

Size range, dia.

Class		0–0.12	0.12–0.24	0.24–0.40	0.40–0.71	0.71–1.19	1.19–1.97	1.97–3.15	3.15–4.73
RC1	Hole	+0.2 / −0	+0.2 / −0	+0.25 / −0	+0.3 / −0	+0.4 / −0	+0.4 / −0	+0.5 / −0	+0.6 / −0
	Shaft	+0.1 / −0.25	−0.15 / −0.3	−0.2 / −0.35	−0.25 / −0.45	−0.3 / −0.55	−0.4 / −0.7	−0.4 / −0.7	−0.5 / −0.9
RC2	Hole	+0.25 / −0	+0.3 / −0	+0.4 / −0	+0.4 / −0	+0.5 / −0	+0.6 / −0	+0.7 / −0	+0.9 / −0
	Shaft	−0.1 / −0.3	−0.15 / −0.35	−0.2 / −0.45	−0.25 / −0.55	−0.3 / −0.7	−0.4 / −0.8	−0.4 / −0.9	−0.5 / −1.1
RC3	Hole	+0.4 / −0	+0.5 / −0	+0.6 / −0	+0.7 / −0	+0.8 / −0	+1.0 / −0	+1.2 / −0	+1.4 / −0
	Shaft	−0.3 / −0.55	−0.4 / −0.7	−0.5 / −0.9	−0.6 / −1.0	−0.8 / −1.3	−1.0 / −1.6	−1.2 / −1.9	−1.4 / −2.3
RC4	Hole	+0.6 / −0	+0.7 / −0	+0.9 / −0	+1.0 / −0	+1.2 / −0	+1.6 / −0	+1.8 / −0	+2.2 / −0
	Shaft	−0.3 / −0.7	−0.4 / −0.9	−0.5 / −1.1	−0.6 / −1.3	−0.8 / −1.6	−1.0 / −2.0	−1.2 / −2.4	−1.4 / −2.8
RC5	Hole	+0.6 / −0	+0.7 / −0	+0.9 / −0	+1.0 / −0	+1.2 / −0	+1.6 / −0	+1.8 / −0	+2.2 / −0
	Shaft	−0.6 / −1.0	−0.8 / −1.3	−1.0 / −1.6	−1.2 / −1.9	−1.6 / −2.4	−2.0 / −3.0	−2.5 / −3.7	−3.0 / −4.4
RC6	Hole	+1.0 / −0	+1.2 / −0	+1.4 / −0	+1.6 / −0	+2.0 / −0	+2.5 / −0	+3.0 / −0	+3.5 / −0
	Shaft	−0.6 / −1.2	−0.8 / −1.5	−1.0 / −1.9	−1.2 / −2.2	−1.6 / −2.8	−2.0 / −3.6	−2.5 / −4.3	−3.0 / −5.2
RC7	Hole	+1.0 / −0	+1.2 / −0	+1.4 / −0	+1.6 / −0	+2.0 / −0	+2.5 / −0	+3.0 / −0	+3.5 / −0
	Shaft	−1.0 / −1.6	−1.2 / −1.9	−1.6 / −2.5	−2.0 / −3.0	−2.5 / −3.7	−3.0 / −4.6	−4.0 / −5.8	−5.0 / −7.2
RC8	Hole	+1.6 / −0	+1.8 / −0	+2.2 / −0	+2.8 / −0	+3.5 / −0	+4.0 / −0	+4.5 / −0	+5.0 / −0
	Shaft	−2.5 / −3.5	−2.8 / −4.0	−3.0 / −4.4	−3.5 / −5.1	−4.5 / −6.5	−5.0 / −7.5	−6.0 / −9.0	−7.0 / −10.5
RC9	Hole	+2.5 / −0	+3.0 / −0	+3.5 / −0	+4.0 / −0	+5.0 / −0	+6.0 / −0	+7.0 / −0	+9.0 / −0
	Shaft	−4.0 / −5.6	−4.5 / −6.0	−5.0 / −7.2	−6.0 / −8.8	−7.0 / −10.5	−8.0 / −12.0	−9.0 / −13.5	−10.0 / −15.0

Table A-15-2 LOCATIONAL CLEARANCE FITS

		Size range, dia.							
Class		0–0.12	0.12–0.24	0.24–0.40	0.40–0.71	0.71–1.19	1.19–1.97	1.97–3.15	3.15–4.73
LC1	Hole	+0.25 / −0	+0.3 / −0	+0.4 / −0	+0.4 / −0	+0.5 / −0	+0.6 / −0	+0.7 / −0	+0.9 / −0
	Shaft	+0 / −0.2	+0 / −0.2	+0 / −0.25	+0 / −0.3	+0 / −0.4	+0 / −0.4	+0 / −0.5	+0 / −0.6
LC2	Hole	+0.4 / −0	+0.5 / −0	+0.6 / −0	+0.7 / −0	+0.8 / −0	+1.0 / −0	+1.2 / −0	+1.4 / −0
	Shaft	+0 / −0.25	+0 / −0.3	+0 / −0.4	+0 / −0.5	+0 / −0.6	+0 / −0.7	+0 / −0.9	+0 / −1.0
LC3	Hole	+0.6 / −0	+0.7 / −0	+0.9 / −0	+1.0 / −0	+1.2 / −0	+1.6 / −0	+1.8 / −0	+2.2 / −0
	Shaft	+0 / −0.4	+0 / −0.5	+0 / −0.6	+0 / −0.7	+0 / −0.8	+0 / −1.0	+0 / −1.2	+0 / −1.4
LC4	Hole	+1.6 / −0	+1.8 / −0	+2.2 / −0	+2.8 / −0	+3.5 / −0	+4.0 / −0	+4.5 / −0	+5.0 / −0
	Shaft	+0 / −1.0	+0 / −1.2	+0 / −1.4	+0 / −1.6	+0 / −2.0	+0 / −2.5	+0 / −3.0	+0 / −3.5
LC5	Hole	+0.4 / −0	+0.5 / −0	+0.6 / −0	+0.7 / −0	+0.8 / −0	+1.0 / −0	+1.2 / −0	+1.4 / −0
	Shaft	−0.1 / −0.35	−0.15 / −0.45	−0.2 / −0.6	−0.25 / −0.65	−0.3 / −0.8	−0.4 / −1.0	−0.4 / −1.1	−0.5 / −1.4
LC6	Hole	+1.0 / −0	+1.2 / −0	+1.4 / −0	+1.6 / −0	+2.0 / −0	+2.5 / −0	+3.0 / −0	+3.5 / −0
	Shaft	−0.3 / −0.9	−0.4 / −1.1	−0.5 / −1.4	−0.6 / −1.6	−0.8 / −2.0	−1.0 / −2.6	−1.2 / −3.0	−1.4 / −3.6
LC7	Hole	+1.6 / −0	+1.8 / −0	+2.2 / −0	+2.8 / −0	+3.5 / −0	+4.0 / −0	+4.5 / −0	+5.0 / −0
	Shaft	−0.6 / −1.6	−0.8 / −2.0	−1.0 / −2.4	−1.2 / −2.8	−1.6 / −3.6	−2.0 / −4.5	−2.5 / −5.5	−3.0 / −6.5
LC8	Hole	+1.6 / −0	+1.8 / −0	+2.2 / −0	+2.8 / −0	+3.5 / −0	+4.0 / −0	+4.5 / −0	+5.0 / −0
	Shaft	−1.0 / −2.0	−1.2 / −2.4	−1.6 / −3.0	−2.0 / −3.6	−2.5 / −4.5	−3.0 / −5.5	−4.0 / −7.0	−5.0 / −8.5
LC9	Hole	+2.5 / −0	+3.0 / −0	+3.5 / −0	+4.0 / −0	+5.0 / −0	+6.0 / −0	+7.0 / −0	+9.0 / −0
	Shaft	−2.5 / −4.6	−2.8 / −4.6	−3.0 / −5.2	−3.5 / −6.3	−4.5 / −8.0	−5.0 / −9.0	−6.0 / −10.5	−7.0 / −12.0
LC10	Hole	+4.0 / −0	+5.0 / −0	+6.0 / −0	+7.0 / −0	+8.0 / −0	+10.0 / −0	+12.0 / −0	+14.0 / −0
	Shaft	−4.0 / −9.5	−4.5 / −9.5	−5.0 / −11.0	−6.0 / −13.0	−7.0 / −15.0	−8.0 / −18.0	−10.0 / −22.0	−11.0 / −25.0
LC11	Hole	+6.0 / −0	+7.0 / −0	+9.0 / −0	+10.0 / −0	+12.0 / −0	+16.0 / −0	+18.0 / −0	+22.0 / −0
	Shaft	−5.0 / −11.0	−6.0 / −13.0	−7.0 / −16.0	−8.0 / −18.0	−10.0 / −22.0	−12.0 / −28.0	−14.0 / −32.0	−16.0 / −38.0

		Size range, dia.							
Class		0–0.12	0.12–0.24	0.24–0.40	0.40–0.71	0.71–1.19	1.19–1.97	1.97–3.15	3.15–4.73
LT1	Hole	+0.4 −0	+0.5 −0	+0.6 −0	+0.7 −0	+0.8 −0	+1.0 −0	+1.2 −0	+1.4 −0
	Shaft	+0.10 −0.10	+0.15 −0.15	+0.2 −0.2	+0.2 −0.2	+0.25 −0.25	+0.3 −0.3	+0.3 −0.3	+0.4 −0.4
LT2	Hole	+0.6 −0	+0.7 −0	+0.9 −0	+1.0 −0	+1.2 −0	+1.6 −0	+1.8 −0	+2.2 −0
	Shaft	+0.2 −0.2	+0.25 −0.25	+0.3 −0.3	+0.35 −0.35	+0.4 −0.4	+0.5 −0.5	+0.6 −0.6	+0.7 −0.7
LT3	Hole			+0.6 −0	+0.7 −0	+0.8 −0	+1.0 −0	+1.2 −0	+1.4 −0
	Shaft			+0.5 +0.1	+0.5 +0.1	+0.6 +0.1	+0.7 +0.1	+0.8 +0.1	+1.0 +0.1
LT4	Hole			+0.9 −0	+1.0 −0	+1.2 −0	+1.6 −0	+1.8 −0	+2.2 −0
	Shaft			+0.7 +0.1	+0.8 +0.1	+0.9 +0.1	+1.1 +0.1	+1.3 +0.1	+1.5 +0.1
LT5	Hole	+0.4 −0	+0.5 −0	+0.6 −0	+0.7 −0	+0.8 −0	+1.0 −0	+1.2 −0	+1.4 −0
	Shaft	+0.5 +0.25	+0.6 +0.3	+0.8 +0.4	+0.9 +0.5	+1.1 +0.6	+1.3 +0.7	+1.5 +0.8	+1.9 +1.0
LT6	Hole	+0.4 −0	+0.5 −0	+0.6 −0	+0.7 −0	+0.8 −0	+1.0 −0	+1.2 −0	+1.4 −0
	Shaft	+0.65 +0.25	+0.8 +0.3	+1.0 +0.4	+1.2 +0.5	+1.4 +0.6	+1.7 +0.7	+2.0 +0.8	+2.4 +1.0

		Size range, dia.							
Class		0–0.12	0.12–0.24	0.24–0.40	0.40–0.71	0.71–1.19	1.19–197	1.97–3.15	3.15–4.73
LN1	Hole	+0.25 −0	+0.3 −0	+0.4 −0	+0.4 −0	+0.5 −0	+0.6 −0	+0.7 −0	+0.9 −0
	Shaft	+0.45 +0.25	+0.5 +0.3	+0.65 +0.4	+0.8 +0.4	+1.0 +0.5	+1.1 +0.6	+1.3 +0.7	+1.6 +1.0
LN2	Hole	+0.4 −0	+0.5 −0	+0.6 −0	+0.7 −0	+0.8 −0	+1.0 −0	+1.2 −0	+1.4 −0
	Shaft	+0.65 +0.4	+0.8 +0.5	+1.0 +0.6	+1.1 +0.7	+1.3 +0.8	+1.6 +1.0	+2.1 +1.4	+2.5 +1.6
LN3	Hole	+0.4 −0	+0.5 −0	+0.6 −0	+0.7 −0	+0.8 −0	+1.0 −0	+1.2 −0	+1.4 −0
	Shaft	+0.75 +0.5	+0.9 +0.6	+1.2 +0.8	+1.4 +1.0	+1.7 +1.2	+2.0 +1.4	+2.3 +1.6	+2.9 +2.0

Table A-15-5   FORCE AND SHRINK FITS

Size range, dia.

Class		0–0.12		0.12–0.24		0.24–0.40		0.40–0.56		0.56–0.71		0.71–0.95		0.95–1.19		1.19–1.58	
FN1	Hole	+0.25	−0	+0.3	−0	+0.4	−0	+0.4	−0	+0.4	−0	+0.5	−0	+0.5	−0	+0.6	−0
	Shaft	+0.5	+0.3	+0.6	+0.4	+0.75	+0.5	+0.8	+0.5	+0.9	+0.6	+1.1	+0.7	+1.2	+0.8	+1.3	+0.9
FN2	Hole	+0.4	−0	+0.5	−0	+0.6	−0	+0.7	−0	+0.7	−0	+0.8	−0	+0.8	−0	+1.0	−0
	Shaft	+0.85	+0.6	+1.0	+0.6	+1.4	+1.0	+1.6	+1.2	+1.6	+1.2	+1.9	+1.4	+1.9	+1.4	+2.4	+1.8
FN3	Hole													+0.8	−0	+1.0	−0
	Shaft													+2.1	+1.6	+2.6	+2.0
FN4	Hole	+0.4	−0	+0.5	−0	+0.6	−0	+0.7	−0	+0.7	−0	+0.8	−0	+0.8	−0	+1.0	−0
	Shaft	+0.95	+0.7	+1.2	+0.9	+1.6	+1.2	+1.8	+1.4	+1.8	+1.4	+2.1	+1.6	+2.3	+1.8	+3.1	+2.5
FN5	Hole	+0.6	−0	+0.7	−0	+0.9	−0	+1.0	−0	+1.0	−0	+1.2	−0	+1.2	−0	+1.6	−0
	Shaft	+1.3	+0.9	+1.7	+1.2	+2.0	+1.4	+2.3	+1.6	+2.5	+1.8	+3.0	+2.2	+3.3	+2.5	+4.0	+3.0

Size range, dia.

Class		1.58–1.97		1.97–2.56		2.56–3.15		3.15–3.94		3.94–4.73		4.73–5.52		5.52–6.30		6.30–7.09	
FN1	Hole	+0.6	−0	+0.7	−0	+0.7	−0	+0.9	−0	+0.9	−0	+1.0	−0	+1.0	−0	+1.0	−0
	Shaft	+1.4	+1.0	+1.8	+1.3	+1.9	+1.4	+2.4	+1.8	+2.6	+2.0	+2.9	+2.2	+3.2	+2.5	+3.5	+2.8
FN2	Hole	+1.0	−0	+1.2	−0	+1.2	−0	+1.4	−0	+1.4	−0	+1.6	−0	+1.6	−0	+1.6	−0
	Shaft	+2.4	+1.8	+2.7	+2.0	+2.9	+2.2	+3.7	+2.8	+3.9	+3.0	+4.5	+3.5	+5.0	+4.0	+5.5	+4.5
FN3	Hole	+1.0	−0	+1.2	−0	+1.2	−0	+1.4	−0	+1.4	−0	+1.6	−0	+1.6	−0	+1.6	−0
	Shaft	+2.8	+2.2	+3.2	+2.5	+3.7	+3.0	+4.4	+3.5	+4.9	+4.0	+6.0	+5.0	+6.0	+5.0	+7.0	+6.0
FN4	Hole	+1.0	−0	+1.2	−0	+1.2	−0	+1.4	−0	+1.4	−0	+1.6	−0	+1.6	−0	+1.6	−0
	Shaft	+3.4	+2.8	+4.2	+3.5	+4.7	+4.0	+5.9	+5.0	+6.9	+6.0	+8.0	+7.0	+8.0	+7.0	+9.0	+8.0
FN5	Hole	+1.6	−0	+1.8	−0	+1.8	−0	+2.2	−0	+2.2	−0	+2.5	−0	+2.5	−0	+2.5	−0
	Shaft	+5.0	+4.0	+6.2	+5.0	+7.2	+6.0	+8.4	+7.0	+9.4	+8.0	+11.6	+10.0	+13.6	+12.0	+13.6	+12.0

Table A-16 AMERICAN STANDARD PIPE

Nominal size, in	Outside diameter, in	Threads per inch	Wall thickness, in Standard No. 40	Extra strong No. 80	Double extra strong
$\frac{1}{8}$	0.405	27	0.070	0.098	
$\frac{1}{4}$	0.540	18	0.090	0.122	
$\frac{3}{8}$	0.675	18	0.093	0.129	
$\frac{1}{2}$	0.840	14	0.111	0.151	0.307
$\frac{3}{4}$	1.050	14	0.115	0.157	0.318
1	1.315	$11\frac{1}{2}$	0.136	0.183	0.369
$1\frac{1}{4}$	1.660	$11\frac{1}{2}$	0.143	0.195	0.393
$1\frac{1}{2}$	1.900	$11\frac{1}{2}$	0.148	0.204	0.411
2	2.375	$11\frac{1}{2}$	0.158	0.223	0.447
$2\frac{1}{2}$	2.875	8	0.208	0.282	0.565
3	3.500	8	0.221	0.306	0.615
$3\frac{1}{2}$	4.000	8	0.231	0.325	
4	4.500	8	0.242	0.344	0.690
5	5.563	8	0.263	0.383	0.768
6	6.625	8	0.286	0.441	0.884
8	8.625	8	0.329	0.510	0.895

Table A-17 MECHANICAL PROPERTIES OF STEELS*

The values shown for hot-rolled (HR) and cold-drawn (CD) steels are *estimated minimum values* which can usually be expected in the size range of $\frac{3}{4}$ to $1\frac{1}{4}$ in. A minimum value is roughly several standard deviations below the arithmetic mean. The values shown for heat-treated steels are so-called *typical values*. A typical value is neither the mean nor the minimum. It can be obtained by careful control of the purchase specifications and the heat treatment, together with continuous inspection and testing. The properties shown in this table are from a variety of sources and are believed to be representative. There are so many variables which affect these properties, however, that their approximate nature must be clearly recognized.

UNS number	AISI number	Processing	Yield strength, kpsi†	Tensile strength, kpsi†	Elongation in 2 in, %	Reduction in area, %	Brinell hardness, H_B
G10100	1010	HR	26	47	28	50	95
		CD	44	53	20	40	105
G10150	1015	HR	27	50	28	50	101
		CD	47	56	18	40	111
G10180	1018	HR	32	58	25	50	116
		CD	54	64	15	40	126
G10180	1112	HR	33	56	25	45	121
		CD	60	78	10	35	167
G10350	1035	HR	39	72	18	40	143
		CD	67	80	12	35	163
		Drawn 800°F	81	110	18	51	220
		Drawn 1000°F	72	103	23	59	201
		Drawn 1200°F	62	91	27	66	180
G10400	1040	HR	42	76	18	40	149
		CD	71	85	12	35	170
		Drawn 1000°F	84	113	23	62	235
G10450	1045	HR	45	82	16	40	163
		CD	77	91	12	35	179
G10500	1050	HR	49	90	15	35	179
		CD	84	100	10	30	197

SAE	UNS	Condition					
2317		Drawn 600°F	180	220	10	30	450
		Drawn 900°F	130	155	18	55	310
		Drawn 1200°F	80	105	28	65	210
		Core‡	107	137	22	52	285
2330		Drawn 400°F	195	221	11	40	425
		Drawn 600°F	171	196	14	49	382
		Drawn 800°F	131	160	18	56	327
		Drawn 1000°F	97	127	23	61	268
		Drawn 1200°F	70	108	27	64	222
2340		Drawn 800°F	164	178	23	53	368
2345		Drawn 800°F	177	188	20	51	388
2350		Drawn 800°F	180	194	17	50	402
3120		Drawn 600°F	145	162	12	45	320
		Drawn 1000°F	91	112	22	68	222
3130		Drawn 600°F	178	210	10	37	404
		Drawn 1000°F	120	137	20	62	276
3140		HR§	64	96	26	56	197
		CD	91	104	17	48	212
3145		Drawn 800°F	157	188	15	50	376
3150		Drawn 800°F	164	195	12	47	380
3240		Drawn 800°F	171	202	12	44	396
3250		Drawn 600°F	211	237	10	40	466
3340		Drawn 600°F	214	243	9	37	477
4130	G41300	Drawn 800°F	183	211	13	47	394
		HR§	60	90	30	45	183
		CD§	87	98	21	52	201
		Drawn 1000°F	133	146	17	60	293
4140	G41400	HR§	63	90	27	58	187
		CD§	90	102	18	50	223
		Drawn 1000°F	131	153	16	45	302
4340	G43400	HR§	69	101	21	45	207
		CD§	99	111	16	42	223
		Drawn 600°F	234	260	12	43	498
		Drawn 1000°F	162	182	15	40	363
4620	G46200	Core‡	89	120	22	55	248

Table A-17 MECHANICAL PROPERTIES OF STEELS* (*continued*)

UNS number	AISI number	Processing	Yield strength, kpsi†	Tensile strength, kpsi†	Elongation in 2 in, %	Reduction in area, %	Brinell hardness, H_B
	4640	Drawn 800°F	94	130	23	66	256
	4650	Drawn 800°F	170	187	13	54	378
		Drawn 800°F	179	198	13	49	410
G15216	52100	HR§	81	100	25	57	192
G61500	6150	HR§	58	91	22	53	183
		Drawn 1000°F	132	155	15	44	302
	8650	HR§	58	99	20	48	197
		Drawn 1000°F	132	155	14	42	311
G87400	8740	HR§	64	95	25	55	190
		CD§	96	107	17	48	223
		Drawn 1000°F	129	152	15	44	302
G92550	9255	HR§	78	115	22	45	223
		Drawn 1000°F	160	180	15	32	352
	9442	Drawn 800°F	180	201	12	43	404
	9840	Drawn 800°F	199	218	12	47	436

* Tabulated in accordance with the Unified Numbering System for Metals and Alloys (UNS), Society of Automotive Engineers, Warrendale, Pa., 1975.

† Multiply strength in kpsi by 6.89 to get the strength in MPa.

‡ Case-hardened, core properties.

§ Annealed.

Table A-18 PROPERTIES OF SOME HIGH-STRENGTH STEELS

AISI number	Processing[a]	Brinell hardness, H_B	Modulus of elasticity, E, Mpsi	Yield strength[b] S_y, kpsi	Ultimate strength S_u, kpsi	Reduction in area, %	True fracture strength, σ_F kpsi	True fracture ductility, ε_F	Straining hardening exponent, n
1045	Q & T 80°F	705	29	265T² 300C	300	2	310T 420C	0.02	0.186
1045	Q & T 360°F	595	30	270	325	41	430/ 395	0.52	0.071
1045	Q & T 500°F	500	30	245	265	51	370/ 330	0.71	0.047
1045	Q & T 600°F	450	30	220	230	55	345/ 305	0.81	0.041
1045	Q & T 720°F	390	30	185	195	59	315/ 270	0.89	0.044
4142	Q & T 80°F	670	29	235T 275C	355	6	375	0.06	0.136
4142	Q & T 400°F	560	30	245	325	27	405/ 385	0.31	0.091
4142	Q & T 600°F	475	30	250	280	35	340/ 315	0.43	0.048
4142	Q & T 700°F	450	30	230	255	42	320/ 290	0.54	0.043

Table A-18   PROPERTIES OF SOME HIGH-STRENGTH STEELS (*continued*)

AISI number	Processing[a]	Brinell hardness, H_B	Modulus of elasticity E, Mpsi	Yield strength[b] S_y, kpsi	Ultimate strength S_u, kpsi	Reduction in area, %	True fracture strength, σ_F kpsi	True fracture ductility, ε_F	Straining hardening exponent, n
4142	Q & T 840°F	380	30	200	205	48	295/ 265	0.66	0.051
4142[e]	Q & D 550°F	475	29	275T 225C	295	20	310/ 300	0.22	0.101T 0.060C
4142	Q & D 650°F	450	29	270T 205C	280	37	330/ 305	0.46	0.016T 0.070C
4142	Q & D 800°F	400	29	210T[f] 175C	225	47	305/ 275	0.63	0.032T 0.085C

Source: Data from R. W. Landgraf, *Cyclic Deformation and Fatigue Behavior of Hardened Steels*, Report no. 320, Dept. of Theoretical and Applied Mechanics, University of Illinois, Urbana, 1968.

[a] AISI 1045: Cold-drawn to 9/16 in rounds from hot-rolled rod.   Austenized 1500°F (oxidizing atmosphere) 20 min, water quenched at 70°F.   AISI 4142: Cold-drawn to 9/16 in rounds from annealed rod.   Austenized at 1500°F (neutral atmosphere), quenched in agitated oil at 180°F.   AISI 4142 Def: Austenized at 1500°F, oil quenched.   Reheated in molten lead, drawn 14 percent through die at reheating temperature to $\frac{5}{8}$ in rods.

[b] 0.2 percent offset method.

[c] $(P_F/A_F)/(P_F/A_F)$ (Bridgman's correction for necking).

[d] T, tension; C, compression.

[e] Deformed 14 percent.

[f] Proportional limit in tension.

Table A-19 MECHANICAL PROPERTIES OF WROUGHT ALUMINUM ALLOYS

These are *typical* properties for sizes of about $\frac{1}{2}$ in. A typical value may be neither the mean nor the minimum. It is a value which can be obtained when the purchase specifications are carefully written and with continuous inspection and testing. The values given for fatigue strength S_f correspond to $50(10)^7$ cycles of completely reversed stress. Aluminum alloys do not have an endurance limit. The yield strength is the 0.2 percent offset value.

UNS alloy number	Temper	Yield strength S_y, kpsi	Tensile strength S_u, kpsi	Shear modulus of rupture S_{su}, kpsi	Fatigue strength S_f, kpsi	Elongation in 2 in, %	Brinell hardness, H_B
A91100	–O	5	13	9.5	5	45	23
	–H12	14	15.5	10	6	25	28
	–H14	20	22	14	9	16	40
	–H16	24	26	15	9.5	14	47
	–H18	27	29	16	10	10	55
A93003	–O	6	16	11	7	40	28
	–H12	17	19	12	8	20	35
	–H14	20	22	14	9	16	40
	–H16	24	26	15	9.5	14	47
	–H18	27	29	16	10	10	55
A93004	–O	10	26	16	14	25	45
	–H32	22	31	17	14.5	17	52
	–H34	27	34	18	15	12	63
	–H36	31	37	20	15.5	9	70
	–H38	34	40	21	16	6	77
A92011	–T3	48	55	32	18	15	95
	–T8	45	59	35	18	12	100
A92014	–O	14	27	18	13	18	45
	–T4	40	62	38	20	20	105
	–T6	60	70	42	18	13	135

Table A-19 MECHANICAL PROPERTIES OF WROUGHT ALUMINUM ALLOYS (*continued*)

UNS alloy number	Temper	Yield strength S_y, kpsi	Tensile strength S_u, kpsi	Shear modulus of rupture S_{su}, kpsi	Fatigue strength S_f, kpsi	Elongation in 2 in, %	Brinell hardness H_B
A92017	—O	10	26	18	13	22	45
	—T4	40	62	38	18	22	105
A92018	—T61	46	61	39	17	12	120
A92024	—O	11	27	18	13	22	47
	—T3	50	70	41	20	16	120
	—T4	48	68	41	20	19	120
	—T36	57	73	42	18	13	130
A95052	—O	13	28	18	17	30	45
	—H32	27	34	20	17.5	18	62
	—H34	31	37	21	18	14	67
	—H36	34	39	23	18.5	10	74
	—H38	36	41	24	19	8	85
A95056	—O	22	42	26	20	35	
	—H18	59	63	34	22	10	
	—H38	50	60	32	22	15	
A96061	—O	8	18	12.5	9	30	30
	—T4	21	35	24	13.5	25	65
	—T6	40	45	30	13.5	17	95
A97075	—T6	72	82	49	24	11	150

Table A-20 MECHANICAL PROPERTIES OF ALUMINUM ALLOY CASTINGS

These are typical properties for $\frac{1}{2}$-in sizes and may be neither the mean nor the minimum, but are properties which can be attained with reasonable care. The fatigue strength is for $50(10)^7$ cycles of reversed stress. Both yield strengths are obtained by the 0.2 percent offset method. Multiply strengths in kpsi by 6.89 to get strength in MPa.

| UNS number and temper | Tension | | | Compressive yield strength, kpsi | Shear modulus of rupture, kpsi | Fatigue strength, kpsi | Brinell hardness H_B |
	Yield strength, kpsi	Ultimate strength, kpsi	Elongation in 2 in, %				
A03190*	18	27	2.0	19	22	10	70
A03190-T6*	24	36	2.0	25	29	10	80
A03330†	19	34	2.0	19	27	14.5	90
A03330-T5†	25	34	1.0	25	27	12	100
A03330-T6†	30	42	1.5	30	33	15	105
A03330-T62†	40	45	1.5	40	36	10	105
A03550-T6*	25	35	3.0	26	28	9	80
A03550-T7*	36	38	0.5	38	28	9	85
A03550-T71*	29	35	1.5	30	26	10	75
A03560-T51*	20	25	2.0	21	20	7.5	60
A03560-T6*	24	33	3.5	25	26	8.5	70
A03560-T7*	30	34	2.0	31	24	9	75

* Sand casting.

† Permanent-mold casting.

Table A-21 TYPICAL PROPERTIES OF GRAY CAST IRON

The American Society for Testing Materials (ASTM) numbering system for gray cast iron is established such that the numbers correspond to the *minimum tensile strength* in kpsi. Thus an ASTM No. 20 cast iron has a minimum tensile strength of 20 kpsi. Note particularly that the tabulations are *typical values*.

ASTM number	Tensile strength S_{ut}, kpsi	Compressive strength S_{uc}, kpsi	Shear modulus of rupture S_{su}, kpsi	Modulus of elasticity, Mpsi		Endurance limit S_e, kpsi	Brinell harness, H_B
				Tension	Torsion		
20	22	83	26	9.6–14	3.9–5.6	10	156
25	26	97	32	11.5–14.8	4.6–6.0	11.5	174
30	31	109	40	13–16.4	5.2–6.6	14	201
35	36.5	124	48.5	14.5–17.2	5.8–6.9	16	212
40	42.5	140	57	16–20	6.4–7.8	18.5	235
50	52.5	164	73	18.8–22.8	7.2–8.0	21.5	262
60	62.5	187.5	88.5	20.4–23.5	7.8–8.5	24.5	302

Table A-22 TYPICAL PROPERTIES OF SOME COPPER-BASE ALLOYS

All yield strengths are by 0.5 percent offset method. Multiply strength in MPa by 0.145 to get strength in kpsi.

UNS number	Alloy name	Form	Temper	Yield strength MPa	Tensile strength MPa	Elongation in 50 mm %	Rockwell hardness
C17000	Beryllium	Rod	Hard	515	790	5	98B
		Rod	Soft	170	415	50	77B
		Sheet	Hard	1000	1240	2	⋯
C21000	Gilding brass	Sheet	Hard	345	385	5	64B
		Sheet	Soft	70	235	45	46F
C22000	Commercial bronze	Sheet	Hard	370	420	5	70B
		Sheet	Soft	70	255	45	53F
		Rod	Hard	380	415	20	60B
		Rod	Soft	70	275	50	55F
C23000	Red brass	Sheet	Hard	395	480	5	77B
		Sheet	Soft	85	275	47	59F
		Rod	Hard	360	395	23	75B
		Rod	Soft	70	275	55	55F
C26000	Cartridge brass	Sheet	Hard	435	525	8	82B
		Sheet	Soft	105	325	62	64F
		Rod	Hard	360	480	30	80B
		Rod	Soft	110	330	65	65F
C27000	Yellow brass	Sheet	Hard	415	510	8	80B
		Sheet	Soft	105	325	62	64F
		Rod	Hard	310	415	25	80B
		Rod	Soft	110	330	65	65F

Table A-22 TYPICAL PROPERTIES OF SOME COPPER-BASE ALLOYS (*continued*)

UNS number	Alloy name	Form	Temper	Yield strength MPa	Tensile strength MPa	Elongation in 50 mm %	Rockwell hardness
C28000	Muntz metal	Sheet	Hard	415	550	10	85B
		Sheet	Soft	145	370	45	80F
		Rod	Hard	380	515	20	80B
		Rod	Soft	145	370	50	80F
		Tube	Hard	380	510	10	80B
		Tube	Soft	160	385	50	82F
C33000	Low-leaded brass	Tube	Hard	415	515	7	80B
		Tube	Soft	105	325	60	64F
C33200	High-leaded brass	Sheet	Hard	415	510	7	80B
		Sheet	Soft	115	340	52	68F
C46200	Naval brass	Sheet	Hard	480	620	5	90B
		Rod	Hard	365	515	20	82B
		Tube	Hard	455	605	18	95B

Table A-23 TYPICAL MECHANICAL PROPERTIES OF WROUGHT STAINLESS STEELS
All yield strengths are obtained using the 0.2 percent offset method.

UNS number	Processing	Yield strength, kpsi	Tensile strength, kpsi	Elongation in 2 in, %	Reduction in area, %	Brinell hardness, H_B
S20100	Annealed	55	155	55		
	¼ hard	75	125	20		
	½ hard	110	150	10		
	¾ hard	135	175	5		
	Full hard	140	185	4		
S20200	Annealed	55	110	55		
	¼ hard	75	125	12		
S30100	Annealed	40	110	60		165
	¼ hard	75	125	25		
	½ hard	110	150	15		
	¾ hard	135	175	12		
	Full hard	140	185	8		
S30200	Annealed	37	90	55	65	155
	¼ hard	75	125	12		
S30300	Annealed	35	90	50	55	160
S30400	Annealed	35	85	55	65	150
S31000	Annealed	40	95	45	65	170
S31400	Annealed	50	100	45	60	170
S41400	Annealed	95	120	17	55	235
	Drawn 400°F	150	200	15	55	415
	Drawn 600°F	145	190	15	55	400
	Drawn 800°F	150	200	16	58	415
	Drawn 1000°F	120	145	20	60	325
	Drawn 1200°F	105	120	20	65	260

Table A-23 TYPICAL MECHANICAL PROPERTIES OF WROUGHT STAINLESS STEELS (*continued*)

UNS number	Processing	Yield strength, kpsi	Tensile strength, kpsi	Elongation in 2 in, %	Reduction in area, %	Brinell hardness H_B
S41600	Annealed	40	75	30	65	155
	Drawn 400°F	145	190	15	55	390
	Drawn 600°F	140	180	15	55	375
	Drawn 800°F	150	195	17	55	390
	Drawn 1000°F	115	145	20	65	300
	Drawn 1200°F	85	110	23	65	225
	Drawn 1400°F	60	90	30	70	180
S43100	Annealed	95	125	20	60	260
	Drawn 400°F	155	205	15	55	415
	Draan 600°F	150	195	15	55	400
	Drawn 800°F	155	205	15	60	415
	Drawn 1200°F	95	125	20	60	260
S50100	Annealed	30	70	28	65	160
S50200	Annealed	30	70	30	75	150

Source: By permission, *Metals Handbook*, 8th ed., vol. 1, American Society for Metals, Metals Park, Ohio, 1961, p. 414.

Table A-24 TYPICAL PROPERTIES OF MAGNESIUM ALLOYS*

Since magnesium does not have an endurance limit, the fatigue strength shown is for $50(10)^7$ cycles of reversed stress.

Magnesium alloy	Tensile yield strength, kpsi	Tensile strength, kpsi	Elonga-tion in 2 in, %	Compres-sive yield strength, kpsi	Brinell hardness H_B	Shear strength, kpsi	Fatigue strength, kpsi
Cast AM 265C	11	27	6	11	48	⋯	11
Cast AM 240-T4	12	35	9	12	52	20	11
Cast AM 260-T6	20	38	3	20	78	22	11.5
Die-cast AM 263	22	34	3	⋯	⋯	⋯	14
Wrought AM 3S	30	40	7	11	40–52	16.7	11
Wrought AM C52S	30	40	17	20	50–71	19	15
Wrought AM C57S	32	44	14	20	55–74	20.5	17
Wrought AM 59S	38	51	9	27	70	22	18

* Courtesy of American Magnesium Corporation.

Table A-25 DECIMAL EQUIVALENTS OF WIRE AND SHEET-METAL GAUGES*

Name of gauge:	American or Brown & Sharpe	Birmingham or Stubs iron wire	United States Standard	Manu-facturers Standard†	Steel wire or Washburn & Moen	Music wire	Stubs steel wire	Twist drill
Principal use:	Nonferrous sheet, wire, and rod	Tubing, ferrous strip, flat wire, and spring steel	Ferrous sheet and plate, 480 lb/ft^3	Ferrous sheet	Ferrous wire except music wire	Music wire	Steel drill rod	Twist drills and drill steel
7/0	⋯	⋯	0.500	⋯	0.490 0			
6/0	0.580 0	⋯	0.468 75	⋯	0.461 5	0.004		
5/0	0.516 5	⋯	0.437 5	⋯	0.430 5	0.005		
4/0	0.460 0	0.454	0.406 25	⋯	0.393 8	0.006		
3/0	0.409 6	0.425	0.375	⋯	0.362 5	0.007		
2/0	0.364 8	0.380	0.343 75	⋯	0.331 0	0.008		
0	0.324 9	0.340	0.312 5	⋯	0.306 5	0.009		
1	0.289 3	0.300	0.281 25	⋯	0.283 0	0.010	0.227	0.228 0
2	0.257 6	0.284	0.265 625	⋯	0.262 5	0.011	0.219	0.221 0
3	0.229 4	0.259	0.25	0.239 1	0.243 7	0.012	0.212	0.213 0
4	0.204 3	0.238	0.234 375	0.224 2	0.225 3	0.013	0.207	0.209 0
5	0.181 9	0.220	0.218 75	0.209 2	0.207 0	0.014	0.204	0.205 5

6	0.162 0	0.203	0.203 125	0.194 3	0.192 0	0.016	0.201	0.204 0
7	0.144 3	0.180	0.187 5	0.179 3	0.177 0	0.018	0.199	0.201 0
8	0.128 5	0.165	0.171 875	0.164 4	0.162 0	0.020	0.197	0.199 0
9	0.114 4	0.148	0.156 25	0.149 5	0.148 3	0.022	0.194	0.196 0
10	0.101 9	0.134	0.140 625	0.134 5	0.135 0	0.024	0.191	0.193 5
11	0.090 74	0.120	0.125	0.119 6	0.120 5	0.026	0.188	0.191 0
12	0.080 81	0.109	0.109 357	0.104 6	0.105 5	0.029	0.185	0.189 0
13	0.071 96	0.095	0.093 75	0.089 7	0.091 5	0.031	0.182	0.185 0
14	0.064 08	0.083	0.078 125	0.074 7	0.080 0	0.033	0.180	0.182 0
15	0.057 07	0.072	0.070 312 5	0.067 3	0.072 0	0.035	0.178	0.180 0
16	0.050 82	0.065	0.062 5	0.059 8	0.062 5	0.037	0.175	0.177 0
17	0.045 26	0.058	0.056 25	0.053 8	0.054 0	0.039	0.172	0.173 0
18	0.040 30	0.049	0.05	0.047 8	0.047 5	0.041	0.168	0.169 5
19	0.035 89	0.042	0.043 75	0.041 8	0.041 0	0.043	0.164	0.166 0
20	0.031 96	0.035	0.037 5	0.035 9	0.034 8	0.045	0.161	0.161 0
21	0.028 46	0.032	0.034 375	0.032 9	0.031 7	0.047	0.157	0.159 0
22	0.025 35	0.028	0.031 25	0.029 9	0.028 6	0.049	0.155	0.157 0
23	0.022 57	0.025	0.028 125	0.026 9	0.025 8	0.051	0.153	0.154 0
24	0.020 10	0.022	0.025	0.023 9	0.023 0	0.055	0.151	0.152 0
25	0.017 90	0.020	0.021 875	0.020 9	0.020 4	0.059	0.148	0.149 5

Table A-25   DECIMAL EQUIVALENTS OF WIRE AND SHEET-METAL GAUGES* *(concluded)*

Name of gauge:	American or Brown & Shape	Birmingham or Stubs iron wire	United States Standard	Manu-facturers Standard†	Steel wire or Washburn & Moen	Music wire	Stubs steel wire	Twist drill
Principal use:	Nonferrous sheet, wire and rod	Tuning, ferrous strip, flat wire, and spring steel	Ferrous sheet and plate, 480 lb/ft³	Ferrous sheet	Ferrous wire except music wire	Music wire	Steel drill rod	Twist drills and drill steel
26	0.015 94	0.018	0.018 75	0.017 9	0.018 1	0.063	0.146	0.147 0
27	0.014 20	0.016	0.017 187 5	0.016 4	0.017 3	0.067	0.143	0.144 0
28	0.012 64	0.014	0.015 625	0.014 9	0.016 2	0.071	0.139	0.140 5
29	0.011 26	0.013	0.014 062 5	0.013 5	0.015 0	0.075	0.134	0.136 0
30	0.010 03	0.012	0.012 5	0.012 0	0.014 0	0.080	0.127	0.128 5
31	0.008 928	0.010	0.010 937 5	0.010 5	0.013 2	0.085	0.120	0.120 0
32	0.007 950	0.009	0.010 156 25	0.009 7	0.012 8	0.090	0.115	0.116 0
33	0.007 080	0.008	0.009 375	0.009 0	0.011 8	0.095	0.112	0.113 0
34	0.006 305	0.007	0.008 593 75	0.008 2	0.010 4	...	0.110	0.111 0
35	0.005 615	0.005	0.007 812 5	0.007 5	0.009 5	...	0.108	0.110 0
36	0.005 000	0.004	0.007 031 25	0.006 7	0.009 0	...	0.106	0.106 5
37	0.004 453	...	0.006 640 625	0.006 4	0.008 5	...	0.103	0.104 0
38	0.003 965	...	0.006 25	0.006 0	0.008 0	...	0.101	0.101 5
39	0.003 531	...	...	...	0.007 5	...	0.099	0.099 5
40	0.003 145	...	...	...	0.007 0	...	0.097	0.098 0

* Reproduced by courtesy of the Reynolds Metal Company. Specify sheet, wire, and plate by stating the gauge number, the gauge name,
and the decimal equivalent in parentheses.

† Reflects present average unit weights of sheet steel.

Table A-26 CHARTS OF THEORETICAL STRESS-CONCENTRATION
FACTORS K_t*

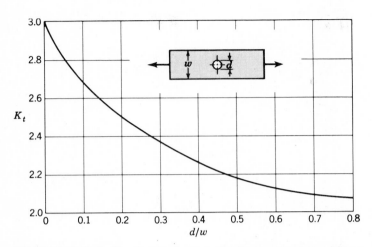

FIGURE A-26-1 Bar in tension or simple compression with a
transverse hole. $\sigma_0 = F/A$, where $A = (w - d)t$, and where t is the
thickness.

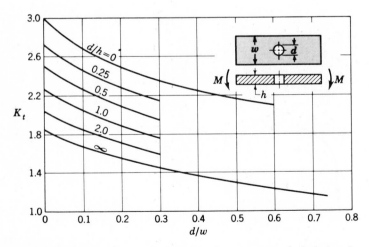

FIGURE A-26-2 Rectangular bar with a transverse hole in bending. $\sigma_0 = Mc/I$, where $I = (w - d)h^3/12$.

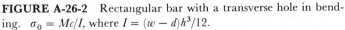

* Unless otherwise stated, these factors are from R. E. Peterson, "Design Factors for Stress Concentration," *Machine Design*, vol. 23, no. 2, February 1951, p. 169, no. 3, March 1951 p.161; no. 5,
May 1951, p. 159; no. 6, June 1951, p. 173; no. 7, July 1951, p. 155; reproduced with the permission
of the author and publisher.

Table A-26 CHARTS OF THEORETICAL STRESS-CONCENTRATION
FACTORS K_t (*continued*)

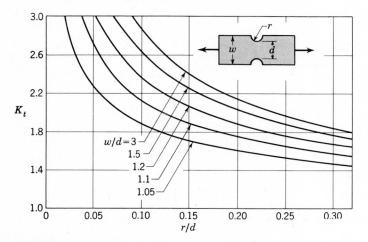

FIGURE A-26-3 Notched rectangular bar in tension or simple
compression. $\sigma_0 = F/A$, where $A = dt$ and t is the thickness.

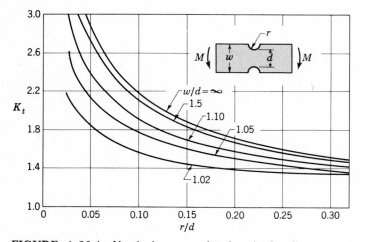

FIGURE A-26-4 Notched rectangular bar in bending. $\sigma_0 = Mc/I$, where $c = d/2$ and $I = td^3/12$. The thickness is t.

Table A-26 CHARTS OF THEORETICAL STRESS-CONCENTRATION FACTORS K_t (*continued*)

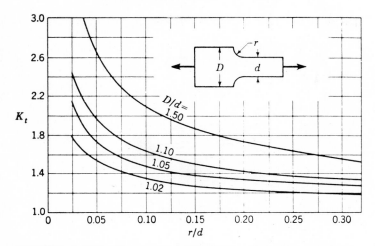

FIGURE A-26-5 Rectangular filleted bar in tension or simple compression. $\sigma_0 = F/A$, where $A = dt$ and t is the thickness.

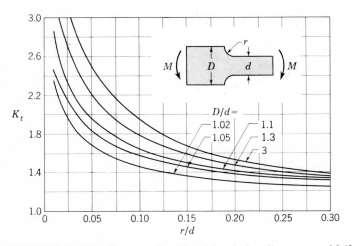

FIGURE A-26-6 Rectangular filleted bar in bending. $\sigma_0 = Mc/I$, where $c = d/2$, $I = td^3/12$, and t is the thickness.

Table A-26 CHARTS OF THEORETICAL STRESS-CONCENTRATION
FACTORS K_t (*continued*)

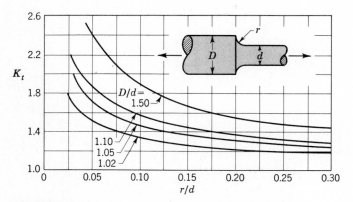

FIGURE A-26-7 Round shaft with shoulder fillet in tension.
$\sigma_0 = F/A$, where $A = \pi d^2/4$.

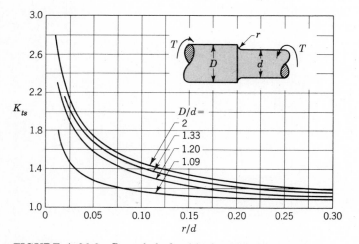

FIGURE A-26-8 Round shaft with shoulder fillet in torsion.
$\tau_0 = Tc/J$, where $c = d/2$ and $J = \pi d^4/32$.

Table A-26   CHARTS OF THEORETICAL STRESS-CONCENTRATION FACTORS K_t (*continued*)

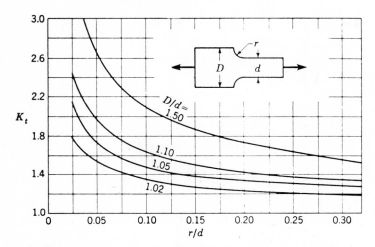

FIGURE A-26-5   Rectangular filleted bar in tension or simple compression. $\sigma_0 = F/A$, where $A = dt$ and t is the thickness.

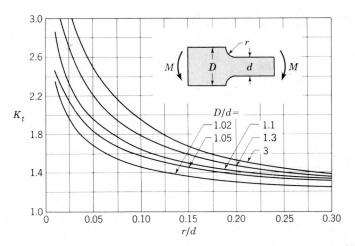

FIGURE A-26-6   Rectangular filleted bar in bending. $\sigma_0 = Mc/I$, where $c = d/2$, $I = td^3/12$, and t is the thickness.

Table A-26 CHARTS OF THEORETICAL STRESS-CONCENTRATION
FACTORS K_t *(continued)*

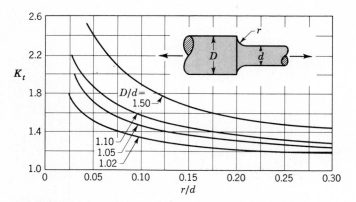

FIGURE A-26-7 Round shaft with shoulder fillet in tension.
$\sigma_0 = F/A$, where $A = \pi d^2/4$.

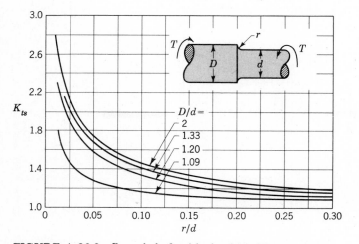

FIGURE A-26-8 Round shaft with shoulder fillet in torsion.
$\tau_0 = Tc/J$, where $c = d/2$ and $J = \pi d^4/32$.

Table A-26   CHARTS OF THEORETICAL STRESS-CONCENTRATION
FACTORS K_t (*continued*)

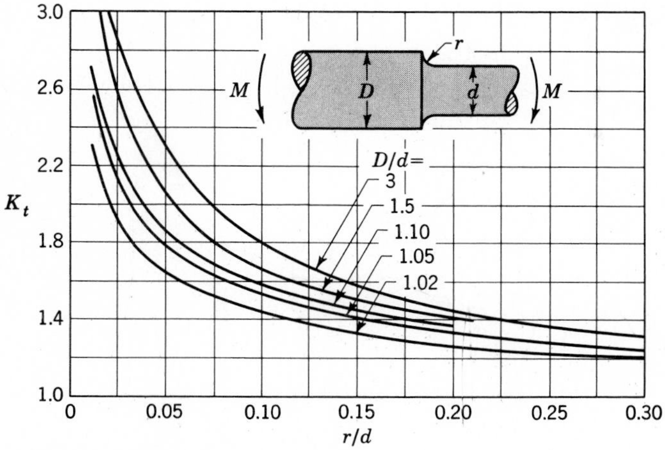

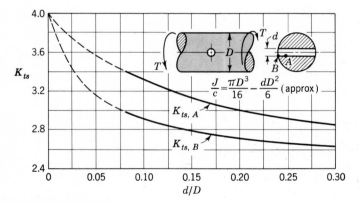

FIGURE A-26-9   Round shaft with shoulder fillet in bending.
$\sigma_0 = Mc/I$, where $c = d/2$ and $I = \pi d^4/64$.

FIGURE A-26-10   Round shaft in torsion with transverse hole.

Table A-26   CHARTS OF THEORETICAL STRESS-CONCENTRATION
FACTORS K_t (*continued*)

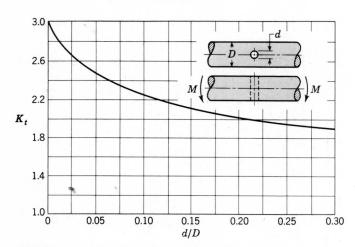

FIGURE A-26-11   Round shaft in bending with a transverse
hole. $\sigma_0 = M/[(\pi D^3/32) - (dD^2/6)]$, approximately.

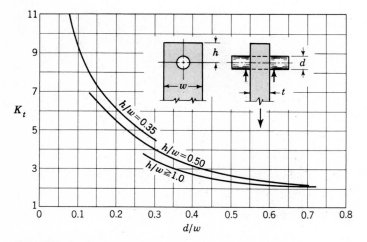

FIGURE A-26-12   Plate loaded in tension by a pin through a
hole. $\sigma_0 = F/A$, where $A = (w - d)t$.   When clearance exists, increase K_t 35 to 50 percent.   (*M. M. Frocht and H. N. Hill, "Stress
Concentration Factors around a Central Circular Hole in a Plate Loaded
through a Pin in Hole," J. Appl. Mechanics, vol. 7, no. 1, March 1940, p.*
A-5.

Table A-26 CHARTS OF THEORETICAL STRESS-CONCENTRATION FACTORS K_t (*continued*)

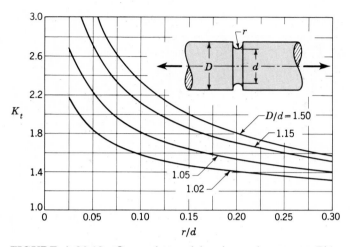

FIGURE A-26-13 Grooved round bar in tension. $\sigma_0 = F/A$, where $A = \pi d^2/4$.

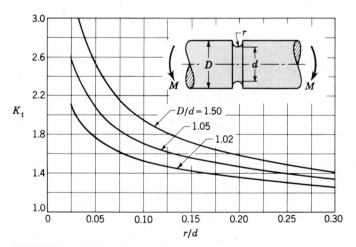

FIGURE A-26-14 Grooved round bar in bending. $\sigma_0 = Mc/I$, where $c = d/2$ and $I = \pi d^4/64$.

Table A-26   CHARTS OF THEORETICAL STRESS-CONCENTRATION
FACTORS K_t (*concluded*)

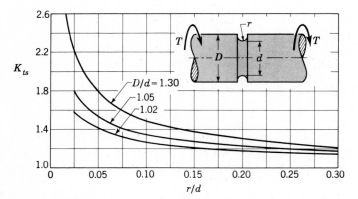

FIGURE A-26-15   Grooved round bar in torsion. $\tau_0 = Tc/J$,
where $c = d/2$ and $J = \pi d^4/32$.

Table A-27   GREEK ALPHABET

Alpha	A	χ		Nu	N	ν	
Beta	B	β		Xi	Ξ	ξ	
Gamma	Γ	γ		Omicron	O	o	
Delta	Δ	δ	∂	Pi	Π	π	
Epsilon	E	ϵ	ε	Rho	P	ρ	
Zeta	Z	ζ		Sigma	Σ	σ	ζ
Eta	H	η		Tau	T	τ	
Theta	Θ	θ	ϑ	Upsilon	Υ		
Iota	I	ι		Phi	Φ	ϕ	φ
Kappa	K	κ	χ	Chi	X	χ	
Lambda	Λ	λ		Psi	Ψ	ψ	
Mu	M	μ		Omega	Ω	ω	

Table A-28 DIMENSIONS OF ROUND-HEAD MACHINE SCREWS (ASA B18.6-1947)

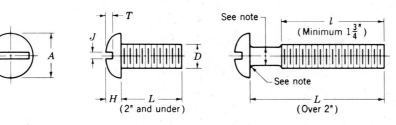

	D	A		H		J		T	
		Head diameter		**Height of head**		**Width of slot**		**Depth of slot**	
Nominal size	Max diameter of screw	Max	Min	Max	Min	Max	Min	Max	Min
0	0.060	0.113	0.099	0.053	0.043	0.023	0.016	0.039	0.029
1	0.073	0.138	0.122	0.061	0.051	0.026	0.019	0.044	0.033
2	0.086	0.162	0.146	0.069	0.059	0.031	0.023	0.048	0.037
3	0.099	0.187	0.169	0.078	0.067	0.035	0.027	0.053	0.040
4	0.112	0.211	0.193	0.086	0.075	0.039	0.031	0.058	0.044
5	0.125	0.236	0.217	0.095	0.083	0.043	0.035	0.063	0.047
6	0.138	0.260	0.240	0.103	0.091	0.048	0.039	0.068	0.051
8	0.164	0.309	0.287	0.120	0.107	0.054	0.045	0.077	0.058
10	0.190	0.359	0.334	0.137	0.123	0.060	0.050	0.087	0.065
12	0.216	0.408	0.382	0.153	0.139	0.067	0.056	0.096	0.072
$\frac{1}{4}$	0.250	0.472	0.443	0.175	0.160	0.075	0.064	0.109	0.082
$\frac{5}{16}$	0.3125	0.590	0.557	0.216	0.198	0.084	0.072	0.132	0.099
$\frac{3}{8}$	0.375	0.708	0.670	0.256	0.237	0.094	0.081	0.155	0.117
$\frac{7}{16}$	0.4375	0.750	0.707	0.328	0.307	0.094	0.081	0.196	0.148
$\frac{1}{2}$	0.500	0.813	0.766	0.355	0.332	0.106	0.091	0.211	0.159
$\frac{9}{16}$	0.5625	0.938	0.887	0.410	0.385	0.118	0.102	0.242	0.183
$\frac{5}{8}$	0.625	1.000	0.944	0.438	0.411	0.133	0.116	0.258	0.195
$\frac{3}{4}$	0.750	1.250	1.185	0.547	0.516	0.149	0.131	0.320	0.242

Notes:

All dimensions are given in inches. Head dimensions for sizes $\frac{7}{16}$ in and larger are in agreement with round-head cap screw dimensions except the minimum values have been decreased to provide tolerances in proportion to balance of table. The diameter of the unthreaded portion of machine screws shall not be less than the minimum pitch diameter nor more than the maximum major diameter of the thread. The radius of the fillet at the base of the head shall not exceed one-half the pitch of the screw thread.

Source: By permission from Vallory H. Laughner and Augustus D. Hargan, *Handbook of Fastening and Joining of Metal Parts.* McGraw-Hill, New York, 1956.

Table A-29   DIMENSIONS OF HEXAGON-HEAD CAP SCREWS (ASA B18.2-1952)

Nominal size or basic major diameter of thread	Body diameter minimum (maximum equal to nominal size)	Width across flats F			Width across corners G		Height H			Radius of fillet R	
		Max	(basic)	Min	Max	Min	Nom	Max	Min	Max	Min
$\frac{1}{4}$	0.2450	$\frac{7}{16}$	0.4375	0.428	0.505	0.488	$\frac{5}{32}$	0.163	0.150	0.023	0.009
$\frac{5}{16}$	0.3065	$\frac{1}{2}$	0.5000	0.489	0.577	0.557	$\frac{13}{64}$	0.211	0.195	0.023	0.009
$\frac{3}{8}$	0.3690	$\frac{9}{16}$	0.5625	0.551	0.650	0.628	$\frac{15}{64}$	0.243	0.226	0.023	0.009
$\frac{7}{16}$	0.4305	$\frac{5}{8}$	0.6250	0.612	0.722	0.698	$\frac{9}{32}$	0.291	0.272	0.023	0.009
$\frac{1}{2}$	0.4930	$\frac{3}{4}$	0.7500	0.736	0.866	0.840	$\frac{5}{16}$	0.323	0.302	0.023	0.009
$\frac{9}{16}$	0.5545	$\frac{13}{16}$	0.8125	0.798	0.938	0.910	$\frac{23}{64}$	0.371	0.348	0.041	0.021
$\frac{5}{8}$	0.6170	$\frac{15}{16}$	0.9375	0.922	1.083	1.051	$\frac{25}{64}$	0.403	0.378	0.041	0.021

Nominal sizes (basic major diameter of thread): 0.2500, 0.3125, 0.3750, 0.4375, 0.5000, 0.5625, 0.6250

3/4	0.7500	0.7410	1 1/8	1.1250	1.100	1.299	1.254	15/32	0.483	0.455	0.041	0.021
7/8	0.8750	0.8660	1 5/16	1.3125	1.285	1.516	1.465	35/64	0.563	0.531	0.062	0.047
1	1.0000	0.9900	1 1/2	1.5000	1.469	1.732	1.675	39/64	0.627	0.591	0.062	0.047
1 1/8	1.1250	1.1140	1 11/16	1.6875	1.631	1.949	1.859	11/16	0.718	0.658	0.125	0.110
1 1/4	1.2500	1.2390	1 7/8	1.8750	1.812	2.165	2.066	25/32	0.813	0.749	0.125	0.110
1 3/8	1.3750	1.3630	2 1/16	2.0625	1.994	2.382	2.273	27/32	0.878	0.810	0.125	0.110
1 1/2	1.5000	1.4880	2 1/4	2.2500	2.175	2.598	2.480	15/16	0.974	0.902	0.125	0.110

Notes:

All dimensions given in inches. *Bold type indicates products unified dimensionally with British and Canadian Standards.* Taper of head (angle between one side and axis) shall not exceed 2 deg; specified width across flats being the largest dimension. Top of head shall be flat and chamfered. Diameter of top circle shall be maximum width across flats within a tolerance of minus 15 percent. Bearing surface shall be flat and either washer-faced or with chamfered corners. Diameters of bearing surface shall be 95 percent of maximum width across flats within a tolerance of plus or minus 5 percent. Bearing surface shall be at right angles to axis of body within a tolerance of 2 deg for sizes up to and including 1 in, and within a tolerance of 1 deg for sizes larger than 1 in. The bearing surface shall be concentric with axis of body within a tolerance of 3 percent of the maximum width across flats. Minimum thread length shall be twice the diameter plus $\frac{1}{4}$ in for lengths up to and including 6 in; twice the diameter plus $\frac{1}{2}$ in for lengths over 6 in. The tolerance shall be plus $\frac{3}{16}$ in or $2\frac{1}{2}$ threads, whichever is greater. On products that are too short for minimum thread lengths the distance from the bearing surface of the head to the first complete thread shall not exceed the length of $2\frac{1}{2}$ threads, as measured with a ring thread gauge, for sizes up to and including 1 in and $3\frac{1}{2}$ threads for sizes larger than 1 in. Threads shall be coarse, fine, or 8-thread series, class 2A for plain (unplated) cap screws. For plated cap screws, the diameters may be increased by the amount of class 2A allowance. Thickness or quality of plating shall be measured or tested on the side of the head. Point shall be flat and chamfered, length of point to first full thread not to exceed $1\frac{1}{2}$ threads. Tolerance on length for sizes up to and including $\frac{3}{4}$ in shall be minus $\frac{1}{32}$ in for lengths up to and including 1 in; minus $\frac{1}{16}$ in for lengths over 1 in to and including 2 in; and minus $\frac{3}{32}$ in for lengths over 2 in to 6 in, inclusive. The tolerance shall be doubled for sizes over $\frac{3}{4}$ in and lengths longer than 6 in. Total runout (eccentricity and angularity) of thread in relation to body for sizes over $\frac{3}{4}$ in shall not exceed 0.010 in for each inch of length when measured in a sleeve gauge; the deviation of shank from a surface plate on which it is rolled shall not exceed 0.0312 in. For sizes over $\frac{3}{4}$ in total runout shall be subject to negotiation. Unless otherwise specified, physical properties of steel cap screws shall correspond to SAE grades 2 or 5. Cap screws may also be made from alloy steel, brass, bronze, corrosion-resisting steel, aluminum alloy, or such other material as specified.

Source: By permission from Vallory H. Laughner and Augustus D. Hargan, *Handbook of Fastening and Joining of Metal Parts*, McGraw-Hill, New York, 1956.

Table A-30   DIMENSIONS OF FINISHED HEXAGON BOLTS (ASA B18.2-1952)

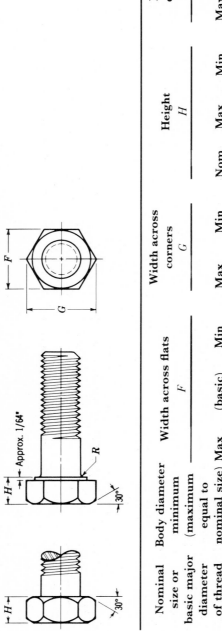

Approx. 1/64"

Nominal size or basic major diameter of thread		Body diameter minimum (maximum equal to nominal size)	Width across flats F			Width across corners G		Height H			Radius of fillet R	
			Max	(basic)	Min	Max	Min	Nom	Max	Min	Max	Min
$\frac{1}{4}$	0.2500	0.2450	$\frac{7}{16}$	0.4375	0.428	0.505	0.488	$\frac{5}{32}$	0.163	0.150	0.023	0.009
$\frac{5}{16}$	0.3125	0.3065	$\frac{1}{2}$	0.5000	0.489	0.577	0.557	$\frac{13}{64}$	0.211	0.195	0.023	0.009
$\frac{3}{8}$	0.3750	0.3690	$\frac{9}{16}$	0.5625	0.551	0.650	0.628	$\frac{15}{64}$	0.243	0.226	0.023	0.009
$\frac{7}{16}$	0.4375	0.4305	$\frac{5}{8}$	0.6250	0.612	0.722	0.698	$\frac{9}{32}$	0.291	0.272	0.023	0.009
$\frac{1}{2}$	0.5000	0.4930	$\frac{3}{4}$	0.7500	0.736	0.866	0.840	$\frac{5}{16}$	0.323	0.302	0.023	0.009
$\frac{9}{16}$	0.5625	0.5545	$\frac{13}{16}$	0.8125	0.798	0.938	0.910	$\frac{23}{64}$	0.371	0.348	0.041	0.021
$\frac{5}{8}$	0.6250	0.6170	$\frac{15}{16}$	0.9375	0.922	1.083	1.051	$\frac{25}{64}$	0.403	0.378	0.041	0.021
$\frac{3}{4}$	0.7500	0.7410	$1\frac{1}{8}$	1.1250	1.100	1.299	1.254	$\frac{15}{32}$	0.483	0.455	0.041	0.021
$\frac{7}{8}$	0.8750	0.8660	$1\frac{5}{16}$	1.3125	1.285	1.516	1.465	$\frac{35}{64}$	0.563	0.531	0.062	0.047
1	1.0000	0.9900	$1\frac{1}{2}$	1.5000	1.469	1.732	1.675	$\frac{39}{64}$	0.627	0.591	0.062	0.047
$1\frac{1}{8}$	1.1250	1.1140	$1\frac{11}{16}$	1.6875	1.631	1.949	1.859	$\frac{11}{16}$	0.718	0.658	0.125	0.110
$1\frac{1}{4}$	1.2500	1.2390	$1\frac{7}{8}$	1.8750	1.812	2.165	2.066	$\frac{25}{32}$	0.813	0.749	0.125	0.110
$1\frac{3}{8}$	1.3750	1.3630	$2\frac{1}{16}$	2.0625	1.994	2.382	2.273	$\frac{27}{32}$	0.878	0.810	0.125	0.110
$1\frac{1}{2}$	1.5000	1.4880	$2\frac{1}{4}$	2.2500	2.175	2.598	2.480	$\frac{15}{16}$	0.974	0.902	0.125	0.110
$1\frac{5}{8}$	1.6250	1.6130	$2\frac{7}{16}$	2.4375	2.356	2.815	2.686	1	1.038	0.962	0.125	0.110

$1\frac{3}{4}$	1.7500	1.7380	$2\frac{5}{8}$	2.6250	2.538	3.031	2.893	$1\frac{3}{32}$	1.134	1.054	0.125	0.110
$1\frac{7}{8}$	1.8750	1.8630	$2\frac{13}{16}$	2.8125	2.719	3.248	3.100	$1\frac{5}{32}$	1.198	1.114	0.125	0.110
2	2.0000	1.9880	3	3.0000	2.900	3.464	3.306	$1\frac{7}{32}$	1.263	1.175	0.125	0.110
$2\frac{1}{4}$	2.2500	2.2380	$3\frac{3}{8}$	3.3750	3.262	3.897	3.719	$1\frac{3}{8}$	1.423	1.327	0.188	0.173
$2\frac{1}{2}$	2.5000	2.4880	$3\frac{3}{4}$	3.7500	3.625	4.330	4.133	$1\frac{17}{32}$	1.583	1.479	0.188	0.173
$2\frac{3}{4}$	2.7500	2.7380	$4\frac{1}{8}$	4.1250	3.988	4.763	4.546	$1\frac{11}{16}$	1.744	1.632	0.188	0.173
3	3.0000	2.9880	$4\frac{1}{2}$	4.5000	4.350	5.196	4.959	$1\frac{7}{8}$	1.935	1.815	0.188	0.173

* In sizes $\frac{1}{4}$ to 1 in, a tolerance of minus 0.050 D may be used when the product is hot-made.

Notes:

All dimensions given in inches. *Bold type indicates products unified dimensionally with British and Canadian Standards.* "Finished" in the title refers to the quality of manufacture and the closeness of tolerance and does not indicate that surfaces are completely machined. Taper of head (angle between one side and axis) shall not exceed 2 deg; specified width across flats being the largest dimension. Top of head shall be flat and chamfered. Diameter of top circle shall be maximum width across flats within a tolerance of minus 15 percent. Bearing surface shall be flat and either washer-faced or with chamfered corners. Diameter of bearing surface shall be 95 percent of maximum width across flats within a tolerance of plus or minus 5 percent. Bearing surface shall be at right angles to axis of body within a tolerance of 2 deg for sizes up to and including 1 in, and within a tolerance of 1 deg for sizes larger than 1 in. The bearing surface shall be concentric with axis of body within a tolerance of 3 percent of the maximum width across flats. Minimum thread length shall be twice the diameter plus $\frac{1}{4}$ in for lengths up to and including 6 in; twice the diameter plus $\frac{1}{2}$ in for lengths over 6 in. The tolerance shall be plus $\frac{3}{16}$ in or $2\frac{1}{2}$ threads, whichever is greater. On products that are too short for minimum thread lengths, the distance from the bearing surface of the head to the first complete thread shall not exceed the length of $2\frac{1}{2}$ threads, as measured with a ring thread gauge, for sizes up to and including 1 in and $3\frac{1}{2}$ threads for sizes larger than 1 in. Threads shall be coarse-, fine-, or 8-thread series, class 2A for plain (unplated) bolts. For plated bolts, the diameters may be increased by the amount of class 2A allowance. Thickness or quality of plating shall be measured or tested on the side of the bolt head. Point shall be flat and chamfered, length of point to first full thread not to exceed $1\frac{1}{2}$ threads. Tolerance on bolt length for sizes up to and including $\frac{3}{4}$ in shall be minus $\frac{3}{32}$ in for lengths up to and including 1 in; minus $\frac{1}{16}$ for lengths over 1 in to and including 2 in; and minus $\frac{3}{32}$ in for lengths over 2 in to 6 in, inclusive. The tolerance shall be doubled for sizes over $\frac{3}{4}$ in and lengths longer than 6 in. Total runout (eccentricity and angularity) of thread in relation to body for sizes up to and including $\frac{3}{4}$ in shall not exceed 0.010 in for each inch of length when measured in a sleeve gauge; the deviation of shank from a surface plate on which it is rolled shall not exceed 0.0312 in. For sizes over $\frac{3}{4}$ in total runout shall be subject to negotiation. Unless otherwise specified, physical properties of steel bolts shall correspond to SAE grades 2 or 5. Bolts may also be made from alloy steel, brass, bronze, corrosion-resisting steel, aluminum alloy, or such other material as specified.

Source: By permission from Vallory H. Laughner and Augustus D. Hargan, *Handbook of Fastening and Joining of Metal Parts,* McGraw-Hill, New York, 1956.

Table A-31 DIMENSIONS OF FINISHED HEXAGON AND HEXAGON JAM NUTS (ASA B18.2-1952)

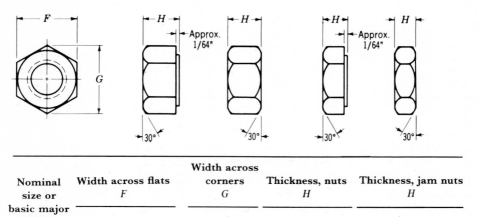

Nominal size or basic major diameter of thread	Width across flats F			Width across corners G		Thickness, nuts H			Thickness, jam nuts H		
	Max	(basic)	Min	Max	Min	Nom	Max	Min	Nom	Max	Min
$\frac{1}{4}$	0.2500	$\frac{7}{16}$ 0.4375	0.428	0.505	0.488	$\frac{7}{32}$	0.226	0.212	$\frac{5}{32}$	0.163	0.150
$\frac{5}{16}$	0.3125	$\frac{1}{2}$ 0.5000	0.489	0.577	0.577	$\frac{17}{64}$	0.273	0.258	$\frac{3}{16}$	0.195	0.180
$\frac{3}{8}$	0.3750	$\frac{9}{16}$ 0.5625	0.551	0.650	0.628	$\frac{21}{64}$	0.337	0.320	$\frac{7}{32}$	0.227	0.210
$\frac{7}{16}$	0.4375	$\frac{11}{16}$ 0.6875	0.675	0.794	0.768	$\frac{3}{8}$	0.385	0.365	$\frac{1}{4}$	0.260	0.240
$\frac{1}{2}$	0.5000	$\frac{3}{4}$ 0.7500	0.736	0.866	0.840	$\frac{7}{16}$	0.448	0.427	$\frac{5}{16}$	0.323	0.302
$\frac{9}{16}$	0.5625	$\frac{7}{8}$ 0.8750	0.861	1.010	0.982	$\frac{31}{64}$	0.496	0.473	$\frac{11}{32}$	0.365	0.323
$\frac{5}{8}$	0.6250	$\frac{15}{16}$ 0.9375	0.922	1.083	1.051	$\frac{35}{64}$	0.559	0.535	$\frac{3}{8}$	0.387	0.363
$\frac{3}{4}$	0.7500	$1\frac{1}{8}$ 1.1250	1.088	1.299	1.240	$\frac{41}{64}$	0.665	0.617	$\frac{27}{64}$	0.446	0.398
$\frac{7}{8}$	0.8750	$1\frac{5}{16}$ 1.3125	1.269	1.516	1.447	$\frac{3}{4}$	0.776	0.724	$\frac{31}{64}$	0.510	0.458
1	1.0000	$1\frac{1}{2}$ 1.5000	1.450	1.732	1.653	$\frac{55}{64}$	0.887	0.831	$\frac{35}{64}$	0.575	0.519
$1\frac{1}{8}$	1.1250	$1\frac{11}{16}$ 1.6875	1.631	1.949	1.859	$\frac{31}{32}$	0.999	0.939	$\frac{39}{64}$	0.639	0.579
$1\frac{1}{4}$	1.2500	$1\frac{7}{8}$ 1.8750	1.812	2.165	2.066	$1\frac{1}{16}$	1.094	1.030	$\frac{23}{32}$	0.751	0.687
$1\frac{3}{8}$	1.3750	$2\frac{1}{16}$ 2.0625	1.994	2.382	2.273	$1\frac{11}{64}$	1.206	1.138	$\frac{25}{32}$	0.815	0.747
$1\frac{1}{2}$	1.5000	$2\frac{1}{4}$ 2.2500	2.175	2.598	2.480	$1\frac{9}{32}$	1.317	1.245	$\frac{27}{32}$	0.880	0.808
$1\frac{5}{8}$	1.6250	$2\frac{7}{16}$ 2.4375	2.356	2.815	2.686	$1\frac{25}{64}$	1.429	1.353	$\frac{29}{32}$	0.944	0.868

Table A-31 DIMENSIONS OF FINISHED HEXAGON AND HEXAGON JAM NUTS
(ASA B18.2-1952) *(continued)*

Nominal size or basic major diameter of thread		Width across flats F			Width across corners G		Thickness, nuts H			Thickness, jam nuts H		
	Max	(basic)	Min	Max	Min	Nom	Max	Min	Nom	Max	Min
$1\frac{3}{4}$	1.7500	$2\frac{5}{8}$	2.6250	2.538	3.031	2.893	$1\frac{1}{2}$	1.540	1.460	$\frac{31}{32}$	1.009	0.929
$1\frac{7}{8}$	1.8750	$2\frac{13}{16}$	2.8125	2.719	3.248	3.100	$1\frac{39}{64}$	1.651	1.567	$1\frac{1}{32}$	1.073	0.989
2	2.0000	3	3.0000	2.900	3.464	3.306	$1\frac{23}{32}$	1.763	1.675	$1\frac{3}{32}$	1.138	1.050
$2\frac{1}{4}$	2.2500	$3\frac{3}{8}$	3.3750	3.262	3.897	3.719	$1\frac{59}{64}$	1.970	1.874	$1\frac{13}{64}$	1.251	1.155
$2\frac{1}{2}$	2.5000	$3\frac{3}{4}$	3.7500	3.625	5.330	4.133	$2\frac{9}{64}$	2.193	2.089	$1\frac{29}{64}$	1.505	1.401
$2\frac{3}{4}$	2.7500	$4\frac{1}{8}$	4.1250	3.988	4.763	4.546	$2\frac{23}{64}$	2.415	2.303	$1\frac{37}{64}$	1.634	1.522
3	3.0000	$4\frac{1}{2}$	4.5000	4.350	5.196	4.959	$2\frac{37}{64}$	2.638	2.518	$1\frac{45}{64}$	1.643	1.643

Notes:

All dimensions given in inches. *Bold type indicates products unified dimensionally with British and Canadian Standards.* "Finished" in the title refers to the quality of manufacture and the closeness of tolerance and does not indicate that surfaces are completely machined. Taper of the sides of nuts (angle between one side and the axis) shall not exceed 2 deg, the specified width across flats being the largest dimension. Tops of nuts shall be flat and chamfered. Diameter of top circle shall be the maximum width across flats within a tolerance of minus 15 percent for washer-faced nuts and within a tolerance of minus 5 percent for double-chamfered nuts. Bearing surface shall be washer-faced or with chamfered corners. Diameter of circle of bearing surface shall be the maximum width across flats within a tolerance of minus 5 percent. Tapped hole shall be counter-sunk $\frac{1}{64}$ in over the major diameter of thread for nuts up to and including $\frac{1}{2}$ in and $\frac{1}{32}$ in over the major diameter of thread for nuts over $\frac{1}{2}$ in size. Bearing surface shall be at right angles to the axis of the threaded hole within a tolerance of 2 deg for $\frac{5}{8}$-in nuts or smaller and 1 deg for nuts larger than $\frac{5}{8}$ in; therefore the maximum total runout of bearing face would equal the tangent of specified angle times the distance across flats. Thread shall be coarse-, fine-, or 8-thread series; class 2B. Suitable material for steel nuts is covered by ASTM A-307; other materials will be as agreed upon by manufacturer and user. Tolerance on width across flats may be increased 0.015 in for hot-formed nuts $\frac{5}{8}$ in and smaller.

Source: By permission from Vallory H. Laughner and Augustus D. Hargan, *Handbook of Fastening and Joining of Metal Parts*, McGraw-Hill, New York, 1956.

Table A-32 MASS AND MASS MOMENTS OF INERTIA OF GEOMETRIC SHAPES

ρ = density, weight/unit volume

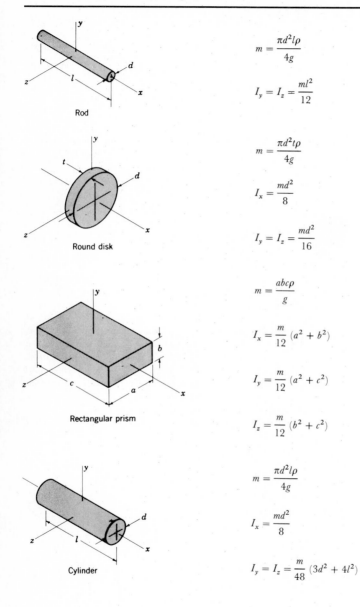

$$m = \frac{\pi d^2 l \rho}{4g}$$

$$I_y = I_z = \frac{ml^2}{12}$$

Rod

$$m = \frac{\pi d^2 t \rho}{4g}$$

$$I_x = \frac{md^2}{8}$$

$$I_y = I_z = \frac{md^2}{16}$$

Round disk

$$m = \frac{abc\rho}{g}$$

$$I_x = \frac{m}{12}(a^2 + b^2)$$

$$I_y = \frac{m}{12}(a^2 + c^2)$$

$$I_z = \frac{m}{12}(b^2 + c^2)$$

Rectangular prism

$$m = \frac{\pi d^2 l \rho}{4g}$$

$$I_x = \frac{md^2}{8}$$

$$I_y = I_z = \frac{m}{48}(3d^2 + 4l^2)$$

Cylinder

Table A-32 MASS AND MASS MOMENTS OF INERTIA OF GEOMETRIC SHAPES (*continued*)

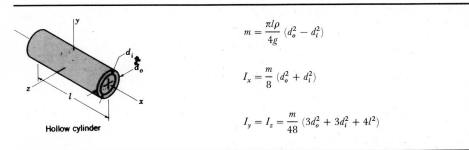

Hollow cylinder

$$m = \frac{\pi l \rho}{4g} (d_o^2 - d_i^2)$$

$$I_x = \frac{m}{8} (d_o^2 + d_i^2)$$

$$I_y = I_z = \frac{m}{48} (3d_o^2 + 3d_i^2 + 4l^2)$$

AREAS UNDER THE STANDARD NORMAL DISTRIBUTION CURVE

$$A = \int_0^z \frac{1}{\sqrt{2\pi}} e^{-z^2/2} \, dz$$

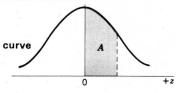

z	0	1	2	3	4	5	6	7	8	9
0.0	0.0000	0.0040	0.0080	0.0120	0.0160	0.0199	0.0239	0.0279	0.0319	0.0359
0.1	0.0398	0.0438	0.0478	0.0517	0.0557	0.0596	0.0636	0.0675	0.0714	0.0754
0.2	0.0793	0.0832	0.0871	0.0910	0.0948	0.0987	0.1026	0.1064	0.1103	0.1141
0.3	0.1179	0.1217	0.1255	0.1293	0.1331	0.1368	0.1406	0.1443	0.1480	0.1517
0.4	0.1554	0.1591	0.1628	0.1664	0.1700	0.1736	0.1772	0.1808	0.1844	0.1879
0.5	0.1915	0.1950	0.1985	0.2019	0.2054	0.2088	0.2123	0.2157	0.2190	0.2224
0.6	0.2258	0.2291	0.2324	0.2357	0.2389	0.2422	0.2454	0.2486	0.2518	0.2549
0.7	0.2580	0.2612	0.2642	0.2673	0.2704	0.2734	0.2764	0.2794	0.2823	0.2852
0.8	0.2881	0.2910	0.2939	0.2967	0.2996	0.3023	0.3051	0.3078	0.3106	0.3133
0.9	0.3159	0.3186	0.3212	0.3238	0.3264	0.3289	0.3315	0.3340	0.3365	0.3389
1.0	0.3413	0.3438	0.3461	0.3485	0.3508	0.3531	0.3554	0.3577	0.3599	0.3621
1.1	0.3643	0.3665	0.3686	0.3708	0.3729	0.3749	0.3770	0.3790	0.3810	0.3830
1.2	0.3849	0.3869	0.3888	0.3907	0.3925	0.3944	0.3962	0.3980	0.3997	0.4015
1.3	0.4032	0.4049	0.4066	0.4082	0.4099	0.4115	0.4131	0.4147	0.4162	0.4177
1.4	0.4192	0.4207	0.4222	0.4236	0.4251	0.4265	0.4279	0.4292	0.4306	0.4319
1.5	0.4332	0.4345	0.4357	0.4370	0.4382	0.4394	0.4406	0.4418	0.4429	0.4441
1.6	0.4452	0.4463	0.4474	0.4484	0.4495	0.4506	0.4515	0.4525	0.4535	0.4545
1.7	0.4554	0.4564	0.4573	0.4582	0.4591	0.4599	0.4608	0.4616	0.4625	0.4633
1.8	0.4641	0.4649	0.4656	0.4664	0.4671	0.4678	0.4686	0.4693	0.4699	0.4706
1.9	0.4713	0.4719	0.4726	0.4732	0.4738	0.4744	0.4750	0.4756	0.4761	0.4767
2.0	0.4772	0.4778	0.4783	0.4788	0.4793	0.4798	0.4803	0.4808	0.4812	0.4817
2.1	0.4821	0.4826	0.4830	0.4834	0.4838	0.4842	0.4846	0.4850	0.4854	0.4857
2.2	0.4861	0.4864	0.4868	0.4871	0.4875	0.4878	0.4881	0.4884	0.4887	0.4890
2.3	0.4893	0.4896	0.4898	0.4901	0.4904	0.4906	0.4909	0.4911	0.4913	0.4916
2.4	0.4918	0.4920	0.4922	0.4925	0.4927	0.4929	0.4931	0.4932	0.4934	0.4936
2.5	0.4938	0.4940	0.4941	0.4943	0.4945	0.4946	0.4948	0.4949	0.4951	0.4952
2.6	0.4953	0.4955	0.4956	0.4957	0.4959	0.4960	0.4961	0.4962	0.4963	0.4964
2.7	0.4965	0.4966	0.4967	0.4968	0.4969	0.4970	0.4971	0.4972	0.4973	0.4974
2.8	0.4974	0.4975	0.4976	0.4977	0.4977	0.4978	0.4979	0.4979	0.4980	0.4981
2.9	0.4981	0.4982	0.4982	0.4983	0.4984	0.4984	0.4985	0.4985	0.4986	0.4986
3.0	0.4987	0.4987	0.4987	0.4988	0.4988	0.4989	0.4989	0.4989	0.4990	0.4990
3.1	0.4990	0.4991	0.4991	0.4991	0.4992	0.4992	0.4992	0.4992	0.4993	0.4993
3.2	0.4993	0.4993	0.4994	0.4994	0.4994	0.4994	0.4994	0.4995	0.4995	0.4995
3.3	0.4995	0.4995	0.4995	0.4996	0.4996	0.4996	0.4996	0.4996	0.4996	0.4997
3.4	0.4997	0.4997	0.4997	0.4997	0.4997	0.4997	0.4997	0.4997	0.4997	0.4998
3.5	0.4998	0.4998	0.4998	0.4998	0.4998	0.4998	0.4998	0.4998	0.4998	0.4998
3.6	0.4998	0.4998	0.4999	0.4999	0.4999	0.4999	0.4999	0.4999	0.4999	0.4999
3.7	0.4999	0.4999	0.4999	0.4999	0.4999	0.4999	0.4999	0.4999	0.4999	0.4999
3.8	0.4999	0.4999	0.4999	0.4999	0.4999	0.4999	0.4999	0.4999	0.4999	0.4999
3.9	0.5000	0.5000	0.5000	0.5000	0.5000	0.5000	0.5000	0.5000	0.5000	0.5000

INDEX